The Early History
of the Earth

BASED ON THE PROCEEDINGS OF A NATO
ADVANCED STUDY INSTITUTE
HELD AT THE UNIVERSITY OF LEICESTER
5–11 APRIL, 1975

Edited by

Brian F. Windley

*Department of Geology,
University of Leicester*

A Wiley–Interscience Publication

JOHN WILEY & SONS

LONDON · NEW YORK · SYDNEY · TORONTO

Library of Congress Cataloging in Publication Data:

NATO Advanced Study Institute, University of Leicester,
  1975.
  The early history of the Earth.

  'A Wiley–Interscience publication.'
  Includes index.

1. Geology, Stratigraphic—Archaen—Congresses.
2. Geology, Stratigraphic—Pre-Cambrian—Congresses.
3. Earth—Origin—Congresses.   4. Geology—Congresses.
I. Windley, Brian F.   II. Title.

QE653.N37 1975   551.7'12   75-26610

ISBN 0 471 01488 5

Set on Linotron Filmsetter and printed in Great Britain
by J. W. Arrowsmith Ltd., Bristol.

Preface

The early history of the Earth was so long and complicated that current students usually work only on one or two aspects of the subject. But the research subjects include field geology, geochemistry, structural geology and tectonics, the evolution of the atmosphere and the oceans, palaeontology, geochronology and metallogeny, so it is not surprising if specialists in one branch commonly have little knowledge of another; and yet many of these fields overlap and knowledge of one should influence and contribute to that of another. The oldest rocks known at present have an age of about 3.8 b.y., but most geologists know little of that pre-geological period before 3.8 b.y. when the core, mantle and protocrust formed. The aim of the NATO Advanced Study Institute, held from 5–11th April, 1975, at Leicester, was to bring together specialists from many fields to produce their latest findings and ideas and to discuss the present state of knowledge on early Earth history in the period 4.5–2.5 b.y. ago. Much of the success of the meeting was due to the fact that 2 hours of every day was devoted to discussion of specific problems. This volume contains the text of papers presented at the meeting with the object of providing an inter-disciplinary approach for the student, teacher and researcher to the problem of how the Earth evolved in its early stages. With such a large subject it is obviously impossible to be comprehensive, but it is hoped that this volume goes some way in providing an integrated compilation. The papers vary from long reviews of major subjects to short reports of recent research. Some are quantitative, some speculative and some highly controversial, but this variation reflects the state of current research in the subject. No one pretends that research into the early history of the Earth has reached an advanced stage of development—far from it; there are not enough constraints to enable one to choose between various alternatives and models in almost every field. I hope that this volume imparts to the student the controversial nature of our knowledge of the Archaean and pre-Archaean.

The idea of having a NATO Advanced Study Institute on the early history of the Earth came from Professor J. V. Smith in 1971 when I was working with him on Archaean rocks in Chicago and when he was involved in the early organization of the 1972 Feldspar NATO ASI at Manchester. I am grateful to him and to Professor J. Sutton (Imperial College, London) for many discussions since then on the organization of the meeting, and to Professor P. C. Sylvester-Bradley of the Department of Geology, Leicester University, who made many useful suggestions and gave much encouragement.

The Institute was attended by 141 scientists from 11 NATO and 67 other countries. The organizers are grateful for a generous grant from the Scientific Affairs Division of NATO which covered the accommodation and partial travel costs of participants from NATO countries, partial support for the field excursions in Scotland, and the organizational expenses of the meeting. The National Science Foundation of the U.S.A. kindly provided travel grants for three research students, and the International Union of Geological Sciences financed two visitors from non-NATO countries. UNESCO gave a grant to enable five delegates from non-NATO countries to attend the first full committee meeting of the International Geological Correlation Project on Archaean Geochemistry, convened by Dr A. Glikson and held during the Conference.

Thanks are due to many people who helped to make the meeting a success, especially: Drs J. Peal and K. Davies, Wardens of Villiers and Gilbert Murray Halls of Residence, for providing excellent facilities; the technical and secretarial staff of the Geology Department of Leicester University for their services; Miss J. Baker and Mr M. Clarke for general assistance throughout the meeting; and Professor J. Watson who led two field excursions, to the Scourian of NW Scotland and of the Outer Hebrides, before and after the meeting. In particular, I am grateful to Judith, my wife, for indispensable administrative and secretarial assistance pre-, syn- and post- the meeting; she also organized the field excursions.

June, 1975   BRIAN F. WINDLEY

Contents

The Early Earth–Moon System

Development of the Earth–Moon System with Implications for the Geology of the Early Earth

J. V. SMITH

Dept. of the Geophysical Sciences, University of Chicago, Chicago, Ill. 60637, U.S.A.

Introduction

The literature on the Earth–Moon system is bedevilled by naïveté and special pleading. This brief review cannot encompass all ideas, and I deliberately select recent accessible papers. For the Moon, I do not reference well-known data, and refer readers to the Proceedings of the Lunar Science Conferences and the journal *The Moon*: controversial ideas and references in other journals are documented more thoroughly. Let me emphasize my predilection for catastrophic processes which lead to molten planets, while recognizing that cooler, less catastrophic processes also occur: also note my psychological bias in favour of the petrologic model for the Moon by Smith et al. (1970a,b) with its crust of plagioclase-rich rocks and basalts, its olivine- and pyroxene-rich mantle, and Fe-rich core (Fig. 1): see also Wood et al. (1970). [The manuscript was prepared before the Sixth Lunar Science Conference, but brief references are made to the preprints. The book *Lunar Science: A Post-Apollo View* by Taylor (1975) appeared during final revision of the manuscript, and quick perusal reveals an excellent survey of the properties of the Moon, together with the Taylor–Jakeš geochemical model of partial melting.]

Prolonged weathering and metamorphism in a water-rich environment together with continuing igneous activity and continental drift have destroyed most of the Earth's early

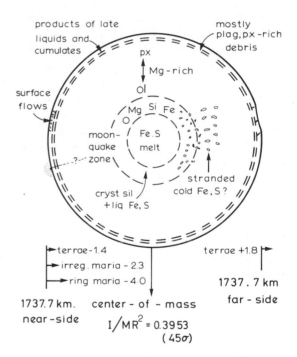

Fig. 1. Model for recent Moon (Smith, 1974) based on model of Smith et al. (1970a,b). Near-side basins are filled with mare basalts whereas far-side ones are unfilled. The moon-quake zone at ~700–1000 km depth is ascribed here to Fe,Ni,S liquid enclosed in solid silicate mantle, but probably would derive from melting of silicate if recent temperature estimates are correct. The postulated core might be as large as 700 km radius if the Fe,Ni,S liquid is rich in sulphide. The displacement of surface features with respect to a mass-centred sphere of mean radius 1737·7 km is shown at the bottom (Kaula et al., 1974)

3

crust. Initial hopes that an inert Moon would supply the missing evidence have been partly justified, but differences in chemical composition and dynamic environment of the two bodies require cautious interpretation. In 1975, chemical and morphological data for the Moon allow serious testing of petrologic and geochemical models but incomplete stratigraphic and geophysical controls lead to severe frustration. Exploration of Mercury, Mars and Venus is mostly confined to morphology and bulk properties, but these clearly favour major differentiation of the silicates, perhaps with separation of a metal-rich core. Theoretical calculations of chemical differentiation in a cooling solar nebula together with dynamic models for accretion and fragmentation can be combined with direct observations on meteorites and telescopic observations of asteroids and comets to provide a basis for the evolving system of planets. All these evidences combine to yield a plausible model in which high temperatures and intense bombardment prevent establishment of a stable crust on Earth before ~ -4 Giga years. In addition, the morphological and age relations on the Moon allow plausible speculations on the transition period in which early crust survived bombardment and remelting.

General Features of Earth and Moon

Established facts for the Earth include its density, seismic properties (which require an irregular crust, mantle and core); magnetism; general chemical properties of the upper crust with probable inferences for the lower crust and upper mantle (ranging from sediments via granitic, basaltic and granulitic to peridoditic and dunitic rocks); time scale running from -3.7 Gy for surviving remnants of the early crust; and evidence of bombardment. The inferred properties of the outer 100–200 km, plus those of the hydrosphere and atmosphere, show that temperatures did not rise high enough and long enough for loss of all the volatile components H_2O, CO_2 and alkalies although some loss can be expected on the basis of the noble gas data (e.g. Ringwood volatilization model: see Ringwood (1975) for

many references). Quite uncertain is how these residual volatiles were originally stored and how they subsequently migrated (e.g. mica was probably stable down to ~ 200 km, and pyroxene could store Na and K at greater depths). Qualitatively, the observed properties of the outer 100–200 km are most easily explained by temperatures reaching the melting of silicates while permitting retention of volatile elements in amphibole, mica, scapolite, liquid, glass, etc. The inferred properties of the middle mantle are consistent with predominance of the $(Mg,Fe)_2SiO_4$ composition, perhaps occurring mainly as a mixture of dense oxides (e.g. stishovite and periclase). Inferences for the lower mantle and core are even more uncertain, but S-bearing Fe-rich material is likely for the latter. The above model provides sufficient framework for present purposes.

Established facts for the Moon include its density (3.34 gm/cm^3); its moment of inertia (~ 0.395 implying near-uniformity but slightly centre-heavy); its asymmetry (centre-of-mass about 2 km closer to Earth than the centre-of-volume); its surface morphology (highlands, irregular and ringed mare basins, craters of all sizes, lineaments and wrinkles); the refractory and reduced nature of, low content of metal-seeking elements in, and ubiquitity of feldspar-rich breccias and basalts in the Apollo samples; the crystallization ages ranging from -3.9 to -3.1 Gy for mare basalts, and some less certain values earlier than -4 Gy for possible fragments of the early crust (e.g. Schaeffer and Husain, 1973; Jessberger et al., 1974); the model ages which indicate that lunar rocks originated mostly from a single differentiated reservoir about $-4\frac{1}{2}$ Gy; its weak seismicity; remanent magnetism of the crustal rocks; mass anomalies, especially positive ones over the centres of basalt-filled ringed basins and negative ones of late craters.

The seismic data are sparse, and mostly confined to travel paths near part of the Earth side. The latest interpretation (Nakamura et al., 1974) distinguishes four and perhaps five zones whose seismic properties are consistent with: I plagioclase-rich crust 50–60 km thick ranging from rubble at the surface to consolidated rock at depth (note that the seismic velocity of a consolidated rock is independent of grain size); II olivine-pyroxene upper mantle 250 km thick (note that olivine has only slightly higher seismic velocity than pyroxene); III 500 km middle mantle with high Poisson's ratio (0·33–0·36); IV lower mantle with high attenuation of shear waves, perhaps resulting from partial melting; V core 170–360 km radius with low P velocity, perhaps resulting from an iron-rich melt. The details are model-dependent, but the evidence for moon quakes at ~ 600-1000 km depth, plus poor transmission of shear waves, definitely favours a hot interior with partial melting. The crust is probably thicker on the far side, perhaps about 100 km thick. Although quite tentative at this time, the easiest interpretation of the seismic data utilizes an olivine-rich mantle and an Fe-rich core. Perturbation of the solar wind yields an estimated profile of the electrical conductivity with depth. The inferred temperature profile requires a mineralogical model and an estimate of the oxidation state of iron. Whatever the details, an olivine or pyroxene mantle with iron in the reduced state allows high temperatures up to and including the melting range of basaltic and Fe,S-rich liquids (Duba and Ringwood, 1973; Schwerer et al., 1974). The simplest explanation of the magnetic properties invokes a dynamo in the early lunar core, but there are many unsolved problems involving dynamics and energy sources (e.g. Sonett and Runcorn, 1973). Many aspects of the internal composition and evolution of the Moon are reviewed by Solomon and Toksöz (1973) and Dainty et al. (1974).

Taking all the data together, a possible model (e.g. Smith and Steele, 1975) utilizes (a) accretion near − 4·5 Gy with only minor later accretion at the surface from impacting bodies: the original bulk composition was dominated by Mg-rich olivine and pyroxene, Ca-rich plagioclase, ~ 5% Fe,S,Ni, minor double-oxides, apatite, etc., thereby requiring the refractory and low-Fe nature to be inherited from the accretion process, (b) total crystal-liquid differentiation near − 4·5 Gy

producing plagioclase-rich asymmetric crust, olivine-rich mantle and Fe,Ni,S liquid core, (c) intense early bombardment down to ~ −4 Gy at which stage the crust had thickened sufficiently to sustain distinct impact basins, (d) intense brecciation and crystal-liquid fractionation of the crust from − 4·5 to − 4·0 Gy, (e) uprising of the mantle under basins, differentiation of the debris and underlying rock, remelting of early cumulates at the crust–mantle interface to produce basalts which flooded the mare basins from − 4 to − 3 Gy, (f) consolidation and shrinkage of the crust and upper mantle plus minor volcanism from − 3 Gy to present (e.g. Muehlberger, 1974), (g) retention of liquid in the lower mantle and core, (h) declining flux of projectiles. [Note that complete simultaneous melting of the Moon is not necessary, and progressive partial melting with complex crystal-liquid fractionation gives an easier interpretation of the temperature profile.] Fig. 2 is a possible model of the *present* Moon.

Important Problems Involving the Moon

Extent of Melting

Many early lunar models utilized a cold Moon in which a molten zone developed near the surface and moved inwards only a short distance. Most present models have accepted the concept of early (~ − 4·5 Gy) melting sufficient to produce an ~ 50 km plagioclase-rich crust, but many models involve melting of only the outer part of the Moon with retention of primary accreted material at the centre (e.g. Taylor and Jakeš, 1974, who consider both possibilities).

Time of Crustal Differentiation and Nature of Bombardment

Taylor (1975) reviews the general features of the early basins. These basins seem to occur randomly over the entire Moon. Faint features indicate partial destruction of earlier basins by later ones. Basins on the near-side contain basalt, while most on the far-side are free of

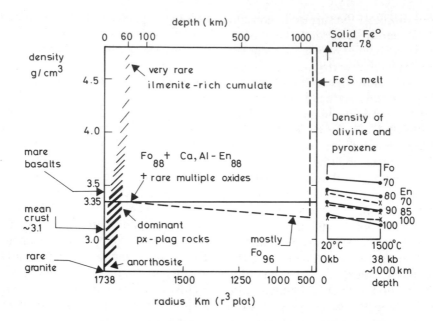

Fig. 2. Possible cross-section of present Moon showing density profile and location of rock types for simplest interpretation of seismic data (Smith and Steele, 1975). The radius is plotted as r^3 to show graphically how the zones contribute to the bulk density. An Fe-rich core could range in density from ~4·5 for FeS liquid to ~7·8 for solid Fe. The densities of olivine and low-Ca pyroxene are shown at the small right-hand diagram. The mantle varies from olivine-rich at the base to olivine plus pyroxene at the top. The lunar crust is dominated by rocks containing plagioclase and pyroxenes but contains some Ba-rich granite and ilmenite-rich cumulate

basalt except for minor quantities. See Stuart-Alexander and Howard (1970), Wilhelms and McCauley (1971) and Wilshire and Jackson (1972) for basin stratigraphy based on surface morphology: note also suggestion of Gargantuan basin (Cadogan, 1974) which covered most of NW quadrant of Moon prior to Imbrium impact. These morphological features require early development of a global crust which was disrupted by impact of large projectiles prior to out-pouring of mare basalts.

The actual time of crustal differentiation and nature of the bombardment are controversial. Tera et al. (1974) and Tera and Wasserburg (1974) interpreted U–Pb data in terms of formation of the lunar crust near −4·42 Gy with widespread metamorphism in a 'terminal lunar cataclysm' at −4·0 to −3·8 Gy which involved blanketing of the Moon's nearside by the debris from basin-forming impacts. Jessberger et al. (1974) used high-resolution Ar data for breccias and mineral separates to suggest that the basins formed over ~10^8 y between −4·0 and −3·9 Gy. Schaeffer and Husain (1974), also using ^{40}Ar–^{39}Ar ages but of small fragments, proposed that the terminal lunar cataclysm results merely from the predominance of ejecta from the Imbrium basin, and proposed that basins developed over some hundreds of millions of years: viz., Nectaris −4·25 ± 0·05, Crisium/Humorum −4·20 to −4·05, Imbrium −3·95 ± 0·05, Orientale −3·85 ± 0·05 Gy. Nunes et al. (1974) claimed that two-stage models are insufficient to explain the U–Pb evolution of many lunar samples; argued for a complex bombardment history from ~ −4·5 to −3·9 Gy with basin chronology similar to that of Schaeffer and Husain;

and suggested that mare basalts retained no Pb from earlier rocks. Further data and polemics are given by Kirsten and Horn (1974), Nunes and Tatsumoto (1975) and Tera and Wasserburg (1975).

Crucial to settling this controversy will be reliable assignment of ejecta to the various basins. McGetchin et al. (1973) developed a semi-empirical formula for ejecta thickness t in terms of the crater radius R and distance from the crater centre r (all in metres) where $t = 0 \cdot 14 R^{0 \cdot 74} (r/R)^{-3 \cdot 0}$.

This model predicts the following thicknesses of Imbrium ejecta: Apollo 15 800 m, Apollo 14 130 m, Apollo 17 100 m and Apollo 16 50 m. Pike (1974) estimated even greater thicknesses. Moore et al. (1974), using the McGetchin et al. model and a spherical Moon, predicted average thicknesses of 0·2 km at 1000 km from the basin centre, falling to 10 metres at 4000 km and rising again to 0·2 km at the antipode.

Morgan et al. (1974) utilized the very low content of siderophile elements in the indigeneous lunar rocks to detect the amount of meteoritic contamination in surface breccias. They calculated the expected thicknesses at the Apollo and Luna sites for 12 basins, and utilized the proportions of less to more volatile siderophile elements to identify six distinct meteoritic components, five of which were tentatively assigned to the bodies responsible for the Imbrium, Serenitatis, Crisium, Nectaris and Humorum or Nubium basins. Using ingenious but model-dependent ideas, they concluded that the bodies consisted of an extinct population of planetesimals or moonlets of roughly chondritic composition, undifferentiated, with 15–40% Fe, and striking at velocities generally less than 8 km/sec. None of the bodies resembled the Earth and Moon in bulk composition, while several appeared to be less refractory and richer in volatiles. Utilizing an estimate that the lunar crust contains no more than 2% of meteoritic material, and that none was lost during impact, Morgan et al. concluded that between 40 and 160 basin-forming objects hit the Moon from the time of formation of the crust ($\sim -4 \cdot 5$ Gy) until the Imbrium impacts at

$-3 \cdot 9$ Gy. This is about twice the number identified by Hartmann and Wood (1971), but early basins could have been obliterated, of course.

Perhaps at this time the most plausible model involves (a) development of the crust, mantle and ? core at $-4 \cdot 5 \pm 0 \cdot 1$ Gy in one major episode, (b) intense bombardment by a variety of projectiles down to $-3 \cdot 9$ Gy, only the late arrivals leaving distinct basins and ejecta blankets. Undoubtedly considerable tolerance exists for varying the timing and nature of the bombardment, and clustering of projectiles is not unlikely especially if a planetisimal disintegrates.

Bulk Chemical Composition

Apart from late meteoritic accretion which is trivial volumetrically, the remarkably uniform partitition coefficients for many geochemical indicators (e.g. Ba, Nb, Hf, Th, vs. REE, Taylor and Jakeš, 1974) appear to rule out significant heterogeneous accretion of the crust, and major loss of volatiles from the surface.

The bulk composition of the Moon is very difficult to estimate since surface rocks are strongly biased by crustal material. Mare basalts have often been interpreted as partial melts of the lunar mantle, but some tricky geochemical problems remain (e.g. Ringwood, 1974). Smith et al. (1970a) argued for a hybrid origin of mare basalts, and Smith and Steele (1975) presented a model for origin of mare basalts at the crust–mantle transition zone (see later). They also argued that minor ultrabasic fragments could be interpreted in terms of an Mg-rich mantle composed principally of olivine. Table 1, column 3, gives their estimate of the bulk composition of the Moon based on a crust 80 km thick of 50% plagioclase An_{93} and 50% pyroxene ($En_{63}Fs_{27}$-$Wo_{10})_{98}$ $(Al_2O_3)_2$, a mantle composed of 80% olivine $Fo_{91}Fa_9$ and 20% pyroxene $(En_{86}Fs_7Wo_7)_{99}$ $(Al_2O_3)_1$, and a hypothetical core with 5·5 wt.% Fe and 1·0 wt.% FeS (Fig. 2). If mare basalts actually derive from the mantle, the olivines and pyroxenes therein must contain more Fe (e.g. $\sim Fa_{15-20}$): this

Table 1. Estimates of chemical composition (wt.%) of Moon and Earth

	1	2	3	4	5	6	7	8
SiO_2	44·9	44·0	42·5	39·8	33·7	45·2	44·6	30·7
TiO_2	0·56	0·3	ne	0·56		0·7	0·2	0·17
Al_2O_3	24·6	8·2	2·6	11·0	26·6	3·5	3·9	3·3
Cr_2O_3	0·10	0·19	ne	0·17		0·4	0·5	0·7
FeO	6·6	10·5	5·6	3·5	2·3	8·0	8·4	4·7
MnO	0·1	0·1	ne	0·04		0·14	0·2	0·08
MgO	8·6	31·0	41·0	28·8	13·1	37·5	37·4	21·9
CaO	14·2	6·0	1·7	8·9	21·6	3·1	3·7	2·7
Na_2O	0·45	0·11	0·05	0·12	1·1	0·57	0·35	0·21
K_2O	0·075	0·012	ne	0·12		0·13	0·12	0·021
Fe			(5·5)	5·6				29·0
FeS			(1·0)	1·1				5·0

See original references for other elements, and for qualifications concerning some listed elements.

1 Lunar crust, 60 km thick, Taylor and Jakeš (1974), Taylor (1975).
2 Whole Moon, geochemical model, Taylor and Jakeš (1974).
3 Whole Moon, simple mineralogic model, Smith and Steele (1975): ne not estimated but values near TiO_2 0·2 Cr_2O_3 0·1 MnO 0·1 K_2O 0·01 would be acceptable: Fe and FeS arbitrary guesses.
4 Whole Moon, solar condensation model, Ganapathy and Anders (1974), similar composition used for phase-equilibria experiments by Hodges and Kushiro (1974): also Ni ~ 0·5.
5 Allende Ca–Al aggregate, Clarke et al. (1970).
6 Pyrolite III, Ringwood (1966a), used as Moon model by Binder (1974). Also Fe_2O_3 0·5 NiO 0·2.
7 Archean pyrolite, Green (1975). Also Fe_2O_3 0·7 NiO 0·3.
8 Whole Earth, solar condensation model, Ganapathy and Anders (1974): also Ni ~ 2.

results in the higher FeO content of the Taylor–Jakeš model (column 2).

Taylor and Jakeš (1974) developed a geochemical model of the Moon using (a) a crust based on orbital remote-sensing data, observed data for various rock types, and selection of a model of 80% anorthositic gabbro (often called highland basalt: note the common use of rock names in a *chemical* rather than a *petrographic* sense) and 20% low-K basalt, (b) an interior whose negative REE anomaly matches the positive one of the crust and whose mineralogic composition is dominated by olivine and pyroxene with properties obeying bulk physical properties and derivation of mare basalt by partial melting. Fig. 3 shows the model developed for − 4·4 Gy. Table 1, columns 1 and 2, shows predicted amounts of major and minor elements in the differentiated part of the bulk Moon and the 60 km crust. Note that this model is basically similar to the one in Figs. 1, 2 and 4, except for the greater depth of origin of the mare basalts, and the concept of a frozen primitive crust to provide high Ni and Mg in crustal rocks.

Ganapathy and Anders (1974) predicted bulk compositions for the Earth and Moon by use of model compositions developed from theoretical studies of condensation of a hypothetical solar nebula (constrained by estimates of pressure, temperature and of the bulk composition of the sun) together with inferences from meteorite texture and chemistry. They used three principal condensates: refractory, early condensate; metallic Ni,Fe; and Mg-rich silicate. During cooling, metal reacts with H_2S to give FeS and with H_2O to give FeO which enters the silicate. Remelting, probably during collisions, leads to loss of volatiles, and reversion of FeS to metal. Finally, a volatile-rich, Mg–Fe silicate component is incorporated. The resulting seven components were adjusted for the Earth and Moon using observed values for certain elements, including U, Th, Mn and K, to give the following:

	Moon (mean of models 3 and 3a)	Earth
Early condensate	0·30	0·09 mass
Metal, remelted	0·05	0·24 fraction
Metal, unremelted	—	0·07
Troilite	0·009	0·05
Silicate, remelted	0·57	0·42
Silicate, unremelted	0·07	0·11
Volatile-rich material	0·0004	0·01
$Mg/(Mg + Fe)$ atomic	0·90	0·89
Fe/Si atomic	0·24	1·26

The resulting bulk compositions are given in Table 1, columns 4 and 8.

Anderson (1973) proposed that the entire Moon is composed of differentiation products of a high-temperature condensate, such as the Ca,Al-rich aggregate from the Allende meteorite (column 5). However, Seitz and Kushiro (1974) and Hodges and Kushiro (1974) found from melting experiments that neither the Allende aggregate nor a 2 : 3 mixture of aggregate and bulk Allende meteorite was suitable since the interior of the Moon would contain considerable clinopyroxene and minor melilite, partial melting of which would yield melts too low in Si and too high in Ca to fit with known lunar rocks. Of course,

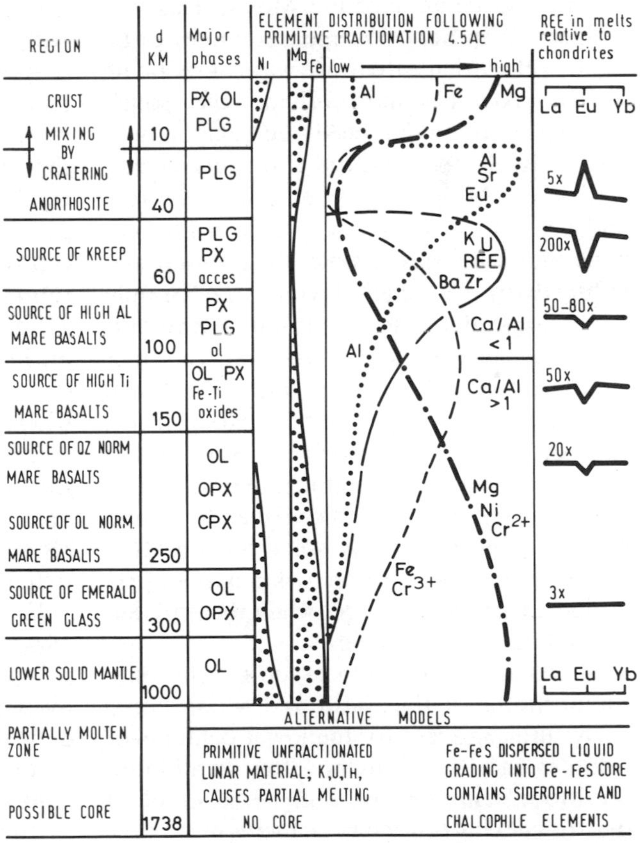

Fig. 3. Geochemical model of Moon proposed by Taylor and Jakeš (1974) for ~ −4·4 Gy. A frozen crust rich in olivine provides the high Ni and Mg and low Al of surface rocks. Melting of the underlying layer down to 1000 km leads to mineralogic zoning similar to that of Figs. 1 and 2. The frozen crust is mixed by impacts with the plagioclase-rich floated cumulate. The upper part of the lithosphere is zoned to provide successively deeper sources for the extrusive rocks ranging from basalts rich in K, REE and P to breccias rich in emerald green glass.

one cannot rule out an Allende-type composition for a hypothetical undifferentiated lunar interior lying inside a differentiated exterior of different bulk composition.

The geochemical and solar-condensation models contain about 4 times as much CaO and Al_2O_3 as the simple mineralogic interpretation of the seismic data. As the only important *observed* Ca,Al-bearing minerals at the lunar surface are anorthite and pyroxene, and the simplest interpretation of the seismic data indicates an extreme upper limit of about 60 km for plagioclase, the easiest way to increase the CaO and Al_2O_3 content is to place more Ca,Al-rich pyroxene in the mantle. However, there is very little diopside-rich pyroxene in the lunar samples, and the phase equilibrium data indicate that the average lunar pyroxene should be low in Ca. It seems difficult to increase the Ca and Al of the simple mineralogic model by more than 2-fold. In the geochemical and solar-condensation models, the Ca and Al contents are estimated from the assumed U content of the Moon. This was deduced from heat-flow measurements to be ~ 60 ppb. Perhaps there are ways of reducing the U content and perhaps the solar-condensation model needs re-examination. Alternatively, the closeness of the seismic velocities of Mg-rich pyroxene and Ca-rich feldspar may allow a rather different mineralogic interpretation of the seismic data. Subjectively, I shall favour the simple mineralogic model as a temporary expedient.

The pyrolite models for the Earth's upper mantle (columns 6 and 7) are fairly close to the mineralogic model for the bulk Moon. Presence of garnet and Ca-rich pyroxene in the Earth (as sampled by peridotite nodules) gives a ready explanation of the higher CaO and Al_2O_3 in the former. A solar-condensation model for the whole Earth (column 8) is fairly similar to a pyrolite model when recalculated to exclude the Fe and FeS.

In conclusion, one can argue for the bulk compositions of the Earth and Moon being fairly similar except for the lower contents of volatiles and Fe,Ni,S in the latter. In addition the Moon's surface has a higher content of silicate-seeking refractory elements and a very low content of metal-seeking refractory elements. These chemical features might be explained by (a) derivation of the Moon and Earth from approximately the same region of the solar nebula, (b) either more efficient collection of Fe-rich metal by the Earth than the Moon, or loss of metal by the Moon, or gain by the Earth, or all three, (c) a high-temperature event for the Moon, but not for the Earth, (d) higher efficiency of separation of metal-seeking elements into Fe-rich bodies (core?) in the Moon. Qualitatively all these features can be explained by disintegrative capture of a pre-Moon by the Earth, followed by rapid accretion of the debris in Earth orbit (Smith, 1974): but other scenarios are possible, including simultaneous accretion.

Relative Position and Orientation of Earth and Moon

The relative position of the Earth and Moon certainly affects the tides, probably affects volcanic activity and convection, and certainly modifies the relative collision rates with incoming bodies.

A liquid or extremely weak satellite in circular orbit around and rotation-locked to a much larger body is gravitationally disrupted inside the Roche distance $d \sim 2.45 \, r_c (\rho_s/\rho_c)^{1/3}$, where r_c is the radius of the central body, and ρ_s and ρ_c are the densities of the satellite and central body. For the Earth (Smith, 1974), the ratio d/r_c is 2·24 for liquid Fe metal (ρ 7·1 gm/cm^3), 2·61 for FeS (4·5), 2·88 for the mean Moon (3.34), and 3.13 for plagioclase (2.6). Satellite rotation reduces the stability whereas mechanical coherence increases it. Accretion of the Moon from a cloud around Earth would be affected by the gravitational gradient from Earth, perhaps with a tendency for an Fe-rich core to accrete before a silicate-rich mantle.

In principle, biological periodicity (Wells, 1963) can yield information on the number of solar days per lunar month. In practice, objective measurement of the periodicity is very difficult, and idiosyncratic environmental and other factors (e.g. Clark, 1974) cause problems. The subject was explored thoroughly in a conference on 'Biological Clocks and

Changes in the Earth's Rotation: Geophysical and Astronomical Consequences' at Newcastle, January, 1974: the proceedings are in press (Rosenberg and Runcorn, 1975).

Tidal energy loss is proportional to the sixth inverse power of the Earth–Moon separation so long as the mechanisms stay unchanged. Dissipation occurs mainly at liquid–solid interfaces (e.g. Wones and Shaw, 1975). Continental break-up with development of shelfs and gulfs should increase the tidal dissipation on Earth, while reduction of the amount of magma upon cooling should reduce the tidal dissipation. Extrapolation backwards on the basis of present-day regression cannot be expected to be linear. Approach of the Moon to the Roche region around the Earth should yield catastrophic heating which should severely modify the surfaces and interiors of both bodies (e.g. Alfvén and Arrhenius, 1969; Wise, 1969, Öpik, 1972; Urey and McDonald, 1971). Early extrapolation of tidal recession predicted that the Moon was at the Roche limit in the middle Precambrian, which is quite unreasonable because of the reliable evidence for stability of many surface features of the Moon since -3 to -4 Gy. The recent prediction by Turcotte et al. (1974) that the Moon approached the Earth at $-2 \cdot 85 \pm 0 \cdot 25$ Gy is effectively based on one datum for a Precambrian stromatolite, and is difficult to accept in the context of the crystallization ages for lunar surface rocks. Turcotte et al. supported their prediction by using evidence for a pulse of high-temperature volcanism on Earth at about $-2 \cdot 8$ Gy, and evidence for life beginning at approximately the same time (with quoted dates of $-3 \cdot 0$ to $-3 \cdot 3$ Gy): however, these can be explained in other ways.

It is just possible that a fortunate combination of accretion circumstances can keep the Moon near the Roche zone for a long period. It is also just possible that the Moon was captured non-catastrophically, but the probability is extremely small, and catastrophic capture is much more likely (see review by Öpik (1972)).

This whole subject is fraught with problems, both observational and theoretical, and I prefer to assume that the Moon was well outside the Roche zone during the entire period for which remnants of stable crust persist on either the Earth or Moon, or both. Thus the Moon would be well outside the Roche zone by the time of formation of the recognizable lunar basins which is between $-3 \cdot 9$ and $\sim -4 \cdot 3$ Gy depending on one's choice of the evidence. Of course, an even earlier time is permissible. Another probable conclusion is that the Moon left the Roche zone very rapidly because of the inverse sixth power of energy dissipation: only a few million years at most should be needed to recede to a distance of (say) ten Earth radii. Perhaps further study of Precambrian stromatolites (e.g. Pannella, 1972) will yield useful information on lunar periodicity and tide heights.

Nature of Mare Basins and Source of Mare Basalts

Many questions are unresolved about mare basins and the source of mare basalts. These questions are closely tied to the chemical zoning of the crust and mantle, the variation of rigidity with depth, and so on. See Taylor (1975) for key references.

The excavation depth of the incoming projectiles is not known, and estimates range from ~ 100 km down to several tens of kilometres. Unfortunately experimental data cannot be reliably extrapolated up to the huge impacts responsible for mare basins. Particularly uncertain are the effects of varying the impact velocity of the projectile and the relative mechanical strengths of the projectile and lunar crust. General background material is given by French and Short (1968) and Baldwin (1963). According to Baldwin (p. 166), increasing the size of a projectile results in a shallower crater (i.e. the depth increases more slowly than the width). The scarcity of ultrabasic material in returned lunar samples suggests that the mare basins were formed entirely in the crust, perhaps limiting the maximum depth to about 50 km. Recent estimates of the thicknesses of ejecta blankets indicate excavation depths of only a few tens of kilometres, especially when account is taken of incorporation of regolith by the ejecta cloud (e.g. Head and Stein, 1975, who suggest a depth of 9 to 20 km for Imbrium). This depth range may be too low to allow explanation of the mascons by mare basalt fill, and perhaps a range from 20 to 50 km is a more reasonable guess.

Basins on the far side (and large, late craters all over the Moon), when unfilled or only partially filled by lava have negative gravity anomalies, which is readily explainable by the unfilled hole. Only partial isostatic adjustment can have occurred, thereby implying a rather rigid crust and mantle. Near-side ringed basins have positive mass concentrations even though their surface is depressed about 3 km below the neighbouring highlands. The older irregular basins have adjusted isostatically but still have surfaces about 1 km below the neighbouring highlands. These data can be explained only if the Moon's outer portion strengthened greatly from the time of formation of the irregular basins ($\sim -4\cdot5$ to $-4\cdot2$ Gy?) to the time of formation of the ringed basins ($\sim 4\cdot2$ to $-3\cdot9$ Gy?) to the time of extrusion of the last mare basalts ($\sim -3\cdot1$ Gy, and perhaps even later).

Models involving a buried dense projectile (e.g. an iron meteorite) cannot explain the negative mascon of unfilled mare basins. Probably the projectiles contain material dominated by chondritic compositions, thereby rendering very difficult distinction from earlier lunar material.

The simplest explanation of isostatic adjustment in the irregular basins is uplifting of a mantle dome together with plutonic and volcanic activity in and around the basin.

No fully convincing explanation exists for the ringed near-side basins, and a complex model utilizing differential strength of the crust and mantle is needed. Perhaps the most popular model invokes a rigid crust which can sustain the basin and a partially molten mantle which supplies mare basalts from some hundreds of kilometres depth into the basin. This model was used by Taylor and Jakeš for their geochemical interpretation (Fig. 3). Presumably, in such a model, the basaltic liquids pass through conduits one to three hundreds of kilometres long without weakening the surrounding rock. Furthermore, the liquid must tend to move sideways towards the basin in order to yield an overall mass concentration.

Fig. 4 shows an alternate model (Smith and Steele, 1975) in which the lavas are produced from the products of the late liquids which lie

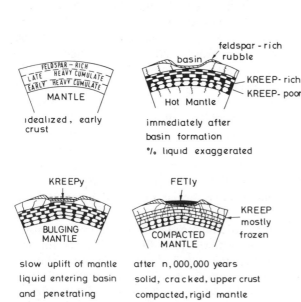

Fig. 4. Model for development of near-side mare basin with positive mascon (Smith and Steele, 1975). Upper left: idealized model of primary differentiation with cumulate and trapped liquid formed between mantle and crust. Upper right: basin and feldspar-rich rubble formed in the the feldspar-rich crust; residual liquid still present in cumulates. Lower left: the mantle is still weak and bulges upwards; liquid (especially KREEP-rich) moves into mare basin and penetrates the crust. Lower right: the mantle has become rigid, and liquid remains only in the lower crust; a few thin flows, some rich in Fe,Ti, terminate the volcanism

between a plagioclase-rich crust and an olivine-pyroxene mantle. These products consist of heavy cumulates enclosing trapped liquid whose content of late-differentiated elements (e.g. K, REE and P) increases upwards. The early mantle loses its liquid as a result of fracturing consequent upon surface impacts, and becomes compacted before the lower crust. Radioactive heating plus preferential absorption of tidal energy at solid–liquid interfaces (e.g. Wones and Shaw, 1975) counterbalances some of the radiation loss from the surface allowing liquid to persist down to about −3 Gy in the lower crust, but not in the highlands (actually very weak microseisms occur even now in the crust). Formation of a basin leads to uplift of the upper mantle which has not yet compacted. The basin begins to fill with the complex products of residual liquid, impact-generated liquid, and partially-melted cumulates. KREEP-rich (granitic) liquids are more viscous than KREEP-poor (basaltic) liquids and tend to form sub-surface bodies and rare surface extrusions. The basaltic liquids tend to reach the surface and flow over large areas. Many varieties of basalts occur, of which the Fe,Ti-rich mare basalts form the final layers in some of the basins. This model has not yet been published in detail and subjected to criticism.

Further complications are (a) the concentric rings (especially well shown by Mare Orientale): these are interpreted as the result of either a giant wave during impact or of late-stage slumping along ring faults (see Taylor, 1975 for summary and arguments in favour of the impact wave); (b) the uncertain thermal regime resulting from impact heating, redistribution of radioactive heat sources, and thermal insulation from the ejecta blanket (Arkani-Hamed, 1973); (c) the unresolved controversies over the interpretation of the phase equilibria of mare basalts (e.g. whether cotectic or near-cotectic behaviour occurs for melting at either low or high pressure) and of the minor and trace element patterns which apparently cannot be explained by a single-stage process (innumerable references in *Proc. Lunar Sci. Conf.*); the much greater extent of mare fill in the near-side than in the far-side basins (Smith et al., 1970b) suggested that crystal–liquid fractionation when the Moon was close to the Earth was affected by the differential gravitational potential resulting in a thicker crust on the far side and a greater concentration of final liquid on the near side. Differential bombardment with transfer of material from the near side to the far side (Wood, 1973) would appear to be ruled out if basins are isotropically distributed over the Moon).

Development of the Earth–Moon System from the Solar Nebula

The fission model for derivation of the Moon from the Earth faces severe dynamical and energy-source problems as does the volatilization–condensation model: however, later impacts on the Moon could have changed the dynamics. Ringwood (1975) and Binder (1975) continue to argue for these models, and a discussion is given in Taylor (1975). See Kaula (1971) and Kaula and Harris (1975) for review of dynamical aspects of lunar origin.

I prefer models involving capture and/or simultaneous accretion, since the dynamic and chemical problems appear to be less severe than for fission. In particular, during the stage of planetary accretion, planetesimals *must* interact either by direct hits or near-misses. There are several mechanisms for capturing a Moon, after which simultaneous accretion would occur for both the Earth and Moon. Non-disintegrative capture is improbable, especially if the approach velocity is high. Consequently, models which explain the refractory nature of the Moon by accretion from an orbit greatly different from that of the Earth are dynamically unlikely (see however, Anderson, 1973, and Cameron, 1973, for arguments in favour of such an origin). Furthermore, oxygen isotope data argue for a related origin for the Earth, Moon and basaltic meteorites, and a different origin for carbonaceous and chondritic meteorites, including the refractory Allende material (Clayton and Mayeda, 1975). Consequently, the following skeletal review deliberately concentrates on models which utilize capture, especially with disintegration. Some pertinent references are: Ruskol (1960, 1963, 1972, 1973) who argues for accretion of the Moon from a cloud of bodies in Earth orbit subject to both augmentation and dissipation by accreting bodies from sun-centred orbits; Öpik (1972) who argues for a near-grazing collision with accretion from a sediment ring of debris; Singer (1972) who argues for non-disintegrative tidal capture of a body accreted close to the Earth's orbit; Alfvén and Arrhenius (1972) who argue for tidal capture of the Moon which swept up the numerous satellites already in Earth orbit; Harris and Kaula (1974) who present a model for co-accretion of the Earth and Moon beginning when the Earth is only 1/10 of its final mass while the Moon stays at about 10 Earth radii during most of its growth; Kaula and Harris (1973) who argue that tidal capture is much less efficient than collision with a satellite or cloud; Wood and Mitler (1974) who argue for preferential capture of part of a differentiated planet giving chemical differentiation; Smith (1974) who listed various ways of obtaining chemical differentiation during disintegrative capture and sequential accretion.

The following framework, utilizing many ideas from Kaula (1975), allows enough tolerances to accommodate existing data on the Earth–Moon system:

I. The solar nebula condenses and coagulates by a variety of processes to produce near-radial swarms of planetesimals. A major process is sequential condensation with falling temperature resulting in early, refractory material rich in Ca,Al and Ti, intermediate separation of Mg-rich silicates, and late separation of volatile components (e.g. Grossman and Larimer, 1974). The temperature decreases away from the Sun (Lewis, 1974) and the solar wind sweeps volatiles away from the Sun, thereby producing a strong chemical trend among the planets. Equilibration during cooling is not complete, leading to only partial oxidation and sulphurization of iron-rich metal. Collisions result in fragmentation and heating. Metal is more coherent than silicate, and plagioclase shock-melts at a lower pressure (~ 400 kb) than olivine and pyroxene (~ 600 kb). These and other mechanical effects (e.g. Kaula, 1974) increase the differentiation of pre-planetary material. The larger is the body the greater is the chance of retaining collision fragments. In the inner solar system, the overall evidence favours domination by Mg-rich silicate and Fe,Ni-rich metal though substantial amounts of Ca,Al,Ti-refractory material and minor amounts of volatiles occur. By chance, a few bodies grow especially large and eat up their neighbours rather like tadpoles growing from frog spawn. Jupiter perturbs bodies in the asteroid belt and inhibits the growth of a planet there, and probably stunts Mars. Interaction of planetesimals and smaller bodies with the planets gives a mixture of direct collisions, non-capture deflections and disintegrative partial or complete capture.

II. The Moon could either develop separately from the Earth with capture after most of the accretion had taken place, or it could be captured early and accrete most of its mass in Earth orbit. Detailed discussion is given in the listed papers. For late capture, the Moon might have differentiated into a plagioclase-rich crust, Mg-silicate mantle and Fe-rich core, thereby permitting processes for chemical separation (e.g. disruption with escape of an Fe-rich core, or preferential sedimentation of Fe-rich debris to the Earth). For early capture, the chemical differentiation between Earth and Moon could be derived from the higher velocity and lower retention efficiency of the Moon when interacting with accreting material (see ideas in Ganapathy and Anders, 1974). Late capture of a fully grown Moon probably would result in a very big heat pulse in the Earth, whereas early capture would allow the Earth and Moon to have separate thermal regimes. Whatever the details, which will probably never be free of controversy, it is possible to envisage the Moon in Earth orbit by $\sim -4 \cdot 5$ Gy fairly close to but outside the Roche limit.

III. The above early stages are probably permanently inaccessible to direct observation, but the later stages are recorded in the

Moon. Using models and plausible speculations it is possible to make inferences on the bombardment history of the early crust of the Earth (see Wetherill, 1975, for review of pre-mare cratering and early solar system history). The sum total of information on the Moon is most easily explained by its being in Earth orbit by about $-4\cdot5$ Gy. The saturation cratering in the highlands may represent the end of the first stage of accretion after which few projectiles remained in Earth-like orbits: if so, the Earth was being bombarded by a similar population. Alternatively, the cratering represents the final stages of accretion by the Moon from a debris cloud in Earth orbit, but the various estimates of the time scale (tens of thousands of years) may be unfavourable.

The heavily cratered surfaces of Mercury, Mars and the Moon can be explained by spraying the entire inner region of the solar system by bodies arising outside Mars, but unfortunately the age relations for the Mercurian and Martian surface are unknown. There are various ways of deriving these projectiles including perturbation by Jupiter of asteroids and comets. The bodies could be supplied sporadically over a long period ($\sim 0\cdot5$ Gy) or in a large cataclysm. For the present purpose, the important feature is that the projectiles would be high-speed (> 10 km/sec) and would have similar impact probabilities on the Earth and Moon (of course, there are many subtle details, including effects resulting from the ~ 2 km/sec velocity of the Moon around the Earth: this velocity depends on the Earth–Moon distance). Consequently, this approach leads to saturation of the terrestrial surface by the effects of projectiles up to ~ 50 km or more across. A significant accretion might occur, perhaps resulting in a change of the surface composition of the Earth if the mantle and core had become fairly stable. Perhaps the higher content of metal- and sulphide-seeking elements in the Earth's crust than in the Moon's crust results from a higher accretion rate of late projectiles on the former (perhaps a significant part of the projectiles hitting the Moon ended up on Earth, together with part of the outer crust since the escape velocity is only ~ 2 km/sec).

IV. If the Earth were too hot to develop a crust, some of the preceding discussion is almost irrelevant to trying to understand the formation of the Earth's crust. The thermal history of the early Earth is very uncertain, but gravitational acceleration during accretion, core formation, and radioactive heating provide more than enough energy to melt the Earth completely within a very short time (e.g. Ringwood, 1966b; Wetherill, 1972; Kaula, 1975). Indeed, the formation of a thick atmosphere and a liquid, near-surface region for a period of $0\cdot n$ Gy seems highly plausible. The initial crust would be very thin and unstable. Of course, the crust must have stabilized sufficiently by $-3\cdot7$ Gy to permit preservation of fragments for which such crystallization ages have been obtained. By this time, the bombardment history of the Moon indicates that no further basin-forming projectiles were available, although smaller ones (up to 5 km across?) were available to produce craters such as Tycho and Copernicus. Perhaps the Sudbury feature records an early impact of such a projectile (Dietz, 1964; Dence, 1972). The next section describes two models by D. H. Green for the early terrestrial crust.

Impact Processes in the Early Crust of the Earth

Although the uncertainties must be borne in mind, the sum total of all the above evidence and model-building points to the Earth having been bombarded by the order of 10^3 to 10^4 large projectiles (order of 10 to 100 km radius) travelling at around 10 to 20 km/sec in the period $-4\frac{1}{2}$ to -4 Gy. Each projectile would cause severe modification of a region hundreds to even thousands of kilometres in diameter, especially as major volcanic activity and lateral movement would occur. Probably material was sprayed from the Moon on to the Earth, but it would have been rapidly lost among the reworked terrestrial crust. Cessation of the bombardment near $-3\cdot9$ Gy would allow $\sim 0\cdot2$ Gy for development of a relatively stable crust, some remnants of which are still preserved though heavily metamorphosed. Perhaps all evidence of early terrestrial basins has been lost for ever, and only speculation remains.

Green (1972), however, suggested that Archean greenstone belts may be the terrestrial equivalent of lunar maria (Fig. 5). In A, a

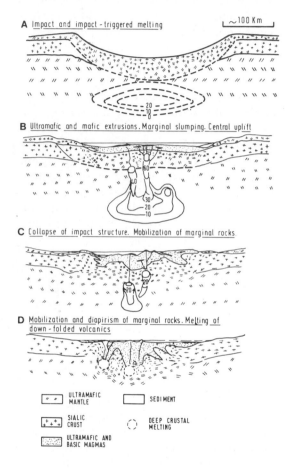

A Impact and impact-triggered melting

~100 Km

B Ultramafic and mafic extrusions. Marginal slumping. Central uplift

C Collapse of impact structure. Mobilization of marginal rocks

D Mobilization and diapirism of marginal rocks. Melting of down-folded volcanics

	ULTRAMAFIC MANTLE		SEDIMENT
	SIALIC CRUST		DEEP CRUSTAL MELTING
	ULTRAMAFIC AND BASIC MAGMAS		

Fig. 5. Model for greenstone belts as terrestrial equivalents of lunar maria (Green, 1972)

crater ~ 300 km diameter and ~ 50 km depth is underlain by the ejecta blanket and brecciated upper mantle. Pressure release causes partial melting below the crater (% shown in A, B, C). Emplacement of ultramafic and mafic magmas, uplift of the mantle and marginal slumping follow (B). Inward collapse and partial melting of the original crust (granitic?) leads to complex interleaving (C) which finally yields to melting of down-folded volcanics,

intrusion of gneiss domes, and remelting of mixed basaltic and 'basement' materials to yield andesitic and dacitic magmas. Green's model was based largely on the Barberton region in South Africa. For this model, the greenstone belts should be erratically distributed on any pre-3 Gy sialic terrain, in response to the uncorrelated supply of projectiles.

Utilizing data from crystallization studies of periodotite magma demonstrating extrusion temperatures of > 1600°C (Green et al., 1975) and exploring the implications of high geothermal gradients often postulated in the Archean, Green (1975) proposed alternatively that greenstone belts result from basaltic material scraped off a subducting peridotitic lithosphere and 'plastered' against 'granitic' nuclei. The Archean 'scraping-off' process resulted from failure of basaltic crust to react to eclogite under high geothermal gradients, and contrasts with modern subduction of basaltic oceanic crust. This model would result in greenstone belts occurring at certain margins of 'granitic' terrains, whereas the former one should result in erratic, crinkled belts *surrounded* by primeval crust. Furthermore, this model is not tied to the pre-3·9 Gy projectiles.

Perhaps both models are valuable, with the former applying to a very early stage before the crust had stabilized sufficiently to leave recognizable geologic features.

Final Comment

Discovery of extensive cratering on Mercury, Mars and probably Venus is leading to comparative studies, with speculations that all the inner planets have been bombarded by the same population of large projectiles (e.g. Wetherill, 1975). In addition, the asymmetry of the inner planets is leading to speculation of a common development of an anisotropic crust (e.g. Lowman, 1975, argues from comparison with the Moon, Mars and Mercury that on Earth the original crust was disrupted by impacts and tectonic fragmentation followed by growth of the ocean basins as the continents thickened by several processes).

Indeed Kaula (1975) reviewed the situation under the Shakespearian title *The Seven Ages of a Planet* comprising condensation, planetesimal interactions, formation, vigorous convection, plate tectonics, terminal volcanism and quiescence. Uniformitarianism and catastrophism have lost their separate identities and become transcended into an integrated philosophy for the development of the solar system. Simultaneously geology, geochemistry, geophysics and solar-system astronomy have lost their separate identities.

Acknowledgements

I thank G. R. Clark II, M. R. House, S. McKerrow and G. D. Rosenberg for supplying information on biological rhythms, and E. Anders, A. G. W. Cameron, R. Ganapathy, W. M. Kaula and G. Wetherill for supplying information on processes in the solar system. NASA is thanked for grant NGL-14-001-171 Res.

References

Alfvén, H., and Arrhenius, G., 1969. 'Two alternatives for the history of the Moon', *Science*, 165, 11–17.

Alfvén, H., and Arrhenius, G., 1972. 'Origin and evolution of the Earth–Moon system', *The Moon*, 5, 210–230.

Anderson, D. L., 1973. 'The composition and origin of the Moon', *Earth Planet. Sci. Lett.*, 18, 301–316.

Arkani-Hamed, J., 1973. 'On the formation of the lunar mascons', *Proc. Fourth Lunar Sci. Conf.*, 3, 2673–2684.

Baldwin, R. B., 1963. *The Measure of the Moon*, University of Chicago Press, Chicago.

Binder, A. B., 1974. 'On the origin of the Moon by rotational fission', *The Moon*, 11, 53–76.

Binder, A. B., 1975. 'On the petrology and structure of a gravitationally differentiated Moon of fission origin', *Lunar Science*, VI, 54–56.

Cadogan, P. H., 1974. 'Oldest and largest lunar basin?', *Nature*, 250, 315–316.

Cameron, A. G. W., 1973. 'Properties of the solar nebula and the origin of the Moon', *The Moon*, 7, 377–383.

Clark, G. R., III, 1974. 'Growth lines in invertebrate skeletons', *Ann. Rev. Earth Planet. Sci.*, 2, 77–99.

Clarke, R., Jarosewich, E., Mason, B., Nelen, J., Gómez, M., and Hyde, J. R., 1970. 'The Allende, Mexico, meteorite shower', *Smithsonian Contrib. Earth Sci.*, No. 5, 1–53.

Clayton, R. N., and Mayeda, T. K., 1975. 'Genetic relations between the Moon and meteorites', *Lunar Sci.*, VI, 155-157.

Dainty, A. M., Toksöz, M. N., Solomon, S. C., Anderson, K. R., and Goins, N. R., 1974. 'Constraints on lunar structure', *Proc. Fifth Lunar Sci. Conf.*, 3, 3091–3114.

Dence, M. R., 1972. 'Meteorite impact craters and the structure of the Sudbury Basin', *Geol. Assoc. Canada* Spec., Paper 10, 7–18.

Dietz, R. S., 1964. 'Sudbury structure as an astrobleme', *J. Geol.*, 72, 412–434.

Duba, A., and Ringwood, A. E., 1973. 'Electrical conductivity, internal temperatures and thermal evolution of the Moon', *The Moon*, 7, 356–376.

French, B. M., and Short, N. M., 1968. Eds. of *Shock Metamorphism of Natural Materials*, Mono Book Corp., Baltimore.

Ganapathy, R., and Anders, E., 1974. 'Bulk compositions of the Moon and Earth, estimated from meteorites', *Proc. Fifth Lunar Sci. Conf.*, 2, 1181–1206. See also *Lunar Science*, V, 254–256.

Green, D. H., 1972. 'Archaean greenstone belts may include terrestrial equivalents of lunar maria?', *Earth Planet. Sci. Lett.*, 15, 263–270.

Green, D. H., 1975. 'Genesis of Archean peridotitic magmas and constraints on Archean geothermal gradients and tectonics', *Geology*, 2, 15–18.

Green, D. H., Nicholls, I. A., Viljoen, M., and Viljoen, R., 1975. 'Experimental demonstration of the existence of peridotitic liquids in earliest Archean magmatism', *Geology*, 2, 11–14.

Grossman, L., and Larimer, J. W., 1974. 'Early chemical history of the solar system', *Rev. Geophys. Space Phys.*, 12, 71–101.

Harris, A. W., and Kaula, W. M., 1974. 'A coaccretional model of satellite formation', Int. Astron. Union Colloq. 28 on *Planetary Satellites*, Ithaca, New York, Aug. 18–20, 1974.

Hartmann, W. K., and Wood, C. A., 1971. 'Moon: origin and evolution of multi-ring basins', *The Moon*, 3, 3–78.

Head, J. W., and Stein, R., 1975. 'Volume of material ejected from major lunar basins: implications for the depth of excavation of lunar samples', *Lunar Science*, VI, 352–354.

Hodges, F. N., and Kushiro, I., 1974. 'Apollo 17 petrology and experimental determination of differentiation sequences in model Moon compositions', *Proc. Fifth Lunar Sci. Conf.*, 1, 505–520.

18

Jessberger, E. K., Huneke, J. C., Podosek, F. A., and Wasserburg, G. J., 1974. 'High resolution argon analysis of neutron-irradiated Apollo 16 rocks and separated minerals', *Proc. Fifth Lunar Sci. Conf.*, **2**, 1419–1449.

Kaula, W. M., 1971. 'Dynamical aspects of lunar origin', *Rev. Geophys. Space Phys.*, **9**, 217–238.

Kaula, W. M., 1974. 'Mechanical processes affecting differentiation of proto-lunar material', *Soviet-American Conf. on Cosmochemistry of the Moon and Planets*, Moscow, June 4–8, 1974. Preprint.

Kaula, W. M., 1975. 'The seven ages of a planet', *Icarus*, submitted.

Kaula, W. M., and Harris, A. W., 1973. 'Dynamically plausible hypotheses of lunar origin', *Nature*, **245**, 367–369.

Kaula, W. M., and Harris, A. W., 1975, 'Dynamics of lunar origin and orbital evolution', *Rev. Geophys. Space Phys.*, **13**, 363–371.

Kaula, W. M., Schubert, G., Lingenfelter, R. E., Sjogren, W. L., and Wollenhaupt, W. R., 1974. 'Apollo laser altimetry and inferences as to lunar structure', *Proc. Fifth Lunar Sci. Conf.*, **3**, 3049–3058.

Kirsten, K., and Horn, P., 1974. 'Chronology of the Taurus–Littrow region III: ages of mare basalts and highland breccias and some remarks about the interpretation of lunar highland rock ages', *Proc. Fifth Lunar Sci. Conf.*, **2**, 1451–1475.

Lewis, J. S., 1974. 'The temperature gradient in the solar nebula', *Science*, **186**, 440–443.

Lowman, P. D., 1975. 'Crustal evolution of the Moon, Mars and Mercury: implications for the origin of continents', *Lunar Science*, **VI**, 521–522.

McGetchin, T. R., Settle, M., and Head, J. W., 1973. 'Radial thickness variation in impact crater ejecta: implications for lunar basin deposits', *Earth Planet. Sci. Lett.*, **20**, 226–236.

Moore, H. J., Hodges, C. A., and Scott, D. H., 1974. 'Multiringed basins—illustrated by Orientale and associated features', *Proc. Fifth Lunar Sci. Conf.*, **1**, 71–100.

Morgan, J. W., Ganapathy, R., Higuchi, H., Krähenbühl, U., and Anders, A., 1974. 'Lunar basins: tentative characterization of projectiles, from meteoritic elements in Apollo 17 boulders', *Proc. Fifth Lunar Sci. Conf.*, **2**, 1703–1736.

Muehlberger, W. R., 1974. 'Structural history of southeastern Mare Serenitatis and adjacent highlands', *Proc. Fifth Lunar Sci. Conf.*, **1**, 101–110.

Nakamura, Y., Latham, G., Lammlein, D., Ewing, M., Duennebier, F., and Dorman, J., 1974. 'Deep lunar interior inferred from recent seismic data', *Geophys. Sci. Lett.*, **1**, 137–140.

Nunes, P. D., and Tatsumoto, M., 1975. 'U–Th–Pb systematics of anorthositic gabbro 78155', *Lunar Science*, **VI**, 607-609.

Nunes, P. D., Tatsumoto, M., and Unruh, D. M., 1974. 'U–Th–Ph systematics of some Apollo 17 lunar samples and implications for a lunar basin excavation chronology', *Proc. Fifth Lunar Sci. Conf.*, **2**, 1487–1514.

Öpik, E. J., 1972. 'Comments on lunar origin', *Irish Astron. J.*, **10**, 190–238.

Pannella, G., 1972. 'Precambrian stromatolites as paleontological clocks', *Proc. 24th Int. Geol. Congr. Montreal*, **1**, 50–57.

Pike, R. J., 1974. 'Ejecta from large craters on the Moon: comments on the geometric model of McGetchin et al.', *Earth Planet. Sci. Lett.*, **23**, 265–271. Discussion M. Settle, J. W. Head and T. R. McGetchin. 271–274.

Ringwood, A. E., 1966a. 'The chemical composition and origin of the Earth'', In Hurley, P. M. (Ed.), *Advances in Earth Sciences*, Cambridge, M.I.T. Press, 287–356.

Ringwood, A. E., 1966b. 'Chemical evolution of the terrestrial planets', *Geochim. Cosmochim. Acta*, **30**, 41–104.

Ringwood, A. E., 1974. 'Minor element chemistry of maria basalts', *Lunar Science*, **V**, 633–635.

Ringwood, A. E., 1975. 'Composition and origin of the Moon', *Lunar Science*, **VI**, 674–676.

Rosenberg, G. D., and Runcorn, S. K. (Eds.), 1975. *Growth Rhythms and History of the Earth's Rotation*, John Wiley, London. In Press.

Ruskol, E. L., 1960. 'The origin of the Moon I. The formation of a swarm of bodies around the Earth', *Soviet Astron.—AJ*, **4**, 657–667.

Ruskol, E. L., 1963. 'The origin of the Moon II. The growth of the Moon in the circumterrestrial swarm of satellites', *Soviet Astron.—AJ*, **7**, 221–227.

Ruskol, E. L., 1972. 'The origin of the Moon III. Some aspects of the dynamics of the circumterrestrial swarm', *Soviet Astron.—AJ*, **15**, 646–654.

Ruskol, E. L., 1973. 'On the model of the accumulation of the Moon compatible with the data on the composition and the age of lunar rocks', *The Moon*, **6**, 190–201.

Schaeffer, O. A., and Husain, L., 1973. 'Early lunar History: ages of 2–4 mm soil fragments from the lunar highlands', *Proc. Fourth Lunar Sci. Conf.*, **2**, 1847–1863.

Schaeffer, O. A., and Husain, L., 1974. 'Chronology of lunar basin formation', *Proc. Fifth Lunar Sci. Conf.*, **2**, 1541–1555.

Schwerer, F. C., Huffman, G. P., Fisher, R. M., and Nagata, T., 1974. 'Electrical conductivity of lunar surface rocks: laboratory measurements and implications for lunar interior temperatures', *Proc. Fifth. Lunar Sci. Conf.*, **3**, 2673–2687.

Seitz, M. G., and Kushiro, I., 1974. 'Melting relations of the Allende meteorite', *Science*, **183**, 954–957.

Singer, S. F., 1972. 'Origin of the Moon by tidal capture and some geophysical consequences', *The Moon*, **5**, 206–209.

Smith, J. V., 1974. 'Origin of Moon by disintegrative capture with chemical differentiation followed by sequential accretion', *Lunar Science*, **V**, 718–720.

Smith, J. V., Anderson, A. T., Newton, R. C., Olsen, E. J., Wyllie, P. J., Crewe, A. V., Isaacson, M. S., and Johnson, D., 1970a. 'Petrologic history of the Moon inferred from petrology, mineralogy and petrogenesis of Apollo 11 rocks', *Proc. Apollo 11 Lunar Sci. Conf.*, **1**, 897–925.

Smith, J. V., Anderson, A. T., Newton, R. C., Olsen, E. J., and Wyllie, P. J., 1970b. 'A petrologic model for the Moon based on petrogenesis, experimental petrology, and physical properties', *J. Geol.*, **78**, 381–405.

Smith, J. V., and Steele, I. M., 1975. 'Comprehensive petrologic model based on total melting of low-volatile ultrabasic Moon', *Lunar Science*, **VI**, 753–755 and preprint of paper to be submitted to *Geochim. Cosmochim. Acta*.

Solomon, S. C., and Toksöz, M. N., 1973. 'Internal constitution and evolution of the Moon', *Phys. Earth Planet. Inter.*, **7**, 15–38.

Sonett, C. P., and Runcorn, S. K., 1973. 'Electromagnetic evidence concerning the lunar interior and its evolution', *The Moon*, **7**, 308–334.

Steele, I. M., and Smith, J. V., 1973. 'Mineralogy and petrology of some Apollo 16 rocks and fines: general petrologic model of Moon', *Proc. Fourth Lunar Sci. Conf.*, **1**, 519–536.

Stuart-Alexander, D. E., and Howard, K. A., 1970. 'Lunar maria and circular basins: a review', *Icarus*, **12**, 440–456.

Taylor, S. R., 1975. *Lunar Science: A Post–Apollo View*, Pergamon, New York.

Taylor, S. R., and Jakeš, P., 1974. 'The geochemical evolution of the Moon', *Proc. Fifth Lunar Sci. Conf.*, **2**, 1287–1305.

Tera, F., and Wasserburg, G. J., 1974. 'U–Th–Pb systematics on lunar rocks and inferences about lunar evolution and the age of the Moon', *Proc. Fifth Lunar Sci. Conf.*, **2**, 1571–1599.

Tera, F., and Wasserburg, G. J., 1975. 'The evolution and history of mare basalts as inferred from U–Th–Pb systematics', *Lunar Science*, **VI**, 807–809.

Tera, F., Papanastassiou, D. A., and Wasserburg, G. J., 1974. 'Isotopic evidence for a terminal lunar cataclysm', *Earth Planet. Sci. Lett.*, **22**, 1–21.

Turcotte, D. L., Nordmann, J. C., and Cisne, J. L., 1974. 'Evolution of the Moon's orbit and the origin of life', *Nature*, **251**, 124–125.

Urey, H., and McDonald, G. J. F., 1971. 'Origin and history of Moon', In *Physics and Astronomy of the Moon*, Z. Kopal (Ed.), 2nd edn, Chap. 6, 213–289. Academic Press, New York.

Wells, J. W., 1963. 'Coral growth and geochronometry', *Nature*, **197**, 948–950.

Wetherill, G. W., 1972. 'The beginning of continental evolution', *Tectonophysics*, **13**, 31–45.

Wetherill, G. W., 1975. 'Pre-mare cratering and early solar system history', *Smithsonian Conf. on Cosmochemistry of Moon and Planets*, J. Pomeroy, Ed., NASA-SP?. Preprint.

Wilhelms, D. E., and McCauley, J. F., 1971. 'Geologic map of the near side of the Moon', *U.S. Geol. Surv. Misc. Geol. Inv. Map I–703*.

Wilshire, H. G., and Jackson, E. D., 1972. 'Petrology and stratigraphy of the Fra Mauro formation at the Apollo 14 site', *U.S. Geol. Surv. Prof. Paper 785*.

Wise, D. U., 1969. 'Origin of the Moon from the earth: some new mechanisms and comparisons', *J. Geophys. Res.*, **74**, 6034–6045.

Wones, D. R., and Shaw, H. R., 1975. 'Tidal dissipation: a possible heat source for mare basalt magmas', *Lunar Science*, **VI**, 878–879.

Wood, J. A., 1973. 'Bombardment as a cause of the lunar asymmetry', *The Moon*, **8**, 73–103.

Wood, J. A., and Mitler, H. E., 1974. 'Origin of the Moon by a modified capture mechanism, *or* half a loaf is better than a whole one', *Lunar Science*, **V**, 851–853.

Wood, J. A., Dickey, J. S. Jr., Marvin, U. B., and Powell, B. N., 1970. 'Lunar anorthosites and a geophysical model of the Moon', *Proc. Apollo 11 Lunar Sci. Conf.*, **1**, 965–988.

Composition of the Core and the Early Chemical History of the Earth

V. Rama Murthy

Department of Geology and Geophysics,
University of Minnesota,
Minneapolis, MN 55455, U.S.A.

Abstract

The process of core formation in the Earth must have occurred within the first 100×10^6 years under conditions that silicate liquidus temperatures were not largely obtained. From a consideration of the *relative* depletion patterns of several volatile elements in the Earth, it is shown that core formation occurred practically simultaneously with the accretion of the Earth by eutectic melting and segregation of Fe–FeS liquids into the core. The lowest melting liquids for any suitable Earth composition will be in the Fe–FeS system; it is hard not to incorporate S into the core irrespective of whether the Earth accreted hot or cold. Thus, the light element in the core required by geophysical data is considered to be sulphur.

The energy released by formation of a core of this composition will raise the temperature of the Earth $\sim 1200°C$ over its base temperature, leading to the major geochemical differentiation reflected in the U–Pb age of the Earth.

Several siderophile and lithophile elements in the outer parts of the Earth can be attributed to planetesimals accreting on the earth *after* the core segregation. Bulk composition of the Earth with Fe–Ni–S core in such a model will not correspond to any particular class of meteorites, but must reflect mixtures of condensates formed over a temperature range in the nebula.

The earliest protocrustal material generated during core–mantle differentiation will almost certainly be liquids of mafic composition, with minor enrichments of various lithophile elements. Subsequent infalling material, which on the basis of siderophile and refractory element abundances appears to be a high temperature condensate, is then efficiently mixed with this mafic protocrustal material. Partial melts produced from such compositions will still be mafic in composition, but trace element trends will be distinctly different. Sialic and/or anorthositic protocrustal material can only be occasionally produced in such a water-rich primitive mantle.

1. Introduction

The origin and chemical composition of the Earth's core is intimately related to the question of the early physical and chemical history of the Earth and its evolution through subsequent geologic time. Recent progress on the

21

geophysical data pertaining to the core, the nature of formation of solid matter in the early solar system and geochemical data on the abundances of some elements in the mantle and crust have made it possible to pursue this question within the context of some reasonable constraints. I will examine here these constraints in order to suggest a model for the origin and chemical composition of the core. Because of the large mass of the Earth's core relative to that of the planet, the process of core formation should result in a global differentiation, and presumably the formation of a proto-crust on the Earth.

Implicit in the present discussion is the idea that the core of the Earth is an *evolutionary feature*, formed as a natural consequence of the initial thermal state of the Earth at or soon after accretion. Clark et al. (1972) have extensively discussed alternative models of inhomogeneous accretion of the Earth in which the core accreted directly from an early high temperature metallic Fe–Ni condensate in the solar nebula. These models require that the planetary mass accreted rapidly, in a period of the order of 10^3–10^4 years. The physical and dynamical problems of accreting planetary masses on such a rapid time scale have never been solved satisfactorily, but the consequences of such rapid accretion, particularly the high temperatures that result in the accreted planet can be examined.

2. Constraints on Core Formation

When did the core of the Earth form? Ideas regarding the time of core formation have ranged from postulates of continuous growth through geologic time (Runcorn, 1962, 1965) to one of rapid formation in an early stage of Earth history (Stacey, 1963; Birch, 1965). The present evidence seems to support the latter point of view.

It is generally accepted that the Earth's magnetic field is produced by fluid motions in the outer core. The remanent magnetism in surficial rocks of the crust records this field. From the presence of such remanence in rocks about $3 \cdot 2 \times 10^9$ years old, Hanks and Anderson (1969) concluded that the core must have

formed in the first billion years of Earth's history and constructed various initial thermal models for the Earth. Remanent magnetism studies on the recently discovered $\sim 3 \cdot 8 \times 10^9$ years old rocks in Greenland and Minnesota (Black et al., 1971; Moorbath et al., 1972; Goldich and Hedge, 1975) now make it possible to narrow this gap between the time of accretion of the earth and the formation of the core.

A more stringent time constraint for the core is provided by a combination of isotopic studies in the U–Th–Pb system and partitioning of Pb and U between the core and the mantle. If the mantle and core are differentiated products of an accreted Earth, as we have assumed, then the estimates of the age of one will necessarily provide the age of the other. Ulrych (1967) has examined the U–Pb isotopic systematics of basalts of several ages to infer that the mantle of the Earth underwent a major differentiation at about $4 \cdot 5 \times 10^9$ years ago. This would suggest that the mantle and, therefore, the core are very early features in the Earth, possibly formed within the first 100×10^6 years of the Earth's history. Oversby and Ringwood (1971) measured the partition of Pb between silicate material and material relevant to core compositions and by relating these measured distribution coefficients to isotopic evolution of lead in the mantle reached the conclusion that the time interval between the accretion of the Earth and formation of the core is less than 100×10^6 years. Thus, it appears that the major structuring of the Earth into the core and the mantle is a very early feature in Earth history.

The initial thermal history of the Earth has been a hotly contested topic in geology for several decades. The earlier ideas of a hot molten earth have been replaced by cold Earth models championed by Urey (1952, 1957). More recently Ringwood (1966a,b) has advocated a model in which a large fraction of the gravitational energy of accretion ($\sim GM^2/R$) has been trapped in the earth. In the case of the earth, the total gravitational energy is of the order of 2×10^{39} ergs. With C_p of $1 \cdot 3 \times 10^7$ ergs g^{-1} deg, and latent heat of fusion of $4 \cdot 2 \times 10^9$ ergs g^{-1}, this energy is enough to raise the temperature of the whole Earth to $2 \cdot 5 \times 10^3$ degrees. However, the critical parameter is the time of accretion which determines the temperature of the accreting planet because a substantial fraction of this energy can be radiated away into space from the surface of the accreting planet. This parameter is not known. The immediately more useful approach appears to be to search for the evidence for a possible high temperature history of the Earth in the abundances of volatile and other trace elements in the upper parts of the Earth.

Geochemical studies of the abundances of volatile and trace elements in the upper mantle and crust reveal some important clues in this regard. Gast (1960) has conclusively demonstrated that the alkalies are deficient in the upper parts of the earth relative to chondrites. Strong depletion (Table 1 and Fig. 1) of the elements S, C, N has been shown by Murthy and Hall (1970). This observation has to be reconciled with the fact that other more volatile elements such as halogens, water and particularly the non-radiogenic rare gases are

Table 1. Volatile element abundance patterns in the Earth

Element	Abundance in crust and mantle[a]	Abundance factor I	Abundance factor II
S	605	$1 \cdot 2 \times 10^{-3}$	6×10^{-3}
C	1590	$2 \cdot 0 \times 10^{-3}$	0·15
N	230	$4 \cdot 7 \times 10^{-3}$	0·29
H (as H$_2$O)	$1 \cdot 2 \times 10^5$	$2 \cdot 2 \times 10^{-2}$	1·3
F	1240	0·34	1·13
Cl	250	0·13	6·5
Br	4	0·19	9·5
I	0·31	0·22	5·5
^{20}Ne	$2 \cdot 7 \times 10^{-4}$	0·12	4·3
^{36}Ar	$5 \cdot 5 \times 10^{-4}$	0·09	2·3
^{84}Kr	$1 \cdot 1 \times 10^{-5}$	0·10	4·6
^{132}Xe	$3 \cdot 9 \times 10^{-7}$	$3 \cdot 2 \times 10^{-3}$	0·15

[a] Data from Murthy and Hall (1970). All abundances are in atoms per 10^6 atoms Si. For hydrogen and xenon, the figures are a minimum; in the case of hydrogen, because of loss from the Earth by escape, and in the case of xenon due to trapping in Earth materials (Fanale and Cannon, 1971). Abundance factor I = (abundance in crust and mantle/abundance in carbonaceous chondrites). Abundance factor II refers to corresponding value for ordinary chondrites.

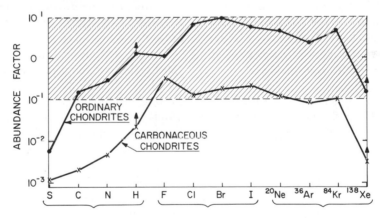

Fig. 1. Abundances of some volatile elements in the mantle of the Earth compared with carbonaceous type-I and ordinary chondrites. Upward arrows for H and Xe indicate that the abundances in each case are a minimum. The abundance factor is defined in Table 1

abundant in approximately chondritic proportions (Anders, 1968; Fanale, 1971). The initial thermal history of the Earth should be recorded in these fractionations.

Clearly, volatilization and loss after the Earth formed at high temperatures cannot be the cause, as it is difficult to conceive of a process which selectively volatilizes the alkalies, S, C and N and retains halogens, water and particularly the rare gases. This leaves us with two choices. Either these patterns are artifacts of cosmochemical fractionations of the material that formed the Earth, or they are due to terrestrial fractionations not involving the high surficial temperatures necessary for volatilization.

Recent work by Grossman (1972) and others have clearly demonstrated that solid matter condensed from the solar nebula at various temperatures. Nebular fractionations for the above volatile elements in diverse types of meteorites (Larimer and Anders, 1967; Anders, 1968) show that the abundances of volatile elements follow each other. The conclusion of Anders (1968) that meteorites are mixtures of condensates formed at various temperatures seems difficult to avoid. If small bodies such as meteorites are mixtures, the conclusion is inescapable that a planet as massive as the Earth must also be a mixture of such condensates rather than one unique type of condensate such as Ringwood (1966a) assumes. In any case, the anomalous depletion patterns of volatiles in the Earth appear to be a terrestrial phenomenon and must be due to a relatively low temperature fractionation in the Earth, since it has been argued above that volatilization from an initial hot Earth was not the cause.

A further clue to the low temperature aspect of the initial Earth is provided by the presence of ^{129}Xe produced from the decay of ^{129}I (half-life = 17×10^6 years) reported by Buolos and Manuel (1971). If this interpretation is correct, this important discovery suggests that the mantle of the Earth had never been totally at temperatures above the liquidus in the first 100×10^6 years of Earth history.

Thus, the two important constraints on the early history of the earth seem to be (1) core formation in the first 100×10^6 years, and (2) below liquidus temperatures of the Earth at accretion.

3. Composition of the Core

Under the constraints discussed above, we have argued elsewhere that the core forms by segregation of Fe–FeS liquids at temperatures substantially below the liquidus temperatures of silicates of the mantle (Murthy and Hall, 1970, 1972). Fe–FeS or analogous systems such as Fe–S–O are characterized by eutectic compositions that melt at substantially lower

temperature than any other solid material in the Earth. (Naldrett, 1969; Brett and Bell, 1969; Usselmann, 1975). It should be noted that the segregation of sulphur into the core along with some C and N will happen irrespective of whether the Earth was at high temperatures initially or formed by cold accretion, because for all reasonable Earth compositions the Fe–FeS liquid is the first liquid formed. It is obvious that any process that requires melting and segregation of iron into the core cannot avoid incorporating sulphur into the core of the Earth. The only exception would be the situation in which S was never present in significant amounts in the material that accreted to form the Earth. However, the trace-element abundance patterns in a wide variety of meteorites (Larimer and Anders, 1967) show that the abundance of S follows that of other volatiles. Thus, in the range of compositions and oxidation-reduction states represented by these bodies, solar nebular processes leading to formation of small solid bodies do not seem to result in a gross fractionation of S from other volatiles. Accordingly, from the presence of other volatiles in the Earth, we assume S was also present.

Present geophysical data on the density of the Earth's core suggest that it is about 10% less dense than Fe–Ni metal at the P, T conditions prevailing in the core regions (Birch, 1964; McQueen and Marsh, 1966). It is also known that the zero pressure high temperature sound speed of the core of 5·10 km/s (Birch, 1952; McQueen and Marsh, 1966) is higher than the corresponding Fe–Ni value of 3·1–3·7 km/s. Therefore, the core must contain about 10–20% of some light element mixed with iron. Many light elements have been considered, but the list can be narrowed down to sulphur and silicon as shown in the detailed discussions of Ringwood (1966b).

Several workers have advocated the presence of silicon in the core, most notably Ringwood (1966a,b). Incorporation of silicon into the core requires a large scale high temperature reduction of the Earth and a disequilibrium segregation of Fe–Si liquid into the core with the production of a massive CO and CO_2

atmosphere, equal to about 1/2 of the Earth's mass in the primitive Earth, which Ringwood argues has been 'blown off' early. The *only suitable* starting composition for the condensates that accreted to form the Earth in Ringwood's model will have to resemble the carbonaceous chondrites in order to satisfy simultaneously the requirements of the reduction process, the proper FeO/FeO + MgO ratios of the mantle and the core : mantle ratio of 31 : 69. It is difficult to imagine a high temperature 'blow off' process which loses S, C, and N preferentially by more than an order of magnitude with respect to rare gases. For example, Table 1 shows that the abundance factors for S and C are of the order of 10^{-3}, whereas those of rare gases are only ~ 0.1, assuming for the moment that the initial composition resembled the carbonaceous chondrites.

In view of these difficulties and the constraints on core formation discussed earlier, I consider sulphur to be the dominant light element in the core.

The composition of the Earth based on a core containing about 10–15% of S and C and a 'pyrolite' (Green and Ringwood, 1963) type of mantle will not correspond to any single class of meteorites (Murthy and Hall, 1972). Appropriate initial compositions can only be represented by a mixture of materials condensed over a wide range of temperatures in the solar nebula. Murthy and Hall (1972) suggested that a suitable mixture of nebular condensates for the Earth in terms of material sampled as meteorites may be about 40% carbonaceous chondrites, 45% ordinary chondrites and 15% of iron meteorites. Purely from geophysical considerations an identical mix has been suggested by Anderson et al. (1972). The compositions of the core and the mantle in such a model of the Earth are shown in Table 2.

An important aspect of core formation with S as the light element is the possibility of segregating substantial quantities of K and other alkali elements into the core, suggested from thermodynamic considerations and evidence from meteorites and metallurgical data (Lewis, 1971; Hall and Murthy, 1971) and by experimental data on the partitioning of K

Table 2. Geochemical model for the Earth

	Ordinary Chondrites	Carbonaceous Chondrites	Iron Meteorites	Mixture[a]
Major Silicates				
SiO_2	51·6	47·2		49·6
MgO	32·3	31·8		32·1
FeO	16·1	21·0		18·3
Total	100·00	100·00		100·00
Metallic Components[b]				
Fe	81·2	63·8	91·5	80·8
S	11·5	36·2	1·3	14·0
Ni	7·3	0·0	7·2	5·2
Total	100·00	100·00	100·00	100·00

[a] Mixture represents 45 : 40 : 15 of ordinary, carbonaceous type I, and iron meteorites (Murthy and Hall, 1972).
[b] Fe and FeS are recomputed as Fe and S. All compositions are reconstituted to represent only major components.

between sulphide melts and K-bearing Fe, Mg silicates (Goettel, 1972).

4. A two-stage History of the Outer Regions of the Earth

A model for the early physical and chemical history of the Earth is presented here based on the following considerations:

1. Core formation occurred in the first 100×10^6 years of the accretion of the Earth under conditions in which iron-rich liquids segregated to the centre of the Earth under equilibrium conditions with the solid silicate mantle.

2. The composition of the core is largely (Fe, Ni) and S with minor amounts of other elements.

3. In the outer few hundred kilometres of the Earth, siderophile elements such as Ni, Co, Cu, Au and Re are present in abundances largely in excess of values expected in an equilibrium segregation of the iron rich core.

4. The crust and upper mantle are also strongly enriched in several lithophile refractory elements, such as Ba, U, Th, and the rare earth elements and depleted in alkalis.

It appears that a self-consistent early history of the Earth can be modelled if it is assumed that core–mantle segregation in an initial Earth was followed by further accretion of some material on to the outer regions of the Earth, and the bulk composition of this late stage material differed radically from the initial accretion material.

Items 1 and 2 have been discussed in great detail in Murthy and Hall (1970, 1972).

The abundances of several strongly siderophile elements in the upper mantle of pyrolitic composition are shown in Table 3; the values are normalized to 10^6 atoms of silicon and corresponding values of cosmic abundances are also shown in this table. Several interesting features emerge from these data. As pointed out by Ringwood (1966a), although several siderophile elements are depleted in the mantle, as would be necessary in any iron-rich core formation, the abundances are still too high for an equilibrium separation of core from the mantle. These excess abundances of several non-volatile siderophile elements in the upper mantle have been used as strong evidence of disequilibrium segregation of the core from the mantle (Ringwood, 1966a).

From recent studies of siderophile elements in lunar surface materials, Morgan et al. (1972) and Ganapathy et al. (1973) observed that the excess siderophile elements in brecciated lunar highland rocks represent the last stages of accretion of metal-bearing planetesimals on the Moon after its initial differentiation. Intense planetesimal flux after the formation of the Moon is well documented (Tera et al., 1974). A similar situation is extremely likely for the Earth. Following Öpik's (1971) model of accretion of planets by sweeping up dust and planetesimals in heliocentric orbits, Ganapathy and Anders (1974) show that for low relative velocities the target radius is proportional to R^2, where R is the radius of the planet. The estimated accretion rate per unit area for the Earth would be a factor of 22 higher than for the Moon (Ganapathy et al., 1970). On the Moon, the siderophile-mixed layer comprises some fraction of the lunar lithosphere of 65 km, and hence we can expect

that in the Earth such accretional material can be mixed to a depth of a few hundred kilometres (Kimura et al., 1974).

From the nature of distribution coefficients for Ni between metal and silicates (Chou et al., 1973) for equilibrium segregation of the core the Ni content of the mantle should correspond to 50–250 ppm. The upper mantle Ni content is 1–2 orders of magnitude greater than this (Ringwood, 1966). From this difference we can estimate the contamination of the upper mantle regions during the terminal stages of accretion that would be necessary. The Ni content of high temperature solar system metal condensates is 5–12 mole % (Grossman, 1972). Most iron meteorites and metal fractions in various chondritic meteorites also encompass this range. Assuming about 7–8 wt.% Ni in such material, the percentage of metal component needed to account for the observed Ni content of the mantle, would be about 2–3% by weight of metal fraction. Such an addition will result in corresponding increases in the other siderophile elements such as Co, Cu, Au and other noble metals. If these siderophile element abundances in the outer parts of the Earth were largely due to addition of late stage component, we should expect that ratios of Ni/Co, Ni/Cu and Ni/Au will be similar to cosmic ratios. Data for the noble metals for the mantle are scanty and Cu appears to be both siderophile and lithophilic in the meteoric

Table 3. Abundances of some nonvolatile siderophile elements in the mantle[a]

Element	Cosmic abundance	U. Mantle abundance
Ni	$4 \cdot 57 \times 10^4$	5040
Co	2300	265
Cu	919	91
Au	0·19	0·02
Wt. Ratios		
Ni/Co	20	20
Ni/Cu	46	55

[a] All abundances normalized to 10^6 atoms of Silicon. Cosmic abundance data from Cameron (1968). Mantle abundances calculated from a mixture 3 parts ultramafic rocks to 1 part basalt from the data of Turekian and Wedepohl (1961) and are somewhat similar to pyrolite abundances of Ringwood (1966a).

component. In contrast, Co strongly follows Ni (Moore et al., 1969). Thus we expect to find nearly cosmic Co/Ni and a ratio of Cu/Ni greater than cosmic abundance. In Table 3 are shown the weight ratio of these elements in the upper mantle and, for comparison, the corresponding ratios of cosmic abundances. The agreement is quite good and supports the present conclusion that a large fraction of the siderophile elements Ni, Co, Cu, Au, Re, etc., in the outer regions of the Earth are due to the postulated terminal accretion of metal.

The concept of a late stage (post-core formation) accretion of metal in the Earth is of some interest in yet another fashion. The fact that the outer parts of the Earth are enriched in several refractory elements has been conclusively demonstrated by Gast (1960). The condensation data for compounds from the solar nebula, such as by Grossman (1972) and others, and the detailed studies of the meteorites and the Moon by Anders and coworkers (Larimer and Anders, 1970; Anders, 1971; Grossman and Larimer, 1974) have stressed the fact that planetary accretion incorporates several types of condensates produced in the solar nebula. A highly detailed set of calculations for the Earth and Moon are given in Ganapathy and Anders (1974). From these studies, it has become obvious that the Earth contains a significant proportion of high temperature refractory condensates of the solar nebula, enriched in Ca, Al and an accompanying complement of refractory lithophile and siderophile trace elements. What is debated is whether these high temperature condensates accreted early or late in the planet's accumulation history (e.g., Gast, 1968; Clark et al., 1972; Anderson and Hanks, 1972).

I suggest that for the Earth the geochemical traits of the crust and upper mantle noted by Gast (1960, 1968) and the accretion of metal at a late stage as noted above indicate that the lithophile refractory condensates were also added, after the core–mantle differentiation in the Earth. Refractory lithophile condensates and metal are the first two high temperature fractions in the solar nebula, and it would be surprising if the earth accreted only one but not the other.

Detailed geochemical calculations and consequences of the present model will be presented elsewhere (Murthy, 1975, in preparation) but for the present it appears that a late-stage accretion of 10–20% by weight of refractory lithophile condensates are sufficient to account satisfactorily for the trace element geochemistry and isotopic evolution of Sr in the Earth's crust and upper mantle.

5. Implications for Early Crustal Evolution

In the model presented above, two discrete stages have been suggested for the evolution of the outer regions of the Earth. The first stage will be dominated by the energetics of the core formation process. When the temperatures in the accreting Earth reach the eutectic melting point of Fe–FeS, the liquid formed will readily sink down because of its high density relative to the silicates. Once the process is initiated, the gravitational energy liberated will further increase the temperature and the rate of sinking of the liquid. Thus, the process of core formation will practically be catastrophic.

The energy released by core formation by iron sinking has been discussed by Urey (1952), Birch (1965) and Flasar and Birch (1973). The energy available for heating has been estimated at about 400 cal/g. In the present model, because of the density difference between Fe-metal and Fe–FeS compositions, the energy released would be approximately 250 cal/g, other factors remaining the same. This is sufficient to cause a temperature increase of $\sim 1200°C$ of the entire Earth over its base temperature, and will lead to a major geochemical differentiation of the entire mantle.

The age of the Earth measured by Patterson (1956) by considerations of the evolution of Pb-isotopes in the Earth on a large scale has been identified with this differentiation (Shimazu, 1965; Ringwood, 1966a). This major thermal event will certainly have resulted in the generation of liquids to produce early protocrustal material, and degas the Earth of a major fraction of its volatiles such as H_2O and rare gases.

For the model that the light element in the core is S, appropriate mixtures of solid matter for a bulk Earth composition are needed. Upon segregation of the core, the primitive mantle composition will be ultramafic. The major oxide compositions are shown in Table 2. Partial melts from this mantle under the circumstances will be dominantly mafic to ultramafic depending upon the degree of melting. The high temperature and high thermal gradients at this period will in general indicate that large fractional melting should have occurred. Under these conditions, the proto-crustal material will nearly resemble the then mantle compositions for the major oxides MgO, FeO, SiO_2. The strongly incompatible elements such as Ba, U, Th, REE, etc., will be only moderately enriched. For example, if the overall behaviour of these elements can be generalized by a distribution coefficient range of $D^{1/s} = 50–100$, then the liquid produced in large scale partial melting of the mantle, say 40–50%, will only show incompatible element enrichments twice or three times the original source.

Thus it appears that the earliest protocrustal material, generated at the time of core formation will necessarily have to be mafic to ultramafic in composition and show about 2–3 times enrichments for the incompatible lithophile trace elements and a strong depletion of alkali elements. It should be emphasized that this would be the case for all models of bulk Earth composition in which the core contains FeS simply because of the dominance of Mg, Fe, Si and O in composition of the Earth and the large thermal energy release during core formation that results in a global differentiation.

The second stage commences with the accretion of Ca, Al and refractory lithophile–enriched material and some metallic Fe–Ni planetesimals. During this stage of accretion, it is doubtful whether any of the materials produced as early crustal material during core formation will be preserved. From the influx of material observed on the Moon after its initial differentiation during the period between 4·6–3·9 × 10^9 years ago, Ganapathy et al. (1970) estimate that this second stage accretion on

the Earth should be mixed in the outer several hundred kilometres of the Earth. However, the outer regions at this time will be highly oxidizing, due to the global degassing of water and other volatiles that occurred during the core-formation period. The siderophile elements brought in through the late-stage accretion will be oxidized and depending on their substitutional properties in the major phases of the mantle, will be incorporated into the silicate lattices or exist as dispersed phases. It can be expected that Ni, Co and Cu will enter silicates while the highly noble elements such as Au and Re will be mixed as dispersed phases due to the absence of any free metal. Thus, the outer regions of the Earth at this stage will be enriched in siderophile and refractory lithophile elements such as Ca, Al, Ba, U, REE and others. Subsequent magmatism in these regions will generate melts that are grossly different in their trace element characteristics and to some extent in the major oxides, particularly CaO and Al_2O_3.

The protocrustal materials generated at this time are more likely to be preserved in the geologic column than the material produced during core formation process. In particular, the high water content and the high temperatures in the mantle may be conducive to the production of anorthositic liquids as shown by Yoder (1968) from experimental work on the anorthite–diopside eutectic at high water pressure and on the anorthite–forsterite–water systems. It is doubtful, however, that the Earth ever had an 'anorthositic' outer layer as does the Moon, because the pressure gradients in the Earth are steeper than in the Moon, and on Earth anorthositic liquids intruded and crystallized at depths > 35 km will undergo subsolidus reactions that will tend to destroy plagioclase-rich assemblages to pyroxene-dominated assemblages. Early crustal anorthosites on the Earth thus depend on emplacement to shallow depths of anorthositic liquids produced in a water-rich upper mantle—either tectonically or due to collisions by planetesimals inducing production of melts and quick eruption to low pressure regimes.

The composition of the protocrustal material from this time must remain speculative but, on the basis of the postulated late stage accretion, it is to be expected that magmatism in the materials must result in melts rich in Ca and Fe among the major elements. It is interesting to find in this context the excess calcium in platform sediments which are products of weathering of old cratonic platforms (Ronov and Yaroshevsky, 1969) and the higher Fe/Mg in early Archean basalts (Glikson, 1971).

The subsequent evolution of the Earth will be governed by a complicated thermal history dominated by U–Th heat production in the outer several hundred kilometres and heat production by ^{40}K in the core which will aid in convective motions in the outer core and the lower mantle. A number of other interesting geophysical and geochemical consequences of the above model will be discussed elsewhere.

Acknowledgements

I am grateful to Professors E. C. Alexander, Clement G. Chase and George Shaw for the benefit of many discussions and suggestions. This work was supported by Grant GA 14950 of the National Science Foundation.

References

Anders, E., 1968. 'Chemical processes in the early solar system, as inferred from meteorites', *Acc. Chem. Res.*, **1**, 289–298.

Anders, E., 1971. 'Meteorites and the early solar system', *Ann. Rev. Astron. Astrophys.*, **9**, *1–34*.

Anderson, D. L., and Hanks, T. C., 1972. 'Formation of the Earth's core', *Nature*, **237**, 387–388.

Anderson, D. L., Sammis, C., and Jordan, T., 1972. 'Composition of the mantle and the core', In *The Nature of the Solid Earth*, E. C. Robertson (Ed.), McGraw-Hill, New York, pp. 41–66.

Birch, F., 1952. 'Elasticity and the composition of the Earth's interior', *J. Geophys. Res.*, **57**, 227–286.

Birch, F., 1964. 'Density and composition of mantle and core', *J. Geophys. Res.*, **69**, 4377–4388.

Birch, F., 1965. 'Energetics of core formation', *J. Geophys. Res.*, **70**, 6217–6221.

Black, L. P., Gale, N. H., Moorbath, S., Pankhurst, R. J., and McGregor, V. R., 1971. 'Isotopic dating of very early Precambrian amphibolite facies gneisses from the Godthaab District, West Greenland', *Earth Planet. Sci. Lett.*, **12**, 245–259.

Brett, R., and Bell, P. M., 1969. 'Melting relations in the Fe-rich portion of the Fe-FeS system at 30 kb pressure', *Earth Planet. Sci. Lett.*, **6**, 479–482.

Buolos, M. S., and Manuel, O. K., 1971. 'The xenon record of extinct radioactivities in the Earth', *Science* **174**, 1334–1336.

Cameron, A. G. W., 1968. 'A new table of abundances of the elements in the solar system', In *Origin and Distribution of the Elements*, L. H. Ahrens (Ed.), Pergamon Press, New York, pp. 125–143.

Chou, C.-L., Baedecker, P. A., and Wasson, J. T., 1973. 'Distribution of Ni, Ga, Ge, and Ir between metal and silicate portions of H-Group chondrites', *Geochim. Cosmochim. Acta*, **37**, 2151–2171.

Clark, S. P., Turekian, K. K., and Grossman, L., 1972. 'Model for the early history of the Earth', In *The Nature of the Solid Earth*, E. C. Robertson (Ed.), McGraw-Hill, New York, pp. 3–18.

Fanale, F. P., 1971. 'A case for catastrophic early degassing of the Earth, *Chem. Geol.*, **8**, 79–105.

Fanale, F. A., and Cannon, W. A., 1971. :Physical adsorption of rare gases on terrigeneous sediments', *Earth Planet. Sci. Lett.*, **11**, 362–368.

Flasar, F. M., and Birch, F., 1973. 'Energetics of core formation: A correction', *J. Geophys. Res.*, **78**, 6101–6103.

Ganapathy, R., and Anders, E., 1974. 'Bulk compositions of the Moon and Earth, estimated from meteorites', *Geochim. Cosmochim. Acta*, Suppl. 5, 1181–1206. Pergamon Press, New York.

Ganapathy, R., Keays, R. R., Laul, J. C., and Anders, E., 1970. 'Trace elements in Apollo-11 lunar rocks: Implications for meteorite influx and origin of the Moon', *Geochim. Cosmochim. Acta*, Suppl. 1, 1117–1142. Pergamon Press. New York.

Ganapathy, R., Morgan, J. W., Krähenbühl, V., and Anders, E., 1973. 'Early intense bombardment of the Moon: Clues from meteoritic elements in Apollo 15 and 16 samples', *Geochim. Cosmochim. Acta*, Suppl. 4, 1239–1261. Pergamon Press, New York.

Gast, P. W., 1960. 'Limitations on the composition of the upper mantle', *J. Geophys. Res.*, **65**, 1287–1297.

Gast, P. W., 1968. 'Upper mantle chemistry and evolution of the Earth's crust, In *The History of the Earth's Crust*, R. A. Phinney (Ed.), Princeton University Press, Princeton, N.J., pp. 15–27.

Glikson, A. Y., 1971. 'Primitive Archean element distribution patterns: Chemical evidence and geotectonic significance', *Earth Planet. Sci. Lett.*, **12**, 309–320.

Goettel, K. A., 1972. 'Partitioning of potassium between silicates and sulfide melts; experiments relevant to Earth's core', *Phys. Earth Planet. Interiors,* **6**, 161–166.

Goldich, S. S., and Hedge, C. E., 1975. '3800-Myr granitic gneiss in south-western Minnesota', *Nature,* **252**, 467–468.

Green, D. H., and Ringwood, A. E., 1963. 'Mineral assemblages in a model mantle composition', *J. Geophys. Res.,* **68**, 937–946.

Grossman, L., 1972. 'Condensation in the primitive solar nebula', *Geochim. Cosmochim. Acta,* **36**, 597–619.

Grossman, L., and Larimer, J. W., 1974. 'Early chemical history of the solar system', *Rev. Geophys. Space Phys.,* **12**, 71–101.

Hall, H. T., and Murthy, V. Rama, 1971. 'The early chemical history of the earth; some critical elemental fractionations', *Earth Planet. Sci. Lett.,* **11**, 239–244.

Hanks, T., and Anderson, D. L., 1969. 'The early thermal history of the Earth', *Phys. Earth Planet. Interiors,* **2**, 19–29.

Kimura, K., Lewis, R. S., and Anders, E., 1974. 'Distribution of gold and rhenium between nickel-iron and silicate melts: Implications for the abundance of siderophile elements on the Earth and the Moon', *Geochim. Cosmochim. Acta,* **38**, 683–701.

Larimer, J. W., and Anders, E., 1967. 'Chemical fractionations in meteorites-II. Abundance patterns and their interpretation', *Geochim. Cosmochim. Acta,* **31**, 1239–1270.

Larimer, J. W., and Anders, E., 1970. 'Chemical fractionations in meteorites-III. Major element fractionations in chondrites', *Geochim. Cosmochim. Acta,* **34**, 367–387.

Lewis, J. S., 1971. 'Consequences of the presence of sulfur in the core of the Earth', *Earth Planet. Sci. Lett.,* **11**, 130–134.

McQueen, R. G., and Marsh, S. P., 1966. 'Shockwave compression of iron-nickel alloys and the Earth's core', *J. Geophys. Res.,* **71**, 1751–1756.

Moorbath, S., O'Nions, R. K., Pankhurst, R. J., Gale, N. H., and McGregor, V. R., 1972. 'Further Rubidium-Strontium age determinations on the very old Precambrian rocks of the Godthaab District, West Greenland', *Nature. Phys. Sci.,* **240**, 78–82.

Moore, C. B., Lewis, C. F., and Nava, D., 1969. 'Superior analyses of iron meteorites', In *Meteorite Research*, P. M. Millman (Ed.), Reidel Publishing Co, Dordrecht, Holland, pp. 738–748.

Morgan, J. W., Laul, J. C., Krähenbühl, U., Ganapathy, R., and Anders, E., 1972 'Major impacts on the Moon: Characterization from trace elements in Apollo 12 and 14 samples', *Geochim. Cosmochim. Acta,* Suppl. 3, pp. 1377–1395. M.I.T. Press, Cambridge, Mass.

Murthy, V. Rama, and Hall, H. T., 1970. 'The chemical composition of the Earth's core: Possibility of sulphur in the core', *Phys. Earth Planet. Interiors,* **2**, 276–282.

Murthy, V. Rama, and Hall, H. T., 1972. 'The origin and composition of the Earth's core', *Phys. Earth Planet. Interiors,* **6**, 123–130.

Naldrett, A. J., 1969. 'A portion of the system Fe–S–O between 900° and 1080°C with its application to sulfide ore magmas', *J. Petrol.,* **10**, 171–201.

Öpik, E. J., 1971. 'Catering on the Moon's Surface', In *Advances in Astronomy and Astrophysics,* Z. Kopal (Ed.), Academic Press, New York, Vol. 8, pp. 107–137.

Oversby, V. M., and Ringwood, A. E., 1971. 'Time of formation of the Earth's core', *Nature,* **234**, 463–465.

Patterson, C., 1956. 'Age of meteorites and the Earth', *Geochim. Cosmochim. Acta,* **10**, 230–235.

Ringwood, A. E., 1966a. 'The chemical composition and the origin of the Earth', In *Advances in Earth Science,* P. M. Hurley (Ed.), M.I.T. Press, Cambridge, Mass., pp. 287–356.

Ringwood, A. E., 1966b. 'Chemical evolution of the terrestrial planets', *Geochim. Cosmochim, Acta,* **30**, 41–104.

Ronov, A. B., and Yaroshevsky, A. A., 1969. 'Chemical composition of the Earth's crust', In *The Earth's Crust and Upper Mantle,* Geophys. Monograph B. P. J. Hart (Ed.), Amer. Geophys. Union, pp. 37–57.

Runcorn, S. K., 1962. 'Convection currents in the Earth's mantle', *Nature,* **195**, 1248–1249.

Runcorn, S. K., 1965. 'Changes in the convection pattern of the Earth's mantle and continental drift; evidence for a cold origin of the earth', In *A Symposium on Continental Drift,* The Royal Society, London, pp. 228–251.

Shimazu, Y., 1965. 'Some aspects of formation processes of planets in relation to their internal constitution', *J. Earth Sci. Nagoya Univ.,* **13**, 59–87.

Stacey, F. D., 1963. 'A new mechanism for convection in the mantle and continental accretion', *Nature,* **197**, 582–583.

Tera, F., Papanastassiou, D. A., and Wasserburg, G. J., 1974. 'Isotopic evidence for a terminal lunar cataclysm', *Earth Planet. Sci. Lett.,* **22**, 1–21.

Turekian, K. K., and Wedepohl, K. H., 1961. 'Distribution of elements in some major units of the Earth's crust', *Geol. Soc. Amer. Bull.,* **72**, 175–192.

Ulrych, T. J., 1967. 'Oceanic basalt leads: A new interpretation and an independent age for the Earth', *Science,* **158**, 252–256.

Urey, H. C., 1952. 'The terrestrial planets', In *The Planets,* Yale University Press, New Haven, Conn., pp. 58–113.

Urey, H. C., 1957. 'Boundary conditions for theories of origin of the Solar System', In *Progress in Physics and Chemistry of the Earth*, L. H. Ahrens, F. Press, K. Rankama, and S. K. Runcorn (Eds.), Pergamon Press, New York, Vol. 2, pp. 46–76.

Usselman, T. J., 1975. 'Experimental approach to the state of the core. Part I. The liquidus relations of the Fe-rich portion of the Fe–Ni–S system from 30–100 kb', *Amer. J. Sci.*, **257**, 278–290.

Yoder, H. S., 1968. 'Experimental studies bearing on the origin of anorthosite', In *N.Y. State Museum and Science Service, Mem. 18*, Isachsen (Ed.), pp. 13–22.

Development of the Early Continental Crust

Part 2: Prearchean, Protoarchean and Later Eras

DENIS M. SHAW

Department of Geology,
McMaster University, Hamilton, Ontario

Abstract

Gravitational energy of accretion and core separation must have produced near-total melting of the outer Earth, in order to (a) expel most of the primordial rare gases, (b) transfer U, Th, REE from early condensates to the surface zone.

Core separation and ensuing cooling must have been completed in 600 to 700 m.y. (Earth age 4550 m.y.) in order for sedimentary rocks dated at 3700–3900 m.y. (Greenland, Minnesota) to develop and be preserved.

The *Prearchean* era comprised the major Earth differentiation when granitophile elements accumulated near the surface: the era terminated in the development of a thin world-wide crust of anorthosite overlying granite and the first condensation of hot acid reducing rains, causing rapid rock breakdown to oxide–carbonate–sulphide–soluble salt assemblages.

Mantle convection aided rupture of crustal flakes along thermal contraction buckles. Anatexis of basic crust, weathering products and subcrustal granite produced light silicic magmas, which reinforced the upper crust. Local thickening localized continental crust nuclei; local thinning localized ocean plate nuclei. Archean greenstone belts acted as weld-sutures, building wider and thicker continental masses, complemented by wider and deeper ocean basins, leading to present-day plate tectonics.

Three models of early sialic crust are developed. The first considers the protocrust (C) to be similar in chemistry to the present-day continental crust. If this is true, then the Sr isotropic data suggest that 2/3 of the continental crust must have been recycled through the upper mantle, so as to exchange radiogenic ^{87}Sr with mantle ^{86}Sr. This amounts to $1·2$ km^3 yr^{-1}, which agrees fairly well with estimates of crustal growth rate at present-day spreading plate boundaries.

The second model proposes C to have the composition inferred from partition coefficient modelling: its composition would be more tonalitic and would necessitate addition of K, Rb, etc., from, and return of Na, Ni, Cr, etc., to the mantle over time, the bulk mass resting constant. Evidence of secular changes in composition, however, is conflicting.

The third model is similar to the second, except that the crustal mass has increased by addition of Si, Al, K, etc., over time. It is not yet possible to substantiate unequivocally any of these models.

Introduction—Constraints on Earth Origin

This section constitutes a review and updating of the discussion in Part 1, emphasizing first the most critical facts.

(a) Volatile elements

The Earth has a nearly complete absence of nonradiogenic RG (rare gases) relative to the stars. Recent studies nevertheless claim that primordial ^{3}He has been detected in mantle rocks (Tolstikhin et al., 1974). In principle the atmospheric stock of such gases could be attributed to solar wind irradiation, but Fanale (1971) shows this source to have been quantitatively trivial.

Other volatile elements such as H, C, N, O, S, F, Cl are relatively more abundant in Earth than the RG. Also Mars has atmospheric CO_2 and some H_2O, Venus has much CO_2 and perhaps H_2SO_4 (Vinogradov et al., 1972) and Jupiter is largely gaseous.

Earth's content of C, H is about what to expect if these were dissolved as CO_2, H_2O in a molten crust (Garrels and Mackenzie, 1971, p. 289).

(b) Radio-isotopes

The U–Th–Pb, Rb–Sr and K–A radiometric systems suggest a common origin for the Earth, Moon and meteorites at about -4500 m.y. Study of extinct radioactivities (e.g. ^{129}I) suggest that nucleosynthesis took place not much earlier (10^7–10^8 yr). Geological history has been documented to -3700 to -3900 m.y. The oldest rocks known to possess paleomagnetism (indicating existence of the core) are > 2700 m.y. old: I am confident that this will soon be extended back by another b.y.

Although meteorite impacts still occur, the period -4600 to -3900 m.y. was one of vastly more frequent arrivals of very large objects at the lunar surface. Mars, Mercury and possibly Venus (Campbell et al., 1970; Goldstein and Rumsey, 1970) are similarly cratered, presumably at the same time: Earth must also have been subject to the same impacting flux (if Earth were in the solar system), which must have been the concluding phase of the period of planetary accretion; it had decreased markedly by -3300 m.y.

(c) Heat

It has not been established whether Earth is heating or cooling, but the emitted surface flux is $\sim 10^4$ less than the solar radiation received. However, Jupiter emits more heat than it receives. Terrestrial heat flux was greater in the early Precambrian by reason of the higher ^{235}U abundance.

Sources of heat liberation in the early Earth include gravitational (accretional) collapse, core separation, solar and lunar tides, continuing radioactivity (U, Th, K), extinct radioactivity (^{26}Al,

etc.). Correct assessment of the relative importance of these is critical to understanding Earth.

Carbonaceous chondrites have never been subjected to heating above 200–300°C (perhaps much less, if they contain ice) and constitute the cosmic cold reservoir of volatile elements.

(d) Accretion

Recent cosmochemical developments have led to widespread acceptance of sequential, temperature-controlled condensation of chemical elements as stable compounds, from nebular gas. Ensuing dust planetesimals coalesced into larger objects (size very controversial) adsorbing low-melting gases (He, Ar, N_2, etc.).

Critical to Earth origin is the conclusion that U, REE condensed early, at high temperature. Thus Wänke et al. (1974) found these elements to be enriched 13 to 20 × over carbonaceous chondrites, in a high-temperature white inclusion in the Allende meteorite: these elements are now, however, in the terrestrial crust and upper mantle.

Many of these facts and others have led to two currently popular models of Earth origin. The first or *homogeneous accretion* theory maintains that nebular condensation preceded accretion. The mechanics of accretion are controversial but led to the establishment of aggregates large enough for gravitational self-collapse. The consequent heating led in turn to metal separation, which sank rapidly to form a core, releasing more heat. Outer parts of Earth were degassed of compounds not sufficiently refractory to remain held by gravitation in the hot protoatmosphere or by solution in molten silicate. Modifications consider fast heating, i.e. times of 10^3–10^5 yr, or slow heating, i.e. times $> 10^8$ yr. Constraints on this choice are:

(i) *Most* but not *all* of the RG must be released;
(ii) The primary Earth differentiation into core, mantle (including crust), atmosphere must take place;
(iii) U, Th, REE must be largely transferred to the surface region;
(iv) Time of core formation by paleomagnetism and necessity for long cooling afterwards.

Hanks and Anderson (1969) have carefully analyzed the possibilities of this theory.

The alternative *inhomogeneous accretion* theory (Turekian and Clark, 1969) was devised to explain some of the more difficult problems with homogeneous accretion, particularly the apparent necessity that a period of early heat liberation would have totally degassed the Earth. In this theory accretion accompanies nebular condensation, the sequence roughly being high-temperature silicates, metallic Fe–Ni, low-temperature silicates, volatile-rich residue. Although this theory is attractive, it is difficult to see why accretion of early planetesimals to a body with a substantial gravitational field would not liberate accretional heat as in the previous model: more troublesome is the necessity to move early condensing U, Th, REE from the core to the surface without liquid (i.e. heating).

The Hanks and Anderson model of homogeneous accretion has been accepted here.

While the metallic core was separating, the rudimentary hot atmosphere consisted largely of H_2, H_2O, N_2 with minor amounts of CO (or CH_4) and S_2 (or H_2S) as shown by Holland (1962). Abundant H_2 would delay crustal cooling by permitting a rate of solar heating higher by as much as 100°C (Sagan and Mullen, 1972): this would be counteracted by CO_2 or NH_3 which are opaque to IR radiation, and by the 60% lower luminosity of the Sun at this time (earlier point on the main sequence curve).

In Part I (Shaw, 1972a) evidence was presented which suggested that

(i) During and after initial accretion, the Earth passed through a period of heating, accompanying core separation, when the mantle was molten;

(ii) The mantle crystallized progressively upwards during convective and radiative cooling, concentrating lithophile elements such as U, Th, K, Rb, REE into a near-surface layer of basaltic composition;

(iii) Consolidation of the residue of the mantle yielded a world-wide proto-crust, the upper part consisting of basic rock, and the lower part forming a 14 km layer, similar to the present-day continental crust;

(iv) The proto-crust contained much of the heat-producing elements, U, Th, K, the remainder being in the upper mantle.

This period, ending with formation of the first solid crust, may be termed the Prearchean era: it is convenient to call the ensuing period the Protoarchean era, until such time that research has positively identified primary crustal rocks.

The present article is concerned with the later evolution of the primary crust, in particular the development of the present pattern of thick (40 km) continental plates, separated by ocean basins. First a qualitative model will be presented, followed by quantitative appraisals.

Later Prearchean Era

The basaltic residual magma overlaid a more-or-less solid peridotite mantle, and was cooling by radiation, conduction and convection. According to Holland (1962), basic magma at 1250°C would be in equilibrium with a gas phase containing $PO_2 \cong 10^{-7}$ bars and consisting mainly of N_2, H_2O, CO_2, Ar with minor CH_4, NH_3, SO_2, CO, He, Ne. Garrels and Mackenzie (1971) argue that much gas could be dissolved in the magma, up to 300 bars H_2O, 45 bars CO_2, 10 bars HCl (at 600°C). These gases were slowly released to form the protoatmosphere, as magma solidification took place.

Rapid radiative cooling would slow down as surface freezing begins. Poorly conducting solids, chiefly olivine, pyroxene and plagioclase separated out and, whereas the first two would sink (carrying down some plagioclase) the lower density of plagioclase would cause it mostly to float. In this way would develop an anorthositic scum, later to become a more rigid crust, overlying liquid magma which in turn overlaid a basic gabbroic cumulate (Fig. 1(a)). The thin crust accumulated in irregular patches, covering eventually the whole Earth—there being no reason for it to behave otherwise. Contraction led to wrinkling, like the scum on cooling boiled milk, roughly parallel to sheet edges.

Below this thin crust, the residual magma solidified within a few kilometres into a granitic layer with a mass of $1 \cdot 89 \times 10^{25}$ g. If this layer covered the whole Earth's surface its thickness would have been 13–14 km (Fig. 1(b)). Since the oceans did not yet exist, it is probable that it did in fact underlie the surface.

As mentioned previously, a high meteorite impact flux was sustained during this Prearchean era. The effect of large impacts was to remelt surface and underlying rocks (Fig. 1(b)), just as on the Moon, helping promote the density inversion of the anorthositic crust for more differentiated rocks below. It should be borne in mind that Earth and Moon must have been much closer at this period (Goldreich, 1972) and the lunar bombardment can not have missed the Earth (see Green, 1972).

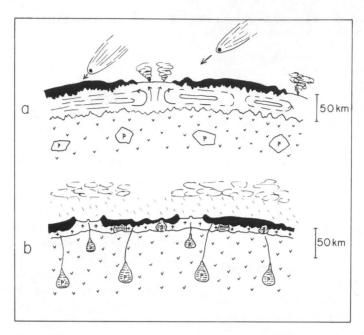

Fig. 1(a). Prearchean era, towards the close. A few kilometres of anorthositic (solid black) accumulate from the early unstable crust, floating in a convecting and degassing residual basalt liquid. Olivine and pyroxenes sink, with some plagioclase, to form gabbroic solid below (v). This passes downwards into eclogite and contains large blocks or lenses of peridotite (p) carried up from the lower mantle. Impacting meteorites arrive frequently at the surface. (b) Protoarchean era. The residual basalt has fractionated and solidified to yield 10–15 km of granite below the anorthositic crust. The surface is pitted with large impact craters which fused and brecciated the crust, allowing the rise of the lighter granite from beneath the anorthosite. Hot acid rains are falling, but the surface is still too hot for water to accumulate and the vapours are weathering the surface rocks. Convective heat rising from below, coupled with pressure release along deep contraction fractures, yield magma reservoirs by melting (horizontal lines) of peridotite (p), gabbro (b) and granite (g)

Although most of the outer Earth is now solid, local hot spots would localize magma pools, ranging in composition from ultramafic to granitic.

Protoarchean Era

As cooling progressed, the initial solid scum underwent successive periods of remelting, volcanism, crystallization and annealing (Fig. 2), building up continental-type crustal nuclei, under a H_2O–CO_2-rich hot atmosphere: oceans were not yet present. The initial crustal configuration of heavier anorthositic overlying lighter silicic material was unstable, and throughout the subsequent reworking these rock-types plus basalt extrusions would tend to invert the stratification. The process would, however, be irregular and haphazard, resulting in a quartz–diorite mixture with abundant large and small blocks of partially assimilated feldspathic amphibolite.

The crust would, however, be unstable, as a result both of tidal and contractional (cooling) stresses. Fracture systems therefore developed and many small polygonal continental nuclei ('ice floes') were bordered by tectonically active zones, which localized heat losses

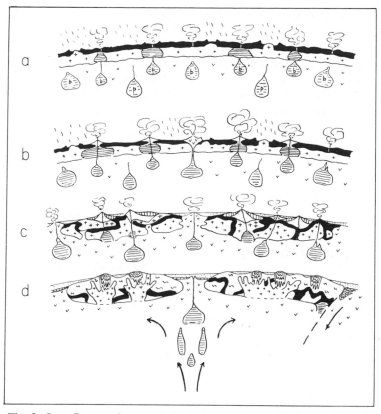

Fig. 2. Late Protoarchean and Archean eras (same symbols as Fig. 1). (a) Cooling contraction fractures the surface, permitting escape of gases and initiating volcanism: the watery proto-atmosphere accumulates and rain falls. (b) Silicic and basaltic volcanism develops, tending to overturn the gravitationally unstable layering so that lighter granitic rock overlies heavier anorthosite. Volcanogenic sediments accumulate on the flanks of volcanoes. (c) The surface regions of the crust are now a mixture of volcanic edifices (unmarked), sediments (▦) and intrusives. Where granitic material predominates, the crust thickens by further addition of remelted granite at the base of the slab, by isostatic compensation. Continental-type crust begins to develop higher surface elevations than adjacent basaltic crust. Water has condensed and covers wide areas (stippled). Continued volcanism begins to build up future greenstone-belt volcanosedimentary piles, including basaltic and ultramafic (komatiite) components, localized by fractures at fairly regularly-spaced intervals. (d) At a later stage the greenstone-belt rocks have been folded and intruded by younger anatectic granites: continental crust older rocks are shown by ($\sim$). The continental slabs are now more coherent and separated by basaltic crust. Water drains off the continents, creating deeper-water seas above the basaltic crust and helping initiate sea-floor spreading above a convection cell. The leading-edge of the right-hand continental nucleus is developing a subduction zone by overriding an oceanic mini-plate

from the interior by vulcanism. Diameters of the nuclei were of the order of 2–3 times their thickness (Fig. 2(c)). The closely spaced distribution of Archean fold-belts has been related by Clifford (1969) to the buckling wavelength of a thin crust under compression (Fig. 3), the points of buckling marking preferred foci of volcanic belt development.

Sub-parallel sets of polygon edges (fractures) in the crust became the sites of linear volcanic centre activity (Fig. 2(a)). Some fractures tapped reservoirs of still unconsolidated silicic magma at shallow depths, others penetrated to deeper pockets of basaltic and ultramafic magma. Contemporaneous silicic ably deeper surface fracture penetration to develop conduits: such deep conduits would be difficult to sustain in the hotter upper mantle of the Protoarchean era.

Partial melting of this kind would lead to a continually more silicic and therefore less-dense crust. Accumulation of silicic volcanic

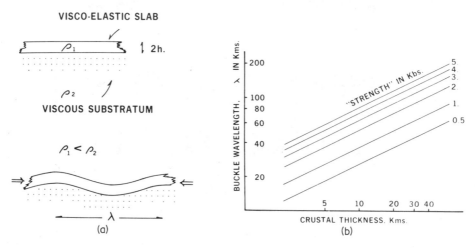

Fig. 3. Buckle wavelength as a function of strength and crustal thickness. Equating the mean greenstone-belt spacing of 53 km in the Superior province to λ in (a), the crustal thickness for strength of 1 to 5 kb in (b) is 4 to 20 km. (Clifford, 1969, Fig. 8)

to ultramafic volcanism thus occurred (Fig. 2(b),(c)). Regeneration of magma was facilitated by upper mantle radioactive heating consequent upon upward accumulation of U, Th, K, during the primary differentiation (see Part 1, Fig. 4). The nature of the magmas produced is a function of P, T, P_{H_2O}, rock composition and the extent of melting. The rocks available would range from peridotite to granite, and extensive melting would produce surface volcanics such as komatiite, tholeiitic basalt, dacite-rhyolite. Small degrees of partial melting of peridotite and gabbro (or eclogite) could produce andesite, dacite, rhyodacite, provided that (a) sufficient H_2O still remained in the upper mantle, and (b) that the melting took place at depths greater than 50 km (Green and Ringwood, 1966; Kushiro, 1970, 1974; Lambert and Wyllie, 1970, 1972). Extensive andesitic magma would be unlikely to form, in the absence of consider-

piles at the surface would tend to displace the anorthosite and basalt downward and would tend to increase the topographic elevation of silicic parts of the crust, relative to areas of mainly basaltic volcanism, by isostatic compensation (Fig. 2(c)). The uniform proto-crust began to separate into elevated, thicker silicic segments separated by lower, thinner, more basic regions.

It is difficult to visualize the character of sediments produced before extensive condensation of water. Hot acid rains would produce an alteration similar to present-day tropical weathering, but with iron remaining unoxidized. Minerals such as silica, alumina, gibbsite, calcite, magnesite, fluorite, siderite, ilmenite, magnetite, pyrite and salts of the alkali metals with halogens, sulphur, nitrogen and boron would be expected. Garrels and Mackenzie (1971) estimate that (at 600°C) at this time the partial pressures of H_2O, CO_2 and

HCl could have been 300, 45 and 10 bars, respectively, condensing to give water containing 1 mole/HCl and 0·5 mole/CO_2: they remark (p. 290) that 'the reactivity of this ocean-atmosphere system would be awesome'.

When surface temperatures dropped to about 100°C, condensation of the primitive oceans would begin near the poles. Surface conditions began to be dominated by solar radiation, and the meteorological-gradational cycles began. Assuming that differences in elevation were small, water would accumulate in shallow depressions, leaching soluble halides, sulphates, carbonates, nitrates and borates, and rapidly becoming saline. Solar ultraviolet radiation penetrated to the Earth's surface in the absence of oxygen (Berkner and Marshall, 1964) and photo-dissociation of water produced an ozone-enriched zone which promoted oxidative weathering, aided by hot-water temperatures. This point must have been reached about 3900 m.y. ago. The possibility of surface temperatures falling below the freezing-point of brine has been proposed by Sagan and Mullen (1972), arising from the reduced luminosity of the Sun: this would be counteracted by the greenhouse effect of abundant CO_2.

During the Protoarchean, continued condensation increased the seawater volume to the present-day values of $1·42 \times 10^{24}$ cm^3: the world-wide thin crust (surface-area $5·101 \times 10^{18}$ cm^2) was gradually inundated, to a mean depth of 2·8 km. For a while the Earth's surface must have been one vast ocean marked here and there by emergent volcanic piles. It would be wrong, however, to equate *either* the volcanoes to modern island-arcs, *or* the subaqueous thin crust to the modern oceanic crust.

Clastic sediments accumulated around volcanic complexes as wholly altered assemblages of the insoluble minerals listed above together with carbonate–halide–sulphate precipitates from seawater (Fig. 2(c) and Table 1). As such sediments become engulfed in later volcanics they slowly descended towards the lower crust, where partial melting led to silicic magma (silica–alumina–alkalis) leaving more refractory Fe–Ca–Mg compounds probably as amphibolite. Thus this process too would tend to increase the silicic nature of the upper crust as time went on.

The acid-leaching regime must have rapidly led to establishment of more neutral waters and consequently less exotic sediments. The ensuing sedimentation regime, however, differed markedly from the present day. The absence of land-masses, highly elevated mountain chains and deep ocean basins prevented development of wide drainage areas complemented by extensive continental shelves. Instead, there were isolated but numerous volcanic complexes, surrounded by narrow sedimentation basins grading out into shallow seas. The ability for sediments to be transported long distances was very restricted, and we can visualize immense marine basins with very little clastic input, adjacent to marginal basins filling rapidly with coarse clastic debris, all subject to vastly higher tides than now because of the proximity of Moon and Earth.

Archean Geological Conditions

In recent years, Archean geology has been intensively re-examined, contrasting both similarities and differences with younger terranes. Emphasis has been given to greenstone belts, some of whose features indicate profound differences with younger mountain-chains and classical (Hall–Kay) geosynclinal concepts.

The salient features of Archean fold-belts presented by Baragar and Goodwin (1969), Goodwin (1968), Clifford (1969), Anhaeusser (1973), Anhaeusser et al. (1969), Hart et al. (1970), Jakeš and White (1971) are as follows:

(a) Elongated, narrow linear character, occurring as sub-parallel sets or swarms across large Precambrian shield areas;
(b) Dominance of basic volcanic rocks and poorly sorted clastic sediments;
(c) Sequential extrusion of ultramafic, basic, intermediate and silicic lavas, this sequence often repeated twice or more;

Table 1. Effects of primary ocean condensation on surface rocks

PROTOATMOSPHERE
N_2 H_2O CO_2 HCl H_2O

Condensation
Leaching

Sialic Crust

Simatic Crust

Residual (Sialic)
SiO_2
Al_2O_3

Solution (Sialic)
Na^+ K^+
Cl^- CO_3^{-2} S^{-2}

Residual (Simatic)
Al_2O_3
SiO_2
Fe, Ti oxides

Solution (Simatic)
Ca^{+2} Mg^{+2} Fe^{+2}
Cl^- CO_3^{-2} S^{-2}

Submergence

Clays
Chert

Clays
Greenalite
Siderite
Dolomite
Chert

(d) K, Rb, Cs, Sr, Ba abundances intermediate between calc-alkalic and oceanic tholeiitic, resembling present-day island arc low-K volcanics;

(e) Strong deformation within many belts;

(f) Low metamorphic grade (greenschist facies) except in contact aureoles of later intrusive plutons;

(g) Tectonically unstable troughs of accumulation, indicating a thin crust (4–21 km, Clifford);

(h) General absence of the voluminous carbonate sediments associated with younger geosynclinal belts;

(i) Uncertain age relations with adjacent sialic cratons.

The low metamorphic grade of most greenstone belts precludes both deep burial and high topographic elevations (from isostatic considerations). Archean mountains could not have been very high. Petrological support for a thin Archean crust has also been given by Hart et al. (1970).

There is no direct evidence of the original nature of the cratonic areas between greenstone belts.

Anhaeusser (1973) carefully examined the evidence from African Archean areas and concluded that the earliest ultramafic and mafic igneous series can nowhere be demonstrated to overlie sialic crust. He interprets these ultramafic–mafic sequences as being remnants of a thin (5–7 km) oceanic crust, which extended over wide areas and was underlain by mantle materials: at this stage of the Earth's evolution no granitic crust existed, but it was slowly built up by island arc volcanism and associated granitic intrusives. The basic character of the first thin crust is also suggested by the enrichment of Ni and Cr in the Fig Tree argillite—one of the oldest South African clastic series (Danchin, 1967).

Anhaeusser's argument, however, gives little consideration to the question of whether the ocean basins existed in Archean times, and consequently whether sea-floor spreading took place. The close-spacing of sub-parallel greenstone belts in many shield areas, and the restricted dimensions of each belt makes a direct comparison with present-day island arcs difficult to maintain, since the latter always occur singly and extend for hundreds of miles, near the contact of an oceanic and continental plate. The present interpretation, whereby greenstone belts are localized at the edges of small thin continental plates or slabs accords better with observation. A corollary is the absence of the oceanic/continental plate dichotomy of the present day.

By contrast Condie (1967) and others have pointed out that among the earliest sedimentary units are greywackes with overall similarity in composition to the present-day continental crust, suggesting a sialic provenance.

These views are all compatible with the argument expressed in the preceding section, where early in the Earth's history a partial overturning of the drowned protocrust began, leaving some areas with mafic surface rock and others with more silicic material. The composition of the rock exposed in any region would strongly influence the sediments derived by erosion.

If it be accepted that widespread seas existed (but ocean basins did not) in early Archean times, then evidence of this should be expected in the sedimentological and biological record. The prevalence of siliceous iron formations (at least 3700 m.y. ago in Greenland: D. Bridgwater, verbal comm.) favoured by deep-acid leaching of continental rocks (Sakamoto, 1950; Moore and Maynard, 1929), and widespread organic activity (Cloud, 1965) supports this view.

The implicit objection in Rubey's (1955) conclusion that the oceans have grown throughout geological time is now less serious than when first published, since plate tectonics theory has provided compelling evidence that much continental and ocean crustal material is recycled through the upper mantle. Present-day volcanic gases are largely recycled surface water, and the degassing required by Rubey's analysis took place during Prearchean times.

If the early seas were richer than now in salts leached from the first rocks, notable enrichments (adsorption) of heavy metals would be expected in early sediments. The abundance of economic deposits of Au, Cu, Zn, Pb in Archean belts resulted from remobilization of such source-rocks. Cameron and Jonasson (1972) suggest that these conditions persisted throughout Archean times, with the main period of Earth out-gassing coming at the end

of the Archean, since the Hg abundance in carbonaceous shales increases from 129 to 513 ppb, between Archean and Proterozoic times, thereafter falling to 42 ppb in the Palaeozoic. Under the present hypothesis, these results would express the slow recycling of early crustal deposits.

Later Crustal Evolution

The process of continental plate thickening, brought about by sialic volcanism, deep partial melting and greenstone-belt suturing, led to increased continental freeboard and consequent draining of water off the highlands. Isostatic balance would favour the accumulation of water over basic rock, leading to development of incipient ocean basins (Fig. 2(d)). These basins were initially small, little more than wide troughs, localized where chance forces drew apart a pair of continental plates.

Assuming as in Part 1, that slow mantle convection continued after the primary earth crystallization, then the top of a mantle convection cell localized a zone of rifting and ocean-crust formation. Spreading then created an ocean basin and continental plates began to behave more coherently, developing the present polarity of craton and ocean plate. Erosional differentiation, accompanied by refusion of sediment and basement along subduction zones, increased continually the continental thickness at the expense of surface-area, building larger ocean basins, The dimensions of plate-margin mobile-belts increased, and their plurality decreased, leading to the marked tectonic style contrast with Archean greenstone belts. The upper limit to this process would be set by the rapid erosion rates of high mountain chains. The analysis of sediment evolution by Li (1972) following Gregor (1970: *in* Garrels and Mackenzie, 1971) suggests that 95% of the sediment mass (and volatiles) were at the surface after the first 1000 m.y. Earth history.

The time of change-over, i.e. the time of oceanic crust development, may well have varied from one area to another. However, there appears much evidence that the early

Table 2. Summary of geochemical data and interpretations from Part 1, Tables 6 and 9

	Initial concentration in whole system c_0	Basalt above 280 km depth — Concentration c_b	Basalt above 280 km depth — % of whole	Residual mantle below	Continental crust fraction — Concentration range in model calculations	Continental crust fraction — Concentration estimate Part 1 Table 1 c_c
K %	0·016	0·104	97·5	0·00047	0·19–3·1	2·3
Rb ppm	0·59	3·84	97·6	0·016	7·4–120	85
Sr ppm	27	154	85·5	4·6	420–4200	450
U 10^{-8} g/g	2·7	(18)*	(97·5)*	0·079		145
Th 10^{-8} g/g	7·8	(51)*	(97·5)*	0·18		600
Pb ppm	0·13	(0·85)*	(97·5)*	0·0038		17·5
La ppm	2·2	15	100	—	190–460	42
Ce ppm	5·4	32	88·5	0·71	120–1100	81
Eu ppm	0·5	1·6 (2·4)^a	48·0(72·0)^a	—	2·1–60	1·5
Yb ppm	0·42	0·11(1·7)^a	3·9(60·7)^a	—	0–34	4·0
Ba ppm	5·5	34·7	94·5	0·35	39–1050	1100
Tl ppm	4·1^b	(27)*	(97·5)*	0·12		600
Ni ppm	1490	194	2·0	1720	0–180	20
Rb/Sr	0·022	0·025		0·0035		0·189
K/Rb	271	278		294	246–258	271
Mass $\div 10^{25}$ g	409·6	61·4		348·2		1·888
Mass fraction	1	0·15		0·85		0·0046

^{a} Spinel lherzolite mantle assemblage, rather than garnet lherzolite.
* Interpolated: see text.
^{b} The Tl abundance in the mantle given in Part 1, Table 1 was over-estimated and should be 1·3 ppb, giving a whole-earth abundance of 4·1 ppb.

Proterozoic depositional troughs were much larger than in the Archean, and developed higher grade metamorphic belts. In the Ontario–Michigan region, the Huronian orogen (folded about 2100 m.y. ago) is the first such 'modern' mountain-belt.

Presumably the mechanical thickness of continental plates has increased over time. Earlier discussion suggested the Protoarchean crust to be 10–20 km thick both in its mechanical and chemical behaviour: i.e. the M-discontinuity lay at the base of the plate. Nowadays the M-discontinuity in a continental plate lies at 40–50 km depth, but the mechanical base of the lithospheric plate is much deeper—100–150 km. This difference probably reflects the waning of the period of high radioactive heat generation following on (a)

accumulation of U, Th, K near the surface, and (b) depletion of ^{235}U.

Composition of the Primary Crust

A summary of the geochemical interpretations from Part 1 is presented in Table 2*. These assume fractional crystallization (and gravitational stratification) of the mantle-crust system, producing first a 280 km surface layer of basaltic composition after 85% solidification. The basaltic layer continued to crystallize, yielding a 0.46% (mass) fraction of continental crust composition near the surface, leaving beneath a depleted basaltic layer amounting to 14·54% of the total mass. Model calculations could

* A similar approach to trace element fractionation models had been made by Prof. A. Masuda, who kindly reminded me that I had (inadvertently) omitted reference to his work (e.g. Geochim. Cosmochim. Acta, **30**, 239–250, 1966; Chem. Geol., **1**, 155–163, 1966).

not be made for some elements (U, Th, Pb) because of a lack of partition coefficients: order-of-magnitude values are given in Table 2 in parentheses, based on similarities in behaviour to K.

Values in column 3 of Table 2 differ slightly for some elements from data in Table 6, Part 1, arising from rounding errors. An arithmetic error occurs on p. 1592, Part 1 where it is stated that '... solidification ... into continental crust would still leave 67% of the U, Th, K ... in the subcontinental mantle': these estimates should actually be 75, 65, 34% for U, Th, K (and corresponding figures for other elements).

The residual material from the basaltic layer after extraction of continental crust material would consist of gabbro (above) and eclogite (below) and, together with peridotite, constituted the upper mantle.

Two hypotheses concerning the sialic part of the primary crust (i.e. ignoring the surface anorthosite cumulate) are presented in Table 3. The first, model R, is that the primary crust had the composition of the present-day continental crust c_c, leaving a residual basaltic layer of composition c_{cr} beneath: this model stipulates no substantial growth of continental mass throughout geological time, and may be called the *recycling model* (see next section).

The second hypothesis, model S, is that the primary sialic crust had a composition c_s determined by fractional crystallization, underlain by a residual basaltic composition c_{sr}. Among the varying fractionation mechanisms discussed in Part 1, the most likely is *continuous equilibration* (because of the long cooling period with high probability of an approach to equilibrium) during crystallization to a garnet lherzolite upper mantle assemblage, so the appropriate concentrations c_s in Table 3 were taken from Table 9, Part 1.

Table 3. Two hypotheses for primary crustal composition

	Model R—primary crust had composition of present continental crust c_c			Model S—primary crust had most probable fractionation model composition c_s		
	c_c	Element % in crust	Residual basaltic layer composition c_{cr}	c_s	Element % in crust	Residual basaltic layer composition c_{sr}
K %	2·3	66·1	0·0346	0·69	19·8	0·0855
Rb ppm	85	66·3	1·27	28	21·8	3·08
Sr ppm	450	7·7	145	470	8·0	144
U 10^{-8} g/g	145	24·7	13·5	(116)*	(19·8)*	14·9
Th 10^{-8} g/g	600	35·4	33·3	(335)*	(19·8)*	42·0
Pb ppm	17·5	61·9	0·318	(5·6)*	(19·8)*	0·70
La ppm	42	8·8	13·8	220	46·0	8·51
Ce ppm	81	6·9	30·3	120	10·2	29·2
Eu ppm	1·5	1·4	1·6(2·4)[a]	2·1(4·3)[a]	1·9	1·58(2·4)[a]
Yb ppm	4·0	4·4	− (1·6)[a]	0·1(2·2)[a]	< 10^{-2}	0·11(1·8)[a]
Ba ppm	1100	92	0·99	190	15·9	29·8
Tl ppb	600	67·3	8·87	(176)*	(19·8)*	22·2
Ni ppm	20	< 10^{-4}	205	180	< 10^{-3}	194
Rb/Sr	0·189		0·009	0·060		0·021
K/Rb	271		272	246		278
K/Ba	21		349	36		29
Ba/Rb	13		0·78	6·80		9·7
Ba/Sr	2·4		0·007	0·40		0·21
Mass fraction	0·0046		0·1454	0·0046		0·1454

[a] Spinel lherzolite mantle assemblage, rather than garnet lherzolite.
* Interpolated: see text.

Since model S accepts an early crust having lower granitophile element abundances (K, Rb, etc.) than now, this model then requires a progressive increase of the granitic components of the crust. But, as set out in Table 3, this must be accomplished by return of gabbrophile elements to the mantle (e.g. Ni, Mg, Fe, Ca), with no net increase in crustal mass, which remains at 0·0046 of the mantle–crust system. This appears reasonable, if the primary equilibrium crystallization of the mantle is accepted as being an efficient process. So although this may be called the *secular growth model* it also involves a crust-mantle reciprocal exchange.

Among major elements the chief difference between the two primary crust models would be in alkalis. The expected values would be:

	Na$_2$O	K$_2$O	
model R	3·2%	2·8%	(Pt. 1, Table 7, col. C and Table 3)
model S	4·5	0·83	

A third hypothesis follows as a corollary of the first two, according to each of which the crustal mass should be a constant. This constant mass is the quantity of residual sialic melt resulting from closed-system equilibrium crystallization of the ultramafic-to-mafic mantle. After this initial equilibration and separation took place, the residual mantle may again be treated as a closed system. A temperature gradient (convection) can lead to partial anatexis and scavenging of residual granitophile materials which, by gravity, could rise to the crust. The third hypothesis (model M) is thus one of slight growth of the mass of the continental crust and progressive depletion of parts of the mantle in crustal elements.

These models are consonant with the earlier discussion. They are summarized in Table 4 and Fig. 4 and will be examined in the next sections.

Isotopic Evidence

A test of the continental evolution models described above and the Rb and Sr abundances calculated in Part 1 (see Table 2) is provided by Sr isotope ratios. Assuming that samples which average out local isotopic irregularities and represent large rock volumes can be obtained, then ^{87}Sr/^{86}Sr ratios should be confined to the area on Fig. 5 between the limiting growth-lines (see Table 5) for continental crust and mantle residue.

This is seen to be the case from the other data shown on Fig. 5, which includes initial ratios for modern ocean water and carbonate sediments (Faure and Powell, 1972), crustal rocks and rocks believed to come from the upper mantle: also included are growth-lines

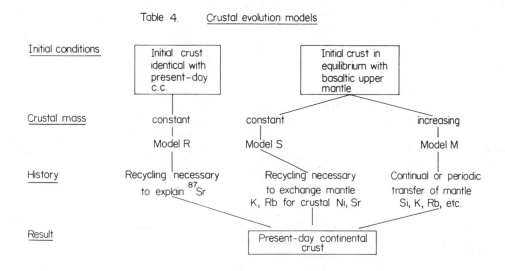

Table 4. Crustal evolution models

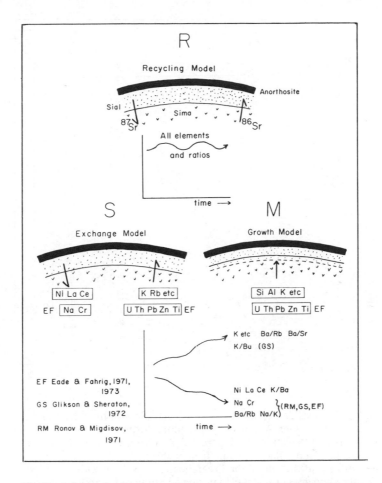

Fig. 4. Schematic diagrams of three crustal models (Table 4) and their expected chemical evolution over time. The trends indicated arise out of this paper except where indicated otherwise (see text)

Table 5. Data for Sr isotope growth lines

Material	Concentration (ppm)			Present* $^{87}Sr/^{86}Sr$	Present† $^{87}Sr/^{86}Sr$
	Rb	Sr	Rb/Sr		
Total mantle + c_c	0·59	27	0·022	0·7031	
15% basalt	3·84	154	0·025	0·7037	0·7036
Model R { crust c_c	85	450	0·189	0·7346	0·7307
Model R { upper mantle c_{cr}	1·27	145	0·009	0·7007	0·7010
Model S { crust c_s	28	470	0·060	0·7103	0·7094
Model S { upper mantle c_{sr}	3·08	144	0·021	0·7030	0·7030

* $(^{87}Sr/^{86}Sr)_0 = 0.6990$. $\lambda_{Rb} = 1.39 \times 10^{-11} \, yr^{-1}$.
† Assuming 0.55×10^9 yr evolution of mantle-crust system to give $^{87}Sr/^{86}Sr = 0.6995$, followed by 4.0×10^9 yr separate evolution.

46

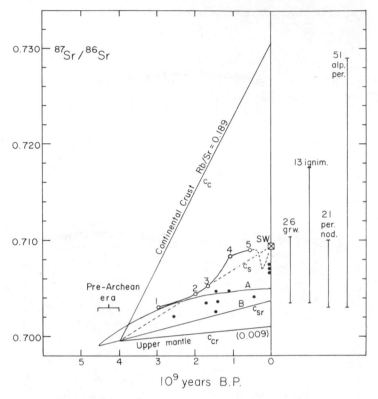

Fig. 5. Possible evolution paths of $^{87}Sr/^{86}Sr$ with time, beginning at
−4·55 By. Lines A and B are the limits for mantle evolution from
Faure and Powell (1972, Figure XII.2). Continental crust and upper
mantle residue lines use data from Tables 2, 3, 5 and commence at
−4·0 By, at the end of the Prearchean era. Solid circles show *initial*
ratios and ages of selected high-quality studies of granites (Peterman
et al., 1967). Open circles are marine limestones, with development
lines to modern seawater taken from Faure and Powell (*op. cit.*). On
the right-hand side are shown data for American Cordilleran
greywackes, Central American Cordilleran ignimbrites, peridotite
nodules in basalts and Alpine-type peridotites (see text)

A and B representing alternative upper mantle evolutionary trends (sub-continental) taken from Faure and Powell (1972).

The growth-line for the present continental crust, however, lies well above these representative continental materials, which would fit better to a line with Rb/Sr = 0·06 (model S) leading to a present-day ratio of 0·7094. Many authors have pointed out this discrepancy and taken it as the starting-point for one or other of two hypotheses, namely either (a) the continental crust has slowly grown in volume by separation from the mantle over geological time (Hurley et al., 1962; Hurley, 1968a,b), or (b) the original crust has undergone extensive

and continual recycling and partial isotopic equilibration with the mantle reservoir of Sr (Peterman et al., 1967; Armstrong, 1968; Armstrong and Hein, 1973). Another possible explanation is, of course, that the element abundances (particularly for Rb) are not correct: this has been carefully analyzed by Hurley (1968a,b) and by Russell and Ozima (1972). Their arguments were used in choosing the abundances used in Table 1 (Part 1) and it is unlikely that this explanation can be accepted.

Evidence favouring recycling is given by Pushkar et al. (1972). Over 10,000 km^3 of late Tertiary ignimbrites occur in Central

America, and have $^{87}Sr/^{86}Sr$ ratios in the range 0·7035–0·7175: associated calc-alkalic lavas centre closely on a ratio of 0·704. Although the authors discuss various explanations of their observations, the most plausible hypothesis would be that the ignimbrites are products of anatexis of older crust, whereas the lavas come from a deeper mantle source.

Additional evidence favouring crustal recycling comes from the $^{87}Sr/^{86}Sr$ ratios of 0·7036–0·7106 in 21 peridotite nodules in alkalic basalts (Leggo and Hutchinson, 1968; Stueber and Murthy, 1966) and ratios of 0·703–0·729 in 51 alpine ultramafic intrusions (various authors; see Faure and Powell, 1972, Table VII.2). These are mostly much higher than the continental basalts ($< 0·706$) and oceanic basalts ($< 0·705$), which nevertheless are believed to come from partial melting of mantle material, from REE fractionation (e.g. Frey, 1969) and other evidence. This material must therefore have been depleted in Rb early in the Earth's history and could be a mixture of deep peridotite and the basalt residue c_{cr}, or less readily the model S residue c_{sr}. The alpine peridotites and nodules (together with their host alkalic basalt magmas) must come from depleted mantle mixed with recycling continental crust.

Recycling also fits well to current views of island arc volcanism, whereby magmas are produced by partial fusion of a descending lithosphere plate along a Benioff subduction zone (see later). The partial fusion will involve oceanic crust, ocean-floor sediments, continental crust (if present, which is usually the case) and mantle rocks. Accepting that such fusion will tend to homogenise the isotopic ratios and that the mantle is effectively a large reservoir of Sr (Table 2) with low values of Rb and of $^{87}Sr/^{86}Sr$ (Table 5), then regenerated crust can have low initial values of $^{87}Sr/^{86}Sr$.

Now consider the apparent rate of recycling of continental crust. If the ratio $^{87}Sr/^{86}Sr$ is S, then for model R continental crust (Table 5) the rate of growth of this ratio is

$$\frac{dS_m}{dt} = \frac{0·7307 - 0·6995}{4 \times 10^9} = 0·0078 \times 10^{-9} \, yr^{-1}.$$

The model S growth rate, using a present-day ratio averaging 0·7094, is

$$\frac{dS_e}{dt} = \frac{0·7094 - 0·6995}{4 \times 10^9} = 0·0025 \times 10^{-9} \, yr^{-1}$$

which is 38% of dS_m/dt.

This suggests that, since the upper mantle acts as a large reservoir of Sr, then 68% of the ^{87}Sr produced in the model R continental crust has undergone recycling. Taking the mass to be $1·89 \times 10^{25}$ g (Part 1) and a mean density of 2·67, this corresponds to recycling of 1·2 km^3 yr^{-1}.

This figure is compatible with estimates obtained from measurements of (a) rate of crustal growth by volcanism (anatectic recycling), and (b) ocean-floor spreading, as follows:

(a) Sapper (1927 in Hess, 1962) estimated volcanic activity to be adding 1 km^3 yr^{-1} to the crust. Jakeš and White (1971) gave figures for volcanic additons to the Japanese and Kurile arcs of $\sim 0·040$ and $\sim 0·078$ km^3 yr^{-1}. The latter figure is from Markhinin (1968 in Dickinson, 1970) whose estimate of the volume of the crust below the Kurile Islands to be $6·5–7·0 \times 10^6$ km^3, or approximately 1/1000 the volume of the whole continental crust: the corresponding estimate for volcanic addition rate to the whole crust could not therefore exceed 78 km^3 yr^{-1} and must be much less, since most of the crust is now stable. A related estimate of continental crustal growth by Hurley and Rand (1969) is 3 km^3 yr^{-1}.

(b) The rate of growth of oceanic crust at median ridges may be equated to the rate of consumption of continental crust by underthrusting. The total length of median ridges may be taken as 64,000 km and oceanic crust is about 5 km thick. If the half-spreading rate now averages 1·3 cm yr^{-1} (Isacks et al., 1968; Le Pichon, 1968) then the growth rate is 8·3 km^3 yr^{-1}.

Both model R and model S accept the necessity of recycling, since the continental crustal mass remains constant. The Sr isotope evidence can therefore be used to support either model, and the return of ^{87}Sr to the mantle by model R is equivalent to extraction of ^{87}Rb by model S. However, the low ^{87}Sr/^{86}Sr ratios of both continental and oceanic basalts and the high ratios of alpine peridotites are less easily accounted for by model S.

Support for model M, whereby continental crustal material is added from primordial mantle, is difficult to find in isotopic data, since the evidence also supports either model R or S, or both.

Isotopic growth in the U–Th–Pb systems is also important here. It will not be examined, however, since uncertainties in the partition coefficients in these elements prevented adequate modelling in Part 1.

Volcanic Geochemistry

The general pattern of volcanic rock geochemistry, as well as that of ^{87}Sr, fits the present models.

Early Archean volcanic sequences are dominated by basalts (Baragar and Goodwin, 1969) which resemble the early island arc tholeiitic series of Jakeš and White (1971) in having low K, Rb, Ba and REE, and chondritic REE abundance patterns. This may be expected if both kinds of rock series come from residual or depleted parts of the mantle (column 9, Table 2). Modern oceanic tholeiites are even more depleted, suggesting that they are products of partial melting of deep residual mantle, brought up by convection: their low ^{87}Sr/^{86}Sr values also attest to this.

Calc-alkalic and alkalic volcanic rock series are minor in Archean terrains, but become more abundant in younger times when the transition to plate tectonics has occurred and near-surface sialic crust can be carried down to great depths by subducting lithospheric plates. Mixing of crustal anatectic melt with mantle-melts will generate the calc-alkalic (later) island arc series (Jakeš and White, *op. cit.*).

Alkalic igneous rocks pose a special problem since they (a) are enriched in K, Rb, Sr, Ba, (b) occur in both oceanic islands and continental rifts, (c) contain peridotite nodules enriched in ^{87}Sr, (d) originate at depths of up to 100 km. Their origin must involve recycled crustal material and low degrees of melting of peridotite: the latter has been substantiated experimentally by Green (e.g. 1970).

Later Scavenging and Secular Evolution

It is clear that crustal recycling would tend to return K, Ni, Sr, etc., to the mantle: there, remelting would send back K, Rb, Ba, Tl to the crust, preferentially retaining Sr, Ni, in accord with the partition coefficients (see Part 1). Anatexis of 'uncontaminated' mantle would also tend to lift K, etc., towards surface regions. These scavenging processes should lead to secular trends in crustal evolution, such as an increase in K, Rb, etc., Rb/Sr and related changes, i.e. models S and M.

Secular chemical trends are extraordinarily difficult to document because of the high noise level of other geological processes: claims have, however, been made by many. For instance, Fahrig and Eade (1968, Eade and Fahrig, 1971, 1973) find that Archean parts of the Canadian shield contain more Na, Ni, Cr and less K, Ti, U, Th, Pb, Zn than Proterozoic segments, and discuss the hypothesis that this might represent a secular change. Glikson (1971) and Glikson and Sheraton (1972) show a secular decrease in Na/K and K/Rb ratios in Kalgoorlie System, Australia, and the Kaapvaal Shield, Transvaal, and suggest that these trends may apply to the evolution of the Archean crust. Other trends are decrease of Ni, Cr, Ba/Rb and increase of K/Ba.

Decrease in igneous Na/K ratios of younger rocks in a particular orogenic belt is of course well known. This gave rise to the 'granite series' concept which has been discussed by many authors (e.g. Read, 1949; Chesworth, 1970), and should add an oscillatory component to any long-term trend for systematic change: such trends have been demonstrated by A. B. Ronov and his associates (e.g. Ronov and Migdisov, 1971) but the sampling has not been very satisfactory for Precambrian shields.

Some relevant data are presented in Tables 6 and 7. Abundances of SiO_2, K and Na/K are given, in order to aid comparisons, since major element abundances are controlled by

Table 6. Element abundances in granitic rocks and gneisses of various ages

	SiO$_2$ %	Na/K	K %	Li ppm	K/Rb	K/Ba	Ba/Rb	Rb/Sr	Ba/Sr	Rb/Tl	K/la^3 Tl
Archean, Western Australia (A, B)											
Pregeosynclinal 1	75	14·0	0·41	—	333	11	31	0·047	1·2	—	—
Pregeosynclinal 2	74	7·3	0·61	—	290	13	22	0·078	1·6	—	—
Syngeosynclinal	71	2·2	1·89	14	282	—	—	0·076	—	—	—
Late kinematic	74	0·8	3·69	161	79	—	—	5·11	—	—	—
Archean, N.W. Scotland (C)											
Hornblende-biotite gneiss	64	4·6	0·76	—	888	10	86	0·013	1·2	—	—
Biotite gneiss	68	3·9	0·94	—	647	9·6	68	0·024	1·6	—	—
Gneiss	61	3·8	0·85	—	773	11	69	0·019	1·3	—	—
Archean (?), English River Belt, Ontario (D)											
Savant — ⌠ 50–60	1·77	1·45	31	204	20	10	0·15	1·55	154	31	
Lake St. Joseph ⎨ 60–70	1·46	1·97	22	209	22	9·8	0·16	1·61	185	39	
section ⌡ >70	0·89	3·07	16	284	31	9·3	0·33	3·03	211	60	
Cedar L. ⌠ 60–70	1·93	1·50	19	255	—	—	0·21	—	133	34	
section ⌡ 70–79	0·74	3·47	8	369	—	—	0·49	—	149	55	
Archean, Louis L. batholith, U.S.A.(K)	66	1·6	2·08	—	248	14	17	0·095	1·67	—	—
Archean greywacke, Wyoming (L)	64	1·4	2·03	—	230	—	—	0·23	—	—	—
Proterozoic, Ontario, Grenville (E)											
Hermon leptites	76	0·5	3·79	26	408	141	2·9	1·3	3·8	—	—
Apsley gneiss (average)	67	1·1	2·57	41	367	47	7·9	0·39	3·1	—	—
Silent L. complex	>69	1·3	2·65	29	363	55	6·6	1·33	8·7	—	—
Paleozoic, Cape Granite, South Africa (F)											
Coarse	70	0·52	4·15	42	160	64	2·5	2·4	5·9	144	23
Medium	74	0·52	4·36	42	163	92	1·8	3·5	6·2	148	24
Fine	76	0·63	4·17	28	137	390	0·35	15	5·4	161	22
Paleozoic, Snowy Mts., Australia (G)											
Granodiorite	68	0·76	2·63	29	217	45	4·9	0·64	3·1	110	24
Gneissic granite	69	0·33	3·48	30	205	59	3·4	1·5	5·2	131	27
Late leucogranite	76	0·69	3·87	29	100	143	0·7	9·2	6·4	185	18
Paleozoic Malsburg pluton, Germany (H)	68	0·68	4·00	—	177	39	4·5	0·75	3·4	—	—
Mezozoic, Sierra Nevada, U.S.A. (J)	68	0·84	2·99	—	253	30	8·5	0·24	2·0	—	—

A. Glikson and Sheraton (1972).
B. O'Beirne *in* A.
C. Sheraton *in* A.
D. Unpublished data, in collaboration with A. M. Goodwin, C. J. Westerman.
E. From Shaw (1972), Breaks and Shaw (1973), Jennings (1969).
F. Kolbe (1966).
G. Kolbe and Taylor (1966).
H. Hahn-Weinheimer and Ackermann (1967).
J. Dodge (1972).
K. Condie and Lo (1971).
L. Condie (1967).

plutonic crystallization processes *in any epoch*. The only apparent secular trends in Table 6 are decrease of Ba/Rb, Na/K and increase of K/Na, Rb/Sr, Ba/Sr: it is also possible that these result from (a) inadequate literature search, (b) granulite facies metamorphism effects, (c) bias towards K, Si rich younger rocks. However, these trends are certainly expected on crystal chemical grounds. The mean values from Table 6 are grouped under approximate ages in Table 5 and two of the ratios are shown in Fig. 6: the trends are clear

Table 7. Mean Values from Tables 3 and 4

Age m.y.	K/Rb	Ba/Rb	K/Ba	Ba/Sr	Locality
~ 4000	246	6·8	36	0·40	Crust ⎫ Model S
	278	9·7	29	0·21	Mantle ⎭
	271	13	21	2·4	Crust ⎫ Model R
	272	0·78	349	0·007	Mantle ⎭
2850	419 (9)	49 (6)	11 (6)	1·4 (6)	⎰ W. Aust., NW ⎱ Scotland, Wyoming
2500	264 (5)	9·7 (3)	24 (3)	2·1 (3)	English River Belt
1000	379 (3)	5·8 (3)	81 (3)	5·2 (3)	Grenville
550	153 (3)	1·6 (3)	180 (3)	5·8 (3)	Cape granite
350	174 (3)	3·0 (3)	82 (3)	4·9 (3)	Snowy Mountains
250	177 (1)	4·5	39	3·4	Malsburg
80	253 (1)	8·5	30	2·0	Sierra Nevada

Number of values in parentheses.

only for Precambrian rocks, and there are insufficient data for younger granites. More high-quality data are needed to test these possible trends.

Conclusions

(1) After nebular condensation the accreting Earth passed through a largely molten stage in order to (a) expel most of the primordial rare gases, (b) transfer REE, U, Th from early condensates to the surface zone. Core separation presumably also occurred at this time, which lasted no more than 600 or 700 m.y.

(2) The Prearchean era ended when a thin solid crust formed across the surface of the whole Earth, and residual magma solidified in the upper mantle. Hot, acid, reducing rains were condensing out of the protoatmosphere and rapidly altering surface rocks to silica–alumina–carbonate–magnetite–pyrite–soluble salt assemblages.

(3) The initial crust, about 4000 m.y. ago, consisted of anorthosite overlying a thin sialic layer. Density contrasts, a heavy meteorite impact flux, contraction fracturing and volcanism, progressively overturned this lithology and developed more sialic surface zones: everything but volcanic peaks had been submerged beneath a universal ocean 2–3 km deep by 3700 m.y. ago.

(4) Volcanism welded crustal nuclei with ultramafic, basaltic and sialic lavas and fragmentals. Anatexis of basic crust, weathering products and subsurface granite formed light silicic magmas which thickened the upper sialic crust in some regions, forming dry land, and contrasting with thinner simatic crust where water deepened and ocean basins developed. As continents emerged, the cooling effect of water over thinner basic crust localized mantle convective upwelling and permitted initiation of seafloor spreading and plate tectonics (probably in Proterozoic times).

(5) The initial crust had a similar mass to the present-day continental crust, but small additions of sialic material may have taken place.

(6) Sr isotopic ratios show that the initial crust must have been extensively recycled. The evidence supports more strongly an initial crust similar to the present day, rather than the alternative of an initial crust richer in Na, Sr, REE and Ni and poorer in K, Rb, Ba. If the former is correct, then 2/3 of the continental crust must have been recycled, in order to exchange radiogenic ^{87}Sr for mantle ^{86}Sr, at a rate of $\sim 1\cdot2\,km^3\,yr^{-1}$. This figure agrees fairly well with estimates of crustal growth at present-day spreading plate boundaries.

(7) If the initial crust did differ in composition from the present-day, then it should be

possible to detect secular chemical changes, either in element abundances or in ratios. None has been certainly identified in the present study, using Na, K, Rb, Ba, Sr, Li, Tl abundances.

(8) Contrasting tectonic and magmatic properties of Archean greenstone belts, relative to younger island arc-fold mountain belts, depend on the absence of a plate tectonic regime in Archean times.

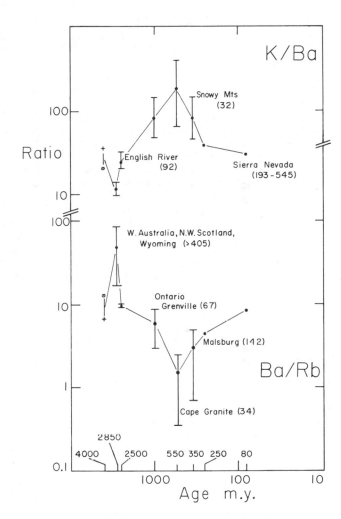

Fig. 6. Relationship of K/Ba and Ba/Rb to age in sialic complexes, as listed in Table 7. Mean values (●) and ranges are shown, the latter for several *average* values, based in each case on the total number of analyses in parentheses. Model R and model S initial crustal values shown by ⊙ and +

References

Anhaeusser, C. R., 1973. 'The evolution of the early Precambrian crust of southern Africa, *Phil. Trans. Roy. Soc. Lond.*, **A273**, 359–388.

Anhaeusser, C. R., Mason, R., Viljoen, M. J., and Viljoen, R. P., 1969. 'A reappraisal of some aspects of Precambrian shield geology', *Geol. Soc. Amer. Bull.*, **80**, 2175–2220.

Armstrong, R. L., 1968. 'A model for the evolution of Sr and Pb isotopes in a dynamic Earth', *Rev. Geophys.*, **6**, 175–199.

Armstrong, R. L., and Hein, S. M., 1973. 'Computer simulation of Pb and Sr isotope evolution of the Earth's crust and upper mantle', *Geochim. Cosmochim. Acta*, **37**, 1–18.

52

Baragar, W. R. A., and Goodwin, A. M., 1969. 'Andesites and Archean volcanism of the Canadian shield', *Proc. Andesite Conf., Bull. 65*, Dept. Geol. and Miner. Industries, State of Oregon, 121–142.

Berkner, L. V., and Marshall, L. C., 1964. 'The history of oxygenic concentration in the Earth's atmosphere', *Faraday Soc. Discussions*, **37**, 122–141.

Breaks, F. W., and Shaw, D. M., 1973. 'The Silent Lake pluton, Ontario: a nodular, sedimentary, intrusive complex', *Lithos*, **6**, 103–122.

Cameron, E. M., and Jonasson, T. R., 1972. 'Mercury in Precambrian shales of the Canadian shield, *Geochim. Cosmochim. Acta*, **36**, 985–1006.

Campbell, D. B., Jurgens, R. F., Dyce, R. B., Harris, F. S., and Pettengill, G. H., 1970. 'Radar interferometric observations of Venus at 70 cm. wavelength', *Science*, **170**, 1090–1091.

Chesworth, W., 1970. 'Anatexie dans une partie de la province de Grenville d'Ontario', *Bull. Soc. geol. de France (7)*, **XII**, No. 5, 875–885.

Clifford, P. M., 1969. 'The Primitive Crust', In *The Primitive Earth* (a symposium), Miami Univ., Oxford, Ohio.

Cloud, P. E., 1965. 'Significance of the Gunflint (Precambrian) microflora', *Science*, **148**, 27–35.

Condie, K. C., 1967. 'Geochemistry of early Precambrian graywackes from Wyoming', *Geochim. Cosmochim. Acta*, **31**, 2135.

Condie, K. C., and Lo, H. H., 1971. 'Trace element geochemistry of the Louis Lake batholith of early Precambrian age, Wyoming', *Geochim. Cosmochim. Acta*, **35**, 1099–1119.

Danchin, R. V., 1967. 'Chromium and nickel in the Fig Tree shale from South Africa', *Science*, **158**, 261–262.

Dickinson, W. R., 1970. 'Relation of andesites, granites and derivative sandstones to arc-trench tectonics', *Rev. Geophys. and Space Physics*, **8**, 813–860.

Dodge, F. C. W., 1972. 'Trace-element contents of some plutonic rocks of the Sierra Nevada batholith', *U.S.G.S. Bull.* 1314-F, F13 p.

Eade, K. E., and Fahrig, W. F., 1971. 'Geochemical evolutionary trends of continental plates—a preliminary study of the Canadian shield', *Geol. Surv. Canada Bull.* 179, 51 p.

Eade, K. E., and Fahrig, W. F., 1973. 'Regional, lithological and temporal variation in the abundances of some trace elements in the Canadian shield', *Geol. Surv. Canada Paper 72–46*, 46 p.

Fahrig, W. F., and Eade, K. E., 1968. 'The chemical evolution of the Canadian shield', *Can. J. Earth Sci.*, **5**, 1247–1252.

Fanale, F. P., 1971. 'A case for catastrophic early degassing of the Earth', *Chem. Geol.*, **8**, 79–105.

Faure, G., and Hurley, P. M., 1963. 'The isotopic composition of Sr in oceanic and continental basalts: application to the origin of igneous rocks', *J. Petrol.*, **4**, 31–50.

Faure, G., and Powell, J. L., 1972. *Strontium Isotope Geology*, Springer-Verlag, 188 pp.

Frey, F. A., 1969. 'Rare earth abundances in a high-temperature peridotite intrusion', *Geochim. Cosmochim. Acta*, **33**, 1429–1448.

Garrels, R. M., and Mackenzie, F. T., 1971. *Evolution of Sedimentary Rocks*, Norton, Inc., New York, 307 pp.

Glikson, A. Y., 1971. 'Primitive Archaean element distribution patterns: chemical evidence and geotectonic significance', *Earth Planet. Sci Lett.*, **12**, 309–320.

Glikson, A. Y., and Sheraton, J. W., 1972. 'Early Precambrian trondhjemitic suites in Western Australia and Northwestern Scotland, and the geochemical evolution of shields', *Earth Planet. Sci. Lett.*, **17**, 227–242.

Goldreich, P., 1972. 'Tides and the earth-moon system', In *Planet Earth*, F. Press and R. Siever (Eds.), W. H. Freeman and Co., pp. 303, 259–268.

Goldstein, R. M., and Rumsey, H., 1970. 'A radar snapshot of Venus', *Science*, **169**, 974–976.

Goodwin, A. M., 1968. 'Evolution of the Canadian shield, *Proc. Geol. Assoc. Canada*, **19**, 1–14.

Green, D. H., 1970. 'The origin of basaltic and nephelinitic magmas', *Trans. Leicester Lit. and Phil. Soc.*, **LXIV**, 26–54.

Green, D. H., 1972. 'Archean greenstone belts may include terrestrial equivalents of lunar maria?' *Earth Planet. Sci. Lett.*, **15**, 263–270.

Green, T. H., and Ringwood, A. E., 1966. Origin of the calc-alkaline igneous rock suite, *Earth Planet. Sci Lett.*, **1**, 307–316.

Hahn-Weinheimer, P., and Ackermann, H., 1967. 'Geochemical investigation of differentiated granite plutons of the Southern Black Forest. II. The zoning of the Malsburg granite pluton as indicated by the elements titanium, zirconium, phosphorus, strontium, barium, rubidium, potassium and sodium', *Geochim. Cosmochim. Acta*, **31**, 2197.

Hanks, T. C., and Anderson, D. L., 1969. 'The early thermal history of the Earth', *Phys. Earth Planet. Interiors*, **2**, 19–29.

Hart, S. R., Brooks, C., Krogh, T. E., Davis, G. L., and Nava, D., 1970. 'Ancient and modern volcanic rocks: a trace element model', *Earth Planet. Sci. Lett.* **10**, 17–28.

Hess, H. H., 1962. 'History of ocean basins', In *Petrologic Studies:* A volume in Honor of A. F. Buddington (Engel, James and Leonard (Eds.)), Geol. Soc. Amer., 599–620.

Holland, H. D., 1962. Model for the evolution of the Earth's atmosphere', *Geol. Soc. Amer. Buddington Vol.*, 447–477.

Hurley, P. M., 1968a. 'Absolute abundance and distribution of Rb, K and Sr in the Earth', *Geochim. Cosmochim. Acta*, **32**, 273–283.

Hurley, P. M., 1968b. Correction to: 'Absolute abundance and distribution of Rb, K and Sr in the Earth', *Geochim. Cosmochim. Acta*, **32**, 1025–1030.

Hurley, P. M., Hughes, H., Faure, G., Fairbairn, H. W., and Pinson, W. H., 1962. 'Radiogenic Sr–87 model of continental formation'. *J. Geophys. Res.*, **67**, 5315–5334.

Hurley, P. M., and Rand, J. R., 1969. 'Pre-drift continental nuclei', *Science*, **164**, 1229–1242.

Isacks, B., Oliver, J., and Sykes, L. R., 1968. 'Seismology and the new global tectonics', *J. Geophys. Res.*, **73**, 5855–5899.

Jakeš, P., and White, A. J. R., 1971 'Composition of island arcs and continental growth', *Earth Planet. Sci. Lett.*, **12**, 224–230.

Jennings, D. S., 1969. *Origin and metamorphism of part of the Hermon Group near Bancroft*, Ontario. Unpub. Ph. D. Thesis, McMaster Univ., Hamilton, Ontario.

Kolbe, P., 1966. 'Geochemical investigation of the Cape Granite, S.W. Cape Province, South Africa', *Trans Geol. Soc. S. Africa*, **69**, 161–199.

Kolbe, P., and Taylor, S. R., 1966. 'Geochemical investigation of the granitic rocks of the Snowy Mountains area, New South Wales', *J. Geol. Soc. Australia*, **13**, 1–25.

Kushiro, I., 1970. 'Systems bearing on melting of the upper mantle under hydrous conditions', *Ann. Rep. Director Geophys. Lab.* for 1968–69, 240–245.

Kushiro, I., 1974. 'Melting of hydrous upper mantle and possible generation of andesitic magma: an approach from synthetic systems', *Earth Planet. Sci. Lett.*, **22**, 294–299.

Lambert, I. B., and Wyllie, P. J., 1970. 'Melting in the deep crust and upper mantle and the nature of the low velocity layer', *Phys. Earth Planet. Interiors*, **3**, 316–322.

Lambert, I. B., and Wyllie, P. J., 1972. 'Melting of gabbro (quartz eclogite) with excess water to 35 kilobars, with geological applications', *J. Geol.*, **80**, 693–708.

Le Pichon, X., 1968. 'Sea-floor spreading and continental drift', *J. Geophys. Res.*, **73**, 3661–3697.

Leggo, P. J., and Hutchinson, R., 1968. 'A Rb-Sr study of ultrabasic xenoliths and their basaltic host rocks from the Massif Central, France', *Earth Planet. Sci. Lett.*, **5**, 71–75.

Li, Y.-H., 1972. 'Geochemical mass balance among lithosphere, hydrosphere and atmosphere', *Amer. J. Sci.*, **272**, 119–137.

Moore, E. S., and Maynard, J. E., 1929. 'Solution, transportation and precipitation of iron and silica', *Econ. Geol.*, **24**, 272–303; 365–402; 506–527.

Peterman, Z. E., Hedge, C. E., Coleman, R. G., and Snaveley, P. D., Jr., 1967. '$^{87}Sr/^{86}Sr$ ratios in some eugeosynclinal sedimentary rocks and their bearing on the origin of granitic magma in orogenic belts', *Earth Planet. Sci. Lett.*, **2**, 433–439.

Pushkar, P., McBirney, A. R., and Kudo, A. M., 1972. 'The isotopic composition of Sr in Central American ignimbrites', *Bull. Volcanol.*, **35**, 265–295.

Read, H. H., 1949. 'A contemplation of time in plutonism', *Q.J.G.S. Lond.*, **105**, 101–156.

Ronov, A. B., and Migdisov, A. A., 1971. 'Geochemical history of the crystalline basement and the sedimentary cover of the Russian and North American platforms', *Sedimentology*, **16**, 137–185.

Rubey, W. W., 1955. 'Development of the hydrosphere and atmosphere with special reference to probable composition of the early atmosphere', *Geol. Soc. Amer. Spec. Paper 62*, 631–650.

Russell, R. D., and Ozima, M., 1971. 'The potassium/rubidium ratio of the Earth', *Geochim. Cosmochim. Acta*, **35**, 679–686.

Sagan, C., and Mullen, G., 1972. 'Earth and Mars: evolution of atmospheres and surface temperatures', *Science*, **177**, 52–55.

Sakamoto, T., 1950. 'The origin of Precambrian banded iron ores, *Amer. J. Sci.*, **248**, 449–474.

Shaw, D. M., 1972a. 'Development of the early continental crust. Part 1. Use of trace element distribution coefficient models for the protoarchean crust', *Can. J. Earth Sci.*, **9**, 1577–1595.

Shaw, D. M., 1972b. 'The origin of the Apsley gneiss, Ontario', *Can. J. Earth Sci.*, **9**, 18–35.

Stueber, A. M., and Murthy, V. R., 1966. 'Sr isotope and alkali element abundances in ultramafic rocks', *Geochim. Cosmochim. Acta*, **30**, 1243–1260.

Tolstikhin, I. N., Mamyrin, B. A., Khabarin, L. B., and Erlikh, E. N., 1974. 'Isotope composition of He in ultrabasic xenoliths from volcanic rocks of Kamchatka', *Earth Planet. Sci. Lett.*, **22**, 75–84.

Turekian, K. K., and Clark, S. P., Jr., 1969. 'Inhomogeneous accumulation of the Earth from the primitive solar nebula', *Earth Planet. Sci. Lett.*, **6**, 346–348.

Vinogradov, A. P., Marov, M. Ya., and Surkov, Yu. A., 1972. 'Examination of the atmosphere of Venus by the Venus 4, Venus 5 and Venus 6 Soviet probes', *Geochem. Internat.*, **3-4**, 241–251.

Wänke, H., Baddenhausen, H., Palme, H., and Spettel, B., 1974. 'On the chemistry of the Allende inclusions and their origin as high temperature condensates', *Earth Planet. Sci. Lett.*, **23**, 1–7.

Crustal Evolution in the Early Earth–Moon System:

Constraints from Rb–Sr Studies

BOR-MING JAHN

The Lunar Science Institute, 3303 NASA Rd. 1, Houston, Texas 77058, U.S.A.

L. E. NYQUIST

TN7, NASA-Johnson Space Center, Houston, Texas 77058, U.S.A.

Abstract

Chronological studies of the lunar samples have revealed several important events: (1) formation of the Moon at about 4·6 b.y. ago, followed by large-scale differentiation within about the first 300 m.y., (2) intense bombardment by interplanetary debris or meteorites about 4·6 to 3·9 b.y. ago which probably culminated at about 3·9–4.0 b.y. ago, (3) eruption of mare basalts 3·9–3·1 b.y. ago, mostly on the nearside of the Moon, (4) a relatively quiescent period followed the last episode of mare flooding.

The lunar Rb–Sr data show that most lunar mare basalts have initial Sr^{87}/Sr^{86} ratios ($= I$) lower than 0·69960, suggesting that the magma sources have evolved in a lunar mantle of very low Rb/Sr ratio ($\sim 0·01$ or less). However, the pattern of I's suggests that the lunar mantle was heterogeneous.

Most of the old crustal segments on the Earth have an imprint of a world-wide thermal event about 2·7 b.y. ago. A few older remnants are found in the Minnesota River Valley ($\sim 3·5$ b.y.), western Greenland ($\sim 3·8$ b.y.) and southern Africa ($\sim 3·5$ b.y.). The available Rb–Sr data of Archean rocks indicates that the minimum I value so far determined is about 0·7004, which is approximately the maximum value for the lunar basalts (KREEP type). The crustal evolution on the Earth probably involved extensive recycling processes in the crust–upper mantle system and transport of crustal material from the lower to the upper mantle. These processes might have effectively kept the Rb/Sr ratio rather constant in much of the upper mantle. However, the upper mantle composition has by no means been completely homogenized, as evidenced from the Rb–Sr data of some Archean rocks and modern oceanic basalts.

The predominance of ages of about 4·0 b.y. for lunar non-mare rocks show this to have been a very active time in lunar history. No rocks older than 4·0 b.y. have been found on the Earth. In the lunar case, the 3·9–4·0 b.y. period appears to be associated with major basin-forming events by meteoritic bombardment. The implications of this interpretation for terrestrial and lunar crustal evolution have not been fully assessed.

I. Introduction

The geological records of early Precambrian strata are very important to the understanding of the early history of the Earth. Few such records of the first billion years of Earth's history have been preserved. Although rocks of 3·5 b.y. or older age are believed to occur in several places on Earth, reliable chronological information is only available for three areas: (1) the Morton and Montevideo gneisses in the Minnesota River Valley (Goldich et al., 1970; Goldich and Hedge, 1974), (2) the quartzofeldspathic gneisses and associated rocks in western Greenland (Moorbath et al., 1972), and (3) the Onverwacht Group of South Africa (Allsopp et al., 1968, 1973; Hurley et al., 1972; Sinha, 1972; Jahn and Shih, 1974). The granitic and associated volcanic-sedimentary piles that occur in the Rhodesia craton may also have been emplaced as early as 3·5 b.y. (Wilson and Harrison, 1973; Robertson, 1973). Younger Late Archean rocks (about 2·5 to 3·0 b.y.) are more widespread and preserved in all shield areas.

The characteristics and evolution of the earliest crust on the Earth have long been debated. The lack of geological records earlier than four billion years ago and the more active tectonic processes in that period as predicted from the Earth's thermal history lead us to conclude that this debate is futile. Perhaps a better way of understanding the Earth's early history is to study extraterrestrial materials which originated in the solar system, i.e. meteorites and returned lunar samples. The information derived from the study of meteorites is always piecemeal, and the origins of various meteorites are problematical. The returned lunar samples from the Apollo and the Luna missions provide a better opportunity to study the earliest history of the solar system and possibly to fill the missing pages of the earliest history of the Earth.

Because the temporal framework of the crustal evolution in the Earth–Moon system can be defined by studying the isotopic systems, most notably, the Rb–Sr and the U–Th–Pb systems, the purposes of this paper are: (1) to examine available high quality Rb–Sr data of lunar and terrestrial rocks older than 2·5 b.y., (2) to compare the similarities and differencies of the isotopic systematics in these rocks of different planetary bodies, and (3) to discuss possible implications of lunar studies to the early history of the Earth.

II. Isotopic Evolution as an Indication of Crustal Evolution

The formation and accumulation of crustal materials is generally accomplished by partial melting of mantle materials and by subsequent differentiation processes. In conjunction with available crystal-liquid distribution coefficients of Rb and Sr during partial melting (Schnetzler and Philpotts, 1970, Philpotts and Schnetzler, 1970; Gast, 1968a; Griffin and Murthy, 1969; Shimizu, 1974; Hart and Brooks, 1974), the patterns of Sr isotopic evolution in the source region (or the mantle) can be used as an indication of the rate of crustal development. The classical treatment of this kind of study, including the use of U–Th–Pb systematics, can be found in Hurley et al. (1962), Patterson (1964), Patterson and Tatsumoto (1964), Wasserburg (1966), Gast (1967, 1968b), Armstrong (1968) and Hurley and Rand (1969). More recently, similar studies were revived by Hart and Brooks (1970) and Russell (1972), Armstrong and Hein (1973), and Russell and Birnie (1974).

Construction of any realistic Sr isotopic evolution model for the upper mantle requires precise isotopic data of mantle-derived rocks. Significant and useful data are very scarce and, therefore, serious mathematical modeling cannot yet be tested. Nevertheless, the following hypothetical evolution patterns serve a qualitative discussion (cf. Hart and Brooks, 1970).

The growth of radiogenic Sr^{87} or the change of Sr^{87}/Sr^{86} ratio(s) in a system (e.g. the upper mantle) is a function of radioactive decay of *in situ* Rb^{87} and of changes in the Rb/Sr ratio if Rb or Sr are added to or taken from the system. Basically, variations of the Rb/Sr ratio as a function of time in the upper mantle due to inward or outward transport of crustal

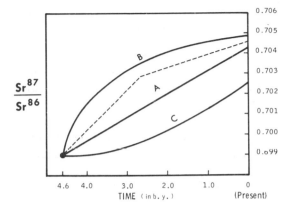

Fig. 1. Hypothetical Sr isotope evolutionary paths for the upper mantle. The change of the Rb/Sr ratio with time, d(Rb/Sr)/dt, for Curve A is 0, for B, <0 and for C, >0. All these three curves are continuous. The broken line for which the Rb/Sr ratio in the second stage is substantially reduced indicates a two-stage discontinuous Sr isotopic evolution

materials can produce three basic continuous evolution patterns (Fig. 1). The slope at any instantaneous point on the curve in Fig. 1 defines a certain ratio of Rb/Sr. Thus curve A represents a constant Rb/Sr ratio in the upper mantle. The implications of these curves are:

Curve A:

(a) The upper mantle is an infinite reservoir for Rb and Sr. Tapping of volcanic liquid from the mantle throughout the geologic history does not change the Rb/Sr ratio of the mantle, or (b) distribution coefficients for Rb and Sr are identical during the partial melting process, i.e. Rb and Sr were removed from the mantle in identical proportions, leaving the Rb/Sr ratio unchanged, or (c) volcanic liquids were derived from different volumes of a homogeneous pristine mantle and each volume was tapped only once from the pristine mantle, or (d) the loss of Rb relative to Sr to the crust was replenished from the lower mantle, or (e) constant mixing of crustal materials with the upper mantle, thus keeping the Rb/Sr ratio in the upper mantle unchanged.

Curve B:

The upper mantle is a finite reservoir for Rb and Sr. Continuous extraction of volcanic liquids not only fractionate Rb and Sr from but also deplete Rb relative to Sr, i.e. Rb was removed preferentially and the Rb/Sr ratio decreases with time.

Curve C:

The Rb/Sr ratio in the mantle increases continuously with time. Since magmatism leading to the formation of sialic crust can only decrease the Rb/Sr ratio in the upper mantle, the only perceptible way of increasing the Rb/Sr ratio is to add Rb preferentially from the lower mantle, a process perhaps similar to rising of deep mantle plumes. This also implies that there is a net gain of crustal material in the crust plus upper mantle regime, e.g. the crust becomes thickened. Alternatively, the return of crustal materials, which were possibly separated from the mantle in the earliest possible time (about 4·5 b.y.) after the Earth's formation, to the upper mantle would also increase the Rb/Sr ratio.

The evolution of Sr^{87}/Sr^{86} in the mantle needs not be continuous as described above. It might have been a multi–stage discontinuous or even only a two-stage discontinuous process (broken line, Fig. 1). We consider all these evolutionary paths plausible and they will be discussed in the context of the available data.

Application of the above models of mantle evolution requires a knowledge of the primordial Sr^{87}/Sr^{86} ratio. This ratio has been more or less directly determined for the Moon and some meteorites. To date it has not been possible to determine the primordial Sr^{87}/Sr^{86} ratio of the Earth directly; however, it is a good assumption that it was the same as for the Moon and meteorites. In fact, several aspects of early solar system history are more clearly preserved in lunar samples than in terrestrial rocks. Consequently, it is instructive to review briefly the lunar data in the following.

III. Lunar Data

(A) *Chemical Characteristics of Lunar Samples*

The lunar crystalline rocks can be divided into two broad categories: (1) mare basalts and

(2) non-mare rocks, consisting of anorthositic rocks, KREEP and VHA basalts, and some rare rock types, such as granitic rock 12013 and dunite 72417, and troctolite 76535. In addition to these crystalline rocks, the lunar samples also include breccias and soils of various composition. In the following we shall outline some of the most important geochemical characteristics of the lunar samples, and then review the necessary Rb–Sr isotopic data.

Mare Basalts. Mare basalts occur exclusively in the lunar basins, but not all lunar basins contain mare basalts (Stuart-Alexander and Howard, 1970). They have been returned by the Apollo 11, 12, 15 and 17 missions. Mare-derived materials have also been found in the regolith fines returned by all lunar missions. Texturally, they resemble terrestrial basalts and appear to have formed by rapid crystallization of lavas. All mare basalts are rather similar in chemical composition, but a few compositional groups can be defined based on some discriminant chemical species, such as Ti, K, etc. Chemical variation within mare basalts has been ascribed to derivation of melt from different source regions (e.g. Taylor and Jakeš, 1974), different degrees of partial melting in a homogeneous source region (Ringwood, 1974a), or it may be due to crystal fractionation of a single parental magma (O'Hara et al., 1970, 1974). The major chemical distinction between mare and most terrestrial basalts are: (1) mare basalts have higher Fe/Mg ratio and Ti content; the lack of ferric iron and the presence of metallic iron indicate a lower f_{O_2} for the lunar environment, (2) mare basalts have unique trace and minor element abundances with enrichment in refractory elements, such as Sr, Ba, Ti, Hf, Zr, U, Th, Y and REE, but with depletion in volatile elements, such as Na, K, Rb and some siderophile elements when compared with chondritic values, (3) mare basalts are characterized by low K/U and very high U^{238}/Pb^{204} ratios as compared with terrestrial rocks. The K/U ratios in chondrites: Terrestrial rocks: lunar samples are about 6×10^4: $1–2 \times 10^4$: $1–3 \times 10^3$. The very high U^{238}/Pb^{204} ratio in lunar samples obviously involved a severe Pb loss in the

source region in pre-lunar history, (4) unlike terrestrial basalts, mare basalts display an outstanding negative Eu anomaly in their REE abundance patterns (Fig. 2(A)), suggesting that they may have derived from a source that contains plagioclase (Gast, 1972) or more likely that the negative Eu anomaly reflects the nature of their source rock rather than due to the role of plagioclase fractionation (Philpotts and Schnetzler, 1970; Schnetzler and Philpotts, 1971; Murthy and Banerjee, 1973; Taylor and Jakeš, 1974; Shih et al., 1975).

Non-Mare Rocks. Non-mare rocks are abundant in the lunar highlands and thus they are very important to our understanding of lunar crustal evolution. Most non-mare rocks contain higher Al_2O_3 than mare basalts. Gast (1973) devided non-mare rocks into anorthosite, VHA and KREEP basalts using the index of Al_2O_3 contents—anorthosite, $Al_2O_3 >$ 25%; VHA basalt, 20–25%; and KREEP, 15–20%. All mare basalts have Al_2O_3 less than 12%. Obviously this index is related to the amount of plagioclase present in individual rocks.

Fragments of anorthositic rocks were identified in almost all landing sites. They range in composition from true anorthosite through gabbroic anorthosite to anorthositic gabbro. All anorthositic rocks exhibit positive Eu anomalies in their REE abundance patterns and their total REE abundances are lower than those of mare basalts (Fig. 2(A)). Because of their very low Rb/Sr ratios, some very pure anorthosites have preserved their primordial Sr composition (i.e. lunar initial Sr^{87}/Sr^{86} ratio) virtually unchanged to the present day.

KREEP Basalts. KREEP materials were first identified as an abundant component in some Apollo-12 soils (Hubbard et al., 1971; Meyer et al., 1971). They have basaltic major element composition but are very unique in minor and trace element characteristics. As indicated by their name they are rich in K, REE and P; their REE abundances are enormously high (Fig. 2(A), about 100–300 X chondritic abundance) with conspicuous negative Eu anomaly (Hubbard and Gast, 1971;

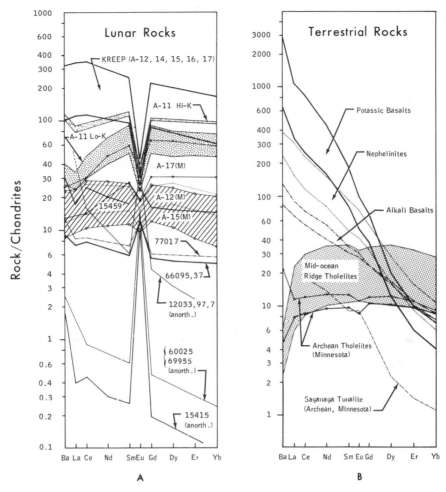

Fig. 2. Ba and REE abundance patterns for various lunar and terrestrial rocks. The difference in the shapes of the patterns is very striking. In the case of lunar rocks, KREEP and mare basalts (A-11, Hi-K, Lo-K, A-17(M), A-12(M), and A-15(M)) have distinct negative Eu anomalies. In contrast, all anorthositic rocks (15459, 77017, 66095, 12033, 60025, 69955 and 15415) have positive Eu anomalies but much lower total REE abundances. VHA basalts or troctolitic rocks (not shown) have REE abundances intermediate of those of KREEP and anorthositic rocks. In the terrestrial case, two Archean tholeiites from Minnesota have REE patterns comparable with the oceanic ridge or island arc tholeiites; and an Archean tonalite shows a strong depletion in heavy REE. Sources: Gast et al. (1970), Hubbard and Gast (1971), Hubbard et al. (1972, 1974), Schnetzler et al. (1972), Laul and Schmitt (1973), Philpotts et al. (1974), Rhodes and Hubbard (1973), Kay et al. (1970), Kay and Gast (1973), Arth and Hanson (1972) and Jahn et al. (1974)

Gast, 1972; Philpotts et al., 1972; Wänke et al., 1972; Haskin et al., 1973; Hubbard et al., 1973a, 1974; Laul and Schmitt, 1973). Terrestrial alkali basalts also have high total REE abundances (Kay and Gast, 1973), but their abundance patterns (slope and lack of Eu anomaly) are totally different from that of KREEP basalts (Fig. 2(A) and (B)). This probably implies different mineralogical controls during partial melting processes under high pressure conditions. For instance, the effect of garnet on REE patterns of melt is quite evident in terrestrial processes but is obscure in lunar processes.

The distribution of KREEP materials is not fully known. Many of the returned samples are brecciated and the majority of the fragments in the Apollo 12 soil are glassy. KREEP materials were found in all landing sites but crystalline KREEP were only returned in Apollo 14, 15 and 16 missions. Orbital gamma-ray data indicate that the surfaces of Mare Ibrium and Oceanus Procellarum are relatively enriched in the KREEP component and that essentially pure KREEP may exist in the regions of the craters Archimedes and Aristarchus in addition to the Fra Mauro region sampled by the Apollo 14 mission (Metzger et al., 1974).

VHA (*Very High Alumina*) *Basalts.* 'VHA basalt' is a chemical classification and its petrogenesis is currently a topic of some debate. The REE, Ba and U contents in VHA basalts are about 1/3 to 1/4 those of KREEP basalts, whereas the relative abundance of these elements is essentially identical with that of KREEP basalts (Hubbard et al., 1973b, 1974). Because of their intermediate chemical characteristics, they are argued by some to be mixtures of the KREEP and the anorthosite components (Dowty et al., 1974a,b; Schonfeld, 1974), although Hubbard et al. (1974) considered the mixing process geologically unsound, and presented arguments favouring an ultimate origin in magmatic processes.

In addition to the above rock types, other materials such as a granitic material (12013), an olivine cumulate (72417) and an olivine–plagioclase cumulate (76535) are also returned, but they constitute insignificant amounts of lunar surface materials.

Nomenclature of non-mare rocks is still in a chaotic state because of their complex origin and the use of different classification criteria. Many non-mare rocks are probably mixtures of a few components which have undergone subsequent impact melting processes (Schonfeld and Meyer, 1972). The so-called ANT suite (Prinz et al., 1973) consists of anorthositic, noritic and troctolitic rocks. All of them have high plagioclase contents (>60%) but low K, REE and P contents. Most of them

exhibit positive Eu anomalies. The so-called highland basalts are in fact synonymous with anorthositic gabbros. KREEP and VHA are terms derived from chemical classification. The same rock (e.g., 14310) which is called KREEP by some may be called Fra Mauro basalt or norite by others; or the same rock (e.g. 62295) may be either called spinel troctolite or VHA basalt. Some spinel troctolite (Prinz et al., 1973) may be also called type A FIIR rock (Delano et al., 1973) or olivine-rich, true spinel-bearing anorthosites (Reid, 1972). Numerous examples of confusing nomenclature can be found in the lunar literature.

In summary, mare basalts are confined in lunar basins and concentrated (>90%) on the near-side of the Moon (Stuart-Alexander and Howard, 1970). Lunar basins also contain some non-mare rocks. On the lunar highlands, anorthositic rocks are the most abundant rock type. From many geophysical and geochemical arguments, none of the returned lunar samples may represent the bulk composition of the Moon. Taylor and Jakeš (1974) proposed that the sources for all lunar samples were differentiated shortly after the formation of the Moon, and the least differentiated sample is probably represented by a green glass (15426). The discussion on the origins of various lunar rocks is beyond the scope of this paper.

(B) Lunar Rb–Sr Data

(1) *The Lunar Initial* Sr^{87}/Sr^{86} ($=I_0$). It is commonly assumed that the Moon as well as the Earth was formed at the same time as other planetary bodies in the solar system. Consequently, the BABI value (basaltic achondrite best initial $= 0 \cdot 69898 \pm 3$, C.I.T. scale) established by Papanastassiou and Wasserburg (1969) has been frequently used as the initial or primordial value for the Earth and the Moon in model age calculations. The most direct determination of the lunar initial Sr^{87}/Sr^{86} ratio ($=I_0$) would be measurement of Sr^{87}/Sr^{86} ratio from a rock or a mineral formed during or immediately after planetary accretion with Rb/Sr = 0. However, in this

Table 1. Summary of Lunar Rb–Sr Age Data

			Age (in b.y.)	$I = (Sr^{87}/Sr^{86})_0$ (C.I.T. scale)	References
(A) Mare basalts:					
Appollo 11	Low-K	(2)	3·63–3·71	0·69906–0·69909	Papanastassiou and Wasserburg (1970)
	High-K	(5)	3·59–3·68	0·69926–0·69939	Papanastassiou and Wasserburg (1970)
Appollo 12		(9)	3·16–3·36	0·69918–0·69957	Papanastassiou and Wasserburg (1971)
Appollo 15		(6)	3·28–3·44	0·69923–0·69937	Papanastassiou and Wasserburg (1973)
Appollo 17		(6)	3·64–3·89	0·69908–0·69924	Evensen et al. (1973)
					Tera et al. (1974)
					Tatsumoto et al. (1973)
					Nunes et al. (1974)
					Nyquist et al. (1974, 1975)
Luna 16		(1)	3·42 ± 0·18	0·69952 ± 5	Papanastassiou and Wasserburg (1972)
(B) Non-mare rocks:					
Appollo 12	12013		4·0	0·7085	Lunatic Asylum (1970)
Appollo 14	14001	KREEP	3·89 ± 0·03	0·70036 ± 6	Papanastassiou and Wasserburg (1971)
	14066	KREEP	4·06 ± 0·14	0·70034 ± 34	Mark et al. (1973)
	14073	KREEP	3·88 ± 0·04	0·70034 ± 4	Papanastassiou and Wasserburg (1971)
	14310	KREEP	3·90 ± 0·07	0·70036 ± 8	Papanastassiou and Wasserburg (1971)
					Compston et al. (1972)
					Murthy et al. (1972)
					Mark et al. (1974)
	14321	KREEP	4·05 ± 0·06	0·70104 ± 33	Mark et al. (1973)
Appollo 15	15434	KREEP	3·91 ± 0·04	0·70058 ± 10	Nyquist et al. (1974)
Appollo 16	62295	VHA	4·00 ± 0·06	0·69957 ± 6	Mark et al. (1974)
	65015	KREEP	3·93 ± 0·02	0·70030 ± 20	WR-Quint. tie line = metamorphic age
			4·42 ± 0·04	0·69917 ± 8	WR-plag. tie line = primary age
					Papanastassiou and Wasserburg (1972)
	68415	Anorth. gabbro	3·84 ± 0·01	0·69920 ± 3	Papanastassiou and Wasserburg (1972)
Appollo 17	72417	Dunite	4·60 ± 0·09	0·69899 ± 5	Albee et al. (1974)
	72275	Pigeonite basalt	4·01 ± 0·04		
	76055	Norite	3·86 ± 0·04	0·69985 ± 6	WR-Quint. tie line
					Tera et al. (1974)
			4·49		WR-plag. tie line
	77135	Troctolite	3·89 ± 0·08	0·69923 ± 4	Nunes et al. (1974)
(C) Anorthositic rocks: Ar^{39}–Ar^{40} dates					
Appollo 15, 16, 17 and Luna 20			3·90–4·20		Husain et al. (1972)
					Schaeffer and Husain (1973, 1974)
					Kirsten et al. (1973)
					Stetler et al. (1973, 1974)
					Turner et al. (1973)
					Podosek et al. (1973)
					Alexander and Kahl (1974)

Notes: (1) C.I.T. Sr scale = JSC scale − 0·00012; NBS/SRM-987 Sr on the C.I.T. scale is 0·71013.
(2) The numbers in parentheses of mare basalt data indicate the number of samples determined by the mineral isochron method.

case one needs independent evidence to assess that the rock or mineral is indeed ancient because it is impossible to determine its age by the Rb–Sr method.

Anorthositic samples returned by Apollo 15, 16 and 17 missions have yielded directly measured Sr^{87}/Sr^{86} ratios slightly lower than BABI (Papanastassiou and Wasserburg, 1972b; Nyquist et al., 1973; Nunes et al., 1974). Based on measurements of the anorthositic portion of breccia 61016, Nyquist et al. (1973, 1974) proposed a lunar initial value, LUNI (Lunar initial, = 0·69903 ± 3 on the JSC scale, or 0·69891 adjusted to the C.I.T. scale). This value is based on an assumed age of the rock of 4·6 b.y., but is insensitive to age uncertainty because of its extremely low Rb/Sr ratio (0·0003–0·0006). It is suggested from interlaboratory comparison of available data that the best estimate for LUNI is 0·69894 ± 4 (on the C.I.T. scale, or 0·69906 on the JSC scale). A detailed discussion of the problem of the lunar initial, I_0, is given by Nyquist (1975).

(2) *Ages of Mare Basalt.* Mare basalts have not been metamorphosed or subjected to any open system behaviour subsequent to lava

extrusion, and the Rb–Sr mineral isochron method has been proven very satisfactory in determination of their crystallization age (Table 1). Most of the ages have been established by the C.I.T. group (Papanastassiou et al., 1970 and their subsequent papers; see references in Table 1). All mare basalts and their mineral constituents have very low Rb/Sr ratios and non-radiogenic Sr compositions placing extreme demands on analytical precision.

Mare basalt ages cannot be determined by the whole-rock isochron method, because of the presence of variable I values—for example, the I values for high-K and low-K Apollo 11 basalts are demonstrably different. There is also a large range in I's for Apollo 12 mare basalts, suggesting heterogeneous sources for these rocks. In addition, very narrow ranges in both Rb/Sr and Sr^{87}/Sr^{86} ratios for all mare basalt whole rock samples ($Sr^{87}/Sr^{86} = 0.6995$ to 0.7000 for Apollo 11 low-K basalts; 0.6997 to 0.7016 for Apollo 12 basalts; 0.7002 to 0.7008 for Apollo 15 mare basalts) present difficulty in achievement of precise age data by the whole-rock isochron method. Careful studies of lunar samples reveal that the seemingly comagmatic rocks in any single landing site are not necessarily derived from a homogeneous source. This suggests that the common practice of constructing whole-rock isochrons for terrestrial rocks may not be strictly valid. Unfortunately, determination of crystallization ages for most old terrestrial rocks has had to rely upon the whole-rock isochron method because most mineral isochrons are the result of later isotopic re-equilibration.

The ages and I values for mare basalts are summarized in Table 1 and also illustrated in Fig. 3. From the radiometric data, mare flooding episodes on the lunar surface extended at least as long as 800 m.y. (from about 3.9 to 3.1 b.y.). However, this may not reflect the entire period of mare flooding, as indicated by comparison to the relative ages of mare units determined by impact crater density (Stuart-Alexander and Howard, 1970; Hartmann, 1972; Boyce et al., 1974). If the crater density technique and the assumptions used in calibrating it to radiometric ages are correct, the mare flooding did not cease until about 1.7 b.y. ago (Boyce et al., 1974). The earliest extrusion of mare basalts was probably after formation of mare basins. If most of the basins were formed within a narrow time span of about 3.9–4.0 b.y. (Tera et al., 1974), then this puts an older limit to the onset of mare flooding. The mechanisms involved in mare flooding are unclear. It was proposed that mare flooding might be induced by meteoritic impacts (Hulme, 1974; Green, 1972), but more likely endogenic volcanic processes played the major role as evidenced from the time relationships of formation of most impact basins and extrusion of mare basalts.

(3) *Ages of Non-mare Rocks.* Rb–Sr age data of non-mare rocks are also included in Table 1. Non-mare rocks were not returned in appreciable amounts until the Apollo 14 mission. Their radiometric ages (both Rb–Sr and Ar^{39}–Ar^{40}) are slightly older than that of the oldest mare basalts (Table 1); and their isotopic systematics are much more complex. Anorthositic rocks, which generally have very low Rb/Sr ratios and very primitive Sr isotopic compositions have not been dated by the Rb–Sr method. However, Ar^{39}–Ar^{40} ages of many anorthositic rocks are about 4.0 b.y. although some ages as old as 4.2 b.y. have been reported (Schaefer and Husain, 1974; Stettler et al, 1974; Kirsten et al., 1974; see the bottom line of Table 1).

Rb–Sr ages of non-mare rocks cluster around 3.9–4.0 b.y. (Table 1). Except for 12013, typical KREEP basalts have the highest range of I values among non-mare rocks (Total range = 0.70019 to 0.70058 except for 14321 reported by Mark et al., 1973). This is about the lowest range for all terrestrial rocks (see Fig. 4 for comparison). Rock 12013, which contains 'lunar granite', is a polymict breccia with highly heterogeneous composition. Because of its complex isotopic systematics, interpretation of its Rb–Sr data is not straightforward. One interpretation (Lunatic Asylum, 1970) is that the 4.5 b.y. whole-rock isochron age represents the time of formation and mixing of rocks of widely different Rb/Sr

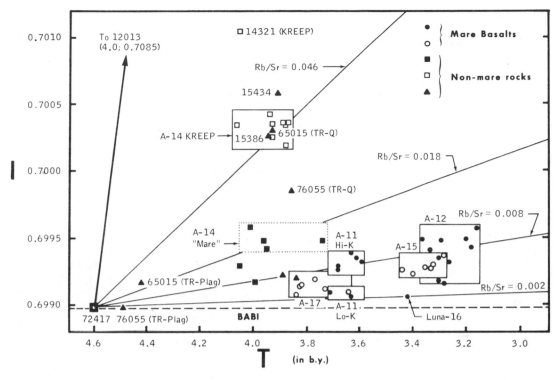

Fig. 3. Sr isotope evolution diagram (or I, T plot) for lunar samples. The majority of the data presented here are from the works of the C.I.T. group. Other data are as reported by workers at the Australian National University, the University of Minnesota, the Johnson Space Center, the University of California at Los Angeles, and U.S. Geological Survey, Denver

ratios; whereas the 4·0 b.y. mineral isochron age represents the time when local Sr isotopic homogenization took place in the polymict breccia. The highly evolved Sr isotopic composition indicates that the rock has evolved in a system with $Rb/Sr \cong 0.4$, higher than average chondritic materials (Fig. 4). No other type of lunar materials have evolved in such a high Rb/Sr system. The antiquity and high I value of this lunar 'granitic' rock together with frequent model ages of 4·3–4·6 b.y. for lunar rock and soils have been cited as strong evidence for an early ($\sim$4·3–4·6 b.y.) lunar differentiation (Papanastassiou and Wasserburg, 1971a). The differentiation of KREEP as the major radiogenic component (or magic component) in lunar soils (Hubbard et al., 1971; Schonfeld and Meyer, 1972) does not materially affect this conclusion.

Non-mare rocks (62295, 76055, 77135, etc.) of composition intermediate of KREEP

and anorthositic rocks have ages about 3·9 to 4·0 b.y., similar to that of KREEP basalts, but their I values generally show an intermediate range (0·69923–0·69985). Rocks 65015 (a KREEP basalt) and 76055 (a noritic breccia) clearly show lack of isotopic homogenization during an impact event about 3·9 b.y. ago. A cumulate rock, 72417 (dunite), returned in Apollo 17 mission, was reported to have an age of 4·60 ± 0·09 b.y. and $I = 0.69899 \pm 5$ (Albee et al., 1974). This age is as old as the solar system and the I value is as primitive as BABI. More analyses on this rock should be performed in order to confirm this important finding.

The frequent occurrence of ages of 3·9 to 4·0 b.y. for non-mare rocks by various radiometric dating methods was cited by Tera et al. (1974) as evidence of a terminal lunar cataclysm during which many of the major lunar basins were formed, e.g. Imbrium and

Serenitatis. Because most of the materials dated were impact-metamorphosed and brecciated, the event(s) at 3·9–4·0 b.y. can hardly be interpreted as due to endogenic magmatic processes.

As the mineral isochron ages for non-mare rocks are probably the result of impact metamorphism, they need not represent the time of their original differentiation. In order to date the time for the formation of KREEP materials, Nyquist et al. (1973, 1974) employed the whole-rock isochron method. Assuming a genetic relationship between anorthositic gabbros and KREEP basalts based on a study of experimental petrology by Walker et al. (1973), whole-rock isotopic data of a series of KREEP and related rocks defined as isochron of $t = 4·26 \pm 0·06$ b.y. and $I = 0·69926 \pm 10$; the age was interpreted as a minimum age of KREEP differentiation (Nyquist et al., 1974). In addition to the whole-rock method, one may use the model age method to estimate the time of KREEP differentiation. With the reasonable assumption that either Rb/Sr fractionation did not occur in impact-produced total melts of KREEP rocks in the 3·9–4·0 b.y. event, or Rb was preferentially increased relative to Sr in impact-produced partial melts, the calculated average model age of $4·30 \pm 0·04$ b.y. should indicate a minimum age of KREEP differentiation (Hubbard and Gast, 1971; Schonfeld and Meyer, 1972). Present evidence suggest that the Kreep differentiation most likely took place at about 4·3–4·5 b.y. ago.

IV. Terrestrial Archean Data

(A) General

The geological settings of terrestrial Archean rocks are very complicated and most of the geochronological studies were done before 1969 when the first Apollo samples were returned. The Rb–Sr data are now more variable than those for lunar samples, both in quality and in characteristics of the isotopic systems. Thus, although there is a much longer history of studying terrestrial rocks than lunar

samples, the availability of Rb–Sr data suitable for discussion of mantle Sr isotopic evolution seems more meagre. More specifically:

(1) Most previous age studies were performed on granitic rocks, because granitic rocks are not only more abundant than basaltic rocks in shield areas, but also possess very radiogenic Sr compositions which do not require high analytical precision to achieve an age determination. Data for comagmatic granitic rocks may define whole-rock isochron ages quite precisely, but they usually do not yield precise I values. Moreover, the origin of granitic rocks is not always clear.

(2) It is possible to determine very precise mineral isochrons for granitic rocks. Unfortunately, all terrestrial Archean granitic rocks have been subjected to one or more later geologic events that reset their isotopic systems. Mineral isochrons almost invariably show younger ages and higher I values than whole-rock isochrons, thus no useful information regarding the mantle Sr composition may be obtained.

(3) Unless exact field relationships are known, the assumption of identical I values for whole rock samples may be erroneous. The common scattering of whole-rock data about an isochron or 'errochron' (Brooks et al., 1972) may in part be attributed to the heterogeneity of original isotopic compositions. The interpretation of such data is ambiguous.

(4) Geological errors, resulting from differential response of mineral or whole-rock system to later events or processes leading to isotopic homogenization, such as chemical weathering, alteration, metamorphism, etc., can complicate the systems and make both age and initial ratio hard to determine. These errors cannot always be avoided even by very careful selection of samples.

(5) In addition to the above geological problems, there are analytical problems. Prior to the introduction of computerized data acquisition systems by Wasserburg et al. (1969) and improved chemical procedures, it

was impossible to measure Sr^{87}/Sr^{86} precisely enough to determine an isochron for low Rb/Sr basaltic rocks. This is one of the reasons that Rb-Sr isochron ages were seldom reported for Archean basalts.

Consequently, the choice of data for discussion of Sr isotope evolution in the Earth's mantle is very limited. Data for modern oceanic basalts are better documented because of recent extensive studies of oceanic basalts in connection with the studies of global tectonics. Furthermore, the geologic problems associated with these young rocks are less severe and can be handled more easily. The data selected for this discussion are listed in Table 2 and illustrated in Fig. 4. The data of oceanic basalts can be easily found in the literature and are not individually included in Table 2, but their ranges of Sr compositions are illustrated in Fig. 4. The (I, T) fields of lunar rocks are also plotted for comparison.

(B) *Terrestrial Rb–Sr Data*

The West Greenland data were obtained for high-grade Amitsoq gneisses (Moorbath et al., 1972). The best estimated I value (0.701) was interpreted either as an 'igneous' initial if a two-stage evolution process was assumed; or as a 'metamorphic' one if the emplacement of parental rocks took place less than 200 m.y. prior to metamorphism (Moorbath et al., 1972). Either way, the data for the Amitsoq gneisses cannot directly reflect the Sr composition of the contemporaneous mantle.

Many isotopic studies have been performed on rocks from the Onverwacht Group and associated granitic intrusions of South Africa, but relatively few data are for basaltic or very low Rb/Sr rocks. Allsopp et al. (1973) reported data for nine basic rocks of the Onverwacht Group and found that the age-corrected I values (assuming $T = 3.2$ b.y.) range from 0.7000 to 0.7058, among them

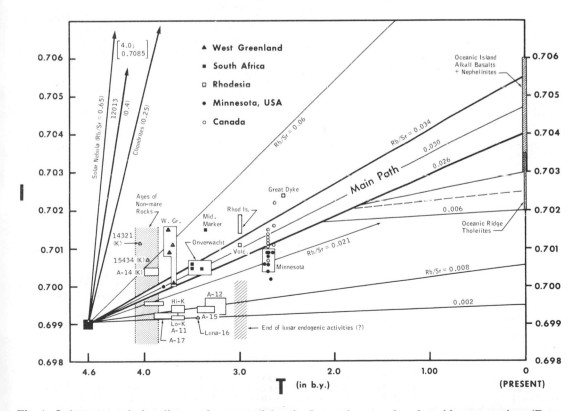

Fig. 4. Sr isotope evolution diagram for terrestrial rocks. Lunar data are also plotted for comparison (Data sources, see Table 2)

Table 2. Age and initial Sr^{87}/Sr^{86} data for terrestrial Archean rocks

Rock unit	t = age in b.y. ± 2	$I = (Sr^{87}/Sr^{86})_0$ ± 2	References	Comments
Minnesota, U.S.A.				
Ely Greenstone	$2 \cdot 69 \pm 0 \cdot 08$	$0 \cdot 70056 \pm 26$	Jahn and Murthy (1975)	I obtained from WR isochron
Newton Lake Fm	$2 \cdot 65 \pm 0 \cdot 11$	$0 \cdot 70086 \pm 24$	Jahn and Murthy (1975)	I obtained from WR isochron
Vermilion Granite	$2 \cdot 70 \pm 0 \cdot 05$	$0 \cdot 70041 \pm 29$	Jahn and Murthy (1975)	I obtained from WR isochron
				Basic data from Peterman et al. (1972) and Jahn and Murthy (1975)
	$2 \cdot 68 \pm 0 \cdot 10$	$0 \cdot 7006 \pm 12$	Peterman et al. (1972)	I obtained from WR isochron
Northern Light Gneiss	$2 \cdot 74 \pm 0 \cdot 10$	$0 \cdot 7007 \pm 4$	Hanson et al. (1971)	I obtained from WR isochron
Saganaga Tonalite	$2 \cdot 71 \pm 0 \cdot 56$	$0 \cdot 7010 \pm 6$	Hanson et al. (1971)	I obtained from WR isochron
Icarus Pluton	$2 \cdot 69 \pm 0 \cdot 48$	$0 \cdot 7100 \pm 14$	Hanson et al. (1971)	I obtained from WR isochron
Giants Range Granite	$2 \cdot 67 \pm 0 \cdot 07$	$0 \cdot 7003 \pm 19$	Prince and Hanson (1972)	I obtained from WR isochron
Morton-Montevideo Gneisses	$3 \cdot 80$	$0 \cdot 700$	Goldich and Hedge (1974)	Both t and I are best values from an eye-ball fit through six data points
Canada				
Pyroxenes, Superior Prov.	$2 \cdot 70$	$0 \cdot 7010$ to $0 \cdot 7014$	Hart and Brooks (1974)	Seven age-corrected I values for primary pyroxenes from two extrusive and three intrusive mafic units
Keewatin + Coutchiching	$2 \cdot 69 \pm 0 \cdot 08$	$0 \cdot 7003 \pm 13$	Hart and Davis (1969)	I obtained from WR isochron
Chibougamau Greenstone belt: Dore Lake Complex	$2 \cdot 72 \pm 0 \cdot 50$	$0 \cdot 7011 \pm 10$	Jones et al. (1974)	I obtained from WR isochron
Anorthosite-1		$0 \cdot 70068 \pm 6$	Jahn (unpublished)	Age-corrected I value; Sr composition measured at JSC; $t = 2 \cdot 70$ b.y.
Anorthosite-2		$0 \cdot 70150 \pm 7$	Jahn (unpublished)	Age-corrected I value; Sr composition measured at JSC; $t = 2 \cdot 70$ b.y.
Chibougamau Pluton	$2 \cdot 56 \pm 0 \cdot 16$	$0 \cdot 7007 \pm 4$	Jones et al. (1974)	I obtained from WR isochron
Meta-meladiorite		$0 \cdot 70133 \pm 6$	Jahn (unpublished)	Age-corrected I value; Sr composition measured at JSC; $t = 2 \cdot 6$ b.y.
Yellowknife:				
SE Granodiorite	$2 \cdot 64 \pm 0 \cdot 08$	$0 \cdot 7011 \pm 16$	Green and Baadsgaard (1971)	I obtained from WR isochron
Metavolcanics	$2 \cdot 63 \pm 0 \cdot 16$	$0 \cdot 7022 \pm 23$	Green et al. (1968)	I obtained from WR isochron
South Africa				
Onverwacht:				
Komati	$3 \cdot 50 \pm 0 \cdot 20$	$0 \cdot 70048 \pm 5$	Jahn and Shih (1974)	I obtained from mineral isochron
Anorthositic norite		$0 \cdot 70061 \pm 6$	Jahn and Shih (1974)	Age-corrected I value ($t = 3 \cdot 5$ b.y.)
Anorthosites and komatiites		$0 \cdot 7000$ to $0 \cdot 7006$	Allsopp et al. (1973)	Age-corrected I values assuming $t = 3 \cdot 2$ b.y.
Middle Marker Horizon	$3 \cdot 36 \pm 0 \cdot 07$	$0 \cdot 7015 \pm 18$	Hurley et al. (1972)	I obtained from WR isochron
Rhodesia:				
Bulawayan	$3 \cdot 0 - 3 \cdot 1$	$0 \cdot 70110 \pm 5$	Jahn (unpublished)	Age-corrected I value assuming $t = 3 \cdot 0$ b.y. Measured = $0 \cdot 70122$
Bulawayan limestones		$0 \cdot 7014$ to $0 \cdot 7019$	Bell and Blenkinsop (1974)	Measured values; uncorrected
Great Dyke	$2 \cdot 53 \pm 0 \cdot 09$	$0 \cdot 7024 \pm 8$	Davies et al. (1970)	I obtained from WR isochron
West Greenland:				
Amitsoq Gneisses:				
Narssaq Area	$3 \cdot 75 \pm 0 \cdot 09$	$0 \cdot 7015 \pm 8$	Moorbath et al. (1972)	All I's obtained from WR isochrons;
Qilangarssuit	$3 \cdot 74 \pm 0 \cdot 10$	$0 \cdot 7009 \pm 11$	Moorbath et al. (1972)	Best age = $3 \cdot 70 - 3 \cdot 75$ b.y.
Praestefjord	$3 \cdot 69 \pm 0 \cdot 23$	$0 \cdot 7001 \pm 17$	Moorbath et al. (1972)	Best $I = 0 \cdot 7010 - 0 \cdot 7015$
Isua Area	$3 \cdot 70 \pm 0 \cdot 14$	$0 \cdot 7011 \pm 20$	Moorbath et al. (1972)	

Note: All I values have been adjusted based on NBS-SRM987 Sr Standard = $0 \cdot 71025$ or E & A Sr Standard = $0 \cdot 7081$.

five fall in the range of 0·7000 to 0·7006. Jahn and Shih (1974) obtained a mineral isochron with $T = 3·50 \pm 0·20$ b.y. and $I = 0·70048 \pm 5(2_\sigma)$ for a basaltic komatiite of the Onverwacht Group. Because of the metamorphic nature of this rock, the I value was considered as an upper limit for the Sr isotopic composition of the mantle beneath that crustal segment. An age-corrected I value (assuming $T = 3·5$ b.y.) for one anorthositic norite is $0·70061 \pm 6$, slightly higher than that of the above basaltic komatiite. It should be noted that less precise data for these two rocks were included in Alsopp et al (1973), and we intended to use the more precise data in the plot of Fig. 4.

V. Rama Murthy (personal communications) measured Sr isotopic compositions of sedimentary barite samples which occur within the Onverwacht Group and found that the lowest measured value was $0·70047 \pm 5$; no correction was made for its I value due to a very low Rb/Sr ratio. This value is identical to our basaltic komatiite initial and also sets an upper limit for the upper mantle value. An age of $3·36 \pm 0·07$ b.y. and $I = 0·7015 \pm 18$ was determined for the Middle Marker Horizon by the whole-rock isochron method (Hurley et al., 1972). The large uncertainty in I is due to a long extrapolation from very radiogenic data points and limits the usefulness of these data for discussion of the Sr composition in the mantle.

A limited number of K–Ar ages and some common Pb data indicate that the Rhodesian craton is perhaps as old as 3·5 b.y. (Wilson and Harrison, 1973; Robertson, 1973). The preliminary data obtained at the Johnson Space Centre indicate that the age of the Midland–Bulawayan Group is about 3·0–3·1 b.y. with an I value of about 0·7011. Our data also show the effects of later thermal disturbances at about 2·7 b.y. and younger. Bell and Blenkinsop (1974) measured Sr^{87}/Sr^{86} ratios in some Bulawayan limestones and they reported a range of 0·7014 to 0·7019 (adjusted to our Sr scale, but not corrected for ages) which agrees with our data within their error limits. They suggested that the Sr isotopic composition in these limestones may merely reflect a localized source provenance, perhaps the volcanic rocks. A whole-rock isochron of $T = 2·53 \pm 0·09$ b.y. and $I = 0·7024 \pm 8$ was reported by Davies et al. (1970) for the Great Dyke of Rhodesia.

Within the Canadian Shield, the granite–greenstone terrain of the Vermilion district, northeastern Minnesota, has been studied extensively. The ages for all volcanic and intrusive rocks are $2·70 \pm 0·05$b.y. (total range) and the I values range from 0·7003 to 0·7010 (Hanson et al., 1971; Prince and Hanson, 1972; Peterman et al., 1972; Jahn and Murthy, 1975). Most recently, Goldich and Hedge (1974) reported whole rock Rb–Sr isotopic data for Montevideo gneisses in the Minnesota River Valley, and suggested that an isochron of $T = 3·8$ b.y. and $I = 0·700$ best suited for most data points. Interestingly, this age is the oldest one yet reported for terrestrial rocks, and is also older than the zircon age reported earlier by the same authors (Goldich et al., 1970). More work seems needed to confirm this significant finding.

Two whole rock isochron ages of $2·64 \pm 0·08$ and $2·63 \pm 0·16$ were reported for Yellowknife granodiorites and metavolcanic rocks, respectively. The associated I values for metavolcanics, however, had a very large uncertainty (Table 2; Green et al., 1968; Green and Baadsgaard, 1971). The Coutchiching meta-sediments and the Keewatin metavolcanics in the Rainy Lake area were dated by Hart and Davis (1969). They obtained a combined whole-rock isochron yielding $T = 2·69 \pm 0·08$ b.y. and $I = 0·7008 \pm 13$ (or 0·7007 adjusted to our scale). Recently Hart and Brooks (1974) analysed seven primary pyroxenes selected from some cumulate pyroxenite layers and segregations in thick ultramafic flows and sills in three different areas of the Canadian Shield. They found that all the age-corrected I values were relatively uniform, ranging from 0·7010 to 0·7014 with an average of (0·7012) and they interpreted these values as possible upper mantle values 2·7 b.y. ago. They further suggested that the modern oceanic island and island arc magmas, which have $Sr^{87}/Sr^{86} = 0·7035 - 0·7060$, may have evolved from the presumably closed mantle system throughout the history of the Earth.

Two meta-anorthositic and one meta-dioritic rocks of very low Rb/Sr ratios (0·0017–0·0055) from the Chibougamau greenstone belt were measured for their Sr isotopic composition. Using the isochron ages reported by Jones et al. (1974) for these rocks, the age-corrected I values are 0·70068 ± 6; 0·70150 ± 7, and 0·70133 ± 6 (Table 2; Jahn, unpublished). The first number resembles the average Minnesota value, while the last two are similar to those reported by Hart and Brooks (1974).

No data from the Aldan Shield, southeastern Siberia, the Baltic Shield, and the Ukranian Shield, the Brazilian Shield, the Indian Shield, or the western Australian Shield are shown in Fig. 4. Arriens et al. (1971) reported a large number of Rb–Sr data for Archean granitic rocks from western Australia, but in most cases the ages were not very precisely determined and the uncertainties in I values were about an order of magnitude larger in average than those selected here.

V. Discussion

All the (I, T) values for lunar samples in Fig. 4 were obtained by the mineral isochron methods, whereas most of those for terrestrial rocks were by the whole rock isochron method with only a few notable exceptions. The quality of I values for lunar samples is generallly far superior to that for terrestrial counterparts for reasons mentioned earlier. Nevertheless, the significant spread of both lunar and terrestrial data (Figs. 3 and 4) suggest that the mantles in both planets have been heterogeneous for a long period of time.

Lunar rocks, particularly the mantle-derived mare basalts, have more primitive initial $^{87}Sr/^{86}Sr$ ratios than do the terrestrial rocks (Fig. 4). This presumably reflects a lower Rb/Sr ratio (by about a factor 5–10) in the lunar mantle than in the Earth's mantle. Although KREEP basalts have I values which are comparable with those of some of the terrestrial rocks, the interpretation of these data is not yet clear. The occurrence of the lunar cataclysm at ~3·9 AE ago suggest that the isotopic systematics of KREEP rocks may have been reset by this event. The $^{87}Sr/^{86}Sr$ ratio in the source of KREEP basalts may have been significantly lower at the time of their original differentiation than the mineral isochron I values of ~0·7005 indicates. A whole-rock isochron for KREEP-related rocks suggests that they may have differentiated ~4·3 AE ago when $I = 0·69926$, a value more consistent with low Rb/Sr in the lunar mantle (cf. Nyquist et al., 1974).

Considering only radiometric age data, the history of lunar magmatism seems to have terminated about 3·1 b.y. ago. However, photogeologic studies of crater densities suggest that lunar magmatisms may have persisted to about 2 b.y. ago (Boyce et al., 1974). The presence of mascons and their relation with mare materials suggest that the lunar crust was rigid enough to support the dense mascons before the cessation of mare flooding. Endogenic lunar evolution probably approached completion 3·0 b.y. ago, and was certainly completed by about 2 b.y. ago. On the other hand, the evolution in the Earth was still in a young stage. It has been suggested that the Archean lithosphere was not even rigid enough to form plates and thus plate tectonic concepts were not applicable to Archean time (Burke and Dewey, 1973). Of course, this is still a point of controversy. However, it is imaginable that during the Archean time, creation of new and destruction of old crustal materials, or mixing and recycling processes between the crust and mantle would likely take place more efficiently than at present. The rarity of lunar rocks with ages in excess of 4·0 b.y. has been widely interpreted as due to resetting of isotopic systems by intense bombardment of the moon in the interval of 4·0–4·6 b.y. ago. The exact nature of the bombardment implied by the isotopic data is under some debate (see, for example, Tera et al., 1974; Schaeffer and Husain, 1974; Hartmann, 1975), nevertheless, it is reasonable to suggest that the Earth also underwent an early heavy bombardment which could have destroyed the pre-existing crustal materials.

Models on lunar crustal evolution are available in the lunar literature (for examples, Smith et al., 1970; Wood et al., 1970; Ringwood and Essene, 1970; Papanastassiou and Wasserburg, 1971; Gast, 1972; Toksöz et al., 1972; Taylor and Jakeš, 1974). As mentioned above, Rb–Sr data suggest that the heterogeneous nature of the lunar mantle was established shortly after the formation of the Moon. The bulk of the Moon was subdivided into many subsystems of different Rb/Sr ratios as a result of major fractionation processes about 4·3–4·6 b.y. ago. The majority of model ages for lunar basalts (both mare and non-mare types) except the Apollo 11 high-K basalts are close to 4·6 b.y.; strongly suggesting that the source regions for these rocks have remained relatively isolated, and that convection in the lunar mantle probably did not operate to any significant extent. Since convection may help homogenize mantle composition (chemical and isotopic), the lack of it would likely result in a discontinuous pattern of Sr isotope evolution, that is, one or more stages of linear growth paths as shown in Figs. 3 and 4.

In contrast, convection in the terrestrial mantle has probably played a significant role in the Earth's tectonic history and crustal evolution. Continuous tapping of partial melts in which the Rb/Sr ratio are expected to be higher than in their source would reduce the Rb/Sr ratio in the residual mantle, and thus Hart and Brooks (1970) proposed a convex curvilinear path for mantle Sr isotopic evolution (similar to Curve B of Fig. 1). The available data, however, indicate that (1) the mantle is not a simple system that can be defined by a single growth curve but instead the mantle in the Archean time was heterogeneous, and (2) mantle convection did not effectively homogenize the chemical and isotopic composition.

It is known that modern oceanic ridge (or floor) basalts have a rather uniform composition with an average Rb/Sr = 0·008 and $Sr^{87}/Sr^{86} = 0·7025$ (Hart, 1971). These two parameters result in a model age of about 8·2 b.y. which is unreasonably older than the age of the Earth. This high model age suggests that a depletion of Rb relative to Sr in the source region took place sometime in the past but significantly after planetary formation. In addition, there are other 'anomalous' chemical characteristics observed in ridge basalts, such as, very low contents of K, Ba and La, and very high K/Rb ratios (500 to 1500). Thus, oceanic ridge basalts cannot have evolved in a single-stage isolated mantle source from the primordial Sr composition 4·6 b.y. ago, because the calculated Rb/Sr ratio for this model evolution is 0·018, a value substantially greater than the inferred value of about 0·006–0·008 for the source of ridge basalts.

On the other hand, all available oceanic alkali basalts and nephelinites have model ages less than 4·6 b.y., but greater than 2·0 b.y. with a mean of 3·0–3·2 b.y. (basic data from the literature). Assuming a single stage evolution, the formula used by Schonfeld and Meyer (1972) allows a model calculation of the enrichment factor of Rb/Sr in the melt relative to the source, i.e.

$$F = \frac{e^{\lambda T_0} - e^{\lambda T}}{e^{\lambda T_m} - e^{\lambda T}}$$

where F = Rb/Sr enrichment factor in the melt; T_0 = age of the Earth or Moon = 4·6 b.y.; T = age of the rocks = 0 for modern oceanic rocks; T_m = model age of rock; and λ = decay constant of Rb^{87} = $1·39 \times 10^{-11}$ yr^{-1}. The average enrichment factor calculated is about 50%, i.e. the Rb/Sr ratio in the alkali basalts and nephelinites was increased by 50% over the source during partial melting (a few percent of partial melting, based on calculation of Kay and Gast, 1973; Gast, 1968). Alternatively, if one assumes that the average Sr^{87}/Sr^{86} ratio = 0·7045 for these rocks, a single stage evolution from $I_0 = 0·699$ would require Rb/Sr = 0·029, which reflects the source value. The average Rb/Sr ratios in the rocks are about 0·040–0·045, therefore an enrichment factor of about 50% is also derived. The above calculations reveal that a single stage evolution for alkali basalts and nephelinites in oceanic environment is reasonable and the source for these rocks may still be quite pristine and not very depleted in lithophile elements. This is consistent with the observation of Peterman and Hedge (1971) who found that Sr^{87}/Sr^{86} ratios in oceanic basalts were in positive correlation with $K_2O/(K_2O + Na_2O)$ ratios and suggested that alkali basalts from many oceanic islands represented the most primitive mantle materials which were least depleted in alkali metals.

As shown in Fig. 4, the roughly linear array of Archean data for the 3·8 b.y. Montevideo gneiss, the 3·5–3·3 b.y. Onverwacht volcanic rocks and barites, the ~3·1–3·0 b.y. Rhodesian metabasalts, the 2·7 b.y. pyroxenes and meta-anorthosites and meta-diorite of Canadian volcanic rocks (but not those of Minnesota) and modern alkali basalts suggests that these rocks might have evolved from a mantle system with a rather constant Rb/Sr ratio = 0·025–0·035. We shall call this evolutionary path the Main Path (Fig. 4). This roughly linear array of data (similar to Curve A of Fig. 1) may be due to: (1) the upper mantle is an infinite reservoir for Rb and Sr in comparison with the volume of volcanic material removed from it, (2) upward transport (thus depletion) of alkali metals from the upper mantle to the crustal regime was compensated by the input of these elements from below, or (3) alkali elements lost to the crustal regime were returned and recycled through the upper mantle. The relative constancy of trace element contents in volcanic rocks of different geologic ages has been explained as a combined result of (2) and (3) above (Jahn et al., 1974).

The (I, T) field of Minnesota samples indicates that their source has evolved in a system of $Rb/Sr \cong 0·02$ if a single stage model is assumed. Although modern ocean ridge basalts and Minnesota data form a linear trend, modern ridge basalts do not have the same parentage as the Minnesota samples because of their very low Rb/Sr ratios. From the data of Minnesota and Canadian samples, it is evident that the Archean mantle 2·7 b.y. ago was distinguishably heterogeneous. It is suggested that the source for Minnesota samples was depleted probably earlier than 3·5 b.y. ago.

The available distribution coefficients for possible mantle minerals (Gast, 1968; Philpotts and Schnetzler, 1970; Schnetzler and Philpotts, 1970; Griffin and Murthy, 1969; Shimizu, 1974; Hart and Brooks, 1974) and the general range of Rb/Sr enrichment factors estimated for lunar mare basalts (see Nyquist, 1975 for discussion) indicate that the most likely range for Rb/Sr enrichment for large degrees of partial melting (about 20%) is about 1·0 to 1·4 (liquid/source). Therefore, the inferred mantle source for oceanic ridge basalts has a Rb/Sr ratio of about 0·006–0·008. Using the average Sr^{87}/Sr^{86} ratio of 0·7025 and Rb/Sr = 0·006 for the source of oceanic ridge basalts, a Sr isotope evolution path may be extrapolated backward in time and is found to intercept the Main Path at about 1·8–2·0 b.y. ago (Fig. 4). If, in an extreme case, no fractionation of Rb/Sr during partial melting is assumed, then the intercept would be found at about 2·7 b.y. ago. This suggests that the source for ocean ridge basalts was isolated about 1·8 to 2·7 b.y. ago following a large scale mantle fractionation leading to severe alkali depletion. New isotopic data by Sun (1973) for oceanic basalts indicate a linear trend in a Pb^{207}/Pb^{204} vs Pb^{206}/Pb^{204} plot. This linear trend may be interpreted as a mixing line or as a secondary isochron. If the secondary isochron interpretation is chosen, the slope which corresponds to an age of 1·8 b.y. suggests that the source for oceanic basalts was indeed fractionated into various μ ($= U^{238}/Pb^{204}$) subsystems. Geologically, the period of 1·8–2·0 b.y. B.P. recorded significant thermal and orogenic activities on the Earth. The status of 1·8 to 2·0 b.y. B.P. period should receive more attention in attempts to explain the terrestrial crustal evolution.

Crustal evolution is obviously controlled by the thermal history of individual planets, which in turn is a function of planetary size and content of heat-producing elements, such as K, U, and Th. The Moon is presently cool and tectonically inactive, but the Earth still retains enough thermal energy to have tectonic activity throughout its geologic history. The chemical and isotopic characteristics of lunar crustal material probably have been frozen since about 2 to 3 b.y. ago, except for modifications which have occurred due to meteoritic bombardment. The effect of these modifications on the interpretation of lunar sample data is somewhat controversial and is an active area of current research. On the Earth, continuous tectonic and other surface activities, such as metamorphism, weathering, mechanical mixing and sedimentary differentiation, etc., constantly modified the chemical and isotopic composition of most crustal and even mantle rocks.

Finally, the following contrasting points on crustal evolution in the early Earth–Moon system are presented: (1) In addition to several episodes of mare flooding (3·9 to 3·1 b.y., based on radiometric data only), crustal evolution on the Moon was heavily influenced by impact processes. Impact processes might have been important in the Earth's early history ($\geq 4\cdot0$ b.y.) but no record of such importance (relative to magmatism) has been preserved. The hypothesis of impact origin for the Onverwacht greenstone belt by Green (1972) remains controversial or untenable if sedimentary layers, such as banded ironstone, barite, cherts and jasperites, within the mafic–ultramafic units of the Onverwacht are to be accounted for. (2) Again considering only the radiometric time scale, the crustal evolution virtually ceased about 3·0 b.y. ago on the Moon, but still continues on the Earth. The crustal segments on the Earth were probably completed prior to 2·5–2·7 b.y. ago; this period is one of the most important episodes of terrestrial crustal evolution and events at that time are recorded in almost all continents. (3) The crustal evolution has been heavily influenced by water and other volatiles on the Earth but not on the Moon. Anorthositic rocks are thought to be the earliest formed 'scum' on the Moon (Wood et al., 1970) and KREEP differentiation likely took place very early in the Moon's history (4·3–4·5 b.y.). In contrast there is indication that the earliest terrestrial sialic crust is plagioclase-rich tonalitic rocks, but the plagioclase is Na-rich and not Ca-rich as occurred on the Moon. Because the bulk composition, the volatile contents and the thermal histories are so different in both planets, does any comparison become futile and meaningless? Or do events recorded in lunar history have implications for terrestrial evolution? Was terrestrial crust-mantle evolution always dominated by endogenic processes? Certainly endogenic processes are of primary importance for both planets, but did the intense bombardment of the Earth–Moon system by planetesimal-size objects during the period of 4·0–4·6 b.y. ago also profoundly affect their evolution? Perhaps a better understanding of crustal evolution in both the Moon and the Earth will evolve from more detailed study of terrestrial Archean rocks and lunar samples.

Acknowledgements

Comments and suggestions by C. Meyer, Jr., of the Johnson Space Center are greatly appreciated. The Lunar Science Institute is operated by the Universities Space Research Association under Contract No. NSR 09-051-001 with the National Aeronautics and Space Administration.

References

Albee, A. L., Chodos, A. A., Dymek, R. F., Gancarz, A. I., Goldman, D. S., Papanastassiou, D. A., and Wasserburg, G. J., 1974. *Lunar Science* V. Pt. 1, 3–5.

Alexander, E. C., and Kahl, S. B., 1974. 'Ar^{40}–Ar^{39} studies of lunar breccias', *Proc. 5th Lunar Sci. Conf.*, **2**, 1353–1373.

Allsopp, H. L., Ulrych, T. J., and Nicolaysen, L. O., 1968. 'Dating of some significant events in the history of the Swaziland System by the Rb–Sr isochron method', *Can. J. Earth Sci.*, **5**, 605–619.

Allsopp, H. L., Viljoen, M. J., and Viljoen, R. P., 1973. 'Strontium isotopic studies of the mafic and felsic rocks of the Onverwacht Group of the Swaziland Sequence', *Geologische Rundschau*, **62**, 902–917.

Armstrong, R. L., 1968. 'A model for the evolution of strontium and lead isotopes in a dynamic Earth', *Rev. Geophysics*, **6**, 175–200.

72

Armstrong, R. L., and Hein, S. M., 1973. 'Computer simulation of Pb and Sr isotope evolution of the Earth's crust and upper mantle', *Geochim. Cosmochim. Acta*, **37**, 1–18.

Arriens, P. A., 1971. 'The Archean geochronology of Australia', *Spec. Publs. Geol. Soc. Aust.*, **3**, 11–23.

Arth, J. G., and Hanson, G. N., 1972. 'Quartz diorites derived by partial melting of eclogite or amphibolite at mantle depths', *Contr. Min. Petrol.*, **37**, 161–174.

Bell, K., and Blenkinsop, J., 1974. 'Sr isotope composition of the Bulawayan limestone' [Abs.], *Geol. Soc. Ann. Meeting Miami Beach*, **6**, p. 650.

Boyce, J. M., Dial, A. L., and Soderblom, L. A., 1974. 'Ages of the lunar nearside light plains and maria', *Proc. 5th Lunar Sci. Conf.*, **1**, 11–23.

Brooks, C., Hart, S. R., and Wendt, I., 1972. 'Realistic use of two-error regression treatments as applied to rubidium-strontium data', *Rev. Geophys. Space Phys.*, **10**, 551–577.

Burke K., and Dewey, J. F., 1973. 'An outline of Precambrian plate development', In Tarling, D. H., and Runcorn, S. K. (Eds.), *Implication of Continental Drift to the Earth Sciences*, 1035–1046.

Compston, W., Vernon, M. J., Berry, H., Rudowski, R., Gray, C. M., and Ware, N., 1972. 'Apollo 14 mineral ages and the thermal history of the Fra Mauro formation', *Proc. 3rd Lunar Sci. Conf.*, **2**, 1487–1501.

Davies, R. D., Allsopp, H. L., Erlank, A. J., and Manton, W. I., 1970. 'Sr-isotopic studies on various layered mafic intrusions in southern Africa', *Spec. Publ. Geol. Soc. S. Afr.*, **1**, 576–593.

Delano, J. W., Bence, A. E., Papike, J. J., and Cameron, D. D., 1973. 'Petrology of the 2–4 mm soil fraction from the Descartes region of the Moon and stratigraphic implications', *Proc. 4th Lunar Sci. Conf.*, **1**, 537–551.

Dowty, E., Prinz, M., and Keil, K., 1974a. ' "Very high alumina basalt": A mixture and not a magma type', *Science*, **183**, 1214–1215.

Dowty, E., Prinz, M., and Keil, K., 1974b. 'Igneous rocks from Apollo 16 rake samples', *Proc. 5th Lunar Sci. Conf.*, **1**, 431–445.

Evensen, N. M., Murthy V. Rama and Coscio, M. R., Jr., 1973. 'Rb-Sr ages of some mare basalts and the isotopic and trace element systematics in lunar fines', *Proc. 4th Lunar Sci. Conf.*, **2**, 1707–1724.

Gast, P. W., 1967. 'Isotope geochemistry of volcanic rocks', In *Basalts, a Treatise on Rocks of Basaltic Composition*, Hess and Poldervaart (Eds.), Interscience, New York Vol. 1, 325–358.

Gast, P. W., 1968a. 'Trace element fractionation and the origin of tholeiitic and alkaline magma types', *Geochim. Cosmochim. Acta*, **32**, 1057–1086.

Gast, P. W., 1968b. 'Upper mantle chemistry and evolution of the earth's crust', In *The History of the Earth's Crust*, R. A. Phinney (Ed.), Princeton University Press, pp. 15–27.

Gast, P. W., 1972. 'The chemical composition and structure of the Moon', *The Moon*, **5**, 121–148.

Gast, P. W., 1973. 'Lunar magmatism in time and space', *Lunar Science* **IV**, LSI, 275–277.

Gast, P. W., Hubbard, N. J., and Weisman, H., 1970. 'Chemical composition and petrogenesis of basalts from Tranquility Base', *Proc. Apollo 11 Lunar Sci. Conf.*, **2**, 1143–1163.

Goldich, S. S., Hedge, C. E., and Stern, T. W., 1970. 'Age of the Morton and Montevideo gneisses and related rocks, southwestern Minnesota', *Geol. Soc. Am. Bull.*, **81**, 3671–3696.

Goldich, S. S., and Hedge, C. E., 1974. '3800-Myr granitic gneiss in south-western Minnesota', *Nature*, **252**, 467–468.

Green, D. C., Baadsgaard, H., and Cumming, G. L., 1968. 'Geochronology of the Yellowknife area, Northwest Territories, Canada, *Can, J. Earth Sci.*, **5**, 725–735.

Green, D. C., and Baadsgaard, H., 1971. 'Temporal evolution and petrogenesis of an Archean crustal segment at Yellowknife, N.W.T. Canada', *Jour. Petrol.*, **12**, 177–217.

Green, D. H., 1972. 'Archean greenstone belts may include terrestrial equivalents of lunar maria?', *Earth Planet. Sci. Lett.*, **15**, 263–270.

Griffin, W. L., and Murthy, V. Rama, 1969. 'Distribution of K, Rb, Sr, and Ba in some minerals relevant to basalt genesis', *Geochim. Cosmochim. Acta*, **33**, 1389–1414.

Hanson, G. N., Goldich, S. S., Arth, J. G., and Yardley, D. H., 1971. 'Age of the early Precambrian rocks of the Saganaga Lake—Northern Light Lake area, Ontario-Minnesota', *Can. J. Earth Sci.*, **8**, 1110–1124.

Hart, S. R., 1971. 'K, Rb, Cs, Sr, and Ba contents and Sr isotope ratios of ocean floor basalts', *Phil. Trans. Roy. Soc. Lond.*, **268**, 573–587.

Hart, S. R., and Brooks, C., 1970. 'Rb-Sr mantle evolution models', *Dept. Terr. Mag.*, Carnegie Inst., Ann. Rept. 1968/1969, 426–429.

Hart, S. R., and Brooks, C., 1974. 'The geochemistry and evolution of early Precambrian mantle', *Ann. Rept. Dept. Terr. Mag.*, Carnegie Inst.

Hart, S. R., and Brooks, C., 1974. 'Clinopyroxene-matrix partitioning of K, Rb, Cs, Sr and Ba', *Geochim. Cosmochim. Acta*, **38**, 1799–1806.

Hart, S. R., and Davis, G. L., 1969. 'Zircon, U-Pb and whole rock Rb-Sr ages and early crustal development near Rainy Lake, Ontario', *Geol. Soc. Amer. Bull.*, **80**, 595–614.

Hartmann, W. K., 1972. 'Paleocratering of the Moon: Review of post-Apollo data', *Astrophys. Space Sci.*, **17**, 48–64.

Hartmann, W. K., 1975. 'Lunar "cataclysm": a misconception?', *Icarus*, **24**, 181–187.

Haskin, L. A., Helmke, P. A., Blanchard, D. P., Jacobs, J. W., and Telander, K., 1973. 'Major and trace element abundances in samples from the lunar highlands', *Proc. 4th Lunar Sci. Conf.*, **2**, 1275–1296.

Hubbard, N. J., and Gast, P. W., 1971. 'Chemical composition and origin of nonmare lunar basalts', *Proc. 2nd Lunar Sci. Conf.*, **2**, 999–1020.

Hubbard, N. J., Gast, P. W., Rhodes, J. M., Bansal, B. M., Wiesmann, H., and Church, S. E., 1972. 'Non-mare basalts: Part II', *Proc. 3rd Lunar Sci. Conf.*, **2**, 1161–1179.

Hubbard, N. J., Meyer, C., Jr., Gast, P. W., and Wiesmann, H., 1971. 'The composition and derivation of Apollo 12 soils', *Earth Planet. Sci. Lett.*, **10**, 341–350.

Hubbard, N. J., Rhodes, J. M., Gast, P. W., Bansal, B. M., Shih, C. Y., Wiesmann, H., and Nyquist, L. E., 1973a. 'Lunar rock types: The role of plagioclase in non-mare and highland rock types', *Proc. 4th Lunar Sci. Conf.*, **2**, 1297–1312.

Hubbard, N. J., Rhodes, J. M., and Gast, P. W., 1973b. Chemistry of very high Al_2O_3 lunar basalts', *Science*, **181**, 339–342.

Hubbard, N. J., Rhodes, J. M., Wiesmann, H., Shih, C., and Bansal, B. M., 1974. 'The chemical definition and interpretation of rock types returned from the non-mare regions of the Moon', *Proc. 5th Lunar Sci. Conf.*, **2**, 1227–1246.

Hulme, G., 1974. 'Generation of magma at lunar impact crater sites', *Nature*, **252**, 556–558.

Hurley, P. M., Hughes, H., Faure, G., Fairbairn, H. W., and Pinson, W. H., 1962. 'Radiogenic Sr-87 model of continental formation', *J. Geophys. Res.*, **67**, 5315–5334.

Hurley, P. M., Pinson, W. H., Jr., Nagy, B., and Teska, T. M., 1972. 'Ancient age of the Middle Marker Horizon: Onverwacht Group, Swaziland Sequence, South Africa', *Earth Planet. Sci. Lett.*, **14**, 360–366.

Hurley, P. M., and Rand, J. R., 1969. 'Pre-drift continental nuclei', *Science*, **164**, 1229–1242.

Husain, L., Schaeffer, O. A., Funkhouser, J., and Sutter, J., (1972). 'The ages of lunar material from Fra Mauro, Hadley Rille, and Spur Crater', *Proc. 3rd Lunar Sci. Conf.*, **2**, 1557–1567.

Jahn, B., and Murthy, V. R., 1975. 'Rb–Sr ages of the Archean rocks from the Vermilion district, northeastern Minnesota', *Geochim. Cosmochim. Acta.* (in press).

Jahn, B., and Shih, C., 1974. 'On the age of the Onverwacht Group, Swaziland Sequence, South Africa', *Geochim. Cosmochim. Acta*, **38**, 873–885.

Jahn, B., Shih, C., and Murthy, V. R., 1974. 'Trace element geochemistry of Archean volcanic rocks', *Geochim. Cosmochim. Acta*, **38**, 611–627.

Jones, L. M., Walker, R. L., and Allard, G. O., 1974. 'The rubidium-strontium whole-rock age of major units of the Chibougamau greenstone belt, Quebec', *Can. J. Earth Sci.*, **11**, 1550–1561.

Kay, R. W., and Gast, P. W., 1973. 'The rare-earth content and origin of alkali-rich basalts', *J. Geology*, **81**, 653–682.

Kay, R., Hubbard, N. J., and Gast, P. W., 1970. 'Chemical characteristics and origin of oceanic ridge volcanic rocks', *J. Geophys. Res.*, **75**, 1585–1613.

Kirsten, K., and Horn, P., 1973. 'Chronology of the Taurus-Littrow region III: Ages of mare basalts and highland breccias and some remarks about the interpretation of lunar highland rock ages', *Proc. 5th Lunar Sci. Conf.*, **2**, 1451–1475.

Laul, J. C., and Schmitt, R. A., 1973. 'Chemical composition of Apollo 15, 16 and 17 samples', *Proc. 4th Lunar Sci. Conf.*, **2**, 1349–1367.

Lunatic Asylum, 1970. 'Mineralogic and isotopic investigations on lunar rock 12013', *Earth Planet. Sci. Lett.*, **9**, 137–163.

Mark, R. K., Cliff, R. A., Lee-Hu, C., and Wetherill, G. W., 1973. 'Rb-Sr studies of lunar breccias and soils', *Proc. 4th Lunar Sci. Conf.*, **2**, 1785–1795.

Mark, R. K., Lee-Hu, C. N., and Wetherill, G. W., 1974. 'Rb-Sr age of lunar igneous rocks 62295 and 14310', *Geochim. Cosmochim. Acta*, **38**, 1643–1648.

Metzger, A. E., Trombka, J. I., Reedy, R. C., and Arnold, J. R., 1974. 'Element concentrations from lunar orbital gamma-ray measurements', *Proc. 5th Lunar Sci. Conf.*, **2**, 1067–1078.

Meyer, C., Jr., Brett, R., Hubbard, N. J., Morrison, D. A., Mckay, D. S., Aitkin, F. K., Taketa, H., and Schonfeld, E., 1971. 'Mineralogy, chemistry of origin of the KREEP component in soil samples from the Ocean of Storms', *Proc. 2nd Lunar Sci. Conf.*, **1**, 393–411.

Moorbath, S., O'Nions, R. K., Parkhurst, R. J., Gale, N. H., and McGregor, V. R., 1972. 'Further Rb-Sr age determinations on the very early Precambrian rocks of the Godthaab District, West Greenland', *Nature (Phys. Sci.)*, **240**, 78–82.

Murthy, V. Rama, and Banerjee, S. K., 1973. 'Lunar evolution: how well do we know it now?', *The Moon*, **7**, 149–171.

Murthy, V. Rama, Evensen, N. M., Jahn, B., and Coscio, M. R., Jr., 1972. 'Apollo 14 and 15 samples: Rb–Sr ages, trace elements, and lunar evolution', *Proc. 3rd Lunar Sci. Conf.*, **2**, 1503–1514.

Nunes, P. D., Knight, R. J., Unruh, D. M., and Tatsumoto, U. S. 1974a. 'The primitive nature of the lunar crust and the problem of initial Pb isotopic compositions of lunar rocks: A Rb-Sr and U-Th-Pb study of Apollo 16 samples', *Lunar Sci. V*, LSI, 559–561.

Nunes, P. D., Tatsumoto, M., and Unruh, D. M., 1974b. 'U-Th-Pb and Rb-Sr systematics of

74

Apollo 17 Boulder 7 from the North Massif of the Taurus-Littrow Valley', *Earth Planet. Sci. Lett.*, **23**, 445–452.

Nyquist, L. E., 1975. 'Lunar Rb-Sr chronology. A review paper to be published in *Phys. and Chem. of the Earth.*

Nyquist, L. E., Bansal, B. M., Wiesmann, H., and Jahn, B. M., 1974. 'Taurus-Littrow chronology: Some constraints on early lunar crustal development', *Proc. 5th Lunar Sci. Conf.*, **2**, 1515–1539.

Nyquist, L. E., Hubbard, N. J., Gast, P. W., Bansal, B. M., Wiesmann, H., and Jahn, B., 1973. 'Rb-Sr systematics for chemically defined Apollo 15 and 16 materials', *Proc. 4th Lunar Sci. Conf.*, **2**, 1823–1846.

Nyquist, L. E., Hubbard, N. J., Gast, P. W., Church, S. E., Bansal, B. M., and Wiesmann, H., 1972. 'Rb-Sr systematics for chemically defined Apollo 14 breccias', *Proc. 3rd Lunar Sci. Conf.*, **2**, 1515–1530.

O'Hara, M. J., Biggar, G. M., Hill, P. G., Jefferies, B., and Humphries, D. J., 1974. 'Plagioclase saturation in lunar high-titanium basalt', *Earth Planet. Sci. Lett.*, **21**, 253–268.

O'Hara, M. J., Biggar, G. M., Richardson, S. W., Ford, C. E., and Jamieson, B. G., 1970. 'The nature of seas, mascons, and the lunar interior in the light of experimental studies', *Proc. Apollo 11 Lunar Sci. Conf.*, **1**, 695–710.

Papanastassiou, D. A., and Wasserburg, G. J., 1969. 'Initial strontium isotopic abundances and the resolution of small time differences in the formation of planetary objects', *Earth Planet. Sci. Lett.*, **5**, 361–376.

Papanastassiou, D. A., and Wasserburg, G. J., 1970. 'Rb-Sr ages from the Ocean of Storms', *Earth Planet. Sci. Lett.*, **8**, 269–278.

Papanastassiou, D. A., and Wasserburg, G. J., 1971a. 'Lunar chronology and evolution from Rb-Sr studies of Apollo 11 and 12 samples', *Earth Planet. Sci. Lett.*, **11**, 37–62.

Papanastassiou, D. A., and Wasserburg, G. J., 1971b. 'Rb-Sr ages of igneous rocks from the Apollo 14 mission and the age of the Fra Mauro Formation', *Earth Planet. Sci. Lett.*, **12**, 36–48.

Papanastassiou, D. A., and Wasserburg, G. J., 1972a. 'Rb-Sr age of a Luna 16 basalt and the model age of lunar soils', *Earth Planet. Sci. Lett.*, **13**, 368–374.

Papanastassiou, D. A., and Wasserburg, G. J., 1972b. 'The Rb-Sr age of a crystalline rock from Apollo 16', *Earth Planet. Sci Lett.*, **16**, 289–298.

Papanastassiou, D. A., and Wasserburg, G. J., 1972c. 'Rb-Sr systematics of Luna 20 and Apollo 16 samples', *Earth Planet. Sci. Lett.*, **17**, 52–63.

Papanastassiou, D. A., and Wasserburg, G. J., 1973. 'Rb-Sr ages and initial strontium in basalts from Apollo 15', *Earth Planet. Sci. Lett.*, **17**, 324–337.

Papanastassiou, D. A., Wasserburg, G. J., and Burnett, D. S., 1970. 'Rb-Sr ages of lunar rocks from the Sea of Tranquility', *Earth Planet. Sci. Lett.*, **8**, 1–19.

Patterson, C., 1964. 'Characteristics of lead isotope evolution on a continental scale in the earth', In H. Craig, S. L. Miller and G. J. Wasserburg (Eds.), North-Holland Co., Amsterdam, pp. 244–268.

Patterson, C., and Tatsumoto, 1964. 'The significance of lead isotopes in detrital feldspar with respect to chemical differentiation within the Earth's mantle', *Geochim. Cosmochim. Acta*, **28**, 1–22.

Peterman, Z. E., Goldich, S. S., Hedge, C. E., and Yardley, D. H., 1972. 'Geochronology of the Rainy Lake region, Minnesota-Ontario', *Geol. Soc. Amer. Memoir*, **135**, 193–215.

Peterman, Z. E., and Hedge, C. E., 1971. 'Related strontium isotopic and chemical variations in oceanic basalts', *Geol. Soc. Amer. Bull.*, **82**, 493–500.

Philpotts, J. A., and Schnetzler, C. C., 1970a. 'Phenocryst-matrix partition coefficients for K, Rb, Sr and Ba with applications to anorthosite and basalt genesis', *Geochim. Cosmochim. Acta*, **34**, 307–322.

Philpotts, J. A., and Schnetzler, C. C., 1970b. 'Apollo 11 lunar samples: K, Rb, Sr, Ba and rare-earth concentration in some rocks and separated phases', *Proc. Apollo 11 Lunar Sci. Conf.*, **2**, 1471–1486.

Philpotts, J. A., Schnetzler, C. C., Nava, D. F., Bottino, M. J., Fullagar, P. D., Thomas, H. H., Schuhmann, S., and Kouns, C. W., 1972. 'Apollo 14: Some geochemical aspects', *Proc. 3rd Lunar Sci. Conf.*, **2**, 1293–1305.

Philpotts, J. A., Schuhmann, S., Kouns, C. W., Lum, R. K. L., and Winzer, S., 1974. 'Origin of Apollo 17 rocks and soils', *Proc. 5th Lunar Sci. Conf.*, **2**, 1255–1267.

Podosek, F. A., Huneke, J. C., Gancarz, A. J., and Wasserburg, G. J., (1973). 'The age and petrography of two Luna 20 fragments and inferences for widespread lunar metamorphism', *Geochim. Cosmochim. Acta*, **37**, 887–904.

Prince, L. A., and Hanson, G. N., 1972. 'Rb-Sr isochron ages for the Giants Range Granite, northeastern Minnesota', *Geol. Soc. Amer. Memoir*, **135**, 217–224.

Prinz, M., Dowty, E., and Keil, K., 1973. 'Mineralogy, petrology and chemistry of lithic fragments from Luna 20 fines: Origin of the cumulate ANT suite and its relationship to high-alumina and mare basalts', *Geochim. Cosmochim. Acta*, **37**, 979–1006.

Reid, J. B., Jr., 1972. 'Olivine-rich, true spinel-bearing anorthosites from Apollo 15 and Luna 20 soils—possible fragments of the earliest formed lunar crust', *The Apollo 15 Lunar Samples*, LSI, 154–156.

Rhodes, J. M., and Hubbard, N. J., 1973. 'Chemistry, classification and petrogenesis of Apollo 15 mare basalts', *Proc. 4th Lunar Sci. Conf.*, **2**, 1127–1148.

Ringwood, A. E., 1972. 'Zonal structure and origin of the Moon', Lunar Sci. III, LSi, 651–653.

Ringwood, A. E., 1974a. 'Minor element chemistry of maria basalts', *Lunar Science* **V**, [Abs.], 5th Lunar Sci. Conf.., LSI, 633–638.

Ringwood, A. E., 1974b. 'Heterogeneous accretion and the lunar crust', *Geochim. Cosmochim. Acta*, **38**, 983–984.

Ringwood, A. E., and Essene, E., 1970. 'Petrogenesis of Apollo 11 basalts, internal constitution and origin of the Moon', *Proc. Apollo 11 Lunar Sci. Conf.*, **1**, 769–799.

Robertson, D. K., 1973. 'A model discussing the early history of the Earth based on a study of lead isotope ratios from veins in some Archean craters of Africa', *Geochim. Cosmochim. Acta*, **37**, 2099–2124.

Russell, R. D., 1972. 'An evolutionary model for lead isotopes in conformable ores and in ocean volcanics', *Rev. Geophys. and Space Phys.*, **10**, 529–549.

Russell, R. D., and Birnie, D. J., 1974. 'A bidirectional mixing model for lead isotope evolution', *Phys. Earth. Planetary Int.*, **8**, 158–166.

Schaeffer, O. A., and Husain, L., 1973. 'Early lunar history: ages of 2 to 4 mm soil fragments from the lunar highlands', *Proc. 4th Lunar Sci. Conf.*, 1847–1863.

Schaeffer, O. A., and Husain, L., 1974. 'Chronology of lunar basin formation', *Proc. 5th Lunar Sci. Conf.*, **2**, 1541–1555.

Schnetzler, C. C., and Philpotts, J. A., 1970. 'Partition coefficients of rare-earth elements between igneous matrix material and rock-forming mineral phenocrysts. Pt. II', *Geochim. Cosmochim. Acta*, **34**, 331–340.

Schnetzler, C. C., and Philpotts, J. A., 1971. 'Alkali, alkaline earth and rare-earth element concentrations in some Apollo 12 soils, rocks and separated phases', *Proc. 2nd Lunar Sci. Conf.*, **2**, 1101–1122.

Schnetzler, C. C., Philpotts, J. A., Nava, D. F., Schuhmann, S., and Thomas, H. H., 1972. 'Geochemistry of Apollo 15 basalt 1555 and soil 15531, *Science*, **175**, 426–428.

Schonfeld, E., 1974. 'The contamination of lunar highland rocks by KREEP: Interpretation by mixing models', *Proc. 5th Lunar Sci. Conf.*, **2**, 1269–1286.

Schonfeld, E., and Meyer, C., Jr., 1972. 'The abundances of components of the lunar soils by a least-squares mixing model and the formation age of KREEP', *Proc. 3rd Lunar Sci. Conf.*, **2**, 1397–1420.

Shih, C., Wiesmann, H. W., and Haskin, L. A., 1975. 'On the origin of high-Ti mare basalts [Abs.], In *Lunar Science.* **VI**, 6th Lunar Sci. Conf., 735–737.

Shimizu, N., 1974. ' An experimental study of the partitioning of K, Rb, Cs, Sr and Ba between clinopyroxene and liquid at high pressures, *Geochim. Cosmochim. Acta*, **38**, 1789–1798.

Sinha, A. K., 1972. 'U–Th–Pb systematics and the age of the Onverwacht Series, South Africa', *Earth Planet. Sci. Lett.*, **16**, 219–227.

Smith, J. V., Anderson, A. T., Newton, R. C., Olsen, E. J., Wyllie, P. J., Crewe, A. V., Isaacson, M. S., and Johnson, D., 1970. 'Petrologic history of the moon inferred from petrography, mineralogy and petrogenesis of Apollo 11 rocks', *Proc. Apollo 11 Lunar Sci. Conf.*, **1**, 897–925.

Stettler, A., Eberhardt, P., Geiss, J., Grögler, N., and Maurer, P., (1973). 'Ar^{39}–Ar^{40} ages and Ar^{37}–Ar^{38} exposure ages of lunar rocks', *Proc. 4th Lunar Sci. Conf.*, **2**, 1865–1888.

Stettler, A., Eberhardt, P., Geiss, J., Grögler, N., and Maurer, P., 1974. 'Sequence of terra rock formation and basaltic lava flows on the Moon', *Lunar Science.* **V**, Pt. II, 738–740.

Stuart-Alexander, D. E., and Howard, K. A., 1970. 'Lunar maria and circular basins—A review', *Icarus*, **12**, 440–456.

Sun, S. S., 1973. 'Lead isotope studies of young volcanic rocks from oceanic islands, mid-ocean ridges, and island arcs', *Ph.D. Thesis*, Columbia University, 139 pp.

Tatsumoto, M., Nunes, P. D., Knight, R. J., Hedge, C. E., and Unruh, D. M., 1973. 'U–Th–Pb, Rb–Sr and K measurements of two Apollo 17 samples' [Abs.], *EOS*, **54**, 614–615.

Taylor, S. R., and Jakeš, P., 1974. 'The geochemical evolution of the Moon', *Proc. 5th Lunar Sci. Conf.*, **2**, 1287–1305.

Tera, F., Papanastassiou, D. A., and Wasserburg, G. J., 1974. 'Isotopic evidence for a terminal lunar cataclysm', *Earth Planet. Sci. Lett.*, **22**, 1–21.

Toksöz, M. N., Solomon, S. C., Minean, J. M., and Johnston, D. H., 1972. 'Thermal evolution of the Moon', *The Moon*, **4**, 190–213.

Turner, G., Cadogan, P. H., and Yonge, C. J., (1973). 'Argon selenochronology', *Proc. 4th Lunar Sci. Conf.*, **2**, 1889–1914.

Walker, D., Grove, T. L., Longhi, J., Stolper, E. M., and Hays, J. F., 1973. 'Origin of lunar feldspathic rocks', *Earth Planet. Sci. Lett.*, **20**, 325–336.

Wänke, H., Baddenhausen, H., Balacescu, A., Teschke, F., Spettel, B., Dreibus, G., Palme, H., Quijano-Rico, M., Kruse, H., Wlotzka, F., and Begemann, F., 1972. 'Multielement analyses of lunar samples and some implication of the results', *Proc. 3rd Lunar Sci. Conf.*, **2**, 1251–1268.

Wasserburg, G. J., 1966. 'Geochronology and isotopic data bearing on development of the continental crust', In Hurley, P. M. (Ed.), *Advances in Earth Sciences*, M.I.T. Press, Cambridge, Mass., pp. 431–459.

Wasserburg, G. J., Papanastassiou, D. A., Nenow, E. V., and Bauman, C. A., 1969. 'A programmable magnetic field mass spectrometer with on-line data processing', *Rev. Sci. Inst.*, **40**, 288.

Wilson, J. F., and Harrison, N. M., 1973. 'Recent K-Ar age determinations on some Rhodesian granites', *Spec. Publ. Geol. Soc. S. Afr.*, **3**, 69–78.

Wood, J. A., Dickey, J. S., Marvin, U. B., and Powell B. N., 1970. Lunar anorthosites and a geophysical model of the Moon', *Proc. Apollo 11 Lunar Sci. Conf.*, **1**, 965–988.

Giant Impacting and the Development of

Continental Crust

ALAN M. GOODWIN

Precambrian Research Group,
University of Toronto, Toronto, Canada.

Introduction

The current high level of interest in Precambrian problems has stimulated a flood of relevant data which, combined with spectacular results of lunar and other space probes, has focused attention upon problems of origin and tectonic development of the early Earth's crust. We may confidently anticipate both accelerated data accumulation and deepening understanding of these problems. In brief, the study of primitive crust is unusually fluid and dynamic with geological and geophysical observations equally relevant as never before. It follows that any current interpretation will be subjected to rapid, rigorous and possibly dramatic testing.

Available Precambrian data indicate the following: (1) most sialic crust is of Archean (>2·5 b.y.) age; and (2) more tenuous but central to the problem, this sialic crust has retained a Pangaean or quasi-Pangaean pattern throughout most if not all Precambrian time. The Precambrian Earth, by implications, featured an asymmetric (one-sided) crescentic distribution of sialic, hence continental crust in relation to the major obversely disposed oceanic crust of which the Pacific Basin is the remnant. By partial analogy with the Moon, terrestrial asymmetry may be related, directly or indirectly, to giant meteorite impacting following core formation but during differentiation of sialic crust.

Precambrian Crust

Distribution

Each continent of the world comprises a Precambrian core with Phanerozoic additions both in the form of cover rocks, e.g. Hudson and Interior Platforms of the Canadian Shield and extensive platform cover of the Siberian Shield, and marginal accretions themselves situated either at continental–oceanic interfaces, e.g. Cordilleras of North and South America or eastern fold belts of Australia, or between converging continental plates, e.g. Alpine and Himalayan fold belts (Fig. 1). The present distribution of continents with their Precambrian cores is manifestly a product of a long, varied history of crustal evolution, including fragmentation and dispersion of continental crust. Fragmentation occurred in response to deep global tectonic processes—at least during the Phanerozoic. To understand the early Precambrian, crustal-spawning processes we must reconstruct the Precambrian fragments and interpret the resulting global patterns. This process of reconstruction becomes increasingly difficult with geological age and is most difficult for the Archean crust whose record is fragmentary and opaque but for which a rationale at least must be sought.

The exposed Precambrian core of a continent is conventionally called a shield. The buried extension—the crypto-shield—may exceed it in area. It is the total Precambrian

77

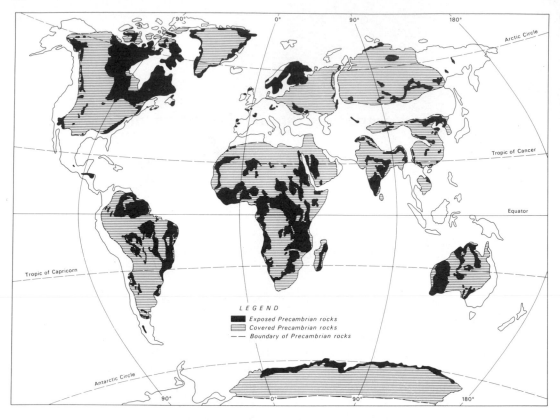

Fig. 1. Distribution of Precambrian rocks in modern continents

crust—exposed and buried—that forms the Precambrian core or stable continental platform. In addition, some Precambrian crust representing former margins of the platforms has been incorporated in Phanerozoic mountain belts, e.g. interior Rocky, Ural, and Himalayan fold belts, to form integral parts of these younger peripheral mobile belts.

Within a particular continent the relative size and disposition of Precambrian core and Phanerozoic accretion reflect in substantial measure the history of fragmentation and resulting direction of plate motion. Thus the Precambrian core relative to the Phanerozoic accretion may be central and predominant (Africa) or offset and more nearly equal in area (Americas, Eurasia, Australia). Most continents contain a fragment only of a pre-existing larger Precambrian crustal mass. For example, in the several continents resulting from the fragmentation of Gondwana, the major Phanerozoic accretions—each situated

respectively west (South America), south (Antarctic), east (Australia) and north (Africa and India) in response to the relative direction of plate motion.

In the northern hemisphere the conventional Precambrian shields arranged from west to east and north to south are: Canadian and Greenland (Laurentian Platform), Baltic and Ukrainian (Russian Platform), Anabar and Aldan (Siberian Platform), Sino-Korean, Indian, Arabian with north-central and western African (as part of African Craton) and Guyanan shields. In the southern hemisphere they are: Brazilian, south-central and southern African, Australian and Antarctic. In addition several micro-shields (e.g. Madagascar), representing island slivers, are present.

The pre-drift reconstruction of Precambrian crust omitting certain internal Phanerozoic (Pan-African) fold belts is shown successively in (1) the Atlantic reconstruction after Bullard et al. (1965) and (2) the Dietz

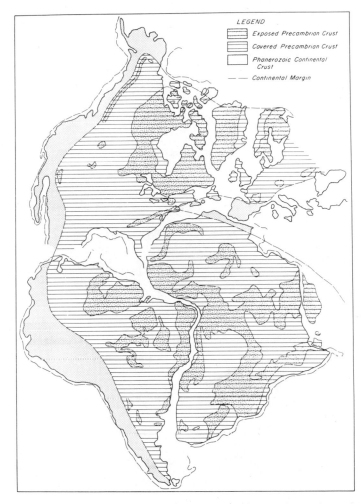

Fig. 2. Distribution of Precambrian and Phanerozoic continental crust in the fit of the continents around the Atlantic (Bullard et al., 1965)

and Holden (1970) fit of Pangaea with modern continental margins shown for ease of reference (Fig. 5). The global continuity of Precambrian crust within Pangaea is the outstanding feature. There is indeed substantial supportive evidence that the northern and southern supercontinents of Laurasia and Gondwana, respectively, were in close proximity if not in fact united as illustrated rather than substantially separated into two independent, widely separated landmasses. 'New' or Phanerozoic crust constitutes mainly the circum-Pacific and the future Mediterranean–Himalayan fold belts (Alpine) of Cenozoic age situated at the leading edges of the modern continents in addition to certain transecting or internal Phanerozoic fold belts, e.g. Urals (not illustrated).

Basins

Attention has been directed to the presence of a number of large basins and/or depressions distributed within the main shield areas of the world. Analysis of the Canadian Shield and its larger enclosing supershield, Laurentia, with Hudson Bay at the geographic centre has led to the interpretation of the central basin as a paleoplume or group of related plumes about which the supershield has grown (Fig. 3) as briefly outlined below.

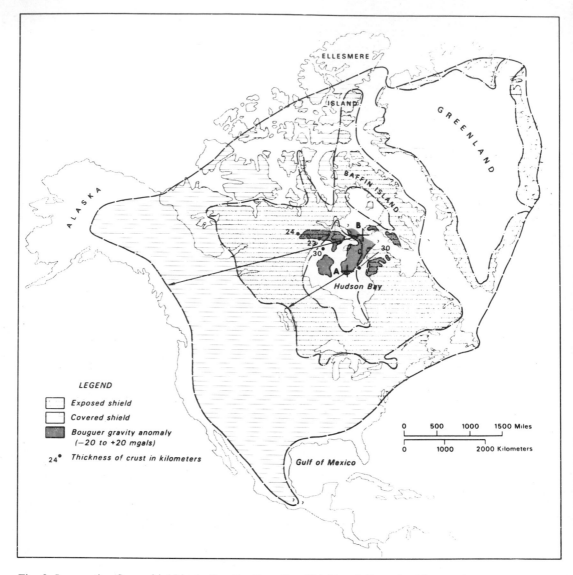

Fig. 3. Laurentian Supershield including the Canadian Shield and Greenland in pre-drift reconstruction

The Canadian Shield is the product of a long, complex history of crustal growth. Analysis of belt patterns by eras indicates that (1) this shield grew from a number of small primitive, possibly interconnected, crustal units or nodules by processes of unit enlargement to form a single continuous shield, and (2) growth occurred about a long-lived crustal anomaly situated at the geographic centre of the shield in Hudson Bay. Three main stages of growth are recognized: (1) in early Precambrian time, enlarging crustal nodules were grouped about a negative crustal anomaly in the Hudson Bay region; (2) in late Precambrian time a single continuous shield features a pronounced positive crustal anomaly in the form of a broad epeirogenic uplift centred on Hudson Bay; and finally (3) from early Phanerozoic time to the present, a fully developed continental shield has featured a pronounced negative crustal anomaly in the form of a central basin in the same Hudson Bay region.

Vertical fluctuations of the enlarging shield are interpreted in terms of mantle plume tectonics. The hypothesis has been elaborated

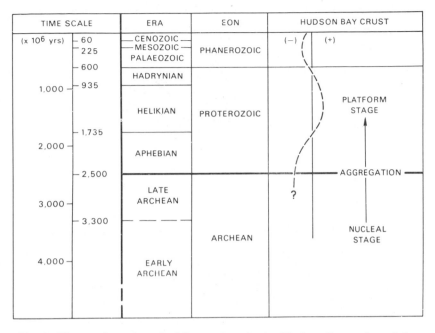

Fig. 4. Time scale and vertical fluctuations in the Hudson Bay region of the
Canadian Shield

(Goodwin, 1974) that Hudson Bay represents a paleoplume. Accordingly, a mantle plume or, more probably, a group of related plumes operating in the Hudson Bay region during Precambrian time was a key element in growth of the shield. Such growth featured: (1) early development of unstable protocontinental nuclei or nodules distributed about Hudson Bay (Archean time, about 4000–2500 m.y.); (2) their enlargment and aggregation to form a number of metastable to stable cratons similarly grouped around Hudson Bay (Aphebian time, about 2500–1700 m.y.); (3) attainment of a continuous shield featuring a broad central plateau over Hudson Bay, the product of epeirogenic uplift with attendant volcanism, rifting, and development of local basins and swells (Helikian and Hadrynian times, about 1700–600 m.y.); and finally (4) collapse of the Hudson Bay region to negative crustal status marking lateral migration of the consolidated shield and supporting lithospheric plate away from its parent plume (about 600 m.y. to present) (Fig. 4). The postulated Hudson Bay plume was larger and longer lived by an order

of magnitude than Cenozoic plumes as presently recognized.

Precambrian crust of the world, reassembled in Pangaea, is interpreted in terms of a number of such supershields (Fig. 5), each of which may have developed by similar, though not necessarily, synchronous growth stages about a central mantle plume or family of mantle plumes, now represented in part by a large central basin. Accordingly, a global array of Precambrian mantle plumes which formed in response to some fundamental thermal event, controlled the distribution of the growing shields and resultant continental crust which had aggregated in recognizable global form by Phanerozoic time at least. A major implication of these relationships is that 'plate tectonics' of Cenozoic-type with its history of fragmentation and global dispersion of continental crust did not operate in Precambrian time on a scale sufficient to destroy the fundamental symmetry of the growing shields. However, this does not preclude the operation of substantial intra-shield plate motion, precursor of Cenozoic plate tectonics, or of local

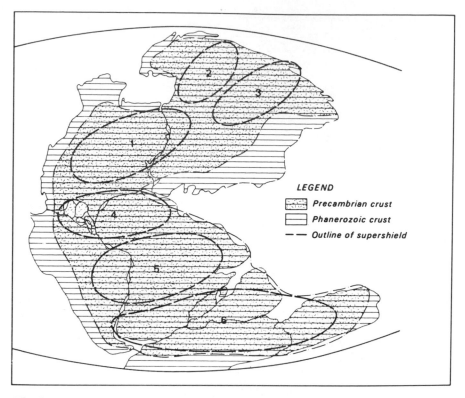

Fig. 5. Distribution of Precambriam crust and possible Supershields in the Pangaean reconstruction

marginal addition of extra-shield crustal segments at any or all stages of shield development.

Possible plume-generated analogues of Hudson Bay may be represented by the Siberian and Russian Platforms and by the Tadoueni, Chad, Congo and Kalahari basins in Africa. This particular array of possibly plume-generated, intra-shield basins which is distributed on the long axis of Precambrian crust in the Pangaean fit, may be a product of deep-seated, long-lived, mantle thermal convection systems.

Archean Crust

All Precambrian shields contain Archean (>2.5 b.y.) crust distributed both as readily recognized ancient nuclei or cratons and as widespread commonly overprinted hence obscure components of post-Archean fold belts. With due allowance for this latter type of overprinted and digested (anatexis) crust, the total quantity of Archean crust greatly exceeds that presently defined on acceptable isotopic basis and, indeed, appears to greatly predominate within the total Precambrian crust.

The distribution of isotopically defined Archean crust is illustrated in reconstructions of Laurasia and Gondwana (Fig. 6(a), (b)). All shields are represented except for the Sino-Korean.

Distribution

Eurasia

Siberian Platform. Defined Archean crust is reported in the Anabar craton (Russian Geological Atlas, Map 196–197, 1964) of the northern Siberian Platform, and from the Aldan shield to the south where the Tcharskaja massif in the west, representing the most ancient block, includes prevalent gneiss and migmatite which yield Pb–U isotopic dates of 2950 m.y. (Tougarinov, 1968, pp. 649–655). In the Aldan River basin

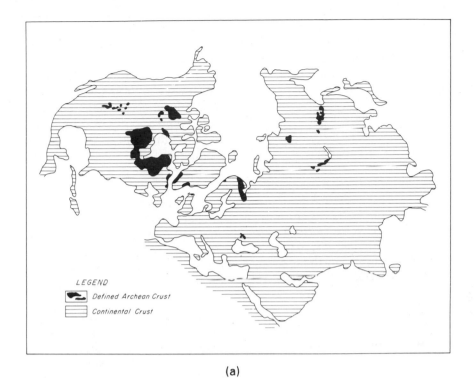

Fig. 6. Distribution of isotopically-defined Archean crust in (a) Laurasia and (b) Gondwana

to the east three thick groups of rocks, Barsalinskaja, Olekminukaja and Kurultinskaja, collectively 17,000 m thick include biotite, biotite–sillimanite–garnet and pyroxene–garnet gneiss.

Russian Platform. In the Ukranian Shield of the southern Russian Platform, Precambrian I division as defined by Semenenko is a stable platform formed by ancient folded 'geosynclinal formations'. The Konkian orogenic cycle operative within the period 3·5–3·1 b.y. and the Aulian cycle, operative within the period 3·1–2·7 b.y. have been recognized (Semenenko et al., 1968, pp. 661–671).

In the Baltic Shield to the north the Saamo–Karelian Zone corresponds to the area of the Karelides, i.e. eastern part of the shield. It contains three subzones of which the northeastern and southwestern subzones show a multistage structure including a lower structural or basement stage in which the rock formations have isotopic ages exceeding 2·6 b.y., and the central or Belomorian subzone composed of Belomorian and granulite formation of 'Archean age', i.e. >2·5 b.y. (Kratz, 1968, pp. 657–660). Preliminary isotopic data from the Kola Peninsula suggest that ages as old as 3·8 b.y. may be present (Pb–Pb whole rock method).

Scotland. Undoubted Archean rocks with an age of 2·8 b.y. are present in the Scourian gneiss of Scotland (Bridgwater et al., 1973b).

North America

Laurentian Supershield. In Southern Greenland Archean ages of about 3·8 b.y. have been obtained by Rb–Sr whole rock isochron and whole rock Pb–Pb isotope measurements on a well-defined supracrustal and gneissic succession. A supracrustal sequence (Isua series) is geological oldest followed by a gneissic phase (Amîtsoq). The supracrustals include conglomerate, carbonates, sandstone, ultramafic sheets, volcanics, quartzite and iron-formation. A younger gneissic series (3·0 b.y.) can be separated from the Amîtsoq by the presence of deformed amphibolite dikes (Ameralik) cutting the latter. Younger events

include high-grade metamorphism at 2·8 b.y. and igneous intrusive activity at 2·6 b.y. (Bridgwater et al., 1973a).

Similar rocks occur in Labrador where the oldest date so far is 3·4 b.y. There is an early supracrustal series (Upernavik) of volcanics and sediments followed by extensive quartzofeldspathic gneiss (Uivak) distinguished from still younger gneiss by the presence of deformed mafic dikes (Saglek) (Bridgwater et al., 1975).

The Canadian Shield includes undoubted Archean rocks of Superior and Slave Provinces plus numerous Archean outliers within the intervening isotopically (K–Ar) younger Churchill provinces, such as the Rankin Inlet–Ennadai greenstone belt situated west of Hudson Bay. In addition Archean rocks to at least 2·9 b.y. are present in the Montana, Wyoming, Dakota region of north-central United States. Also Archean gneiss of age 3·8 b.y. has been dated in the Minnesota River Valley (Goldich and Hedge, 1974). Elsewhere the oldest currently dated Archean rocks in the Canadian Shield are granitic gneisses of the Nain Province on the east coast of Labrador with an age of 3·4 b.y. (Bridgwater et al., 1975) and the English River Gneiss Belt of western Ontario of minimum age 3·008 b.y. (Krogh et al., 1975).

South America

Venezuela. The oldest rocks found so far in South America are in a high-grade metamorphic sequence in the Imataca complex of Venezuela in the Guyanan Shield. An east–northeast-trending belt of quartz–feldspar gneiss, amphibolite, amphibole–pyroxene gneiss, iron-formation and migmatite includes plentiful granulite. The age range indicated by whole rock Rb–Sr isochron plots is 2·7–3·4 b.y. The Kanuko Group in Guyana and Venezuela contains banded biotite and biotite–garnet gneiss in amphibolite and granulite facies, hypersthene gneiss, and charnockites. These rocks which are at least 2·0 b.y. old by the Rb–Sr whole rock, isochron method, may be equivalent to the Imataca complex, the two forming the rim of the Venezuela–Guyana basin. Possible Archean correlatives are the Coereni and Falawatra groups in Surinam and the Ile de Cayenne basement in Surinam (Hurley and Rand, 1947, pp. 391–410).

Brazil. In Brazil to the south, three possible Archean nuclei are present namely the Guapore craton located on the Central Brazil Shield, the Sao Luiz cratonic area, a narrow feature near the northern coast between Belern and Sao Luiz, and the Sao Francisco craton, located southeast of the Parnaibo basin. A fourth has been proposed, the cratonic area of Rio de la Plata, located in southern Uruguay and north-eastern Argentina. However, to date, the only isotopically defined Archean rocks lie in the eastern Brazilian Shield in the Quadrilatera Ferrifero (Almeida et al., 1974, pp. 411–446) and in Rio de Janeiro and Minas Gerais states (Cordani et al., 1973).

Africa

Southern Africa. As reviewed by Kröner (in press) the geochronological record of the oldest rocks in southern Africa comes from the Kaapvaal and Rhodesian cratons and from the island of Malagasy. The best documented area of the Kaapvaal Craton is the Barberton greenstone belt which includes the Swaziland Sequence divided into a lower Onverwacht Group, a middle Fig Tree Group, and an upper Moodies Group. In the Rhodesian greenstone terrain the corresponding threefold division is also recognized as the Sebakwian, Bulawayan and Shamvaian Groups, respectively. In Swaziland Rb–Sr ages from the Ancient Gneiss Complex vary between 3·4–3·1 b.y. (Hunter, 1974) and the older tonalitic gneiss of Rhodesia has a Rb–Sr age not significantly older than 3·5 b.y. (Hickman, 1974). Basalts from the lower Onverwacht Group yield an age of $3·5 \pm 0·2$ b.y. (Jahn and Shih, 1974). In Malagasy the Androyan, Graphite and Vohibory sequences (Besairie, 1967) may be remnants of highly metamorphosed volcanic rocks which were intruded by granite dated at 3020 m.y. (Hottin, 1970).

There is widespread geochronological evidence for a major tectonogenetic event at 2·6–2·7 b.y. which has affected most of the older rocks. This is logically related to the similar widespread African tectonothermal event to the west including Limpopo deformation called by Kröner the Limpopo–Liberian tectonogenetic cycle.

To the south and west of the Kaapvaal craton the Namaqua–Natal Metamorphic Complex includes rocks correlated by Joubert (1971) and Vazner (1974) with the Kheis Group which is considered to have been deformed and metamorphosed prior to 2·5 b.y. (Vazner, 1974) and is related to the Limpopo event further east.

South-Central Africa. A large part of north-eastern Zambia is underlain by the Bangweulu Block which consists of granitic gneiss and metasupracrustals. Although no reliable age determinations have been reported from these rocks, detrital monazite from quartzite in the Irumi Hills near Mkushi was dated at 2·7 b.y. (Cahen and Snelling, 1966, p. 111) and is most probably derived from the Bangweulu basement (Kröner, in press, p. 7).

East Africa. The basement complex of Tanzania comprises an Archean granite–greenstone terrain commonly known as the Tanzanian Shield and a gneissic terrain to the east which was probably a part of the Archean complex but was extensively reworked during the Pan African tectonism (600 m.y.). Relics of charnockitic granulites yield ages of 3080 m.y. Some granites are reported to be definitely older than 2·7 b.y. (Kröner, in press, p. 11); and the Masaba biotite–adamellite yielded a whole rock Rb–Sr age of 2930 ± 80 m.y. (Old and Rex, 1971, p. 356).

The basement gneiss of southern and central Uganda consists of amphibolite-grade metamorphic rocks which were affected 2·6 b.y. ago. Since the Aruan (pre-2·6 b.y.) sequence rests upon older Archean rocks in north-western Uganda there is no doubt that an Archean link existed between the shield area of Tanzania and western Kenya and the Archean of Zaire via the Ugandan basement complex (Kröner, in press, p. 11).

Central Africa. The West Nile metamorphic complex of Uganda includes the high grade pre-2·9 b.y. Granulite Group or Watnian, the pre-2·6 b.y. Western Grey Gneiss or Aruan, and the Eastern grey Gneiss or Miriam. Leggo (1974) demonstrated with detailed Rb–Sr and Pb–U geochronology that the Watnian granulites were metamorphosed 2·9 b.y. ago. The Aruan gneisses of north and central Uganda were evidently involved in the Limpopo–Liberian cycle, as defined by Kröner, 2·6–2·7 b.y. ago. The West Nile Complex has been correlated with the Bomu Complex further west which consists of amphibole gneiss and migmatite, itself overlain by the Ganguan sequence which is cut by quartz veins for which Cahen and Snelling (1966, p. 51) reported two common lead model ages of 3480±100 m.y.

These complexes are regarded by Cahen and Snelling (1966) as part of the large North Congo Shield presumed in existence more than 3·5 b.y. ago.

The Cameroun basement to the west yields Rb–Sr biotite ages of ±2300 m.y. from the Ebolowa and similar granites. However, Pb–Pb zircon ages of 2·5 and 3·2 b.y. (Cahen and Snelling, 1966, p. 137) are noted by Kröner as evidence of early Precambrian age.

In the Congo Republic a quartz vein cutting altered granite near Lamberene yielded a Pb–Pb age of 2540±70 m.y.; a common lead age of 2930 m.y. was obtained from granite in the Efot region near the Gabon–Equatorial Guinea border.

The Kasai shield contains a basement complex of which the oldest rocks belong to the Upper Luanyi Gneiss with associated pegmatites dated at 3270±170 m.y. (Cahen and Snelling, 1966). Associated with the Luanyi gneisses is a broad belt of charnockites and mafic intrusions dated at ±2880 m.y. (Pb–U zircon ages). The major part of the Kasai shield is composed of the Dibaya Granite–Migmatite Complex which experienced severe metamorphism at ±2720 m.y., obviously part of the continent-wide Limpopo–Liberian diastrophism as defined by Kröner. Basement rocks in central Angola which resemble the Dibaya Complex of Kasai and the Upper Zambezi rocks of eastern Angola were metamorphosed 2·5–2·8 b.y. ago.

West Africa. Detailed work in West Africa indicates the presence of extensive metamorphic complexes which suffered granulite facies metamorphism 3·0 b.y. ago. In the Reguibat Ridge of northern Mauritania and western Algeria high-grade Archean basement complexes are overlain unconformably by Birrimian-type metasupracrustal sequences (Kröner, in press, p. 19). The Amsaga 'Series' of the Atar region was metamorphosed 3·0 b.y. ago and an age of 2·8 b.y. was obtained in the Tiris 'Series'. The adjoining Zagoride belt of the Anti-Atlas region of Morocco consists of schist and mafic intrusions metamorphosed 2·5–2·7 b.y. ago (Choubert et al., 1974).

Hurley et al. (1971) reported ages varying from 2650 to 2820 m.y. from Liberia and Sierra Leone on the basis of which they distinguish a major tectonothermal event at 2700±200 m.y. and define the Liberian province which was also subjected to an earlier granulite metamorphism at ±3·0 b.y. as supported by Ph–U and Pb–Pb ages of 2890–3020 m.y. from the Sula Mountains of Sierra Leone (Choubert and Faure-Muret, 1971).

North-Central Africa. Farther east in the Hoggar, high-grade gneiss and charnockite of the Ouzzalien in the inlier of In Ouzzal (Caby, 1970) were metamorphosed 3·3 b.y. ago. Kröner also cites indications of Archean ages from the central (Aleksod massif) and north-eastern (Oumelalen massif) part of the Hoggar.

India

The Dharwar Craton of southern India contains a complex assemblage of granitic gneiss, migmatite and charnockite associated with numerous greenstone belts. The Peninsular Gneiss, a polymetamorphic granitic complex is as old as 2950±120 m.y. (Pichamuthu, 1974, p. 344). The amphibolites of the Kolar belt and the hornblende schist of the Hutti area have given ages of 2900±200 m.y. and 3295±200 m.y., respectively.

The craton displays increasing metamorphic grade to the south. Thus the greenstone belts are preferentially concentrated in the northern half of the craton and the charnockite–khondalite terrain in the southern half.

The oldest reported age for the Peninsular Gneisses is that from a gneiss pebble in Kaldurga conglomerate dated by the Rb–Sr isochron method at 3250 ± 150 m.y. (Radhakrishna, 1974, p. 446, Table 1). Other ages and lithologies listed are charnockite, Satnur—3100 m.y.; 'Oldest' gneisses, S. India—3065 ± 75 m.y.; granite, N. Mysore—2990 ± 120 m.y.; Peninsular gneiss, Bangalore—2950 ± 110 m.y.; Patna, Kolar—2920 ± 100 m.y.; amphibolite, Kolar—2900 ± 200 m.y. Thus widespread Archean ages are clearly demonstrated.

Australia

Archean crust as known is confined to the Yilgarn and Pilbara cratons in western Australia. Ages up to 3·1 b.y. have been determined in the Pilbara craton in addition to widespread ages of 2·8 b.y. in the greenstone belts of the Yilgarn craton. In the Warrawoona Succession of the Pilbara craton (De Laeter and Blockley, 1972) plagioclase-rich granites provide ages of 3·15–2·9 b.y. The thick tholeiitic sequences in Western Australia are interpreted as original Archean crust first developed during rifting of an unidentified pre-existing crust.

Antarctica

Archean crust including high grade granulitic terrain is widespread in northern Antarctica. Few reliable age dates have been provided so far as known. However, Halpern reports Rb–Sr dates of about 3000 m.y. (Grikurov et al., 1972) and in all likelihood plentiful old Archean ages will be forthcoming as isotopic dating proceeds.

Pangaean Distribution

The outline of Precambrian crust as known is shown in the Dietz and Holden (1970) reconstruction of Pangaea together with the distribution of defined Archean crust as outlined above (Fig. 7). An outstanding feature is

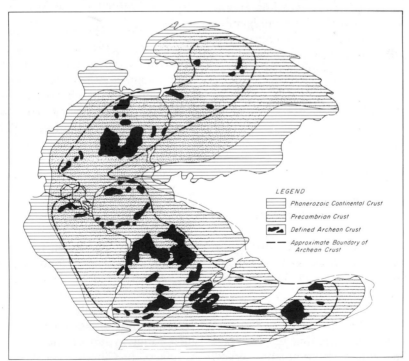

Fig. 7. Location of Precambrian crust, defined Archean crust, and approximate boundary of total Archean crust in the Pangaean reconstruction

the widespread distribution of widely separated Archean portions which extend in assembled Pangaean form from eastern Siberia through northern Europe and North America via South America and Africa to India, Antarctica and Australia.

Unquestionably a substantial additional amount of Archean crust remains to be defined by conventional isotopic methods. But an even greater amount is considered to be present in metamorphically overprinted crust much of which may defy conventional isotopic dating by presently known techniques. This, together with original Archean crust which has experienced complete or substantial regeneration to appear as post-Archean intrusions, must be allowed for in estimating total Archean distribution limits. In this regard, some guidance is provided by the Canadian and African shields wherein defined Archean crust is sufficiently widespread to support the interpretation that Archean crust originally extended over the complete areas of the Precambrian shields with the exception of certain off-shield and marginal geosynclinal accumulations, e.g. Coronation geosyncline of the Canadian Shield. On this basis a boundary line drawn around all defined Archean portions in the Pangaean reconstruction is considered to· provide a reasonable estimate of the distribution pattern of primary Archean crust. Indeed the estimated boundaries probably err on the conservative side.

Accordingly, Archean crust is considered to have been distributed over the main part of the total Precambrian reconstruction. An important implication is that the major part of the present sialic crust is of Archean age. However, this does not rule out substantial post-Archean sialic accretion.

Thus the conclusion is reached that major Archean sialic crust of asymmetric, crescentic distribution is represented in the Pangaean reconstruction.

Pre-Permian Distribution

How far back in geologic time may we trace this particular pattern of continental crust? This important point is highly contentious at present due to limited, partly conflicting data.

However, a significant and growing body of data supports the interpretation of a Pangaean or quasi-Pangaean presence throughout much if not all Precambrian time.

Early Phanerozoic–Late Precambrian. According to the Pangaean reconstruction, Precambrian crust was continuous through the two supercontinents of Laurasia and Gondwana, themselves united to form the single continental mass of Pangaea. The arms of Laurasia and Gondwana unite to form the Pangaean crescent which embraces the 'gaping gore' (Carey) containing the Tethyan part of Panthalassa, the universal ocean, including the future Mediterranean Sea. Other possible interpretations allow for some separation and independent motion not only of Gondwana and Laurasia (Brock, 1973; King, 1973; Piper, 1973) but also of Laurentian and Siberian portions of Laurasia (Creer, 1973). However, common to most interpretations so far offered is a general crescentic array of continental crust in late Precambrian–early Phanerozoic time whether in single continuous or constrained fragmental form.

Precambrian. Despite the uncertainties in reconstructing Precambrian sialic crust by eras imposed by the severe limitations of present data a number of pertinent Precambrian patterns have been discerned or proposed. Specific widespread elements of apparent order have been recognized in Pangaean crust which collectively support long-lived, *in situ*, continental growth with gross preservation of the global pattern and, concurrently, militate against repeated indiscriminate fragmentation, widespread global dispersion and unrelated regrouping.

(1) Hurley has shown tentatively that the distribution of geochronologic provinces in the reconstructed supercontinent of Laurasia appears to be concentric with an inner concentration of roughly equidimensional ancient cratons and a peripheral distribution of increasingly younger mobile belts with the youngest provinces almost completely circumscribing the older land mass (Hurley, 1970, p. 189). Hurley suggests that roughly

equidimensional masses of ancient continental crust received adjacent portions of successively younger blocks in a process of continuing coalescence or *in situ* growth (Hurley, 1970, p. 195).

Whereas the effects of regional metamorphic overprinting as known impose severe limits on the validity of this interpretation in detail, the data do indicate some discernible overall coherent growth pattern in the Laurasian landmass which is suggestive of long-lived consistent *in situ* growth rather than repeated indiscriminate, unconstrained fragmentation, dispersion and regrouping of component blocks.

(2) A similar pattern is apparent in the reconstructed supercontinent of Gondwana (for example, King, 1973, Fig. 5). Kröner (in press) contends that new structural, geochronological and paleomagnetic data from Africa provide strong evidence for the existence of large continental plates since at least the early Proterozoic which were later transected by linear mobile belts and thus partly destroyed. Many phenomena indicate processes involving little or no relative motion of crustal plates or crustal shortening between them suggesting that large-scale dispersive movements of major continental fragments were uncommon or absent in the early and middle Precambrian.

(3) Along similar lines Engel and Kelm (1972) concluded from studies of the nature and patterns of orogenic belts formed from the early Archean to the Recent that the post-Permian, large-scale dispersive movements of major continental fragments are uncommon, possibly unique in the geological record. They stress the accordant tectonic patterns that result when the Americas, Africa, India and Australia are reclustered into the pre-Triassic Gondwana and western Laurasia defined by many lines of evidence. In this mega-continent, especially in Gondwana, fold belts of Paleozoic, Proterozoic and Archean ages are juxtaposed into subparallel, coherent entities. In general, the fold patterns decrease in age towards the periphery. They conclude that these first-order features seem to require

a consistent interrelation and orientation of southern Gondwana at least, with remarkably uniform regional force fields in the mantle under and seaward beyond Gondwana throughout at least 3·5 b.y. of geologic time. The further accordance, along great circles or parallels of the Archean fold belts of Permian North America with those in Permian Gondwana suggests either a single mega-Archean orogenic system, or two complementary subparallel systems which have undergone little differential reorientation throughout the entire Proterozoic and Paleozoic (Engel and Kelm, 1972).

(4) Dearnley (1966) has shown that a reconstruction of Archean tectonic lineaments, using the drifting continents model, yields a convergent pattern with trends meeting at the poles and diverging towards the equator. He interpreted this pattern in terms of a two-cell convection current system. Such a convergent pattern is not in accord with unrestrained fragmentation and dispersion.

(5) Goodwin (1973a and b) has shown that the world-wide distribution pattern of early Proterozoic sedimentary banded iron-formation about 2000 m.y. ago in age, which is the major single interval of iron deposition in Earth history, is generally parallel or subparallel to that of modern oceanic ridges. The iron-formations may be related in origin and pattern to some type of incipient early Proterozoic plate motions. The apparent subparallelism between the distribution of early Proterozoic iron-formations and the modern plate boundaries supports the contention of the stability of pre-Permian continental nuclei and imposes severe restrictions on the degree of pre-Permian dispersion, disorientation and reassemblage of Precambrian nuclei.

(6) In similar vein Herz (1969) has shown that most anorthosites lie in two principal belts when plotted on a pre-drift continental reconstruction. Anorthosite ages in the belts cluster around 1300 ± 200 m.y. and range from 1100 to 1700 m.y. Again, such an element of consistency would be difficult to reconcile with random motion of plate fragments.

(7) Additional constraint is imposed by Precambrian paleo-magnetic data. For example, Irving and Lapointe (1975) conclude that interior Laurentia which includes all but some possible peripheral parts, e.g. the Grenville segment of the Canadian Shield seems to have remained predominantly in the northern hemisphere and does not appear to have moved randomly over the surface of the Earth. Similarly, Ueno et al. (1975) conclude from paleomagnetic data that the relative position of the Laurentian (Canadian and Greenland) and Baltic Shields before and after the Caledonian orogeny were very similar with the clear implication that prior to this orogenic cycle these two regions were near together.

While far from conclusive, these data and interpretations are the basis for the following working hypothesis according to which sialic crust was distributed in Pangaean or quasi-Pangaean form throughout much, if not most, Precambrian time. This implies a consistent asymmetric, crescentic distribution of terrestrial sialic crust with due allowance for possible intermittent constrained crustal fragmentation and limited dispersion followed by regrouping within the general Pangaean framework.

One possible explanation of such a terrestrial distribution pattern is provided by analogy with lunar maria, the product of giant meteorite impacting, as examined below.

Lunar Maria

Major features of the lunar surface are the large basins or maria each approximately

Fig. 8. The full Moon showing general distribution pattern of major lunar maria. (From Thomas A. Mutch, *Geology of the Moon*, rev. ed. (copyright © 1972 by Princeton University Press) Fig. III-5, p. 41. Reprinted by permission of Princeton University Press)

200 km in diameter, the direct consequence of giant meteorite or asteroid impacting. Lunar maria are confined to the near side of the Moon where they are distributed in a crudely crescentic or horsehoe pattern (Fig. 8). The giant impacting as dated occurred 4·2–3·8 b.y. ago following development of the lunar crust to an average thickness of 65 km. Since the time of giant impacting, the maria have been infilled with basalt lava flows hence their dark colour in contrast to the lighter coloured anorthositic 'highlands'. Apart from this activity plus constant bombardment by conventional meteorites, the lunar surface has been essentially frozen since that time (Strangway and Sharpe, 1975).

Eleven major mare arranged in crude chronological order from oldest to youngest with ages as known in brackets are as follows: Mare Fecunditas, Smythii, Tranquilitatus, Nubium, Procellarum, Serenitatis, Humorum (4·13–4·20 b.y.), Nectaris (4·20 b.y.), Crisium (4·13 b.y.), Imbrium (4·00 b.y.) and Orientale (3·85 ± 0·05 b.y.) (Strangway and Sharpe, 1975, p. 25). Thus the indicated time interval of impacting is in the order of 350 m.y. There is no apparent impact pattern with the possible exception of the linear alignment of Crisium, Serenitatis and Imbrium. On this basis the overall impact pattern presumably represents many superimposed infalls. However, more data are required to define the age and pattern of individual impacts and lines of impact.

The lunar crust is about 60 km thick on the front side of the Moon. It has been inferred that the upper 20–25 km is basaltic whereas the underlying material is essentially anorthosite or gabbroic anorthosite. The time of development of the crust is uncertain but has been inferred at 4·4–4·3 b.y. (Hubbard and Gast, 1971). Strangway and Sharpe (1975) note the presence of a dunite clast dated at 4·6 b.y. implying some very early differentiation. They further note that most lines of evidence point to a planetary-wide melting event at or near the time of origin of the Moon.

The mare basalts are extensively formed and developed on the front side of the Moon. They were apparently developed in the time interval 3·8 to 3·2 b.y., long after the differentiation that formed the lunar crust and following the major meteorite impacts. The basalt cover is very thin over much of its exposed surface. Strangway and Sharpe (1975) develop the hypothesis that the disturbance of the lunar crust by giant impacting in a cooling crust is the cause of the basaltic magma generation some hundreds of millions of years after the causative impact. As such they would represent a second generation of differentiation in the upper mantle. However, Wood (1973, p. 99) and others consider it unlikely that lava generation was directly related to the act of basin excavation.

A number of models have been proposed to account for the frontside–backside asymmetry. For example, Wood (1973) suggests that the thickness of the crust may vary systematically with the light anorthosite material being thicker on the backside than on the frontside. Alternatively, meteorite bombardment has been proposed as the cause of lunar asymmetry (Wood, 1973, p. 99).

Whatever the reason it is clear that the Moon is asymmetric in geographic, geophysical and chemical terms (Wood, 1973, p. 72) and it experienced major impacting 4·2–3·8 b.y. ago.

Postulated Terrestrial Impacts

It is altogether probable that planet Earth experienced mega-impacting synchronous with lunar impacts, namely 4·2–3·8 b.y. ago. If anything, the terrestrial impacting meteorites or asteroids would have been larger and denser (possibly by an order of magnitude) commensurate with the greater size and density, hence gravitational attraction, of Earth. In contrast to the comparative 'deep freeze' status of the lunar crust which notwithstanding may have developed a deep-seated, long-lived thermal response as noted above, the effects of terrestrial mega-impacting of planet Earth with its comparatively thin crust and hot convecting interior were, by inference, even more dramatically profound and long-lived. As with lunar counterparts, the postulated terrestrial mega-impacts may have been asymmetric (one-sided) and of accumulative crescentic or horseshoe shape. It is postulated that mega-impacting on Earth which succeeded the major differentiation-inducing, high temperature, core-forming event, initiated or alternatively guided pre-existing deep-seated, long-lived global thermal convection systems including deep mantle plumes which by successive stages of sialic differentiation have produced existing continental crust. Pangaea, the resulting crescent-shaped landmass is, by inference, not only a direct outgrowth but also a faithful reproduction of the terrestrial mega-impact pattern.

Proposed Crustal Evolution

A highly simplified three-stage development of sialic crust is offered for consideration as follows:

1. Accretion and core-form-   4·6–4·2 b.y.
   ing event

2. Giant impacting   4·2–3·8 b.y.

3. Accelerated sialic differ-   3·8–2·8 b.y.
   entiation and develop-
   ment of sialic crust

Accretion and Core-forming Event (4·6–4·2 b.y. ago)

Following accretion of planet Earth the principal event of terrestrial evolution was undoubtedly the segregation of the core which comprises one-third of the Earth's mass. Birch (1965) simplified the energetics of core formation as follows: (1) undifferentiated Earth is assumed to comprise a homogeneous mixture of core and mantle material; (2) after a period of radioactive heating, melting of iron begins at relatively shallow depth; (3) the molten-iron layer becomes unstable and descends while the mantle material, mostly solid, rises; and (4) separation is completed in a time so short that radioactive heating, with an approximate allowance for thermal expansion, is 400 cal/g. This is equivalent, according to Birch, to a mean rise of temperature of about 1600°.

The thermal consequences of core formation are such that it must antedate the oldest known surface rocks at 3·8 b.y. Indeed, Hanks and Anderson (1969) prefer a model wherein

Earth accreted in 200,000 years or less (4·5 b.y. ago), and that large-scale differentiation of Earth upon accretion seems most likely. The temperatures in the deep interior reflect residual accretionary energy, including that of core-formation, and are not primarily a result of radioactive heating.

Accordingly it is likely that accretion and core-formation were early events in Earth history, about 4·5 b.y. ago, with the possibility of exceptionally long-lived thermal consequences to the present.

Giant Impacting (4·2–3·8 b.y. ago)

It is proposed that Earth experienced mega-impacting synchronous with lunar impacts, namely 4·2–3·8 b.y. ago. Terrestrial impacting meteorites or asteroids were probably larger and denser commensurate with the Earth's greater gravitational attraction. As with lunar counterparts, terrestrial mega-impacts or the consequences of mega-impacting may have been asymmetric (one-sided) and of cumulative crescentic or horseshoe-shape. Possible causes of the postulated terrestrial asymmetry as with existing lunar asymmetry are speculative and uncertain.

Mega-impacting of planet Earth with its comparatively thin crust and hot, convecting interior shortly following the major high temperature, differentiation-inducing, core-forming event had profound consequences upon the course of subsequent differentiation including the production of sialic crust. It is proposed that mega-impacting initiated or alternatively guided existing deep-seated, long-lived global convection systems including deep mantle plumes which played a fundamental role in the ongoing process of mantle differentiation.

Accelerated Sialic Differentiation (3·8–2·8 b.y. ago)

One major consequence of upper mantle differentiation was the production of sialic, or protocontinental crust. Using the lunar model (Strangway and Sharpe, 1975) it is proposed that the thermal time-lag following terrestrial impact lasted for many hundreds of millions of years. Furthermore, it is proposed that a deep thermal convection system operated during this long period of time. As a result upper mantle and median mantle material, rich in lithophile or protosialic elements, underwent accelerated sialic differentiation resulting in selective asymmetric concentration of sialic crust at the Earth's surface. Major sialic differentiation occurred throughout this long interval with periodic pervasive sialic emplacement in the crust, culminating and eventually phasing out with major sialic addition marking global cratonization at and near the end of Archean time (2·8–2·7 b.y.).

Pangaea, the resulting crescent-shaped land-mass is, by inference, not only an outgrowth but also a possible reproduction of the terrestrial mega-impact pattern. The postulated mega-impact sites, in direct physical form, would have succumbed shortly after impact to the rapidly convecting mantle–crust systems. However, in so far as the ensuing deep-seated sial-inducing global convection systems were maintained over this long time interval, it is conceivable that the sialic continents in gross Pangaean, hence global, asymmetric form were developed in consequence of the impacting. At this stage major emphasis is placed upon crustal symmetry of Moon and Earth in relation to impacting with subordinate emphasis only upon the intriguing, yet possibly fortuitous, crescentic patterns present in both bodies. In conclusion, final emphasis is placed upon the crustal asymmetries of Moon and Earth and their postulated respective relationships, direct or indirect, to giant meteorite impacting.

Acknowledgements

I gratefully acknowledge valuable discussions with numerous earth scientists including many referenced in the text on the broad subject of generation of the early Earth's crust. However, interpretative aspects are my own responsibility unless otherwise stated. I acknowledge with particular thanks a valuable preprint on Precambrian geotectonic evolution of Africa provided by Dr A. Kröner. All illustrations with the exception of Fig. 8 were drafted by F. Jurgeneit, Department of Geology, University of Toronto.

References

Almeida F. F. M., de, Amoral, G., Cordani, U. G., and Kawashita, K., 1974. 'The Precambrian evolution of the South American Margin south of the Amazon River? In *The Ocean Basins and Margins*, A. E. M. Nairn and F. G. Stehli (Eds.), Plenum Press, New York.

Besairie, H., 1967. 'The Precambrian of Madagascar', In *The Precambrian*, K. Rankama (Ed.), Vol. 3, 133–142, Interscience, London.

Birch, F., 1965. 'Energetics of core formation', *J. Geophys. Res.*, **70**, No. 24, 6217–6221.

Bridgwater, D., Escher, A., Jackson, G. D., Taylor, F. C., and Windley, B. F., 1973a. 'Development of the Precambrian Shield in West Greenland, Labrador and Baffin Island', *Amer. Assoc. Petrol. Geol. Mem.*, 19, *Arctic Geology*, 99–116.

Bridgwater, D., Watson, J., and Windley, B. F., 1973b. 'The Archaean craton of the North Atlantic region', *Phil. Trans. Roy. Soc. Lond.*, **A273**, 493–512.

Bridgwater, D., Collerson, K. D., Hurst, R. W., and Jesseau, C. W., 1975. 'Field characters of the early Precambrian rocks from Saglek, coast of Labrador', *Geol. Surv. Canada Pap.*, **75–1**, Pt. A, 287–296.

Brock, A., 1973. 'Precambrian polar wander and continental drift of Africa and North America: some comments', In *Implications of Continental Drift to the Earth Sciences*, D. H. Tarling and S. K. Runcorn (Eds.), 33–36, Academic Press, London.

Bullard, E. C., Everett, J. E., and Smith, A. G., 1965. 'The fit of the continents around the Atlantic', *Phil. Trans. Roy. Soc. Lond.*, **A258**, 4 pp.

Caby, R., 1970. 'La chaine pharusienne dans le nord-ouest de l'Ahaggar (Sahara central, Algerie); sa place dans l'orugenese du Precambrien superieur en Afrique', *D.Sc. Thesis*, Univ. Montpellier, 336 pp.

94

Cahen, L., and Snelling, N. J. 1966. *The Geochronology of Equatorial Africa*, 195 pp. North-Holland Publ., Amsterdam.

Choubert, G., and Faure-Muret, A., 1971. 'Bouclier eburneen (ou liberio-iovoirien)', In *Tectonique de l'Afrique, UNESCO Earth Sciences* 6, 185–200, Paris.

Choubert, G., Faure-Muret, A., Hassenforder, B., and Jeanette, D., 1974. 'Nouvelle interpretation du Precambrien ancien de l'Anti-Atlas (Maroc)', *C. R. Acad. Paris*, **D278**, 2095–2098.

Cordani, U. G., Delhal, J., and Ledent, D., 1973. 'Orogeneses superposee dans le Precambrian du Brasil Sud-Oriental (Etats de Rio de Janiero et de Minas Gerais)', *Rev. Brasileira de Geociencias*, **3**, 1.

Creer, K. M., 1973. 'On the arrangement of the landmasses and the configuration of the geomagnetic field during the Phanerozoic', In *Implications of Continental Drift to the Earth Sciences*, D. H. Tarling and S. K. Runcorn (Eds.), Academic Press, London, 47–76.

Dearnley, R., 1966. 'Orogenic fold belts and a hypothesis of Earth evolution', In *Physics and Chemistry of the Earth*, L. H. Ahrens, F. Press, S. K. Runcorn and H. C. Vrey (Eds.), Pergamon Press, London, Chap. VII.

De Laeter, J. R., and Blockley, J. G., 1972. 'Granite ages within the Archean Pilbara block, Western Australia', *J. Geol. Soc. Australia*, **19**, 363–370.

Dietz, R. S., and Holden, J. C., 1970. 'Reconstruction of Pangaea: break-up and dispersion of continents, Permian to Present', *J. Geophys. Res.*, **75**, 4939–4956.

Engel, A. E. J., and Kelm, D. L., 1972. 'Pre-Permian global tectonics: a tectonic test', *Geol. Soc. Amer. Bull.*, **83**, 2325–2340.

Goldich, S. S., and Hedge, C. E., 1974. '3800-Myr granite gneiss in south-western Minnesota', *Nature*, **252**, No. 5483, 467–468.

Goodwin, A. M., 1973a. 'Archean iron-formations and tectonic basins of the Canadian Shield', *Econ. Geol.*, **68**, No. 7, 915–933.

Goodwin, A. M., 1973b. 'Plate tectonics and evolution of Precambrian crust', In *Implications of Continental Drift to the Earth Sciences*, D. H. Tarling and S. K. Runcorn (Eds.), Academic Press, London, 1047–1069.

Goodwin, A. M., 1974a. 'The most ancient continental margins', In *The Geology of Continental Margins*, C. A. Burk and C. L. Drake (Eds.), Springer Verlag, New York, 767–780.

Goodwin, A. M., 1974b. 'Precambian belts, plumes, and shield development', *Amer. J. Sci.*, **274**, 987–1028.

Grikurov, G. E., Ravich, M. G., and Soloviev, D. S., 1972. 'Tectonics of Antarctica', In *Antarctic Geology and Geophysics*, R. J. Adie (Ed.), Int.

Un. Geol. Sci. Ser. B, No. 1, 457–468, Universitetsforlaget, Oslo.

Hanks, T. C., and Anderson, D. C., 1969. 'The early thermal history of the Earth', *Phys. Earth Planet. Interior*, **2**, 19–29.

Herz, N., 1969. 'Anorthosite belts, continental drift, and the anorthosite event', *Science*, **164**, 944–947.

Hickman, M. H., 1974. '3500-Myr-old granite in southern Africa', *Nature*, **251**, 295–296.

Hottin, G., 1970. 'Geochronolgie et stratigraphie Malagaches—essai d'interpretation', *Serv. Geol. Malagasy, Docum. Bur. Geol.*, **182**, 21 pp.

Hubbard, N. J., and Gast, P. W., 1971. 'Chemical composition and origin of non-mare lunar basalts', *Geochim. Cosmochim. Acta*, Suppl. 2, **2**.

Hunter, D. R., 1974. 'Crustal development in the Kaapvaal Craton I, The Archaean', *Precambrian Res.*, **1**, 259–294.

Hurley, P. M., 1970. 'Distribution of age provinces in Laurasia', *Earth Planet. Sci. Lett.*, **8**, 189–196.

Hurley, P. M., Leo, G. W., White, R. H., and Fairbairn, H. W., 1971. 'Liberian age province (about 2700 m.y.) and adjacent provinces in Liberia and Sierra Leone', *Geol. Soc. Amer. Bull.*, **82**, 3483–3490.

Hurley, P. M., and Rand, J. R., 1974. 'Outline of Precambrian chronology in lands bordering the South Atlantic, exclusive of Brazil', In *The Ocean Basins and Margins*, A. E. M. Nairn and F. G. Stehli (Eds.), Plenum Press, New York.

International Tectonic Map of Africa, ASGA/UNESCO 1968 Co-ordinator: G. Choubert. 1:5,000,000.

Irving, E. and Lapointe, P. L., 1975. 'Paleomagnetism of Precambrian rock of Laurentia', *Geoscience Canada*, June issue.

Jahn, B., and Shih, C., 1974. 'On the age of the Onverwacht Group, Swaziland Sequence, South Africa', *Geochim. Cosmochim. Acta*, **38**, 873–885.

Joubert, P., 1971. 'The regional tectonism of the gneisses of part of Namaqualand', *Precambrian Res. Unit, Univ. Cape Town, Bull.*, 10, 220 pp.

King, L., 1973. 'An improved reconstruction of Gondwanaland', In *Implications of Continental Drift to the Earth Sciences*, D. H. Tarling and S. K. Runcorn (Eds.), Academic Press, London, 851–864.

Kratz, K. O., Gerling, E. K., and Lobach-Zhuchenko, S. B., 1968. 'The isotope geology of the Precambrian of the Baltic Shield', *Can. J. Earth Sci.*, **5**, 657–660.

Krogh, T. E., Harris, N. B. W., Davis, G. L., and Goodwin, A. M., 1975. 'Archean rocks in the Eastern Lac Seul region of the English River gneiss belt, northwestern Ontario', *Can. J. Earth Sci.*,

Kröner, A. (in press). 'The Precambrian geotectonic evolution of Africa: plate accretion versus plate destruction', *Geol. Soc. Amer. Bull.*

Leggo, P. J., 1974. 'A geochronological study of the basement complex of Uganda', *J. Geol. Soc. Lond.*, **130**, 263–277.

Mutch, T. A., 1972. *Geology of the Moon*, 391 pp. Princeton University Press, Princeton, New Jersey.

Old, R. A., and Rex, D. C., 1971. 'Rubidium and strontium age determination of some pre-Cambrian granitic rocks, SE Uganda', *Geol. Mag.*, **108**, 353–360.

Pichamuthu, C. S., 1974. 'The Dharwar Craton', In *The Geochemistry of the Precambrian rocks of India. J. Geol. Soc. India*, **15**, No. 4, 339–346.

Piper, J. D. A., 1973. 'Geological interpretation of paleomagnetic results from the African Precambrian', In *Implications of Continental Drift to the Earth Sciences*, D. H. Tarling and S. K. Runcorn (Eds.), Academic Press, London, 19–32.

Radhakrishna, B. P., 1974. 'Peninsular gneissic complex of the Dharwar Craton—a suggested model for its evolution', In *The Geochemistry of the Precambrian rocks of India. J. Geol. Soc. India*, **15**, No. 4, 439–454.

Semenenko, N. P., Scherbak, A. P., Vinogradov, A. P., Tougarinov, A. I., Eliseeva, G. D., Cotlovskay, F. I., and Dimidenko, S. G., 1968. 'Geochronology of the Ukrainian Precambrian', *Can. J. Earth Sci.*, **5**, 661–671.

Strangway, D. W., and Sharpe, H. N., (in press). 'A model of lunar evolution', *The Moon*.

Tougarinov, A. I., 1968. 'Geochronology of the Aldan Shield, southeastern Siberia', *Can J. Earth Sci.*, **5**, 649–655.

Ueno, H., Irving, E., and McNutt, R. H., 1975. 'Paleomagnetism of the Whitestone anorthosite and diorite, the Grenville polar track, and relative motions of the Laurentian and Baltic Shields', *Can. J. Earth. Sci.*, **12**, 209–226.

Vazner, V., 1974. 'The tectonic development of the Namaqua mobile belt and its foreland in parts of the northern Cape', *Precambrian Res. Unit, Univ. Cape Town, Bull.*, 14, 201 pp.

Wood, J., 1973. 'Bombardment as a cause of the lunar assymetry', *The Moon*, **8**, 73–103.

General Archaean Tectonics

Tectonic Relationships in the Archaean

J. Sutton

Dept. of Geology, Imperial College, London, England.

Stable and Unstable Archaean Crust

The instability of the Archaean crust has been over-emphasized. It now seems likely that long periods of time separated the successive groups of greenstone belts and associated tonalites. Some Archaean sedimentary sequences unconnected with greenstone belts can be compared with Phanerozoic shelf deposits. No doubt the parallel should not be pushed too far but there undoubtedly is an Archaean association of sediments marked by orthoquartzites, limestones and on occasion by shales which contrasts with the badly sorted sediments often found within the predominantly volcanic greenstone successions. Too much should not be made of this contrast; the presence, for example, of cherts in many greenstone belts indicates relative stability and an absence of clastic deposition during parts of the time such belts were evolving. Moreover, the possible shelf sea Archaean orthoquartzite–limestone associations are usually accompanied by igneous rocks. These are typically ultrabasic and basic types including peridotites, gabbros and anorthosites. In addition to lavas, dyke swarms, ring complexes, sills, laccoliths, a variety of differentiated igneous bodies occur in this association. These suggest, perhaps, a fractured crust under tension.

As the dating of these rocks improves, they will no doubt be broken down into groups of different ages, but as of now the picture they present is of a crust where sedimentation was interrupted from time to time by fracturing and the intrusion or extrusion of basic and ultrabasic magmas. Locally, cross-cutting relations can still be detected, indicating that the entire crust has not been so intensely deformed as to obliterate the original relations of the rocks. The presence of recognizable unconformities serves to make the same point. In this respect these possible Archaean shelf sediments and associated basic and ultrabasic volcanics contrast with the greenstone and tonalite associations in which deformation normally has destroyed all primary discontinuities.

One may suspect, but not at present prove, that another contrast exists. The tonalites associated with greenstone belts are in many regions now known to be of the same age as the accompanying basic volcanics, and like them probably derived from the mantle. In contrast the orthoquartzite–limestone assemblages, which are accompanied by basic and ultrabasic activity some of which, since it includes volcanic products is roughly contemporaneous, are not always accompanied by contemporaneous acid rocks. This is well shown in the south east of Rhodesia (Wilson, 1968) and in much of the South African Kaapvaal craton where Archaean sediments unconformably overlie very old greenstone tonalite complexes over 3 b.y. old (Hunter, 1974). These late Archaean deposits occur with dyke swarms, volcanic successions and intrusive complexes such as the Rhodesian Mashaba complex with the remarkable

Archaean ring intrusion described by Wilson (1968), but appear to lack widespread contemporary granites or tonalites. There was, however, certainly some production of granite as, for example, in the basement below the Witwatersrand succession. It is unusual to find Archaean supracrustals so well preserved as in Southern Africa and more commonly such rocks occur as in north west Scotland or Greenland as relics in gneissose complexes. Geochronometry at the present time is not usually able to show what time interval, if any, separated the formation of the gneissose complexes from the accumulation of the supracrustals: when this becomes possible we shall learn how much Archaean crust originated under conditions where widespread granitic or tonalitic intrusion accompanied vulcanicity and sedimentation and how much was formed from possible stable shelf sequences accompanied by basic and ultrabasic rocks but which lack widespread contemporary granites.

We can sum up the position by postulating two Archaean extremes:

(a) The tonalite–greenstone association with badly sorted sediments.
(b) The basic–ultrabasic–hypabyssal and volcanic association linked with well sorted orthoquartzites, limestones and shales of possible shelf sea provenance.

The first association is linked with widespread crustal deformation and tonalitic activity. The second association which typically retains some original features, was accompanied by only limited contemporary granitic activity and developed in faulted rather than intensely folded terrain.

The Repetition of Greenstone Belts

Two facts about the timing of greenstone belts are now well documented. Belts can accumulate in 50–100 m.y. and are accompanied by tonalites and granites formed over the same interval of time as the volcanic activity (e.g. Green and Baadsgaard, 1971).

Most known greenstone belts formed between 3·7 b.y. and 2·5 b.y. ago. Still older belts may be revealed; in a few regions belts continued to form until 2 b.y. ago.

Up to the present not more than three generations of greenstone belts have been identified in any one region. Even if the true position is that twice as many groups of greenstone belts evolved, the total time this would have taken, c. 500 m.y. in all, represents less than half of the 1200 m.y. period over which we have some understanding of Archaean geology (3·7 b.y.–2·5 b.y.).

The question arises as to what happened during the intervals between successive greenstone belts? One way of tackling this problem is to examine another but possibly related question. Through the Archaean, the maximum size of successive greenstone belts, as they are preserved today, apparently became progressively larger. This may be a misleading effect due to greater erosion earlier in the Archaean having left only small remnants of the older belts or it may on the other hand reflect a real evolutionary change. What is more certain is that the clusters of contemporary greenstone belts and accompanying tonalites occupied progressively smaller regions as Archaean time advanced. We appear to see a record of two contrasting changes. Successive individual belts became larger, but the area accompanied by contemporary clusters of belts decreased through the Archaean until they ceased to be a feature of the crust early in the Proterozoic about 2 b.y. ago.

Because of the decrease in size of individual clusters of greenstone belts and associated tonalites, it is possible to examine late Archaean crust around the margins of such terrains. Earlier in the Archaean clusters either overlapped or were so extensive that the entire crust was marked by greenstones and tonalites.

By examining Archaean crust outside the greenstone–tonalite complexes we can learn about other types of change which developed in Archaean time and we may get a clue as to the kind of event which may have occurred between successive phases of greenstone and tonalite production.

Nature of the Archaean Crust outside Greenstone–tonalite Complexes

The first indications of an early Archaean or Katarchaean crust were isolated age determinations made 20 to 15 years ago which suggested the existence of a crust at least 3·5 b.y. old (e.g. Catanzaro, 1963; Slawson et al., 1963). Several were determinations on galena which provided indirect evidence of very old source rocks. These were sufficient to lead to several proposals for major subdivisions of the Precambrian which in effect distinguished older and younger Precambrian. In 1967 I suggested a Katarchaean sequence of events among which I tentatively included the as yet undated Malene group which Mac-Gregor had already suggested might be over 3·6 b.y. in age. Lead isotope distribution in Rhodesia investigated by Robertson (1973) indicated in his view two major events at about 3·5 b.y. and 2·7 b.y. in that country. Recent work by Moorbath and colleagues (see this volume) has clarified the Rhodesian chronology and has shown that the greenstone–tonalite associations, widely developed in Southern and Central Africa prior to 3 b.y. and which include rocks as old as 3·6 b.y., were followed by a younger group of greenstones about 2·7 b.y. ago. By that time a striking contrast has developed between parts of that area.

In South Africa and in south east Rhodesia a succession of sedimentary basins developed ranging in age from the Pongola basin c. 3 b.y. (Hunter, 1974) to the youngest members of the Witwatersrand succession which straddle the Archaean–Proterozoic boundary c. 2·5 b.y. in age. Among these successions are the sediments and volcanics which in south east Rhodesia unconformably overlie the older tonalite gneiss near Shabani (Wilson, 1968; Morrison and Wilson, 1971) and the Messina Formation and its possible equivalents within the Limpopo belt. In contrast, in western and northern Rhodesia further greenstones and tonalites formed about 2·7 b.y. ago (see Moorbath, this volume). The sedimentary successions south of these younger greenstone belts were accompanied by basic and ultrabasic igneous activity and were deformed almost everywhere to some degree. But outside the Limpopo belt it would be true to say that the cover and basement remained clearly identifiable, so that the younger Archaean sequences could be readily recognized and even within the highly deformed Limpopo belt the Messina Formation could be mapped, though uncertainty remained as to its relation to adjoining gneisses.

The development of domes of basement below the Witwatersrand basin during the accumulation of the sedimentary cover is well known and indicates a limited instability below this Archaean basin. Despite this circumstance there would be general agreement that the relatively flat-lying sediments of the Witwatersrand basin and of the still older Pongola basin reflected rather different conditions from those which produced the tightly folded greenstone belts, concordant as they typically are, with the contemporary tonalites and surrounding gneisses. In that situation a regional transfer of acid material from mantle to crust took place in contrast to the limited granitic activity within the South African basins. Dolerite dykes systems have been identified below the Messina formation and in the older Rhodesian tonalites below the younger Archaean deposits near Shabani (Wilson, 1968). Anorthosites and gabbros occur in the Limpopo belt associated with the Messina group, and at Mashaba, Wilson has described a remarkable igneous complex which includes an ultrabasic ring complex, dyke swarms and inclined sheets several tens of kilometres long emplaced along gently dipping Archaean fault planes (Wilson, 1968).

The anorthosite–gabbro association is now well known in a region extending from Angola north eastwards through the Limpopo belt to Malagasy and thence into India and also in a belt extending from North America through Greenland into Northern Europe (Windley, 1973). Both belts cross regions which contain areas of Archaean rocks which lack young Archaean greenstone belts but which flank areas in which such young greenstone belts (c. 2·7 b.y. or younger) occur. It would appear that the 3 b.y. old anorthosite sequences either formed in intervals of time between successive greenstone belts, or were intruded in regions where the production of greenstones had come to an end. Of considerable interest in this connection are districts around two of the younger greenstone–tonalite complexes, namely the Superior Province in Canada, and the Yilgarn block in Western Australia (Wilson, 1958), where anorthosites and gabbros have been reported in each instance in association with belts of altered supracrustals.

Are these outcrops representative of rocks formed between successive greenstone complexes or should they be regarded as marginal phenomena developed around greenstone–tonalite complexes?

Much attention has been paid to the relations of Archaean granulites and the greenstone–tonalite complexes. Lambert and Heier, using an argument based on measured heat flow and the known abundance of heat-producing elements in West Australian tonalites, argued that the concentration of radioactive heat-producing elements must fall off with depth (Lambert, 1971). Windley and Bridgwater (1971) suggested from their knowledge of the North Atlantic craton a closely similar model in which granulite-rich terrain occurred at depth below the greenstones and tonalites. It could also be argued that granulites occur around the margins of greenstone complexes as, for example, in the Wheat Belt in S.W. Australia and the Pikwitonei province of the Canadian Shield. Where age determinations support such a hypothesis, one might suggest that such granulites represent an extensive thermal effect around the greenstone/tonalite-rich regions. However, in several instances the ages suggest a time difference between granulites and neighbouring greenstone complexes with an indication that the granulites are rather older than the youngest greenstone–tonalite complexes. Accordingly, the granulite metamorphism and the rocks affected by it may be due to events occurring between the times two greenstone–tonalite complexes formed.

What is becoming apparent is that towards the end of the Archaean, areas of newly-formed greenstone–tonalite terrain were becoming restricted and that around the complexes in which these rocks were forming, was a network of belts marked by other phenomena which included granulite metamorphism, the local intrusion of anorthosite and in some instances occurrence of the shelf sea orthoquartzite–limestone association with its attendant basic and ultrabasic rocks. It seems likely that this network provided the basis on which the younger systems of linear mobile belts were to devlelop later in the Precambrian.

A special case of such a mobile belt is provided by the Limpopo belt recognized for some time as one of the earliest linear belts in the crust. It was first visualized as separating two rather similar stable blocks, the Rhodesian and Kaapvaal cratons. It is now seen as separating two rather different blocks, in one of which greenstone and tonalite formation continued until at least 2·7 b.y. ago (Moorbath, this volume), while in the other, as we have seen, sedimentary basins formed on a wide scale from about 3 b.y. onwards. Coward et al. (1970) have pointed out the importance of transcurrent movements in the Limpopo belt. These may take up the differential displacements that occurred as the region to the north was intensely deformed when the younger greenstone belts of the Rhodesian Midlands, e.g. Que Que area, were deformed about 2·7 b.y. ago, while correspondingly intense deformation was not occurring to the south of the Limpopo belt in the Kaapvaal craton. On this view the Limpopo belt would be an asymmetrical structure separating more intensely and less intensely deformed blocks during late Archaean times. Such a transcurrent structure anticipates the many long belts of transcurrent movement and high strain which are a feature of the Proterozoic crust.

Movement of Material Between Crust and Mantle

It is interesting to note that some of the earliest (but not the first) examples of $^{87}Sr/^{86}Sr$ ratios from crustal rocks which are distinctly higher than the values to be expected in contemporary mantle came from pegmatites and granites about 2·5–2·4 b.y. old (Arriens, 1971; Penner and Clark, 1971). These occur as recognizably post-tectonic bodies. In other words, they occur in crust to some degree stabilized. The accumulation of ^{87}Sr relative to ^{86}Sr indicates that the rocks may contain or have been derived from, older crustal material which has been isolated for long enough to build up an anomalously high ^{87}Sr/^{86}Sr ratio. Still older occurrences of this anomaly are few but are coming to light. They should prove one of the best means of locating very old stabilized crust which was then subject to reworking. The rarity of such examples in analyzed Archaean rocks (Moorbath, 1975) as compared with the more numerous instances in which crust and mantle ^{87}Sr/^{86}Sr values are indistinguishable, suggest that Archaean tectonics permitted an exchange of material between crust and mantle on a scale which did not persist throughout the Proterozoic and Phanerozoic (see Armstrong, 1968).

It will be a matter of some interest to establish whether long-term variations in the extent of the transfer of matter from mantle to crust can be established within the Archaean. The lead isotope data recorded by Robertson (1973) suggest that such might be the case.

Conclusions

1. Greenstone belts and related tonalites or granites were formed repeatedly during the Archaean but they probably developed over periods which in total occupied less than half Archaean time.

2. There are records of sediments often including orthoquartzites and limestones, accompanied by basic and ultrabasic activity but not by the extensive intrusion of acid rocks which suggest another type of Archaean assemblage. It appears to reflect a more stable crust, possibly a shelf sea. Some local granites developed under these conditions, but there was not the extensive introduction of granites or tonalites which radically altered the crust in the greenstone terrains.

3. Individual greenstone belts may have varied in size during the Archaean. The complexes of tonalites and clusters of greenstone belts taken as a whole become smaller.

4. Around the restricted younger Archaean greenstone–tonalite complexes one can find regions marked by granulite metamorphism affecting, at least in some instances, rocks of the orthoquartzite–limestone–basic–ultrabasic assemblages. These late Archaean areas of high-grade metamorphism became the site of many of the later Precambrian mobile belts.

5. Certain examples of such young Precambrian mobile belts, for example the Limpopo belt, lie between an area in which younger greenstone belts developed and an Archaean region which lacks such rocks. Trans-current movements in such late Archaean mobile belts may be accommodation structures accompanying the intense deformation of the neighbouring block in which young Archaean greenstones were evolving and undergoing deformation.

6. Since the late Archaean mobile belts influenced later Precambrian tectonics and often became the site of younger mobile belts, their recognition has been more difficult than that of the greenstone–tonalite association.

7. There is evidence that rocks over 2·4 b.y. in age could accumulate ^{87}Sr to build up anomalously high ^{87}Sr/^{86}Sr ratios, but in general this did not happen as commonly as in post-Archaean rocks. Since early Archaean crustal rocks including granites undoubtedly existed, there must have been a more effective way of cycling material between crust and mantle during the Archaean than is provided by plate tectonics as developed since late Proterozoic times.

8. The reason for the repeated development of greenstone belts in short periods (50–100 m.y.) over the 1200 m.y. of Archaean time for which we have a record (from 3·7 b.y. onwards) is not understood. It may be connected with phenomena similar to those which later produced the abrupt

changes in direction of polar wander paths and the variations in the amount of granite (Gastil, 1960) and in the quantity of material transferred from mantle to crust, known to have occurred during the Proterozoic and Phanerozoic.

References

Armstrong, R. L., 1968. 'A model for the evolution of strontium and lead isotopes in a dynamic Earth, *Rev. Geophys.*, **6**, 175–199.

Arriens, P. A., 1971. 'The Archaean Geochronology of Australia', *Spec. Publ. Geol. Soc. Aust.*, **3**, –11–24.

Catanzaro, E. J., 1963. 'Zircon ages in southwestern Minnesota', *J. Geophysics Res.*, **68**, 2045–2046.

Coward, M. P., Francis, P. W., Graham, R. H., and Watson, J., 1970. 'Large-scale Laxfordian structures of the Outer Hebrides in relation to those of the Scottish mainland', *Tectonophysics*, **10**, 425–435.

Coward, M. P., Graham, R. H., James, P. R., and Wakefield, J., 1973. 'A structural interpretation of the northern margin of the Limpopo orogenic belt, Southern Africa', *Phil. Trans. Roy. Soc.*, **A273**, 487–492.

Gastil, G., 1960. 'Continents and mobile belts in the light of mineral dating', *Int. geol. Cong. 21st Sess.*, Pt 9, 162–169.

Green, D. C., and Baadsgaard, H., 1971.'Temporal evolution and petrogenesis of an Archaean crustal segment at Yellowknife N.V.T., Canada', *J. Petrol.*, **12**, 177–217.

Hunter, D. R., 1974. 'Crustal development in, the Kaapvaal craton I, The Archaean', *Precambrian Research*, **1**, 259–326.

Lambert, I. B., 1971. 'The composition and evolution of the deep continental crust', *Spec. Publ. Geol. Soc. Aust.*, **3**, 419–428.

Moorbath, S., 1975. 'Evolution of Precambrian crust from strontium isotope evidence', *Nature*, **254**, 395–398.

Morrison, E. R., and Wilson, J. F., 1971. *Symposium on the granites, gneisses and related rocks.* Excursion guide book, Geol. Soc. S. Africa, Rhodesian Branch.

Penner, A. P., and Clark, G. S., 1971. 'Rb–Sr age determinations from the Bird River area, south eastern Manitoba', *Geol. Ass. Canada*, Sp. paper 9,105–109.

Robertson, D. K., 1973. 'A model discussing the early history of the earth based on the study of lead isotope ratios from veins in some Archaean cratons of Africa', *Geochim. Cosmochim. Acta.*, **37**, 2099.

Slawson, W. F., Kanaseowich, E. R., Ostie, R. G. and Farquhar, R. M., 1963. 'Age of the North American crust', *Nature*, **200**, 413–414.

Sutton, J., 1967. 'The extension of the Geological Record into the Precambrian', *Proc. Geol. Ass.*, **78**, 493–534.

Wilson, A. F., 1958. 'Advances in the knowledge of the structure and petrology of the Precambrian rocks of south-western Australia, *J. Roy. Soc. West. Aust.*, **41**, 57–83.

Wilson, J. F., 1968. 'The geology of the country around Mashaba', *Bull. geol. Surv. Rhod.*, 62.

Windley, B. F., 1973. 'Archaean anorthosites: a review'. *Spec. Publ. Geol. Soc. S. Afr.*, **3**, 319–332.

Windley, B. F., and Bridgwater, D., 1971. 'The evolution of Archaean Low-and-High-grade terrains', *Spec. Publ. Geol. Soc. Aust.*, **3**, 33–46.

New Tectonic Models for the Evolution of Archaean Continents and Oceans

BRIAN F. WINDLEY

Geology Department, The University, Leicester, England.

Introduction

There is currently a proliferation of new data and ideas on the Archaean, with considerable disagreement as to the role of 'plate tectonics' in the construction of the early continents. One point does seem to stand out above all others: it is increasingly apparent that Archaean rocks, their associations and chemistry, are remarkably similar to modern examples in continental margins and island arcs, and an appeal to a primitive form of microcontinental collision and arc formation in the Archaean is therefore increasingly convincing. The aim of this paper is to review some of the current trends of thought about Archaean tectonic evolution and, in particular, to interweave some recently published data and models with those published in this volume.

It is still convenient to discuss the situation in terms of low-grade (greenstone belt-granite) and high-grade (gneissic) terrains, although it is true that their differences may not be so great as once suggested. Recent age determinations have shown that both these terrains went through at least two stages of growth which commonly fall in the periods 3800–3600 m.y. and 3100–2600 m.y. (Hawkesworth et al., 1975; Moorbath, this volume). The majority of the main greenstone belts in Canada, Rhodesia and West Australia, and the larger part of the high-grade region in Greenland were formed in the second period. Unconformities between younger greenstone belts and older gneisses are uncommon, but are clearly seen in the Belingwe area, Rhodesia (Bickle et al., in press) and the Cross Lake area, Manitoba (Ermanovics, this volume).

From strontium isotope evidence Moorbath (1975, this volume) points out that the majority of Archaean volcanic and plutonic rocks represent juvenile additions to the continental crust and that within any one region the time interval between the extraction of juvenile igneous material and the regional metamorphism is less than 50–100 m.y. The isotope evidence demonstrates that the majority of Archaean igneous rocks (volcanics, granites and gneisses) were not derived by partial melting of older sialic material, but were derived directly from the upper mantle. The evidence can be used as substantial support for some kind of plate tectonic activity in the Archaean, particularly when one considers the fact that the Archaean calcalkaline volcanics, tonalites and granodiorites concerned have geochemical characteristics similar to those of equivalent rocks in modern island arcs and continental margins which *are* situated at plate boundaries and were derived from mantle sources.

Regions Dominated by Low-grade Greenstone belts

Some Archaean regions, like Rhodesia and the Superior Province in Canada, have a high proportion of well-preserved supracrustal rocks which are largely in a greenschist facies state. The remainder of the regions are made up about equally of older gneissic rocks and coeval to slightly younger tonalite and K-rich granite plutons. Much attention has naturally been focused on the mode of formation of these low-grade greenstone belts and, not surprisingly, a marked division of opinion has arisen between classical static models and those more actualistic interpretations which involve some degree of primitive plate interaction.

(a) Classical concepts

Knowledge of the evolution of greenstone belts was advanced considerably by the reappraisal of Anhaeusser et al. (1969), following which Glikson proposed the well-publicized (1970, 1971, 1972) down-sagging basin model, which has been modified to take into account some subduction (Anhaeusser, 1973; Glikson, in press) and meteorite impact (Glikson, this volume). The early Glikson models took no account of the gneiss–granulite regions but the later models interpret them as the deeper crustal levels of the greenstone belts—a less than convincing argument because the high-grade regions contain their own distinctive shallow-water shelf-type sedimentary orthoquartzite–carbonate association (Sylvester-Bradley, 1975; Sutton, this volume), which is totally unlike the sedimentary assemblage of the greenstone belts. The only conclusion that satisfies these relations is that these parts of the high-grade complexes formed at the Earth's surface (they are often intercalated with meta-volcanics) and in a different tectonic setting to the greenstone belts.

A prerequisite for the deep downsagging model of Glikson is that the basins should be filled with vast volcano-sedimentary deposits, commonly more than 10 km thick. However, recent reconsiderations make such original thicknesses unlikely for some key greenstone belts. Coward et al. (this volume) show that the successions in the Rhodesian belts have been subject to extensive thrusting and stratigraphic duplication, and Burke et al. (this volume) suggest that there are few tectonically unrepeated sections in greenstone belts.

A more fundamental problem with the Glikson classical model (e.g. 1970 and in press) is that the greenstone belts are considered to have developed from some form of primitive and extensive oceanic crust and thus pre-date the formation of sialic nuclei. But the most recent isotopic data are totally inconsistent with such an idea. Clearly the majority of the main greenstone belts in Rhodesia, W. Australia and Canada formed in the late Archaean in the period 3100–2600 m.y. ago, long after the development of sialic crust that is now known in Canada, the U.S.A., Greenland and Rhodesia.

As an alternative, a rift valley-small ocean basin model was proposed by Windley (1973a). This was consistent with relationships considered important at that time, in particular with the evidence from the Abitibi belt (Goodwin and Ridler, 1970) and with the age relationships with older gneissic 'basement'. It formed the basis for the tectonic model for the belts of the Kaapvaal craton by Hunter (1974), but in the light of the most recent information I would consider it outdated. For one reason, it does not specify where in the continental environment the rift is situated, and secondly, it does not provide an adequate mechanism for the generation of the intrusive tonalites.

(b) Actualistic Concepts

Plate tectonic models for the evolution of the Archaean crust have been proposed by several authors including Talbot (1973) and Rutland (1973). One reason for believing that some kind of, perhaps primitive, plate activity was responsible for the generation of Archaean igneous rocks is the fact that it is increasingly evident that the vast majority of the rocks are little different in major and trace element chemistry to equivalent Mesozoic–Tertiary rocks which clearly did form only as a direct result of plate tectonics (White et al., 1971; Engel et al., 1974; Jahn et al., 1974).

So far most specialists have compared greenstone belts with modern island arcs, but a new approach considers them as Archaean analogues of late Phanerozoic marginal basins that evolved in extensional back-arc environments, like those in the western Pacific (Karig, 1971). The marginal basin model is proposed for the first time in some detail in two papers in this volume (Burke, Dewey and Kidd; Tarney, Dalziel and de Wit).

Examples of fossil marginal basins that are remarkably similar to greenstone belts are the Rocas Verdes Complex in S. Chile (Dalziel et al., 1974; Tarney et al., this volume), Smartville terrane, N.W. Sierra Nevada (Moores, 1975), Round Pond area, South-Central Newfoundland, and two in western Newfoundland which noticeably contain basaltic komatiites (Burke et al., this volume), and the Klamath Mountains, California.

Like the greenstone belts (Litherland, 1973) the marginal basin rocks occur as mega-xenoliths or elongate discontinuous strips within engulfing 'granitic' rocks; they both typically have early ultramafic–mafic volcanic successions followed by calc–alkaline volcanic suites and volcanoclastic and clastic sediments (Glikson, this volume). Fossil marginal basin rocks have commonly been involved in extensive inter-thrusting with basement rocks in a manner not dissimilar to that described by Coward et al. (this volume) from greenstone belts in Rhodesia.

In proposing a fossil marginal basin (back-arc) origin for greenstone belts one obvious question arises: what rocks were forming in the adjacent arc environment and do we see them in Archaean terrains today?

Regions Dominated by High-grade Gneisses

The predominant rock type in these regions is tonalitic-granodioritic gneiss which occupies 80–85% of the present exposure surface (e.g. Kalsbeek, this volume). Within the gneisses are a variety of metamorphosed sediments, volcanics and layered igneous complexes, but a major problem is the origin of the gneisses themselves; the principal current ideas on their source material vary from andesitic–dacitic volcanics (Bowes et al.,

1971; Viswanathan, 1974) to modified earlier gneisses (Tarney et al., 1972), to andesitic magma that crystallized under plutonic conditions, to dioritic rocks by a crustal underplating process (Holland and Lambert, 1975).

A novel approach to the origin of these rocks is to regard them as the Archaean equivalents of the tonalite–granodiorite batholiths that occur along the main arc axis of modern continental margins like the Andes and western North America (Tarney, this volume; Smith and Windley, in prep., Tarney and Windley, in prep.). The geochemical comparisons are made by Tarney (this volume) and a plutonic origin for the gneisses is favoured by the discovery in West Greenland and Labrador that in low strain areas they can be seen to be derived by deformation of homogenous 'granitic' rocks (McGregor, 1973; Collerson et al., this volume).

Support for the plutonic continental margin origin comes from the group of calcic anorthosite–gabbro–ultramafic rocks that occur in layered igneous complexes as remnants within the gneisses on several continents (Windley, 1973b), because remarkably similar rocks occur in relict complexes within the mesozonal tonalitic–granodioritic batholiths along the eastern Pacific (Murray, 1972; Nishimori, 1974, 1975). These have An_{80-100} plagioclases in anorthositic gabbros, cumulate textures and layering, primary igneous low-Si amphiboles, and enrichments in Fe—all very similar, for example, to the Fiskenaesset complex, West Greenland (Windley and Smith, 1974). Nishimori considers these rocks to represent the hydrous refractory residue of a basaltic–andesitic magma parental to the calc–alkaline tonalite–granodiorite suite that forms the Peninsular Range batholith of California. In view of the remarkable similarities with the modern examples in terms of composition and mode of occurrence, it would seem likely that the Archaean rocks had a broadly similar origin in relation to some primitive form of plate movements.

At this point the actual field occurence of these Archaean rocks should be noted. In the Fiskenaesset region of West Greenland the anorthosite complex was intruded into basic volcanics (Myers, this volume) and then engulfed by enormous quantities of tonalites and granodiorites, now represented by plagioclase-rich gneisses which occupy more than 50% of the region (Kalsbeek, this volume) (in Peru the main batholith has reached the surface and engulfed volcanic rocks). No basement gneisses have been recognized and thus the amphibolite–anorthosite suite occurs as enormous sheet-like rafts within the intruding calc–alkaline rocks (Kalsbeek and Myers, 1973).

Often, however, the layered complexes are not bordered by, or intruded into, supracrustal rocks; they just occur as vast relics floating in the younger engulfing granitic rocks. Clearly the early fractionated residue could either rise to a high level of the crust to be intruded into supracrustal rocks, or else it could remain at some depth; in either case they would be engulfed by their own more highly fractionated granitic suite.

In some regions rocks belong to the layered complexes occur in the gneisses as inclusions on a 1–100 m scale, but there is little evidence that their break-up was caused by regional deformation. A more promising approach might be to consider that much of the break-up of the complexes was caused by intrusion of the tonalite-granodiorite magma and that this was then accentuated by some deformation when the plutonic rocks were converted to gneisses and subsequently folded and refolded.

Barker and Peterman (1974) suggested that high-grade regions in the Transvaal and Rhodesian cratons consist largely of a bimodal gneiss–amphibolite association formed by recrystallization of dacites and tholeiites derived by partial melting of a downgoing slab of hydrous oceanic lithosphere. However, I would suggest that in the North Atlantic craton, the Limpopo belt in southern Africa and the Peninsular Gneiss region of southern India there is a trimodal association of predominateiy tonalitic–granodioritic gneisses, amphibolites and anorthosite–gabbro–ultramafic complexes, the unmetamorphosed equivalents of which can be found along modern continental margins.

Archaean Tectonic Environments

If the ideas expressed by Burke et al. (this volume), Tarney (this volume), Tarney et al. (this volume) and in this paper are correct, then one should look towards primitive forms of microcontinental collision and arc formation as a means of explaining Archaean development rather than unique forms of tectonics, such as downsagging basins (Glikson, 1971), basalt underplating (Fyfe, 1974), subcrustal accretion (Holland and Lambert, 1975) and skimming orogeny (Talbot, 1973). An appeal for comparison with modern tectonic environments has a certain advantage in that the hypothesis can be tested by comparative study of equivalent modern and ancient rock suites.

The Archaean environment suggested by Tarney and Windley (in prep.) is one in which parallel arcs and marginal basins gave rise to tonalitic–granodioritic belts and greenstone belts, respectively, the arcs being the site of high heat flow and subsequent granulite-grade metamorphism. An essential characteristic of the model is that the so-called high- and low-grade terrains developed coevally and laterally, rather than coevally and vertically as envisaged by Glikson (in press and this volume).

There seems very little liklihood that any high-grade gneissic regions are simply highly metamorphosed greenstone belts as the proportion of rock types is so different: in the one, granitic rocks often amount to no more than 10–20%, in the other they reach 80–85%. This argument applies at least as long as the granitic gneisses are largely derived from pre-existing plutonic granitic rocks.

Komatiites are now known in the Barberton, Canadian and West Australian greenstone belts, as well as in the Mesozoic ophiolite complex in Cyprus, the Appalachian fossil margin basins in Newfoundland, and the high-grade meta-volcanics in West Greenland (Rivalenti, this volume); but their geotectonic significance is far from clear as they also occur in Archaean sediments that lie unconformably on a gneissic basement in Rhodesia (Bickle et al., in press) and in an Archaean stable cratonic sedimentary suite in the Keewatin district of Canada (Schau, 1975). Certainly they cannot be used as a sole indication of an oceanic crustal environment, nor as a means of concluding that enclosing gneisses and granulites are derived by recrystallization of a greenstone belt volcanic suite (Viswanathan, 1974).

It should not be surprising to find some metamorphosed oceanic-type tholeiites or fossil marginal basin rocks within the high-grade gneissic regions, in spite of the fact that it is suggested here that the latter represent an arc-type development. For example, K-poor oceanic-type meta-tholeiites occur as amphibolite layers in some high-grade gnessic regions (Friend, 1975), and it is clear in this case that the bordering gneisses were formed from younger intrusive granitic rocks. In South Chile the fossil marginal basins are, in places, bordered on both sides and engulfed by plutons and batholithic rocks (Tarney et al., this volume). As Burke et al. (this volume) emphasize, the complications of development in such continental margins may be immense due to continuous or episodic oceanward migration of the subduction zone, complexities associated with the opening and closing of the marginal basins and subsequent arc–arc or arc–microcontinental collisions.

Burke et al. (this volume) suggest that high-grade regions can be explained solely as reactivated basements, but one cannot escape the fact that there are areas with apparently no basement which are dominated by intrusive granitic rocks—these cannot be explained just in terms of basement reactivation, and the isotopic evidence of Moorbath (1975 and this volume) supports this.

Greenland Tectonics

A relevant question that may be asked at this point concerns the tectonic relationship between the older (Amîtsoq gneisses) and younger (Malene Group) rock units in the Godthaab district of West Greenland. According to Bridgwater et al. (1974) and Myers (this volume) the Amîtsoq gneisses and Malene supracrustals (largely basic volcanics) were probably formed in different parts of the crust (continental and oceanic) and were driven together by subhorizontal thrusting to form alternating slices of basaltic and granitic crust above subduction zones. However, in a particularly important paper, Chadwick and Coe (this volume) point out that in some areas the Malene supracrustal rocks are clearly cut by abundant amphibolite dykes, which are of the same lithological types and have undergone the same deformation and migmatisation as the Ameralik dykes in the gneisses. They therefore conclude, very reasonably, that it is most likely that the dykes in the Malene Group are also Ameralik dykes; the main implications of this are that the Malene Group supracrustal rocks were deposited as a cover on a basement of Amitsoq gneisses and that the unconformity has been removed by deformation. Lateral thrusting must therefore have been important, but not on such a large scale as to move the rock units very far, and so Chadwick and Coe (this volume) are of the opinion that there is no evidence to support the hypothetical large-scale crustal interleaving model of Bridgwater et al. (1974). The fact that an unconformable cover sequence can be converted by deformation to lie conformably on its basement-gneisses is no surprise when one considers the excellent demonstration of such tectonic relations in the Ivigtut area of S.W. Greenland (Henriksen, 1969).

It is interesting that Collerson et al. (this volume) compare the anorthosites in Labrador with those in West Greenland and the Cross Lake area of Manitoba. Those at Cross Lake clearly lie unconformably on a gneissic basement (Ermanovics, this volume); can it not be, therefore, that the Cross Lake sequence represents an undeformed equivalent of the Godthaabsfjord sequence?

There is a recent change in emphasis in the interpretation of the tectonic style of some Archaean regions. Not more than a couple of years ago greenstone belts were commonly considered to have undergone a vertical style of deformation. However, Coward et al. (this volume) demonstrate the existence of large-scale thrust and nappe structures in the

110

Rhodesia craton somewhat similar to those suggested by Bridgwater et al. (1974) for the high-grade region of West Greenland. Burke et al. (this volume) likewise place emphasis on the presence of thrust structures in greenstone belts and a tectonic development more by horizontal shortening than vertical stretching or downsagging. Whilst making analogies with Archaean environments it may be significant to note that the major thrust belts in modern continental margins occur in the gap between the arc and the marginal basin; more extensive thrusting in the Archaean, a reflection of a more prominent horizontal tectonic regime, would have easily inter-thrusted the arc gneisses with the marginal basin supracrustals.

References

Anhaeusser, C. R., 1973. 'The evolution of the early Precambrian crust of southern Africa', *Phil Trans. R. Soc. Lond.*, **A273**, 359–388.

Anhaeusser, C. R., Mason, R., Viljoen, M. J., and Viljoen, R. P., 1969. 'A reappraisal of some aspects of Precambrian shield geology', *Bull. Geol. Soc. Amer.*, **80**, 2175–2200.

Barker, F., and Peterman, Z. E., 1974. 'Bimodal tholeiitic-dacite magmatism and the Early Precambrian crust', *Precambrian Res.*, **1**, 1–12.

Bickle, M. J., Martin, A., and Nisbet, E. G., (In press). 'Basaltic and peridotitic komatiites and stromatolites above a basal unconformity in the Belingwe greenstone belt, Rhodesia', *Geology*.

Bowes, D. R., Barooah, B. C., and Khoury, S. G., 1971. 'Original nature of Archaean rocks of North-west Scotland, *Geol. Soc. Aust. Sp. Publ.*, **3**, 77–92.

Bridgwater, D., McGregor, V. R., and Myers, J. S., 1974. 'A horizontal tectonic regime in the Archaean of Greenland and its implications for early crustal thickening', *Precambrian Res.*, **1**, 179–197.

Dalziel, I. W. D., de Wit, M. J., and Palmer, K. F., 1974. 'A fossil marginal basin in the southern Andes', *Nature*, **250**, 291–294.

Engel, A. E. J., Itson, S. P., Engel, C. G., Stickney, D. M., and Cray, E. J., Jr., 1974. 'Crustal evolution and global tectonics: a petrogenic view', *Bull. Geol. Soc. Amer.*, **85**, 843–858.

Friend, C. R. L., 1975. 'The geology and geochemistry of the Pre-Ketilidian basement complex in the Ravns Storø area, Fiskenaesset region, southern West Greenland', Unpubl. Ph.D. thesis, Univ. London, England.

Fyfe, W. S., 1974. 'Archaean tectonics', *Nature*, **249**, p. 338.

Glikson, A. Y., 1970. 'Geosynclinal evolution and geochemical affinities of early Precambrian systems', *Tectonophysics*, **9**, 397–433.

Glikson, A. Y., 1971. 'Primitive Archaean element distribution patterns: chemical evidence and geotectonic signficance', *Earth Planet. Sci. Lett.*, **12**, 309–320.

Glikson, A. Y., 1972. 'Early Precambrian evidence of a primitive ocean crust and island nuclei of sodic granite', *Bull. Geol. Soc. Amer.*, **83**, 3323–3344.

Glikson, A. Y., (In press). 'Archaean to Early Proterozoic shield tectonics: relevance of plate tectonics', *Proc. Geol. Ass. Can.*,

Goodwin, A. M., and Ridler, R. H., 1970. 'The Abitibi orogenic belt', In *Geol. Surv. Can.* Pap. 70–40 (Ed. A. J. Baer), 1–30.

Hawkesworth, C. J., Moorbath, S., O'Nions, R. K., and Wilson, J. F., 1975. 'Age relationships between greenstone belts and "granites" in the Rhodesian Archaean craton', *Earth Planet. Sci. Lett.*, **25**, 251–262.

Henriksen, N., 1969. 'Boundary relations between Precambrian fold belts in the Invigtut area, Southwest Greenland', In *Geol. Ass. Can. Sp.* Pap. 5, 143–154.

Holland, J. G., and Lambert, R. St. J., 1975. 'The chemistry and origin of the Lewisian gneisses of the Scottish mainland: the Scourie and Inver assemblages and sub-crustal accretion', *Precambrian Res.*, **2**, 161–188.

Hunter, D. R., 1974. 'Crustal development in the Kaapvaal craton. I. The Archaean', *Precambrian Res.*, **1**, 259–294.

Jahn, B.-M., Shih, C. Y., and Murthy, V. R., 1974. 'Trace element geochemistry of Archaean volcanic rocks', *Geoch. Cosmoch. Acta*, **38**, 611–627.

Kalsbeek, F., and Myers, J. S., 1973. 'Geology of the Fiskenaesset region', In *Grønlands Geol. Unders. Rapp.*, **51**, 5–18.

Karig, D. E., 1971. 'Origin and development of marginal basins in the western Pacific', *J. Geophys. Res.*, **76**, 2542–2561.

Litherland, M., 1973. 'Uniformitarian approach to Archaean "schist relics" ', *Nature*, **242**, 125–127.

McGregor, V. R., 1973. 'The early Precambrian gneisses of the Godthaab district, west Greenland', *Phil. Trans. R. Soc. Lond.*, **A273**, 343–358.

Moorbath, S., 1975. 'Evolution of Precambrian crust from strontium isotopic evidence', *Nature*, **254**, 395–398.

Moores, E. M., 1975. 'The Smartville terrane, northwestern Sierra Nevada, a major pre-Late Jurassic ophiolite complex', *Geol. Soc. Amer., Abstr. with Prog.*, **7**, p. 352.

Murray, C. G., 1972. 'Zoned ultramafic complexes of the Alaskan type: feeder pipes of andesitic volcanoes', *Geol. Soc. Amer. Mem.*, **132**, 313–335.

Nishimori, R. K., 1974. Cumulate anorthositic gabbros and peridotites and their relation to the origin of the calc-alkaline trend of the Peninsular Range batholith. *Geol. Soc. Amer.*, Abstr. with Prog., **6**, 229–230.

Nishimori, R. K., 1975. 'Fractional crystallisation modeling of the Peninsular Range batholith based on the Peninsular Range gabbros', *Geol. Soc. Amer.*, Abstr. with Prog., **7**, 356–357.

Rutland, R.W., 1973. 'Tectonic evolution of the continental crust of Australia', In *Implications of Continental Drift to the Earth Sciences*, D. H. Tarling and S. K. Runcorn (Eds.), Academic Press, London, Vol. 2, 1011–1033.

Schau, M., 1975. 'Meta-komatiites in a stable Archaean environment', *Geol. Soc. Amer.*, Abstr. with Prog., **7**, p. 849.

Sylvester-Bradley, P. C., 1975. 'The search for Protolife', *Phil. Trans. R. Soc. Lond.*, **189**, 213–233.

Talbot C. J., 1973. 'A plate tectonic model for the Archaean crust', *Phil. Trans R. Soc. Lond.*, **A273**, 413–428.

Tarney, J., Skinner, A. C., and Sheraton, J. W., 1972. 'A geochemical comparison of major Archaean gneiss units from Northwest Scotland and East Greenland', *Proc. 24th Int. Geol. Congr. Montreal*, **1**, 162–174.

Viswanathan, S., 1974. 'Contemporary trends in geochemical studies of early Precambrian greenstone-granite complexes', *J. Geol. Soc. India*, **15**, 347–379.

White, A. J. R., Jakeš, P., and Christie, D. M., 1971. 'Composition of greenstones and the hypothesis of sea-floor spreading', In *Symp. on Archaean rocks*, Ed. J. E. Glover. Geol. Soc. Aust. Sp. Publ. **3**, 47–56.

Windley, B. F., 1973a. 'Crustal development in the Precambrian', *Phil. Trans. R. Soc. Lond.*, **A273**, 321–341.

Windley, B. F., 1973b. 'Archaean anorthosites; a review with the Fiskenaesset complex, West Greenland as a model for interpretation', *Geol. Soc. S. Afr. Sp. Publ.*, **3**, 312–332.

Windley, B. F., and Bridgwater, D., 1971. The evolution of Archaean low- and high-grade terrains, *Geol. Soc. Aust. Sp. Publ.*, **3**, 33–46.

Windley, B. F., and Smith, J. V., 1974. 'The Fiskenaesset complex, West Greenland. Pt 2: General mineral chemistry from Qeqertarssuatsiaq', *Bull. Grønlands Geol. Unders.*, **108**, 1–54 (also *Meddr. Grønland* **196**).

Dominance of Horizontal Movements, Arc and Microcontinental Collisions During the Later Permobile Regime

KEVIN BURKE, JOHN F. DEWEY, and W. S. F. KIDD

Department of Geological Sciences, State University of New York at Albany, 1400 Washington Avenue, Albany, New York 12222

Abstract

In the latter part of the permobile regime ($\simeq 3\cdot5$–$2\cdot5$ b.y. ago), at a time when many of the oldest rocks now preserved were formed, most of the present continental mass had differentiated and the Earth's surface can be envisaged as rather less than one-third covered with continental material accreted by arc-amalgamation but only locally differentiated into granitic and granulitic fractions by Tibetan-type continent–continent collision processes. The continental masses were not significantly thinner than today's continents but were up to a hundred times smaller in area and hence much more numerous.

The remaining two-thirds of the Earth's surface was oceanic and the amount of water in the oceans about the same as today. Because heat-generation rates were perhaps three times higher than now, thermal gradients were rather steeper than at present and additional thermal energy was dissipated through greater ridge activity. Ridges either spread faster or the total length of ridge was longer than today or, more probably, both. The total length of subduction zone now roughly matches the total ridge length and this was presumably also the case 3 b.y. ago.

The lenticular style of the Superior Province is the characteristic signature of the rapid horizontal movements and consequent arc and microcontinental collisions of this time. Greenstone terrains such as the Superior Province preserve few obvious signs of the torsionally rigid behaviour at rupture and collision which became general during the Proterozoic. For this reason we characterize the deformation, although it involved sea-floor spreading, subduction and continental drift, as permobile deformation and not as plate tectonics.

Collisional processes affecting fossiliferous strata of Phanerozoic times generally disrupt the ophiolite, island arc, continental margin and exogeosynclinical successions involved and collision zones such as the Alps and Himalayas are recognized as areas of great tectonic complexity. Students of greenstone belts, who do their stratigraphy without fossil control, commonly report undisturbed successions. It may be that there are unrecognized tectonic boundaries within greenstone belts, recognition of which would help to resolve some of the supposed differences between these terrains and arc rocks involved in later collisions.

113

Introduction

Although, in the early history of the Earth, fractionation of anorthosite and dominance of structural evolution by impact and related eruption of basalt are likely to have characterized the surface development of the planet, no material produced in this phase appears to have survived in a recognizable state. By the end of the permobile regime (Burke and Dewey, 1973) conditions were much closer to those of today. We here suggest that these conditions imply a dominance of horizontal motion of a rigid lithosphere (boundary conduction layer in the case of the oceans) with, as now, rates and amounts of horizontal motion exceeding those of vertical motion by one hundred to one thousand times.

The Permobile Ocean: Faster Mantle Fractionation

Jakeš (1973), among others, has shown that the average composition of continental crust resembles that of island arcs and has suggested that continents have been made by a two-stage fractionation of mantle material, the first stage producing ocean floor igneous rocks at ridges and the second island arc igneous rocks above sinking inclined slabs of oceanic lithosphere. Continents form from arcs by collision following horizontal motion. In the first stage partial melting of mantle material to produce basalt and leave depleted pyrolite involves only vertical movement but once these rocks have been differentiated and emplaced horizontal motion comes into operation as the new lithosphere is carried away from the spreading ridge. Cooling and thickening of ageing lithosphere in this process is important in the convective dissipation of heat generated within the Earth. Because the thickness and elevation of cooling lithosphere change as the square root of the lithosphere's age (Parker and Oldenburg, 1973) more heat is dissipated by faster spreading ridges than by slow.

Fig. 1 shows that much more heat was being generated by radioactive decay in the Earth 2·5 b.y. ago than now. Many authors (for example, Goodwin, 1973) have emphasized

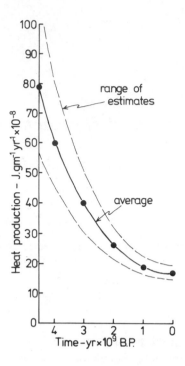

Fig. 1. Total heat production through geological time by U, Th, and K. (After Lee, 1967)

the close resemblance of the rocks formed in the latter part of the permobile regime and preserved in greenstone belts to those formed at ridges and above sinking slabs today. It is evident (Jakeš, 1973) from this resemblance that the two-stage fractionation of mantle material at ridges and above descending slabs was then operating much as today. Because two or three times as much heat was being generated and because there is no evidence of the operation of any heat-dissipating process peculiar to ancient times, we infer that ridge processes were operating two or three times as fast as today. This greater intensity of operation could have been achieved either by greater average spreading rates or by greater total ridge length, or by both. Assuming that the volume of the Earth was about the same as now, there would have been either a greater length of subduction zone to consume the extra ocean floor or subduction would have worked faster. Since there is some evidence (see, for example, Forsythe and Uyeda, in press) that there is a general average for sub-duction rates at about 6 cm/yr (w.r.t. one of the slowly moving reference frames), a substantially greater length of subduction zone seems more likely.

The picture that emerges is that of an ocean in late permobile time with a great length of fast spreading ridge and a large number of long trenches, the two processes producing at a great rate ocean floor and arc rocks very like those of today. There seems little likelihood that oceanic thermal gradients were much greater in those times. Partial melting of pyrolite to make basalt presumably took place at temperatures and pressures similar to those of today and the greater amount of heat coming from within the Earth was mainly dissipated by the faster convective removal of basalt and depleted pyrolite at the faster spreading ridges. Ultramafic lavas appear to be rather more common in the Archaean than in later rocks and their occurrence perhaps indicates local melting at higher temperatures of the ancient pyrolite. Since even in the Archaean the proportion of these ultramafic lavas within basaltic piles is low, it seems unlikely that thermal gradients at ridges were much steeper than now. We suggest, as did Moores (1973), that Archaean oceanic crust is likely to have been somewhat thicker than present oceanic crust, both because the mantle was then likely to have been a little less depleted, and because a slightly higher percentage of partial melting occurred due to slightly higher thermal gradients. In contrast, the Archaean oceanic lithosphere (and continental lithosphere) is likely to have been rather thinner than at present.

Because all oceans soon 'self-destruct' (Atwater's term), it is impossible to make direct observation of extensive areas of old oceanic material and inferences about ancient ocean can only be made indirectly or from the dismembered slivers of oceanic and arc rocks to be found in greenstone belts. In contrast, continents are 'non-geodegradable' (Dietz's term) and once made, persist, although subject to episodic internal fractionation (Dewey and Burke, 1973) due to continent–continent collision. For this reason we do not consider here

the idea that the Earth's surface was once covered with a greater area of continent.

Ancient Continents

Inspection of Fig. 1 shows that very roughly twice as much heat was generated by radioactive decay between 4·5 and 2·5 b.y. ago as has been generated since. If this heat was the sole source of formation of the continents through two-stage fractionation and continental accretion through arc collision, then about two-thirds of the continental mass would have been produced by 2·5 b.y. ago. However, the Earth was already inhomogeneous at planetary accretion and additional heat was generated in its early history from short-lived radioisotopes, and from gravitational potential energy during accretion and subsequent core formation. Major impacts must have provided additional energy, at least partly dissipated through mantle fractionation, over a relatively long time range, mainly between 4·5 and 3·9 b.y. ago if terrestrial history paralleled that of the Moon. Because these processes made available an unknown, but probably large, amount of additional heat to promote mantle fractionation, it seems likely that more than two-thirds of the present continental mass was produced under the permobile regime and that by its end ($\cong 2·5$ b.y. ago) much, if not nearly all, of the present continental mass was in existence.

Other lines of evidence supporting this view include: (1) the increasingly widespread recognition of very old continental rocks (Hart and Goldich, 1975), (2) the arguments of Armstrong (1968) and Wise (1974) indicating that much of the present oceanic water volume was present by 2·5 b.y. ago and that continents stood then with their present freeboard and (3) evidence of recycling of strontium isotopes (Armstrong, 1968).

Radioactive isotopes concentrated in the continents at the end of the permobile phase were themselves generating approximately twice as much heat as now and for this reason continental thermal behaviour must have been somewhat different. However, there appears no strong evidence of persistence of extreme thermal gradients or very different tectonic behaviour within the ancient continental masses because there is little sign of widespread partial melting, other than that due to continent–continent collision.

The only geological evidence that clearly shows anomalous behaviour of continental lithosphere prior to about 2·5 b.y. ago is the extensive and very thick shelf sediment sequence of the Witwatersrand Triad (2·8–2·4 b.y.). This includes a huge thickness of silicic and mafic volcanics (Whiteside, 1970) also not like anything seen in younger shelf sequences. The Huronian (2·4–2·2 b.y.) is of more normal thickness and lithology over most of its extent, except in its southern-most part (Frarey and Roscoe, 1970). This southern area is bounded by faults (Card, 1970) and there are minor amounts of basalt confined to the base of the very thick (>10 km) sequence of shallow water sediments. It is possible that this part of the Huronian occupies an east–west trending aulacogen, opening eastward into the more highly deformed and metamorphosed turbidite-bearing sequence of the 'Southern Province,' which perhaps represents the continental slope/rise, outer shelf and underlying graben facies deposits of a rifted continental margin.

Partial melting of continental crust averaging a diorite–granodiorite or andesitic composition causes differentiation into a residual anhydrous anorthosite and granulite rich fraction and a hydrous granitic fraction (as was shown experimentally by Green (1969), and as is well displayed on a relatively small scale in such hot-spots where minor amounts of continental crust have melted; e.g. the White Mountains and the Jos plateau). Continental crust, in which neither the high level granitic nor the deep seated granulitic and anorthositic rocks are abundant and in which there is a marked preponderance of granodioritic (andesitic) bulk compositions, is unlikely to have been partially melted and hence is unlikely to have been subjected to exceptional thermal gradients. Although both granulitic and anorthositic rocks are known among the rocks formed in premobile times, they are restricted in their distribution. We do not know of any 'granitic' Archaean terrain, including those with granulites and anorthosites, whose properties (e.g. Bridgwater et al., 1974) do not resemble in most major aspects those of younger terrains of basement reactivation, which are clearly associated with major continental collisions. The preponderance of granodioritic compositions among the continental rocks formed in late premobile times (the greenstone–granodiorite terrains) we take as evidence that there has been relatively little partial melting and internal fractionation after arc accretion except in areas that have suffered post-premobile time collisions.

Although a great proportion of existing continent appears to have formed by the end of permobile times, there are no very large areas of old continent preserved intact. This is because there was, in permobile time, a great total length of plate boundary so that any piece of continent stood a large risk of being fragmented by development of new plate boundaries. Continued convergence after collision, the process that produces the Tibetan and Central Asian environments with accompanying *in situ* continental fractionation (Dewey and Burke, 1973; Molnar and Tapponier, in press), was relatively uncommon in permobile times perhaps for a similar reason. As there was a great total length of plate boundary, motion may have been more readily transferred to other boundaries after collision. Also, break-up by rifting and opening of oceans and Tibetan reactivation following continental collisions since 2·5 b.y. ago must have fragmented, and modified beyond recognition respectively, a significant proportion of continental crust originally formed in the Archaean.

Archean Tectonics—The Style of the Superior Province

General Considerations

A basaltic dyke swarm not younger than 3 b.y. and perhaps as old as 3·7 b.y. cuts the oldest rocks yet recognized on earth (McGregor, 1973). This fact clearly shows that the upper lithosphere of that time behaved *instantaneously* in a brittle manner under stresses similar to those responsible for younger dyke swarms of which most in the Proterozoic and Phanerozoic are directly connected to rifting and the opening of oceans.

A general and sustained surface temperature of less than 100°C since 03·4 b.y. ago is necessitated by the presence of thick sections of that and younger ages containing pillow lavas. This observation, combined with reasonable estimates of heat generation (Fig. 1) and a likely distribution of the heat-producing elements also requires that there was a rigid lithosphere of significant thickness (~20–40 km) by 3·4 b.y. ago. Because mantle heat production was at least twice the present rate at the end of the permobile regime, strain rates due to first-order lithospheric tectonics cannot have been less than at present. Therefore such tectonics must have operated within the brittle regime from the top of the lithosphere to a depth perhaps slightly less than, but essentially the same as, that today. We therefore infer: (1) that a rigid lithosphere (boundary conduction layer in the case of the oceans) has been present since at least 3·7 b.y. ago; (2) that lithospheric relative movement has been by ocean-floor spreading, transform faulting, and subduction at least since this time, even though there was probably a significantly greater length of plate boundary, faster average relative plate motion, and many small 'plates' prior to 2·5 b.y. ago; and (3) that continental fragments behaved in a similar manner to those of later times, and that the plastic compressive deformations seen within them are due to orogenic events *localized* in time and space and due to subduction and/or collision-related processes, which may involve the *local* and *temporary* occurrence of greatly elevated thermal gradients and silicic magmas. Sequences dominantly composed of mafic rocks over 10 km thick all in low greenschist metamorphic facies (e.g. Glikson, 1972), whether composed of several thrust sheets or not, clearly demonstrate that equilibrium thermal gradients at least in these areas were not generally extreme in Archean times (not above 25°C/km).

Alkaline volcanic and/or plutonic rocks seem to be absent from rocks formed prior to 02·5 b.y. ago (with one exception noted below). This absence of alkaline magmas indicates that in environments now occupied by alkalic magmas (hot spots with or without rifting/spreading, in both continents and oceans), the percentage of mantle partial melt was relatively high (>10%) and perhaps that this melting took place at shallower depths (less than 40 km) and may suggest that a thinner lithosphere was present prior to 2·5 b.y. ago.

No aulacogens (failed rift arms) have been recognized prior to about 2·2 b.y. ago (2·4 b.y. ago if the southernmost part of the Huronian is in an aulacogen). We suggest that the faster pace of plate motion and the thinner lithosphere prior to 2·5 b.y. ago resulted in very few failures of rifts to evolve into oceans, and therefore that there are no Archean aulacogens preserved in the small sample of essentially unmodified Archean crust remaining for inspection. Two types of Archean crustal terrain seem to be present. One is a 'granitic,' 'gneissic' or 'continental' terrain, and which has been most thoroughly studied in West Greenland (Bridgwater, McGregor and Myers, 1974). The other kind is the greenstone–granodiorite, or 'oceanic' terrain. We consider the Superior Province the best example of such a terrain because its area greatly exceeds that of other greenstone–granodiorite terrains and because it has been so thoroughly studied. It is evident that the tectonic environment responsible for greenstone–granodiorite terrains such as the Superior must be compatible with plate tectonics because the age of the Birrimian greenstone–granodiorite terrain of West Africa (~2·1–1·8 b.y.) is the same age and younger than the age of opening (~2·15 b.y.) of the ocean now represented by the Labrador Trough–Circum–Ungava–Nelson Front suture, which gives clear evidence of the operation of plate tectonics from this time onward (Burke and Dewey, 1972, 1973).

*Island-arc Complex Assemblages and
  Tectonics Compared with Greenstone–
  Granodiorite Terrains*

Several authors (e.g. Engel, 1968; Folins-bee et al., 1968; Anhaeusser et al., 1968; Jakeš and Gill, 1970; Burke and Dewey, 1972) have put forward arguments suggesting that the granodiorite–greenstone terrains represent island-arc rocks. Present-day island arcs and related surroundings contain five significantly different tectonic environments that all might be represented somewhere in the granodiorite–greenstone terrains if, as is likely, the analogy is valid. These environments are (Fig. 2A): (a) remnant arc; (b) marginal basin, including oceanic crust formed by back-arc spreading (Karig, 1971); (c) island arc; (d) arc/trench gap; and (e) trench mélange. Environments (a), (c) and (d) include oceanic crust on which the arc was built (Fig. 2B), and (e) includes pieces of oceanic crust and overlying pelagic sediment in thrust slices and as tectonic blocks in mélange. Upon arc–arc or arc–microcontinent collision the arc/trench gap and trench mélange are extremely likely, judging by their rarity in Phanerozoic orogenic belts, to be destroyed by being telescoped, sutured out, overthrust and eroded, perhaps because they are founded on essentially unmodified oceanic crust, which is readily subducted and/or obducted. Arcs involved in collisions tend to be uplifted, the calc–alkaline volcanic and volcaniclastic pile being eroded off and exposing the granodioritic plutonic terrain underneath. An example of this behaviour includes the Ladakh 'granite' (mostly granodiorite) in the Himalayas, which represents the roots of a volcanic arc that collided with the Asian continent in the Cretaceous. It is surrounded on both sides by narrow belts of ophiolitic rocks, including on the southern side 'flysch' derived from the arc and its erosion (Gansser, 1964). Another example is the Mesozoic batholithic belt of southern Chile, where the only evidence of the previously overlying volcanic rocks is now found in volcaniclastic sediments overlying the ophiolite complex floor of a Cretaceous marginal basin to the east (Dalziel et al., 1974). This marginal basin opened by spreading behind the arc and closed by collision of the arc with South America, deforming the oceanic crust and overlying sediments in upright folds and steep tectonic slide zones (Dalziel et al., 1974), and forming a greenstone belt (Tarney, Dalziel, and de Wit, this volume). Even without collisions, volcanic arcs tend to end up being represented by their plutonic roots, for example the Sierra Nevada batholith complex of California. These examples illustrate that it is more likely that the volcanics and volcaniclastics preserved in greenstone belts are, where they are narrow anastomosing belts separated from one another by large areas of granodioritic material (such as in the Kenora area of the Superior Province), more likely to represent marginal basins, and possibly arc/trench gaps and 'main' ocean sutures than they are to represent the island arc itself. However, where most of the outcrop area is greenstone with little granodiorite, as in the Abitibi area of the Superior Province, the arc volcanics, as might be predicted, seem to be well preserved (Goodwin and Ridler, 1970). It must be emphasized that significant quantities of marginal basin and arc/trench gap rocks are also likely to be present in this and similar areas.

Although the bulk of lava flows in any one marginal basin–arc–arc/trench gap complex may be largely confined to the arc, and oceanic crust of various ages, they are not wholly restricted to the arc. Arc/trench gaps at any one time are defined by the trench and by the volcanic front which is controlled by the depth to the descending slab (Fig. 2A). If, during a time interval, the dip of the descending slab increases or decreases, volcanism will occur in what was the arc/trench gap or marginal basin, respectively. Therefore, in an area representing an arc complex built over a significant time period the distinction between arc and arc/trench gap and between arc and marginal basin may not be clear-cut. Significant amounts of basaltic pillow lava are known within the mafic volcaniclastic sediment sequences overlying the oceanic crust of two originally very narrow (<50 km) 'fossil' marginal basins in western Newfoundland (Upadhyay et al., 1971; Kidd, 1974). These may represent a temporary shallowing of the

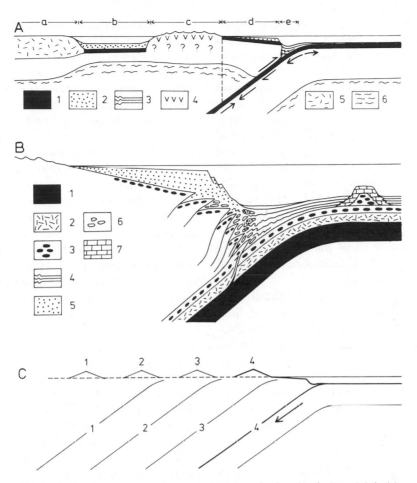

Fig. 2. A. Section across convergent plate margin showing (left to right): (a) remnant arc of microcontinent; (b) marginal basin; (c) volcanic arc; (d) arc/trench gap; (e) trench; together with oceanic lithosphere and subduction zone. 1—oceanic crust; 2—volcaniclastic sediments; 3—pelagic sediments; 4—arc volcanics; 5—remnant arc or microcontinental crust; 6—asthenosphere

B. Arc/trench gap-trench environment. 1—oceanic mantle (depleted harzburgite-dunite); 2—gabbro; 3—dykes and pillow lava; 4—pelagic sediments; 5—volcaniclastic sediments; 6—trench mélange; 7—carbonate platform and detritus on seamount

C. Proposed model for evolution of parts of convergent plate margins (e.g., Klamath Mountains, California). Previous sites of subduction 1, 2, 3 relate to previous arcs 1, 2, 3. Active arc 4 due to active subduction zone 4. Sites of subducted zones 1, 2, and 3 are not occupied by slabs after subduction moves to younger site

dip of a subducted slab, or perhaps some other process peculiar to marginal basins. It is important that the ophiolite complex pillow lavas (oceanic crust) of these two 'fossil' marginal basins, but not the lavas in the sediments above, appear to be basaltic komatiites (Strong, in prep.; Kidd, 1974).

If, in addition to these processes and the opening and closing of marginal basins, arc complexes are constructed with continuous or episodic oceanward migration of the subduction zone, such as has been suggested for the Klamath Mountains in California (Fig. 2C), horrendous complexities of tectonic and magmatic chronology are likely, quite apart from additional complexities superimposed during arc–arc or arc–microcontinent collisions.

As well as granodioritic plutonic complexes representing the roots of arcs, granodioritic to granitic plutonism during and immediately following collision is probable, due to partial melting of arc or microcontinental crust, or of the underlying mantle forming the rest of the lithosphere, by overthickening during collision (Dewey and Kidd, 1974). This process may have contributed a large proportion of the plutons in granodiorite–greenstone terrains, including most or all of the widely reported plutons that are syn-kinematic to some of the deformation sequence.

Two examples of alkaline volcanics and sills are reported from greenstone terrains. Both are in the Abitibi belt of the Superior Province, one near Kirkland Lake (Ridler, 1970), the other near Timmins (Goodwin, 1972). These may be island arc rocks, as suggested by Cooke and Moorhouse (1969) or, if there are more tectonic boundaries present than have presently been recognized, Archean seamounts washed into a trench complex (as is portrayed imminent in Fig. 2B).

Relatively small blocks mostly composed of quartzo-feldspathic rocks occur within the dominantly greenstone–granodiorite terrain of the Superior Province, the English River Gneiss Belt and the Quetico block being the best known examples (Goodwin, 1972). It is reported that the rocks are largely 'paragneiss', derived from quartzo-feldspathic sediments. We suggest that it is more likely that they will be found, after more detailed study avoiding use of the word paragneiss, to consist mostly of extremely deformed plutonic rocks, such as reported for the 'gneissic' terrain of West Greenland by Bridgwater et al. (1974). The boundaries of the two blocks mentioned are reported as zones of intense faulting and mylonitization (Goodwin, 1972). In some areas, particularly on the northern side of the English River Belt in Manitoba (Wilson et al., 1972), and to a lesser extent on the northern side of the Quetico block, tectonic slivers of serpentinite, and serpentinized ultramafic

rock, are clearly associated with these zones, giving them a similar character to oceanic suture zones of younger orogenic belts. We suggest that these two blocks, and others like them, are microcontinents in terms of their present relationships to the granodiorite–greenstone terrains. Remnant arcs, and inactive arc complexes, can also be regarded for tectonic purposes as microcontinents with respect to active convergent plate boundaries and their active arcs. Such microcontinents may in some cases be identifiable by containing a mixture of plutonic rocks of two greatly different ages, one from its arc history, and a younger from its collision with an active arc that will only give ages near and including the younger age range. Such a relationship was suggested to be common in the Superior greenstone–granodiorite terrain by A. C. Lawson (1913) on the basis of clasts of granitic rock in greenstone belt conglomerates. Lithologic associations very similar to those in greenstone–granodiorite terrains, and deformed in a similar style, can be found locally within Phanerozoic orogenic belts. One example is the Round Pond area of south-central Newfoundland (Fig. 3A). This area, which includes a well-developed ophiolite complex, was deformed and intruded by silicic magmas during the Acadian continental collision, which was relatively weak in the salient of central Newfoundland (Dewey and Kidd, 1974). The 'triangular syncline' morphology, the scale, and the lithological assemblage in the Round Pond area are very similar to those in Archean greenstone–granodiorite terrains, illustrated for comparison in Fig. 3B by the Bulawayan greenstone-belt (Amm, 1940). The steep metamorphic gradients shown by the narrow dynamo-thermal aureoles of the silicic plutons in the Round Pond area are a particular point of resemblance to the Archean terrains. An even better example is found in the northwestern Sierra Nevada of California (Fig. 4); the resemblance of this area to Archean greenstone–granodiorite terrains is astounding. The ultramafic and gabbroic rocks in this area are almost all or perhaps entirely variably dismembered ophiolite complexes including, just to the

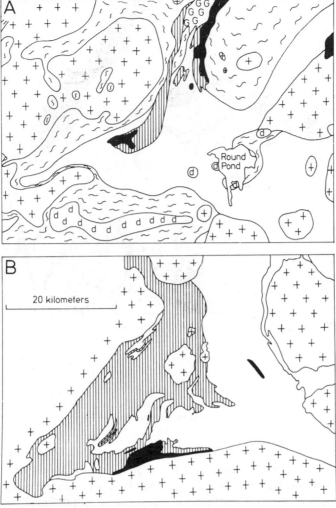

Fig. 3. Outline geological maps at the same scale to illustrate similarities in gross morphology, rock types and structural relationships between part of a Phanerozoic orogenic belt and part of an Archean greenstone belt. A. Round Pond area, central Newfoundland (Williams, 1970). B. Bulawayan greenstone belt, Rhodesia (Amm, 1940). Black ultramafic rocks; G—gabbro; white—sediments; vertical lines—mafic volcanics (mainly basalt and metabasalt); broken wavy lines—high grade metamorphic rocks (probably dynamo-thermal aureoles around 'granites'; crosses—'granites'; d—diorite

south of the area shown, one with well preserved sheeted dykes (R. G. W. Kidd, pers. comm.; Moores, 1975). Tectonic boundaries and slides are abundant (E. Nisbet, pers. comm., in prep.). The ophiolite complexes and the mafic volcanics and volcaniclastics are probably at least partly of marginal basin origin.

It is important to recognize that the apparently swirly and plastic deformation and intrusion features seen in the Phanerozoic analogues cited for the Archean greenstone–granodiorite terrains are the sum of a complex series of events. In each case these occurred during a significant time period in an area where the *instantaneous* situation at any time

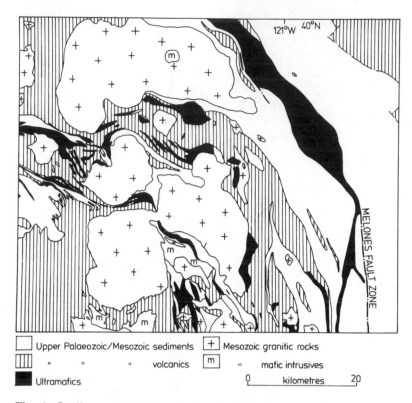

Fig. 4. Outline geological map of part of Northwestern Sierra Nevada, California (Burnett and Jenkins, 1962) to illustrate similarity to Archean greenstone belts

was either relative plate motion at sharply defined boundaries, or, during short-lived orogenic episodes, significant horizontal shortening with *consequent* upward motion of material to compensate. These episodes may be due either to collision across a marginal basin or a 'main' ocean, or to other (at present) poorly understood compressive orogenic episodes within and behind Andean and island arc complexes. We see no reason to suppose anything different for the Archean terrains.

An extensive gneissic terrain containing granulite-facies rocks forms the part of the Superior Province in the Ungava Peninsula. The radiometric ages (2·5–2·4 b.y.) and the nature of the rocks (Stevenson, 1968) show this to be an area of basement reactivation, which we interpret as due to convergence after a continental collision about 2·5 b.y. ago. This type of orogeny, where internal continental fractionation takes place (Burke and Dewey, 1970, 1973), seems to have been relatively

uncommon in Archean time, as shown by the large areas of granodiorite–greenstone terrain that have survived. We suggest that it was the general incompleteness, or weakly developed nature of collisions in the Archean that has lead to widespread preservation of the greenstone belt material only locally preserved in younger orogenic belts.

The supposed 'vertical' tectonic style reported from greenstone belts seems to be based on the alleged control of deformation by the silicic plutons, the alleged simple synclinal structures of the belts and, perhaps, on the subvertical elongation lineation commonly observed. The latter is a necessary consequence of horizontal shortening across a steep cleavage, because vertical displacement of plastically deforming material is usually easier than lateral displacement. The supposed simple synclinal structure of greenstone belts is a myth (Ramsay, 1963), discussed below. The steep inclination of cleavage and the overall steep structure of greenstone belts is quite compatible with collisional tectonics, being observed in Phanerozoic orogenic belts where incomplete collision has occurred (Dewey and Kidd, 1974). It is also found in very complete collisions in tightly sutured zones such as the Indus suture (Gansser, 1964). Although the Indus suture, and others like it, represent oceans driven out by low to moderately dipping subduction and thrusting, the suture zone itself becomes oversteepened in the later stages of the collisional process. This effect is also demonstrable in the Baie Verte Lineament (Kidd, 1974) on a scale more compatible with an individual greenstone belt. Ramsay (1963) has shown that the silicic diapirs do not significantly control the deformation in the Barberton greenstone belt. It is difficult to see how a strong regional cleavage and associated horizontal shortening could be impressed on relatively hard, cold volcanics and sediments by soft, mushy granitoid diapirs. An externally applied horizontal compressive stress, such as is available during arc/arc or arc/ microcontinent collision, is mechanically more reasonable, and is more compatible with the behaviour of the planet on which the major horizontal shortening clearly documented by Bridgwater et al. (1974) for the Archean of West Greenland occurred.

Ophiolite Complexes and Archean Oceanic Crust and Mantle

No examples of oceanic crust and mantle similar to fully developed ophiolite complexes have as yet been reported from greenstone belts. Our suggestion that Archean oceanic crust was thicker than present oceanic crust leads to the idea that possible examples of tectonic slices of Archean oceanic crust may not often reach down to the depleted harzburgite-dunite that is a diagnostic and characteristic of their Phanerozoic counterparts. We also suggest that the somewhat higher heatflow and perhaps less depleted mantle in Archean times resulted in a significantly greater thickness of essentially ultramafic cumulates and a proportionately lesser thickness of gabbro than is seen in Phanerozoic ophiolite complexes. As Moores

(1973) pointed out the stratiform anorthosites of the Early Precambrian (e.g. Fiskenaesset and the Limpopo) may also represent the upper part of the cumulate gabbro of the thick Archean oceanic crust. Many potential examples of Archean oceanic crust (regarding ultramafic cumulates as crust for brevity) are stated to be sills. For example, the Bird River 'sill' in Manitoba consists of cumulate ultramafics and cumulate gabbro (Wilson et al., 1972). The relationships of the 'sill' to surrounding rocks are not well known, perhaps because it has been assumed to be a sill. It is stated (Wilson et al., 1972) that the contacts with volcanics or sediments are nowhere seen in outcrop, except where they are faulted. It seems entirely possible that they could be wholly tectonic, or perhaps, in some places, that there might be a gradation into diabase, sheeted dykes and pillow lava. In considering this, and many other examples of large supposed sills in the Archean greenstone belts that might have been chosen for discussion, we urge that the possibility of their being samples of oceanic crust (and perhaps mantle) be very carefully considered, if only because of the rich reward of information, particularly geochemical information, that would ensue from the positive identification as Archean oceanic crust. We do not intend to deny the existence of large genuine differentiated sills in greenstone belts, nor do we suggest that the Bird River Sill is definitely not a sill. We only wish to point out that there is reasonable doubt about the external relationships of many of these bodies that needs to be clarified by detailed mapping. Many Phanerozoic ophiolite complexes, now known to be variably dismembered and tectonically bounded slices of oceanic crust and mantle, were originally described as intrusive sills, despite the essential absence of significant metamorphic aureoles and the presence, usually overlooked, of narrow tectonic boundaries with large displacements. One example includes the Bay of Islands Complex (Smith, 1958). A more dismembered example, the Mings Bight and related Baie Verte Lineament Ophiolite Complexes (Bird et al., 1971; Dewey and Bird, 1971) were also originally described as concordant lenticular ultramafic and gabbro sills (Watson, 1943). Two common mistakes are to regard plutonic rocks as necessarily intrusive and in isolation from surrounding mafic volcanics. The fact that layering in them is coplanar with bedding in surrounding volcanics is immaterial; this is the usual case for ophiolite complexes of Phanerozoic orogenic belts in all stages of tectonic dismemberment. Many examples of supposed sills very similar to the cumulate ultramafic and gabbroic parts of ophiolite complexes are known in Archean greenstone belts, and we suggest that many of them are good candidates for samples of Archean oceanic crust, perhaps including, in some instances, depleted non-cumulate upper mantle as well.

These possible ocean floor fragments in greenstone belts with associated mafic volcanics and volcaniclastics closely resemble those dismembered ophiolites interpreted as being of marginal basin origin in Phanerozoic mountain belts (e.g. Bird et al., 1971; Dalziel et al., 1974). We suggested above that the contents of marginal basins are very likely to compose a large proportion of Archean greenstone belts. We suggest that there are two reasons why any Archean oceanic crust that may be identified will be more likely to represent marginal basin rather than main ocean crust. First, the permobile ocean with many scattered microcontinents and arcs appears likely to have had many more areas in which young ocean floor was close to arc and microcontinental sediment sources than in later times. Secondly, marginal basin ocean floor has a greater chance of being young, and hence thin and hot, near to a subduction-resistant arc or remnant arc (microcontinent) than does main ocean floor, and therefore has a greater chance of being obducted and hence preserved.

The Paradox of Greenstone Belt Stratigraphy

The environments represented in greenstone belts closely resemble those found in later orogenic belts where intense tectonism is the rule and although both intensive and extensive characters of the permobile earth indicate a dominance of horizontal motion, many publications on greenstone belt geology (e.g. Glikson, 1972, Fig. 4) report very thick (>10 km) tectonically unrepeated sections, or report such sections containing a few tectonic slides and then describe a stratigraphic sequence as if they were not present (e.g. Viljoen and Viljoen, 1970). We would like to suggest that tectonic boundaries, especially thrusts and tectonic slides concordant with bedding, are likely to be present, although as yet generally unidentified, throughout these sections. Map patterns in the Bulawayan greenstone belt (Amm, 1940) support the notion that such repetition is important. Indeed, the term 'schist belt' formerly used in Africa to describe what are now known as greenstone belts, even though it is not appropriate to the general metamorphic grade, does describe accurately, and speaks volumes for the general tectonic condition and state of deformation of the rocks contained within them. With the absence of fossil evidence the problem of sorting out stratigraphic sequences is acute, and we list briefly the following cautionary examples for those who insist that greenstone belt sequences are very thick and do not contain major tectonic repetitions and/or excisions. The examples are essentially instances where apparently homoclinal sequences have been shown from fossil evidence to consist of numerous parallel thrust sheets, or from the discovery of narrow high strain zones (mylonite or tectonic slide zones) throughout the sequence.

(1) East of the Green Mountains in Vermont an east-facing regular stratigraphic sequence is claimed from later Precambrian arkoses to Middle Ordovician volcanics and sediments (e.g. Cady, 1968). It is certain that the Taconic allochthon was expelled from this zone making the possibility of a continuous stratigraphic sequence negligible. The zone is riddled with detached isoclinal fold hinges and narrow mylonite (or slide) zones of high strain.

(2) In Newfoundland, the Bay of Islands ophiolite allochthon together with a series of underlying thrust sheets was long regarded as

a continuous upward facing stratigraphic sequence lying conformably above a Cambro–Ordovician platform carbonate sequence (Smith, 1958), as indeed was the Taconic allochthon of New York State. The discovery of fossils in the allochthons led to a re-evaluation of the sections showing them to be a stacked sequence of thrust sheets (Rodgers and Neale, 1963; Bird, 1969).

(3) The Scardroy and other Lewisian inliers in the Moines were long regarded as parts of the Moine 'stratigraphy' except by Kennedy (1955, especially the discussion following) until Clough's basal conglomerate in Glenelg was finally accepted as the unconformable base of the Moines, and Lewisian radiometric ages were determined.

(4) The stacking of parallel *decken* of Schrattenkalk in the Glaurus valley was recognized by careful lithologic and faunal work and the recognition of the significance of the thin *Lochseitenkalk* bands as calcareous 'mylonites' (Trümpy, 1969).

(5) As a final example, there is the awful lesson of the Col de Genet section in the French Alps, where, until foraminifera were collected, a continuous stratigraphic sequence was 'recognized' from Jurassic carbonates through a *schistes lustré* section with 'interstratified' ophiolites and flysch. Once fossil repetition was detected, very thin mylonite zones were found.

It is emphasized that addition (or removal) of strata several kilometres thick, or juxtaposition of sequences originally deposited a hundred or more kilometres apart, may take place along mylonite (slide) zones as little as a few centimetres thick and without causing any anomalous deformation outside them. Also, once one slide zone is recognized in any succession, it is almost certain that many are present.

Most of the examples cited above are from relatively flat-lying overthrust- and nappe-type situations. Narrow tectonic slides with large displacement are not confined to such terrains. For instance, a stack of nappes may be deformed into tight folds, and will then not only contain the earlier tectonic slides between nappes but also new ones coplanar with the axial surfaces of the folds. Simpler terrains affected by fairly upright large-scale tight folding, a single steep axial surface cleavage, and a moderately to steeply pitching elongation lineation are well known to contain and be bounded by large tectonic slides, for example the Croagh Patrick Syncline (Dewey, 1967) and the Baie Verte Lineament (Kidd, 1974; Dewey and Bird, 1971). It is interesting that both these examples are of rocks deposited in small Appalachian/Caledonian marginal basins, and deformed during weak continental collision (Caledonian, Acadian) by closure of the basins.

The greenstone belt that has been most adequately mapped and studied from a structural point of view is the Barberton Mountain Land. In a germinal, but conspicuously and unjustly ignored paper, Ramsay (1963) showed that the majority of rocks in this belt are strongly and complexly deformed in a manner no different from parts of Phanerozoic orogenic belts. He demonstrated that there are at least three distinct and regionally significant deformation episodes recognizable; that they all involved *horizontal* shortening as their principal strain component; and that the granodioritic plutons in his map area were intruded during the second deformational episode, are affected by it, and do not cause significant deformation themselves. His mapping also shows the presence of major tectonic slides formed during the second, cleavage-producing deformation. Inspection of the maps of Viljoen and Viljoen (1970, Fig. 7) shows that these second episode tectonic slides are abundant throughout the Barberton greenstone belt. Their maps also show (Viljoen and Viljoen, 1970, Fig. 2 and Fig. 7) that there is at least one major tectonic slide present that was formed during the first deformation (at the base of the Komati Formation) and their description of 'sheared and talcose' rocks just below the Middle Marker horizon may indicate the presence of another early tectonic slide. This terrain therefore resembles the hypothetical example given above where a stack of nappes is tightly folded with development of new tectonic slides. It also is stated (Viljoen and Viljoen, 1970) that the uppermost unit (Swartkoppie) of the Onvervacht sequence is everywhere separated from the other units by 'faults' (second episode tectonic slides), and yet it is shown with the other units in a continuous stratigraphic column. The Barberton Mountain Land seems to be regarded unofficially as the type example of a greenstone belt, and rightly so in view of the detailed mapping available and the relatively good exposure. As this belt has the structural properties listed above, it is highly unlikely that other greenstone belts are very different.

Wood (1966) stated that the dominant structure in Rhodesian greenstone belts is a steeply inclined cleavage across which up to 90% shortening has occurred. In the Superior Province a steep cleavage is the rule, and narrow zones of high strain, which are probably tectonic slides, are common (Goodwin, 1972). In particular, the Kirkland Lake–Larder Lake 'break' (Ridler, 1970; Goodwin, 1972) greatly resembles major tectonic slides in younger terrains. It contains, at the Kerr-Addison mine, a magnesite–quartz–fuchsite rock tectonically derived from ultramafic rock, and in this respect closely resembles major tectonic slides, found in some Phanerozoic terrains, that contain this rock type and tectonic slivers of ophiolite-derived harzburgite and dunite. Examples are found in Newfoundland (Kidd, 1974; Dewey and Bird, 1971) and California (including the area of Fig. 4). The magnesite–quartz–fuchsite rock was described as carbonate iron-formation by Ridler (1970), despite the negligible iron content, and illustrates the problems encountered in interpreting supposed stratigraphic sections when structural effects have not been considered. Major tectonic slide zones derived from mafic or felsic plutonics, volcanics or volcaniclastics may very easily be mistaken as merely strongly cleaved fine-grained mafic or felsic volcaniclastics if a worker is unfamiliar with the properties of tectonic slides. Archean greenstone belts in Western Australia are also complexly and strongly deformed and tectonic boundaries are common (P. F. Williams, pers.

comm.). Supposed very thick, intact stratigraphic sequences described from there (e.g. Glikson, 1972) are therefore also unlikely to be real, especially in view of the generally poor outcrop.

We do not wish to suggest that partial stratigraphic sections cannot be obtained from greenstone belts. We do, however, suggest that major tectonic slides are common, and that stratigraphic sequences presently described from greenstone belts, especially those more than about 10 km thick, are likely to consist of several tectonically stacked sequences. If this is the case, correlation between tectonic slices is likely to be difficult or impossible in the enforced absence of fossils, unless very distinctive marker bands are present. We emphasize the fact that young untectonized volcanic stratigraphy is very heterogeneous and complex on both small and large scales, and that the few basic lithologies involved make even gross lithologic correlation very difficult merely in Neogene examples. We also suggest that it is very important to state where a thickness measurement was made, because significant changes occur during deformation, with shortening of 50 to 90% across cleavage and thickening of two to five times in fold hinges being commonly reported from greenstone belts (Wood, 1966).

Once the stratigraphy of greenstone belts is recognized as the product of tectonic stacking, many of the anomalous features associated with the belts cease to be anomalous. Thus apparent vast thicknesses become readily understandable and repeated cycles from ultramafic to mafic cease to be hard to explain. We suggest that the long-recognized close resemblance of the rocks of greenstone belts to those of oceans, marginal basins and arcs is real, and that they do mark the places where oceanic crust has been removed, just as do the suture zones of younger orogenic belts.

Conclusions

The characteristics of ancient ocean and continent outlined here have been inferred by assuming that rocks similar to those forming today were made by similar processes, and by allowing for the effects of faster heat generation in the past. Because much of the heat generated in the Earth today is dissipated in making oceanic lithosphere at spreading ridges, in ageing it on ocean floors, in partly melting descending slabs of oceanic lithosphere below island arcs, and in emplacing the igneous products of this melting, we have inferred that these processes operated more effectively during the permobile regime prior to 2·5 b.y. ago. The picture outlined by considering these intensive properties is of an Earth at the end of the permobile regime covered much as today with about one-third continent and two-thirds ocean, and the volume of water produced from the mantle concurrently with lithosphere being also similar to the present (Wise, 1974). We suggest that the length of plate boundary was, however, much greater, and this explains many of the differences. The extensive characters of exposed Archean terrains are consistent with this picture and permit its refinement.

We suggest that models based on erroneous views of the structure of greenstone–granodiorite terrains are unlikely to be realistic. Strong and Stevens (1974) state that the greenstone belts have been deformed by vertically acting forces, which is clearly incorrect (Ramsay, 1963), and they imply a random pattern of Archean magmatism, a point which has yet to be demonstrated. Their model is based on the intersection of higher Archean geotherms with the peridotite solidus. The present asthenosphere is thought to be a zone of partial melting resulting from the same process (Wyllie, 1971) and the effect of a higher Archean geotherm would be to increase its thickness and to reduce slightly the thickness of the lithosphere, thereby probably facilitating lithospheric relative motion. Present lithospheric relative motion leads to ordered magmatic sequences with calc–alkaline magmatic products in arcs cutting and overlying tholeiites of the oceanic crust. Archean sequences resemble those of the present day and imply that magmatism in the Archean was also not random. Possible convective instability in a thicker asthenosphere with a somewhat thinner lithosphere would presumably give rise to 'hot spots' like those of the present day, except that their composition would probably be tholeiitic rather than alkalic.

We therefore infer that the rocks in the granodiorite–greenstone terrains were made at spreading ridges (and perhaps 'hot spots') in oceans and marginal basins, and in and around arcs above descending slabs of oceanic lithosphere. We emphasize the potential importance of marginal basins and oceanic crust in the greenstone assemblages. We propose that the deformation and some of the plutonic activity occurred due to horizontal shortening during collision of arcs and microcontinents, and we suggest that the greenstone-belt sequences are very much more tectonized than has generally been recognized.

References

Amm, F. L., 1940. 'The geology of the country around Bulawayo', *Bull. Geol. Surv. S. Rhodesia*, **35**, 307 pp.

Anhaeusser, C. R., Roering, C., Viljoen, M. J., and Viljoen, R. P., 1968. 'The Barberton Mountain Land: a model of the elements and evolution of an Archean fold belt', *Trans. Geol. Soc. S. Africa*, **71** (Annex), 225–253.

Armstrong, R. L., 1968. 'A model for the evolution of strontium and lead isotopes in a dynamic Earth', *Rev. Geophys.*, **6**, 175–200.

Bird, J. M., 1969. 'Middle Ordovician gravity sliding-Taconic region', In Kay, M. (Ed.), *North Atlantic–Geology and Continental Drift*, Amer. Assoc. Petrol. Geologists Mem. 12, pp. 670–686.

Bird, J. M., Dewey, J. F., and Kidd, W. S. F., 1971. 'Proto-Atlantic ocean crust and mantle: Appalachian/Caledonian ophiolites', *Nature Phys. Sci.*, **231**, 28–31.

Bridgwater, D., McGregor, V. R., and Myers, J. S., 1974. 'A horizontal tectonic regime in the Archean of Greenland and its implications for early crustal thickening', *Precambrian Res.*, **1**, 179–197.

Burke, K. C. A., and Dewey, J. F., 1972. 'Orogeny in Africa', In Dessauvagie, T. F., and Whiteman, A. J., Eds., *African Geology:* Univ. Ibadan Press, pp. 583–608.

Burke, K. C. A., and Dewey, J. F., 1973. 'An outline of Precambrian plate development', pp.

1035–1045, In Tarling, D. H., and Runcorn, S. K., *Implications of Continental Drift to the Earth Sciences,* Vol. 2, Academic Press, London.

Burnett, J. L., and Jenkins, C. W., 1962. *Chico sheet, Geological Map of California, 1:250,000,* California Division of Mines and Geology, Department of Conservation, San Francisco.

Cady, W. M., 1968. 'The lateral transition from the miogeosynclinal to the eugeosynclinal zone in northwestern New England and adjacent Quebec', pp. 151–161, In Zen, E-an, White, W. S., and Hadley, J. B. (Eds.), *Studies of Applachian Geology: Northern and Maritime,* Wiley–Interscience, New York.

Card, K. D., 1970. 'Comment on "The Huronian Supergroup north of Lake Huron (Frarey and Roscoe, 1970)" ', p. 157, In Baer, A. J. (Ed.), *Basins and Geosynclines of the Canadian Shield,* Geol. Surv. Canada Paper 70–140.

Cooke, D. L., and Moorhouse, W. W., 1969. 'Timiskaming volcanism in the Kirkland Lake area, Ontario, Canada', *Can. J. Earth Sci.,* **6**, 117–132.

Dalziel, I. W. D., de Wit, M. J., and Palmer, K. F., 1974. 'A fossil marginal basin in the southern Andes', *Nature,* **250**, 291–294.

Dewey, J. F., 1967. 'The structural and metamorphic history of the lower Palaeozoic rocks of central Murrisk, Co. Mayo, Eire', *Quart. J. Geol. Soc. London,* **123**, 125–155.

Dewey, J. F., and Bird, J. M., 1971. 'Origin and emplacement of the ophiolite suite: Applachian ophiolites in Newfoundland', *J. Geophys. Res.,* **76**, 3179–3206.

Dewey, J. F., and Burke, K. C. A., 1973. 'Tibetan, Variscan, and Precambrian basement reactivation: products of continental collision', *J. Geol.,* **81**, 683–692.

Dewey, J. F., and Kidd, W. S. F., 1974. 'Continental collisions in the Appalachian-Caledonian orogenic belt: variations related to complete and incomplete suturing', *Geology,* **2**, 543–546.

Engel, A. E. J., 1968. 'The Barberton Mountain Land: clues to the differentiation of the Earth', *Trans. Geol. Soc. S. Africa,* **71** (Annex), 255–270.

Folinsbee, R. E., Baadsgaard, H., Cumming, G. L., and Green, D. C., 1968. 'A very ancient island arc', pp. 441–448, In *The Crust and Upper Mantle of the Pacific Area.* Geophys. Monograph **12**. Amer. Geophys. Union, Washington.

Forsythe, D., and Uyeda, S., 1975. 'On the relative importance of driving forces of plate motion', *J. Geophys. Res.* (in press)

Frarey, M. J., and Roscoe, S. M., 1970. 'The Huronian Supergroup north of Lake Huron', pp. 143–157, In Baer, A. J. (Ed.), *Basins and Geosynclines of the Canadian Shield,* Geol. Surv. Canada Paper 70-40.

Gansser, A., 1964. *The Geology of the Himalayas,* Wiley–Interscience, New York. 289 pp.

Glikson, A. Y., 1972. 'Petrology and geochemistry of metamorphosed Archean ophiolites, Kalgoorlie-Coolgardie, Western Australia', pp. 121–189, In *Bull. 125,* Bureau of Mineral Resources, Geology and Geophysics, Canberra.

Goodwin, A. M. (coordinator), 1972. 'The Superior Province', pp. 527–624, In Price, R. A., and Douglas, R. J. W. (Eds.), *Variations in Tectonic Styles in Canada,* Geol. Assoc. Canada Spec. Paper 11.

Goodwin, A. M., 1973. 'Plate tectonics and evolution of Precambrian crust', pp. 1047–1069, In Tarling, D. H., and Runcorn, S. K. (Eds.), *Implications of Continental Drift to the Earth Sciences,* Vol. 2, Academic Press, London.

Goodwin, A. M., and Ridler, R. H., 1970. 'The Abitibi orogenic belt', pp. 1–24, In Baer, A. J. (Ed.), *Basins and Geosynclines of the Canadian Shield,* Geol. Surv. Canada Paper 70–40.

Green, T. H., 1969. 'High-pressure experimental studies on the origin of anorthosite', *Can. J. Earth Sci.,* **6**, 427–440.

Hart, S. R., and Goldich, S. S., 1975. 'Most ancient known rocks may be found in all Earth's shields', *Geotimes,* **20** (3), 22–24.

Jakeš, P., 1973. 'Geochemistry of continental growth', pp. 999–1009, In Tarling, D. H., and Runcorn, S. K. (Eds.), *Implications of Continental Drift to the Earth Sciences,* Academic Press, London, Vol. 2.

Jakeš, P., and Gill, J., 1970. 'Rare earth elements and the island arc tholeiitic series', *Earth Planet. Sci. Lett.* **9**, 17–28.

Karig, D. E., 1971. 'Origin and development of marginal basins in the western Pacific', *J. Geophys. Res.,* **76**, 2542–2561.

Kennedy, W. Q., 1955. 'The tectonics of the Morar anticline and the problem of the north-west Caledonian front', *Quart. J. Geol. Soc. London,* **110**, 357–382.

Kidd, W. S. F., 1974. 'Evolution of the Baie Verte Lineament, Burlington Peninsula, Newfoundland', Unpublished Ph.D. thesis, University of Cambridge.

Lawson, A. C., 1913. 'The Archean geology of Rainy Lake restudied', *Geol. Surv. Canada Mem.,* 40, 115 pp.

Lee, W. H. K., 1967. 'Thermal history of the Earth', Unpublished Ph.D. thesis, University of California at Los Angeles.

McGregor, V. R., 1973. 'The early Precambrian gneisses of Godthaab district, West Greenland', *Phil. Trans. Roy. Soc. London,* **A273**, 343–358.

Molnar, P., and Tapponier, P., 1975. 'Tectonics of Asia: consequences and implications of a continental collision', *Science,* in press.

Moores, E. M., 1973. 'Plate tectonic significance of Alpine peridotite types', pp. 963–975, In Tarling, D. H. and Runcorn, S. K., *Implications of Continental Drift to the Earth Sciences,* Academic Press, London, Vol. 2.

Moores, E. M., 1975. 'The Smartville terrain, northwestern Sierra Nevada, a major pre-late Jurassic ophiolite complex', Abstract, *Geol. Soc. Amer., Abstracts with Programs*, **7**, p. 352.

Parker, R. L., and Oldenburg, D. W., 1973. 'Thermal model of ocean ridges', *Nature Phys. Sci.*, 137–139.

Ramsay, J. G., 1963. 'Structural investigations in the Barberton Mountain Land, eastern Transvaal', *Trans. Geol. Soc. S. Africa*, **66**, 353–398.

Ridler, R. H., 1970. 'Relationship of mineralization to volcanic stratigraphy in the Kirkland-Larder Lakes area, Ontario', *Geol. Assoc. Canada, Proc.* **21**, 33–42.

Rodgers, J., and Neale, E. R. W., 1963. 'Possible "Taconic" klippen in western Newfoundland', *Amer. J. Sci.*, **261**, 713–730.

Smith, C. H., 1958. 'Bay of Islands igneous complex, western Newfoundland', *Geol. Surv. Canada Mem. 290*, 132 pp.

Stevenson, I. M., 1968. 'A geological reconnaissance of Leaf River map-area, New Quebec and Northwest Territories. *Geol. Surv. Canada Mem. 356*, 112 pp.

Strong, D. F., and Stevens, R. K., 1974. 'Possible thermal explanation of contrasting Archean and Proterozoic geological regimes', *Nature*, **249**, 545–546.

Trümpy, R., 1969. 'Die helvetischen decken der Ostschweiz: versuch einer palinspastichen Korrelation und ansätze zu einer Kinematischen analyse', *Eclog. Geol. Helv.*, **62**, 105–138.

Upadhyay, H. D., Dewey, J. F., and Neale, E. R.

W., 1971. 'The Betts Cove ophiolite complex, Newfoundland: Appalachian oceanic crust and mantle', *Geol. Assoc. Canada Proc.*, **24**, 27–34.

Viljoen, M. J., and Viljoen, R. P., 1970. 'Archean vulcanicity and continental evolution in the Barberton region, Transvaal', pp. 27–39, In Clifford, T. N., and Gass, I. G. (Eds.), *African Magmatism and Tectonics*, Oliver and Boyd, Edinburgh.

Watson, K. de P., 1943. 'Mafic and ultramafic rocks of the Baie Verte area, Newfoundland', *J. Geol.*, **51**, 116–130.

Whiteside, H. C. M., 1970. 'Volcanic rocks of the Witwatersrand Triad', pp. 73–88, In Clifford, T. N., and Gass, I. G., Eds. *African Magmatism and Tectonics*, Oliver and Boyd, Edinburgh.

Williams, H., 1970. *Red Indian Lake (East Half) Newfoundland*, Geol. Surv. Canada Map 1196A.

Wilson, H. D. B., Brisbin, W. C., McRitchie, W. D., and Davies, J. C., 1972. 'Archean geology and metallogenesis of the western part of the Canadian Shield', *Excursion guidebook A33-C33*, 24th Int. Geol. Congress, Montreal.

Wise, D. U., 1974. 'Continental margins, freeboard and the volumes of continents and oceans through time', pp. 45–58, In Burk, C. A., and Drake, C. L., *The Geology of Continental Margins*, Springer-Verlag, New York.

Wood, D. S., 1966. 'The Rhodesian basement', *10th Ann. Rept. Inst. African Geol.*, Univ. Leeds, pp. 18–19.

Wyllie, P. J., 1971. *The Dynamic Earth*, Wiley, New York, 416 pp.

Marginal Basin 'Rocas Verdes' Complex from S. Chile: A Model for Archaean Greenstone Belt Formation

J. TARNEY

Department of Geological Sciences, University of Birmingham, England.

I. W. D. DALZIEL

Lamont-Doherty Geological Observatory of Columbia University, U.S.A.

M. J. DE WIT

Department of Geological Sciences, Columbia University, U.S.A.

Abstract

The typical Archaean greenstone-belt association—synclinal successions of ultramafic and mafic lavas, minor andesitic and silicic lavas, clastic sediments, cherts, ironstones, etc., separated and surrounded by diapiric gneiss domes—has few direct counterparts in the Proterozoic or Phanerozoic. The character of the volcanics is oceanic or island arc and the metamorphism low grade. The intervening gneisses or diapiric batholiths are generally of calc–alkaline character.

In the earliest Cretaceous a marginal basin opened up along the western margin of the southern S. American continent behind an andesitic island arc. The crust of the basin consists of an ophiolitic suite of minor picrite, gabbro, sheeted dykes, pillow lavas, chert, andesitic volcaniclastics and sediments, with some silicic volcanics and intrusives. By the mid-Cretaceous the basin started closing again, deforming the marginal basin rocks into an elongate synform bordered by the batholiths of the Pacific margin. The metamorphic grade is low and the deformation heterogeneous. The overall geotectonic situation parallels in general that of a typical Archaean greenstone belt.

Similar marginal basins are common in the Phanerozoic record. An outline of the characteristics of marginal basins, using modern-day examples from the Western Pacific and an overview of the mechanisms by which marginal basins are believed to form, reveals that marginal basins provide an actualistic counterpart to Archaean greenstone belts. Taking into account the higher geothermal gradient in the Archaean, with its influence on lithospheric plate size and thickness, plate subduction and nature of the crust–mantle interface, a reasonable geotectonic model for Archaean greenstone belts is possible which is also reconcilable with the geochemical features of the associated rock types.

131

Introduction

Any model for Archaean greenstone belts is subject to a number of constraints. It is the recognition of these constraints that has posed the greatest difficulty in constructing successful models. Indeed there has been a tendency to neglect constraints and assume that geotectonic conditions were so fundamentally different in the Archaean that extreme non-uniformitarian models are possible. We tend towards a more moderate view, namely that near-uniformitarian models should not be discarded until shown to be completely unworkable. This is not to claim, however, that conditions in the Archaean parallel exactly those of today: a major difference, amplified below, would be the higher geothermal gradient in the Early Precambrian, with its influence on 'plate' motions, 'plate' size and thickness, crustal and mantle behaviour and rates of crustal generation and continental growth.

Reviews of previous models for Archaean greenstone belts have been given by Windley (1973, this volume) and Hunter (1974). The models tend to fall into two main categories: (a) those which regard the basic–ultrabasic lavas as either primitive ocean floor or island arc volcanic sequences subsequently underplated by granitic material direct from the mantle (Glikson, 1971; Anhaeusser, 1973) and (b) those which regard the greenstone-belt sequences as being erupted and deposited on a rifted pre-existing granitic crust (Windley, 1973; Rutland, 1973; Hunter, 1974). More extreme models such as the lunar-analogous meteorite impact theory of Green (1972) are now generally discounted. A major difficulty with models of the first type is the problem of generating vast quantities of the tonalitic magma which underplates the greenstone belts by essentially a single-stage melting process. On the other hand, models of the second type are consistent with a hiatus of several hundred million years between basement gneisses and younger greenstone belts in Rhodesia (Hawkesworth et al., 1975) and in the case of younger greenstone belts with the proven existence of considerable older continental crust elsewhere. However, there are other difficulties, such as explaining the occurrence of lavas with oceanic geochemistry in a tectonic environment which, in the Proterozoic and Phanerozoic, produces volcanics of alkaline type. Also the geochemistry of the tonalitic plutons which invade the greenstone belts is more consistent with a continental subduction environment than a continental rift situation. Strontium isotope data (Moorbath, 1975) would seem to rule out significant melting of the underlying granitic crust, too, in the production of such plutons.

In striving to provide a more comprehensive model for greenstone-belt formation we would suggest the following as important factors to be considered:

1. Most greenstone belts appear to have formed in the period 3·4–2·7 b.y. ago. While in some regions a demonstrable time-lag of several hundred million years exists between the greenstone-belt sequences and the older intervening gneissic areas, the relationship in other regions remains to be elucidated, and the possibility must be allowed that they may be almost contemporaneous.

2. The overall dimensions; synclinal form, either cuspate (older) or linear (younger) and often multiple; the low grade of metamorphism; base metal mineralization; associated intrusion of batholiths; low to moderate deformation, but often zonally intense (Coward and James, 1974).

3. The compositions of the peridotitic and basaltic komatiite lavas in the lower parts of greenstone-belt sequences are distinctive (Brooks and Hart, 1974) but their geochemistry is consistent with melting of mantle peridotite (Arth and Hanson, 1975). The high liquidus temperature ($\sim$1650°C) of peridotitic komatiite implies high degrees of partial melting ($\sim$70%) of such peridotitic mantle (Green et al., 1975) in keeping with postulated higher geothermal gradients in the Archaean (Green, 1975). The general geochemistry of the greenstone basalts has been compared with modern ocean ridge tholeiites (Glikson,

1971) although lithophile element concentrations tend to be rather higher, element ratios such as K/Rb not so extreme (Hart et al., 1970) and rare earth patterns not so light-RE depleted (Jahn et al., 1974; Arth and Hanson, 1975). Others have argued that such geochemical features are closer to modern island arc tholeiites (cf. Anhaeusser, 1973; Jahn et al., 1974) although most arc tholeiites have much lower Cr and Ni contents (Jakeš and White, 1972). Caution must be exercised in the use of geochemical data on low-grade metamorphosed basaltic rocks as independent evidence of magmatic affinity or tectonic environment (Kay and Senechal, 1975), although some elements and element ratios appear to be diagnostic (Pearce, 1975). Lavas at higher stratigraphic levels in greenstone-belt sequences may include more intermediate and felsic types (Baragar and Goodwin, 1969) with distinctly more calc–alkaline trace element abundance levels and more highly fractionated rare earth patterns (Jahn et al., 1974; Condie, this volume).

4. The thick clastic sedimentary sequences in greenstone belts indicate derivation from a mixed volcanic–granitic terrain (Condie et al., 1970; Condie, this volume). On the other hand, the association of cherts and pillow lavas is typical of ophiolite complexes of oceanic origin.

5. The geochemical features of the intervening gneiss belts and the invading tonalitic to potassic granite plutons, with their generally high lithophile element contents and fractionated light-REE enriched patterns, are most similar to calc–alkaline igneous plutons produced at Andean-type continental margins (cf. Tarney, this volume). Strontium isotope data preclude significant remelting of older continental crust in the production of these rocks, at least on present evidence (Arth and Hanson, 1975; Moorbath, 1975), but their overall geochemical characteristics are consistent with an origin through partial melting of

amphibolite/eclogite/peridotite material (Arth and Hanson, 1975). Large areas of Archaean gneisses and plutons, even in the high grade terrains, seem to have similar geochemical characteristics (Tarney, this volume) and low initial $^{87}Sr/^{86}Sr$ ratios (Moorbath, 1975) suggesting a major period of crustal generation from the mantle environment *while* the greenstone belts were forming. The generation of *large volumes* of calc–alkaline, dominantly tonalitic, magma is most easily accomplished at continental margins by the two-stage model of Ringwood (1974). Direct single-stage melting of hydrous peridotitic mantle is unlikely to produce the required trace element levels (Tarney, this volume) found in Archaean gneisses, nor the major element compositions either (cf. Nicholls and Ringwood, 1973).

Previous models for greenstone belts find difficulty in satisfying *all* these constraints. We believe, however, that a marginal basin model, which must necessarily incorporate a tectonic situation near a continental margin, satisfies most of them and provides a more satisfactory alternative.

Marginal Basins

The term marginal basin as used in the Western Pacific is a purely geographical one. Karig (1971a) suggested the term for semi-isolated basins or series of basins lying behind volcanic island arc systems. The marginal basins in the Western Pacific (Karig, 1974; Coleman, 1975) are equidimensional to linear in outline, have broadly similar geological and geophysical characteristics, and differ from one another primarily because of variation in age and sediment distribution.

Seismic refraction profiling and bottom sampling have revealed that the crust beneath marginal basins is similar to that of normal oceanic lithosphere (Karig, 1971a). Morphological studies and sediment distribution patterns have indicated that most of the Western Pacific marginal basins were probably created by crustal extension due to periodic splitting of arc systems, whereby the active arc migrates away from the Asiatic continental margin, leaving behind remnant or non-active arcs (Karig, 1971a, 1974). The fundamental mechanism for the opening of marginal basins is, however, still a controversial topic. Karig (1971a) favours the extensional mechanism to be in response to large mantle diapirs rising behind the active arcs. Whether the actual mechanism is such an active one, or whether it is a more passive phenomenon merely due to release of horizontal compressive stresses across the arcs, as preferred by others (Scholz et al., 1971; Kelleher and McCann, 1975) is still not known, but perhaps both occur. Alternatively, it has been suggested that in some cases, the subduction of an active spreading ridge may have resulted in back arc spreading (Uyeda and Miyashiro, 1974). However, as a universal mechanism for marginal basin development, such a process would seem rather unlikely: the supply of ridges able to be subducted, if indeed such features can be subducted (Kelleher and McCann, 1975), would be somewhat limited.

Generally in the Western Pacific, back-arc spreading has given rise to poorly defined magnetic anomalies of low amplitude. However, in the South Atlantic, active fast spreading behind the South Sandwich Island Arc over the last 8 m.y. has produced a series of well-defined magnetic anomalies which can be correlated confidently with the geomagnetic reversal time scale (Barker, 1972). A similar magnetic lineation pattern has recently been described in the Shikoku marginal basin in the Western Pacific (Watts and Weissel, 1975). This suggests that the generation of oceanic crust in at least some marginal basins is similar to the accretional processes at mid-oceanic ridges (Cann, 1974). Whatever the final cause of back-arc extension, there is no doubt that it is a widespread feature associated with subduction and tectonic processes in arc systems—processes which, moreover, are just as likely to have occurred in the Archaean.

Crustal extension is not the only way to form marginal basins. In cases where a volcanic arc develops directly on oceanic crust, away from the continental margin, original oceanic crust is trapped between a newly developed subduc-

tion zone and a continent or previously extensional marginal basin. Magnetic lineations at high angle to the island arc in the Western Philippine basin (Uyeda and Miyashiro, 1974) suggest that some basins are underlain by entrapped oceanic crust.

The regional variation in thickness and type of accumulated sediment in marginal basins of the Western Pacific is a function of tectonic position and age of the basin (Karig, 1971a). Although sediment cover in active marginal basins isolated from continents is thin, as might be expected, sedimentation rates are very high. Coarse wedges or volcanic aprons accumulate at rates exceeding 100 metres/m.y., decreasing with distance from the volcanoes. Inactive marginal basins separated by remnant arcs and lying behind the active basins, show similar basement morphology to the active basins, but thick sedimentary cover obscures the basement in most older basins. Basins which lie adjacent to continental margins tend to have thicker accumulations of terrigenous sediments. The sediment cover ranges from a few hundred metres in basins protected from continental sources (i.e. Philippine Sea) to several kilometres of terrigenous turbites where a continent forms a basin flank (i.e. Japan Sea).

Along the Western Pacific margin, marginal basin formation has been operational since at least the early Tertiary (Karig, 1971a; Packham and Falvey, 1971; Brothers and Blake, 1973). Presently, active (opening) marginal basins, inactive marginal basins (Karig, 1971a, 1974; Uyeda and Miyashiro, 1974), closing marginal basins (Karig, 1973) and recently closed (deformed) marginal basins or remnants thereof (Bain, 1973; Johnson and Molnar, 1972), occur geographically and tectonically adjacent to one another. They are even interspersed (i.e. Solomon Islands, Hackman, 1973; Yap Islands, Shiraki, 1971) but separated by the remnant or active arcs.

The closure of marginal basins is generally linked to arc polarity reversals (subduction flips; Karig, 1974) and may be a direct result of minor collisions of bouyant features, such as aseismic ridges, with the arcs (Kelleher and McCann, 1975). Arc polarity reversals appear to have been common throughout the

Cenozoic, but temporally and spatially irregularly developed, both across and along the Pacific belt. Such events have led to some very complexly deformed and/or obducted zones, such as for example between the Philippines and the Ryuku Islands (Taiwan and Luzon; Karig, 1973), New Caledonia (Brothers and Blake, 1973) and New Guinea (Davies and Smith, 1971; Johnson and Molnar, 1972). Further closing of present basins must produce very complicated overall tectonic patterns of linear interspersed basaltic and calc–alkaline belts. The modern-day situation along the Western Pacific margin has interesting application in terms of multiple Archaean greenstone belts (see below).

The marginal basins in the Western Pacific can thus basically be divided into three types, a division based on their tectonic position and their mode of origin:

1. Trapped in origin, by the development of an island arc away from a continental margin on earlier oceanic crust (West Philippine basin, Aleutian basin).

2. Extensional in origin, flanked on the continental side by an inactive calc–alkaline volcanic chain (remnant arc) and on the oceanic side by an active or inactive volcanic chain. (Parce Vela basin, Shikoku basin, Tonga-Kermadec basin).

3. Extensional in origin flanked on the inner side by a continental margin, and on the outer side by a volcanic arc (i.e. Japan Sea, Sea of Okhotsk, South China Sea, Okinawa Trough).

While many marginal basins have opened up behind oceanic island arcs, the Japanese example also illustrates that back-arc spreading can strip off a narrow sliver of continental margin (Sugimura and Uyeda, 1973). In the South Atlantic too, the long narrow strip of the Antarctic Peninsula may have arisen by the same process, possibly during the Upper Mesozoic, producing a marginal basin which eventually widened into the Weddell Sea (Barker and Griffiths, 1972; Dalziel, 1974). In the same area, the South Shetland Islands represent a narrow strip of Palaeozoic to Tertiary continental crust separated from the Antarctic Peninsula itself by Bransfield Strait.

The latter appears to be a small ocean basin which has opened relatively recently behind the South Shetlands Arc in response to subduction linked to the spreading centre in western Drake Passage (Barker, 1975). Some of these basins have remnant blocks of continental crust (i.e. Yamato Rise in Sea of Japan, Sugimura and Uyeda, 1973). Early crustal extension prior to the production of oceanic crust may be preceded by extensive silicic volcanism (Sugimura and Uyeda, 1973). Volcano-tectonic rift zones with associated silicic volcanism are found in tectonic positions of active interarc basins in some continental trench-arc systems (i.e. Taupo volcanic zone, New Zealand, Karig, 1971).

The recognition that there are different types of marginal basins is of obvious importance. The fact that arcs can form on continental crust or on oceanic crust (and whether such arcs become mature) must be of geochemical concern. Substantial geochemical differences might be expected between continental margin and interoceanic marginal basins, and also their respective arc volcanics (cf. Dickinson, 1975).

In all the above cases the intervening ocean basins between the continental island arcs and the continent itself are not exposed. In Southern Chile, however, a marginal basin analogous to the Japan Sea situation began to develop along the Pacific margin of southern South America in the late Jurassic–early Cretaceous. Continued expansion of the marginal basin was interrupted during the mid-Cretaceous, and uplift associated with deformation along the Pacific margin was accomplished by the late Cretaceous, preserving the relationships. Not inappropriately perhaps, the marginal basin rocks have long been known as the 'rocas verdes'.

'Rocas Verdes' Marginal Basin Complex, S. Chile

Narrow, almost en-echelon, discontinuous lenses of mafic igneous rocks define the spine of the Andean Cordillera from Cape Horn at 56° S to at least as far north as 51° S, a distance of more than 700 km, with a width narrowing from 100 to 30 km northward (Fig. 1). These

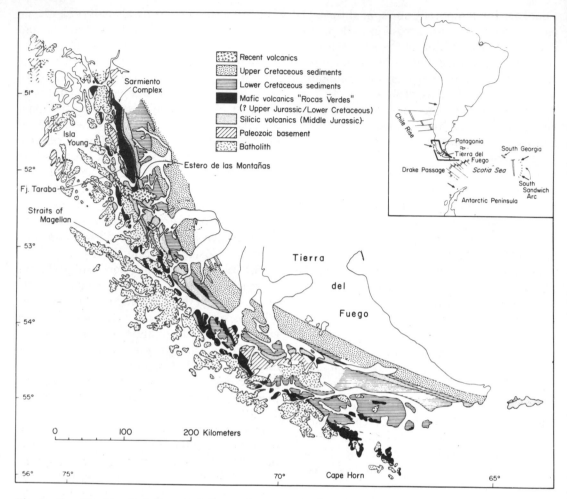

Fig. 1. Simplified 'geological map of the southernmost Andes showing the outcrop of the marginal basin 'rocas verdes' mafic volcanics in relation to the pre-existing Palaeozoic basement, the later sediments and Cordilleran batholith (reproduced, with permission, from *Nature*, 250, p. 291 (1974))

rocks commonly termed 'rocas verdes', are flanked on the Pacific side by the Patagonian Batholith, a vast continuous granodiorite body averaging 60 km in width. The batholith shows clear evidence of magmatic origin (Krank, 1932, Stewart and Suarez, 1971, Dalziel and Cortés, 1972) with features typical of circum-Pacific batholiths (Cobbing and Pitcher, 1972). Increased volumes of monzonite and adamellite on the eastern side are evidence of transverse petro-chemical variation. Minor gabbroic plutons are not uncommon.

The batholith is intruded into and contains large xenoliths of Palaeozoic metasediments and middle-to-late Jurassic silicic volcanics both readily recognizable as parts of the South America continental 'basement' that underlies the Cretaceous shelf sediments to the east of the 'rocas verdes' (see below).

Radiometric dating (Halpern, 1973) indicates three distinct climaxes of magmatic activity, in the late Jurassic–early Cretaceous (155–120 m.y.), late Cretaceous (100–75 m.y.) and later Tertiary (50–10 m.y.). A number of small Miocene plutons occur in the Pre-cordillera, east of the main batholith, indicating eastward migration of magmatic activity with time.

On the Atlantic side of the 'rocas verdes', the middle-upper Jurassic sequence of silicic

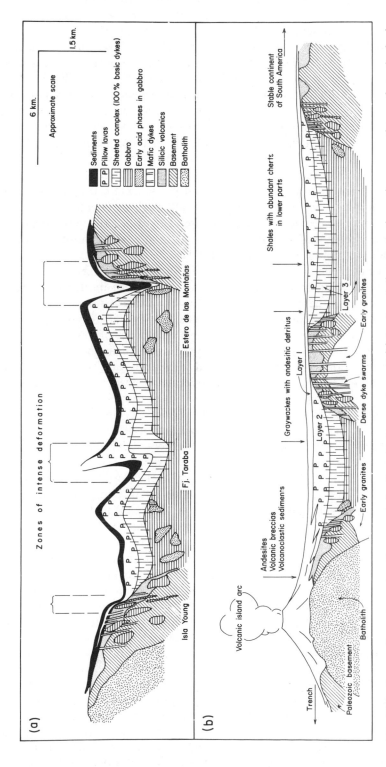

Fig. 2(a). Composite east–west section across the Sarmiento complex and (b) Restored diagrammatic section across the southern Andes in the early Cretaceous (before basin closure) showing tectonic relationship of the marginal basin complex to the pre-existing Palaeozoic basement and earlier granitic batholiths and the contemporaneous calc–alkaline andesitic volcanism of the (continental-based) island arc to the west (reproduced, with permission, from *Nature*, 250, p. 292 (1974))

volcanics rest unconformably, with locally thick conglomerates at their base, on Palaeozoic metamorphic basement. Pre-'rocas verdes' granitic plutons cut these basement schists and gneisses. The silicic volcanics are widespread, extending in the sub-surface as far as the Atlantic continental margin.

The 'rocas verdes' are overlain by a thick sequence of lower-middle Cretaceous sediments, which in the east consist of black mudstones of shelf facies, thickening to 2 km towards the west. Within the cordillera, a marked facies and further thickness change occurs. Thick greywackes become frequent and thin cherts appear interbedded with the mudstones. Abundant immature andesitic volcanic detritus in the greywackes suggests the presence of an active volcanic source during sedimentation. In the southern and widest part of the 'rocas verdes' belt, almost directly against the batholith (which in this region frequently has andesitic dykes) the greywackes thicken and are interbedded with andesite lavas, pyroxene tuffs, and andesitic breccias and conglomerates with boulders several metres across. The sequence exceeds 3 km in thickness. The variation in facies and thickness, and independent current indications (Dott et al., 1975) indicate that the volcanic source lay on the Pacific side of the sedimentary basin.

The dating indicates coeval magmatic activity of the batholith and the andesitic volcanicity. Dalziel et al. (1974) have thus interpreted the coastal batholith as a plutonic root of an overlying calc–alkaline volcanic chain (cf. Hamilton, 1969; Cobbing and Pitcher, 1972), reviving earlier suggestions of Krank (1932) of a comagmatic origin for the batholith and the andesitic volcanics.

In detail, the central belt of 'rocas verdes' consists of large lenses (i.e. Sarmiento Complex, up to at least 120 km long and 20 km wide and more than 3000 m thick) of mafic rocks (Fig. 1). The lenses are sub-parallel to the Cordillera. They intrude and separate blocks of the Palaeozoic basement and the overlying middle-upper Jurassic silicic volcanics (Fig. 2a).

The mafic rocks represent the upper parts of ophiolite assemblages. A thick intrusive unit, consisting mainly of gabbro (>1 km in thickness) that may show cumulative layering, is cut by, and passes upwards into, a thin sheeted dyke complex (<300 m thick). At still higher levels the sheeted complex is overlain by a thick (>2 km) extrusive unit consisting of near horizontal pillow lavas, pillow breccias and tuffs (de Wit and Stern, in prep.). The extrusives are in turn overlain by, and intercalated with, minor amounts of chert and jasper. Base metal showings are frequent throughout the belt, although so far only in one case (Mina Cutter Cove) of sufficient tonnage and high enough grade to be mined (Stewart et al., 1971).

The intrusive unit has steeply dipping igneous contacts with the continental blocks. In places the extrusive units overlap onto these blocks, usually being separated from them by a variable sequence of clastic (often volcaniclastic) sediments including conglomerates, mafic and silicic tuffs, and lavas. The extrusives dip moderately inwards (Fig. 2a). Thus the mafic lenses are funnel-shaped in cross-section, have normal igneous contacts and are autochthonous.

The sheeted dyke complexes are composed of 100% dykes. They strike consistently subparallel to the Cordillera and swarms of parallel mafic dykes and gabbro sills are present outside the main mafic intrusive complexes, along the entire belt, cutting the rocks of the continental blocks (metamorphic basement, silicic volcanics, and early pre-'rocas verdes' granites). Where large volumes of these intrusives are present in the continental blocks, especially near the margins of the mafic lenses, remelting and assimilation of the older sialic material is frequently evident.

The absence of extensive flysch-like sediment layers within and below the mafic extrusives suggests that the latter were extruded very rapidly during a major pulse of magmatism in the early Cretaceous.

The spine of the Cordillera thus consists of huge strips of mafic crust separated by elongate blocks of older continental crust and this whole belt must therefore represent a zone of extension and intrusion of basaltic magma behind an active andesitic volcanic chain.

The volcanic chain and Patagonian batholith had previously been emplaced into South American continental crust, as in the present central Andes, supplying this continent with volcanics and detritus. This then suggests that subduction of the floor of the Pacific Ocean beneath this part of South America was initiated by the late Jurassic (Dalziel et al., 1974). Extension moved the volcanic chain towards the Pacific, relative to South America, producing a volcanic island arc based on continental crust and separated from the stable continent by a marginal basin whose floor had geophysical characteristics partly 'oceanic' and partly 'continental'. This geotectonic setting is analogous to the present Japan Sea area (Figs. 2a, 3, Dalziel et al., 1974).

The present outcrop pattern of the 'rocas verdes', is partly a manifestation of the closure and uplift of the marginal basin during the mid- to late-Cretaceous, when the volcanic arc was translated back towards the continental margin (Dalziel et al., 1974). The resultant movement caused the deformation and uplift of the marginal basin floor and its margins, accentuating its synformal shape, although the rocks are still clearly autochtonous. The deformation was regional and essentially of single phase; the strain states are very heterogeneously distributed across the belt. The sediments overlying the 'rocas verdes' may display a strongly developed cleavage (generally upright) and open to isoclinal folds in wide zones, whilst elsewhere they are flat-lying and essentially undeformed. Furthermore, pillow lavas directly below intensely deformed sediments may show no apparent sign of deformation at all (de Wit, 1975). The 'rocas verdes' complex is itself deformed in zones of intense, and locally even polyphase, deformation, commonly concentrated along major lithology contrasts at the margins of the basin. Large areas along and across the marginal basin show no obvious signs of deformation. Deformation generally dies off rapidly outward away from the margins of the basin. Extensive

major subvertical shear zones are present in the batholith basin and continent.

Middle to upper greenschist facies mineral assemblages generally prevail throughout the 'rocas verdes' with only locally contact metamorphism (cordierite–anthophyllite) along the batholith side. The area centred around Cordillera Darwin is an exception. Here, the rocks of the basin and specifically its eastern margin are polyphase deformed and locally amphibolite–facies metamorphic assemblages are present (kyanite, staurolite, chloritoid; de Wit, in prep.). Minor eastward thrusting of the rocks occurs and significant remobilization and deformation of the Palaeozoic basement and pre-'rocas verdes' granites is sometimes evident (Dalziel and Cortes, 1972; de Wit, 1975). Dalziel et al. (1974) have indicated that this may be a manifestation of local reversed subduction within this part of the marginal basin during closure.

Throughout the belt, late granitic–granodioritic plutons cut the deformed marginal basin rocks in places extensively, engulfing mega-xenoliths of previously deformed 'rocas verdes' (i.e. just north of the Strait of Magellan). The volcanic arc had evidently moved considerably eastward even after the closure of the marginal basin.

Whatever the fundamental causes behind the opening and closing of the marginal basin, it is clear that there are many obvious similarities between the 'rocas verdes' marginal basin complex and a typical Archaean greenstone belt:

(a) At the borders of the basin the ophiolites abut against and overlap on to a pre-existing continental crust consisting of older deformed Palaeozoic basement with younger calc–alkaline plutons and associated silicic to intermediate volcanics.

(b) The overall form of the basin complex is synformal and elongated parallel to the continental margin. The dimensions are similar to a typical greenstone belt. The occurrence of pillow lavas and cherts, a low grade of regional metamorphism, local contact and higher grade metamorphism (Coward and James, 1974) and sulphide mineralization also adds to the comparison.

(c) The general deformation pattern and style of the 'rocas verdes' and adjacent rocks, associated with basin closure and uplift, is comparable with that seen in many greenstone belts (cf. Coward and James, 1974).

(d) The dominant rock type—greywackes with a high proportion of volcanogenic detritus—and shales, are equivalent. Although the sedimentary sequence is attenuated compared with, say, the Barberton greenstone sequence (Anhaeusser, 1973), this may be attributed to several factors (tectonic environment during opening, premature basin closure, etc.).

(e) The volcanic sequence in the 'rocas verdes' is dominantly basaltic and tholeiitic in character, although there are minor picritic intrusions locally (Suarez, pers. comm.). Peridotitic komatiite lavas are of course absent, but this can be directly attributed to the higher geothermal gradients in the Archaean (Green, 1975). Little geochemical data is available on marginal basin volcanics. Hart et al. (1972) have analyzed dredged basalts from behind the Marianas arc. Relevant trace element data are shown in Table 1. Also shown are preliminary data for the 'rocas verdes' complex at Sarmiento (Saunders et al., 1975a), and for fresh dredged basalts from the back arc spreading centre in the Scotia Sea (Saunders et al., 1975b). Typical trace element values for Archaean basalts, modern ocean ridge basalts and island arc tholeiites are also included for comparison. Archaean basalts differ from arc tholeiites in their markedly higher Cr and Ni contents and from ocean ridge basalts by their higher content of lithophile elements, K, Rb and Ba. Marginal basin basalts from all three localities, however, appear to have the high Cr and Ni values typical of ridge basalts coupled with the higher lithophile element contents of arc tholeiites. REE patterns of the Scotia Sea basalts also tend to be slightly light-REE enriched rather than depleted (Saunders et al., 1975) thus strengthening the comparison with Archaean basalts (cf. Jahn et al., 1974; Arth and Hanson, 1975).

(f) Many Superior Province greenstone belts are characterized by a greater proportion of andesitic and felsic lavas, (Baragar and Goodwin, 1969) or bimodal basalt–felsic lavas (Hubregtse, this volume). More salic

Table 1. Typical values of selected trace elements in Archaean greenstone basalts compared with suggested values for fresh ocean ridge basalts, island arc tholeiites and preliminary data for marginal basin basalts.

	Cr	Ni	Rb	Sr	Y	Ba	
Archaean basalts							
Australia	407	150	3	97	17	—	Glikson (1971)
Australia	400	116	10	102	22	—	Glikson (1971)
Canada	296	108	—	196	—	85	Baragar and Goodwin (1969)
Canada	175	85	—	124	15	85	Baragar and Goodwin (1969)
Vermilion	—	—	5	150	—	70	Arth and Hanson (1975)
Mean	~320	~120	6·4	153	~18	98	Jahn et al. (1974)
Ocean Ridge Basalt	297	110	1·1	135	43	10·5	cf. Jahn et al. (1974)
Island Arc Tholeiite	50	30	5	200	~19	75	Jakeš and White (1972)
Marginal Basin Basalts							
Marianas	231	69	4	208	—	34	Hart et al. (1972)
Scotia Sea	270	72	5	195	29	60	Saunders et al. (1975b)
Sarmiento	197	80	5	150	27	90	Saunders et al. (1975a)

lavas also characterize the upper parts of the Barberton belt (Anhaeusser, 1973). Andesitic and felsic lavas, however, are the dominant volcanic product in the Andes north of the Chile Rise, and silicic volcanicity is common in Japan throughout its evolution (Sugimura and Uyeda, 1973). Extensive andesitic lavas and detritus only appear in the western part of the 'rocas verdes' complex, however. The absence of a significant andesitic component within a marginal basin itself may be due to its tectonic environment, and life-span, or the ending of subduction.

(g) The Cordilleran batholith disrupts the 'rocas verdes', predominantly on its western margin, but also on the east, situations paralleling that in many greenstone belts. In some places the disruption has left the 'rocas verdes' as mega-xenoliths in a granitic terrain (cf. Litherland, 1973). Although there is as yet little geochemical data on the batholith in S. Chile, the geochemistry of equivalent batholiths of Mesozoic age down the western margin of the Americas is well known. The range of the bulk compositions and trace element abundances is similar to that in the diapiric plutons invading Archaean greenstone belts (cf. Arth and Hanson, 1975) and in many cases they have the low $^{87}Sr/^{86}Sr$ ratios indicating mantle or lower crustal derivation (Faure and Powell, 1972).

Marginal Basin Model for Archaean Greenstone Belts

The significant point of the model is that all the major rock types found in Archaean greenstone belts—together with their distinctive geochemistries—are present in the 'rocas verdes' marginal basin complex in a similar geotectonic and chronological situation. The model is almost uniformitarian and combines the oceanic character of the greenstone basaltic sequences implied in the Glikson and Anhaeusser models with the rifting and pre-existing continental crust of the Windley and Hunter models.

Marginal basins, whether intra-oceanic or continental margin, are a common feature of the present plate tectonic regime, and are increasingly being recognized throughout the Phanerozoic: in the Alpine belt (Boccaletti and Guazzone, 1974), in California (Schweickert and Cowen, 1974, 1975) and also in the Appalachian–Caledonian belt (Bird et al., 1971; Kidd, 1974; Bursnall and de Wit, 1975). They are invariably associated with regions of (continental) crustal generation. Their apparent absence from the Proterozoic may be due to the fact that they are difficult to recognize or that the Proterozoic is primarily a period of crustal re-working rather than crustal generation. Nevertheless, if subduction occurred in the Archaean, we would suggest that marginal basin formation is just as likely as in the Phanerozoic.

While a marginal basin model is sufficiently flexible to account for many features of greenstone belts, there are two areas where the 'rocas verdes' model needs to be modified in order to account more comprehensively for Archaean greenstone belts:

(a) The mafic igneous rocks are in the form of a gabbro—sheeted dyke—pillow lava sequence (i.e. typical ophiolite), whilst greenstone-belt volcanics invariably occur as a lava sequence.

(b) The 'rocas verdes' complex adequately mirrors a single greenstone belt, and while more than one marginal basin may form at a continental margin (see above), accounting for multiple sub-parallel greenstone belts such as those of the Canadian Superior Province and S.W. Australian Yilgarn block is more difficult.

These difficulties can be overcome by considering the effect of a higher geothermal gradient in the Archaean and its influence on crustal behaviour and then on the probable nature of convection and the rate of crustal generation.

To form an ophiolite suite in a continental marginal basin such as the 'rocas verdes' implies rifting apart of a strip of continent

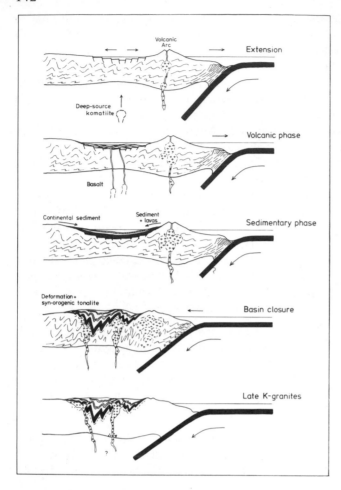

Fig. 3. Suggested development of an Archaean greenstone belt according to the 'rocas verdes' model, with back-arc extention producing a greater component of crustal thinning rather than crustal rifting. Magma production exceeds that required for simple extension. Sediment source is mixed volcanigenic/continental and sequence may include calc–alkaline andesitic and salic lavas from the adjacent volcanic arc. Later movement of arc towards continent produces deformation and synclinal form of greenstone belt. Andean-type tonalitic to granitic plutons (derived from mantle by two- or three-stage process; cf. Ringwood, 1974) may be syn- to post-tectonic, with compositions dependent upon depth of melting of subducted oceanic crust

during back-arc extension. Under present-day conditions this is inevitable because the crust is still rigidly welded to the underlying lithosphere. However, with a higher geothermal gradient in the Archaean temperatures at the base of the crust may have been high enough for the crust to behave in a much more ductile fashion. There is ample evidence of this in terms of the extreme deformation and horizontal flattening seen in Archaean lower crustal gneisses (Sheraton et al., 1973).

Thus in the Archaean, back-arc tension would give rise to a greater component of crustal thinning prior to any crustal fracturing, with the greenstone sequence being laid down on older crust and the lavas being extruded sub-aerially or more likely as pillowed flows as the basin deepened enough to become a marginal sea. The intrusive components, such as gabbros and sheeted dykes, typical of the Phanerozoic ophiolites, are thus of much less significance in the Archaean greenstone belts. This is readily explicable if there was less rifting (cf. Moores and Vine, 1971; de Wit and Stern, in prep.). If the rate of magma supply is great relative to extension, any sheeted dyke complex will be thin or non-existent relative to the extrusive and plutonic units. The enhanced rate of magma supply in Archaean greenstone belts is reflected of course in the large proportion of more peridotitic magmas, particularly in the lower parts of greenstone lava sequences.

Erosion of the volcanic products of the adjacent island arc and the uplifted older crust on either side would provide the extensive clastic sediments to fill the basin. Continued magma generation from remelted hydrous crust and mantle during the subduction process would provide the calc–alkaline batholith in a sequence becoming progressively more potassic as in the 'rocas verdes'. Ending of the greenstone sequence and closure of the basin might be determined by a number of complex factors not yet fully understood for present-day examples.

Accounting for the multiple sub-parallel linear greenstone belts of the S.W. Australian Yilgarn block and the Superior/Slave Provinces of Canada in terms of 'rocas verdes' marginal basin model requires further discussion.

Back-arc extension might be transmitted for a larger distance behind the subduction zone, and occur in a series of nodes similar to that envisaged by Rutland (1973). While a very shallow subducting plate might accomplish this (cf. Uyeda and Miyashiro, 1974), it is not easily reconcilable with the data of Green et al. (1975) which suggests derivation of peridotitic komatiite liquids from depths of ~200 km.

The multiple linear greenstone belt terrains are among those where it is most difficult to demonstrate a significant time interval between the greenstone belts and the intervening granitic/gneiss terrains (Jahn et al., 1974; Arth and Hanson, 1975). The very low initial $^{87}Sr/^{86}Sr$ ratios of the latter together with similar low ratios for many other granitic gneiss terrains in the Archaean (Moorbath, 1975) implies high rates of generation of continental crust from the mantle during the Archaean (3·8–2·6 b.y.). Generation of so much dominantly tonalitic crustal material with geochemical characteristics similar to the calc–alkaline magmas of continental margins is most easily achieved by the two stage 'conveyor-belt' process of Ringwood (1974), but implies more rapid convection, possibly in smaller convection cells. In this situation, basaltic material in the down-going slab would not be as cool as in present-day subduction zones, and water might also be retained more effectively. The result would be that remelting of ocean crust would take place at shallower depths and be constrained at high pH_2O, thus producing dominantly dacitic (tonalitic) magmas (cf. Nicholls and Ringwood, 1973; Arth and Hanson, 1975).

Applying the 'rocas verdes' marginal basin model to conditions of accelerated lateral crustal growth might lead to the following sequence of events. Subduction of oceanic crust at the continental margin produces a back-arc greenstone sequence, as in Fig. 3. Continued crustal growth by lateral accretion causes the continental arc and trench to migrate leading eventually to termination of the greenstone cycle. Development of a second or even multiple marginal greenstone basins would be expected as the trench/arc continues to migrate, in direct analogy with the present situation in the western Pacific where the marginal basins develop discontinuously in stepwise fashion parallel to one another (cf. Karig, 1971b, 1974). These could eventually be welded together by subsequent closure, forming larger continental cratons and preserving the basins (cf. Talbot, 1973).

The later diapiric batholiths invading the greenstone belts could be derived by the same subduction process or by remelting of the newly generated crustal material: the latter would still be consistent with Sr isotope evidence.

If this model is applicable, some progressive age differences might be expected between adjacent greenstone belts. So far little geochronological work has been done in this regard apart from the zircon dates on granite gneisses from part of the Superior Province greenstone terrain (Krogh and Davis, 1972) which suggests a 200 m.y. across-strike age difference, younging southwards.

As a final point we note the apparent restriction of greenstone belts to the period 3·4–2·7 b.y. There must be an underlying reason for this. A significant event on many continents at 2·9–2·6 b.y. was a widespread episode of granulite facies metamorphism affecting deeper crustal levels. Granulite terrains are typically more basic than the average crust and are notably low in the mobile lithophile elements K, Rb, Th and U, and hence more refractory and rigid (Tarney et al., 1972). If, in applying the ('rocas verdes') marginal basin model, a rather ductile lower crust is an essential feature in producing the crustal thinning necessary for an en-sialic marginal basin, then clearly this is less likely to take place subsequent to this granulite facies event. Later marginal basins would develop by crustal rifting rather than by crustal thinning, and evidence of such (wider) marginal basins is more likely to be destroyed during arc polarity reversal rather than preserved, except in special cases such as the 'rocas verdes'.

Acknowledgements

The model developed in this paper, and the ideas therein, arose directly from studies on marginal basins in the Scotia Sea and Southern Chile supported by the Natural Environment Research Council (U.K.) and the Office of Polar Programs of the National Science Foundation (U.S.A.). We thank P. F. Barker, S. D. Weaver and A. E. Wright for helpful criticism of the manuscript.

References

Anhaeusser, C. R., 1973. 'The evolution of the early Precambrian crust of southern Africa', *Phil. Trans. R. Soc. Lond.*, **A273**, 359–388.

Arth, J. G., and Hanson, G. N., 1975. 'Geochemistry and origin of the early Precambian crust of northeastern Minnesota', *Geochim. Cosmochim. Acta*, **30**, 325–362.

Bain, J. H. C., 1973. 'A summary of the main structural elements of Papua, New Guinea', In P. J. Coleman (ed.), *The Western Pacific: Island Arcs, Marginal Seas, Geochemistry*, Univ. W. Aust. Press, Perth, 147–161.

Baragar, W. R. A., and Goodwin, A. M., 1969. 'Andesites and Archaean volcanism of the Canadian Shield', *Oregon State Dept. Geology and Mineral Industries Bull.*, **65**, 121–142.

Barker, P. F., 1972. 'A spreading centre in the east Scotia Sea', *Earth Planet. Sci. Lett.*, **15**, 123–132.

Barker, P. F., 1975. 'The tectonic framework of Cenozoic Volcanism in the Scotia Sea region: A Review', *Bull. Volc.* (in press).

Barker, P. F., and Griffiths, D. H., 1972. 'The evolution of the Scotia Ridge and Scotia Sea', *Phil. Trans. R. Soc. Lond.*, **A271**, 151–183.

Bird, J. M., Dewey, J. F., and Kidd, W. S. F., 1971. 'Proto-Atlantic oceanic crust and mantle: Appalachian-Caledonian Ophiolites', *Nature Phys. Sci.*, **231**, 28–31.

Boccaletti, M., and Guazzone, G., 1974. 'Remnant arcs and marginal basins in the Cenozoic development of the Mediterranean', *Nature, London*, **252**, 18–21.

Brooks, C., and Hart, S. R., 1974. 'On the significance of komatiite', *Geology*, **2**, 107–110.

Brothers, R. N., and Blake, M. C., 1973. 'Tertiary plate tectonics and high pressure metamorphism in New Caledonia', *Tectonophysics*, **17**, 337–358.

Bursnall, J., and de Wit, M. J., 1975. 'Timing and development of the Orthotectonic Zone in the Appalachian Orogen of Northwest Newfoundland', *Can. J. Earth Sci.* (in press).

Cann, J. R., 1974. 'A model for oceanic crustal structure developed', *Geophys. J. R. Astr. Soc.*, **39**, 169–187.

Cobbing, E. J., and Pitcher, W. S., 1972. 'The coastal batholith of central Peru', *Jl geol. Soc. Lond.*, **128**, 421–460.

Coleman, P. J., 1975. 'On island arcs', *Earth-Sci. Rev.*, **11**, 47–80.

Condie, K. C., Macke, J. E., and Reimer, T. O., 1970. 'Petrology and geochemistry of early Precambrian greywackes from the Fig Tree Group, South Africa', *Geol Soc. Am. Bull.*, **81**, 2759–2776.

Coward, M. P., and James, P. R., 1974. 'The deformation patterns of two Archaean greenstone belts in Rhodesia and Botswana', *Precambrian Res.*, **1**, 235–258.

Dalziel, I. W. D., 1974. 'Evolution of the margins of the Scotia Sea', In Burk, C. A., and Drake, C. L. (Eds.), *The Geology of Continental Margins*, Springer-Verlag, New York, 567–579.

Dalziel, I. W. D., and Cortes, R., 1972. 'Tectonic style of the southernmost Andes and the Antarctandes', *Rept 24th Int. geol. Congress, Montreal*, **3**, 316–327.

Dalziel, I. W. D., de Wit, M. J., and Palmer, K. F., 1974. 'Fossil marginal basin in the southern Andes', *Nature, London*, **250**, 291–294.

Davies, H. L., and Smith, I. E., 1971. 'Geology of Eastern Papua', *Geol. Soc. Am. Bull.*, **82**, 3299–3312.

de Wit, M. J., 1975. 'The geology of the central Cordillera, Southern Chile; the anatomy of a Mesozoic marginal basin' (in preparation).

de Wit, M. J., and Stern, C. R., 1975. 'Pillow Talk' (in preparation).

Dickinson, W. R., 1975. 'Potash-depth (K-h) relations in continental margin and intra-oceanic magmatic arcs', *Geology*, **3**, 53–56.

Dott, R. H., Bruhn, R. L., de Wit, M. J., and Winn, R., 1975. 'Age and provenance of the Tekenika beds' (in preparation).

Faure, G., and Powell, J. L., 1972. *Strontium Isotope Geology*, Springer-Verlag, Berlin.

Glikson, A. Y., 1971. 'Primitive Archaean element distribution patterns: chemical evidence and geotectonic significance', *Earth Planet. Sci. Lett.*, **12**, 309–320.

Green, D. H., 1972. 'Archaean greenstone belts may include terrestrial equivalents of lunar maria?', *Earth Planet. Sci. Lett.*, **15**, 263–270.

Green, D. H., 1975. 'Genesis of Archaean peridotitic magmas and constraints on Archaean geothermal gradients and tectonics', *Geology*, **3**, 15–18.

Green, D. H., Nicholls, I. A., Viljoen, M. J., and Viljoen, R. P., 1975. 'Experimental demonstration of the existence of peridotitic liquids in earliest Archaean magmatism', *Geology*, **3**, 11–14.

Hackman, B. D., 1973. 'The Solomon Island Fractured Arc', In P. J. Coleman (ed.), *The Western Pacific, Island Arcs, Marginal Seas, Geochemistry*, Univ. of W. Aust. Press, Perth, 179–191.

Halpern, M., 1973. 'Regional geochronology of Chile south of 50°S latitude', *Geol. Soc. Am. Bull.*, **84**, 2407–2422.

Hamilton, W., 1969. 'The Volcanic Central Andes: a model for the Cretaceous batholiths and tectonics of North America', *Oregon State Dept Geology and Mineral Industries Bull.*, **65**, 175–183.

Hart, S. R., Brooks, C., Krogh, T. E., Davis, G. L., and Nava, D., 1970. 'Ancient and modern volcanic rocks: a trace element model', *Earth Planet. Sci. Lett.*, **10**, 17–28.

Hart, S. R., Glassley, W. E., and Karig, D. E., 1972. 'Basalts and sea floor spreading behind the Mariana island arc', *Earth Planet. Sci. Lett.*, **15**, 12–18.

Hawkesworth, C. J., Moorbath, S., O'Nions, R. K., and Wilson, J. F., 1975. 'Age relationships between greenstone belts and "granites" in the Rhodesian Archaean craton', *Earth Planet. Sci. Lett.*, **25**, 251–262.

Hunter, D. R., 1974. 'Crustal development in the Kaapvaal Craton, I. The Archaean', *Precambrian Res.*, **1**, 259–294.

Jahn, B., Shih, C., and Murthy, V. R., 1974. 'Trace element geochemistry of Archaean volcanic rocks', *Geochim. Cosmochim. Acta*, **38**, 611–627.

Jakeš, P., and White, A. J. R., 1972. 'Major and trace element abundances in volcanic rocks of orogenic areas', *Geol. Soc. Am. Bull.*, **83**, 29–40.

Johnson, T., and Molnar, P., 1972. 'Focal Mechanisms and Plate tectonics of the Southwest Pacific', *J. Geophys. Res.*, **77**, 5000–5032.

Karig, D. E., 1971a. 'Origin and development of marginal basins in the western Pacific', *J. Geophys. Res.*, **76**, 2542–2561.

Karig, D. E., 1971b. 'Structural history of the Mariana island arc system', *Geol. Soc. Am. Bull.*, **82**, 323–344.

Karig, D. E., 1973. 'Plate convergence between the Philippines and the Ryukyu islands', *Marine Geology*, **14**, 153–168.

Karig, D. E., 1974. 'Evolution of arc systems in the western Pacific', *An. Rev. Earth Planet. Sci.*, **2**, 51–75.

Kay, R. W., and Senechal, R. G., 1975. 'Limitations on the use of chemical data in establishing the magmatic affinities of the Troodos ophiolite complex, Cyprus' (in press).

Kelleher, J., and McCann, W., 1975. 'Buoyant zones, great earthquakes and dynamic boundaries of subduction', *J. Geophys. Res.* (in press).

Kidd, W. S. F., 1974. Unpubl. Ph.D. thesis, Cambridge Univ., U.K.

Krank, E. H., 1932. 'Geological investigations in the Cordillera of Tierra del Fuego', *Acta Geographica, Helsingfors*, **4**, 231.

Krogh, T. E., and Davis, G. L., 1972. 'Zircon U–Pb ages of Archaean metavolcanic rocks in the Canadian Shield', *Carnegie Inst. Washington Yearbook*, **70**, 241–242.

Litherland, M., 1973. 'Uniformitarian approach to Archaean "Schist Relics"', *Nature, Phys. Sci.*, **242**, 125–127.

Moorbath, S. M., 1975. 'Evolution of Precambrian crust from strontium isotopic evidence', *Nature, London*, **254**, 395–398.

Moores, E. M., and Vine, F. J., 1971. 'The Troodos Massif, Cyprus and other ophiolites as oceanic crust: evaluation and implications', *Phil. Trans. R. Soc. Lond.*, **A268**, 443–466.

Nicholls, I. A., and Ringwood, A. E., 1973. 'Effect of water on olivine stability in tholeiites and the production of silica-saturated magmas in the island arc environment', *J. Geol.*, **81**, 285–300.

Packham, G. H., and Falvey, D. A., 1971. 'An hypothesis for the formation of marginal seas in the western Pacific', *Tectonophysics*, **11**, 79–109.

Pearce, J. A., 1975. 'Basalt geochemistry used to investigate past tectonic environments on Cyprus', *Tectonophysics*, **25**, 41–67.

Ringwood, A. E., 1974. 'The petrological evolution of island arc systems', *Jl geol. Soc. Lond.*, **130**, 183–204.

146

Rutland, R. W. R., 1973. 'Tectonic evolution of the continental crust of Australia', In D. H. Tarling and S. K. Runcorn (Eds.), *Implications of Continental Drift to the Earth Sciences,* Academic Press, London, 2, 1011–1034.

Saunders, A. D., Tarney, J., Dalziel, I. W. D., and de Wit, M. J., 1975a. 'Geochemistry of the Sarmiento ophiolite complex, S. Chile' (in preparation).

Saunders, A. D., Tarney, J., O'Nions, K., Pankhurst, R. J., and Barker, P. F., 1975b. 'Geochemistry of basalts from the "back-arc" spreading centre in the Scotia Sea' (in preparation).

Scholz, C. H., Barazangi, M., and Sbar, M. L., 1971. 'Late Cenozoic evolution of the Great Basin, western United States, as an ensialic interarc basin', *Geol. Soc. Am. Bull.,* 82, 2979–2990.

Schweickert, R. A., and Cowan, D. S., 1974. 'Pre-Titonian magmatic arcs and subduction zones, Sierra Nevada, California', *Geol. Soc. Am.,* Abs. with Programs, 6, 3, 251–252.

Schweickert, R. A., and Cowan, D. S., 1975. 'Early Mesozoic tectonic evolution of the western Sierra Nevada, California', *Bull. Am. Geol. Soc.,* 86 (in press).

Sheraton, J. W., Tarney, J., Wheatley, T. J., and Wright, A. E., 1973. 'The structural history of the Assynt district', In R. G. Park and J. Tarney (Eds.). *The Early Precambrian of Scotland and Related Rocks of Greenland,* University of Keele, 31–43.

Shiraki, K., 1971. 'Metamorphic basement rocks of Yap Islands, western Pacific: possible oceanic crust beneath an island arc', *Earth Planet. Sci. Letters,* 13, 167–174.

Stewart, T. W., and Suarez, M., 1971. 'Preliminary studies of stratigraphy and structure in the Patagonian Cordillera between 51° and 53°S', *1st South Am. Geol. Cong., Lima* (abstract).

Stewart, T. W., Cruzat, A., Page, B., Suarez, M., and Stambuk, V., 1971. 'Estudio geologico economico de la Cordillera Patagonica entre los paralelos 51° y 53°30'S, Provincia de Magalanes', Unpubl. Report Instituto de investigaciones geol., Satiago, Chile.

Sugimura, A., and Uyeda, S., 1973. *Island Arcs, Japan and its Environs,* Elsevier, Amsterdam, 247 pp.

Talbot, C. J., 1973. A plate tectonic model for the Archaean crust', *Phil. Trans. R. Soc. Lond.,* A273, 413–427.

Tarney, J., Skinner, A. C., and Sheraton, J. W., 1972. 'A geochemical comparison of major Archaean gneiss units from North-West Scotland and East Greenland', *Rept. 24th Int. geol. Congress, Montreal,* 1, 162–174.

Uyeda, S., and Miyashiro, A., 1974. 'Plate tectonics and the Japanese Islands: a synthesis', *Geol. Soc. Am. Bull.,* 85, 1159–1170.

Watts, A. B., and Weissel, J. K., 1975. 'Tectonic history of the Shikoku marginal basin', *Earth Planet. Sci. Lett.,* 25, 239–250.

Windley, B. F., 1973. 'Crustal development in the Precambrian', *Phil. Trans,. R. Soc. Lond.,* A273, 321–341.

Early Precambrian Granulites—Greenstones, Transform Mobile Belts and Ridge-rifts on Early Crust?

M. B. KATZ

School of Applied Geology, University of New South Wales, Kensington, N.S.W., Australia.

Abstract

In most Archaean terrains there is a characteristic intimate association of granulite mobile belts and greenstones. Other important constituents of these terrains are gneisses and granites. This relationship between greenstones and granulites indicates a close genetic tie between low- and high-grade rocks, referred to by some observers as the 'cratonic paradox'. Age dating shows that the granite gneiss may be greater than 3000 m.y.; while the granulites and greenstones appear to be somewhat younger, about 2700 m.y. This suggests that the granite-gneiss forms an older basement associated with later, almost contemporaneously developed granulites and greenstone belts.

The greenstones are widely interpreted as representing the upwelling of mantle material through early crustal breaks and rifts. In contrast, the granulites form conspicuously linear mobile belts characterized by high-temperature metamorphism, igneous activity and intense structural deformation and dislocation. It is proposed that these mobile belts reflect early zones of weakness of essentially transform nature. The thin, hot sialic protocrust, now represented by the older granites and gneisses, was deformed by mantle influences, but was competent enough to undergo fundamental fracturing. These early fractures penetrated the thin crust, localized mantle activity and were the sites of high geothermal gradients and severe dislocation over a relatively wide zone. Movement and displacement along these early transform zones caused high-angle rifting, crustal fragmentation and formation of oceanic crust. Thus the high-grade metamorphic mobile transform zone and rift-oceanic crust zone were forming at about the same time, as can be seen in present-day transform–ridge relationships. The early crust, however, behaved in a different manner and transform zones were then wider and geologically more active.

It is proposed that these early transform zones are the precursors of the high-grade granulite mobile belts of the Archaean. The apparent paradoxical relationships of the granulites to the greenstones can be explained by use of a plate tectonic transform-ridge model. Such relationships are illustrated by examples taken from the classic Archaean terrains of South India, Southwest Australia, South Africa and Canada.

Introduction

In recent years much attention has been devoted to the fundamental problems of the evolution and origin of early Precambrian basement, cratons and mobile belts, especially the relationships among granite–gneiss terrains, high-grade metamorphic belts and granulites, and greenstone belts (Anhaeusser et al., 1969; Engel and Kelm, 1972; Windley, 1973; Sutton, 1973). Some workers have envisaged the formation of greenstone belts on a relatively thin crust, predating the formation of high-grade metamorphic belts (Anhaeusser et al., 1969). Others have shown that the granulite–gneisses are basement to the greenstone belts (Windley, 1973). Recent work in Greenland has revealed a primordial granite-gneiss basement underlying both granulite facies metamorphic rocks and greenstone-type sequences (McGregor, 1973). The relative ages of granulite facies metamorphism and the greenstone belts are so similar that there may be no significant differences. Age data from Archaean terrains of Southwest Australia (Arriens, 1971; Compston and Arriens, 1968) and South India (Crawford, 1969) suggest that the oldest rocks are granite-gneisses, and reconizable granulites and greenstones are somewhat younger but with no significant overall age difference. This supports the formation of a granite-gneiss basement, with subsequent, almost contemporaneous, evolution of granulite and greenstone belts.

The granulites contrast with the greenstones in terms of composition, igneous and metamorphic characteristics, structure and tectonic setting, and relative mobility (Anhaeusser et al., 1969). There is some suggestion that the gneisses and granites formed the earliest protocrust, which was subsequently deformed into greenstone belts and granulite facies mobile belts (Sutton, 1973). The greenstone belts, which are well defined in the literature, are widely interpreted as representing upwelling of mantle material through early crustal breaks and rifts (Windley, 1973; Sutton, 1973). In contrast, the granulite belts are considered to be true mobile belts, characterized by high-grade (high-temperature) metamorphism, igneous activity and intense structural deformation and dislocation.

Granulite Facies Mobile Belts

The granulites seem to occur along linear belts that can be traced through the proposed Precambrian supercontinent (Katz, 1974; Piper, 1974). Such belts are thought to reflect fundamental processes operative early in the evolution of the Earth's crust. The dry, completely recrystallized character of the granulites ensures their resistance to subsequent geological events and favours survival of their original isotopic age, mineralogy, textures and even gross structural characteristics (Watson, 1973). Thus they serve as excellent indicators of the tectonic–metamorphic conditions of the early Precambrian and are also valuable correlation units.

The granulite belts are characterized by a high-temperature, low-pressure metamorphic facies series (Saggerson, 1973), although there is some evidence for paired metamorphic belts (Katz, 1972). Most belts show strong linearity with steep dips and parallel alignments of all structural elements. Major strike slip dislocations are often present, along with shear and cataclastic zones (Watson, 1973). The textures of many granulites indicate extreme deformation and recrystallization. Platy quartz occurs in some granulites, others are uniformly recrystallized and exhibit equigranular granoblastic textures (Eskola, 1952). Associated with these belts are later igneous rocks such as anorthosites and granites, and alkali rocks suites, the intrusion of which are controlled by major lineaments (Bridgwater and Windley, 1973).

Plate Tectonics

The role of plate tectonics in the evolution of early Precambrian granulite facies mobile belts has been considered by some workers. The role of subduction has been presented on the basis of possible paired-belt arrangements (Katz, 1972), and presence of possible oceanic crust (Garson and Livingstone, 1973), but recent palaeomagnetic data implies a Precambrian supercontinent which would preclude large scale subduction processes (Piper, 1974). The predominant ensialic setting of the granulite belts and lack of well-developed ophiolite, along with their structural continuity suggests an *in situ* origin for many such belts (Briden, 1973). The model that is currently receiving most attention relates these high-temperature granulite facies mobile belts to early lineament-rift zones which were loci of high geothermal gradients and igneous activity (Windley, 1973). The role of lineament control on the formation of these belts has been emphasized, and it has been shown that such lineaments exert perennial control from the early Precambrian initiation of the mobile belts up to present-day plate tectonic activity (Katz, 1975; McConnell, 1972; Sundaram et al., 1964). The exact nature of the lineaments has not been defined. They may be normal faults bounding linear rifts, thrust and subduction zones or transform fault zones. The evidence from granulite facies belts, which largely preserve their metamorphic and tectonic entity, indicate that these belts are zones of high temperature and high strain, a combination that cannot be easily explained using a simple rift model. Also it is evident that these belts are related to, and contemporaneous with, the greenstones. It is proposed that these mobile belts reflect zones of essentially transform nature. The early, thin sialic protocrust was deformed by mantle influences but competent enough to undergo relatively brittle fracturing. Such early fundamental fractures penetrated the thin, hot crust and localized mantle activity. They were the sites of high geothermal gradients and severe dislocation over a relatively wide zone. Relicts of the original crust, or their remobilized equivalents tended to survive in areas outside the main zones of metamorphism and deformation.

Movement and displacement along protocrustal transform zones caused rifting at high angles to the trend of zones (Wilson, 1965). The rifts developed in tensional domains of transform movement and were the sites of formation of oceanic crust. Thus the high-grade metamorphic, mobile transform zones and rift-greenstone, oceanic crust zones were evolving at about the same time. An example relating granite gneiss protocrust to mobile granulite belt and greenstone belt from a portion of a Gondwanic early Precambrian terrain extending from Sri Lanka, South India through East Antarctica to S.W. Australia (Katz, 1974) is considered below.

Gondwanic Belt—A proto-transform zone

This granulite belt (Fig. 1), was the site of an early transform zone, possibly related to fundamental regmatic lineament patterns thought to exist in the early crust (Badgley, 1965). This belt was a relatively wide 'ductile' zone of transform movement for which a possible present-day analogue would be the broad transform zone proposed for the San Andreas fault system in Western U.S.A. (Atwater, 1970). Within any such transform zone there would be complex structural and tectonic patterns. Along faults parallel and oblique to the overall transform direction there would be strike-slip movement and compressional and tensional forces related to the movements. Such a transform zone would have all the properties of a mobile belt with tectonism, igneous activity and metamorphism (Elders et al., 1972).

The role of faulting and large-scale transcurrent movements within such mobile belts has been described by Sutton and Watson (1959), Mason (1973) and McConnell (1969). A system of faults in South India (Grady, 1971) may reflect such an ancient fundamental transform zone. The Darling fault of S.W., Australia (Johnstone et al., 1973) may be an extension of the same zone (Fig. 1). Within this belt the small-scale structures are linear and parallel to the faults, implying extreme ductile shearing and attenuation. At shallow levels, zones of tension would promote block faulting, tilting and development of sedimentary accumulations from protocrustal material. The sediments would be later involved in the mobility of the belt and metamorphosed to rocks such as khondalites, quartzites and marbles. Charnockites may have formed at deeper

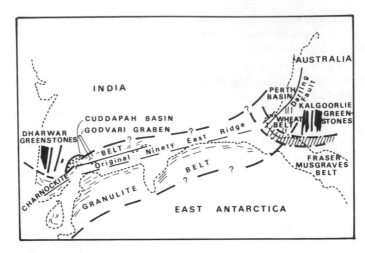

Fig. 1. Reconstruction of Gondwanaland showing 'Granulite Belt' (charnockites of India and East Antarctica and the Wheat Belt of Southwest Australia) in relation to associated major faults, other tectonic elements, and the 'Greenstones' (Dharwars of India and Kalgoorlie System of Western Australia) (after Narayanaswami, 1970, Grady, 1971, Johnstone et al., 1973 and Katz, 1974)

levels in zones of tension that are probable sites of mantle protuberances and high thermal gradients. All these tectonic, sedimentational and metamorphic events are responses to major transform mobility.

Rifts and spreading ridges were developed at high angles to the transform directions and in such areas, oceanic crust, the precursors of the greenstones, were emplaced. In South India such greenstones are represented by the Dharwars, and in S.W. Australia by the Kalgoorlie System. The Dharwar greenstones–charnockitic granulite mobile belt relationship (Fig. 1) is interpreted as a true high-angle ridge–transform relationship in the early Precambrian. The relationship between the Kalgoorlie System greenstones and the adjacent sub-parallel Wheat Belt granulites is not so clear. The Wheat Belt granulites have been considered as the infracrustal roots of the greenstones (Glikson and Lambert, 1973). According to the scheme outlined above, the Kalgoorlie greenstones–Wheat Belt granulites relationship should be similar to the Dharwar greenstones–Charnockitic relationship. However, both of these areas have been affected by subsequent events which may have obscured the original relationships.

Speculative Schematic Evolution of the Gondwanic Belt and Other Related Belts

It is postulated that, at an early stage in the Earths' history fundamental fractures and lineaments were formed in the protocrust (Fig. 2A). These zones of weakness reflected mantle influences and served to control the position of wide, relatively ductile transform zones which developed in the granite-gneiss crust. These fracture-transform zones may have been boundaries to proto-plates. Movement and displacement along these transforms would initiate tectonism, sedimentation, metamorphism and igneous activity. These zones reached their maximum mobility at about 2600 m.y. and were major discontinuities and deep-seated zones within the ancient crust (>3000 m.y.) Contemporaneous with the period of maximum mobility, rift zones and spreading ridges formed at high angles, penetrated the thin sialic crust and permitted the generation of mantle-derived oceanic crust (Fig. 2B).

The fundamental dextral-type transforms were never completely stabilized. Since the development of the Dharwar–Kalgoorlie greenstones and Charnockitic and Wheat

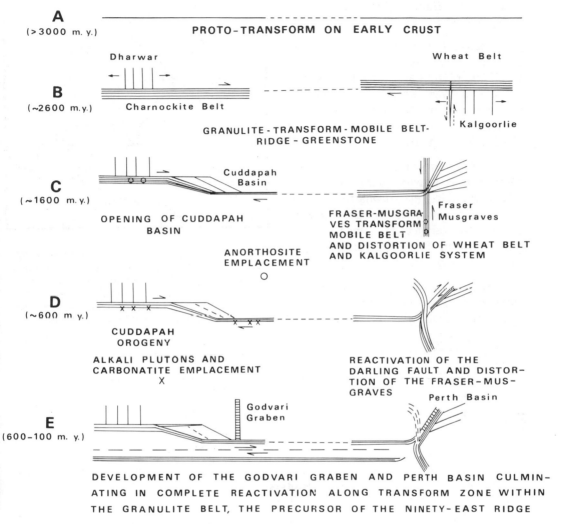

SOUTH INDIA   SOUTHWEST AUSTRALIA

A (>3000 m. y.)   PROTO-TRANSFORM ON EARLY CRUST

Dharwar   Wheat Belt

B (~2600 m.y.)   Charnockite Belt

GRANULITE - TRANSFORM - MOBILE BELT-
RIDGE - GREENSTONE   Kalgoorlie

C (~1600 m.y.)   Cuddapah Basin

OPENING OF CUDDAPAH BASIN

FRASER-MUSGRAVES TRANSFORM MOBILE BELT AND DISTORTION OF WHEAT BELT AND KALGOORLIE SYSTEM   Fraser Musgraves

ANORTHOSITE EMPLACEMENT

D (~600 m.y.)

CUDDAPAH OROGENY

ALKALI PLUTONS AND CARBONATITE EMPLACEMENT

REACTIVATION OF THE DARLING FAULT AND DISTORTION OF THE FRASER-MUSGRAVES

Perth Basin

E (600-100 m. y.)   Godvari Graben

DEVELOPMENT OF THE GODVARI GRABEN AND PERTH BASIN CULMINATING IN COMPLETE REACTIVATION ALONG TRANSFORM ZONE WITHIN THE GRANULITE BELT, THE PRECURSOR OF THE NINETY-EAST RIDGE

Fig. 2. Schematic evolution of major mobile belts, greenstones and tectonic elements of the Gondwanaland of India and Southwest Australia (see text for details)

Belt granulites at about 2600 m.y. in older granite and gneiss protocrustal relicts (> 3000 m.y.) and remobilized equivalents, at least four other episodes of mobility are known to have complicated the original pattern.

(i) The first episode (Fig. 2C) at about 1600–1400 m.y. is recognized in South India as the Eastern Ghats orogeny (Crawford, 1969), and in S.W. Australia as the Musgraves–Fraser event (Compston and Arriens, 1968). In S. India a

major deflection along this Eastern Ghats belt caused the opening of the Cuddapah basin, while in S.W. Australia a new transform-mobile belt formed along the inside edge of the Kalgoorlie greenstone belt. This new sinistral transform zone, orthogonal to the older transform was the precursor of the Frazer–Musgraves mobile belt. Movement along the latter caused displacement and rotation of the Kalgoorlie System and its repositioning up against the Wheat Belt. At this time

152

anorthosite intrusions were being emplaced along major faults and lineaments both in the Eastern Ghats and in the Musgraves (Fig. 2C).

(ii) The second episode (Fig. 2D) at about 600 m.y. was accompanied by intrusion of alkalic plutons and carbonatites in S. India (Narayanaswami, 1970) and the rejuvenation of the west part of the Wheat Belt transform, especially the Darling Fault (Crawford, 1969).

(iii) The third episode (Fig. 2E) was Phanerozoic activity along transforms which includes the formation of the Gondwana rifts such as the Godavari graben (Sundaram et al., 1964) and later rifts such as the Perth Basin (Johnstone et al., 1973) (Fig. 2E).

(iv) The fourth and latest episode was an intense reactivation during the Early Cretaceous that led to the break-up and

Fig. 3. Present position of the Ninetyeast Ridge showing apparent control of younger oceanic transforms by major continental faults and lineaments in India and Western Australia (after Sclater and Fisher, 1974, reproduced with permission of The Geological Society of America)

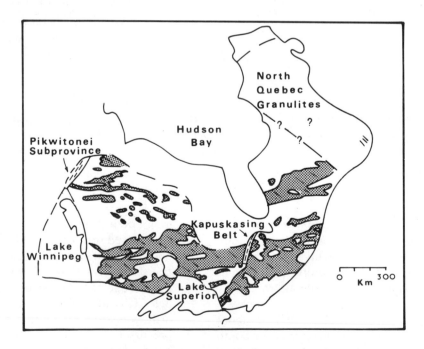

Fig. 4. Greenstone–granulite relationships in the Superior Province of Canada. Both the Pikwitonei and Kapuskasing granulites are zones of strong faulting and igneous activity (after Goodwin et al., 1972, reproduced with permission of Geological Association of Canada)

fragmentation of Gondwanaland. Strong dextral transform movements led to the development of the now dormant Ninetyeast ridge (Fig. 1), which played a major role in the early evolution of the Indian Ocean (Sclater and Fisher, 1974). Other major faults and lineaments recognized in South India can be shown to be the continental extension of major contemporary transforms on the Indian Ocean Ridge and served to control their position (Fig. 3). Similar oceanic transforms may be expected on the extension of the Darling Fault to the south (Fig. 3). The break-up and fragmentation of Australia from Antarctica also follows old transform directions of the Fraser–Musgraves mobile belt.

Similar relationships between high-grade granulite mobile belts of supposed transform origin and greenstone belts, of Archaean age, may be recognised between the Superior Province greenstone and granulites of Canada, and the South African and Rhodesian greenstone belts and the granulites of the Limpopo Mobile Belt. In the Superior Province east–west trending greenstone belts in Northern Quebec make contact with a large area of granulites (Goodwin et al., 1972). Although the details of this relationship have not been defined, it appears that this granulite terrain strikes NNE at high angles to the greenstone trends and may represent an Archaean transform mobile belt (Fig. 4). Two other smaller granulite belts cut across the Superior Province greenstones; one forming the northwest

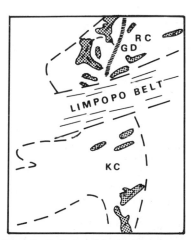

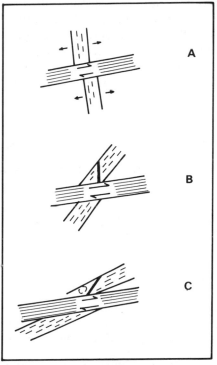

Fig. 5. Relationship between granulite Limpopo Belt and greenstones of the Rhodesian craton (RC) and Kaapvaal craton (KC), also showing the position of the Great Dyke (GD). Schematic evolution of the greenstones, granulites and the Great Dyke is illustrated in A, B and C.

A: Limpopo granulite transform with greenstone ridge (>2700 m.y.).
B: Reactivation of the Limpopo transform with dextral rotation of greenstones and emplacement of Great Dyke (2700 m.y.).
C: Further reactivation of Limpopo transform with dextral rotation of greenstones and Great Dyke. Remobilization of granites in zones of attenuation (2000 m.y.) (after Bliss and Stidolph, 1969)

154

boundary of the Province known as the Pik-witonei subprovince and the other known as the Kapuskasing Belt. Both terrains are associated with major dislocation and igneous intrusion and may represent other zones of transform movement along an Archaean proto-ridge now represented by the Superior greenstones (Fig. 4).

Perhaps the most spectacular example of a transform mobile belt is the Limpopo Belt. This granulite facies belt is an extensive ENE linear zone separating two NW trending greenstone terrains; the Rhodesian craton to the north and the Kaapvaal craton to the south (Mason, 1973) (Fig. 5). This belt is charac-terized by extreme deformation, shear dislo-cation and igneous activity. In particular the role of transcurrent movement in the evolu-tion of this belt has been emphasized. It is proposed that the Limpopo Belt is another transform mobile belt and during its initiation (>2700 m.y.) developed greenstone belts at high angles (Fig. 5A). An important dextral reactivation of this transform zone

(2700 m.y.) rotated the older greenstone belts to a NW orientation, and at the same time, limited orthogonal tension and rifting permit-ted emplacement of the Great Dyke (Bliss and Stidolph, 1969) (Fig. 5B). Repeated move-ment at about 2000 m.y. caused further dextral rotation of structural elements in the surrounding cratons (Fig. 5C).

The apparent paradoxical association of greenstone belts and granulite facies belts in many Archaean terrains can thus be shown to be largely a result of fundamental transform movement in the early crust. The role of transform tectonics in later Precambrian and Phanerozoic mobile belts may not be as evi-dent. As the crust thickened and geothermal gradients were reduced other tectonic processes became more active, with more important roles played by sea-floor spreading and subduction. However, the early transform directions reflected by high-grade Precambrian mobile belts asserted a strong control on subsequent tectonic and plate–tectonic process.

Acknowledgements

I wish to thank H. G. Golding for his careful review of the paper and E. Scheibner for his helpful comments.

References

Anhaeusser, C. R., Mason, R., Viljoen, M. J., and Viljoen, R. P., 1969. 'A reappraisal of some aspects of Precambrian shield geology', *Bull. geol. Soc. Am.*, **80**, 2175–2200.

Arriens, P. A., 1971. 'The Archaean geochronol-ogy of Australia', *Spec. Publ. geol. Soc. Aust.*, **3**, 11–23.

Atwater, T., 1970. 'Implications of plate tectonics for the Cenozoic tectonic evolution of Western North America', *Bull. geol. Soc. Am.*, **81**, 3513–3536.

Badgley, P. C., 1965. *Structural and Tectonic Princi-ples*, Harper and Row, New York, 521 pp.

Bliss, N. W., and Stidolph, P. A., 1969. 'A review of the Rhodesian basement complex', *Spec. Publ. geol. Soc. S. Afr.*, **2**, 305–334.

Briden, J. C., 1973. 'Applicability of plate tectonics to pre-Mesozoic time', *Nature*, **244**, 400–405.

Bridgwater, D., and Windley, B. F., 1973. 'Anorthosites, post-orogenic granites, acid volcanic rocks and crustal development in the North Atlantic shield during the Mid-Proterozoic', *Spec. Publ. geol. Soc. S. Afr.*, **3**, 307–332.

Compston, W., and Arriens, P. A., 1968. 'The Precambrian geochronology of Australia', *Can. J. Earth Sci.*, **5**, 561–584.

Crawford, A. R., 1969. 'India, Ceylon and Pakis-tan. New age data and comparisons with Australia', *Nature*, **233**, 380–384.

Elders, W. A., Rex, R. W., Meidav, T., Robinson, P. T., and Biehler, S., 1972. 'Crustal spreading in Southern California', *Science*, **178**, 15–24.

Engel, A. E. J., and Kelm, D. L., 1972. 'Pre-Permian global tectonics: A tectonic test', *Bull. geol. Soc. Am.*, **83**, 2325–2340.

Eskola, P., 1952. 'On the granulites of Lapland', *Amer. J. Sci., Bowen Volume*, 133–171.

Garson, M. S., and Livingstone, A., 1973. 'Is the South Harris complex in North Scotland a Precambrian overthrust slice of oceanic crust and island arc?', *Nature. Phys. Sci.*, **243**, 74–76.

Glikson, A. Y., and Lambert, I. B., 1973. 'Relations in space and time between major Precambrian shield units: An interpretation of Western Australian data', *Earth Planet. Sci. Lett.*, **20**, 395–403.

Goodwin, A. M., Ambrose, J. W., Ayers, L. D., Clifford, P. M., Currie, K. C., Ermanovics, I. M., Fahrig, W. F., Gibb, R. A., Hall, D. H., Innes, M. J. S., Irvine, T. N., Maclaren, A. S., Norris, A. W., and Pettijohn, F. J., 1972. 'The Superior Province', *Geol. Assoc. Canada, Spec. Pap.*, **11**, 527–624.

Grady, J. C., 1971. 'Deep main faults in South India', *J. geol. Soc. India.*, **12**, 56–62.

Johnstone, M. H., Lowry, D. C., and Quilty, P. G., 1973. 'The geology of southwestern Australia—a review', *J. Roy. Soc. W. Austr.*, **56**, 5–15.

Katz, M. B., 1972. 'Paired metamorphic belts of the Gondwanaland Precambrian and plate tectonics', *Nature*, **239**, 271–273.

Katz, M. B., 1974. 'Paired metamorphic belts in Precambrian granulite facies terrains of Gondwanaland', *Geology*, **2**, 237–241.

Katz, M. B., 1975. 'Precambrian granulite facies belts, lineaments and plate tectonics', *Proc. 1st Int. Conf. New Basement Tectonics*, **1**, (in press).

Mason, R., 1973. 'The Limpopo mobile belt–Southern Africa', *Phil. Trans. R. Soc. London.*, **273**, 463–485.

McConnell, R. B., 1969. 'Fundamental fault zones in the Guiana and West African shields in relation to presumed axes of Atlantic spreading', *Bull. geol. Soc. Am.*, **80**, 1775–1782.

McConnell, R. B., 1972. 'Geological development of the rift system of eastern Africa', *Bull. geol. Soc. Am.*, **83**, 2549–2472.

McGregor, V. R., 1973. 'The early Precambrian gneisses of the Godthaab district, West Greenland', *Phil. Trans. R. Soc. Lond.*, **273**, 243–258.

Narayanaswami, S., 1970. 'Tectonic setting and manifestation of the upper mantle in the Precambrian rocks of South India', *Proc. 2nd Sympos. Upper Mantle Proj.*, **5**, 377–402.

Piper, J. D. A., 1974. 'Proterozoic crustal distribution, mobile belts and apparent polar movements', *Nature*, **251**, 381–384.

Sclater, J. G., and Fisher, R. L., 1974. 'Evolution of the east central Indian Ocean with emphasis on the tectonic setting of the Ninetyeast ridge', *Bull. geol. Soc. Am.*, **85**, 683–702.

Saggerson, E. P., 1973. 'Metamorphic facies series in Africa: a contrast', *Spec. Publ. Soc. S. Afr.*, **3**, 227–234.

Sundaram, R. K., Swami Nath, J., and Venkatesh, V., 1964. 'Tectonics of Penninsular India', *Int. Geol. Cong. 22nd Session, New Delhi*, Part IV, 539–556.

Sutton, J., 1973. 'Some changes in continental structure since early Precambrian time', in Tarling, D. H., and Runcorn, S. K. (Eds.), *Implications of Continental Drift to the Earth Sciences*, 1071–1081.

Sutton, J., and Watson, J. V., 1959. 'Metamorphism in deep-seated zones of transcurrent movement at Kungwe Bay, Tanganyika Territory', *J. Geol.*, **67**, 1–13.

Watson, J. V., 1973. 'Effects of reworking on high-grade gneiss complexes', *Phil. Trans. R. Soc. Lond.*, **273**, 443–455.

Wilson, J. J., 1965. 'A new class of faults and their bearing on continental drift', *Nature*, **207**, 343–347.

Windley, B. F., 1973. 'Crustal development in the Precambrian', *Phil. Trans. R. Soc. Lond.*, **273**, 321–341.

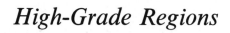
High-Grade Regions

Chemical Composition and Origin of Archean Granulites and Charnockites

K. S. HEIER

Norges geologiske undersøkelse, Box 3006, 7001 Trondheim, Norway.

Abstract

Some granulite facies terrains of Archean age are discussed. It is argued that sufficient evidence does not exist today to separate Archean and Proterozoic granulites on chemical grounds.

As a first step towards the solution of this problem, the following questions must be answered.

(1) Is the mode of Archean granulite facies metamorphism radically different from that of younger terrains?

(2) Are the geochemical characteristics recognized in medium- to high-pressure granulite facies rocks characteristic of the chemistry of the early protocrust rather than a product of high-grade regional metamorphism?

It is suggested that a 'working group' should be established with the mandate to attempt a correlation of Norwegian, Scottish and Greenland-Archean granulites.

Introduction

The topic of this syposium is *The Early History of the Earth*, and the time interval to be discussed is the one between 2·5 b.y. ago and the time of Earth formation approximately 4·5 b.y. ago. This represents only slightly less than half Earth history. The Archean is hereby defined as rocks older than 2·5 b.y.

A literature survey concerning the age boundary between the Archean and the Proterozoic was presented by Rankama (1970). The boundary falls within two groups, viz., a low-age group ranging from about 2100 Ma to about 1700 Ma, and a high-age group ranging from about 3000 Ma to about 2300 Ma. Most of the low ages have been proposed by Precambrian stratigraphers in the U.S.S.R. The high ages are prefered by most stratigraphers in Canada, India and Australia, with some found to be acceptable in the U.S.S.R. Thus there is no consensus as to the exact age of this boundary.

I have been requested to restrict myself to a discussion of granulites and charnockites formed within the Archean. This is not meant to be a discussion of granulite facies rock geochemistry and origin in general but rather of the particulars of the very oldest of such rocks. For a general review of granulite facies rocks and their origin I can refer to an earlier paper (Heier, 1973). I presume that since Archean granulites and charnockites have been singled out it implies that they may be

159

different from similar rocks of Proterozoic or even younger ages. There is, I believe, some evidence for a chemical distinction between Archean ultramafics and younger ones, and thus presumably if such rocks are metamorphosed to granulite facies, they would display a similar chemical disparity. However, as far as I know, there are no obviously significant differences between the granulites of the Archean and the Proterozoic. The only way to determine this would be to have a good look at a large number of chemical data from well-dated rocks. Such data are not yet available. It is hoped that isotopic data may give some evidence of differences between them; however, regional heterogeneities in mantle composition could easily obscure any age-related differences.

In my attempt to make a worthwhile contribution to this topic I am grateful for advice rendered by my colleagues Dr R. L. Oliver (Adelaide), Dr A. Glikson (Canberra), Professor A. F. Wilson (Brisbane) and Dr R. K. O'Nions (Oxford). In spite of their valuable advice and my own considerations, this paper cannot be more than a mere introduction to the subject.

For instance, are Archean granulites more dominantly of the low pressure–high temperature type than granulites of younger ages? I do not think that this question can be fully answered today but it should be easier to discuss than the question of chemical pecularities.

One difficulty naturally arises from the mere problem of obtaining reliable dates from old metamorphic terrains. In a number of areas granulite facies rocks of Archean age occur in polymetamorphic terrains, and one is invariably faced with the duplicate problem of ascertaining that the 'age' obtained is real and that the rock chemistry has not been severely altered during subsequent metamorphic episodes. The analytical understanding of metamorphic age patterns is therefore brought to its fullest complexity in this discussion.

The first step towards an understanding of the problem that one could make is to map out the different areas of granulite facies rocks in the world. This was done by Oliver (1969) and Heier (1973). Their maps show the world-wide distribution of granulite facies rocks; however, which of these are of Archean or of younger age is uncertain. I would like to take the opportunity here to ask for information concerning this question, and that this information be forwarded to me for later publication. The information I would like is shown in Table 1.

If a compilation of this sort could result from this meeting and volume I will consider my contribution not wasted and, indeed, worthwhile. I will now describe some fairly well-authenticated areas of Archean granulites.

Lofoten–Vesterålen, North Norway

I will first turn your attention to the area I am most familiar with myself; the

Table 1. Information on granulite facies rocks

Area:	Dated by: (author/methods)	Petrology by:	Geochemistry by:	Information given by:

Comments

Lofoten–Vesterålen area in North Norway. A number of papers have described the geology, geochronology, petrology and chemistry of the area. The most recent ones are by Devaraju and Heier (1974), Griffin et al. (1974) and Taylor (1975). These papers give references to most of the existing literature.

The paper by Taylor (1975) concentrates on the geochronology of the area, and it is shown that whole rock lead model ages of 3.460 ± 70 m.y. (3.5 b.y.) occur within a restricted area of 'migmatitic gneisses'. The gneisses represent part of a sequence that extends all over the Lofoten–Vesterålen area. There is ample geological and structural evidence to demonstrate that the sequence contains the oldest rocks in the area.

Rb/Sr age data by Taylor (1975) on the same rocks indicate an 'age' of $2300 + 150$ m.y. and an initial Sr-isotope ratio of $0.7126 + 0.0011$. This is a higher value than has previously been reported from any Precambrian granulites, but is in general accord with the higher age indicated by the Pb/Pb data.

The earlier dating by Heier and Compston (1969) did not recognize the subdivision of the gneisses into two sequences; an older 'migmatitic' sequence, and a younger 'supracrustal' sequence.

Heier and Compston (1969) gave an age interpretation based on the same geological information, but using more Rb/Sr age data than was available to Heier (1960). Taylor (1975) has combined his own and earlier age data with the new geological maps which have become available during the past 5 years as the result of mapping carried out under the supervision of Dr W. L. Griffin.

At present the general consensus of the geological history is:

(1) Formation of the precursors of the Vikan gneisses probably as intermediate igneous rocks. (Pre- 3.460 ± 70 m.y.)
(2) High-grade metamorphism culminating in partial melting and migmatite formation at Vikan. (3.460 ± 70 m.y.)
(3) Granulite facies metamorphism, depletion of Rb (2.300 ± 150 m.y.)
(4) Uplift of Vikan gneisses to the surface. Deposition of a supracrustal sequence consisting of sediments (mainly chemical precipitates) and volcanics.
(5) Deep burial dehydration and second granulite facies metamorphism, accompanied by deformation and intrusive activity. (1.900–1.700 m.y.)

With respect to the chemical composition of the gneisses, the older group represents a migmatitic sequence for which it is not possible to demonstrate a supracrustal origin unequivocally, while the younger group represents a set of paragneisses (Griffin, pers. com.). Both sequences have mineral compositions representative of medium- to high-pressure granulite facies rocks (but are also possibly represented by amphibolite facies rocks further to the north east). The granulite facies rocks of both age groups are characterized by high K/Rb ratios, low Rb/Sr, K/U and Th/U ratios. REE pattens of the rocks are inconclusive.

As far as additional isotopic geochemistry is concerned, one interesting aspect is the indicated lack of U (relative to Pb) depletion in the old Langøy rocks. Though this characteristic cannot be said to be convincingly demonstrated by the Norwegian granulites (where frequently the opposite seems to be true), the Greenland granulites generally exhibit this characteristic.

Before we leave the topic of the 3.5 b.y. old Norwegian granulites, I must emphasize that not only are they the oldest rocks in Europe, but they also occupy a very interesting position along the continental margin of Northern Europe.

The Lewisian of Scotland

The recent papers by Holland and Lambert (1973) and Drury (1974) give most of the references to earlier work and the most up-to-date review of our problem in this area. While the paper by Drury discusses mainly chemical effects imposed by retrogressive metamorphism of early granulite facies complexes,

Holland and Lambert give a broad discussion of direct interest to the problem we discuss here.

A general map of the area is given by Holland and Lambert (1973, Fig. 1). Three petrochemical assemblages can be recognized. The Scourie assemblage is supposed to represent a primitive intermediate crust which is characterized by a depletion in Rb, U, Th and which fractionated under 'meta-plutonic' conditions more than 2600 m.y. ago. The Scourie and Inver assemblages are, apart from differences in volatile constituents, essentially isochemical. What is most interesting in this area, without going into details, is the relation between these units and the Laxfordian, and a comparison of these relations with the Archean gneiss sequence of the Lofoten–Vesterålen island discussed above . It appears that here in the mainland of Scotland we also have an old protocrust as represented by the Scourian gneisses, mixed with a younger set of gneisses, the Laxfordian, which can be regarded as a set of paragneisses consisting of supracrustal metasediments whose formation substantially post-dates the initial formation of the Scourie assemblage. As far as I can understand it, the present consensus of opinion is that the deposition of the Laxfordian sediments and their subsequent metamorphism took place in the time interval of 2200–2600 m.y. ago. The Laxfordian metamorphism is, however, one of amphibolite facies rather than granulite facies.

One of the important points raised in the paper by Holland and Lambert is whether the primitive intermediate-siliceous crust was depleted in Rb, U and Th or whether this is a secondary effect caused by granulite facies metamorphism. This is a point well worth taking up for discussion in this symposium, but I don't think it can be answered definitely yet. The evidence today seems to indicate that it is rather a general characteristic of granulite facies rocks irrespective of age. Though outside the scope of this symposium, it should be added that most of the younger events recognized in the Lewisian of mainland Scotland are also observed in North Norway.

West Greenland

Moving from Norway and Scotland into Greenland the problem becomes even more complex. The oldest reliable rock dates on Earth were reported from this area by Black et al. (1971), Moorbath et al. (1972) and Pankhurst et al. (1973). The oldest ages refer to amphibolite facies rocks and are recorded as 3700 to 3750 m.y. old with a fairly low initial $^{87}Sr/^{86}Sr$ ratio of 0·7010–0·7015. It should be noted, however, that these are Rb/Sr ages. The high initial $^{87}Sr/^{86}Sr$ ratios of 0·7126 reported from the Lofoten–Vesterålen area were on rocks which indicate Rb/Sr ages of approximately 2·300 m.y. but where the whole rock Pb model ages were about 3·500 m.y. Thus these provinces in Greenland and Norway may well be contemporaneous.

The granulite facies metamorphism of the Fiskenæsset area appears to be younger. Pb/Pb ages of this metamorphism are calculated around 2·900 m.y. (Black et al., 1973). The geochemistry of the Fiskenæsset complex granulite facies rocks is well described in a paper by Windley and Smith (1974). According to these authors the Fiskenæsset complex is an igneous intrusion and thus not directly comparable with the rocks described above from Norway and Scotland. In order to be fitted into a general pattern the Greenland rocks need more study in particular with respect to correlation between geologic field interrelations, petrologic descriptions, systematic dating and geochemistry. These rocks will be discussed in more detail in this symposium, and it appears premature in a 'presymposium' paper which is given here to go into too much detail.

I propose that a 'working group' be established after this meeting with the object of making an attempt to draw together the data from Norway, Scotland and Greenland into one story.

Australia

The Australian continent is well known for its granulite facies terranes. A number of these areas are of the high temperature/low pressure type and of Proterozoic age. Lambert and Heier (1968a,b) published on the geochemistry of granulite facies rocks from different parts of the continent. Wilson (written communication) confirms that 'the only Archean granulites yet recognized in Australia are those found in the Wheat Belt of Western Australia'. He further states that 'as far as I know there are no significant differences between the granulites of the Archean and the Proterozoic'. The same view is taken by Glikson who, in a written communication, states 'As you know, with the notable exception of Scotland, few chemical data are published on well-dated Archean granulites, as contrasted to Proterozoic granulites. Thus, little was added to the knowledge of the chemistry (particularly trace elements) of the Wheat Belt granulites following yours and Lambert's work, although unpublished theses in the University of Western Australia include some excellent petrological studies of these rocks'.

Other Provinces

I am thankful to Dr Glikson for having turned my attention to the Vol. 15, No. 4 issue of the *Journal of the Geological Society of India*. This issue contains several papers on the Geochemistry of the Precambrian rocks of India, and one of the papers includes information on the chemistry of charnockites. He also turned my attention to an abstract by Heuter and Bocher in the recent conference on 'Geology and Geochemistry of the Earliest Precambrian rocks (Minnesota)' where 3·4 b.y. amphibolite facies gneisses were reported. I don't know if there are granulite facies rocks in this area since the abstract has not yet become available to me.

Conclusion

If the purpose of my contribution here was to point out pecularities of Archean granulites as compared with granulites of younger ages, I have failed. At present I cannot see that there is any defined chemical distinction between granulite facies rocks of various age groups. However, the exercise of thinking through the problem has helped me to raise, rather than to conclude, the following two questions:

(1) Is the mode of Archean granulite facies metamorphism radically different from that of younger terranes?
(2) Are the geochemical characteristics recognized in medium- to high-pressure granulite facies rocks characteristic of the early protocrust or are they a product of high-grade regional metamorphism?

I further want to make the suggestion that (i) the information asked for in Table 1 is distributed and returned to the author for later publication; (ii) a discussion of 'style of granulite facies metamorphism with time' (e.g. PT relationships) is encouraged; (iii) a 'working group' is established to attempt a correlation of Norwegian, Scottish and Greenland-Archean granulites.

References

Black, L. P., Gale, N. H., Moorbath, S., Pankhurst, R. J., and McGregor, V. R., 1971. 'Isotopic dating of very early Precambrain amphibolite facies gneisses from the Godthaab district, West Greenland', *Earth Planet. Sci. Lett.*, **12**, 245.

Black, L. P., Moorbath, S., Pankhurst, R. J., and Windley, B. F., 1973. '207Pb/206Pb whole rock age of the Archean granulite facies metamorphic event in West Greenland', *Nature, Phy. Sci.*, **244**, 50–53.

Devaraju, T. C., and Heier, K. S., 1974. 'Precambrian rocks on Hadseløy, Lofoten-Vesterålen', *Norg. Geol. Unders., Bull.*, **312**, 26, 31–58.

Drury, S. A., 1974. 'Chemical changes during retrogressive metamorphism of Lewisian granulite facies rocks from Coll and Tiree', *Scott. J. Geol.*, **10**, 237–256.

Griffin, W. L., Heier, K. S., Taylor, P. N., and Weigand, P. W., 1974. 'General geology, age and chemistry of the Raftsund mangerite intrusion, Lofoten-Vesterålen', *Norg. Geol. Unders. Bull.*, **312**, 1–30.

Heier, K. S., 1960. 'Petrology and geochemistry of highgrade metamorphic and igneous rocks in Langøy, Northern Norway', *Norg. Geol. Unders.*, **207**, pp. 246.

164

Heier, K. S., 1973. 'Geochemistry of granulite facies rocks and problems of their origin', *Phil. Trans. R. Soc. Lond.*, **A273**, 429–442.

Heier, K. S., and Compston, W., 1969. 'Interpretation of Rb–Sr age patterns in high-grade metamorphic rocks, North Norway', *Norsk Geol. Tidsskr.*, **49**, 257–283.

Holland, J. G., and Lambert, R. St. J., 1973. 'Comparative major element geochemistry of the Lewisian of the mainland of Scotland', In *The Early Precambrian of Scotland and Related Rocks of Greenland*, R. G. Park and J. Tarney (Eds.), University of Keele, **1973**, 51–62.

Lambert, I. B., and Heier, K. S., 1968a. 'Chemical investigations of deep-seated rocks in the Australian Shield', *Lithos.*, **1**, 30–53.

Lambert, I. B., and Heier, K. S., 1968b. 'Estimates of crustal abundances of thorium, uranium and potassium', *Chem. Geol.*, **3**, 233–238.

Moorbath, S., O'Nions, R. K., Pankhurst, R. J., Gale, N. H., and M. Gregor, V. R., 1972. 'Further rubidium–strontium age determinations on the very early Precambriam rocks of the Godthaab District, West Greenland', *Nature, Phys. Sci.*, **240**, 78.

Oliver, R. L., 1969. 'Some observations on the distribution and nature of granulite-facies terrains', *Geol. Soc. Australia, Special Publ.*, No. 2, 259–268.

Pankhurst, R. J., Moorbath, S., Rex, D. C., and Turner, G., 1973. 'Mineral age patterns in ca. 3700 m.y. old rocks from West Greenland', *Earth Planet. Sci. Lett.*, **20**, 157–170.

Rankama, K., 1970. 'Proterozoic, Archean and other weeds in the Precambrian rock garden', *Bull. Geol. Soc. Finland*, 42, 211–222.

Taylor, P. N., 1975. 'An early Precambrian age for migmatitic gneisses from Vikan in Bø, Vesterålen, North Norway', *Earth Planet. Sci. Lett.* (in press).

Windley, B. F., and Smith, J. W., 1974. 'The Fiskenæsset complex, West Greenland, Part II. General mineral chemistry from Qeqertarssuatsiaq', *Grønland Geol. Unders. Bull.*, **108**, pp. 54 (also. *Meddr. Grønland, 196*, 4).

The Early Precambrian Gneiss Complex of Greenland

JOHN S. MYERS

Geological Survey of Greenland,
Øster Vodgade 10, DK 1350, Copenhagen, Denmark.

The main features of the early Precambrian complex are summarized with emphasis on the Isua–Godthåb–Fiskenæsset region of West Greenland. The account is essentially a synopsis of a more detailed description of these rocks by Bridgwater et al. (in press).

Early Precambrian rocks which have been stable since 2600 m.y. ago outcrop for 600 km on the west coast and 200 km on the east coast of southern Greenland (Fig. 1). To the north and south, similar rocks were reworked in the Nagssugtoqidian and Ketilidian mobile belts during the Proterozoic era.

More than 80% of the early Precambrian complex consists of quartzo-feldspathic gneisses considered to be derived from intrusive granitic rocks emplaced during at least two major episodes of plutonism about 3700 and 3100 m.y. ago. Supracrustal rocks, mainly amphibolites derived from volcanics, with smaller amounts of semi-pelitic gneisses, quartzite and calcareous rocks, make up 15% of the complex, and layered basic igneous complexes dominated by leucogabbro and anorthosite make up the remaining 5%.

Table 1. Sequence of major events in the Godthåb–Isua and Fiskenæsset regions

Godthåb–Isua	Fiskenæsset
Qôrqut granite, 2600 m.y.	
F_3, end of granulite and amphibolite facies metamorphism, 2850 m.y.	
F_2	
	Ilivertalik augen granite 2850 m.y.
Intrusion and deformation (F_1) of calc–alkaline Nûk gneisses, 3040 m.y.	Major intrusion and deformation (F_1) of calc–alkaline gneisses
Layered anorthosite complexes, > 3040 m.y.	
Supracrustal amphibolites and minor metasediments, > 3040 m.y.	
– – – – – – – – Relative age unknown – – – – – – – – Ameralik dykes (D_2), > 3040 m.y.	
Intrusion and deformation (D_1) of calc–alkaline Amîtsoq gneisses 3750 m.y.	
Isua supracrustals 3750 m.y.	
? Older granitic crust	

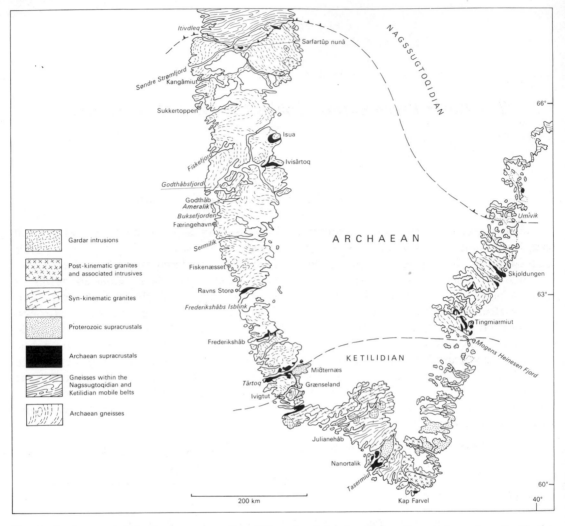

Fig. 1. Outline map of the Archaean complex of Greenland showing the location of places described in the
text

1. Isua Supracrustal Rocks and Gneisses

The oldest known rocks are quartzo-feldspathic gneisses and a sequence of metavolcanics and quartz-banded ironstones which occur at Isua (Fig. 1, Table 1), (Allaart, this volume). The ironstones have given a Pb/Pb isochron age of 3760 ± 70 m.y. (Moorbath et al., 1973), and some of the gneisses may be older than the Isua supracrustals (James, in press), although they have given a Rb/Sr whole rock isochron age of 3700 ± 140 m.y. (Moorbath et al., 1972). Both the gneisses and supracrustal rocks at Isua are cut by two groups of metadolerite dykes (Fig. 2), correlated with the Ameralik dykes near Godthåb (Bridgwater and McGregor, 1974).

2. Amîtsoq and Nûk Gneisses

The Ameralik dykes near Godthåb (Fig. 3) were used by McGregor (1973) to distinguish between two groups of gneisses: Amîtsoq gneisses which are older than the dykes, and Nûk gneisses which intrude and disrupt the dykes and Amîtsoq gneisses. This division was substantiated by Rb/Sr whole rock isochrons

Fig. 2. Isua supracrustal rocks (right foreground) and gneisses cut by two groups of Ameralik dykes (after Bridgwater et al., in press)

which showed ages of 3700–3750 m.y. for the Amîtsoq gneisses (Moorbath et al., 1972) and 3040 ± 50 m.y. for the Nûk gneisses (Pankhurst et al., 1973). Beyond the Godthåb–Isua region, the quartzo-feldspathic gneisses can in some places be divided into different intrusive groups based on cross-cutting and compositional relationships. Many of these gneisses are cut by amphibolite dykes, but the dykes appear to be less numerous than in the Godthåb–Isua region. It has not yet been possible to correlate the Amîtsoq and Nûk gneisses and Ameralik dykes of the Godthåb–Isua region with similar rocks elsewhere because the rocks lack distinctive lithological features, and relatively little isotopic work has been done elsewhere in the early Precambrian complex.

3. Amphibolites

The most abundant supracrustal rocks in the early Precambrian complex are amphibolites. They occur throughout the complex as strips and trains of fragments in the quartzo-feldspathic gneisses. In the Godthåb region the major amphibolite units were called Malene supracrustals by McGregor (1973)

and are veined and fragmented by the Nûk gneisses. Contacts between Malene supracrustals and the Amîtsoq gneisses are generally tectonic and the relative age of these units is unknown. Elsewhere, amphibolite units similar to the Malene supracrustals are extensively veined by gneisses (Fig. 4).

Most of the amphibolites are hornblende–plagioclase rocks with a granoblastic texture. They are generally massive or show poorly defined discontinuous compositional banding. In a few places where they are least deformed, they contain relict pillow lava, pillow lava breccia and agglomerate structures (Fig. 5), which in the Fiskenæsset region are cut by large numbers of thin amphibolite dykes and sills. In some places they contain thicker gabbroic and ultramafic sills with rhythmic igneous layering. All stages can be seen in the conversion of the pillow lava and agglomerate structures by deformation and recrystallization into amphibolite with lenses of leucocratic amphibolite and poorly defined compositional layering of the type which is widespread throughout the Fiskenæsset region (Fig. 6). A large proportion of the amphibolites thus appears to be derived from basic sub-aqueous volcanic rocks.

168

Fig. 3. Ameralik dyke cutting pegmatite-banded Amîtsoq gneiss south of Godthåb (after McGregor, 1973)

4. Metasedimentary Rocks

Small amounts of metasedimentary rocks occur throughout the early Precambrian complex. Most of these rocks occur in association with the amphibolite units and appear to have originally accumulated with the volcanic rocks. The most abundant metasedimentary rocks are rusty brown weathering quartz–plagioclase rocks with abundant biotite and various amounts of potash feldspar, garnet, staurolite, cordierite, sillimanite and amphiboles. Other rock types include quartz–cordierite gneisses, quartz–plagioclase–anthophyllite gneiss, and quartz-rich gneisses containing various amounts of garnet, sillimanite and magnetite.

5. Anorthosites

Metamorphosed anorthosites and associated gabbroic rocks form one of the most distinctive rock units in the early Precambrian gneiss complex. They occur throughout the complex as concordant layers in amphibolites and as layers and trains of inclusions in quartzo–feldspathic gneisses, and they provide one of the best marker horizons for tracing out regional structures (Fig. 7). They extensively preserve igneous textures and in some places igneous minerals and way-up structures (Figs. 8 and 9).

These rocks are most abundant and best preserved in the Fiskenæsset region where they also contain chromite layers (Ghisler, in

Fig. 4. Amphibolite fragments in gneiss derived from intrusive granodiorite, north-east of Fiskenæsset

press). This feature led Windley (1969) to conclude that they represent a stratiform basic igneous complex which he called the Fiskenæsset complex. Original igneous contact relationships are rarely seen but show that the anorthosite complex intruded basic volcanic rocks (Escher and Myers, 1975).

6. Structures

The main structural feature of the early Precambrian complex of Greenland, which is apparent from the scale of regional maps to single outcrops, is the predominance of layered rocks. On a large scale the layering is mostly the result of interleaving of different rock units of different age and geological provenance by a combination of magmatic injection and tectonic transport (Bridgwater et al., 1974) (Fig. 10). On a small scale much of the layering is the result of intense deformation of original inhomogeneities in the rocks such as pillow lava structures (Figs. 5 and 6),

large igneous crystals, inclusions, and cross-cutting veins and dykes. Some layering also formed by rapid changes in the degree of deformation from place to place and within one outcrop (Fig. 11). Once zones of intense deformation were established they tended to be reworked during subsequent deformation episodes.

The oldest known deformation structures are strongly marked planar fabrics in the supracrustal rocks at Isua and in some of the surrounding gneisses (D_1). The fabric generally parallels the bedding of the supracrustal rocks, it is accompanied by a steeply plunging linear component, and was produced by simple shear with vertical displacement of the gneisses to the north by 7 km (James, in press). This deformation may be contemporary with the dates of about 3760–3700 m.y. obtained from these rocks by Moorbath et al. (1972, 1973). These fabrics are refolded by tight and isoclinal folds with steep axial surfaces (D_2) associated with intrusion of the Ameralik

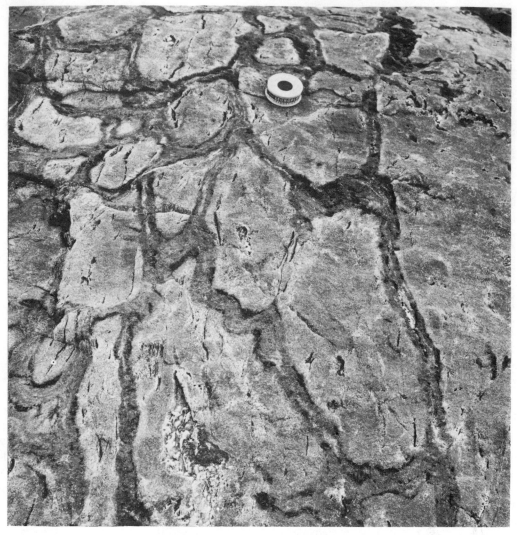

Fig. 5. Relatively little deformed pillow lava structures, Bjørneøen, Godthåbsfjord (after Bridgwater et al., in press)

dykes (James, in press). The relation between the structures at Isua and those elsewhere in the early Precambrian complex is unknown.

The oldest major structures recognized in most of the early Precambrian complex are recumbent isoclinal folds or nappes, and thrusts (F_1). These accompanied and followed the intrusion of the Nûk gneisses 3040 m.y. ago in the Godthåb region and of similar granitic intrusions elsewhere. Much of the granitic material was intruded as sheets associated with thrusting. This can most clearly be seen in the Fiskenæsset region where the layered anorthosite complex provides primary structures and a distinct stratigraphy by which the thrust movements can locally be measured. Deformation continued after the intrusion of enormous amounts of granitic sheets and major recumbent isoclinal folds were formed, probably as nappes. Deformation fabrics from this episode are strongly marked and are mostly planar. The layered nature of the early Precambrian gneiss complex was accentuated on both large and small scales, and over wide areas the layering was sub-horizontal.

Fig. 6. Amphibolite with discontinuous melanocratic and leucocratic layers derived by strong deformation from pillow lava structures, north-east of Fiskenæsset

The flat-lying layering was refolded twice by folds with steep axial surfaces at high angles to each other, and dome-and-basin interference patterns were formed in a large part of the early Precambrian complex. These structures are well developed in the Fiskenæsset region (Fig. 7). The first of these folds (F_2) are tight or isoclinal and were accompanied by the development of very strongly marked planar fabrics. In much of the Fiskenæsset region the axial surfaces of the F_2 folds trend east–west and dip steeply southwards. The second group of folds (F_3) have steep axial surfaces which trend NW–SE. The folds generally have one steep attenuated limb and a flat-lying, less deformed limb. Planar fabrics developed locally in the steep attenuated fold limbs, and a lineation was widely developed by small-scale crinckling of thin layering. Throughout the Fiskenæsset region this lineation plunges gently south-eastwards whereas in the Godthåb region to the north, a similar lineation plunges moderately south-westwards. It is not yet known whether this variation is an original feature or was the result of later

reorientation. The deformation of this episode was most intense in a broad linear belt which trends NE–SW from Godthåbsfjord to the region of Buksefjorden, in which the main layering of the gneisses became sub-parallel with the belt.

7. Early Precambrian Crustal Evolution

A granitic crust was probably already in existence before the oldest known supracrustal rocks were deposited at Isua. These supracrustal rocks are mainly intermediate and basic volcanic rocks, sills, volcanoclastic sediments and chemical precipitates. They were intruded by a group of heterogeneous tonalites and granodiorites which form much of the Amîtsoq gneisses.

Near Godthåb the Amîtsoq gneisses include rocks derived from homogeneous porphyritic granite which are considered by McGregor (in Bridgwater et al., in press) to be the oldest component of the Amîtsoq gneisses and to represent remnants of the Earth's protocrust. Chadwick et al. (1974), however, note that

172

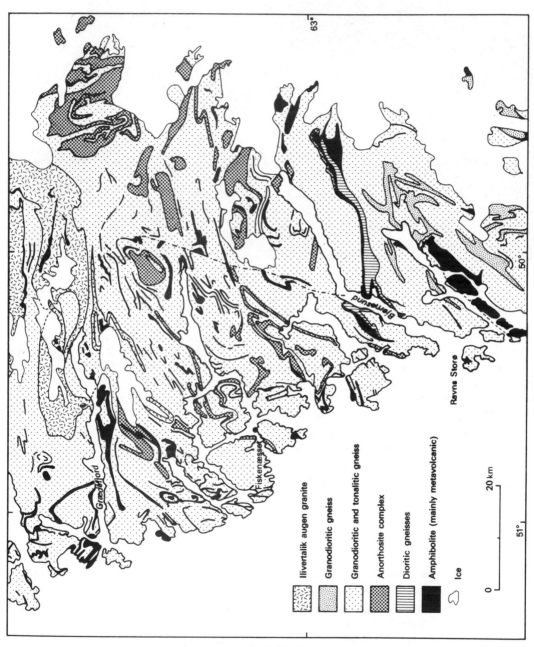

Iivertalik augen granite

Granodioritic gneiss

Granodioritic and tonalitic gneiss

Anorthosite complex

Dioritic gneisses

Amphibolite (mainly metavolcanic)

Ice

0   20 km

Fig. 7. Simplified map of the Fiskenæsset region, after published and unpublished GGU reports

Grædfjord

Fiskenæsset

Bjørnesund

Ravns Storø

63°

50

51°

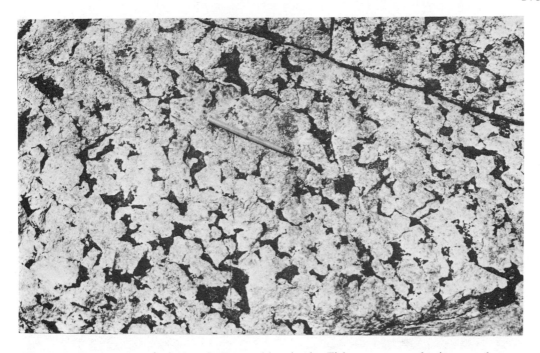

Fig. 8. Relict cumulus plagioclase in leucogabbro in the Fiskenæsset anorthosite complex at Majorqap qâva, north-east of Fiskenæsset

Fig. 9. Composition-graded layers in uniform gabbro in the Fiskenæsset anorthosite complex at Majorqap qâva, north-east of Fiskenæsset

Fig. 10. Units of metavolcanic amphibolite (A), intruded by a sheet of anorthosite (An), and by quartzo-feldspathic gneiss (Gn), east of Fiskenæsset

Fig. 11. Leucogabbro schist, below, derived by deformation from leucogabbro similar to that above with relict igneous texture. At Majorqap qâva, north-east of Fiskenæsset

these granites cut folded banding of Amîtsoq gneiss and regard these granites as one of the youngest components of the Amîtsoq gneisses.

All these rocks are older than about 3700 m.y. when they were deformed and metamorphosed. They are cut by a dense swarm of basic dykes, the Ameralik dykes, which at Isua were associated with regional deformation.

Most of the amphibolites of the early Precambrian complex appear to have been derived from sub-aqueous basic and intermediate volcanic rocks older than 3040 m.y. They were once thought to have accumulated upon older granitic crust such as the Amîtsoq gneisses (McGregor, 1973; Windley et al., 1973), but this hypothesis is now in doubt (Kalsbeek and Myers, 1973). An alternative hypothesis is that they originated in a different place to the Amîtsoq gneisses, as either oceanic crust or on a thin continental crust, and were tectonically interleaved (Bridgwater et al., 1974).

The amphibolites were intruded by basic dykes, basic and ultramafic sills, and layered sills of leucogabbro, gabbro and anorthosite with minor amounts of ultramafic rocks. These intrusions may have been emplaced during the same magmatic episode in which the metavolcanic amphibolites were erupted.

All these rocks were intruded by large volumes of calc–alkaline magmas, mainly as sheets, about 3040 m.y. ago. Intrusion was associated with deformation, mainly thrusting, and together this led to considerable thickening of the granitic crust (Bridgwater et al., 1974).

The earliest nappe-like folds (F_1) may have formed under the influence of gravity during isostatic uplift of this thickened granitic crust. They may have been refolded (F_2) whilst still ductile as the nappes piled up against more rigid crust. As uplift waned and the pile of granitic sheets and older rocks stiffened, it was deformed by steep shear zones (F_3) which were associated with dehydration of the lower part of the early Precambrian complex 2800 m.y. ago, and led to final stabilization of this portion of continental crust.

Acknowledgements

This paper is published with the permission of the Director of the Geological Survey of Greenland.

References

Bridgwater, D., and McGregor, V. R., 1974. 'Fieldwork on the very early Precambrian rocks of the Isua area, southern West Greenland', *Rapp. Grønlands geol. Unders.*, **65**, 49–54.

Bridgwater, D., McGregor, V. R., and Myers, J. S., 1974. 'A horizontal tectonic regime in the Archaean of Greenland and its implications for early crustal thickening', *Precambrian Res.*, **1**, 179–197.

Bridgwater, D., Keto, L., McGregor, V. R., and Myers, J. S., in press, 'Archaean gneiss complex of Greenland', In *Geology of Greenland*, Watt, W. S., and Escher, A. E. (Eds.), Grønlands geol. Unders., Copenhagen.

Chadwick, B., Coe, K., Gibbs, A. D., Sharpe, M. R., and Wells, P. R. A., 1974. 'Mapping of the Precambrian basement in the Buksefjorden region, southern West Greenland', *Rapp. Grønlands geol. Unders.*, **65**, 54–60.

Escher, J. C., and Myers, V. S., 1975. 'Early Precambrian volcanics and anorthosite in the Fiskenæsset region, southern West Greenland', *Rapp. Grønlands geol. Unders*, **75**.

Ghisler, M., in press, *The Geology, Mineralogy and Geochemistry of the Pre-orogenic Archaean Stratiform Chromite Deposits at Fiskenæsset, West Greenland*, Borntraeger Verlag, Berlin.

James, P. R., in press, 'Deformation of the Isua block, West Greenland: a remnant of the earliest stable continental crust', *Can. J. Earth Sci.*

Kalsbeek, F., and Myers, J. S., 1973. 'The geology of the Fiskenæsset region', *Rapp. Grønlands geol. Unders.*, **51**, 5–18.

McGregor, V. R., 1973. 'The early Precambrian gneisses of the Godthåb district, West Greenland', *Phil. Trans. R. Soc. Lond.*, **A273**, 343–358.

176

Moorbath, S., O'Nions, R. K., Pankhurst, R. J., Gale, N. H., and McGregor, V. R., 1972. 'Further rubidium–strontium age determinations on the very early Precambrian rocks of the Godthåb district, West Greenland', *Nature phys. sci.*, **240**, 78–82.

Moorbath, S., O'Nions, R. K., and Pankhurst, R. J., 1973. 'Early Precambrian age for the Isua iron formation, West Greenland', *Nature, Lond.*, **245**, 138–139.

Pankhurst, R. J., Moorbath, S., and McGregor, V. R., 1973. 'Late event in the geological evolu-tion of the Godthåb district, West Greenland', *Nature phys. sci.*, **243**, 24–26.

Windley, B. F., 1969. 'Evolution of the early Pre-cambrian basement complex of southern West Greenland', *Spec. Pap. geol. Ass. Canada*, **5**, 155–161.

Windley, B. F., Herd, R. K. and Bowden, A. A., 1973. 'The Fiskenæsset complex, West Green-land, Part I. A preliminary study of the strati-graphy, petrology, and whole-rock chemistry from Qeqertarssuatsiaq', *Bull. Grønlands geol. Unders.*, **106** (Also *Meddr Grønland*, 196, **2**), 80 pp.

The Pre-3760 m.y. Old Supracrustal Rocks of the Isua Area, Central West Greenland, and the Associated Occurrence of Quartz-banded Ironstone

JAN H. ALLAART

The Geological Survey of Greenland, Copenhagen, Denmark.

Abstract

The Isua supracrustal belt forms an incomplete oval-shaped arc between 12 and 25 km in diameter around a dome-shaped area of gneiss. The distribution of the lithologies does not show a bilateral symmetry and the belt does not appear to be a rim syncline surrounding a dome structure. It is more probable that an early phase of deformation produced a major fault slice comprising the supracrustal belt which was buckled along a north-east axis by a later phase of deformation. At least two generations of isoclinal folds have been recognized. The axes of both generations are almost parallel and plunge steeply to the south-east.

There is a great variety of rock types. The most common is an iron-rich amphibolite with tholeiitic affinities. An important unit consists of siliceous and carbonate-bearing biotite–muscovite schists which contain large numbers of strongly stretched fragments of metarhyolite, some of which are rich in potassium. These schists are undoubtedly of volcanic origin and might represent submarine or subaerial ash flows. Conformable bodies of talc schist with relics of dunite are common.

Biotite–garnet schists are abundant in many places and locally contain staurolite and graphite. Quartz-banded ironstones are of widespread occurrence. Calc–silicate rocks and lenses of marbles of considerable size occur in a few places and small intercalations of carbonate-bearing rock occur throughout the supracrustal belt. In the north-east a horizon of conglomerate occurs with deformed cigar-shaped pebbles of quartzite.

The metamorphic grade is mainly low pressure amphibolite facies, but retrograde metamorphism to greenschist facies has occurred in many places.

A large body of basic amphibole–chlorite–albite schist with garbenschiefer texture intruded the supracrustal rocks during declining metamorphic conditions.

Introduction

The supracrustal rocks at Isua, central West Greenland, which contain a major deposit of banded ironstone were discovered in 1966 by the geologists of Kryolitselskabet Øresund A/S and mapped in great detail (Keto, 1970). A Pb/Pb whole rock isochron age of

3760 m.y. for the banded ironstone has been obtained by Moorbath et al. (1973). The supracrustal belt consists of an incomplete oval-shaped arc around a dome-shaped area of gneiss (Fig. 1). This gneiss has given a Rb/Sr whole rock isochron age of 3700 ± 140 m.y. (Moorbath et al., 1972). To the north-west the arc of supracrustal rocks is cut off by a NE–SW trending fault. In the eastern part, the schistosity dips steeply to the south-east with steeply south to south-east plunging lineations and fold axes. In the western half, dips are vertical or steep to the south-east, and linear elements plunge steeply to the south-east.

In the Isua area several sets of well-preserved metadolerites cross-cut the supracrustals and gneisses (Fig. 2). Towards the south-west these dykes can be seen to be progressively broken up by folding and migmatization. These dykes are correlated with the Ameralik dyke relics in the Amîtsoq gneisses around Godthåb.

Relations to Surrounding Gneisses

In general the supracrustal rocks are remarkably little affected by the mobilization phenomena seen in the surrounding gneisses. Only locally do thin veins of gneiss penetrate the schists and amphibolites. Close to the contacts thicker light-coloured pegmatite sills and lenses occur in the dark-coloured supracrustal rocks.

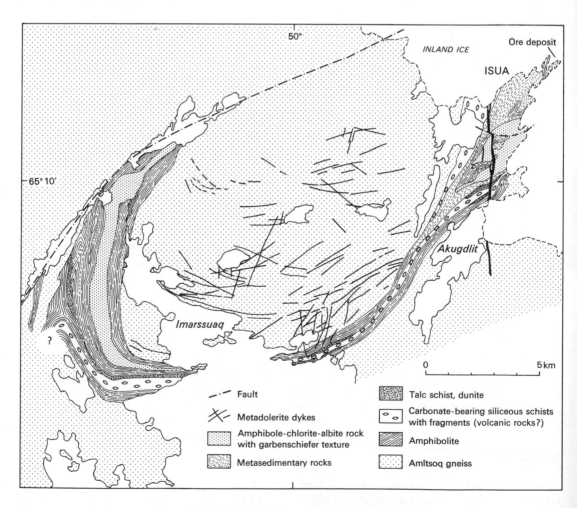

Fig. 1. Map of the Isua supracrustal belt

Fig. 2. Two sets of metadolerites (Ameralik dykes?) in the gneisses. There is a radial set which runs diagonally and a tangential set (parallel to horizontal margin of the picture) of which many dykes are boudinaged

Lithostratigraphy

The slight degree of migmatization suggested that there might be a good chance of obtaining clear-cut information about the internal disposition of the belt and whether it really formed a synform as had been suggested by Keto (1970) and Bridgwater and McGregor (1974). Unfortunately it became clear that the distribution of the lithologies does not show a general bilateral symmetry, although in the eastern part of the belt, north of the lake Akugdlit (see Fig. 1) where the belt reaches its maximum breadth of 3 km, a repetition of one of the units may occur. Here the following succession has been mapped from the south-eastern outer margin to the north-western inner margin of the belt: (1) Marginal amphibolites with occasional thin layers of garnet–biotite schist and magnetite-bearing banded quartzite. This unit is about 350 m wide. (2) Siliceous and carbonate-bearing biotite–muscovite schists (100–400 m wide) which in places are regularly banded with layers of large concentrations of strongly attenuated fragments. Most of the fragments consist of potassium-rich acid metavolcanic rocks. (3) In the succeeding unit, which ranges between 100 m and a few kilometres in width, a small proportion of amphibolites, arranged asymmetrically, occur with metasediments: carbonate rocks, carbonate–silicate rocks, garnet–biotite schists, garnet–staurolite micaschists, biotite–hornblende schists and quartzites which are magnetite-bearing and finely banded. Quartzites increase in abundance towards the main iron ore body which also makes up part of this unit near the margin of the Inland Ice. This unit also contains an important conglomerate horizon with deformed cigar-shaped pebbles in a matrix of banded, siliceous garnet–muscovite–biotite schist. Along the north-western side of unit 3 a 100–150 m broad zone of garnet-rich schists occurs. It has not been observed along the south-eastern side of the unit. The relations in unit 3 are partly obscured by a body of amphibole–chlorite–albite schist with garbenschiefer texture, which has a very irregular form with discordant contacts mainly intruding unit 3. (4) Siliceous and carbonate-bearing biotite–muscovite schists, with horizons of stretched acid metavolcanic fragments. These schists may constitute a repetition of unit 2, but this is impossible to prove conclusively.

Towards the south-west the belt quickly thins out to a width of about 800 m. Units 1, 2 and 3 continue towards the lake Imarssuaq, but unit 4 wedges out. Close to this lake unit 3 contains a much greater proportion of amphibolite than to the north-east. The transition where amphibolites start to predominate over metasediments occurs halfway between the lake and the Inland Ice.

Concordant zones of talc–chlorite schists with relics of dunite can be followed for several kilometres. East of Imarssuaq lake they are mainly concentrated along both sides of unit 2, the siliceous and carbonate-bearing schists with attenuated acid metavolcanic fragments.

West of lake Imarssuaq units 1, 2 and 3 reappear. Unit 3 is the only one which can be followed continuously to the north-western end of the arc where it is cut off by the NE–SW trending fault. In this part of the area. L. Keto (pers. comm. 1975) reported a great proportion of metasedimentary rocks. In the westernmost part of the belt there are complications with respect to units 1 and 2, where they appear to bend around to the south-west. The detailed relations have yet to be mapped out. At the inner side of unit 3 another broad concordant body occurs of the same amphibole–chlorite–albite schist with garbenschiefer texture as in the north-west. This body is followed by a zone of amphibolites with some bands of quartzites and associated banded ironstone, garnet–biotite schists and concordant ultrabasic bodies.

As the above description shows, a bilateral symmetry is not developed in the supracrustal belt and if this supracrustal belt is not a type of rim syncline around a gneiss dome, then the question arises as to how the present form originated.

Structural Development

At least two phases of deformation have been recognized in the supracrustal belt. James (1975) suggests that during the first the supracrustal belt formed an east-west trending fault slice and the 'inner' gneiss block to the north moved upwards about seven kilometres relative to the 'outer' gneiss block to the south. The first deformation phase produced a schistosity parallel to the original banding in the metasediments and a lineation and attenuation of the fragments in the acid to intermediate metavolcanics and the pebbles in the intraformational conglomerates. In different localities the fragments in the acid to intermediate metavolcanics show clear differences in degree of flattening. At a locality close to the east coast of Imarssuaq lake the matrix of these rocks is richer in biotite and carbonate than normal, and here the degree of flattening is clearly less than elsewhere in more siliceous parts of the unit. This difference can probably be explained by a slightly greater difference in competency between fragments and matrix.

It is in the most strongly flattened fragments that the folding produced by the second phase of deformation is especially clearly visible. At a few localities medium-scale isoclinal folds have been observed which deform the early schistosity. James (1975) has suggested that the buckling of the east-west trending supracrustal belt was produced during this second phase, and that a major fold hinge occurs in the north-east where the supracrustal belt bends to the south under the Inland Ice. Suggestions of a major fold hinge have also been observed in the westernmost part of the supracrustal belt and will be investigated during the 1975 summer season.

Petrography

Igneous Rocks

Amphibolites. Amphibolites are the most common rocks in the supracrustal belt. They are in general fine-grained and homogeneous. Relics of pillow lavas have as yet not been observed by the present author, although geologists of Kyrolitselskabet Øresund A/S have reported pillow structures in a small occurrence of amphibolite in the north-east close to the Inland Ice (L. Keto, pers. comm., 1975). Usually the amphibolites show a pronounced lineation which is most prominent close to the contacts with the surrounding gneisses. The freshest examples consist of about two-thirds of green hornblende and one-third of basic plagioclase (bytownite) and quartz. Alteration products of the plagioclase are clinozoisite–epidote and scapolite. Titanite is a widespread accessory. Chemically (see Table 1) the amphibolites have a basaltic composition. The total iron content varies between 10 and 15%, while the K_2O percentage is always greater than 0·2. The rocks show similarities with slightly oversaturated tholeiites, but they have a rather high iron content.

biotite, muscovite, microcline, plagioclase and in many places also calcite are the main constituents. The anorthite content of the plagioclase is variable. In the carbonate-rich samples the plagioclase is very basic, and in the carbonate-poor samples plagioclase is andesine. Microcline is usually not very conspicuous in the schists. The potassium is evidently mainly concentrated in the micas. Scapolite, epidote and clinozoisite are present in some places replacing plagioclase. Tour-

Table 1. Chemical Analyses of Igneous Rocks

	Amphibolites		Acid to intermediate metavolcanic rocks Matrix			Fragments			Amphibole–Albite –Chlorite Schists	
GGU nr	173269	172975	172904	172956	158451	172914	172954	172909	172952	173273
SiO_2	51·32	49·88	49·63	61·07	67·73	52·42	69·47	70·30	49·39	47·97
TiO_2	1·31	0·90	0·28	0·55	0·47	0·35	0·50	0·45	0·33	0·28
Al_2O_3	13·84	13·21	8·18	12·86	15·14	9·44	15·31	14·12	17·40	14·72
Fe_2O_3	3·01	2·80	0·97	0·93		0·80	0·64	0·52	0·71	1·41
FeO	11·05	10·30	4·25	3·94	3·05	4·49	2·14	0·98	8·08	8·68
MnO	0·23	0·29	0·48	0·17	0·07	0·43	0·06	0·06	0·24	0·22
MgO	4·38	8·20	5·39	2·68	1·77	4·69	1·25	1·18	11·65	15·95
CaO	10·53	10·83	12·66	5·02	5·88	11·32	3·17	1·22	7·98	6·35
Na_2O	2·08	2·13	0·04	0·19	2·43	0·00	3·14	0·13	0·69	1·03
K_2O	0·38	0·49	4·09	7·05	1·67	4·24	2·34	8·09	0·16	0·11
P_2O_5	0·18	0·16	0·21	0·22	0·18	0·22	0·14	0·13	0·17	0·13
CO_2			12·20	3·55		10·10		1·10		
H_2O^+	0·84	1·06	0·92	1·00	0·89	0·92	0·86	1·10	3·28	2·67
	99·15	100·25	99·30	99·23	99·29	99·42	99·02	99·38	100·03	99·52

Intercalations of metasedimentary rocks such as garnet–biotite schists, pure quartzites, quartz-banded ironstone and calc–silicate rocks, occur throughout the belt. These are rarely found close to the contacts with the gneisses, but become progressively more common away from it.

Siliceous and Carbonate-bearing Biotite–Muscovite Schists with Meta-rhyolitic Fragments. This unit is homogeneous and intercalations of other rock types are extremely rare. Just east of the lake Imarssuaq the northern half of the unit has a darker colour than the southern half. The more melanocratic part is richer in carbonate and biotite and the more leucocratic part is more siliceous. Quartz,

maline, apatite and opaque ore are accessories.

The siliceous and carbonate-bearing biotite–muscovite schists are in places regularly banded with layers containing large concentrations of strongly attenuated fragments (see Fig. 3). Most of the fragments consist of potassium-rich, acid metavolcanic rock in which phenocryts of quartz and microline can be recognized. Many of the quartz phenocrysts have an ellipsoidal shape. Well-developed microcline phenocrysts have been observed, but most of them have been recrystallized and replaced by muscovite aggregates.

Matrix and fragments vary in composition between dacitic and rhyolitic (see Table 1), and the potassium content reaches a maximum

Fig. 3. Concentration of deformed fragments of metarhyolite in the siliceous and carbonate-bearing biotite–muscovite schists. The width of the photograph is 1·70 m

of 10% while the percentage of CO_2 occasionally exceeds 10%. Some of the fragments are more siliceous than the matrix, but there is a clear overlap in composition.

It is difficult to evaluate the origin of these rocks. They have been interpreted as sedimentary rocks (Bridgwater and McGregor, 1974) and the metavolcanic fragments as conglomerate pebbles. Another interpretation (Keto, 1970) is that they are greywackes. In the author's opinion both interpretations are improbable for the following reasons. (1) This unit forms a very homogeneous series. There is no trace of alternating beds of quartzitic and micaceous layers. If these rocks really were sedimentary rocks a considerable degree of heterogeneity would have been expected within the series. (2) The great overlap in composition between matrix and fragments rather suggests that these rocks are of volcanic

origin. However, the continuity of this unit over 25 to 30 km makes it probable that the schists represent a series of subaerial or submarine ash flows while the fragments might represent agglomeratic material derived from acid lavas near the crater of a volcano. If these rocks originated as submarine ash flows the high content of carbonate could be readily explained, although the high content of potassium cannot be accounted for and might be the result of some kind of alkali metasomatism.

Talc Schists and Dunites. There are numerous conformable bodies of ultrabasic rocks and many of these are a few kilometres long and a few hundred metres wide. The most important ones occur along the contacts of different units. In general the ultrabasic bodies consist of talc schists with chlorite, anthophyllite and carbonate as common constituents. In the field these rocks show a brownish colour with lilac and pink tinges. Fresh ultrabasics consist of dunite and show a typical red-brown weathering colour. They occur as relics within the talc schist bodies which can easily be recognized as red-brown ridges. Olivine is the only main constituent and opaque ore is an accessory. Usually the olivine is partly altered into serpentine. Carbonate is often present interstitially, and in many cases occurs in considerable quantities. The carbonate is usually associated with completely fresh olivine and it appears that the carbonate is not a retrogressive mineral. Probably a process of carbonitization has taken place during an early stage of the history of the supracrustal belt.

The dunites have a very high Mg content and very low contents of Al_2O_3, CaO, Na_2O and K_2O, showing a highly refractive character. It is not known whether these bodies represent thrust slices from the mantle or intrusions. However, numerous bodies of fuchsite-bearing quartzites are closely associated with the ultrabasic bodies and the fuchsite is undoubtedly the result of metasomatism from the ultrabasics.

Amphibole–Albite–Chlorite Schists with Garbenschiefer Texture. These schists occur in three places in the supracrustal belt. Close to the Inland Ice in the north-easternmost part of the belt an irregular body occurs in the metasedimentary unit. West of lake Imarssuaq a thick and conformable body of this rock type occurs in the metasedimentary unit. A small occurrence is situated east of the big lake close to the northern contact of the belt.

Prisms of amphibole are arranged at random within the schistosity planes producing a texture similar to that called garbenschiefer. In the westernmost part of the belt, where the body is about 800 m broad, mineral banding, marked by differences in the concentration of the amphibole, is commonly displayed. In the bands which are not so rich in amphibole a clearly developed porphyritic texture can be recognized. Phenocrysts of amphibole and albite occur in a matrix of fine-grained unoriented albite grains and chorite. The albite aggregate and the chlorite occur in alternating layers. Occasionally large crystals of biotite have been observed. The porphyritic texture of this rock type is interpreted as truly magmatic.

This unusual rock with its hydrous minerals and its greenschist facies mineral assemblage is thought to have intruded the supracrustal rocks at a late stage of their metamorphic history during declining temperature and under hydrothermal conditions, possibly between the two phases of deformation or contemporaneously with the second deformation phase.

The chemical composition of these schists is variable. The silica content is usually close to 50% and most of the analyses show a high alumina content. The rocks are poor in alkalis, especially potassium and contain no carbonate.

Metasedimentary Rocks

Metasedimentary rocks are evenly distributed throughout the supracrustal belt, but only predominate over other rock types in the north-eastern part between the two zones of acid metavolcanic rocks and the Inland Ice.

Quartzites and Banded Ironstones. The quartz-banded ironstones (see Fig. 4) are the most spectacular rock type. They usually show a brown weathering colour and are very useful marker horizons, even though in many places they are only a few metres or less thick. Banded quartzites are closely associated with the banded ironstones and occur also independently. These leucocratic rocks are most abundant in the north-east, close to the Inland Ice; locally they are magnetite-bearing. West of lake Imarssuaq, west of the broad intrusion with garbenschiefer texture the leucocratic quartzites are regularly interbanded with conspicuous reddish-brown carbonate rocks.

The quartzites and banded ironstones are considered to have originated as chemical sediments, but all the original chert has been recrystallized as a result of the regional metamorphism. The mineral assemblage is very simple: quartz, cummingtonitic amphibole, carbonate, magnetite.

184

Fig. 4. Refolded isoclines of banded ironstone. This structure occurs in the core of a larger D_2 fold

The Main Iron Ore Body. This body is only partly exposed in the north-east of the supracrustal belt in a nunataq along the margin of the Inland Ice. Most of the body is covered by ice; it is situated outside the main end moraine of the Inland Ice and is covered by an accumulation of secondary ice up to 100 m in thickness with very low stream velocities. Kryolitselskabet Øresund A/S estimate the ore reserves at two billion tons (Keto, 1970). Geophysical evidence makes it conceivable that an originally continuous zone of ore with a north-east trend was faulted and dextrally displaced towards the south. A considerable enrichment of the ore has taken place along the faults. In the exposed iron ore (see Fig. 5) bands of magnetite and quartzite alternate regularly, defining fold structures and a strongly developed lineation. According to Professor Preston E. Cloud (pers. comm., 1974) the Isua banded ironstones are closely similar to the late Archaean and Proterozoic banded ironstone from many other parts of the world.

Intraformational Conglomerates with Quartzite Pebbles. In some localities in the north-eastern part of the belt conglomerates are well developed (see Fig. 6), and are interbanded with carbonate-bearing rocks. The pebbles consist exclusively of quartzite presumably derived from chert deposits occurring somewhere in the neighbourhood. These rocks are clearly banded and contain a very simple mineral assemblage. They are biotite–muscovite schists in which either quartz or carbonate predominate. The pebbles occur usually in the more quartz-rich beds, and these often also contain garnet. Tourmaline and magnetite are accessories.

Carbonate–Silicate Rocks and Carbonate Rocks. In the low area south-west of the conglomerate developments, carbonate–silicate rocks are abundant and constitute a zone which can be followed over several kilometres. They are rather coarse-grained rocks. Two main types can be distinguished:

Fig. 5. Banded ironstone from the main ore body, showing parallel folding

(a) Garnet-bearing biotite–plagioclase schists, always with very basic plagioclase, and with tourmaline as a common accessory;

(b) Carbonate–amphibole rock often with biotite and quartz; secondary minerals include epidote–clinozoisite and chlorite.

In some places in the supracrustal belt banded rocks occur in which biotitc-plagioclase–quartz bands alternate with hornblende–plagioclase bands. The plagioclase is always very basic. These rocks might represent metamosphosed marls. They also contain occasionally considerable quantities of magnetite arranged in bands.

Pure carbonate rocks occur in many places and form in general thin layers regularly interbanded with other metasedimentary rocks (see Fig. 7). They always show a red-brown colour and are ankeritic in composition (E. C. Perry, pers. comm., 1975). In a few places there are larger lenticular bodies of carbonate rock, several hundred metres in length and up to 50 m broad. Kryolitselskabet's investiga-tions (L. Kcto, pers. comm., 1975) show that these are dolomitic.

Biotite-rich Schists. Biotite-rich schists occur as 1–2 m thick zones in the amphibolite, and also in the areas where metasedimentary rocks are common. These are black or dark in colour and consist of biotite, basic plagioclase, quartz, a varying quantity of carbonate, garnet, ilmenite and magnetitie. Occasionally they contain a considerable quantity of hornblende. Tourmaline is a common accessory. Most of the biotite has abundant pleochroic haloes, the radioactive mineral being mainly allanite.

Pelitic Rocks. Biotite–muscovite schists with porphyroblasts of garnet and staurolite have been found in a few localities. They also contain oligoclase–andesine but no carbonate. A small quantity of graphite (about 1%) has been found in one of the specimens. Tourmaline is a conspicuous accessory. In the chemical analyses of two staurolite-bearing

Fig. 6. Conglomerate with quartzite pebbles in a matrix of siliceous and garnet-bearing muscovite–biotite schists

schists carbonate is not present, but like representatives of other sedimentary rocks they contain an exceptionally high quantity of iron (see Table 2). It would appear that the sea contained more iron in solution during this early stage of the Earth's history than at present.

The distribution of the metasedimentary rocks varies within the supracrustal belt. In general they appear to be in the minority with respect to the amphibolites and the acid metavolcanic rocks, and only in the north-eastern part of the belt in unit 3 do sedimentary rocks predominate over other rock types.

Grade of Metamorphism

The grade of metamorphism throughout the Isua supracrustal belt is low-pressure amphibolite facies. This is shown by the occurrence of basic plagioclase (labradorite–bytownite) in the amphibolites and other rock

Fig. 7. Disharmonically folded ankeritic carbonate rock, occurring in association with the schists of Fig. 6

types which contain carbonate and plagioclase. Staurolite occurs in rocks of pelitic origin. Relics of talc have not been found in rocks which contain ankeritic carbonates. Carbonate is never seen in association with quartz. In all cases micas and amphiboles are associated with carbonate and quartz. Diopside appears to be absent from all supracrustal rocks.

A retrograde metamorphism to greenschist facies is widespread in the supracrustal belt and especially well marked in the neighbourhood of the NW–SE trending fault in the north-west.

Conclusions and Problems

There are two problems which at the present stage of investigations in and around the Isua area cannot be solved. These are the following: (1) Is the Isua supracrustal belt just a fragment of the upper part of an originally

Table 2. Chemical analysis of metasedimentary rocks

GGU nr	17299	172977	173019	173153	173164
SiO_2	46·72	57·75	55·10	47·40	52·68
TiO_2	1·50	1·97	0·60	1·31	1·22
Al_2O_3	21·08	21·31	16·69	14·41	13·02
Fe_2O_3	4·94	3·99	7·33	2·09	2·53
FeO	11·57	5·86	9·56	11·76	11·42
MnO	0·18	0·11	0·31	0·27	0·23
MgO	2·67	1·60	3·23	3·52	5·35
CaO	1·39	1·20	1·36	7·50	4·94
Na_2O	0·55	1·04	1·80	0·48	1·06
K_2O	5·86	2·99	2·45	4·60	3·89
P_2O_5	0·22	0·12	0·09	0·22	0·18
CO_2				4·90	2·05
H_2O^+	2·05	1·66	1·26	1·18	1·08
	98·73	99·60	99·78	99·64	99·65

172999 Tourmaline-bearing garnet–muscovite–biotite schist.
172977 Garnet–staurolite schist.
173019 Garnet–staurolite–biotite schist (graphite-bearing).
173153 Garnet–biotite schist (black and carbonate-bearing).
173164 Hornblende-bearing biotite schist (black in colour).

more extensive succession, or was it originally only a very small supracrustal belt which has remained almost intact? In the author's opinion the first alternative is the most probable, but only future mapping in the region around the Isua supracrustal belt will demonstrate whether other fragments of a possibly more extensive succession still survive. (2) The low degree of migmatization of the supracrustal rocks by the gneisses suggests that the supracrustals were deposited on a basement of granitic rocks which has been partly reactivated, and not on ocean floor. However, if the Isua belt is a high-level fragment of an original more extensive succession, the Isua area itself would not be the most favourable site to look for depositional basement relations. Such features might be preserved in the neighbourhood of possible relics from the deepest parts of the supracrustal succession, which may be revealed during the mapping programme to be carried out in the region around Isua in the years to come.

There are few more points and questions which might be raised concerning the supracrustal belt itself.

(a) It appears that basic and acid metavol-canic rocks are well represented in the Isua belt, while andesitic rocks are rare or absent. Is there really an andesitic gap in the volcanic series of Isua?

(b) The common high potassium content of the acid metavolcanic rock unit is difficult to understand. More systematic sampling of the unit is needed before attempts can be made to solve this problem.

(c) It appears that graphite is widespread in the metasedimentary rocks, even in the banded ironstones. However, it always occurs in very small quantities. If the occurrence of graphite is an indication of life, this would imply that the quantity of living matter was also very small during the deposition of the Isua supracrustals. Are small quantities of bacteria necessary for the production of such major occurrences of banded ironstones as that of Isua?

(d) The occurrence of the intraformational conglomerates might imply the presence of streaming water during the deposition of the supracrustals and probably also the existence of delta-like environments.

Acknowledgements

This paper is published with the permission of the Director of the Geological Survey of Greenland. The geologists of Kryolitselskabet Øresund A/S, especially L. Keto, are kindly acknowledged for their assistance and interest in the work and for many inspiring discussions.

References

Bridgwater, D., and McGregor, V. R., 1974. 'Field work on the very early Precambrian rocks of the Isua area, southern West Greenland', *Rapp. Grønlands geol. Unders.*, **65**, 49–54.

James, P. R., 1975. 'Deformation of the Isua supracrustal belt, Southern West Greenland', *Rapp. Grønlands geol. Unders.*, 75.

Keto, L., 1970. 'Isua, a major iron ore discovery in Greenland', *Kryolitselskabet Øresund A/S.*, 1–13.

Moorbath, S., O'Nions, R. K., Pankhurst, R. J., Gale, N. H., and McGregor, V. M., 1972. 'Further rubidium–strontium age determinations on the very early Precambrian rocks of the Godthaab district, West Greenland', *Nature Phys. Sci.*, **240**, 78–82.

Moorbath, S., O'Nions, R. K., and Pankhurst, R. J., 1973. 'Early Archaean age for the Isua iron foundation, West Greenland', *Nature*, **245**, 138–139.

Amîtsoq Gneiss Geochemistry: Preliminary Observations

RICHARD ST. JOHN LAMBERT

Department of Geology, University of Alberta, Edmonton, Alberta, Canada.

JAMES GRENVILLE HOLLAND

Department of Geological Sciences, Science Laboratories, South Road, Durham, England.

Abstract

A small suite of Amîtsoq gneisses, almost all from the augen-gneiss sub-unit, has been analysed for major and trace elements. The chemistry of these leucocratic gneisses is very unusual in that their feldspar-related mineralogy and chemistry is similar to that in calc–alkali igneous suites, but their iron–magnesium ratios lie on an iron-enriched or tholeiitic trend. Their trace-element chemistry is characterized by very low Th and U and low Pb and Rb; Y declines as Ca declines; and most samples so far analyzed are comparatively deficient in heavy REE. Hydrous fractionation of a reduced uppermost mantle containing garnet and hornblende is a possible mode of origin for these rocks.

Introduction

This paper presents data obtained at the Universities of Alberta and Durham, and by Dr B. Mason at the Smithsonian Institution, on sixteen samples of Amitsoq Gneiss, ten collected by V. R. McGregor in 1967 and six in 1972. The first ten samples were comparatively small (1–2 kg) but the second set (listed on Table 3) were very large (50–100 kg) and are regarded as highly representative. These six samples have been studied isotopically and petrologically in great detail (Baadsgaard, 1973; Baadsgaard, Lambert and Krupicka, in preparation) and Mason has made a chemical study of all the minerals in two of them.

The Amitsoq gneisses are among the oldest rocks known: using present uranium decay constants and an ^{87}Rb half-life of $4 \cdot 9 \times 10^{10} a$, the most probable age of formation is 3600 Ma (Moorbath et al., 1972; Baadsgaard, 1973). They have undergone a protracted history of isotopic age re-setting (Pankhurst et al., 1973). Their structure and general characteristics have been described by McGregor (1973), who regards them as of igneous origin. O'Nions and Pankhurst (1974) consider that the gneisses, while igneous, cannot have formed from a single cogenetic suite of igneous rocks, and that some at least formed by partial melting of eclogite formed from pre-existing basaltic crust.

Major Element Chemistry

The sixteen analyses are given in Tables 1–3, with norms and some petrographic data

191

Table 1. Amîtsoq gneiss analyses (XRF): Normal series

	155820	110860	110858	110857	110857 dupl.	155819
SiO_2	59·45	63·29	65·25	65·50	65·87	67·27
Al_2O_3	12·42	14·20	14·89	15·30	15·25	13·86
Fe_2O_3	12·56	8·67	6·55	6·00	5·98	6·36
MgO	1·86	1·40	1·26	1·24	1·11	1·05
CaO	6·22	4·06	3·50	4·03	3·98	3·36
Na_2O	3·02	3·21	3·44	4·41	4·22	2·71*
K_2O	1·52	3·57	3·83	2·32	2·38	4·18
TiO_2	1·94	1·07	0·85	0·80	0·82	0·82
MnO	0·16	0·12	0·09	0·08	0·08	0·08
S	0·11	0·03	0·02	0·02	0·02	0·02
P_2O_5	0·74	0·37	0·32	0·29	0·29	0·27
Rb ppm	31	68	63	47	47	89
Sr	289	234	250	300	300	190
Y	60	37	27	25	28	54
Zr	324	505	377	282	325	370
Nb	15	13	11	9	9	15
Ba	476	804	844	756	792	694
Ni	2	8	9	7	4	4
Cu	23	15	19	14	15	18
Zn	164	121	92	82	84	87
$U^†$	0·16	—	—	—	—	0·53
$Th^†$	0·39	—	—	—	—	0·94

	155817	155818	110986	110987	110866	110869
SiO_2	67·79	70·31	70·39	72·13	72·89	77·90
Al_2O_3	13·51	13·80	14·46	14·04	14·99	11·85
Fe_2O_3	6·60	4·16	3·01	2·72	0·78	1·39
MgO	1·13	0·74	0·59	0·57	0·18	0·26
CaO	3·30	2·07	1·84	1·82	1·19	1·64
Na_2O	3·46	3·26	3·18	3·43	3·28	3·63
K_2O	2·86	4·88	5·87	4·76	6·48	3·01
TiO_2	0·96	0·54	0·42	0·36	0·15	0·23
MnO	0·08	0·06	0·04	0·04	0·01	0·03
S	0·01	0·02	0·01	0·01	0·01	0·01
P_2O_5	0·31	0·17	0·18	0·12	0·05	0·03
Rb ppm	80	109	136	125	157	93
Sr	191	152	238	150	250	128
Y	33	21	11	11	5	3
Zr	389	314	312	252	44	137
Nb	18	11	8	5	2	12
Ba	618	541	1100	584	1402	500
Ni	4	2	<2	7	<2	<2
Cu	12	8	6	7	6	47
Zn	101	59	45	43	8	21
$U^†$	0·40	0·40	—	—	—	2·19
$Th^†$	0·66	1·49	—	—	—	2·14

* Significantly lower than the wet chemistry figure (3·52).
† Analyst R. A. Burwash, delayed neutron method.

on Tables 4 and 5, and some additional trace element data on Table 6. The first two tables give results obtained by J. G. Holland at the University of Durham using a Philips PW 1212 automatic XRF spectrometer, the results being calibrated against international standards with mass absorption corrections made by an iterative procedure. Six analyses were made by A. Stelmach at the Department of Geology, University of Alberta, using orthodox wet chemical methods (Table 3 and 110999 on Table 2). Rb and Sr obtained by XRF were internally calibrated using Baadsgaard's results. Agreement between the two sets is broadly satisfactory, except for Na$_2$O in 155819 and perhaps for small systematic differences in TiO$_2$ and K$_2$O. The U and Th were obtained by R. A. Burwash using the delayed-neutron method.

With some exceptional features as discussed in the following the analyses show that the augen-gneisses and potassic members of the intrusive granite sheet (Table 1) conform to an igneous pattern, at first sight most closely resembling a weakly calcic calc–alkali series

with an alkali–lime index just above 60. However, a sub-set (Table 2) is very poor in K, Rb, Y, Zr, Nb, Ba and Cu, and rich in Ca and Sr compared with the others: these are plagioclase-rich rocks, two grey gneisses and the others plagioclase-rich units in the intrusive granite sheet. In this K-poor set, primary potash feldspar is absent and biotite is comparatively scarce. Features which are common to all rocks analyzed so far are: very high Fe$_2$O$_3$/MgO (except in 110988X, 110999) for a given SiO$_2$ content; high SiO$_2$/Al$_2$O$_3$ and high modal quartz, especially in the lower-SiO$_2$ rocks; very low U and Th despite K/Rb of 270–505 (with a lower value in the K-poor series (125–405); and in the augen gneisses, Y declining as Ca declines (Lambert and Holland, 1974).

The chemistry of these gneisses is more closely examined in Fig. 1–5: they show that there may be some reasons for regarding the samples analyzed as more heterogeneous in origin than has so far been implied.

Their normative characteristics in Q–Ab–Or (Fig. 1) show that, if igneous and closed to

Table 2. Amîtsoq gneiss analyses (XRF): K-poor series

	110988X	110999*	110802	110865	110988Y
SiO$_2$	65·60	65·69	69·72	76·00	76·15
Al$_2$O$_3$	15·51	16·79	15·65	11·92	12·59
Fe$_2$O$_3$	5·07	4·57	2·90	2·93	1·81
MgO	2·10	1·94	0·94	0·57	0·59
CaO	4·74	4·65	4·02	3·78	4·07
Na$_2$O	4·72	4·47	4·81	3·52	3·87
K$_2$O	1·42	1·15	1·23	0·76	0·60
TiO$_2$	0·54	0·50	0·51	0·41	0·21
MnO	0·08	0·07	0·02	0·05	0·02
S	0·01	—	0·02	0·01	0·01
P$_2$O$_5$	0·21	0·17	0·17	0·05	0·09
Rb ppm	29	28	39	51	29
Sr	415	401	340	265	270
Y	13	—	2	13	3
Zr	101	—	127	297	85
Nb	5	—	4	5	2
Ba	198	—	553	192	113
Ni	28	—	7	3	4
Cu	9	—	4	6	4
Zn	83	—	49	30	26
U	—	0·14	—	—	—
Th	—	0·28	—	—	—

* Wet chem. analysis, recalc.; Rb, Sr by I.D.

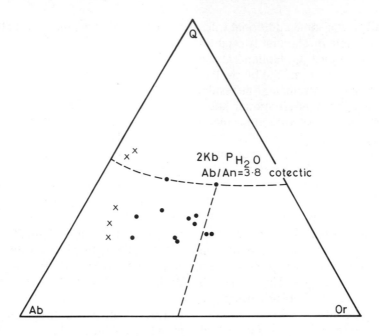

Fig. 1. Normative Q–Ab–Or plot for Amîtsoq gneisses

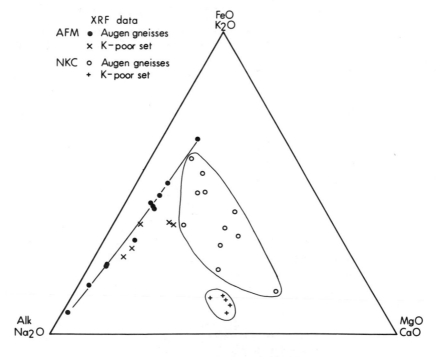

Fig. 2. Alkalies–FeO–MgO and Na₂O–K₂O–CaO plots of Amîtsoq gneisses
(% by wt.)

alkali migration, they come from several series which have undergone separate fractionation patterns under possibly widely varying pressures (load and/or water). Further, the Q-rich rocks 110869, 110865 and 110988Y appear to have either originated by the partial melting of sediments or by igneous fractionation under exceptionally high P_{H_2O}, or have achieved their unusual composition by metasomatic gain of silica. Either the second or third possibility is preferred on the grounds that these rocks come from an intrusive sheet which also contains typical granite compositions.

The patterns exhibited by these rocks on the molecular ratio diagram (Fig. 2) add further to complexity of interpretation. The AFM ratios are truly exceptional for a series of Archean gneisses: indeed they are unique (see AFM plots in Lambert, 1975, this volume and further data (in press, V. R. McGregor)). Among igneous series the AFM ratios in the augen-gneisses compare only with a tholeiitic Fe-enriched fractionation pattern. However, the generally high Ca and Sr, the trend of Ca/Y and the REE patterns (Fig. 4 and O'Nions and Pankhurst, 1974), are most unlike the kinds of chemical trends which we associate with pyroxene–plagioclase fractionation under oxygen and/or hydrogen closed-system conditions. Thus the pattern exhibited by the augen-gneisses should not be regarded as tholeiitic on the sole grounds of its AFM character. The molecular plot of Na_2O-K_2O-CaO shows clearly the calcic nature of these rocks and the very distinctive nature of the K-poor set.

Trace Element Chemistry

Variation of trace elements in the Amîtsoq gneisses is illustrated in Fig. 3, where K/Rb, K/Sr, K/Ba and Ca/Y are plotted against normative Q + Or + Ab. Each set of data for the augen-gneisses appears to show a break at Q + Or + Ab = 72, on Table 1 corresponding to the break between 110857 and 155819. The major elements support this division when Al_2O_3 and Na_2O trends are considered, and of the trace elements not shown on Fig. 3, Zr and Nb also show a break in their variation against SiO_2 at this point. There is thus a suggestion that these gneisses comprise not just two series, but at least three, characterized by:

(a) moderate to high K_2O, SiO_2 <66%, Rb and Y relatively low and Sr and Ba relatively high;
(b) moderate to high K_2O, SiO_2 <66% and Rb, Y, Sr and Ba contrasting to group (a); and
(c) the K-poor set.

Table 3. Amîtsoq gneiss: Chemical analyses

	155820	110999	155819	155817	155818	110869
SiO_2	58·04	65·29	66·99	67·94	70·46	78·69
Al_2O_3	12·85	16·69	14·37	13·87	13·72	11·95
Fe_2O_3	2·78	1·32	1·00	1·02	0·97	0·14
FeO	9·02	2·94	4·28	4·40	2·63	0·95
MgO	2·02	1·93	0·96	1·04	0·59	0·24
CaO	6·04	4·63	3·38	3·45	2·15	1·56
Na_2O	3·39	4·45	3·52	3·56	3·12	3·25
K_2O	1·37	1·15	4·04	2·47	4·45	2·91
TiO_2	2·38	0·50	0·93	1·08	0·61	0·23
MnO	0·18	0·07	0·07	0·07	0·06	0·02
P_2O_5	0·78	0·17	0·29	0·33	0·19	0·04
H_2O tot	0·92	0·60	0·46	0·80	0·80	0·26
	99·77	99·74	100·31	100·03	99·75	100·24

Analyst A. Stelmach.

Table 4. CIPW Norms and Modes of Amîtsoq gneisses, normal series

	155820	110860	110858	110857	110857 dupl.	155819
Q	17·6	17·1	19·1	18·9	20·1	24·3
Or	9·0	21·3	22·6	13·7	14·2	24·8
Ab	25·6	27·2	28·9	37·4	35·6	22·9
An	16·1	13·8	14·2	15·2	15·8	13·4
Cor	—	—	—	—	—	—
Di	12·9	5·7	2·8	4·0	3·5	2·8
Hyp	11·5	10·2	8·9	7·5	7·5	8·3
Ol	—	—	—	—	—	—
Mt*	3·6	2·5	1·9	1·7	1·7	1·9
Ilm	3·7	2·1	1·6	1·5	1·5	1·5
Norm An%	39	34	33	29	31	37
Modal Analyses[†]						
Quartz	3	2	2	2		2
Plagioclase	1	1	3	1		1
Microcline	g	3	1	3		3
Hornblende	2	5	c	b		a
Biotite	4	4	4	4		4
Epidote	f	c	a	a		b
Allanite	e	e	e	e		e
Sphene	d	a	b	c		c
Apatite	a	d	d	d		d
Muscovite	—	—	—	—		—
Oxides	b	b	—	f		f
Zircon	h	f	f	g		g
Calcite	c	—	—	—		—

	155817	155818	110986	110987	110866	110869
Q	25·6	25·5	23·3	28·0	25·1	41·0
Or	17·0	28·9	35·0	28·4	38·2	17·8
Ab	29·4	27·6	27·1	28·8	27·7	30·7
An	12·9	8·6	7·7	8·7	6·8	7·2
Cor	—	—	—	—	0·3	—
Di	3·0	1·5	1·0	0·2	—	0·8
Hyp	8·4	5·6	4·2	4·3	1·2	1·6
Ol	—	—	—	—	—	—
Mt*	1·9	1·2	0·9	0·8	0·2	0·4
Ilm	1·8	1·0	0·8	0·8	0·3	0·4
Norm An%	31	24	22	23	20	19
Modal Analyses[†]						
Quartz	2	3	2	2	2	1
Plagioclase	1	2	3	3	3	3
Microcline	4	1	1	1	1	2
Hornblende	b	c	—	—	—	—
Biotite	3	4	4	4	a	a
Epidote	a	a	a	a	b	c
Allanite	e	f	d	d	—	—
Sphene	c	b	b	b	—	—
Apatite	d	d	c	c	d	d
Muscovite	—	—	—	—	—	b
Oxides	f	e	—	—	c	e
Zircon	g	h	e	e	—	f
Calcite	—	g	—	—	—	—

* Assuming $Fe_2O_3 = 0·25$ FeO (from Stelmach's analyses, allowing for minor late oxidation).
[†] See footnote to Table 5.

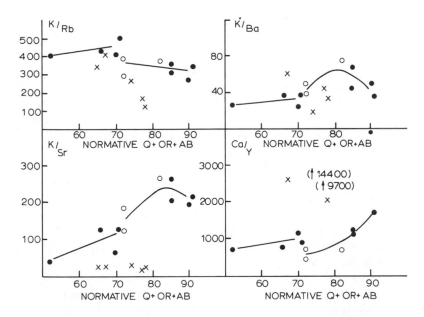

Fig. 3. K/Rb, K/Ba, K/Sr and Ca/Y for Amîtsoq gneisses. Small dots are low SiO$_2$ members of the normal series; open circles the high SiO$_2$ members; large dots are the granites from the intrusive sheet and crosses the K-poor set. Two of the latter plot well off-scale on Ca/Y as indicated by arrows

Table 5. CIPW Norms and modes of Amîtsoq gneisses, K-poor series

	110988X	110999	110802	110865	110988Y
Q	18·3	20·3	26·3	43·3	42·0
Or	8·4	6·8	7·3	4·5	3·6
Ab	40·0	37·9	40·8	29·8	32·7
An	16·9	22·4	17·5	14·5	15·2
Di	5·5	0·6	2·1	3·7	4·2
Hyp	8·3	9·7	4·3	2·7	1·4
Mt	1·5	1·3	0·8	0·9	0·5
Ilm	1·0	1·0	1·0	0·8	0·4
Norm An%	30	37	30	32	32
Modal Analyses					
Quartz	—	2	2	2	2
Plagioclase	—	1	1	1	1
Microcline	—	c	a	d	d
Hornblende	—	4	—	e	b
Biotite	—	3	3	a	a
Epidote	—	a	b	c	e
Allanite	—	e	e	—	—
Sphene	—	b	c	b	c
Apatite	—	d	d	f	f
Muscovite	—	—	—	—	—
Oxides	—	f	—	—	—
Zircon	—	g	f	g	g

Major mineral abundances given in order of abundance 1 to 5, 1 being the most abundant. Minor minerals (<5%) given in order a to h, 'a' being the most abundant.

With the quantity of data available and the lack of information on possible metasomatic exchanges, the division into (a) and (b) cannot yet be regarded as more than a suggestion, but it does conform with O'Nions and Pankhurst's conclusion regarding heterogeneity.

The Rb, Sr, Ba and Y data show, besides the possibility of a geochemical break in the 'normal' series, that some anomalous trends are present in these LIL elements. K/Rb is moderately high and, except in the K-poor set, is much higher than would be anticipated in a granodiorite–granite series. In the K-poor set, K/Rb appears to be very low for such plagioclase-rich rocks. K/Sr and K/Ba appear to show an unusual decline in the highly siliceous rocks, opposite to that which might be expected in a calc–alkali series (Nockolds and Allen, 1953). The decline in Y relative to Ca is one of the most extreme known, and is indicative of extreme garnet and/or hornblende fractionation (Lambert and Holland, 1974). The K-poor set has some of the lowest Y contents known in such rocks.

The additional REE data are given in Fig. 4. 155819 is nearest to a heavy REE fractionated (depleted) version of 155739 (a granite, see O'Nions and Pankhurst, 1974). Both have similar total REE and a negative Eu anomaly, which is otherwise absent in these rocks. 155819 has a similar mineralogy to 155739, but has a small quantity of residual hornblende

(see Baadsgaard, Lambert and Krupicka, in prep., for mineralogy). The Ce_N/Yb_N ratios for 155819, 155739 are 15 and 6, respectively, however, so that 155819 parallels O'Nions and Pankhurst's sample 155735 to a certain extent, rather than 155739.

155818 has appreciable heavy REE depletion and a positive Eu anomaly, and has unusually high Yb compared with Ho and Er. It is fairly similar to O'Nions and Pankhurst's sample 155805, a granodiorite, but is slightly less fractionated (heavy REE depleted) if Ce_N/Yb_N is used as a guide. Ce_N/Er_N, however, are close to 5·5 and 6·5 in 155818 and 155805, respectively, indicating a broadly similar pattern.

These results therefore confirm the work of O'Nions and Pankhurst in demonstrating significant heavy REE depletion in some of the Amîtsoq gneisses. The most depleted samples all appear to be relatively poor in hornblende, in so far as the necessary information is available. We therefore wonder if a multi-stage history, involving interfacing in the mantle with garnet, followed by strong hornblende (? tschermakite) fractionation (removing heavy REE, Y and MgO but enriching the magma strongly in SiO_2) might not produce the required chemical characters and be consistent with suggestions of extreme water pressures. The close comparison of the Bonsall tonalite with 155818 (except for Eu) may be

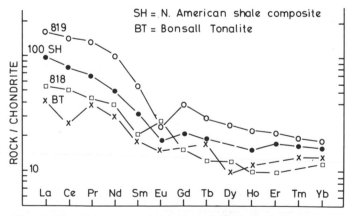

Fig. 4. Chondrite-normalized plot of REE in 155818 and 155819

Table 6. Additional trace element analyses

	Cs	Hf	Pb (ppm)
155818	0·83	5·1	12
155819	1·0	8·2	15

	La	Ce	Pr	Nd	Sm	Eu	Gd	Tb	Dy	Ho	Er	Tm	Yb
155818	16	42	5·0	22	4·2	1·9	4·4	0·57	3·5	0·66	1·9	<0·3	2·1
155819	48	120	16	58	11	1·8	12	1·4	7·7	1·7	4·5	0·55	3·5

Analyst, B. Mason.

significant in this regard, the Bonsall tonalite being a well-known member of the calc–alkaline Southern California batholith. The analogy between 155819 and average North American shale (Fig. 4) is not so easy to interpret, however.

The total variance in the group of analyses as a whole, even though identified as having several sub-groups, is very striking (Fig. 5) and casts a little light on the fractionation problem, howsoever caused. The factors 6 and 7 account for 71% of the variance in an R- mode promax oblique primary pattern matrix: they are clearly to be identified as a plagioclase v. alkali feldspar factor, and a quartz v. ferromagnesian factor, respectively. Of particular interest is the close association of Mn, Nb, P, Ti, Y, Zn and Zr with Fe. Mineralogically, this is due at least in part to the uniformity of mineralogy in these rocks: the only Fe-rich minerals are biotite and hornblende, so it is almost inevitable that Mn, Nb, Ti and Zn correlate with Fe. However, P, Zr and to a lesser extent Y indicate some broader control at work, the most obvious being simultaneous crystal fractionation of the relatively dense apatite and zircon with ferromagnesians. This is consistent with the igneous hypothesis and with the postulate of deep-seated origin under hydrous conditions, which will necessarily produce maximum opportunity for strong fractionation, as the magma ascends great distances and cools over a predictable large liquidus–solidus interval.

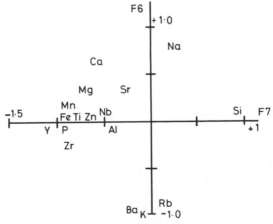

Fig. 5. R-mode factor analysis: factor loadings for each element against the two dominant factors. The elements not shown (S, Cu, Ni) have low loadings of no statistical significance

The other trace elements not yet considered, Cs, Hf, Pb, Th and U, create other problems, but consideration of the first three is deferred until more data are available. The abundances of Th and U can only be called most unusual. K/U ratios range from 11,000 to 100,000 with mean K/mean U = 38,000, and mean Th/mean U = 1·5. Low Th and U are in general accord with the Pb isotope data already published (Oxford Isotope Geol. Laboratory and McGregor, 1971). O'Nions and Pankhurst (1974) consider that the isotopic evidence indicates that these rocks have remained closed as a whole to Rb, Sr, U and Th since 3600 Ma ago: if so, the U and Th abundances can only be due to events of that or greater age. However, Moorbath (pers. comm.) considers that recent U loss has occurred, basing his argument on failure of the U/Pb systematics to yield reasonable age patterns. Baadsgaard (pers. comm.) finds that both U and Pb are well below expected abundances in all these rocks and minerals, and is therefore of the opinion that recent loss is not significant. On balance, we conclude that low U, Th, Pb and perhaps Rb is a fundamental characteristic of these gneisses. One possible mechanism for such a pattern has been discussed by Heier (1973), but no evidence for a granulite facies event affecting these rocks is otherwise known.

We are therefore led to consider the possibility that the source region of these rocks had, in today's terms, an anomalous composition. The alternative, of extreme fractionation (eclogitic or otherwise), does not seem feasible, for such a process is usually regarded as one which produces concentrations of incompatible elements in the low-temperature liquid phase. Speculatively, we would note that the effective absence of U and Th from many of these rocks, the unusual scarcity of Mg compared with Fe, the low level of Y in the siliceous differentiates and considerable heavy REE depletion, accord with the possibility that at this stage of Earth's development, U, Th, Mg, Y and heavy REE were, relative to today, concentrated in the mantle. Lunar, meteoritic and theoretical studies suggest that these elements are concentrated in early formed fractions of accreting planets (Grossman, 1972, 1973; Anderson, 1974). However, Ti is at a normal level in these rocks compared with averages of calc–alkaline igneous rocks (Nockolds, 1954) whereas if the mantle was still at an early stage of differentiation 3600 Ma ago, Ti should be concentrated there.

Whatever the ultimate source of these rocks was, however, and whatever its composition, the AFM trend and otherwise calc–alkaline nature may perhaps be explained by hydrous differentiation under unusually low f_{O_2} conditions. Normally, hydrous magmas will tend to be comparatively oxidizing, so the question is raised whether the source region of these rocks had an appreciable carbon (CO_2, C or CH_4) content.

To date, therefore, there are many options open concerning the origin of these gneisses: by no means all have been mentioned or discussed in the foregoing.

Acknowledgements

The authors wish to thank Dr V. R. McGregor for his stimulating discussions of these rocks and for supplying the entire collection of samples. Thanks are also due to Dr B. Mason for generously donating his geochemical data, to Dr R. A. Burwash for allowing us to use his Th and U data, and to Dr H. Baadsgaard and Dr J. Krupicka for their contributions by way of discussion and provision of unpublished isotopic and petrographic data. This work has been carried out as part of research programmes financed by the National Research Council of Canada and the Natural Environment Research Council, U.K., assistance which is gratefully acknowledged. The paper is published by permission of the Director of the Greenland Geological Survey.

Appendix: Localities of Analyzed Specimens

110802 Not known
110857-8 SE corner of Qilangârssuit, 63°51·2'N 51°37·9'W
110860 As last, 63°51·1'N 51°37·8'W
110865-6, 110869*, 110987-8 As last 63°50·9'N 51°37·6'W

* The specimen identified as 110869 in this report and in Baadsgaard, 1973, was collected and noted as 110986. Baadsgaard and the authors are certain, however, that their usage is consistent. The localities of McGregor's specimens 110869 and 110986 are, coincidentally, identical.

110999 E coast of Angissorssuaq, 63°55·8'N 51°44·1'W
155817 Coast between Narssaq and Kanajorssuit, 64°0·8'N 51°39·0'W (same locality and rock type as 155702, O'Nions and Pankhurst, 1974)
155818 As last, 64°1·2'N 51°38·5'W
155819 As last, 64°0·5'N 51°39·0'W
155820 As last, 64°0·6'N 51°39·1'W

References

Anderson, D. L., 1974. 'The interior of the Moon', *Physics Today*, **27**, 44–49.

Baadsgaard, H., 1973. 'U–Th–Pb dates on zircons from the early Precambrian Amitsoq gneisses, Godthaab district, West Greenland', *Earth Planet Sci. Lett.*, **19**, 22–28.

Grossman, L., 1972. 'Condensation in the primitive solar nebula', *Geochim. Cosmochim. Acta*, **36**, 597–619.

Grossman, L., 1973. 'Refractory trace elements in Ca-Al-rich inclusions in the Allende meteorite', *Geochim. Cosmochim. Acta*, **37**, 1119–1140.

Heier, K. S., 1973. 'Geochemistry of granulite facies rocks and problems of their origin', *Phil. Trans. R. Soc. Lond.*, **A273**, 429–442.

Lambert, R. St. J., and Holland, J. G., 1974. 'Yttrium geochemistry applied to petrogenesis utilising calcium–yttrium relationships in rocks and minerals', *Geochim. Cosmochim. Acta*, **38**, 1393–1414.

McGregor, V. R., 1973. 'The early Precambrian gneisses of the Godthåb district, West Greenland', *Phil. Trans. R. Soc. Lond.*, **A273**, 343–358.

Moorbath, S., O'Nions, R. K., Pankhurst, R. J.,

Gale, N. H., and McGregor, V. R., 1972. 'Further rubidium–strontium age determinations on the very early Precambrian rocks of the Godthaab district, West Greenland', *Nature Phys. Sci.*, **240**, 78–82.

Nockolds, S. R., 1954. 'Average chemical composition of some igneous rocks', *Bull. geol. Soc. Amer.*, **65**, 1007–1032.

Nockolds, S. R., and Allen, R., 1953. 'The geochemistry of some igneous rock series', *Geochim. Cosmochim. Acta*, **44**, 105–142.

O'Nions, R. K., and Pankhurst, R. J., 1974. 'Rare element distribution in Archaean gneisses and anorthosites, Godthåb area, West Greenland', *Earth Planet Sci. Lett.*, **22**, 328–338.

Oxford Isotope Geology Laboratory and McGregor, V. R., 1971. 'Isotopic dating of very early Precambrian amphibolite facies gneisses from the Godthaab district, West Greenland', *Earth Planet. Sci. Lett.*, **12**, 245–259.

Pankhurst, R. J., Moorbath, S., Rex, D. C., and Turner, G., 1973. 'Mineral age patterns in ca. 3700 m.y. old rocks from West Greenland', *Earth Planet. Sci. Lett.*, **20**, 157–170.

New Evidence Relating to Archaean Events in Southern West Greenland

B. CHADWICK and K. COE

Department of Geology, University of Exeter.

Introduction

In 1972 the University of Exeter began a field mapping and research programme in the Buksefjorden region, southern West Greenland, as a joint project with the Geological Survey of Greenland. The field work will be completed in 1977, but already more than 60% of the 2100 km² area has been mapped at scales of 1 : 10,000 and 1 : 20,000. Lithological and structural control has been established along the coast extending over 40 km from north to south of the region, but chronological interpretation and correlation with areas to the north (Godthåbsfjord; McGregor, 1973) and south (Fiskenaesset region; Kalsbeek and Myers, 1973) are at present uncertain.

The purpose of this paper is to present some of the problems of interpretation and regional correlation in order to sound a note of caution against drawing premature conclusions, including generalizations such as early crustal thickening (Bridgwater, McGregor and Myers, 1974), and to highlight the misleading restrictions imposed by the use of limited isotope data.

Reconnaissance by Noe-Nygaard and Ramberg (1961) and Berthelsen (1955) established the significance of certain minor intrusions and hinted at the antiquity of the gneisses in the Godthåb district, but it was McGregor (1968) who first recognized in the Godthåbsfjord area (which includes northwest Buksefjorden) an older (Amîtsoq) and younger (Nûk) group of gneisses, which he suggested were separated in time by intrusion of basic dykes, the Ameralik dykes, and the formation of the Malene supracrustal rocks. Work in the Isua region, about 100 km northeast of Godthåb, (Bridgwater and McGregor, 1974) added to the stratigraphy and chronology set out by McGregor in 1973 which forms the basis of Table 1. The position of suites of various intrusions including basic amphibolite dykes younger than the Ameralik dykes found in the south of the Buksefjorden region is also included. The scheme of tectonic events recognized by the Exeter group is given as well in order to place the following discussion in context.

All place names referred to in the Buksefjorden region can be located on the Danish Geodetik Institute 1 : 250,000 map 63 V 1.

Amîtsoq Gneisses

This term has been used consistently in the sense of McGregor (1973; Black et al., 1971) and his definition with its stipulation that gneisses of this suite have been injected by Ameralik dykes has been followed. Two distinct groups of Amîtsoq gneiss occur in the Buksefjorden region: an older group of mainly leucocratic, migmatitic tonalite gneisses with intrafolial folds, intruded by a younger group of microcline augen gneisses which in areas of

203

Table 1

	Ages × 10^6 years	
16 Slight regional metamorphism	Ca 1600 >2000	I Minor faulting (16)
15 Emplacement of basic dykes	<2600	H Emplacement of basic dykes (15)
14 Emplacement of Qôrqut granite and related pegmatites	Ca2600	G Emplacement of granite sheets mainly in N. Buksefjorden. Waning deformation (11) (12) & (14)
13 Shearing and thrusting with syntectonic emplacement of Kangâmiut dykes		
12 High-grade metamorphism outlasting 10 and 11	2800–2900	F Maintained high-grade metamorphism. Deformation producing upright and fanned folds with coaxial lineation, dome and basin interference, lobe and cusp structure and straight belts (10)
11 Emplacement of syn- and late-tectonic granites		
10 Formation of nappes followed by dome and basin interference patterns, partly modified by transcurrent movements in linear belts	2800–3040	E Injection of minor basic intrusions. Regional deformation; thrust nappes at maintained high grade metamorphism; emplacement of Sermilik gneisses; injection of minor intrusions (10)
9 Intrusion of syn- and late-tectonic calk-alkali rocks (Nûk gneisses) as sub-concordant sheets	3040	D Migmatization of all earlier rocks; Intrusion of granitic suite with anatexis of Amîtsoq gneiss. Anatexis and granite injection continues throughout E and F (9)
8 Emplacement of stratiform anorthosites and related sheets		
7 Major thrusting intercalating Amîtsoq gneiss and Malene supracrustal rocks		C Intrusion of basic sills, emplacement of stratiform anorthosites; intrusion of Ameralik dykes; extrusion of lavas, tuffs, etc. (5) (6) & (8)
6 Formation of Malene supracrustal rocks		
5 Intrusion of Ameralik dyke swarm		
		Deposition of pelitic and other sediments (Malene) (6)
4 Deformation of Amîtsoq gneiss and Isua supracrustals		B Injection of parents of Amîtsoq augen gneiss. Deformation and migmatization (3)
3 Intrusion of syn- and late-tectonic granites (Amîtsoq)	Ca 3750	Intrusion of granitic material (parent of foliated Amîtsoq gneiss) (3)
2 Deposition of Isua supracrustal rocks		A Formation of parent material of the pre-Amîtsoq rocks (2)
1 Formation of very old granitic crust	>3750	

Left hand column—abbreviated from Bridgwater, McGregor and Myers (1974).
Right hand column—Chronological Succession for Buksefjorden region, with the nearest equivalent shown in brackets.
Some of the material used for calculating ages in the central column comes from the Buksefjorden region.

low deformation resemble porphyritic granite (Plate 1A). Xenolithic inclusions of the older foliated gneiss appear in the augen gneiss locally and inclusions of pre-Amîtsoq lithotypes occur in both varieties.

Black et al. (1971) obtained whole rock isochrons of 3980 ± 170 m.y. (Rb/Sr; initial ratio $^{87}Sr/^{86}Sr = 0.699$) and 3620 ± 100 m.y. (Pb/Pb) from 18 samples of Amîtsoq gneisses from the Godthåbsfjord region which included 14 samples of older and younger Amîtsoq gneiss and pre-Amîtsoq 'anorthositic gneiss' from northwest Buksefjorden. In a later paper Moorbath et al. (1972) present additional evidence of the antiquity of a further 40 or so samples of Amîtsoq gneiss from the same area. Once more, older and

younger Amîtsoq gneisses that are recognized in the Buksefjorden region fall on the same Rb/Sr isochrons (3750 ± 90 m.y.; 3740 ± 100 m.y.; 3690 ± 230 m.y.) and have similar initial ratios $^{87}Sr/^{86}Sr$ about 0.701. Zircons separated from six samples of older migmatitic and younger augen granite gneiss of the Amîtsoq suite in north-west Buksefjorden give intersections on the concordia curve at 3650 ± 50 m.y. and 3648 ± 85 m.y. (Baadsgaard, 1973). Interpretations of the isotope data are ambiguous, the ages and initial ratios being regarded by Black et al. (1971) and Moorbath et al. (1972) as indicators of either an igneous or a metamorphic event with the igneous parents of the Amîtsoq gneisses being intruded up to, but not more than, 200 m.y.

205

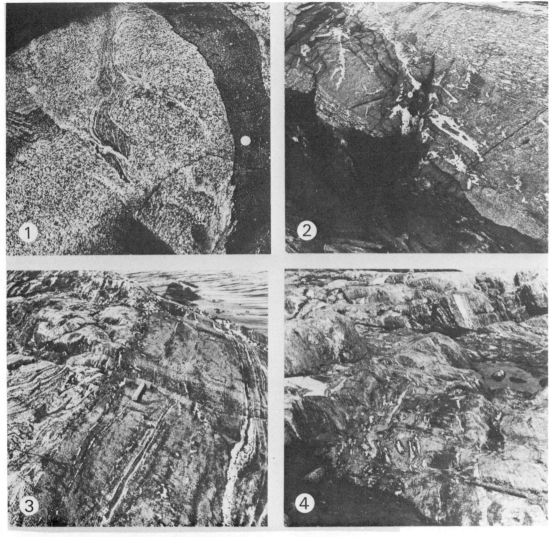

Fig. 1. Ameralik dyke cutting Amîtsoq augen gneiss with xenolithic inclusions of older foliated Amîtsoq gneiss, West Qilángârssuit

Fig. 2. Intersecting Ameralik dykes, Qilángârssuit

Fig. 3. Discordant amphibolite dyke in Malene banded amphibolite, north Qilángârssuit

Fig. 4. Discordant amphibolite dyke in Malene garnetiferous gneiss, Simiútat

Fig. 5. Differentiated ultramafic-metagabbroic sheet in Malene garnetiferous gneiss, Simiútat

earlier. They suggest this metamorphic age must be earlier than Ameralik dyke injection.

The isotope data obtained so far mark events common to all the Amîtsoq gneisses and might be reconciled with the field evidence if the formation of older migmatized gneiss (K-poor) and injection of younger porphyritic granite gneiss (K-rich) took place within a period represented by the error of the isochron, although it is difficult to reconcile the data with pre-Ameralik dyke metamorphism affecting all the Amîtsoq gneisses. It is also remarkable that the Amîtsoq gneisses should be regarded as having remained as closed systems with respect to Rb and Sr on the whole rock scale (Moorbath et al., 1972) but possibly open systems with respect to U–Th–Pb (Baadsgaard, 1973) during multiphase regional anatexis, injection and deformation that took place after intrusion of the Ameralik dykes.

Additional problems arise when attempts are made to apply 'Amîtsoq gneiss' as a lithological term because it covers not only rocks of widely variable composition but also significantly different ages. Furthermore, the term may only be used where Ameralik dykes are present. In the absence of Ameralik dykes, isotope evidence or perhaps certain trace element abundances may be the only criteria available for identifying these particular gneisses.

Ameralik Dykes and Malene Supracrustal Rocks

McGregor (1968, 1973) introduced the term Ameralik dyke to describe certain small bodies of variably deformed and migmatized amphibolites which are abundant in the Amîtsoq gneisses in the Godthåb district. The dykes were used by McGregor to make a fundamental distinction between the Amîtsoq gneiss cut by, or containing, the dykes and the Nûk gneiss where the dykes are absent. McGregor (1973) also recorded his belief that the Ameralik dykes do not occur in the Malene supracrustal rocks.

Ameralik dykes in that part of the Buksefjorden region mapped so far are concentrated in the north-west, especially on the Narssaq peninsula and the Qilángârssuit-Simiútat skaergaard. The dykes are abundant but all have undergone severe disruption by heterogeneous deformation and migmatization: most dykes, even those a few metres thick, extend only 5–100 m, and only rarely show discordances greater than 30° with the older migmatitic foliation of the host Amîtsoq banded gneisses. It has not been possible to discern any original boundary to the Ameralik dyke swarm: detection of such a boundary may in any event be unlikely in view of subsequent deformation and swamping by granite injections and remobilization. Recognition of Amîtsoq gneiss without Ameralik dykes may also be impossible in the absence of isotope evidence.

Seven different Ameralik dyke lithotypes have been recognized in the Amîtsoq gneisses, the differences being based solely on field criteria such as composition, presence or absence of plagioclase inclusions and degree of migmatization. Original igneous minerals and textures have been completely replaced and obliterated by metamorphism, although chills, igneous banding and megacrystic aggregates of plagioclase survive in some dykes. Cross-cutting relations in a few outcrops (Plate 1B) show that there are at least three different ages of dyke, although the difference in ages may not be significantly large. All the dyke lithotypes have been affected by the same deformation and migmatitic events and it seems likely that they may all be related to one period of basic intrusion. Major element chemistry of Ameralik dykes in the Amîtsoq gneisses of the Buksefjorden region indicates a range of composition from tholeiitic to alkali basalts with $Na_2O + K_2O$ less than 5% and SiO_2 c. 45–54%. Average oxide values for 21 dykes include TiO_2 0·89%, Al_2O_3 12·89%, total iron 13·30%, MnO 0·23%, MgO 8·51% and CaO 10·47%. High alumina basalt types are rare.

Some of these Ameralik dyke lithotypes occur abundantly as thin sheets with discordances up to 30° in certain parts of the Malene supracrustal rocks on the Qilángârssuit–Simiútat group of islands (Plate 1C, 1D), the dykes and supracrustal rocks having undergone the same deformation and migmatization as Ameralik dykes in the adjacent Amîtsoq gneisses. The Malene rocks comprise various garnet–sillimanite, quartz–cordierite–antho-phyllite–staurolite and pale mica–sillimanite gneisses and banded, diopsidic metavolcanic amphibolites locally with abundant epidote and carbonate. Some parts of the gneisses and banded amphibolites contain concordant sheets of a metagabbroic amphibolite and ultramafic rocks, locally with igneous differen-tiation banding (Plate 1E).

Lithological similarities with Ameralik dykes in the same area lead us to suggest that the dykes in the Malene rocks be regarded as members of the Ameralik dyke suite, although the possibility remains that some of the dykes in the supracrustal rocks (and *some* in the Amîtsoq gneisses) could be of younger age and part of the group of discordant amphibo-lites related to Nûk gneisses found in the south of the Buksefjorden region, and described below.

Two significant implications arise if the dis-cordant amphibolites in the Malene supra-crustal rocks are Ameralik dykes: (i) the Malene supracrustal rocks were deposited on a sialic basement of Amîtsoq gneisses contain-ing inclusions of an older supracrustal se-quence represented by the pre-Amîtsoq suite, and (ii) if the thrusting model of Bridgwater, McGregor and Myers (1974) is correct, then the dykes were intruded after the thrusting. If the second case holds then it follows that the Ameralik dykes can have no genetic relation to the Malene metavolcanic amphibolites although they could still be related to the Malene metagabbroic amphibolites. How-ever, in our view there is no positive evidence to support the Bridgwater, McGregor and Myers (1974) model, any interleaving of Malene and Amîtsoq sheets being attributable to post-dyke deformation.

Accepting that the Malene rocks were deposited as a cover on a basement of Amîtsoq gneisses it follows that:

(i) The Ameralik dykes in the basement and cover may be cogenetic with the metavol-canic banded amphibolites and could be regarded as feeders to extrusive rocks and intrusive sheets such as the metagabbroic amphibolites.

(ii) The garnetiferous gneisses which contain locally abundant dykes and metagabbroic sheets may in part be older than the basic igneous activity.

(iii) The extensive, quartz-rich, cordierite-anthophyllite gneisses where dykes are absent may be younger than the basic igne-ous activity. The unusually high SiO_2, Al_2O_3, MgO content indicated by the mineralogy suggests these rocks may also be metavolcanic.

Malene supracrustal rocks on the numerous small islands west of Qilángârssuit show a variety of contact relations with variably mig-matized leucocratic orthogneisses. Some of these gneisses contain Ameralik dykes but in many instances ages of the gneisses are uncer-tain because they lack dykes, although many are lithologically identical to Amîtsoq banded gneisses. The leucocratic gneisses generally occur as broadly concordant sheets 1–50 m thick and are traceable along strike for up to 500 m, a distribution that might be attributed to post-dyke interleaving of basement and cover slices (schuppen), not igneous intrusion. On the other hand, leucocratic orthogneisses on islands directly west of Qilángârssuit show gradational metasomatic boundaries with Malene supracrustal gneisses, microcline and muscovite in the supracrustal rocks increasing in proportion as the orthogneiss is approached. Ameralik dykes are absent and the orthogneiss may be of Nûk affinity. It is cut by microgranites and aplo-pegmatites believed to be part of the Qôrqut complex of the Godthåbsfjord region, whilst further south Amîtsoq gneiss (Ameralik dykes present) con-tains xenolithic strips and pods of Malene supracrustal gneisses close to the contact with similar gneisses that form the east of the island Querssuaq. The relations strongly suggest that the Amîtsoq gneiss intrudes the Malene rocks.

The relations outlined above have impor-tant implications because some of the Malene supracrustal rocks could be regarded as pre-Amîtsoq, although the intrusive relation seen in Querssuaq may have been derived by post-dyke anatectic reworking of the Amîtsoq basement to the Malene supracrustal rocks so that they became locally disrupted in a manner similar to that of parts of the Ameralik dykes within more extensive outcrops of Amîtsoq gneiss. Although the possibility of a pre-Amîtsoq affinity for some of the Malene rocks remains, we conclude at this stage of our investigations that there are representatives of two supracrustal sequences in the Buksefjorden region, i.e. the older pre-Amîtsoq suite and the younger Malene supracrustal rocks.

Nûk Gneisses

The term Nûk gneiss was first used in print by Black et al. (1971). McGregor's use indicates that members of the suite are younger than Ameralik dykes and Amîtsoq gneisses and younger than the Malene supracrustal rocks. An injection origin is declared. The parents were injected over a long period of time but were mostly syntectonic intrusions that 'migmatized the older rocks, including the earlier phases of the intrusive suite. The last phases were discrete plutons' (McGregor, 1973). Isotope studies on both Nûk and Amîtsoq rocks have led to positive statements (Pankhurst et al., 1973) that the parents of the Nûk could not be derived from Amîtsoq gneiss, although McGregor (*op. cit.*, p. 352) admits that 'it is conceivable that some might be derived from acid volcanic rocks within the supracrustal sequence'. Pankhurst et al. (1973) suggest that the parents of the Nûk were derived from a source region near a lower crust–upper mantle interface, and Black et al. (1973) indicated that 'significant addition of crustal material from upper mantle sources' took place within the general period ~3000–2800 m.y. It will be shown that the term Nûk has become largely unusable, being on the one hand too rigidly defined with insistence on the intrusive origin and on the other too loosely defined encompassing a very wide range of lithotypes developed over a long period of time and, in our opinion, with various geneses. The negative aspect of the definition, i.e. the absence of Ameralik dykes, is clearly unfortunate and was open to criticism even before the discovery of later amphibolite dykes. In a petrographic sense the term is imprecise and covers the same range of lithotypes as does Amîtsoq.

Mapping in the areas around Buksefjorden and especially south and east of the head of the fjord has shown the Nûk rocks to be multi-phase injections, migmatized and deformed. The earliest phases post-date amphibolites presumed to be of the Malene succession, masses of which occur either as agmatitic xenoliths or forming parts of a banded succession. In later phases xenoliths of amphibolites are less abundant and the gneisses are progressively more leucocratic. The earliest phase is widespread in occurrence and characteristically well-banded with prominent hornblende bands, although it is lithologically variable, particularly with regard to its fabric. Commonly it passes over a few metres into a more homogeneous paler gneiss and thence into leucocratic granitic rock which is isotropic or only slightly foliated. Such relations clearly indicate progressive homogenization with associated remobilization. On the other hand, xenoliths, some very sharp edged, some with diffuse margins, of one gneiss type occur in the others (Plate 2A, 2B). Assessing the age relations on intersections indicates clearly that the last phase was a trondhjemitic rock injected in discordant sheets, although lithologically identical material has been derived from the banded gneiss.

In relation to periods of deformation the picture is less clear, with trondhjemite sheets sometimes cutting late folds in the banded gneiss, almost always showing a fabric parallel to the axial surface of the late folds, but also occasionally almost undeformed.

Syntectonic injection is confirmed by relations with small-scale structures. A protoclastic fabric is present in middle and late phases, whilst several periods of highly ductile deformation are also preserved. Granite formation continued through the major deformation into and after the waning stages. In the areas of greatest granite complexity (Plate 2C) the metamorphic grade did not exceed amphibolite facies but there is a gradation to the south where granulite facies was reached. Very similar relations have already been

Fig. 6. Nûk gneisses: sharp-edged xenolithic inclusion of early phase in later, east Buksefjorden

Fig. 7. Nûk gneisses: diffuse-edged xenolithic inclusion of early phase in later, east Buksefjorden

Fig. 8. Nûk gneisses: early gneiss showing high degree of mobility, east Buksefjorden

Fig. 9. Intra-Nûk amphibolite dykes, north of Alángordlia

Fig. 10. Intra-Nûk amphibolite dyke cutting augen and foliated gneiss, north of Alágordlia

described (Chadwick et al., 1974) from Amitsuarssugssuaq in the south part of the Buksefjorden region but in that case the presence of Ameralik dykes indicates that Amîtsoq gneiss as well as early phases of Nûk gneiss are migmatized and incorporated in the later stages.

It is in the eastern part of the ground, south of Eqaluit valley, that the regional metamorphic grade reached granulite facies. In the southern part of the tract extending from Alángordlia at least as far south as Sermilik syntectonic gneisses of various lithologies are found. These, now referred to as the Sermilik Gneisses, include distinctive augen gneiss. Although outcrop is almost complete, connecting these units with the lithologically comparable Ilivertalik granite of the Midgaard region, unpublished isotope studies from the two areas give conflicting results. In the Sermilik area the chronology has been precisely determined (see Table 1). Protoclastic gneisses were injected during a period of isoclinal folding and metamorphosed to granulite facies during a subsequent period of asymmetric folding. In the granulite facies areas two suites of amphibolite dykes occur. These are separable on lithological grounds into relatively thin (0·3 m) sheets of pale amphibolite, and thicker sheets (average 12 m but reaching 100 m in width) of more massive black amphibolites. Members of the first group (Plate 2D, 2E) pre-date the intrusion of the augen gneiss but are discordant to early stages of the Nûk gneiss (and to members of the Malene supracrustal amphibolites) whilst some members of the second group cut the augen gneiss. Both suites have been raised to granulite facies metamorphism. Thus basalt sheet injection took place in the Sermilik–Alángordlia area almost at the height of the metamorphism and at a time of granite activity with high mobility a short distance to the north. Lithologically the dykes are very similar to some of the Ameralik dykes, distinguishable from them by virtue of fortuitous relations exposed at the western end of Alángordlia, where the host for the thin, pale amphibolite sheets cuts and contains xenolithic inclusions of Amîtsoq gneiss with Ameralik dykes.

Qôrqut Granite

The precisely defined Qôrqut Granite (McGregor, 1973) has its main outcrop north of the ground being investigated, but there are extensive sheets of acid material, mostly pegmatitic in the north and north-west. These have been equated with the Qôrqut because of their petrography (being granites ss) and because of their occurrence in horizontal sheets. Nevertheless, in parts of the area around the mouth of Buksefjorden even the pegmatite phases are strongly deformed and the granitic masses logically regarded as Qôrqut pre-date certain of the phases related to Nûk sequences. Clearly the evolution of the Nûk was protracted beginning with granitic injection and proceeding, no doubt in a non-uniform manner over a wide area, throughout a period of high-grade metamorphism and several episodes of deformation. No major breaks occur in the evolution between Nûk and Qôrqut, and a division cannot be made in relation to periods of deformation. One reason why this has not been clearly stated before derives from the published isotopic data which are as follows:

For the Nûk parent 3050 m.y. (Pankhurst et al., 1973).

For the metamorphism of the Nûk rocks at granulite facies 2850 m.y. (Black et al., 1973).

For the injection of the Qôrqut granite 2600 m.y. (Black et al., 1973).

Without wishing to suggest that these dates are in any way *wrong* for the samples from which they were calculated, it is clear that acceptance at face value has not been wholly useful in understanding the geology of the area.

Conclusions

Considerable progress has been made in the last 5 years in understanding the geology of the central West Greenland block. Our state of knowledge is, however, still too uncertain for us to support some of the evolutionary models that have been suggested. The cautionary nature of the present contribution justifies listing the principal outstanding problems.

1. The true relationship between Amîtsoq gneiss and the Malene supracrustal rocks.
2. The relationship between the Ameralik dykes and the Malene rocks.
3. The justification for a chronological subdivision of the Malene rocks.
4. Correlations between (parts of) the Malenes and the Ravn Storø succession (Andersen and Friend, 1973).
5. The possibility of the existence of amphibolites and other supracrustals related neither to Malenes nor Ravn Storø.
6. The explanation of the isochrons for the

Amîtsoq gneisses in view of their extremely complex history.
7. The significance of the initial Sr 87/86 for the Nûk gneisses.
8. The distinction (especially in the field) between Nûk and Amîtsoq gneiss and the exacerbation of this problem if it can be proved that Amîtsoq gneiss post-dates part of the Malene succession.
9. Distinctions between Nûk and Qôrqut granite.
10. The significance of isotope data with respect to the crystallization of the anorthosite bodies.

Acknowledgements

This contribution is based on field studies in progress by the authors and A. D. Gibbs, M. R. Sharpe, J. G. Stainforth, K. J. Vines and P. R. A. Wells. The project is supported jointly by N.E.R.C. and the Geological Survey of Greenland, and we wish to thank the Director, K. Ellitsgaard-Rasmussen, for his continued interest and permission to publish.

References

Andersen, L. S., and Friend, C., 1973. 'Structure of the Ravns Storø amphibolite belt in the Fiskenaesset region', *Rapp. Grønlands geol. Unders.*, **51**, 37–40.

Baadsgaard, H., 1973. U–Th–Pb dates on Zircons from the early Precambrian Amîtsoq gneisses, Godthaab district, West Greenland', *Earth Planet. Sci. Lett.*, **19**, 22–28.

Berthelsen, A., 1955. 'Structural Studies in the Pre-Cambrian of Western Greenland', *Meddr. Grønland, Bd.*, **135**, Nr. 6.

Black, L. P., Gale, N. H., Moorbath, S., Pankhurst, R. J., and McGregor, V. R., 1971. 'Isotopic dating of very early Precambrian amphibolite facies gneisses from the Godthaab District, West Greenland', *Earth Planet. Sci. Lett.*, **12**, 245–259.

Black, L. P., Moorbath, S., Pankhurst, R. J., and Windley, B. F., 1973. '207Pb/206Pb whole rock age of the Archaean Granulite Facies Metamorphic Event in West Greenland', *Nature Phys. Sci.*, **244**, 50–53.

Bridgwater, D., and McGregor, V. R., 1974. 'Field work on the very early Precambrian rocks of the Isua Area, Southern West Greenland', *Rapp. Grønlands geol. Unders.*, **65**, 49–54.

Bridgwater, D., McGregor, V. R., and Myers, J. S., 1974. 'A horizontal tectonic regime in the Archaean of Greenland and its implications for early crustal thickening', *Precambrian Research*, **1**, 179–197.

Chadwick, B., Coe, K., Gibbs, A. D., Sharpe, M. R., and Wells, P. R. A., 1974. 'Field evidence relating to the origin of ~3000 m.y. gneisses in Southern West Greenland', *Nature*, **249**, 136–137.

Kalsbeek, F., and Myers, J. S., 1973. 'The geology of the Fiskenaesset region', *Rapp. Grønlands geol. Unders.*, **51**, 5–18.

McGregor, V. R., 1968. 'Field Evidence of very old Precambrian rocks in the Godthåb Area, West Greenland', *Rapp. Grønlands geol. Unders.*, **15**, 31–35.

McGregor, V. R., 1973. 'The early Precambrian gneisses of the Godthaab district, West Greenland', *Phil. Trans. R. Soc. Lond.*, **A293**, 343–358.

Moorbath, S., O'Nions, R. K., Pankhurst, R. J., Gale, N. H., and McGregor, V. R., 1972. 'Further rubidium-strontium age determinations of the very early Precambrian rocks of the Godthaab district, West Greenland', *Nature Phys. Sci.*, **240**, 78–82.

Noe-Nygaard, A., and Ramberg, H., 1961. 'Geological Reconnaissance Map of the Country between latitudes 69°N and 63°45 N, West Greenland', *Meddr. Grønland Bd.*, **123**, Nr. 5.

Pankhurst, R. J., Moorbath, S., and McGregor, V. R., 1973. 'Late Event in the Geological Evolution of the Godthaab District, West Greenland', *Nature*, **243**, 24–26.

7

Geochemistry of Metavolcanic Amphibolites from South-West Greenland

Giorgio Rivalenti

Instituto di Mineralogia e Petrologia, Università di Modena, Italy.

Summary

Available data on amphibolites and ultramafics (pyroxenites and hornblendites) occurring in supracrustal sequences of high-grade Archaean terrains of Greenland are examined. When metamorphism is isochemical, the amphibolites, which split normatively into tholeiitic and alkali olivine basalts, are similar to basalts of oceanic domains; minor differences consist in lower Ti and higher Cr, Ni and Fe for comparable Mg content. Chemically, ultramafics can be classified as komatiites. The association of komatiites, tholeiites and alkali olivine basalts is interpreted as deriving from partial melting of little differentiated Archaean mantle under low to intermediate pressure and steep geothermal gradient. The three magmas underwent low-pressure fractionation dominated by olivine during their ascent. Similarities with greenstone belts are pointed out.

Introduction

According to Windley and Bridgwater (1971) and Windley (1973) the Archaean terrains may be divided in two types: those dominated by greenstone belts and granites and those dominated by high-grade facies rocks. While the ultramafic–mafic sequence of greenstone belts has received much attention in recent years, owing to its bearing on the evolution of the early crust, data on mafic rocks of the high-grade terrains are scarce. The principal object of this paper is to give a geochemical contribution to the knowledge and interpretation of the mafic rocks of the Archaean craton of Greenland, dominated mainly by terrains of the second type (Windley and Bridgwater, 1971; Bridgwater et al., 1973).

The Archaean history of Greenland is summarized by Bridgwater et al. (1973), who show that in the time span from 3800 to 2600 m.y. amphibolites are met either in older discordant dykes cutting the early granitic basement (McGregor, 1968, 1969, 1973) or, intermixed with metasediments, in younger supracrustal units overlying the granitic basement or in migmatitic gneisses that might represent transformed supracrustal sequences or basement. Anorthosites and associated igneous bodies postdate the supracrustal sequence (Windley et al., 1973). High-grade metamorphism is dated at about 2800 m.y.

In this paper the amphibolites in the Archaean basement of Fiskenaesset, Frederikshåb and Neria (between Frederikshåb and Ivigtut) are examined.

213

Geological and Petrographic Features

The main geological and petrographic features of the Fiskenaesset, Frederikshåb and Neria areas are reported by Jensen (1966, 1968), Andrews (1968), Dawes (1968, 1970), Higgins (1968), Rivalenti and Rossi (1972), Kalsbeek (1970), Kalsbeek and Myers (1973), Myers (1973), Hopgood (1973), Andersen and Friend (1973) and in a number of unpublished field reports of the geologists of the Geological Survey of Greenland.

In the Frederikshåb and Fiskenaesset areas the amphibolites occur in horizons intercalated either with metasediments (pelites, semipelites and graywackes) or with gneisses, which are, at least in part, migmatized metasediments. In the Neria area amphibolites form layers and agmatites in migmatitic gneisses interpreted by Kalsbeek as isochemically metamorphosed metasediments. Primary structures in the high-grade terrains are rare, although relict pillow structures and primary textural features have been recognized (Dawes, 1970; Myers, 1973; Kalsbeek and Myers, 1973). In the occurrences examined, therefore, amphibolites appear to be the mafic extrusives of a supracrustal sequence. In many bodies some stratigraphy is recognizable, starting with ultramafic lenses and layers on one side of the amphibolite body and passing to normal amphibolite on the other side. Ultramafics consist of lenses of peridotites or bands of pyroxenites and hornblendites. The structure of single amphibolite layers is generally massive or banded, single bands reaching a thickness up to several metres. The metamorphic grade is of granulite and amphibolite facies in the Fiskenaesset region, amphibolite facies (in part retrogressed from granulite; Rivalenti and Rossi, 1972) in the Frederikshåb area and amphibolite facies in the Neria area.

Chemistry

The average chemical composition of amphibolites from Fiskenaesset, Frederikshåb and Neria are reported in Table 1. Average composition of Fiskenaesset ultramafics (pyroxenites and hornblendites only) is reported in Table 2. A study of the correlations between the various major and trace elements has shown that, except for water, K and Rb, the Fiskenaesset rocks have not suffered important chemical metamorphic alterations; water is higher in amphibolite facies (average 1·40%) than in granulite facies (average 1·07%), the K/Rb ratio is 592 in granulites and 281 in amphibolite facies rocks (Rivalenti and Rossi, 1975). A plot of K versus Rb shows that the granulite facies and amphibolite facies samples can be completely discriminated (Fig. 2 in Rivalenti and Rossi, 1975), which demonstrates that one or both of the elements have undergone some mobilization. The mutual relationships of all the other elements, as shown by a correlation matrix (Table 5 in Rivalenti and Rossi, 1975), and their similar behaviour in granulite and amphibolite facies have lead to the conclusion that they behave as relatively immobile during metamorphism. The Frederikshåb amphibolites, although of igneous origin, probably owe their chemical variability in part to metamorphic differentiation (Rivalenti, 1970) and those of Neria are considered by Kalsbeek and Leake (1970) as the product of isochemical metamorphism of basaltic tuffs in part mixed with sedimentary material.

In spite of the different processes affecting the amphibolites in the three areas, the averages of Table 1 reveal that they are remarkably similar. In particular, the similarity is much greater between the Fiskenaesset and Frederikshåb amphibolites, except for Ni and Cr, than with the Neria amphibolites, that differ from the others mainly for a sensibly higher SiO_2 and K_2O content. The Fiskenaesset and Neria amphibolites are also very similar to the average Archaean metabasalts of Canada reported by Goodwin (1968).

Normatively, the amphibolites of Fiskenaesset split into two groups: tholeiitic and alkali olivine basalts. Those of Frederikshåb and Neria all have a tholeiitic composition.

Chemical Affinities

Although the average composition of the three areas in question is similar, it is safer to

Table 1 Averages for Archaean amphibolites of Greenland compared with Archaean amphibolites of Canada and with oceanic and island arc tholeiites. A = 22 tholeiitic amphibolites from Fiskenaesset (Rivalenti and Rossi, 1975); B = 22 amphibolites from Fiskenaesset with alkali olivine basalt composition (Rivalenti and Rossi, 1975); C = 23 tholeiitic amphibolites from Frederikshåb (Rivalenti, 1970; Rivalenti and Rossi, 1972); D = 54 Neria tholeiitic amphibolites (Kalsbeek and Leake, 1970); E = 162 Canadian Archaean amphibolites (Goodwin, 1968); F = range in oceanic basalts (data from: Engel et al., 1965; Manson, 1967; Aumento, 1969; Jakeš and Gill, 1970; Shido et al., 1971: only abyssal tholeiites with $1.50 < Fe/Mg > 1.69$; Hallberg and Williams, 1972; Dmitriev and Bougault, 1973); G = range in island arc tholeiites (data from: Jakeš and Gill, 1970; Jakeš and White, 1971; Hallberg and Williams, 1972).

	(A)	(B)	(C)	(D)	(E)	(F)	(G)
SiO_2	49·50	47·61	50·16	51·05	48·9	49·3–50·5	45–70
TiO_2	1·12	0·98	1·10	1·03	1·06	1·4–2·4	0·5–1·5
Al_2O_3	14·43	14·65	14·53	14·13	14·5	14·6–17·0	14–19
Fe_2O_3	3·24	2·73	2·55	4·23	2·14	0·9–3·2	1·6–8·8
FeO	8·39	8·94	10·27	8·04	9·03	6·8–9·7	1·8–7·0
MnO	0·20	0·20	0·21	0·20	0·21	0·13–0·20	0·12–0·23
MgO	6·90	7·00	6·72	7·63	6·27	7·1–8·6	0·6–6·7
CaO	10·55	11·71	10·48	8·95	8·74	10·6–11·7	2·6–11·7
Na_2O	2·78	3·31	1·92	2·55	2·51	2·2–2·8	2·4–6·7
K_2O	0·46	0·54	0·36	1·03	0·45	0·12–0·53	0·4–1·2
P_2O_5	0·14	0·10	0·04	0·25	0·07	0·15–0·26	0·06–0·58
H_2O^+	1·37	1·03	1·32		3·34	0·32–0·79	1·7–4·3
Cr	366	342	242	310		296–400	2–50
Ni	141	123	83	217		97–170	6–30
Rb	11	11				0·2–5	3–10
Sr	121	122				70–130	100–200
Zr	80	62				95–115	54–103
Ba	20	15		140		14–66	50–279
CaO/Al_2O_3	0·73	0·80	0·72	0·70	0·60	0·69–0·73	0·21–0·74
tot FeO/tot FeO + MgO	0·62	0·62	0·65	0·61	0·64	0·54–0·61	0·59–0·84
Na_2O/K_2O	6·04	6·13	5·33	2·48	5·58	4·24	3–6

restrict the comparison with recent basaltic groups to the isochemically metamorphosed rocks of Fiskenaesset and use the others with caution.

Tholeiitic and Alkali Olivine Basalts

The Fiskenaesset alkali olivine basalts differ from the associated tholeiites only on account of their lower silica and higher sodium content; as regards the concentrations of all the other elements, including incompatible and transitional elements, they are similar. The low content of incompatible elements of the alkalic group makes it different from any actual alkali basalt series. Both alkali and tholeiitic basalts follow an iron-enrichment trend (Fig. 1), which excludes affinities with calc–alkalic suites. The general low content of incompatible elements and high Cr, Ni, Mg and Fe suggests affinities with basalts of oceanic or island arc domains. The comparison reported in Table 1 shows that for most elements the affinities are greater with oceanic basalts. Oceanic affinities are also indicated by the discrimination diagrams of Pearce and Cann (1973): in spite of extrapolation to metamorphosed Archaean rocks discrimination fields constructed with actual basalts, most amphibolites plot in the ocean-floor basalt field (Fig. 2). Some differences with recent oceanic tholeiites do, however, occur, the Fiskenaesset rocks having a lower TiO_2 and Na_2O/K_2O ratio, more similar to island arc values. A low titanium content seems a characteristic common to Archaean mafic suites (Glikson, 1971, 1972; Hallberg and Williams, 1972). The low Na_2O/K_2O ratio should not be considered significant, for

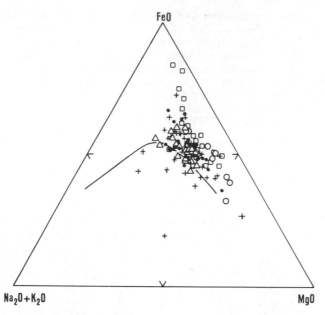

Fig. 1. (MgO)–(total FeO)–(Na$_2$O + K$_2$O) plot showing the iron enrichment trend of the amphibolites. Dots = Fiskenaesset tholeiites; triangles = Fiskenaesset alkali olivine basalts; squares = Frederikshåb amphibolites; crosses = Neria amphibolites: circles = Fiskenaesset ultramafics. The line of separation between tholeiitic and other magma types is plotted according to Irvine and Baragar, 1971

potassium has probably varied as a result of metamorphic alteration (Rivalenti and Rossi, 1975). Moreover, oceanic basalts with a Na$_2$O/K$_2$O ratio as low as that of the Fiskenaesset rocks do occur (Manson, 1967). Other minor differences, which do not, however, affect their oceanic affinity and which seem peculiar to Archaean oceanic rocks (Glikson, 1971, 1972), are that, for any given MgO content, the Greenland rocks are richer in iron, Ni and Cr than actual oceanic basalts.

The presence of alkali basalts does not contrast with oceanic similarities, for such a feature is known in oceanic environments (Aumento et al., 1971).

The main difference between the Frederikshåb amphibolites and those from the Fiskenaesset region is that the former are lacking in alkalis and have a lower Cr and Ni content; however, the Cr and Ni values still fall within the range for oceanic basalts and are considerably higher than those found for island arc basalts.

The oceanic character of the Neria amphibolites, if such it was, is actually masked by the chemical alteration due to the postulated mixing with sediments, which increases at least potassium well beyond the content of oceanic rocks. They differ, however, from calc–alkali and shoshonitic island arc suites on account of their iron-enrichment trend, their high Cr and Ni and their lower Ba and potassium for comparable Na$_2$O/K$_2$O values, respectively.

Pyroxenites and Hornblendites

These are characterized by: high CaO, MgO, Cr and Ni; low TiO$_2$, P$_2$O$_5$, Sr, Zr and Ba, equal, or slightly lower, than in tholeiitic or alkali amphibolites; a CaO/Al$_2$O$_3$ ratio bigger than 1. Rocks with these characteristics have been described in the greenstone belts of South Africa by Viljoen and Viljoen (1969a,b,c), who consider them as a new class of rocks classified as peridotitic and basaltic komatiites, and in the greenstone belts of

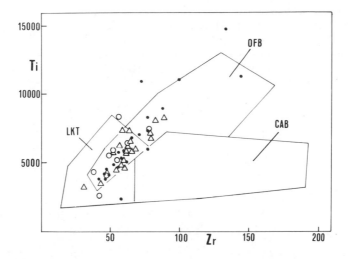

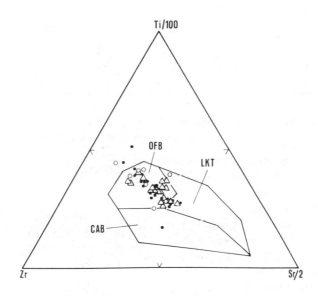

Fig. 2. Discrimination diagrams according to Pearce and Cann (1973). OFB = ocean floor basalts: LKT = low potassium tholeiites; CAB = calc alkali basalts. Other symbols as in Fig. 1

Australia (Williams, 1972) and Canada (Pyke et al., 1973). They differ, however, from typical komatiites with similar magnesium content in that they have lower silica (Table 2). They differ more markedly from picritic basalts, as is evident from a comparison with Clarke's (1970) analyses, which are lower in iron and higher in TiO_2, Zr, Ba and Sr, and with the picrites of the Lesser Antilles island arc (Sigurdsson et al., 1973), which, for a comparable MgO content, have much higher TiO_2, Sr, Zr, Ba and almost double the Al_2O_3 content (Table 2). The ultramafics in question have the greatest affinity, therefore, with komatiites, although the Greenland rocks should be considered an anomalous group.

Table 2 Average of 9 Fiskenaesset pyroxenites and hornblendites (A), compared with (B) South African komatiites with similar MgO (Viljoen and Viljoen, 1969a), with (C) Baffin Island picrites (Clarke, 1970) and (D) Lesser Antilles island arc picrites (Sigurdsson et al., 1973)

	(A)	(B)	(B)	(C)	(C)	(D)
SiO_2	47·81	52·73	52·22	47·5	46·8	45·19
TiO_2	0·92	0·85	0·56	0·97	1·51	0·96
Al_2O_3	7·94	9·83	5·42	13·8	12·9	16·67
Fe_2O_3	2·99	1·23	0·98	3·3	2·9	2·20
FeO	10·11	9·70	8·88	6·7	8·0	6·85
MnO	0·25	0·22	0·22	0·17	0·17	0·17
MgO	11·72	10·10	15·25	11·8	11·2	11·61
CaO	12·83	9·99	12·83	11·7	11·6	11·78
Na_2O	2·15	2·65	1·21	1·53	2·00	3·27
K_2O	0·44	0·46	0·09	0·10	0·25	0·95
P_2O_5	0·09	0·06	0·05	0·10	0·18	0·34
H_2O^+	1·63	1·93	2·05	2·20	2·25	
Cr	1035			890	753	766
Ni	341			314	393	355
Rb	16			2	5	26
Sr	84			161	286	756
Zr	53			66	98	126
Ba	14			56	95	592
CaO/Al_2O_3	1·62	102	2·37	0·85	0·90	0·71
tot FeO/tot FeO+MgO	0·52	0·51	0·39	0·47	0·49	0·43
Na_2O/K_2O	4·89	5·76	13·44	15·30	8·00	3·44

On the Primitive Character of the Magmas

Rocks with the chemical characteristics described above are generally considered 'primitive' in that they should represent direct melts of the mantle (Engel et al., 1965; Gale, 1973). If, however, the $F = (FeO)/(FeO+MgO)$ ratio is taken as a differentiation index, the three groups show a wide range of variation, the komatiites ranging between 0·41 and 0·70, the alkali olivine basalts between 0·53 and 0·74 and the tholeiites between 0·41 and 0·94. Fig. 3 plots the F ratio against Al_2O_3, CaO and MgO. It is evident that the fractionation trend of the three groups is dominated by olivine, which causes a marked depletion of MgO from the magma. The possibility of an orthopyroxene fractionation instead of olivine is ruled out by a silica versus F plot (Fig. not shown) where silica remains constant at increasing F. In the komatiites olivine fractionation is accompanied by clinopyroxene fractionation, which reduces CaO in the residue with respect to Al_2O_3, thus causing alumina enrichment. In the alkali olivine basalts and in the tholeiites

olivine fractionation may also be accompanied by fractionation of some clinopyroxene and plagioclase, which on the one hand causes a slight decrease in CaO and on the other keeps Al_2O_3 constant. Such a feature can be interpreted as a low-pressure fractionation trend for both tholeiites and alkali basalts (see Green and Ringwood, 1967). The komatiite trend can also be interpreted as a low-pressure fractionation trend, but starting from a different parental magma, possibly similar to the peridotitic komatiites of Viljoen and Viljoen (1969a).

The iron-enrichment trend and the constancy of silica as differentiation increases (Fig. not shown) indicate that differentiation occurred under low or decreasing oxygen fugacity (Osborn, 1959; Hamilton and Anderson, 1967).

Hypothesis as to the Origin of the Magma Suites

Whatever hypothesis is put forward, it must explain the following points:

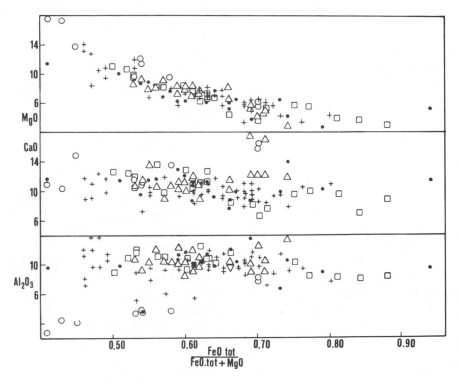

Fig. 3. Plot of total FeO/total FeO + MgO against Al_2O_3, CaO and MgO. Symbols as in Fig. 1. See text for discussion

(1) The existence of the three magma types, each of which shows a low-pressure fractionation trend, and their close association;

(2) The higher Cr and Ni and slightly lower incompatible element content of the komatiites;

(3) The similar content of incompatible and coherent trace elements in alkali basalts and tholeiites.

It is well known (Yoder and Tilley, 1962) that tholeiites and alkali olivine basalts cannot be related to each other by low-pressure fractionation because of the impossibility of straddling the thermal divide diopside–olivine–plagioclase below 8 kb. Above 8 kb the thermal divide disappears and there are several mechanisms which enable residual liquids to be derived at one side of the olivine-gabbro plane from parental liquids at the other (Yoder and Tilley, 1962; Tilley and Yoder, 1964; Green and Ringwood, 1964, 1967;

Kushiro, 1965; O'Hara and Yoder, 1967). The derivation of alkali olivine basalts from olivine tholeiites is determined by the fractionation of aluminous orthopyroxene or of orthopyroxene + subcalcic clinopyroxene (Green and Ringwood, 1967), which would result in a decrease of silica, and increase of alkalis, alumina and calcium in the residue. According to the distribution coefficients reported by Gast (1968), such a process would enrich the residual liquid at least in Ni and incompatible elements, although not to the extent to which the actual alkali basalts are enriched, in respect of the parent olivine tholeiite. As this clearly does not happen in the present case, tholeiites and alkali olivine basalts are unlikely to be genetically related by fractionation at high pressure.

Tholeiites and alkali olivine basalts can also be produced by different amounts of partial melting from pyrolitic mantle at different pressures (Green and Ringwood, 1967) or of garnet–lherzolite mantle with eclogite (high P

and *T*) or lherzolite and harzburgite (lower *P* and *T*) fractionation (O'Hara, 1968). None of these processes, except for the high-pressure eclogite fractionation, would produce a marked enrichment of incompatible elements and they are thus able to afford an explanation for the similarity in trace element content of alkali olivine and tholeiitic basalts. In such a scheme, the production of these magmas can be envisaged in a low to intermediate pressure range, in different thermal environments (O'Hara, 1968).

Komatiites have been said to derive from 80% partial melting of pyrolitic mantle under adiabatic conditions (Green 1972), from partial melting of Archaean undifferentiated mantle (Viljoen and Viljoen, 1969c), from partial melting of mantle in mid-oceanic environments at less than 5 kb (Gale, 1973), from partial melting of a chemically layered mantle (Cawthorn and Strong, 1974) and from partial melting of a four-phase lherzolite mantle above 30 kb (McIver and Lenthall, 1974). Although a mechanism of meteoritic impact as proposed by Green (1972) seems improbable because of the lack of any impact structure in the fields examined, it does not seem possible to escape the postulation of a strong decompression, to emplace the komatiitic magma as extrusives, and of a high geothermal gradient, to keep it at the liquidus at relatively shallow depth.

The similar content of incompatible elements in the three magmas may be explained if the enrichment factor for these elements in magmas produced by partial melting of the mantle is taken into account. It is known (O'Hara, 1968) that the enrichment factor of incompatible elements in the melt is at its maximum during the initial stages of melting, decreases abruptly as melting increases, remains roughly constant as melting proceeds (i.e. for more than 20% partial melts in a garnet–lherzolite model, according to O'Hara, 1968). Thus no significant variation of incompatible elements is to be expected in the present magmas, which are formed by at least 20% and more partial melting, except, as is the case, for a slightly lower content in most advanced partial melting products, i.e. komatiites.

It is beyond the scope of this paper to draw inferences regarding the Archaean mantle, and in any event the data are as yet too inadequate to make such inferences reasonable, but it is worth pointing out that the particular abundance of Fe, Ni and Cr and the low Ti of the Archaean amphibolites, at least of those of Fiskenaesset and Neria, are a characteristic found in other Archaean volcanic suites in other parts of the world (Glikson, 1972; Hallberg, 1972) and this points to the existence of a less differentiated Archaean mantle.

From the foregoing, a hypothesis explaining the three magmas formed at the same sites may be proposed. It has been shown that they may derive from partial melting of mantle in a relatively narrow, low to intermediate pressure range under a steep geothermal gradient. The mechanism that is most likely to be able to produce such a feature consists, as proposed by Windley (1973) for greenstone belts, in the formation of a proto-oceanic ridge system determined by rifting of the thin proto-crust by diapiric upwelling of the mantle (Hunter, 1974). The magnitude and duration of the mantle plume will control the amount of distension and the type of magma produced. Initial thinning and rifting of the proto-crust has adiabatically emplaced the mantle high enough to produce the extrusive komatiites, while the magmas representing less extensive mantle partial melts might have been formed in the following less critical conditions. Supporting this hypothesis is the presence of ultramafics on one side only of the amphibolites, probably at the bottom, and the geochemical evidence already discussed. It is unfortunate that the reconnaissance sampling was not careful enough to permit the reconstruction of the stratigraphy relative to the alkali olivine and tholeiitic rocks, for their relationship might give some indications as to the sequence of phenomena determining the association.

If the amphibolites of the supracrustal terrains formed according to the model presented, i.e. in downsagging basins of the early protocrust, then the mafic dykes, now discordant amphibolites that cut the early basement rocks of Greenland (Dawes, 1970; McGregor,

1973), might represent the feeders, through the rift faults, where the ascending magma underwent the low-pressure olivine fractionation, and should therefore be genetically related to the supracrustal amphibolites.

Conclusions

The chemical characteristics of the amphibolites of high-grade supracrustal sequences, at least where metamorphic alteration has not masked the original characteristics, indicate affinities with oceanic environments. It is tentatively suggested that ultramafic–mafic sequences of these terrains were formed in proto–oceanic ridge zones by mantle upwelling and rifting of the early proto-crust under a high geothermal gradient.

The ultramafic–mafic association and geochemical characteristics of the high-grade supracrustals are very similar to those found in greenstone belts (Anhaeusser et al., 1969; Goodwin and Shklanka, 1967; Viljoen and Viljoen, 1969a,b,c; Glikson, 1971, 1972). Such a similarity may suggest that greenstone belts and high-grade-supracrustal sequences differ in the post-formation history and not in their genesis.

Acknowledgements

Thanks are due to the Director of the Geological Survey of Greenland (GGU), Dr K. Ellitsgaard-Rasmussen, for permission to publish this paper. A grant of C.N.R. (Rome) is also acknowledged.

References

Andersen, L. S., and Friend, C., 1973. 'Structure of the Ravns Storø amphibolite belt in the Fiskenaesset region', *Rapp. Grønlands geol. Unders.*, **51**, 37–40.

Andrews, J. R., 1968. 'The structural chronology of the Nigerdlikasik area, Frederikshåb District', *Rapp. Grønlands geol. Unders.*, **15**, 45–48.

Anhaeusser, C. R., Mason, R., Viljoen, M. J., and Viljoen, R. P., 1969. 'A reappraisal of• some aspects of Precambrian shield geology, *Bull. Geol. Soc. Am.*, **80**, 2175–2200.

Aumento, F., 1969. 'Diorites from the Mid-Atlantic Ridge at 45°N', *Science*, **165**, 1112–1113.

Aumento, F., Loncarevic, B. D., and Ross, D. I., 1971. 'Hudson Geotraverse: geology of the Mid-Atlantic Ridge at 45°N', *Phil. Trans. R. Soc. Lond.*, **A268**, 623–650.

Bridgwater, D., Watson., and Windley, B. F., 1973. 'The Archaean craton of the North Atlantic region', *Phil. Trans. R. Soc. Lond.*, **A273**, 493–519.

Cawthorn, R. G., and Strong, D. F., 1974. 'The petrogenesis of komatiites as evidence for a layered Upper Mantle', *Earth Planet. Sci. Lett.*, **23**, 369–375.

Clarke, D. B., 1970. 'Tertiary basalts of Baffin Bay: possible primary magma from the mantle', *Contr. Mineral and Petrol.*, **25**, 203–224.

Dawes, P. R., 1968. 'Geological investigations on Dalagers Nunatakker, with special reference to metamorphosed basic dykes', *Rapp. Grønlands geol. Unders.*, **15**, 35–40.

Dawes, P. R., 1970. 'Bedrock geology of the nunataks and semi-nunataks in the Frederikshåbs Isblink area of southern West Greenland', *Rapp. Grønlands geol. Unders.*, **29**, 1–60.

Dmitriev, L., and Bougault, H., 1973. 'The petrological interpretation of basalt chemistry', *Deep Sea Drilling Project*, University of California, San Diego, 30–50.

Engel, A. E. J., Engel, C. G., and Havens, R. G., 1965. 'Chemical characteristics of Oceanic basalts and the Upper Mantle', *Bull. Geol. Soc. Am.*, **76**, 719–734.

Gale, G. H., 1973. 'Paleozoic basaltic komatiite and ocean-floor type basalts from Northeastern Newfoundland', *Earth Planet. Sci. Lett.*, **18**, 22–28.

Gast, P. W., 1968. 'Trace element fractionation and the origin of tholeiitic and alkaline magma types', *Geochim. Cosmochin. Acta*, **32**, 1057–1086.

Gill, J. B., 1970. 'Geochemistry of the Viti Levu, Fiji, and its evolution as an island arc', *Contr. Mineral. and Petrol*, **27**, 179–203.

Glikson, A. Y., 1971. 'Primitive Archaean element distribution patterns: chemical evidence and geotectonic significance', *Earth Planet. Sci. Lett.*, **12**, 309–320.

Glikson, A. Y., 1972. 'Early Precambrian evidence of a primitive ocean crust and island arc nuclei of sodic granite', *Bull. Geol. Soc. Am.*, **83**, 3323–3344.

222

Goodwin, A. M., 1968. 'Archaean protocontinental growth and early history of the Canadian Shield', *Internat. Geol. Congr. 23rd*, **1**, 69–89.

Goodwin, A. M., and Shklanka, R., 1967. 'Archaean volcano-tectonic basins: form and pattern', *Can. J. Earth Sci.*, **4**, 777–795.

Green, D. H., 1972. 'Archaean greenstone belts may include terrestrial equivalents of Lunar Maria?', *Earth Planet. Sci. Lett.*, **15**, 263–270.

Green, D. H., and Ringwood, A. E., 1964. 'Fractionation of basalt magmas at high pressures', *Nature*, **201**, 1276–1279.

Green, D. H., and Ringwood, A. E., 1967. 'The genesis of basaltic magmas', *Contr. Mineral. and Petrol.*, **15**, 103–190.

Hallberg, J. A., 1972. 'Geochemistry of the Archaean volcanic belts in the Eastern Goldfields of Western Australia', *J. Petrol.*, **13**, 45–56.

Hallberg, J. A., and Williams, D. A. C., 1972. 'Archaean mafic and ultramafic rock association in the Eastern Goldfields region, Western Australia', *Earth Planet. Sci. Lett.*, **15**, 191–200.

Hamilton, D. L., and Anderson, G. M., 1967. 'Effects of water and oxygen pressures on the crystallization of basaltic magmas', In Hess, H. H., and Poldervaart, A. (Eds.), *Basalts— The Poldervaart Treatise on Rocks of Basaltic Composition*, John Wiley, New York, Vol. 1, 445–482.

Higgins, A. K., 1968. 'The Tartoq Group on Nunaquqortoq in the Itlerdlak area, South-West Greenland', *Rapp. Grønlands geol. Unders.*, **17**, 1–17.

Hopgood, A. M., 1973. 'The pre-Ketilidian history of the gneisses North-West of Frederikshåbs Isblink, Fiskenaesset region', *Rapp. Grønlands geol. Unders.*, **51**, 54–59.

Hunter, D. R., 1974. 'Crustal development in the Kaapvaal Craton, I. The Archaean', *Precambrian Research*, **1**, 259–294.

Irvine, T. N., and Baragar, W. R. A., 1971. 'A guide to the chemical classification of the common volcanic rocks', *Can. J. Earth Sci.*, **8**, 523–548.

Jakeš, P., and Gill, J. B., 1970. 'Rare Earth elements and the island arc tholeiitic series', *Earth Planet. Sci. Letters*, **9**, 17–28.

Jakeš, P., and White, A. J. R., 1971. 'Composition of island arcs and continental growth', *Earth Planet. Sci. Lett.*, **12**, 224–230.

Jensen, S. B., 1966. 'Field work in the Frederikshåb area', *Rapp. Grønlands geol. Unders.*, **11**, 32–35.

Jensen, S. B., 1968. 'Field work in the Frederikshåb area', *Rapp. Grønlands geol. Unders.*, **15**, 40–44.

Kalsbeek, F., 1970. 'The petrography and origin of the gneisses, amphibolites and migmatites in the Qasigialik area, South-West Greenland', *Bull. Grønlands geol. Unders.*, **83**, 70 pp.

Kalsbeek, F., and Leake, B. E., 1970. 'The chemistry and origin of some basement amphibolites between Ivigtut and Frederikshåb, South-West Greenland', *Bull. Grønlands geol. Unders.*, **90**, 36 pp.

Kalsbeek, F., and Myers, J. S., 1973. 'The geology of the Fiskenaesset region', *Rapp. Grønlands geol. Unders.*, **51**, 5–18.

Kushiro, I., 1965. 'The liquidus relations in the system forsterite-$CaAl_2SiO_6$-silica and forsterite-nepheline-silica at high pressures', *Carnegie Inst. Wash. Yearbook*, **64**, 103–104.

Manson, V., 1967. 'Geochemistry of basaltic rocks: major elements', In Hess, H. H., and Poldervaart, A. (Eds.), *Basalts—The Poldervaart Treatise on Rocks of Basaltic Composition*, John Wiley, New York, Vol. 1, 215–269.

McGregor, V. R., 1968. 'Field evidence of very old Precambrian rocks in the Godthåb area, West Greenland', *Rapp. Grønlands geol. Unders.*, **15**, 31–35.

McGregor, V. R., 1969. 'Early Precambrian geology of the Godthåb area, *Rapp. Grønlands geol. Unders.*, **19**, 28–30.

McGregor, V. R., 1973. 'The early Precambrian gneisses of the Godthåb District, West Greenland', *Phil. Trans. R. Soc. Lond.*, **A273**, 343–358.

McIver, J. R., and Lenthall, D. H., 1974. 'Mafic and ultramafic extrusives of the Barberton Mountain Land in terms of CMAS system', *Precambrian Research*, **1**, 327–343.

Myers, J. S., 1973. 'Field evidence concerning the origin of early Precambrian gneisses and amphibolites in part of the Fiskenaesset region', *Rapp. Grønlands geol. Unders.*, **51**, 19–22.

O'Hara, M. J., 1968. 'The bearing of phase equilibria studies in synthetic and natural systems on the origin and evolution of basic and ultrabasic rocks', *Earth-Sci. Rev.*, **4**, 69–133.

O'Hara, M. J., and Yoder, H. S., 1967. 'Formation and fractionation of basic magmas at high pressures', *Scot. J. Geol.*, **3**, 67–117.

Osborn, E. F., 1959. 'Role of oxygen pressure in the crystallization and differentiation of basaltic magma', *Amer. J. Sci.*, **257**, 609–647.

Pearce, J. A., and Cann, J. R., 1973. 'Tectonic setting of basic volcanic rocks determined using trace element analyses', *Earth Planet. Sci. Lett.*, **19**, 290–300.

Pyke, D. R., Naldrett, A. J., and Eckstrand, O. R., 1973. 'Archaean ultramafic flows in Munro Township, Ontario', *Bull. Geol. Soc. Am.*, **84**, 955–970.

Rivalenti, G., 1970. 'Genetical problems of banded amphibolites in the Frederikshåb District, South West Greenland', *Atti Soc. Tosc. Sci. Nat. Mem.*, **A77**, 342–357.

Rivalenti, G., and Rossi, A., 1972. 'The geology and petrology of the Precambrian rocks to the North-East of the fjord Qagssit, Frederikshåb District, South-West Greenland', *Bull. Grønlands geol. Unders.*, **103**, 98 pp.

Rivalenti, G., and Rossi, A., 1975. 'Geochemistry of Precambrian amphibolites in an area near Fiskenaesset, South West Greenland', *Bull. Soc. Geol. It.*, in press.

Shido, F., Miyashiro, A., and Ewing, M., 1971. 'Crystallization of abyssal tholeiites', *Contr. Mineral. and Petrol.*, **31**, 251–266.

Sigurdsson, H., Tomblin, J. F., Brown, G. M., Holland, J. G., and Arculus, R. J., 1973. 'Strongly undersatured magmas in the Lesser Antilles Island Arc', *Earth Planet. Sci. Lett.*, **18**, 285–295.

Tilley, C. E., and Yoder, H. S., 1964. 'Pyroxene fractionation in mafic magma at high pressures and its bearing on basalt genesis', *Carnegie Inst. Wash. Yearbook*, **63**, 114–121.

Viljoen, M. J., and Viljoen, R. P., 1969a. 'The geology and geochemistry of the lower Onverwacht group and a proposed new class of igneous rocks', In *Upper Mantle Project*, Geol. Soc. S. Africa Spec. Pub., **2**, 55–86.

Viljoen, M. J., and Viljoen, R. P., 1969b. 'Evidence of the existence of a mobile extrusive peridotitic magma from the Komati Formation of the Onverwacht Group', In *Upper Mantle Project*, Geol. Soc. S. Africa Spec. Pub., **2**, 87–112.

Viljoen, R. P., and Viljoen, M. J., 1969c. 'Evidence for the composition of the primitive mantle and its products of partial melting from a study of the rocks of the Barberton Mountain Land', In *Upper Mantle Project*, Geol. Soc. S. Africa Spec. Pub., **2**, 275–296.

Williams, D. A. C., 1972. 'Archaean ultramafic, mafic and associated rocks, Mt Monger, Western Australia', *J. Geol. Soc. Australia*, **19**, 163–188.

Windley, B. F., 1973. 'Crustal development in the Precambrian', *Phil. Trans. R. Soc. Lond.*, **A273**, 321–341.

Windley, B. F., and Bridgwater, D., 1971. The Evolution of Archaean Low- and High-grade Terrains, Spec. Pub. Geol. Soc. Australia, **3**, 33–46.

Windley, B. F., Herd, R. K., and Bowden, A. A., 1973. 'The Fiskenaesset complex, West Greenland. Part I. A preliminary study of the stratigraphy, petrology, and whole rock chemistry from Qeqertarssuatsiaq', *Bull. Grønlands geol. Unders.*, **106**, 80 pp.

Yoder, H. S., and Tilley, C. E., 1962. 'Origin of basalt magmas: an experimental study of natural and synthetic rock systems', *J. Petrol.*, **3**, 342–532.

Metamorphism of Archaean Rocks of West Greenland

F. KALSBEEK

The Geological Survey of Greenland, Copenhagen, Denmark.

The Archaean of West Greenland consists mainly of migmatitic gneisses and amphibolites with local major and minor occurrences of anorthositic and associated rocks. One of the most striking features of the rocks in most outcrops is their extreme inhomogeneity. The gneisses commonly are irregularly pegmatite banded, often they contain inclusions of darker gneiss and amphibolite, and the amphibolites and anorthositic rocks are generally stronger migmatized. The gneisses themselves are variable in composition grading from tonalitic to granitic types, but these differences cannot generally be seen in the field, the rocks being too fine-grained and fresh.

To obtain a quantitative impression of the distribution of the different rock types throughout the Fiskenæsset region, which is currently being mapped by the Geological Survey of Greenland (GGU), a collection of 300 rock samples, taken on the basis of a grid covering *c.* 4000 km^2, was studied. The grid localities were visited by helicopter, two samples being taken at each of 150 grid points. To promote unbiased sampling, location of the grid points in the field was left to the pilot. It was attempted to obtain two representative samples at each locality, but due to the inhomogeneity of the rocks the choice of representative samples was generally a matter of personal judgement. The collection can therefore at best be regarded as representative for the region, but not as a true random sample.

Thin sections of the 300 samples were studied, and the distribution of the three major rock types (Table 1) agrees well with earlier estimates based on the study of sand

Table 1. Distribution of major rock types, exclusive late dolerite dykes, in the Fiskenæsset region

	Vol. %
Granitic gneiss (K-feldspar $\geq 20\%$)	16·0
K-feldspar gneiss ($10\% \leq K\text{-fsp} < 20\%$)	9·0
Plagioclase gneiss (K-fsp $< 10\%$, hbl. $< 10\%$)	52·0
Hornblende gneiss hornblende $\geq 10\%$)	5·3
Quartz-poor gneiss (quartz $\leq 10\%$)	2·0
Total gneisses	84·3
Amphibolites	9·3
Anorthositic and associated rocks	5·0
Other rocks (mylonite, albitite, dolerite)	1·3
	99·9

226

samples (Kalsbeek et al., 1974). The most common gneiss type 'plagioclase gneiss' has a tonalitic to granodioritic composition, generally with less than 10% (visual estimate) of microcline. More granitic gneisses also occur and microline-rich gneisses (>20% microline) seem to be more common than gneisses with 10–20% microline. The mineralogical composition of the gneisses is shown in Fig. 1.

Biotite is the most common mafic mineral in the gneisses, while hornblende occurs in about a third of the samples. Hypersthene and diopside occur in a number of gneiss samples mainly in the north-western part of the region,

but some preliminary results were obtained. The information that has been used is listed below.

Hypersthene is restricted to the north-western part of the region and a minor area just north-west of the head of Bjørnesund. It occurs both in gneisses and in amphibolites.

Diopside occurs irregularly in a large area in the northern part of the region; it is found in amphibolites, anorthositic rocks and in some of the gneisses. In the southern, lower grade, part of the terrain, diopside has only been found in one amphibolite very near to

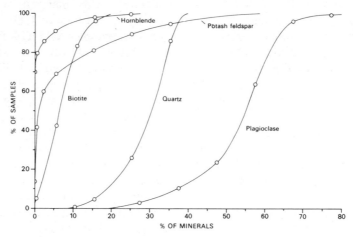

Fig. 1. Mineralogical composition of gneisses in the Fiskenæsset region. The indicated percentages of samples have *less* than the indicated amounts of the different minerals. Based on visual estimates for 253 gneiss samples

whereas muscovite and epidote are found as late porphyroblasts in many gneiss samples from the southern part of the terrain. Garnet occurs in six of the gneiss samples, and sillimanite has been found in only one sample. Other metamorphic indicator minerals such as kyanite, cordierite and staurolite have been found in the area but do not occur in the grid collection.

Metamorphic Zoning

On the basis of the 300 samples from the grid collection it was attempted to define zones of different metamorphic grade. Due to the scarcity of suitable rock compositions to provide index minerals, this proved to be difficult,

Frederikshåbs Isblink. It is possible that the metamorphic grade increases again further south-east. Abundant hypersthene has been found in sands collected at the front of Frederikshåbs Isblink (Kalsbeek et al., *op. cit.*), and Dawes (1970) has found hypersthene in metamorphosed basic dykes on Dalagers Nunatakker.

Hornblende varies in colour from brownish-green in the north-western higher grade terrain to distinctly bluish-green in the southern lower grade terrain. In a number of cases brownish-green hornblende occurs as cores in bluish-green hornblende, indicating a decreasing grade of metamorphism. Although these colour descriptions are subjective, they are confirmed by repeated observations. Brownish-green hornblende surrounded by bluish-green hornblende has also been found near Frederikshåbs Isblink.

Epidote occurs especially in the gneisses in the southern part of the region. Epidote occurs both as a clearly secondary mineral in altered rocks (generally replacing biotite and often inheriting trains of inclusions parallel with the cleavage of the original biotite crystals) and as porphyroblasts (often poikiloblasts) without a secondary appearance. The latter may occur in completely fresh rocks in which the plagioclase (An_{25-30}) does not visibly change composition towards bordering epidote crystals. In these fresh rocks the epidote is also often associated with the biotite, and the impression is gained that these rocks may have formed by progressive metamorphism of altered gneisses. Epidote does not normally occur in the amphibolites; where it occurs it is clearly secondary.

Muscovite occurs in the same general area as the epidote, but unlike the epidote it is restricted to gneisses without hornblende. Muscovite also occurs both in strongly altered rocks as a clearly secondary mineral, and in fresh rocks as porphyroblasts (poikiloblasts)

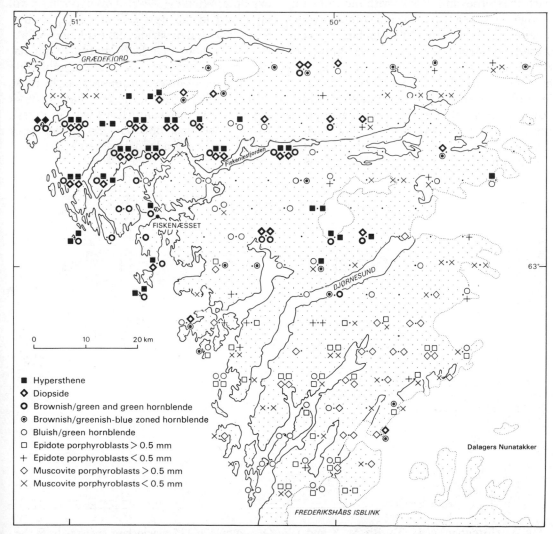

- ■ Hypersthene
- ◆ Diopside
- ◉ Brownish/green and green hornblende
- ◎ Brownish/greenish-blue zoned hornblende
- ○ Bluish/green hornblende
- □ Epidote porphyroblasts > 0.5 mm
- + Epidote porphyroblasts < 0.5 mm
- ◇ Muscovite porphyroblasts > 0.5 mm
- × Muscovite porphyroblasts < 0.5 mm

Fig. 2. Distribution of some characteristic metamorphic indicator minerals in the Fiskenæsset region

228

up to *c*. 1 mm in length. Both the muscovite and the epidote porphyroblasts commonly show myrmekitic rims.

It was hoped that the distribution of these minerals plotted on the map (Fig. 2) would define a regular metamorphic zoning. This is hardly the case. Hypersthene-bearing rocks (hornblende–granulite facies) occupy a sickle-shaped area in the north-western part of the region with one arm running southwards along the coast and one arm passing the inner part of Fiskenæsfjorden. A smaller area of hypersthene-bearing rocks, seemingly in line with this eastern arm, occurs north of the inner part of Bjørnesund. Rocks with diopside and (brown)–green hornblende but without hypersthene (high-rank amphibolite facies) occur scattered throughout the area north of Bjørnesund. These rocks do not define a regular zone around the area of hornblende granulite facies rocks. Rocks with epidote and muscovite porphyroblasts and with bluish-green hornblende (low-rank amphibolite facies), define a coherent area in the southern part of the region. Small epidote and muscovite por-

phyroblasts (<0.5 mm), and bluish-green hornblende, occur scattered throughout the rest of the region, but rarely in hypersthene-bearing rocks. This distribution of the accessory opaque minerals and sphene is clearly related to the distribution of the main minerals, sphene occurring especially in the lower grade part of the region and opaque minerals in the higher grade part. Rocks belonging to all three metamorphic grades may be completely fresh and apparently consist of stable paragen-eses.

Earlier, it was assumed that much of the Archaean area of West Greenland had been downgraded from granulite facies (see, for example, Windley, 1969). This hypothesis will be tested in the next section of this paper. Since field and experimental evidence (Wink-ler, 1967) shows that the formation of migma-tites requires at least high rank amphibolite facies, it seems indeed probable that the rocks in the southern part of the area have been of higher grade originally (Kalsbeek, in press). In thin sections, however, no evidence of earlier higher grade parageneses has been found.

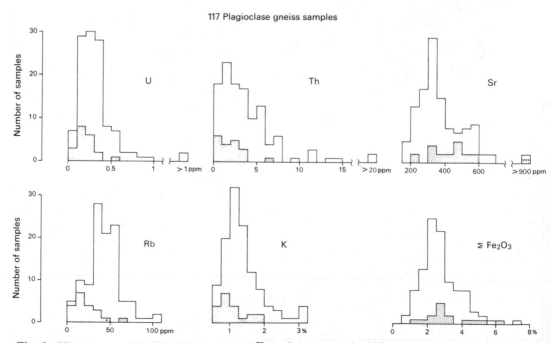

Fig. 3. Histograms of U, Th, Rb, Sr, K and $\sum$ Fe$_2$O$_3$ contents in 117 plagioclase gneisses from the Fiskenæsset region. White: all samples. Hatched: hypersthene-bearing gneisses alone. Th was not detected in 13 samples; of these 7 were arbitrarily put in the 0–1 and 6 in the 1–2 ppm column

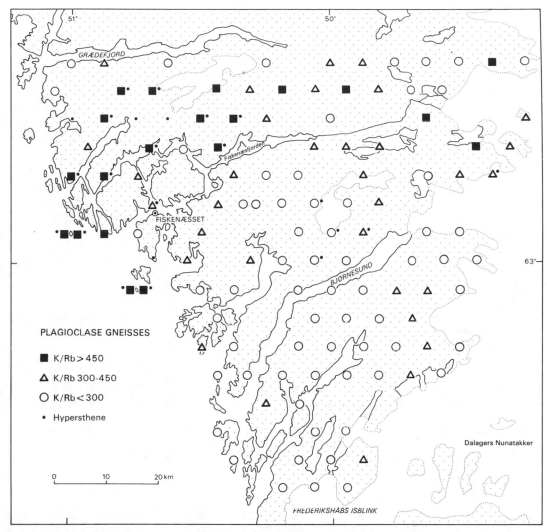

Fig. 4. Distribution of K/Rb ratios for 117 plagioclase gneisses from the Fiskenæsset region. Note the correlation of K/Rb with metamorphic grade as shown in Fig. 2. Heavy dots indicate samples with hypersthene

Chemistry

It is known that granulite facies rocks are often poor in such elements as Rb, U and Th (Lambert and Heier, 1967, 1968; Heier, 1973; Tarney et al., 1972), and retrograded granulite facies rocks retain these features (Green, et al., 1972; Tarney et al., *op. cit.*). This was used in the present case to see which rocks may have been in granulite facies during earlier periods of metamorphism. For convenience, the investigation was restricted to the plagioclase gneisses (K-feldspar < 10%) since these form the largest rock group of

restricted major element composition. Rb, Sr and Fe were measured twice in 117 plagioclase gneisses by X-ray fluorescence spectrography, and are accurate to about 5–10% relative. U, Th and K were analyzed by γ-spectrography, and partly delayed neutron counting following neutron activation, on all samples. Results for U and K are accurate to a few per cent, for Th to about 10–20%.

The Rb content of the plagioclase gneisses varies from *c.* 1 ppm to > 100 ppm (Fig. 3). K/Rb varies from normal values of *c.* 250 to

230

values > 1000. There is a clear correlation of K/Rb with the metamorphic grade of the rocks, most hypersthene-bearing gneisses having K/Rb > 450. The epidote- and muscovite-bearing rocks of the southern part generally have K/Rb < 300 (Fig. 4), and this seems to indicate that these rocks have not been in granulite facies. Samples from the area of hypersthene-bearing rocks just north of the inner part of Bjørnesund (Fig. 2) do not have high K/Rb values; perhaps granulite facies conditions here lasted only for a shorter period. Scattered samples from the northern part of the region have K/Rb > 450 but do not contain hypersthene; perhaps some of

these rocks have earlier been in granulite facies.

U and Th contents (Fig. 3) are low throughout the area and there is only a loose correlation between the metamorphic grade and the U content of the samples (Fig. 5). However, 18 out of 19 hypersthene gneisses have less than 0·35 ppm U. A few samples near Frederikshåns Isblink, where there were indications of increasing metamorphic grade, have U < 0·20 ppm.

Th/U ratios are generally high (Fig. 6) indicating depletion of U relative to Th also in the southern part of the region where K/Rb ratios are normal.

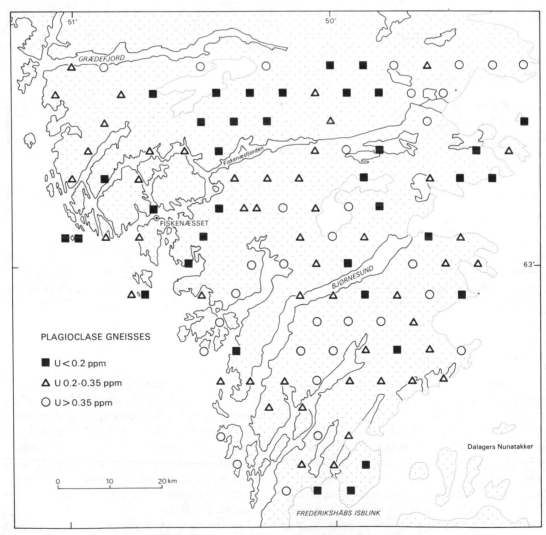

Fig. 5. Distribution of U contents of 117 plagioclase gneisses from the Fiskenæsset region

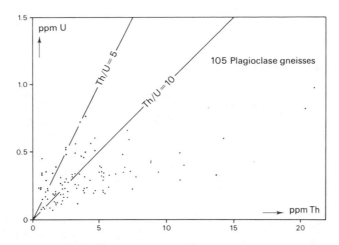

Fig. 6. Relationship between U and Th contents in 104 plagioclase gneisses. Note poor correlation between Th and U and high Th/U ratios. In 13 samples Th was not detected

Because of the very inhomogeneous nature of the rocks, an attempt has been made to make mineralogical and chemical estimates of the average bed rock in the Fiskenæsset region with the help of sand samples. Due to the cold climate and sparse vegetation, fluvial sands prove to be very useful in this respect (Kalsbeek et al., 1974). By a study of sand samples, it has earlier been shown (Kalsbeek, 1974) that the granulite facies rocks from the Nordland region (north of Godthåb) have about six times less U than the amphibolite facies rocks of the Frederikshåb district. For Rb there is also a very marked difference, the Nordland sands having an average of 10 ppm Rb, and the Frederikshåb sands 52 ppm (average K/Rb ratios are, respectively, 871 and 292). This is in line with the results from the Fiskenæsset region and confirms that the rocks of the Frederikshåb district probably were never in granulite facies.

Cause of the Low U, Th and Rb Contents

Lambert and Heier (1967, 1968) have suggested that depletion of U, Th and Rb in the lower crust might be due mainly to partial anatexis. Tarney et al. (1972) thought that these elements were removed by upward permeating volatiles resulting from mantle degassing during the early Precambrian.

In the present case an anatectic model seems able to explain the facts adequately. In this model, gneisses like those in the southern part of the Fiskenæsset region, with normal K/Rb ratios, are regarded as the original rocks from which granitic melts were derived by partial melting, leaving hypersthene gneisses as a residue. If this model were correct, (1), granulite facies metamorphism and the development of granitic rocks would be complementary processes. Moreover, (2), the model can be tested with major element analyses of the different rock types in the area.

1(a) Granitic rocks in the Archaean of West Greenland seem to be associated in space and time with the granulite facies metamorphism. Mappable granite bodies are rare in the southern part of the Archaean block, a stretch of 250 km between Ivigtut and Fiskenæsset where there are also no signs of granulite facies metamorphism. Large outcrops of granitic rocks (Ilivertalik granite; Kalsbeek and Myers, 1973) occur in the northern part of the Fiskenæsset region where granulite facies rocks also occur.

(b) K-rich gneisses seem to be more common north of Bjørnesund than further south (20% versus 10% of the gneiss samples in the grid collection having $K > 2 \cdot 5 \%$; Fig. 7). This difference is not statistically significant, there

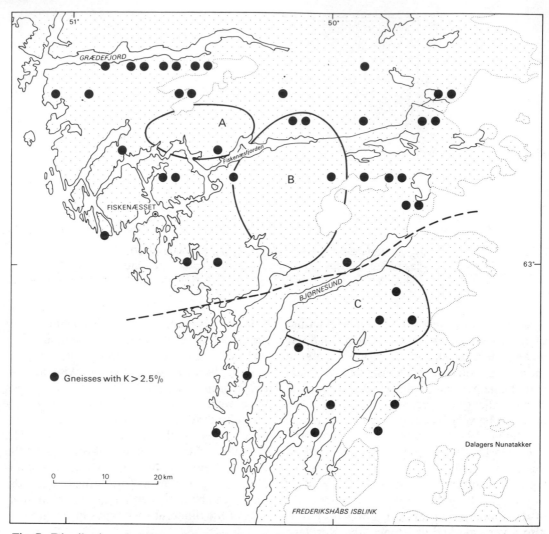

Fig. 7. Distribution of gneiss samples with K > 2·5%. Analyses of composite sand samples from areas A, B and C are given in Table 3

being a probability of *c.* 20% that the two samples were drawn from one population).

(c) The age of the Ilivertalik granite (*c.* 2800 m.y.: Pidgeon et al., in press) agrees with that of the granulite facies metamorphism (Black et al., 1973).

(2) To some extent the major element composition of the rocks supports the anatectic model. Approximately 25% granite could be derived by partial melting of average amphibolite facies gneiss, leaving behind a residue of average hypersthene gneiss composition.

(a) Hypersthene-bearing plagioclase gneisses are on average impoverished in Si and K and enriched in Fe, Mg and Ca compared with non-hypersthene bearing plagioclase gneisses (Table 2, a and b). The differences are not large and their significance can only be proven with large numbers of samples. For two groups of nine samples each (Table 2, a and b), a ranking test demonstrated only the significance of the difference in Ti. Working with a larger group of samples (Fig. 3) the significance of the differences in K and Fe also became evident. Comparable differences between granulite facies and amphibolite

Table 2. Average chemical composition of various gneiss types from the Fiskenæsset region

	a	b	c	d		s
SiO_2	66·50	67·81	68·09	67·67	71·19	2·90
TiO_2	0·49	0·36	0·35	0·44	0·27	0·16
Al_2O_3	16·12	16·09	15·98	15·85	15·04	1·21
Fe_2O_3	4·13	3·29	3·25	3·60	2·01	1·25
MnO	0·06	0·04	0·04	0·05	0·03$_5$	0·04
MgO	1·82	1·42	1·32	1·52	0·62	0·44
CaO	4·49	4·04	3·95	3·87	1·99	0·56
Na_2O	4·66	4·58	4·61	4·55	4·21	0·50
K_2O	1·15	1·36	1·52 (1·91)	1·80	3·74	1·03
H_2O	0·55	0·69	0·74	0·54	0·50	0·28
P_2O_5	0·17	0·14	0·13	0·15	0·09	0·03

a. Average of 9 hypersthene-bearing plagioclase gneisses.
b. Average of 9 epidote/muscovite-bearing plagioclase gneisses.
c. Average of 13 gneisses south of Bjørnesund. K_2O between parentheses is the average for all 93 gneiss samples south of Bjørnesund.
d. Calculated mixture of 75% (a) and 25% (e).
e. Average of 10 granitic gneisses with standard deviations.

facies rocks occur in many other areas and have been noted in Greenland and by Ramberg (1951).

The granitic gneisses in the area (Table 2, e) are obviously richer in SiO_2 and K_2O than the plagioclase gneisses. Twenty-five per cent of average granitic gneiss, combined with 75% average hypersthene gneiss, gives a composition (Table 2, d) which very closely agrees with average gneiss from the area south of Bjørnesund (Table 2, c). In this last average, apart from biotite–plagioclase gneisses, a few samples of gneiss with more K-feldspar and hornblende gneiss were also included.

(b) Although the hypersthene gneisses on average are poorer in K_2O than the non-hypersthene-bearing gneisses, this difference seems to be cancelled out by the more common presence of granitic rocks in the same areas. Average fluvial sand from the granulite facies terrain does not contain less K_2O than sands from the amphibolite facies terrain (Table 3).

Because of the inhomogeneity of the rocks and the resulting sampling problem, and since different gneiss types cannot readily be differentiated in the field, a quantitative treatment of the gross major element composition of the rocks, as presented above, should be treated with due caution.

Table 3. Chemical composition of composite sand samples from areas A, B and C (Fig. 7) in the Fiskenæsset region. Recalculated to 100% water-free. (From Kalsbeek et al., 1974)

	a	b	c
SiO_2	69·7	69·3	69·6
TiO_2	0·3	0·3	0·4
Al_2O_3	15·2	15·3	15·1
Fe_2O_3	0·8	1·0	2·2
FeO	2·2	1·9	1·4
MnO	0·0	0·0	0·0
MgO	1·9	2·0	1·6
CaO	4·1	4·5	3·9
Na_2O	4·3	3·9	4·2
K_2O	1·4	1·7	1·4
P_2O_5	0·1	0·1	0·0

a. Average of 19 sands rich in hypersthene.
b. Average of 24 sands with some hypersthene.
c. Average of 14 sands without hypersthene.

Discussion

(1) The differences in K/Rb ratios and U contents between the granulite facies rocks (north-western part of Fiskenæsset region and the Nordland area) and amphibolite facies rocks (southern part of Fiskenæsset region and Frederikshåb region) are so clear that it

234

now seems nearly certain that these last areas have never been in granulite facies.

(2) There is at least circumstantial evidence that anatectic processes have been operative in causing the chemical differences between the granulite facies rocks and the amphibolite facies rocks.

(3) There is evidence that U depletion may have taken place before K/Rb ratios were severely changed (as in the southern part of the Fiskenæsset region).

A few other points which relate to this study may be raised.

(4) A thin Archaean crust (of the order of 10 km) and steep thermal gradients as suggested, amongst others, by Bridgwater and Fyfe (1974) seem to be difficult to reconcile with the rather monotonous metamorphic grade of the Archaean rocks of south-west Greenland. Over stretches of hundreds of kilometres neither greenschist facies nor granulite facies rocks may occur. All rocks have been hot enough to become migmatites but not so hot that massive remelting occurs. U, Th and K contents are low and decrease with increasing grade of metamorphism so that overall heat generation must have been very low at depth already 3000 m.y. ago.

There is little direct mineralogical evidence regarding thermal gradients during the metamorphism of the Archaean rocks from Greenland. Both kyanite and cordierite, indicating respectively fairly high and fairly low pressure, have been reported. Cordierite seems to be more common in areas where intrusive granitic rocks are prominent. Herd (1972), on the basis of mineral parageneses in different rocks, estimates that pressures during metamorphism were of the order of 7–9 kb. Platt and Myers (in press), on the basis of a study of corona structures formed between olivine and plagioclase in rocks associated with the anorthosite complex, conclude to pressures of c. 9 kb. On balance, the evidence does not seem to favour an exceptionally thin crust or exceptionally steep thermal gradients.

(5) It has been argued on the basis of initial Sr^{87}/Sr^{86} ratios (Pankhurst et al., 1973) that younger rocks (the 'Nûk' gneisses) in the Archaean block cannot have been derived by partial melting of much older rocks (the 'Amîtsoq' gneisses). Field evidence elsewhere (Chadwick et al., 1974) seems to indicate that this is not always so. The granulite facies gneisses studied are commonly so poor in Rb that Sr/Rb ratios $\gg 20$ become normal. If younger granitoid rocks formed by partial melting of such rocks, initial $^{87}Sr/^{86}Sr$ ratios would no longer be a safe guide as to their origin.

Acknowledgements

The U, Th and K analyses used in this paper were made at the Danish Atomic Energy Commissions Research Establishment, Risø, under the direction of civilingeniør Leif Løvborg. Rb, Sr and Fe were analyzed by the writer in the X-ray fluorescence laboratory of the Petrological Institute, University of Copenhagen under the direction of Dr John Bailey. The analyses shown in Table 2 were made by Dr G. Hornung, University of Leeds; those of Table 3 by civilingeniør Ib Sørensen, GGU. Sand samples from the Nordland area were collected by geologists of Kryolitselskabet Øresund A.S., Copenhagen, and those from the Frederikshåb area by cand.mag. Stig Bak Jensen, GGU. The writer gratefully acknowledges the help and cooperation of these institutions and persons. The publication of this paper was permitted by the Director of the Geological Survey of Greenland.

References

Black, L. P., Moorbath, S., Pankhurst, R. J., and Windley, B. F., 1973. '^{207}Pb/^{206}Pb whole rock age of the Archaen granulite facies metamorphic event in West Greenland', *Nature (Phys. Sci)*, **244**, 50–53.

Bridgwater, D., and Fyfe, W. S., 1974. 'The pre-3 b.y. crust: Fact–fiction–fantasy', *Geosci. Canada*, **1**, 7–11.

Chadwick, B., Coe, K., Gibbs, A. D., Sharpe, M. R., and Wells, P. R. A., 1974. 'Field evidence relating to the origin of 3,000 Myr gneisses in southern West Greenland', *Nature (phys. Sci.)*, **249**, 136–137.

Dawes, P. R., 1970. 'Bedrock geology of the nunataks and semi-nunataks in the Frederikshåbs Isblink area of southern West Greenland', *Rapp. Grønlands geol. Unders.*, **29**, 60 pp.

Green, T. H., Brunfelt, A. O., and Heier, K. S., 1972. 'Rare-earth element distribution and K/Rb ratios of granulites, mangerites and anorthosites, Lofoten-Vesteraalen, Norway', *Geochim. Cosmochim. Acta*, **36**, 241–257.

Heier, K. S., 1973. 'Geochemistry of granulite facies rocks and problems of their origin', *Phil. Trans. Roy. Soc. Lond.*, **A273**, 420–442.

Herd, R. K., 1972. 'The petrology of the sapphirine-bearing and associated rocks of the Fiskenæsset Complex, West Greenland', *Ph.D Thesis*, Univ. London.

Kalsbeek, F., 1974. 'U, Th and K contents and metamorphism of Archaean rocks from South-West Greenland', *Bull. geol. Soc. Denmark*, **23**, 124–129.

Kalsbeek, F., in press. 'Metamorphism in the Fiskenæsset region' In 2nd Progress report on the geology of the Fiskenæsset region, South-West Greenland, *Rapp. Grønlands geol. Unders.*

Kalsbeek, F., and Myers, J. S., 1973. 'The geology of the Fiskenæsset region', *Rapp. Grønlands geol. Unders.*, **51**, 5–18.

Kalsbeek, F., Ghisler, M., and Thomsen, B., 1974. 'Sand analysis as a method of estimating bedrock compositions in Greenland, illustrated by fluvial sands from the Fiskenæsset region', *Bull. Grønlands geol. Unders.*, **111**, 32 pp.

Lambert, I. B., and Heier, K. S., 1967. 'The vertical distribution of uranium, thorium and potassium in the Continental Crust', *Geochim. Cosmochim. Acta*, **31**, 377–390.

Lambert, I. B., and Heier, K. S., 1968. 'Geochemical investigations of deep-seated rocks in the Australian shield', *Lithos*, **1**, 30–53.

Pankhurst, R. J., Moorbath, S., and McGregor, V. R., 1973. 'Late event in the geological evolution of the Godthaab district, west Greenland', *Nature (phys. Sci.)*, **243**, 24–26.

Pidgeon, R. T., Aftalion, M., and Kalsbeek, F., in press. 'The age of the Ilivertalik granite', In 2nd Progress report on the geology of the Fiskenæsset region, South-West Greenland', *Rapp. Grønlands geol. Unders.*

Platt, R. G., and Myers, J. S., in press. 'Corona structures from the Fiskenæsset Complex, West Greenland', *Progr. Experimental Petr.*, **3**.

Ramberg, H., 1951. 'Remarks on the average chemical composition of granulite facies and amphibolite to epidote–amphibolite facies gneisses in West Greenland', *Medd. Dansk geol. Foren.*, **12**, 27–34.

Tarney, J., Skinner, A. C., and Sheraton, J. W., 1972. 'A geochemical comparison of major Archaean gneiss units from Northwest Scotland and East Greenland', *24th IGC, 1972*, **1**, 162–174.

Windley, B. F., 1969. 'Evolution of the early Precambrian basement complex of southern West Greenland', *Spec. pap. geol. Ass. Can.*, **5**, 155–161.

Winkler, H. G. F., 1967. *Petrogenesis of Metamorphic Rocks* (2nd edition). Berlin: Springer-Verlag.

Crustal Development of the Archaean Gneiss Complex:

Eastern Labrador

K. D. Collerson

Dept. of Geology, Memorial University, St. John's, Newfoundland, Canada.

C. W. Jesseau

Dept. of Geology, Memorial University, St. John's, Newfoundland, Canada.

D. Bridgwater

Geological Survey of Greenland, on exchange Geological Survey of Canada, Ottawa, Canada.

Abstract

Archaean gneisses that outcrop in the Nain Province, eastern Labrador, form the westernmost extension of the disrupted North Atlantic Craton. They are bounded on the west and south by Proterozoic mobile zones. Best exposures are in the northern and southern regions of the Province, where gneisses have not been cut by Proterozoic anorthosites.

The gneiss complex consists principally of quartzo–feldspathic rocks, thought to be derived from granodioritic parents. Subordinate amounts of metasedimentary and meta-igneous supracrustal units intruded by differentiated meta-basic igneous rocks are interlayered with quartzo-feldspathic gneisses. Regional metamorphism is predominantly amphibolite facies except in the north where granulite facies rocks occur.

In the north, geological relations suggest that the gneisses have undergone a long and complex history of development, similar to that established in the Godthaab area of West Greenland. The oldest rocks (the Uivak gneisses) are a composite group of tonalitic migmatites and intrusive granitic (*sl.*) gneisses characterized by the presence of deformed mafic dykes (Saglek dykes). Uivak gneisses are tectonically interlayered with a cover sequence (the Upernavik supracrustals) consisting predominantly of pelites and amphibolites together with ultramafic pods and layered meta-basic bodies. Saglek dykes have not been recognized cutting the supracrustal suite. Two generations of younger granite gneisses occur as deformed sheets intrusive into both the Uivak gneisses and the Upernavik supracrustals. The final plutonic episode in the area is the emplacement of granite sheets into shear belts.

Archaean rocks in the southern part of the Province bear a close resemblance to those in the north except that they may contain remnants of comparatively well-preserved greenstone belts and they have been affected more strongly by a late-stage development of granites which agmatized the earlier gneiss complex.

238

Introduction

The Archaen gneiss complex exposed in the Nain Province, eastern Labrador (Fig. 1) is part of the disrupted North Atlantic Craton (Bridgwater et al., 1973c). It is dominated by large areas of quartzo-feldspathic gneiss of a variety of ages interlayered with subordinate although geologically important belts of supracrustal rocks derived from sedimentary and volcanic parents (Fig. 2). Differentiated metabasic–anorthositic bodies cut early granite gneisses and the supracrustal sequences.

This paper summarizes the present state of knowledge regarding the geological evolution of Archaean gneisses in eastern Labrador. Its principal object is to provide a stronger basis for the geological comparison of Labrador with other Archaean high-grade gneiss terrains.

We wish to dedicate this review to E. P. (Pep) Wheeler 2nd (1900–1974) in memory of his outstanding contribution to and comprehension of the geology of Labrador.

Geological Setting

Archaean gneisses crop out in eastern Labrador in an attenuated roughly triangular shaped area approximately 600 km long, reaching a maximum width of nearly 100 km inland from Hopedale (Fig. 1). The gneiss complex consists of intensely deformed high-grade gneisses with northerly to north-easterly

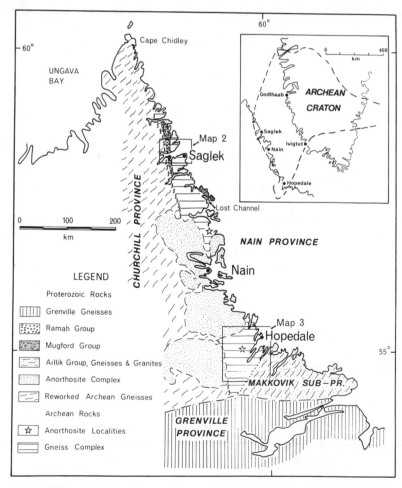

Fig. 1. Map showing the major geological divisions in Labrador and localities mentioned in the text (after Bridgwater et al., 1973c, and Greene, 1974)

Fig. 2. Interlayered gently dipping units of Uivak gneiss and younger granitic gneisses with Upernavik supracrustals at Cape Uivak (relief approximately 600 m)

striking structural trends. Metamorphism is predominantly amphibolite facies except in the north where granulite facies rocks occur.

The Archaean rocks are unconformably overlain by two early Proterozoic sequences (Fig. 1): the Ramah Group (Daly, 1902; Taylor, 1969; Knight, 1973) and the Mugford Group (Daly, 1902; Taylor, 1970; Barton, 1975) which are believed to be equivalent to the Vallen and Sortis Groups in south-western Greenland (Knight, 1974).

The Archaean block is bounded by linear mobile zones which were active during the Proterozoic (Hudsonian Orogeny; Stockwell, 1964). Rocks of the Churchill Province to the immediate west and north of the Nain Province consist predominantly of re-worked Archaean gneisses and subordinate amounts of metamorphosed early Proterozoic sediments and volcanics. The boundary zone between the two Provinces is a belt of sheared rocks marked by intense ductile and brittle thrusting (Morgan, 1974). In the southern part of the Nain Province, the Makkovik Sub-Province (Taylor, 1971), Archaen rocks are unconformably overlain and largely obscured by the English River Greenstone sequence and by the Aillik, Moran and Bruce River Groups (Gandhi et al., 1969; Sutton et al., 1971; Sutton, 1972; Smyth et al., 1975). As in the Churchill Province, the basement and cover have also been deformed and metamorphosed during the Hudsonian Orogeny (Gandhi et al., 1969).

The Churchill Province and Makkovik Sub-Province are regarded as being largely equivalent to the Nagssugtoqidian and Ketilidian mobile belts, respectively, in Greenland (Bridgwater et al., 1973a,b).

Proterozoic post-tectonic plutonic rocks associated with the Nain anorthosite-adamellite complex (Wheeler, 1960) intrude and disrupt the continuity of outcrop within the Archaean gneiss complex, producing marked contact aureoles which range up to granulite facies (Wheeler, 1960; Berg, 1974; Speer, 1975). Best exposures of Archaen rocks are therefore north and south of the Nain anorthosite–adamellite complex.

Field Characteristics of the Archaean Gneiss Complex

Saglek Area

Geological relationships in the Saglek Area (Fig. 3) suggest that the gneisses have undergone a long and complex history of development similar to the Godthaab region of West Greenland (McGregor, 1973).

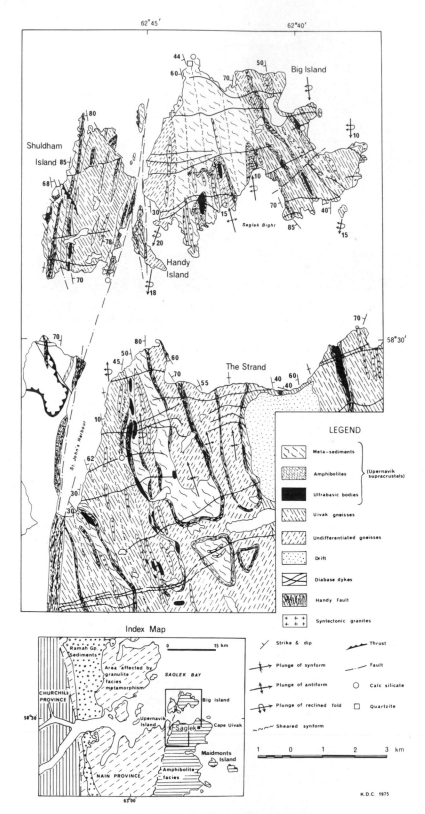

Fig. 3. Geological map of Saglek area

The oldest rocks in the area, the Uivak gneisses, yield a whole-rock Rb/Sr isochron of 3622 ± 72 m.y. with an initial ratio of 0.7014 ± 0.0008 (Hurst et al., 1975). The Uivak gneisses are a composite group of quartzofeldspathic rocks made up of at least two igneous suites. The older group of gneisses (Uivak I gneisses) are typically leucocratic, fine- to medium-grained biotite-bearing layered rocks with septa of 'tonalitic' grey gneiss interlayered with coarser-grained concordant pegmatites 1–2 cm across, containing a higher concentration of potash feldspar. The Uivak I grey gneisses locally contain metasedimentary and basic igneous inclusions of quartzitic gneiss with graphitic garnet–sillimanite layers and hornblendite.

Porphyritic granodioritic gneisses (Uivak II gneisses) with well-developed feldspar augen, high iron content, and a moderately high mafic mineral content pass laterally into strongly foliated gneisses not distinguishable with certainty in the field from the normal grey gneisses. On the east shore of Saglek Bight the grey Uivak I gneisses are intruded by deformed sheets of a variable granitic suite ranging from quartz monzonite to ferro-diorite (Fig. 4) and by layers of iron-rich hornblendite, all of which are intruded by amphibolite dykes (Saglek dykes).

The Saglek dykes generally occur as deformed concordant layers and disrupted pods within the Uivak gneisses. Their original discordant nature is locally preserved (Fig. 5). The groundmass of the dykes is generally fine-grained and granoblastic, containing dispersed plagioclase megacrysts or aggregates of megacrysts and anorthositic xenoliths (Fig. 6). Distribution of the plagioclase megacrysts is highly irregular, commonly occurring along one margin of the dykes or concentrated in pockets in which up to 60% of the total rock is made up of megacrysts and xenoliths.

The Upernavik supracrustal rocks (Fig. 7) consist of a group of meta-sediments ranging from pelites to quartzites and marbles. They occur with a varied group of meta-igneous rocks including amphibolites, ultamafic pods and layered basic bodies. The main meta-sedimentary units preserved are pelitic schist and gneiss, commonly with cordierite, anthophyllite and sillimanite. The composition varies considerably from pelitic biotite-rich units (sometimes with graphite), to semi-pelitic quartz- and feldspar-rich units. Garnetiferous and silmanite-bearing quartzites

Fig. 4. Intrusive relations between banded leucocratic Uivak I gneisses and a melanocratic monzonite phase of the Uivak II gneisses, Saglek Bight. Note the thin Saglek dyke, upper right-hand corner

242

Fig. 5. Outcrop of Uivak I gneisses containing deformed and rotated basic dykes (Saglek dykes). The dyke in the left foreground has a pronounced apophysis into the gneisses. Saglek Bight

Fig. 6. Saglek dyke with local accumulation of anorthosite inclusions and plagioclase megacrysts. Saglek Bight. Hammer for scale

are locally interbanded with the pelitic units. Calc–silicate-rich layers and lenses (diopside–amphibole rocks) and forsterite and phlogopite marbles are a minor although important association in the meta-sedimentary sequence.

The basic rocks within the supracrustal succession were derived from both volcanics and stratified basic igneous bodies. The meta-volcanics are generally fine-grained schistose amphibolites. Some units show small-scale layering with alternating dark amphibole-rich units grading into more felsic rocks, possibly representing graded volcanic ash units or intensely deformed altered pillows (Fig. 8).

Layered basic igneous bodies found at many localities within the supracrustal sequence locally cut across earlier layering in the metasediments. The layered bodies vary from a few metres thick to 30–50 m in width. They show an overall variation in the relative amount of feldspar and mafic material from top to bottom, and local rhythmic variations internally. Some contain distinctive units within the layered sequences; garnet-bearing leuconorites occur near the presumed upper contacts, while layers very rich in ilmenite occur locally lower in the sequence.

Fig. 7. Pelitic and semi-pelitic gneisses with strong compositional banding due to variation in the amounts of garnet, biotite and felsics. (Upernavik supracrustal meta-sediment) Upernavik Island

Fig. 8. Upernavik supracrustal amphibolite showing fine-scale compositional banding and disruption by later granitic intrusion. Maidmonts Island

Ultramafic masses (meta-peridotites) occur as pods and layers in the gneiss complex, ranging up to several hundred metres in width and traceable along strike for up to 1 km. They commonly occur adjacent to the meta-volcanic amphibolites although units of tremolite schists do occur interbanded with the amphibolites.

Younger quartzo-feldspathic gneisses derived largely by re-working of the Uivak gneisses at 3133 ± 156 m.y. (Rb/Sr whole-rock age) with an initial ratio of 0.7063 ± 0.0012 (Bridgwater et al., 1975; Hurst et al., 1975) occur as deformed sheets, intrusive into both the Uivak gneisses and the Upernavik supracrustals (Fig. 9). Although they occupy a similar position in the sequence of events in the Saglek area as the Nûk gneisses in Godthaabsfjord, West Greenland (McGregor, 1973), they do not form such a well defined calc–alkaline suite. Locally the younger gneisses form discrete sheets which are concordant to the regional structures but which locally transgress earlier fabrics. Generally, however, the younger quartzo-feldspathic gneisses occur as migmatitic veins and stringers breaking up the earlier gneisses and supracrustals. Many are apparently formed by local remobilization of the earlier rocks and there is a gradation between older gneisses with well-preserved structures cut by Saglek dykes and areas of migmatites in which the neosome becomes dominant leaving scattered remnants of older gneiss and Saglek dyke.

The late syntectonic granites differ from all earlier granitic gneisses in the area in that they possess a relatively simple fabric which shows evidence of only one main phase of deformation. Most outcrops of the late syntectonic granites have a sugary, commonly slightly pink appearance in the field.

Other relatively minor rock types include pale grey granite sheets emplaced along shear belts, several generations of diabase dykes, and the formation of mylonite and pseudotachylite related to the major movements along the Churchill mobile zone.

Hunt River Area

The Archaean gneisses in the southern part of Labrador are characterized by three thick north-easterly trending, predominantly mafic supracrustal belts which are infolded into polyformational quartzo-feldspathic gneisses (Taylor, 1970; Greene, 1974). The southern-most belt, the Ugjoktok Bay supracrustal rocks described by Sutton (1972), bears many similarities to the Hunt River Belt (Fig. 10).

Fig. 9. Unit of extensively migmatized Uivak II gneiss (quartz monzonite) intruded by a more homogeneous sheet of younger granitic gneiss. Maidmonts Island.

The Hunt River Belt, the northermost supracrustal sequence, comprises chiefly fine- to medium-grained, massive to schistose amphibolites, a series of ultramafic lenses, and minor meta-sediments. The amphibolites, which include garnetiferous and pyroxeniferous varieties and thinly banded felsic-enriched varieties, represent highly deformed basic volcanics. The ultramafics include locally discordant units of hornblendite, tremolite schist, and a series of lenses of serpentinite. In certain of the serpentinite lenses preserved harzburgite patches retain coarse textures involving elongate olivine and interstitial enstatite. These textures may well represent igneous cumulate textures. The ultramafics appear to have originated as a series of intrusive sheets into the volcanic sequence. Meta-sediments are represented by a variety of garnet–staurolite–cordierite-bearing schists, biotite–garnet–rusty units and grey, feldspar–biotite schists.

The Hunt River Belt is infolded into a complex of polyformational quartzo-feldspathic gneisses. A well-banded tonalitic to dioritic gneiss is the most extensive gneiss type. The strong compositional banding developed during a period of intense deformation which included re-working of the supracrustals. The mafic component of the banded gneisses varies in scale from small semi-continuous alternating bands enriched in hornblende several centimetres thick to relatively large amphibolite lenses up to 250 m across. The larger amphibolitic lenses, some with ultramafic pods, are regarded as schlieren of meta-volcanics related to the supracrustals. Intrafolial folds within the dominant gneissic banding provide evidence for an even earlier gneissic terrain. Rare zones of low intensity deformation within the banded gneisses expose a weakly banded to homogeneous hornblende-bearing quartzo-feldspathic gneiss containing deformed amphibolite pods which are clearly discordant to the gneissic foliation. These mafic pods closely resemble deformed dykes.

Other gneissic rock types distinguished in the area include:

(1) Homogeneous granodioritic gneisses which form fairly extensive units intrusive into the banded gneisses. They have a strong planar schistosity defined mainly by anastomosing flakes of biotite and locally by a weakly developed gneissic banding.

(2) Mixed gneisses transitional between the banded and the homogeneous varieties, characterized by relatively well-banded granodioritic and hornblende–biotite-rich units intercalated with the banded and homogeneous gneisses on a scale of tens of metres.

(3) Agmatitic gneisses. The development of the agmatitic gneiss broadly coincides with a major phase of folding in the area. Large amounts of coarse pegmatitic leucosome were injected syntectonically and axial planar to these megascopic folds. In Fig. 10 the area of extensive leucosome emplacement is represented by a single well-defined linear unit. However, the regional effects of this event are quite widespread.

Contacts between the Hunt River Belt and the gneissic terrain are to a great extent fault controlled and are commonly characterized by the presence of coarse-grained pink pegmatitic units.

Elsewhere, the gneisses are separated from the Hunt River Belt by a quartzofeldspathic rock unit approximately 5 m thick. In places this unit is fine-grained, white and quartzitic-looking. It is variable in composition along strike, becoming enriched in potassium feldspar and pink in colour. In other places hornblende is present in sufficient amounts to form a well-banded pink gneissic rock. This unit may represent an intrusive rock emplaced along the margins of the Belt, but the highly variable nature of the unit makes its origin uncertain.

A small body of intensely deformed anorthositic gabbro outcrops in the banded gneisses south of the Hunt River Belt (Fig. 10). It is composed of intricately layered units of calcic plagioclase (An_{70}) and green pleochroic hornblende. Although field relations are difficult to establish, the body appears to be conformable to the surrounding banded gneisses. Similar field relations are described by Wiener (1975) from a gabbro–anorthosite complex at Tessiuyakh Bay, north of Nain (Fig. 1). These Archaean anorthosite occurrences in Labrador are regarded as being broadly comparable with examples described by Windley (1969) and Bridgwater et al. (1973c) from West Greenland, and Ermanovics and Davison (this volume) from the Cross Lake area in Manitoba.

Minor rock types in the area include elliptical-shaped bodies of quartz monzodiorite, which are intrusive into the amphibolite belt, and post-tectonic diabase dykes.

Structural Development

The structural development of the Archaean gneisses is complex and heterogeneous. In the northern part of the Archaean block, major rock units show evidence of both flattening and intense rotational shearing. The regional intercalation of units is interpreted to be the result of thrusting (cf. Bridgwater et al., 1974) accompanied by folding and transposition of layering parallel to the axial planes of the folds. This interpretation is supported in part by the occurrence of numerous rootles intrafolial folds in both the Uivak gneisses and the Upernavik supracrustals. In the south, except for the preservation of rare intrafolial folds, the intensity of thrusting and transposition has almost completely obliterated the original character of the early gneisses. Structural relations are believed to reflect greater intensity of deformation and higher mobility during early tectonism of this region.

Early folds and associated planar and linear structural elements are re-oriented and refoliated during several later periods of folding deformation. In the north, the later deformations resulted in the formation of gently plunging, north–south trending, reclined to recumbent, antiforms and intensely sheared synforms which are characterized by axial planar syntectonic emplacement of granitic sheets (Fig. 3). However, in the south the later stages of folding were accompanied by extensive syntectonic development of agmatite in the gneisses.

Widespread deformation of the Archaean gneiss complex concluded with relatively open folding which produced intricate megascopic interference patterns.

Brittle deformation resulting in the development of fault zones, net-like patches of pseudotachylite and shear belts continued into the Proterozoic particularly near the margins of the Archaean block.

Metamorphism

The metamorphic grade of most of the gneiss complex and, in particular, where key chronological relationships have been established, is amphibolite facies. Common mineral associations include:

1. Biotite, garnet, staurolite, cordierite, fibrolite, sillimanite, graphite, and felsics in pelitic and semi-pelitic schists.
2. Blue-green to green hornblende, actinolite, cummingtonite, diopside, garnet, biotite and plagioclase (An_{30-40}) in amphibolites.
3. Calcite, phlogopite and forsterite in marbles.
4. Quartz, sillimanite, and rare detrital? feldspar in impure quartzites.
5. Olivine, tremolite, hornblende, brucite, calcite, talc, and fuchsite in the ultramafics.
6. Quartz, plagioclase, microline, biotite, muscovite, garnet and allanite in the quartzo-feldspathic gneisses.

In the Saglek area west of the Handy Fault (Fig. 3), deeper structural levels are exposed and regional metamorphism is transitional between amphibolite and granulite facies. Orthopyroxene ($\pm$ garnet) is a common phase in rocks of basic composition, particularly in the differentiated mafic intrusives and mafic volcanics of the Upernavik supracrustals. A significant but rare assemblage is quartz, clinopyroxene and garnet in mafic rocks. Pelitic and semi-pelitic gneisses commonly contain garnet, biotite, cordierite, sillimanite, gedrite, corundum, hercynite and graphite. Relics of kyanite surrounded and almost completely pseudomorphed by white mica are rarely observed. However, their presence does

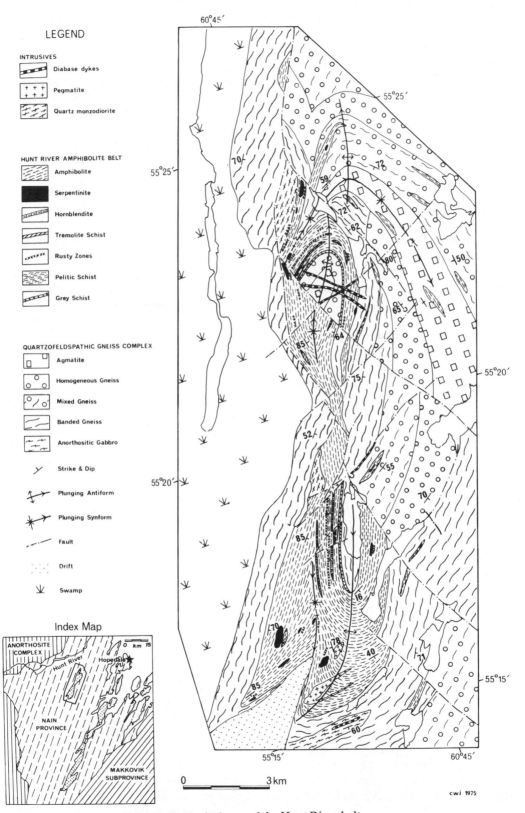

Fig. 10. Geological map of the Hunt River belt

serve to place some constraints on the ambient geothermal gradient during the metamorphism of the area.

In this area considerable evidence also exists for the restricted development of a later period of static granulite facies metamorphism. This event is marked by the emplacement of orthopyroxene-bearing quartzofeldspathic pegmatites into the pre-existing metamorphites, resulting in the production of thin contact (granulite facies) aureoles in the intruded lithologies.

Widespread partial retrogression of the granulite facies rocks to the west of the Handy Fault is observed and appears to be related to shearing and the emplacement of late tectonic granite sheets during the last phases of Archaean igneous activity. Further partial retrogression of both the granulite facies and the amphibolite facies gneisses occurred throughout the complex. This resulted typically in the formation of greenschist facies alteration products and is interpreted as related to the effects of late Archaean and Proterozoic movements along planes parallel to the boundary of the Churchill Province to the west.

Geochemistry

Mean compositions of representative groups of Archaean rocks from the Saglek area and the Hunt River area are presented in Tables 1 and 2. The Uivak gneisses, which are subdivided on the basis of field and petrographic evidence into two groups, show distinctive geochemical differences (Table 1). The earlier Uivak I grey gneisses are a moderately homogeneous suite of tonalities to granodiorites with a fairly narrow range in chemistry. In general they are more sodic and less potassic than the Uivak II gneisses of approximately equivalent silica content. In addition the Uivak II gneisses show a marked degree of iron enrichment relative to the Uivak I tonalites (Fig. 11A).

The younger migmatitic 3100 m.y. gneisses (Table 1) are distinctly more potassic than the older Uivak I gneisses and plot in a similar although more extensive field when represented on a FMA ternary diagram (Fig. 11B).

The average compositions of the metavolcanics, meta-ultramafics, meta-sediments and basic intrusives are presented in Table 2. The garnetiferous and pyroxene-bearing amphibolites are chemically distinct and indicate a bulk composition control of the mineralogy. The garnetiferous amphibolites in particular show anomalous trace element contents.

When compared with other basic volcanic rocks (Table 3) it is readily apparent that the Labrador amphibolites are lower in titanium and have lower K/Rb ratios (believed to be a metamorphic effect) than modern oceanic tholeiites. Mordern island arc tholeiites have lower nickel and chromium contents than the Labrador supracrustal amphibolites. The amphibolites appear to compare most favourably with meta-basalts associated with ultramafics in the lower parts of volcanic cycles from Archaean greenstone belts (cf. Analysis 7, Table 3).

Pelitic meta-sediments from the Upernavik supracrustals in the Saglek area and the Hunt River amphibolite belt have extremely high values of nickel and chromium (Table 2). It would appear unlikely that sediments with such anomalous trace element concentrations would have been derived from a sialic provenance without the influence of significant ultramafic contribution.

Conclusions

The Archaen gneiss complex of eastern Labrador is composed of an intimate tectonic intercalation of polyformational quartzofeldspathic ortho-gneiss and subordinate amounts of meta-volcanic and meta-sedimentary supracrustals. A tentative sequence for the development of the gneiss complex in the Nain Province is presented in Table 4.

The character of the earliest formed gneisses is best known in the Saglek area where a 3622 m.y. old suite of meta-igneous granitic rocks have been discovered. The

Table 1. Mean composition of Archaean ortho-gneisses from Labrador

	1	2	3	4	5	6	7	8
SiO_2	69.89	67.66	62.80	50.85	47.20	72.14	69.10	62.53
TiO_2	0.34	0.64	1.33	0.91	1.42	0.22	0.32	0.84
Al_2O_3	15.46	14.00	12.70	15.88	11.90	14.83	15.87	16.73
Fe_2O_3	0.43	0.50	---	---	---	0.23	0.07	0.75
FeO	1.78	3.90	10.26	10.44	14.40	1.22	1.75	4.33
MnO	0.03	0.06	0.11	0.17	0.31	0.03	0.03	0.06
MgO	0.78	1.14	1.48	6.18	7.52	0.40	0.69	2.05
CaO	2.44	2.79	4.03	7.20	9.92	1.65	2.54	3.64
Na_2O	5.10	4.56	3.30	3.58	2.75	4.14	4.83	4.53
K_2O	2.68	3.62	2.13	2.74	1.17	4.62	3.21	3.06
P_2O_5	0.12	0.19	0.43	0.20	0.16	0.08	0.10	0.20
L.O.I.	0.56	---	---	---	---	0.41	0.30	0.84
TOTAL	99.61	99.06	98.57	98.15	96.75	99.97	98.81	99.56
Zr (ppm)	169	211	194	44	---	166	110	319
Sr	500	156	188	105	---	274	412	471
Rb	109	170	94	15	---	146	60	136
Zn	54	63	52	136	---	43	44	84
Cu	8	12	41	92	---	4	9	29
Ba	504	644	702	97	---	1332	420	1223
Ni	21	23	24	136	---	21	29	59
Cr	17	9	15	466	---	13	21	128
K/Rb	200	165	185	537	---	284	266	182
Rb/Sr	0.255	1.090	0.500	0.143	---	0.559	0.144	0.286
NO. OF ANALYSES	6	5	2	4	1	8	3	3

1. *Vivak I Gneiss.*
2. *Vivak II Gneiss - Quartz Monzonite.*
3. *Vivak II Gneiss - Monzonite.*
4. *Vivak II Gneiss - Diorite.*
5. *Vivak II Gneiss - Hornblendite.*
6. *Migmatitic Gneisses with greater than 70% SiO_2.*
7. *Migmatitic Gneisses with between 65% - 70% SiO_2.*
8. *Migmatitic Gneisses with less than 65% SiO_2.*

Major element analyses computed at the Geological Survey of Canada.
Trace element analyses computed at the Geology Department of Memorial University of Newfoundland.

extent of these old rocks is not known for certain but work by Hurst (1974) and Barton (1975b) suggests that rocks of similar age extend at least as far south as Lost Channel, 100 km south of Saglek. The presence of these old rocks over a comparatively large area in Labrador supports the idea that a major part of the North Atlantic craton may have consisted of these early gneisses which have been re-worked by later Archaen events. The

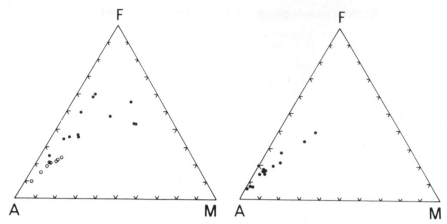

Fig. 11. FMA (total Fe as FeO : MgO : Na₂O + K₂O) ternary plot for Saglek gneisses. (A) Uivak gneisses: open circles—Uivak I gneisses; dots—Uivak II gneisses. (B) Younger migmatitic gneisses

Table 2. Mean composition of meta-volcanics, meta-ultamafics, meta-sediments and basic intrusives from Labrador

	1	2	3	4	5	6	7	8	9	10	11	12	13	14	15	16	17	18
SiO₂	49.07	50.59	49.07	54.00	60.30	73.70	38.30	38.83	45.15	46.78	49.60	49.10	47.87	48.86	49.20	53.57	54.36	77.00
TiO₂	1.03	0.68	0.81	1.53	0.84	0.76	0.02	0.07	0.44	0.43	0.12	1.20	1.30	0.62	0.21	0.75	0.97	0.34
Al₂O₃	14.52	12.08	14.79	12.96	18.70	13.50	0.26	0.95	4.08	7.93	5.20	14.13	14.18	17.22	26.60	19.33	20.30	11.85
Fe₂O₃	2.04	1.20	1.58	3.08	0.80	N.M.	5.60	4.86	1.78	1.19	1.40	2.73	3.61	1.00	0.69	3.17	5.83	N.D.
FeO	10.28	10.87	8.95	11.69	3.60	1.39	7.90	5.13	6.44	8.66	6.30	10.33	8.94	8.36	2.45	8.93	5.20	2.30
MnO	0.23	0.21	0.22	0.19	0.08	0.02	0.23	0.10	0.21	1.70	0.15	0.21	0.17	0.17	0.07	0.14	0.18	0.07
MgO	7.22	8.56	6.47	3.81	2.76	0.57	41.90	38.02	19.66	21.91	29.80	6.53	7.89	6.90	1.55	6.16	4.83	1.24
CaO	10.22	11.00	13.74	7.66	4.58	3.83	0.16	0.61	13.51	4.78	5.40	10.27	10.30	11.10	11.03	1.74	1.70	1.87
Na₂O	2.08	2.16	1.69	2.11	5.00	4.15	0.28	0.09	0.36	0.10	0.85	3.14	2.24	2.35	3.31	2.01	0.07	3.11
K₂O	0.39	0.74	0.38	0.40	1.73	0.78	0.02	0.01	0.07	0.02	0.09	0.80	0.81	0.71	1.13	2.51	2.23	1.9
P₂O₅	0.27	0.07	0.24	0.15	0.28	0.12	0.04	N.D.	0.10	0.09	0.03	0.12	0.36	0.06	0.42	0.05	0.07	0.0
L.O.I.	2.18	1.37	2.22	1.75	2.20	N.M.	3.70	11.88	6.38	5.70	1.50	1.03	1.82	1.17	2.14	0.73	2.44	0.3
TOTAL	99.53	99.53	100.16	99.33	100.87	98.82	98.41	100.55	98.18	99.29	100.44	99.59	99.49	99.02	98.80	99.09	98.18	100.0
Zr (ppm)	55	45	44	60	144	86	14	17	27	25	17	62	54	43	35	67	25	
Sr	134	79	121	71	293	69	10	24	51	25	18	112	140	143	214	66	49	2
Rb	23	26	23	109	108	37	1	14	13	13	4	15	24	21	118	97	4	
Zn	93	89	83	123	48	117	59	46	52	126	40	148	87	93	48	128	45	
Cu	87	71	83	51	80	49	1	16	440	26	6	90	116	27	28	51	186	
Ba	116	115	86	134	629	429	5	8	32	23	27	109	104	487	129	480	22	6
Ni	100	147	107	53	94	72	2028	2064	510	424	1178	83	117	132	47	577	1546	
Cr	337	599	505	29	215	97	3809	3132	2715	2285	2646	221	443	359	66	1687	1114	
K/Rb	141	280	137	31	133	175	332	59	45	130	187	531	280	422	80	213	463	
Rb/Sr	0.172	0.462	0.190	1.535	0.368	0.536	0.100	0.583	0.255	0.520	0.222	0.149	0.171	0.144	0.551	1.676	0.082	0.2
NO. OF ANALYSES	14	6	6	4	1	1	1	10	2	1	1	4	6	5	1	3	1	

1. *Amphibolite, Hunt River Belt.*   2. *Amphibolite, Saglek Area.*   3. *Diopside Amphibolite, Hunt River Belt.*   4. *Garnet Amphibolite, Hunt River Belt.*
5. *Meta-Dacite? Saglek Area.*   6. *Meta-Rhyolite? Saglek Area.*   7. *Meta-Peridotite, Saglek Area.*   8. *Meta-Peridotites, Hunt River Belt.*   9. *Hornblend Schists, Hunt River Belt.*   10. *Tremolite Schist, Hunt River Belt.*   11. *Opx, Cpx, Ol, Spinel Rock, Saglek Area.*   12. *Saglek Dykes (Amphibclis Saglek Area.*   13. *Amphibolites in Gneisses, Hunt River Area.*   14. *Meta-Gabbros, Saglek Area.*   15. *Gabbroic Anorthosite, Hunt River Belt.*   16. *Pel Schists, Saglek Area.*   17. *Pelitic Schist, Hunt River Belt.*   18. *Psammitic Schists, Saglek Area.*

Table 3. Comparison of Labrador meta-volcanics with other volcanic rocks

	1	2	3	4	5	6	7	8	9
SiO_2	49.07	50.59	49.07	54.00	49.90	49.20	51.20	48.80	51.60
TiO_2	1.03	0.68	0.81	1.53	1.51	1.39	0.96	0.91	0.80
Al_2O_3	14.52	12.08	14.79	12.96	17.20	15.80	15.20	15.20	15.90
FeO	12.12	11.95	10.37	14.46	8.70	9.20	1.07	1.09	9.50
MnO	0.23	0.21	0.22	0.19	0.17	0.16	0.22	0.20	---
MgO	7.22	8.56	6.47	3.81	7.20	8.50	6.40	6.13	6.70
CaO	10.22	11.00	13.74	7.66	11.80	11.10	10.70	9.20	11.70
Na_2O	2.08	2.16	1.69	2.11	2.70	2.70	2.80	2.10	2.40
K_2O	0.39	0.74	0.38	0.40	0.16	0.26	0.20	0.31	0.44
P_2O_5	0.27	0.07	0.24	0.15	---	---	---	---	0.11
L.O.I.	2.18	1.37	2.22	1.75	---	---	1.20	5.24	0.45
Zr (ppm)	55	45	44	60	95	100	---	140	70
Sr	134	79	121	71	130	123	102	196	200
Rb	23	26	23	109	1	---	10	---	5
Zn	93	89	83	123	---	---	57	---	---
Cu	87	71	83	51	77	87	111	106	---
Ba	116	115	86	134	14	12	---	85	---
Ni	100	147	107	53	97	123	116	108	30
Cr	337	599	505	29	297	296	400	296	50
K/Rb	141	280	137	31	1300	---	166	---	---
Rb/Sr	0.172	0.462	0.190	1.535	0.007	---	0.098	---	---

1. Mean of 14 amphibolites (Hunt River Belt).
2. Mean of 6 amphibolites (Saglek area).
3. Mean of 6 diopside amphibolites (Hunt River Belt).
4. Mean of 4 garnet amphibolites (Hunt River Belt).
5. Mean of 10 oceanic tholeiites (Engel et al., 1965).
6. Mean of 98 oceanic tholeiites (Cann, 1971).
7. Mean of 85 meta-basalts - ultrabasic-basic suite, Norseman, W. Aust. (Glikson, 1971).
8. Mean of 45 meta-basalts - calc alkaline suite, Superior Province (Baragar and Goodwin, 1969).
9. Average basalt from island arc tholeiite series (Jakes and White, 1971).

effects of late granite injection and consequent loss of regional stratigraphic control is particularly marked in the southern part of the Archaen block (Hunt River Belt) where rocks of similar general aspect to those outcropping at Saglek are found as relics within a network of late to post tectonic granite veins.

The meta-volcanic supracrustal rocks do not correlate exactly with other known volcanic terrains. This may be due in part to metamorphism, or may reflect a unique volcanic environment, preceding the formation (or concentration) of the sial to form major continental land masses and preceding the comparatively stable conditions under which the major greenstone belts were deposited.

Table 4. Tentative Chronology for the Nain Province

17. Proterozoic deformation.	8. Emplacement of basic sheets and ultramafic bodies into the supracrustals.
16. Deposition of early Proterozoic sediments and volcanics.	7. Deposition of Upernavik supracrustals.
15. Intrusion of diabase dykes.	6. Intrusion of basic dykes (Saglek dykes).
14. Widespread development of agmatite and intrusion of granite and pegmatite into shear belts.	5. Deformation–formation of Uivak gneisses (3600 m.y.).
13. Emplacement of syntectonic granite sheets.	4. Intrusion of 'granitic suite' (parent of Uivak II gneisses).
12. Granulite facies metamorphism.	3. Deformation–metamorphism.
11. Reactivation of earlier gneisses (3100 m.y.).	2. Intrusion of 'granite' (parent of Uivak I gneisses).
10. Intrusion of anorthosites.	1. Deposition of pre-Uivak gneiss supracrustals.
9. Intercalation of Uivak gneisses and Upernavik supracrustals.	

Acknowledgements

This research work was supported by the National Research Council of Canada (Grant A8694 to ·K.D.C.). Additional financial support to cover field work in southern Labrador was received by K.D.C. in 1973 from DREE, through the Department of Mines and Energy (St. John's, Newfoundland). Support for D.B. in 1974 was provided by the Geological Survey of Canada.

We are sincerely grateful to R. W. Hurst and Dr G. W. Wetherill for unpublished geochronological details on rocks from Saglek collected by R. W. H. in 1974, and to Dr I. Ermanovics for critically reading the manuscript.

References

Baragar, W. R. A., and Goodwin, A. M., 1969. 'Andesites and Archaean volcanism of the Canadian Shield', In *Proceedings of the Andesite Conference, Oregon Dept. Geol. and Min. Industries, Bull.*, **65**, 121–142.

Barton, J. M., 1975a. 'The Mugford Group volcanics, Labrador', *Can. J. Earth Sci.*, (in press).

Barton, J. M., 1975b. 'Rb–Sr isotopic systematics and chemistry of the 3·6 b.y. Hebron Gneiss, Labrador', *Earth Planet. Sci. Lett.*, (in press).

Berg, J. H., 1974. 'Mineral assemblages in the contact aureole of the Hettasch intrusion and some P-T estimates for the emplacement of the Nain Complex, coastal Labrador', *Geol. Assoc. Can., Prog., Abstr., St. John's, Newfoundland*, 8–9.

Bridgwater, D., Escher, A., Jackson, G. D., Taylor, F. C., and Windley, B. F., 1973a. 'Development of the Precambrian Shield in West Greenland, Labrador and Baffin Island', *Am. Assoc. Petrol. Geol., Mem.*, **19**, 99–116.

Bridgwater, D., Escher, A., and Watterson, J., 1973b. 'Tectonic displacements and thermal activity in two contrasting Proterozoic mobile belts from Greenland', *Phil. Trans. R. Soc. Lond.*, **A273**, 513–533.

Bridgwater, D., Watson, J., and Windley, B. F., 1973c. 'The Archaean craton of the North Atlantic region', *Phil. Trans. R. Soc. Lond.*, **A273**, 493–512.

Bridgwater, D., McGregor, V. R., and Myers, J. S., 1974. 'A horizontal · tectonic regime in the Archaean of Greenland and its implications for early crustal thickening', *Precambrian Res.*, **1**, 179–197.

Bridgwater, D., Collerson, K. D., Hurst, R. W., and Jesseau, C. W., 1975. 'Field characters of the early Precambrian rocks from Saglek, coast of Labrador', *Geol. Surv. Can.*, Paper 75–1, Part A, 287–296.

Cann, J. R., 1971. 'Major element variations in ocean floor basalts', *Phil. Trans. R. Soc. Lond.*, **A268**, 495.

Daly, R. A., 1902. 'The geology of the northeast coast of Labrador', *Bull. Mus. Comp. Zool., Harvard Univ.*, **38**, geol. ser., 5, 205–270.

Engel, A. E. J., Engel, C. G., and Havens, R. G., 1965. 'Chemical characteristics of oceanic basalts and upper mantle', *Geol. Soc. Amer. Bull.*, 76, 719–734.

Gandhi, S. S., Grasty, R. L., and Grieve, R. A. F., 1969. 'The geology and geochronology of the Makkovik Bay area, Labrador', *Can. J. Earth Sci.*, **6**, 89–97.

Glikson, A. Y., 1971. 'Primitive Archaean element distribution patterns: chemical evidence and geotectonic significance', *Earth Planet. Sci. Lett.*, **12**, 309–321.

Greene, B. A., 1974. 'An outline of the geology of Labrador', *Dept. Mines and Energy, Nfld., Info. Circ.*, **15**, 64.

Hurst, R. W., 1974. 'The early Archaean of coastal Labrador', In *The Nain Anorthosite Project: Field Report 1973*, S. A. Morse (Ed.), Geol. Dept., U. Mass., Amherst, Mass., **13**, 29–32.

Hurst, R. W., Bridgwater, D., Collerson, K. D., and Wetherill, G. W., 1975. 'Rb/Sr systematics in very early Archaean gneisses from Saglek Fiord, Labrador', *Earth Planet. Sci. Lett.*, (in press).

Jakeš, P., and White, A. J. R., 1971. 'Composition of island arcs and continental growth', *Earth Planet. Sci. Lett.*, **12**, 224–230.

Knight, I., 1973. 'The Ramah Group between Nachvak Fiord and Bears Gut, Labrador', In *Report of activities, April to October*, 1972, Geol Surv. Can., **73–1**, 156–161.

Knight, I., 1974. 'The Ramah Group–Mugford Group–Snyder Group—a correlation in Labrador and a possible link between the Labrador trough and western Greenland', *Geol. Assoc. Can., Prog., Abstr., St. John's, Newfoundland*, 50.

McGregor, V. R., 1973. 'The early Precambrian gneisses of the Godthaab District, West Greenland', *Phil. Trans. R. Soc. Lond.*, **A273**, 343–358.

Morgan, W. C., 1974. 'The Nain–Churchill boundary in the Torngat Mountains, northern Labrador: an Archaean–Proterozoic structural province contact in the Canadian Precambrian Shield', *Geol. Soc. Amer. Abst.*, **6**, 876–877.

Smyth, W. R., Marten, B. E., and Ryan, B. A., 1975. 'Geological mapping in the central mineral belt, Labrador: redefinition of the Croteau Group', *Nfld. Dept. Mines and Energy, Report of Activities*, 1974 (in press).

Speer, J. A., 1975. 'The contact metamorphic aureole of the Kiglapait intrusion', In *The Nain Anorthosite Project: Field Report 1974*, S. A. Morse (Ed.), Geol. Dept., U. Mass., Amherst, Mass., **17**, 16–26.

Stockwell, C. H., 1964. 'Fourth report on structural provinces, orogenies and time classification of the Canadian Precambrian Shield', In Lowden, J. A., Stockwell, C. H., Tipper, H. W., and Wanless, R. K., *Age Determinations and Geological Studies.* Geol. Surv. Can. Paper **64–17**, 1–21.

Sutton, J. S., Marten, B. F., and Clark, A. M. S., 1971. 'Structural history of the Kaipokok Bay area, Labrador', *Proc. Geol. Assoc. Can.*, **24**, 103–106.

Sutton, J. S., 1972. 'The Precambrian gneisses and supracrustal rocks of the western shore of Kaipokok Bay, Labrador, Newfoundland', *Can. J. Earth Sci.*, **9**, 1677–1692.

Taylor, F. C., 1969. 'Reconnaissance geology of a part of the Precambrian Shield, northeastern Quebec and northern Labrador', *Geol. Surv. Can.*, **68–43**, 13 pp.

Taylor, F. C., 1970. 'Reconnaissance geology of a part of the Precambrian Shield, northeastern Quebec and northern Labrador', *Geol. Surv. Can.*, **70–24**, 10 pp.

Taylor, F. C., 1971. 'A revision of Precambrian structural provinces in northeastern Quebec and northern Labrador', *Can. J. Earth Sci.*, **8**, 579–584.

Wheeler, E. P., 2nd, 1960. 'Anorthosite-adamellite complex of Nain, Labrador', *Geol. Soc. Amer. Bull.*, **71**, 1755–1762.

Wiener, R. W., 1975. 'An Archaean gabbro-anorthosite complex, Tessiuyakh Bay', In *The Nain Anorthosite Project: Field Report 1974*, S. A. Morse (Ed.), Geol. Dept., U. Mass., Amherst, Mass., **17**, 7–15.

Windley, B. F., 1969. 'Anorthosites of southern West Greenland', *Am. Assoc. Petrol. Geol., Mem.* **13**2, 899–915.

Greenstone Belts

Stratigraphy and Evolution of Primary and Secondary Greenstones: Significance of Data from Shields of the Southern Hemisphere

A. Y. GLIKSON

*Bureau of Mineral Resources Geology and Geophysics,
Canberra, A.C.T., Australia.*

Abstract

Aspects of the stratigraphy of Archaean greenstone belts in Western Australia, the Transvaal, Rhodesia, and India are reviewed. Early Precambrian volcanic sequences are classified in terms of two stratigraphically and petrogenetically distinct assemblages, termed primary greenstones and secondary greenstones. Primary greenstones consist of mafic–ultramafic volcanic sequences, including an acid volcanic component. The widespread occurrence of such ultramafic and mafic enclaves in the earliest granites suggests that these rocks are relicts of a once extensive ultramafic–mafic crust which either dates back to a supposed meteorite bombardment phase about 4 b.y. ago, and/or represents continuing generation of oceanic crust throughout the Archaean. The secondary greenstones consist of a bimodal mafic–felsic volcanic assemblage and/or of basalt–andesite–rhyolite cycles. The bimodal suite is commonly accompanied by an ultramafic component. The secondary greenstone are thought to have evolved within linear troughs developed in partly cratonized regions, where primary greenstones were earlier intruded by sodic granites. Both primary and secondary greenstones have been identified in India, in the Swaziland System, and in Rhodesia. In Western Australia at least three and possibly more volcanic-sedimentary cycles can be traced, all containing an ultramafic component; the mafic–felsic bimodal suite is well developed. In India, ultramafic volcanic enclaves in the Dharwar gneiss dome appear to predate the Dharwar greenstone belts, which include cross-bedded quartzite at their base. Because the principal greenstone belts of the Superior and Slave provinces in places overlie granites, and have only a minor ultramafic component, they compare with secondary greenstones. Mafic enclaves in 3 b.y. old gneisses in Manitoba and the Slave Province may represent pre-Keewatin greenstone cycles. The greenstone belt is regarded as the geological entity whose evolution effected a transformation from oceanic crust into sialic shield—a process which took place in different parts of the Earth at different times, and which was terminated by a global thermal event about 2·6 b.y. ago.

1. Early and Current Concepts on Greenstone Belts

The identification of primary volcanic, sedimentary, and geochemical features in low-grade metamorphosed Archaean sequences has resulted in the accumulation of data placing constraints on models of early crustal evolution. Before the application of the results of isotopic dating, the study of the Precambrian was inherently beset by assumptions such as, for example, that increasing metamorphic grade denotes increasing age, or that the crust must have been completely recycled at some stage, or that a world-wide sial formed when the Earth first cooled. It has been widely assumed, and indeed still is by some, that Precambrian igneous metamorphic suites are partly obliterated counterparts of orogenic series analogous to Phanerozoic ones, as for example 'Precambrian provinces represent a long series of orogenic cycles, each of which, though having distinctive peculiarities of its own, is essentially the same kind as the later Caledonian, Hercynian and Alpine cycles' (Holmes, 1948). Similar uniformitarian views have survived the advent of the plate tectonics theory, and modern literature on the Precambrian is characterized by attempts at correlating structural, geophysical, igneous, and geochemical lineaments and boundaries with possible early plate boundaries (Gibb and Walcott, 1971; Katz, 1972; Condie, 1972; Thorpe, 1972; Davidson, 1973; Chase and Gilmer, 1973; Talbot, 1973).

In particular, Precambrian geologists were impressed by the broad stratigraphic similarities between lithological assemblages of Archaean greenstone belts and those of island arc-trench systems (Gill, 1961; Wilson et al., 1965; Folinsbee et al., 1968; Goodwin, 1968, 1971; Green and Baadsgaard, 1971; White et al., 1971; Anhaeusser, 1973), or Alpine-type geosynclines (Pettijohn, 1943; Anhaeusser et al., 1968; Glikson, 1968, 1970; Weber, 1971; Srinivasan and Sreenivas, 1972). The latter comparisons are based on successive occurrence in greenstone belts of a eugeosynclinal-type volcanic association, a flysch-like sequence of turbidites, and in turn a molasse-like conglomerate-rich assemblage in ascending stratigraphic order. Similarities between the Archaean bimodal mafic-felsic volcanic suite and the Alpine spilite–keratophyre suite were also referred to in support of the latter analogy.

Notwithstanding uniformitarian comparisons, many petrological, geochemical, structural, and stratigraphic data from Archaean greenstone belts appear to be incompatible in detail with those of modern tectonic domains (Table 1). MacGregor (1951) drew attention to the unique granite–greenstone patterns of the Rhodesian shield, which he interpreted in terms of differential subsidence of narrow segments of a mafic volcanic layer which overlay a granitic basement. This model was developed further by Talbot (1968) and Fyfe (1973), who regarded the diapiric granites as mantled gneiss domes arising from anatexis and remobilization of sial. Recently it was suggested that Archaean greenstone belts may have evolved in rifted zones developed between diverging sial plates (Windley, 1973). The differences between the granite–greenstone system in the eastern Transvaal and younger orogenic belts were stressed by Anhaeusser et al. (1969). In particular, the recognition of a basal ultramafic–mafic extrusive assemblage in the Swaziland System (Viljoen and Viljoen, 1969a, 1971) has shown that fundamental distinctions exist between Archaean and younger volcanic activity. That important geochemical differences exist between Archaean and younger volcanic and plutonic suites was demonstrated by studies in North America (Wilson et al., 1965; Baragar, 1966, 1968; Baragar and Goodwin, 1969; Hart et al., 1970; Condie et al., 1970; Goodwin, 1971; Condie and Lo, 1971; Arth and Hanson, 1972; Condie, 1972, 1973; Condie and Baragar, 1974; Jahn et al., 1974), South Africa (Viljoen and Viljoen, 1969a, 1969c; Hunter, 1974; Anhaeusser, 1971; Glikson and Taylor, unpublished results), India (Naqvi and Hussain, 1973a, 1973b; Viswanathan, 1974; Sreenivas and Srinivasan, 1974; Naqvi et al., 1974) and Australia (O'Beirne, 1968; Glikson, 1968, 1970, 1971; Glikson and Sheraton, 1972; Hallberg, 1972; Hallberg and

Williams, 1972; Williams and Hallberg, 1973).

The divergence of concepts on the origin and evolution of Archaean greenstone belts stems not only from differences in the emphasis placed upon either their similarities to, or their differences from, younger systems, but also to a large extent from marked differences between the greenstone belts themselves. Bearing in mind that the time span during which greenstone belts are known to have evolved (3.8–2.6 b.y.) exceeds 1·2 b.y., it would have been surprising indeed had not this been the case. It is the aim of this paper to compare the elements of stratigraphy of Archaean volcanic successions in Western Australia, Southern India, Transvaal, and Rhodesia, and to consider environmental models with reference to crustal evolution in the early Precambrian.

Table 1. A comparison of some principal features of Archaean greenstone belts, Proterozoic terrains, Alpine geosynclines and island arcs

	Archaean greenstone belts	Proterozoic volcanic sequences
Geotectonic setting	Outliers within intrusive granites, up to several hundred km in length	Continental shields: platform cover, or within inter-cratonic mobile belts
Intrusive assemblages	Subvolcanic mafic to ultramafic sills; tonalitic, trondhjemitic and granodioritic early plutons; adamellitic and syenitic late plutons	Large mafic dykes and layered lopoliths; granodiorites, granites and alkali granites, rapakivi granites
Volcanic assemblages	Bimodal mafic–felsic suites, abundant ultramafic volcanics (komatiites), basalt–andesite–rhyolite cycles (in Canada). Pillowed basalts common. Alkaline volcanics rare, but occur at high stratigraphic levels. Acid volcanic lenses and pyroclastics are common	Abundant mafic and acid volcanics, often in acid to mafic sequence. Pillowed basalts not very common. Alkaline volcanics rare
Chemistry of volcanics	Low-K oceanic-type tholeiites very common; Na-rich rhyolites very common; Flat REE patterns in tholeiites but highly fractionated in acid volcanics	Continental-type K-normal tholeiites predominate. Acid volcanics are mostly K-rich. Andesites are relatively minor
Sedimentary assemblages	Greywacke-shale association predominates; carbonaceous shales; chert; banded iron formations; conglomerates (both polymictic and monomictic); very minor carbonates and quartzites. Pure shale units are rare	Abundant quartzite (cross bedded); feldspathic sandstones, shales and carbonates; banded iron formations, and locally abundant greywackes present, but less common than quartzose sandstones
Successions	Ultramafic–mafic volcanics, mafic–felsic volcanics, greywacke–slate units, polymictic conglomerates in ascending stratigraphic order	No systematic pattern generally predominates; often mafic volcanics overlie acid volcanics. However, some Proterozoic basins have geosynclinal-like successions (i.e. Coronation geosyncline and Labrador trough)
Typical structures and metamorphism	Synclinal outliers downfolded and downfaulted between granites; anticlines commonly accompanied by major faults; common strike faults; mainly gravity tectonics; mainly low pressure greenschist facies metamorphism	Either little deformed platform cover, or strongly folded and faulted (including rifted) sequences in mobile belts. Metamorphism: up to greenschist in platforms, up to granulite (mostly low pressure) in mobile belts

Table 1 continued on page 260

Table 1—*continued*

	Alpine geosynclines	Island arc and Cordillera chains
Geotectonic setting	Along continent–continent ocean-closure collision sutures; several thousand km in length (i.e. Alpine–Himalaya belt)	Ensimatic, parallel to continent–ocean boundaries, or justaposed with sial–sima boundaries. Several thousand km in length (i.e. circum Pacific belt)
Intrusive assemblages	Ophiolite complexes; large serpentinite and gabbro sheets (part of upper mantle), granodiorite and granite	Early trondhjemite and granodiorite; late shoshonitic intrusions and porphyries
Volcanic assemblages	Ophiolite complexes: spilites, low-K tholeiites, quartz keratophyres, less commonly andesites	Andesites, island-arc tholeiites, high-alumina tholeiites, calcalkaline basalts, shoshonitic volcanics, abundant pyroclastic deposits
Chemistry of volcanics	Spilitized low-K oceanic-type tholeiites, Na-rich acid volcanics (quartz keratophyres)	Island arc tholeiite series and calc-alkaline volcanics are Al-rich, highly depleted in ferromagnesian trace elements, and with fractionated REE patterns for both mafic and acid volcanics
Sedimentary assemblages	Flysch assemblage–turbidite greywacke, shale and carbonates; chert and radiolarian chert; Molasse, impure sandstone, lithic sandstone and conglomerates. Boulder breccias and olistostromes (tectonic slumps)	Volcanogenic lithic greywackes, redeposited lithic and crystal tuffs, shales, abundant foraminiferal carbonates, boulder conglomerates and breccias
Successions	Ophiolite complexes (including the spilite–keratophyre association), to flysch sequences, to molasse	Oceanic tholeiite substratum; mixed island arc tholeiite–calcalkaline volcanic–shoshonite sequences, with turbidites and carbonate at variable stratigraphic levels
Typical structures and metamorphism	Highly compressional tectonic features common: thrusts, nappes, tectonic mélange. Foreland troughs. High-pressure metamorphism may occur	Strong vertical faulting, broad folding within fault blocks, tectonic slides and slumps. Rift zones. High-pressure metamorphism may occur

2. Volcanic Stratigraphy of Greenstone Belts

Kalgoorlie System Yilgarn Shield, Western Australia

Horwitz and Sofoulis (1965) classified the stratigraphy of greenstone belts in the Kalgoorlie–Norseman area in terms of two major volcanic-sedimentary sequences separated by an unconformity which is accompanied by a conglomerate. This unconformity was also mapped in the Coolgardie–Kurrawang area, where a type-section of the Kalgoorlie System includes three ultramafic-mafic assemblages, termed Coolgardie, Mount Robinson, and Red Lake greenstones (Glik-son, 1968, 1970; Kriewaldt, 1969) (Fig. 1). McCall (1969) and students of the University of Western Australia have studied the upper part of this succession in the Lake Lefroy area. Williams (1970, 1973), on the basis of regional mapping east and north of Kalgoorlie, established a stratigraphic column consisting of three volcanic-sedimentary associations, each including a mafic–ultramafic volcanic succession overlain conformably by acid volcanic rocks and derived sediments (Fig. 1). These cycles are separated from one another by pronounced unconformities and/or stratigraphically consistent chert layers. A detailed synthesis of stratigraphic information from the Kalgoorlie–Norseman area carried out by

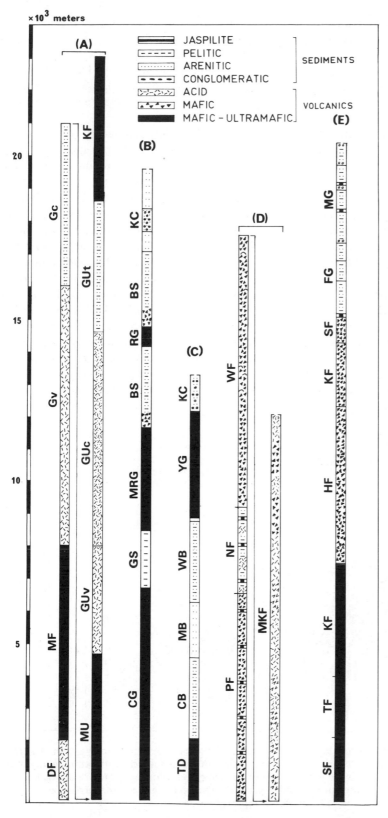

Fig. 1. Generalized composite stratigraphic columns in the Kalgoorlie System, Western Australia, and the Barberton Mountain Land. (A) East and north of Kalgoorlie (Williams, 1970; Williams et al., 1973): DF—Dewtop Formation; MF—Morelands Formation; Gv—Gindalbie Formation, acid volcanics; Gc—Gindalbie Formation, clastic sediments; Mu—Mulgabbie Formation; GUv—Gundockerta Formation, acid volcanics; GUc—Gundockerta Formation, acid volcaniclastics; GUt—Gundockerta Formation, turbidites; KF—Kalpini Formation. (B) Coolgardie–Kurrawang area (Glikson, 1968): CG—Coolgardie greenstones; GS—Gunga siltstones; MRG—Mount Robinson greenstones; BS—Brown Lake sediments; RG—Red Lake greenstones; BS—Black Flag sediments; KC—Kurrawang conglomerate. (C) West Lake Lefroy area (McCall, 1969): TD—Town Dam greenstones; CB—Causeway Beds; MB—Mandila Beds; WB—Wanda Wanda Beds; YG—Yilmia greenstones; KC—Kurrawang conglomerate. (D) Norseman area (Doepel, 1965): PF—Penneshaw Formation; NF—Noganyer Formation; WF—Woolyennyer Formation; MKF—Mount Kirk Formation. (E) Barberton Mountain Land (Anhaeusser, 1973): SF—Sandspruit Formation; TF—Theespruit Formation; KF—Komati Formation; HF—Hooggenoeg Formation; KF—Kromberg Formation; SF—Swartkoppie Formation; FG—Fig Tree Group; MG—Moodies Group. Lithological symbols are as for Figs 2 and 4

Gemuts and Theron (1975) confirmed this subdivision. Durney (1972) documented an unconformity about 400 km north of Kalgoorlie, where older greenstones and an associated granodiorite underlie younger greenstones which are in turn intruded by younger granite. Rb–Sr isotopic determinations assign an age of 2·7 b.y. to the older granodiorite (Roddick et al., 1973), whereas the younger granites of the Kalgoorlie region were dated as *c.* 2·6 b.y. old (Turek and Compston, 1971). Possibly, cycles 1 and 2 of Williams (1970) are even older than the *c.* 2·8 b.y. age assigned to some migmatitic gneisses in the Kalgoorlie area (de Laeter and Blockley, 1972), whereas the age of cycle 3 is restricted by ages of the older and younger granites to the 2·7–2·6 b.y. range. The older granites consist of sodium-rich gneissic types (Glikson and Sheraton, 1972), and intrude in anticlinal positions where they are enclosed by rocks of the first and second cycles. The younger granites include adamellites and porphyritic types, and may also intrude units of the third cycle (Williams et al., 1973; Gemuts and Theron, 1973); the latter

granites have not been affected by regional metamorphism (Glikson, 1968), which is dated in the Kalgoorlie area as *c.* 2·67 b.y. (Turek and Compston, 1971). Stratigraphic columns of the Kalgoorlie System and an interpretation of field relationships are presented in Figs 1 and 2.

The stratigraphy outlined above pertains to the linear north-northwest-striking greenstone belts of the Wiluna–Kalgoorlie–Norseman Zone. This area, defined as the Kalgoorlie subprovince, is flanked on the west and on the east by the Southern Cross and Laverton subprovinces, which differ from the Kalgoorlie subprovince in several important ways (Williams, 1973). Only or mainly one greenstone cycle is recognized in the Southern Cross and Laverton terrains, and has been correlated by Williams (1973) with cycle 1 of the Kalgoorlie subprovince. Basal parts of cycle 2 are found in synclinal positions in the Laverton subprovince. The exposure of only the lower stratigraphic sections in the Laverton and Southern Cross terrains is consistent with their denudation to a deeper level and

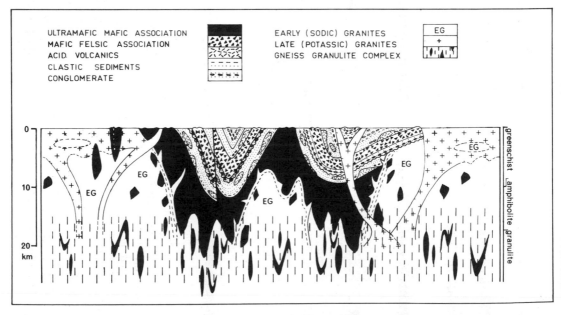

Fig. 2. Interpretation of field relationships between major Archaean rock units in Western Australia, India and South Africa—Rhodesia. No detailed comparison with the observed stratigraphy in any of the above systems is intended, and the figure is meant to portray the general concept of the relations between primary greenstones, secondary greenstones, older granites, younger granites, and high-grade gneiss–granulite complexes

with an interpretation of the Kalgoorlie sub-province as a downfaulted or rifted zone. It is also in accord with the average eastward tilting of the Yilgarn shield in the Perth–Coolgardie cross-section (Glikson and Lambert 1975; Mathur, 1973). It follows from the stratigraphic correlation of Williams (1973) that volcanics of the 1st cycle may have originally extended over most of the Yilgarn shield area, possibly as an almost continuous layer. However, this layer does not necessarily represent the earliest volcanic crust of this region, and evidence for the existence of a yet older volcanic cycle has been found in the Edjudina area north-northeast of Kalgoorlie (Williams, et. al. 1971).

The mafic–ultramafic assemblages display both lateral and vertical changes in the proportions of the various components—which include metamorphosed pillowed to massive tholeiite, dolerite, gabbro, porphyritic gabbro, high-Mg volcanics, peridotitic volcanics, differentiated ultramafic and mafic sills and lenses (Williams and Hallberg, 1973), chert, banded iron formations, and minor extrusive and intrusive acid rocks. The lowermost greenstone cycle (cycle 1) includes small-scale mafic to felsic calc–alkaline cycles which include andesite (Williams, 1973) and acid tuffs and rhyolitic flows (Bye, 1971), and are typically capped by banded iron formations. Banded iron formations are more common in the Southern Cross and Laverton subprovinces than in the Kalgoorlie subprovince. Ultramafic rocks, on the other hand, are more abundant in the Kalgoorlie subprovince (Williams, 1973). In contrast to cycle 1, cycles 2 and 3 display abrupt alternations of ultramafic–mafic and acid volcanic–clastic associations. Another difference between the cycles is the exclusive occurrence of oligomictic volcanically-derived conglomerates above the greenstones of cycle 1, and the appearance of polymictic conglomerates at upper parts of cycles 2 and 3. The uppermost conglomerates include pebbles of jaspilite and sodic granite (Glikson, 1968), signifying uplift and denudation of cycle 1 rocks in neighbouring areas at late stages in the evolution of the Kalgoorlie System.

Dharwar System, southern India

Recent reviews of the stratigraphy of the Dharwar greenstone belts and of the relationships between these successions and the surrounding granite-gneiss terrain (Pichamuthu, 1967, 1970a, 1970b, 1974; Srinivasan and Sreenivas, 1968, 1972; Radhakrishna, 1974) show that the Dharwar System includes at least three cycles, and that ultramafic-mafic enclaves in the gneisses may represent relics of pre-Dharwar volcanic rocks. The latter are exemplified by the Nuggihalli ultramafic zone, which trends northwest between Mysore and Dodguni, and has previously been regarded as intrusive, and by mafic–ultramafic rocks in the Kolar and Bababudan greenstone belts. This interpretation is supported by the pre-3·1 b.y. age of the Kolar rocks (Naqvi et al., 1974).

The pre-Dharwar volcanics include low-K tholeiites, high-Mg volcanics, and serpentinized ultramafic rocks (Viswanathan, 1974; Naqvi et al., 1974). The extrusive origin of some of these rocks is evidenced by pillow structures and quench textures. Chemical analyses of some of these rocks compare with those of 'basaltic komatiites' described from the lower part of the Onverwacht Group in South Africa (Viswanathan, 1974). The mafic and ultramafic enclaves in the Peninsular Gneiss display well pronounced thermal metamorphic aureoles along their contacts with the gneiss, attesting to the intrusive nature of the latter. An age of about 3·3 b.y. was recorded from hornblende schist in the Hutti gold field defining the pre-Dharwar mafic–ultramafic volcanics as the oldest formations identified in Southern India (Radhakrishna, 1967, 1974).

No unequivocal intrusive relations have been found between the Peninsular gneisses and the Dharwar volcanic-sedimentary belts themselves; i.e. the Chitaldrug, Shimoga, and Gadag belts; the arguments in favour of such relations are based on structural discordances between the gneisses and the greenstones, and on thermal aureoles in inclusions, which may, however, be of pre-Dharwar age (see Pichamuthu, 1967; 1970a; 1970b; 1974). Those who favour unconformable relationships between the Peninsular Gneiss and the

Dharwar belts point to the occurrence in the latter of a basal orthoquartzite–carbonate association (Srinivasan and Sreenivas, 1968, 1972), and to the occurrence of granite-derived arenites and conglomerates. It must be borne in mind, however, that the term 'Peninsular gneiss' pertains to a wide range of acid plutonic and metamorphic rocks dated within the 2·6–3·1 b.y. interval (Crawford, 1969), and that probably older rocks occur within this complex, as suggested by the 3250 ± 50 m.y. age of gneiss pebbles from a conglomerate near Kaldurga (Venkatasubramanian and Narayanaswamy, 1974). It is thus clear that the Peninsular gneiss contains pre-Dharwar phases, syn-Dharwar phase (i.e. Champion gneiss) and post-Dharwar phases (i.e. in the Closepet granite–gneiss suite).

Aswanatharayana (1968) suggested that three thermal episodes affected the Dharwar System, about 3 b.y., 2·6–2·3 b.y., and 2 b.y. ago. It is possible, however, that the oldest phase pertains to what Radhakrishna (1967, 1974) considers as pre-Dharwar relicts, and the youngest phase to age resets which post-date the development of the greenstone belt. The stratigraphy of the Dharwar System, as correlated by Srinivasan and Sreenivas (1972) in the Bababudan, Shimoga, Chitaldrug and Kolar belts, is (from top to bottom):

Red beds, shales, silts and sandstones (G.R. Series).
  Local unconformity
  Closepet granites (c. 2·4 b.y., Crawford, 1969). Also younger phases of the Peninsular Gneiss.

Greywackes (Rainbennur greywackes, Shimoga belt).

Jaspilite, banded iron formations, pyritic chert (Shimoga, Chitaldrug and Kolar Belts).

Champion gneiss.

Basalt–keratophyre association, including ultrabasics in the Kolar belt (Grey Trap Rainbennur Series).

Grits, greywackes and conglomerates.
  Unconformity

Orthoquartzite, carbonate, oligomictic conglomerates (Dodguni Series)

Ultramafic rocks
Unconformity

Banded iron formations, quartzite, conglomerate, basalt-felsite association (confined to Bababudan belt; correlation with other belts is uncertain).
Major unconformity

Older phases of Peninsular gneiss and ultramafic enclaves.

It is possible that the ultramafics underlying the Dodguni series are stratigraphic equivalents of the Nuggihalli ultramafics and other enclaves in the Peninsular gneiss. Further geochronological studies are required to elucidate the relationships between the different granite phases and the Dharwar cycles.

Swaziland System

Kaapvaal Shield, Transvaal. Regional mapping in the Barberton Mountain Land area (Visser et al., 1956) has documented what has been shown to be the best exposed greenstone belt in Africa, and even world-wide. Detailed investigations in parts of this area (Anhaeusser et al., 1968, 1969; Viljoen and Viljoen, 1969a,b,c,d,e,f) have shown that the Jamestown Complex, previously regarded as intrusive, is in fact a sequence of ultramafic and mafic extrusive and hypabyssal rocks comprising tholeiites, dolerites, high-Mg basalts, peridotitic flows, minor acid volcanic and pyroclastic rocks, chert, and limestone—stratigraphically termed Onverwacht Group.

The lower part of the Onverwacht Group consists principally of ultramafic to mafic flows and sills, and is divided into three formations, which in ascending stratigraphic order include the Sandspruit, Theespruit, and Komati Formations—a subdivision made possible thanks to the distinctive lithology of the Theespruit Formation, which includes acid tuff and chert intercalations (Viljoen and Viljoen, 1969a). The Komati Formation is overlain by the Middle Marker, a thin but stratigraphically consistent unit of quartz keratophyre, chert, and carbonates—interpreted as a major discontinuity possibly reflecting movements related to the emplacement of the 'ancient tonalites' (Viljoen and Viljoen, 1969a). Isotopic dating of diapiric tonalite and trondhjemite which intrude the lower part of the Onverwacht group gave ages in the range 3·4–3·2 b.y. (Allsopp et al., 1968; Oosthuyzen, 1970). Rb–Sr isotopic work on greenstones indicated an age of at least 3·5 b.y. (Jahn and Shih, 1974).

The upper part of the Onverwacht Group is compositionally distinct from the lower part: it consists of numerous mafic–felsic cycles, which include pillowed or massive tholeiitic basalts capped by acid volcanic assemblages accompanied by chert and carbonate. Andesitic and ultramafic volcanics occur on a minor scale. The age of this sequence is limited to the 3·3–3·0 b.y. range, as defined by the age of the Middle Marker unit (Hurley et al., 1972) and of the intrusive Nelspruit migmatite (Allsopp et al., 1968), respectively. The upper part of the Onverwacht Group is unconformably overlain by a sequence of greywacke, slate, and jaspilite of the Fig Tree Group, which is in turn succeeded by siltstone, quartzite, and polymictic conglomerates of the Moodies Group. The latter includes minor trachytic flows (Anhaeusser et al., 1969). Both the Fig Tree and Moodies Group include detritus derived from the Onverwacht Group and from yet unidentified granites (Anhaeusser et al., 1969; Hunter, 1974; Glikson and Taylor, in prep.).

Rhodesian Greenstone Belts

The earliest greenstones in the Rhodesian shield are enclaves of mafic and ultramafic rocks and banded iron formations scattered in gneissic tonalites dated at c. 3·3 b.y. (Bliss and Stidolph, 1969; Wilson, 1973). These xenoliths are particularly abundant in the Selukwe area, where they have been referred to as 'pre-Sebakwian' by Stowe (1968). However, the enclaves appear to be distinguishable from the Sebakwian Group rocks in their type areas between Selukwe and Gwelo only or mainly by their higher metamorphic grade (pyroxene hornfels facies). The type Sebak-wian Group rocks themselves, originally termed 'magnesian series' by MacGregor (1932), include serpentinized ultramafics, talc schist, banded iron formations, and chert. Some of the ultramafics are of intrusive origin (Harrison, 1968). However, Viljoen and Viljoen (1969g) drew attention to stratigraphic and lithological similarities between the Sebakwian Group and the lower part of the Onverwacht Group. Thus, little-disturbed enclaves of a stratigraphic column analogous to that of the Theespruit–Komati succession occur near Que Que within the Rhodesdale tonalite near its intrusive contact with the Midlands greenstone belt. Wilson (1973) also remarks on the volcanic origin of at least some of the Sebakwian rocks.

The Rhodesdale pluton is dated as c. 3·3 b.y., and is structurally and petrologically similar to the 'ancient tonalites' of the Barberton Mountain Land. A critical outcrop on the Sebakwe River north of Que Que contains a conglomerate which unconformably overlies talc schists of the Sebakwian Group, and includes pebbles of gneiss similar to that of the Rhodesdale pluton. The conglomerate is overlain by the Bulawayan Group, which consists of mafic–acid volcanic cycles analogous to those of the upper sequence of the Onverwacht Group (Bliss and Stidolph, 1969; Viljoen and Viljoen, 1969g). Similar unconformities were recorded along the northeastern boundary of the Shangani batholith. These observations are significant with respect to the origin of the Middle Marker unit in the Barberton Mountain Land, suggesting that the upper part of the Onverwacht Group may have postdated the emplacement of the early tonalites (Viljoen and Viljoen, 1969a).

The Bulawayan Group includes a lower mafic assemblage accompanied by minor ultramafic flows and intercalated chert. Available chemical analyses (Phaup, 1973) indicate the common occurrence of low-K tholeiites. Upper stratigraphic levels show a progressive increase in andesitic to rhyolitic volcanic associations, stratigraphically defined as the Maliami River Formation (Harrison, 1968). These rocks grade laterally and vertically into derived clastic sediments, which were not always mapped from sediments of the overlying Shamwaian Group. The latter group consists of quartz-mica schist, phyllite, banded iron formations, and conglomerate (Bliss and Stidolph, 1969; Wilson, 1973).

3. Significance of Archaean Ultramafic–Mafic Volcanics

Shields of the southern hemisphere abound in mafic–ultramafic relics of what appear to have been extensive volcanic layers older than the earliest granites dated in those areas. Thus, cycle 1 of the Kalgoorlie System, the Nuggihalli ultramafics and other enclaves in the Peninsular gneiss, the lower Onverwacht Group and the Sebakwian Group, represent the earliest units in their respective areas. The shield-wide scale on which these rocks occur, including their high-grade equivalents within Archaean granulite terrains (Wilson, 1971; Pichamuthu, 1970b; Sen, 1974; Radhakrishna, 1974; Glikson and Lambert, in prep.), the lack of evidence for pre-existing granites or sialic crust in their vicinity (Viljoen and Viljoen, 1969a; Glikson, 1972), and the inability of isotopic methods to place older age limits on them, render these rocks the nearest equivalents to 'vestiges of the beginning' in Archaean granite–greenstone terrains.

This is not to say that the exposed ultramafic–mafic units are necessarily representative of the oldest volcanic cycles within their respective areas. As indicated above, cycle 1 in the Kalgoorlie System may well overlie still older cycles (Williams et al., 1973). Likewise, greenstone cycles of the Pilbara system in north-western Australia may well be older than cycle 1, as suggested by the $c.$ 3·1 b.y. age of the intrusive granodiorites (de Laeter and Blockley, 1972). The existence of volcanic cycles older than the Onverwacht Group and Sebakwian Group is likewise probable. Nor is it intended to imply that the earliest recognizable mafic–ultramafic units are relics of the primordial Earth's crust, for it is envisaged that the crustal layer which formed at the time when the Earth cooled must have been largely destroyed by the major meteorite impact phase recorded at 4·0–

3·9 b.y. on the Moon, and which must have affected the Earth (Green, 1972). It is not impossible, however, that relics of the pre-impact crust have been preserved in some regions of the Earth; such rocks can be expected to display extensive brecciation, shatter-cone fractures, vitrification, and high-pressure mineralogical transformations (for example, see Milton et al., 1972). It can be expected that these impacts triggered worldwide volcanic activity arising from deep fracturing of the lithosphere, and consequent rise and adiabatic partial melting of mantle diapirs. It is conceivable that some of the earliest ultramafic–mafic enclaves in Archaean shields indeed represent vestiges of impact-induced volcanism (Green, 1972). On the other hand, because komatiite-type volcanics are known from sequences 2·7–2·6 b.y. old in Western Australia and Canada (Nesbitt, 1971; Hallberg and Williams, 1972; Pyke et al., 1973), where they may overlie thick sedimentary units, they cannot be regarded as the exclusive product of impact events, but rather as the result of endogeneous processes normal in the Archaean.

As suggested by Viljoen and Viljoen (1969a; 1969b) and supported by petrochemical calculations (McIver and Lenthall, 1974; Cawthorn and Strong, 1974), komatiite-type volcanics represent the products of between 30% melting of pyrolite (i.e. low-K tholeiites) and near-complete melting of pyrolite (i.e. extrusive peridotites of the Sandspruit Formation). The constant composition of the low-K tholeiites, as contrasted to the variable chemistry of the high-Mg basalts (Hallberg and Williams, 1972), can perhaps be interpreted in terms of a coexistence of mobile and stable source regions, i.e. partial melting in rising diapirs and the low-velocity zone, respectively. Although high-Mg basalts are known from post-Archaean sequences (Cox et al., 1965; Dallwitz, 1968; Gale, 1973; McIver, 1972), the komatiite suite is best developed in the Archaean. Available data, however, do not allow meaningful chemical comparisons to be made between the Archaean komatiites and more recent analogues. On the other hand, the major-element characteristics of

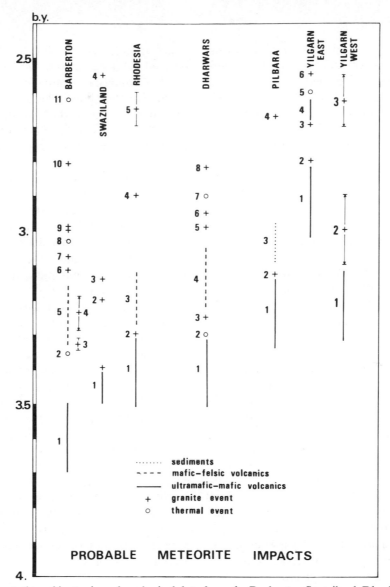

Fig. 3. Interpretation of isotopic and geological data from the Barberton Swaziland, Rhodesia, Dharwars, Pilbara, and Yilgarn terrains. *Barberton*: 1—lower part of Onverwacht Group; 2—Middle Marker; 3—'ancient tonalites'; 4—Dalmein pluton; 5—upper part of Onverwacht Group; 6—Bosmanskop pluton; 7—Hood granite; 8—pegmatite intruded into Fig Tree Group; 9—Nelspruit migmatite and Lochiel granite; 10—Mpageni granite; 11—metamorphism of felsic lavas of the upper part of the Onverwacht Group. *Swaziland*: 1—amphibolites of the 'ancient gneiss complex', intruded by tonalites (Sr87/86 initial = 0·7006, Hunter, 1974); 2—younger gneiss (Sr87/86 initial = 0·7048); 3—younger gneiss (Sr87/86 initial = 0·7022–0·7060); 4—Mbabane pluton. *Rhodesia*: 1—Sebakwian Group; 2—Rhodesdale granite (ancient tonalites); 3—Bulawayan Group; 4—granites; 5—granites. *Dharwars*: 1—Nuggihalli ultramafics and other ultramafic-mafic enclaves in the Peninsular gneiss; 2—hornblende schist in Hutti mines (metamorphic age); 3—gneiss pebbles near Kaldurga; 4—Dharwar greenstone belts; 5—Peninsular gneiss in northern Mysore; 6—Peninsular gneiss; 7—Amphibolite, Kolar greenstone belt; 8—Peninsular gneiss. *Pilbara*: 1—Warrawoona Succession, including at least two greenstone cycles; 2—older granodiorites; 3—Mosquito Creek Succession; 4—adamellites, in part tin-bearing. *Yilgarn east* (*Kalgoorlie System*): 1—greenstone cycles nos. 1 and 2—Menangina gneiss; 3—older granodiorite, Jones Creek; 4—greenstone cycle No. 3; 5—regional metamorphism; 6—younger granites (adamellites). *Yilgarn west* (*Wheat Belt*): 1—basic granulite and amphibolite enclaves in gneiss; 2—older low-Rb gneisses; 3—younger high-Rb granites. Sources of the data are given in the text. See p. 266 for discussion of early meteorite impacts

Archaean low-K tholeiites are mainly consistent with those of modern low-K oceanic tholeiites, allowing detailed consideration of more subtle variations (Glikson, 1970; 1971).

Chemical data from the Kalgoorlie System, from the lowermost and pre-Dharwar volcanics, and from the lower Onverwacht Group and the Sebakwian Groups, indicate that tholeiitic basalts associated with the komatiite suite are much depleted in K, Rb, Cs, Sr, Ba, Zr, Hf, U, and Th (Hallberg, 1972; Naqvi et al., 1974; Viljoen and Viljoen, 1969a; Phaup, 1973). Zr–Y–Ti plots fall into the field of oceanic tholeiites as delineated by Pearce and Cann (1971, 1973) and (Hallberg and Williams, 1972). Chondrite-normalized rare-earth patterns of Archaean mafic rocks from Western Australia and the Transvaal show near-flat curves and no La and Ce depletion (White et al., 1971; Glikson and Taylor, in prep.) The lack of such depletion was referred to by White et al. (1971) in support of the island-arc analogy; however, oceanic tholeiites showing such patterns are also known (Schilling, 1971).

The Fe/Fe+Mg ratios of the Archaean tholeiites are generally high, like those of olivine-depleted oceanic tholeiites (Miyashiro et al., 1969; Shido et al., 1971) and of island arc tholeiites (Jakeš and Gill, 1970). However, in contrast to the latter, Ni, Cr, and Co abundances in Archaean tholeiites are similar to those of modern oceanic tholeiites, suggesting little fractionation of olivine. The high Fe/Fe+Mg ratios may therefore be a characteristic of the primary undifferentiated magma, and may have potential implications for mantle composition and core segregation (Glikson, 1971, 1972). The low to intermediate Al levels of the Archaean tholeiites (about 14–15% Al_2O_3) preclude flotation and concentration of plagioclase, and also suggest a shallow level of magma equilibration, as the breakdown of plagioclase under pressures greater than 5 kb (Green and Ringwood, 1967) would lead to alumina enrichment. The Archaean tholeiites include both olivine-normative and quartz-normative varieties, and according to Green's (1971, p. 713) basalt petrogenetic grid, the maximum depth from which the olivine tholeiites could have been derived is about 60 km, whereas shallower levels of equilibration are indicated for quartz-normative tholeiites. The scarcity of undersaturated basalts in the Archaean (with few exceptions stratigraphically above the mafic–ultramafic suites; McIver and Lenthall, 1974), renders it unlikely that the depth of partial melting exceeded 60 km, or that partial melting within the depth range of 15–60 km was under 25% (see Green, 1971). It is therefore likely that the bulk of the magma equilibrated at depths shallower than 15 km, where evolution of nepheline-normative melts is arrested, and where little magmatic fractionation of olivine took place.

These considerations suggest that the low-K tholeiites evolved within environments to which modern mid-ocean ridges or back-arc spreading centres offer the closest analogues. This interpretation is in accord with the nature of the sediments intercalated with the early mafic–ultramafic units, which include chert, jaspilite, and graphitic slate. It is not inconsistent with the occurrence of acid volcanic and pyroclastic rocks above mafic–ultramafic cycles, as such are also known from modern oceanic domains (Bonatti and Arrhenius, 1970). However, the abundance of the associated acid volcanics, as well as the predominance of the komatiite suite, must be regarded as temporally unique features of Archaean volcanism.

4. Significance of Archaean Mafic–Felsic and Calc–Alkaline Volcanics

In contrast to the ultramafic–mafic suites considered in the preceding section, which evolved in environments showing no evidence of pre-existing or proximal sialic crust, stratigraphically younger greenstones, such as those of cycle 3 of the Kalgoorlie System, the Dharwar System, the upper part of the Onverwacht Group, and the Bulawayan Group, rest on post-granite unconformities and/or include granite-derived arenites and conglomerates. With the possible exception of the Kalgoorlie System, these greenstones include only a minor proportion of ultramafic material. The predominant volcanic assemblage is bimodal, consisting of tholeiitic basalts and dacitic to rhyolitic volcanic assemblages which form discrete cycles in this ascending order. Andesites were reported from the Dharwar, Bulawayan, and upper Onverwacht Groups, but they appear to be less abundant than either mafic or felsic rocks, and are rare in cycles 2 and 3 rocks of the Kalgoorlie System (Viljoen and Viljoen, 1969c; Bliss and Stidolph, 1969; Phaup, 1973; Hallberg, 1971; Williams, 1970, 1973).

In searching for modern analogues of the Archaean mafic–felsic and calc–alkaline volcanic assemblages, there can be little doubt that island arc-trench systems offer the closest resemblances. This comparison has been particularly favoured by Canadian geologists who pointed out the cyclic calc–alkaline nature of the volcanicity, the abundance of andesites and pyroclastic extrusives, the thick sequences of turbidites, and the evidence for older granites or sial in the vicinity of the greenstone belts (Gill, 1961; Wilson et al., 1965; Folinsbee et al., 1968; Goodwin, 1968; 1971; Baragar and Goodwin, 1969; Green and Baadsgaard, 1971; Jahn et al., 1974). The abundance of andesites (Baragar and Goodwin, 1969), the occurrence of high-Al basalts at Noranda (Baragar, 1968), and of volcanics showing shoshonitic affinities in north-western Ontario (Smith and Longstaffe, 1974), are all compatible with the island-arc interpretation. A basal conglomerate containing granitic pebbles and units of cross-bedded orthoquartzites underlie a thick basalt–andesite succession at Goose Lake, Manitoba, testifying to proximal or underlying granite basement (F. J. Elbers, pers. comm., 1974). Similar characteristics are displayed by volcanics of the upper part of the Onverwacht Group and the Bulawayan Group, Dharwar volcanics, and in part cycle 3 of the Kalgoorlie System (Williams, 1973). However, the Archaean sequences also contain ultramafic extrusives unknown in island arcs. Moreover, pending further geochemical studies it appears from available data that the Archaean assemblages display a pronounced mafic–felsic polarity, andesites being relatively uncommon in the Kalgoorlie System (Hallberg, 1972), the upper part of the

Onverwacht Group (Viljoen and Viljoen, 1969), the Bulawayan Group (Bliss and Stidolph, 1969; Phaup, 1973), and the Dharwar System (Naqvi et al., 1974).

Further considerations of the greenstone belt–island-arc analogy indicate differences in the chemistry of the volcanic rocks, in particular of the tholeiitic basalts. Whereas the bulk of island arc mafic volcanics are high-Al tholeiites and calc–alkaline basalts, most Archaean tholeiites have low to intermediate Al contents, with a few exceptions such as at Noranda (Baragar, 1968). $Fe/Fe+Mg$ ratios of tholeiites from Canadian Archaean calc–alkaline suites are equally high or higher than in island-arc basalts (Glikson, 1971). Ni, Cr, and Co abundances in Archaean rocks are higher by factors of two or three than in basalts of island arcs (compare Baragar and Goodwin, 1969; Hallberg, 1972; Naqvi and Hussain, 1973a,b; Taylor et al. 1969; Gill, 1970; and De Long, 1974). The latter features militate against olivine fractionation, which is considered an important process in island-arc petrogenesis. Nor is magnetite separation, regarded by Osborn (1962) as responsible for the development of the calc–alkaline suite, consistent with the high $Fe/Fe+Mg$ ratios in the Archaean rocks; thus low oxygen fugacities in Archaean magmas is indicated, in agreement with the low Fe_2O_3/FeO ratios of these rocks (Goodwin, 1968; Glikson, 1968; Hallberg, 1972). Condie and Baragar (1974) showed that tholeiites from the Abitibi belt have negative Eu anomalies similar to those of lunar highland volcanics (Gast, 1972), possibly confirming very low oxygen fugacities and retention of Eu in Fe^{++} sites of residual or refractory pyroxene.

Structural considerations also suggest that fundamental differences exist between Archaean greenstone belts and modern island arcs. The typical arcuate granite–greenstone pattern has not been observed in island arcs, although its existence at deeper crustal levels cannot be precluded. Whereas the distribution of arc-trench systems is controlled by compressive continent-ocean plate boundaries, no evidence is at hand for continental environments in the Archaean—i.e. no platform

270

volcanic-sedimentary sequences of the type common in the Lower Proterozoic are known. Instead, Viljoen and Viljoen (1969a) suggest that the upper Onverwacht Group has formed within narrow depositories which postdated the emplacement of the 'ancient tonalites'. This is supported by thickness variations of volcanic and sedimentary units across the strike of greenstone belts. It is envisaged that the upper greenstone sequence evolved within

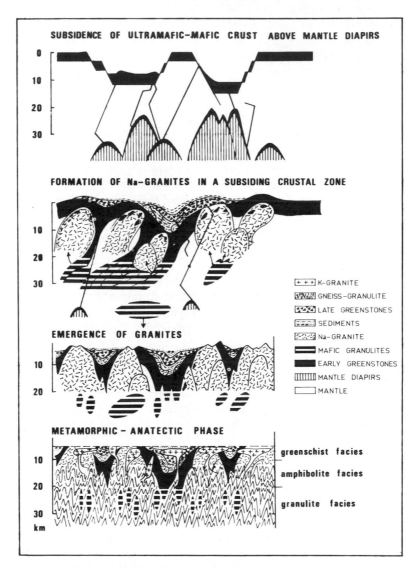

Fig. 4. A model of the evolution of Archaean greenstone belts and cratons (after Glikson and Lambert, in prep.). 1—Rifting of mafic–ultramafic crust and partial melting of subjacent mantle diapirs. 2—Partial melting of eclogite and/or amphibolite of subsiding crustal segments gives rise to sodic melts and granite diapirs. 3—differential vertical movements of the granitic diapirs and intervening segments of ultramafic–mafic crust account for the development of linear troughs in which the secondary greenstones and associated sediments accumulate. 4—A world-wide anatectic–metamorphic episode at c. 2·6 b.y. gives rise to younger granites, which rise toward and spread at high crustal levels (Viljoen and Viljoen, 1969d, 1969e; Glikson and Sheraton, 1972; Glikson and Lambert, 1973)

subsiding or rifted zones developed above the tonalite-intruded and partly cratonized early ultramafic–mafic crust (Glikson and Lambert, in prep.) (Fig. 4). The occurrence of ultramafic volcanics within upper greenstone sequences suggests that mantle diapirs continued to form concomitantly with the development of mafic–felsic and calc–alkaline suites.

The bimodal volcanic suite of greenstone belts has been interpreted in terms of progressive partial melting of subducted mafic crust (Barker and Petermann, 1974). However, such a mechanism should have resulted in acid to basic volcanic cycles, whereas cycles of greenstone belts move from basic to acid. Moreover, no experimental data exist to explain the compositional polarity in terms of partial melting processes. Whereas a mafic–felsic gap has been suggested on theoretical grounds by Yoder (1973), the experimental studies of Green and Ringwood (1968), which show that acid melts can be produced by partial melting of eclogite ($pH_2O = pload$), do not imply a compositional gap upon further melting. That the acid volcanics of the Onverwacht Group have formed under pressure exceeding 9 kb is indicated by the occurrence of resorbed quartz phenocrysts in keratophyres, and by REE patterns which suggest garnet fractionation (Glikson and Taylor, in prep.). Thus, the scarcity of andesites in bimodal volcanic suites may reflect consistent gravity fractionation of garnet from hydrous melts.

5. 'Primary' and 'Secondary' Greenstones

The preceding discussion suggests that no single mode of origin is applicable to the different greenstone suites—there are greenstones and greenstones. Whereas a variety of assemblage types showing different ratios of volcanic to hypabyssal rocks, ultramafic to mafic rocks, and mafic to felsic rocks occur, Archaean volcanic suites lend themselves to a broad division on the basis of their relationships to the encompassing granites, the nature of associated sediments, and evidence on original lateral extent, as follows:

'Primary' greenstones—Mostly mafic–ultramafic volcanic cycles, minor mafic–felsic and intermediate volcanics; chemical sediments and volcanically derived clastics; no evidence exists for exposure of pre-existing granites within or near the volcanic domain. These rocks occur at low stratigraphic levels within greenstone belts, and are widespread as inclusions in the earliest intrusive granites.

'Secondary' greenstones—Mostly mafic–felsic or calc–alkaline volcanic cycles; variable ultramafic component; volcanically derived and granite-derived clastic sediments; unconformably overlie granites and 'primary' greenstone cycles. Original extent restricted to linear troughs.

The field and geochronological relations between primary and secondary greenstones and the enveloping granites are portrayed in Figs. 2 and 3.

There are exceptions to this classification. For example, the occurrence of calc–alkaline volcanic units within cycle 1 of the Kalgoorlie System is inconsistent with the proposed scheme. Also, the proportion of ultramafic rocks differs in the different areas; for example, in the Kalgoorlie System cycle 3 abounds in chlorite–tremolite high-Mg meta-basalts and cogenetic sills (Doepel, 1965; Williams, 1970). As is evidenced by the multiplicity of cycles in the Kalgoorlie System, it is not true that only two major greenstone suites occur in any particular shield. Nor is it intended to imply contemporaneity of either 'primary' or 'secondary' greenstone suites in different shields. For example, geochronological data clearly suggest that the uppermost cycle in the Yilgarn Shield (2·7–2·6 b.y.) and the upper part of the Onverwacht Group (3·2–3·0 b.y.) are not contemporaneous (Fig. 3). On the other hand, no older Rb–Sr age limits can be placed on the lowermost cycles of the Kalgoorlie System, the lower part of the Onverwacht Group, the Sebakwian Group, and the Nuggihalli ultramafics, unless assumptions are made about the original (pre-metamorphic) Rb/Sr ratios in these rocks.

It will be interesting to test the above classification in the Canadian, North Atlantic, Kola, Siberian, and Ukrainian shields. Lithologically and geochemically the Canadian

Keewatin–Abitibi and Yellowknife green-stone belts appear to be akin to secondary greenstones. However, mafic–ultraamfic inclusions in 3 b.y. old gneisses in eastern Manitoba may be derived from still earlier volcanic cycles (Ermanovics, 1974). Further-more, the stratigraphic relationships of pil-lowed and quench-textured volcanics in Ontario and the Abitibi belts are not clear (see Pyke et al., 1973), as is the case with quench-textured ultramafics of the Thompson belt in Manitoba (J. J. Hubregtse, pers. comm., 1974).

Because of the synclinal structure of green-stone belts and the occurrence of primary greenstone at low stratigraphic levels, these rocks can be expected to be more abundant than secondary greenstones at relatively deeper crustal levels of Precambrian Shields—i.e. in association with higher-grade meta-morphic zones (Glikson and Lambert, in prep.). Crustal tilting of the Yilgarn shield eastward (Mathur, 1973), the Dharwar Shield northward (Pichamuthu, 1967), the Kaapvaal Shield westward, and the Superior Shield east-ward (Bell, 1971) are reflected by seismic data, basement-cover relations, and considerations based on metamorphic grade. In each of these instances the metamorphic grade increases toward the more deeply eroded uplifted part of the shield. Mafic and ultramafic enclaves abound in the gneiss–granulite terrain of the Wheat Belt in Western Australia (Wilson, 1958, 1971). The Nilgiris granulites in south-ern India abound in mafic enclaves (Pichamuthu, 1970b; Radhakrishna, 1974), and the Nuggihalli ultramafics are likewise associated with deeper crustal levels of the Dharwar Craton. Hunter (1970) described ultramafic and mafic enclaves from the 'ancient gneiss complex' in Swaziland, and remarked on the scarcity of meta-sediments. Possibly the 'ancient gneiss complex' can be interpreted in terms of the coeval roots of the Swaziland System, where synclinal keels and enclaves of Onverwacht Group rocks were intruded by the 'ancient tonalites' under amphibolite-facies conditions.

6. Conclusions

(1) Archaean greenstone suites in shields of the southern hemisphere can be classified in terms of two broad categories, denoted as 'primary' greenstones and 'secondary' green-stones.

(2) Primary greenstones are principally, although not without exception, of mafic–ultramafic composition, and appear to have evolved as extensive volcanic layers within ensimatic oceanic ridge-type environments.

(3) Secondary greenstones, which postdate primary greenstones within any individual area, are principally bimodal mafic–felsic or mafic–intermediate felsic including a rela-tively minor ultramafic component. This vol-canism postdated granites, and has been con-fined to linear troughs.

(4) Each shield contains both 'primary' and 'secondary' greenstones; however, whether or not this classification is applicable in the Yil-garn and Pilbara Shields should be assessed by further stratigraphically controlled geo-chemical studies.

(5) Primary greenstone relics abound in Archaean gneiss–granulite terrains, which are interpreted as coeval roots of granite–greenstone terrains.

(6) Field evidence from Archaean terrains and igneous geochemical data indicate evolu-tion from ensimatic crustal environments to partly cratonized and fully cratonized domains, a trend supported by geochemical and isotopic data from Archaean sediments (Veizer, 1973; Naqvi and Hussain, 1972; Jakeš and Taylor, 1974).

Acknowledgements

I wish to thank W. B. Dallwitz, J. Ferguson, and J. W. Sheraton for their comments and criticism. Special thanks are due to B. F. Windley for inviting me to present this review paper at the conference on The Early History of the Earth.

References

Allsopp, H. L., Ulrych, J. J., and Nicolaysen, L. O., 1968. 'Dating some significant events in the history of the Swaziland System by the Rb–Sr isochron method', *Can. J. Earth Sci.*, **5**, 605–619.

Anhaeusser, C. R., 1971. 'Cyclic volcanicity and sedimentation in the evolutionary development of Archaean greenstone belts of shield areas', *Geol. Soc. Aust. spec. publ.*, **3**, 57–70.

Anhaeusser, C. R., 1973. 'The evolution of the early Precambrian crust of Southern Africa', *Phil. Trans. R. Soc. Lond.*, **A273**, 359–388.

Anhaeusser, C. R., Mason, R., Viljoen, M. J., and Viljoen, R. P., 1969. 'A reappraisal of some aspects of Precambrian shield geology', *Bull. geol. Soc. Am.*, **80**, 2175–2200.

Anhaeusser, C. R., Roering, C., Viljoen, M. J., and Viljoen, R. P., 1968. 'The Barberton Mountain Land: A model of the elements and evolution of an Archaean fold belt', *Trans. geol. Soc. S. Afr.*, **71** (Annex.), 225–253.

Arth, J. G., and Hanson, G. N., 1972. 'Quartz diorites derived by partial melting of eclogite or amphibolite at mantle depths', *Contr. Miner. Petrol.*, **37**, 161–174.

Aswathanarayana, U., 1968. 'Metamorphic chronology of the Precambrian provinces of south India', *Can. J. Earth Sci.*, **5**, 591–600.

Baragar, W. R. A., 1968. 'Major element geochemistry of the Noranda volcanic belt, Ontario', *Can. J. Earth Sci.*, **4**, 773–790.

Baragar, W. R. A., 1966. 'Geochemistry of the Yellowknife volcanic rocks', *Can. J. Earth Sci.*, **3**, 9–30.

Bargar, W. R. A., and Goodwin, A. M., 1969. 'Andesites and Archaean volcanism in the Canadian Shield', *Oregon Dep. Geol. Min. Industr.*, 121–142.

Barker, F., and Peterman, Z. E., 1974. 'Bimodal tholeiitic-dacitic magmatism and the early Precambrian crust', *Precambrian Res.*, **1**, 1–12.

Bell, C. K., 1971. 'Boundary geology, upper Nelson River area, Manitoba and northwestern Ontario', *Geol. Assoc. Canada Sp. Paper* 9, 11–39.

Bliss, N. W., and Stidolph, P. A., 1969. 'A review of the Rhodesian Basement Complex', *Trans. Geol. Soc. S. Afr. spec. Publ.*, **2**, 305–333.

Bonatti, E., and Arrhenius, G., 1970. 'Acidic rocks on the Pacific Ocean floor', In *The Sea*, A. E. Maxwell (Ed.), Wiley-Interscience, New York, 445–464.

Bye, S. M., 1971. 'The acid volcanic rocks of the Madra area, Yilgarn Goldfield, Western Australia' (Abstract), *Geol. Soc. Aust. spec. Publ.* 3, J. E. Glover (Ed.), 151.

Cawthorn, R. G., and Strong, D. F., 1974. 'The petrogenesis of komatiites and related rocks as evidence for a layered upper mantle', *Earth Planet Sci. Lett.*, **23**, 369–375.

Chase, C. G., and Gilmer, T. H., 1973. 'Precambrian plate tectonics: The mid-continent gravity high', *Earth Planet. Sci. Lett.*, **21**, 70–78.

Condie, K. C., 1972. 'A plate tectonics evolutionary model of the South Pass Archaean greenstone belt, Southwestern Wyoming', *Int. geol. Cong.*, *24th Sess.*, **1**, 104–114.

Condie, K. C., 1973. 'Archaean magmatism and crustal thickening', *Bull. geol. Soc. Am.*, **84**, 2981–2992.

Condie, K. C., and Baragar, W. R. A., 1974. 'Rare-earth element distributions in volcanic rocks from Archaean greenstone belts', *Contr. Miner. Petrol.*, **45**, 237–246.

Condie, K. C., and Lo, H. H., 1971. 'Trace element geochemistry of the Louis Lake batholith of early Precambrian age, Wyoming', *Geochim. Cosmochim. Acta*, **35**, 1099–1119.

Condie, K. C., Macke, J. E., and Reimer, T. O., 1970 'Petrology and geochemistry of early Precambrian graywackes from the Fig Tree Group, South Africa', *Bull. geol. Soc. Am.*, **81**, 2759–2776.

Cox, K. C., Johnson, R. L., Monkman, L. J., Stillman, C. J., Vail, J. R., and Wood, D. N., 1965. 'The geology of the Nuanetsi igneous province', *Phil. Trans. R. Soc. Lond.*, **A257**, 71–218.

Crawford, A. R., 1969. 'Reconnaissance Rb-Sr dating of the Precambrian rocks of southern India', *J. geol. Soc. India*, **10**, 117–166.

Dallwitz, W. B., 1968. 'Chemical composition and genesis of clinoenstatite-bearing volcanic rocks from Cape Vogel, Papua: at discussion', *Int. geol. Cong. 23rd Sess.*, **2**, 229–242.

Davidson, D., 1973. 'Plate tectonics model for the Musgrave Block-Amadeus Basin Complex of Central Australia', *Nature*, **245**, 21–23.

De Laeter, J. R., and Blockley, J. G., 1972. 'Granite ages within the Archaean Pilbara block, Western Australia', *J. geol. Soc. Aust.*, **19**, 363–370.

De Long, S. E., 1974. 'Distribution of Rb, Sr and Ni in igneous rocks, central and western Aleutian Islands, Alaska', *Geochim. Cosmochim. Acta*, **38**, 245–266.

Doepel, J. J. G., 1965. 'The geology of the Yilmia area, west of Lake Lefroy, Western Australia', *Thesis* (unpubl.), Univ. W. Aust.

Durney, D. W., 1972. 'A major unconformity in the Archaean, Jones Creek, Western Australia', *J. geol. Soc. Aust.*, **19**, 251–259.

Ermanovics, I. F., 1974. 'Evidence for early Precambrian Mg-rich crust and pre-Kenoran sialic crust in northwestern Superior Province of the Canadian Shield', *Geol. Surv. Canada* (inpubl.).

Folinsbee, R. E., Baadsgaard, H., Cumming, G. L., and Green, D. C., 1968. 'A very ancient island arc', In *The Crust and Upper Mantle of the Pacific Area*, L. Knopoff et al. (Eds.), *Am. geophys. Un. Monogr.*, **12**, 441–448.

Fyfe, W. S., 1973. 'The generation of batholiths', *Tectonophysics*, **17**, 273–283.

Gale, G. H., 1973. 'Paleozoic basaltic komatiite and ocean floor type basalts from northwestern Newfoundland', *Earth Planet. Sci. Lett.*, **18**, 22–28.

Gast, P. W., 1972. 'The chemical composition and structure of the moon', *Moon*, **5**, 121–148.

Gemuts, I., and Theron, A. C., 1973. 'The stratigraphy and mineralization of the Archaean between Norseman and Coolgardie, Western Australia', In *Economic geology of Australia and New Zealand*, by C. L. Knight, A.I.M.M. (Ed.), in press.

Gibb, R. A., and Walcott, R. I., 1971. 'A Precambrian suture in the Canadian shield', *Earth Planet. Sci. Lett.*, **10**, 417–422.

Gill, J. E., 1961. 'The origin of continents', *Trans. Roy. Soc. Canada*, **55**, 103–113.

Gill, J. B., 1970. 'Geochemistry of Viti-Levu, Fiji, and its evolution as an island arc', *Contr. Miner. Petrol.*, **27**, 179–203.

Glikson, A. Y., 1968. 'The Archaean geosynclinal succession between Coolgardie and Kurrawang, near Kalgoorlie, Western Australia', *Ph.D. Thesis*, Univ, W. Aust. (unpubl.).

Glikson, A. Y., 1970. 'Geosynclinal evolution and geochemical affinities of early Precambrian systems', *Tectonophysics*, **9**, 397–433.

Glikson, A. Y., 1971. 'Primitive Archaean element distribution patterns: Chemical evidence and geotectonic significance', *Earth Planet. Sci. Lett.*, **12**, 309–320.

Glikson, A. Y., 1972. 'Early Precambrian evidence of a primitive ocean crust and island nuclei of sodic granite', *Bull. Geol. Soc. Am.*, **83**, 3323–3344.

Glikson, A. Y., and Lambert, I. B., 1973. 'Relations in space and time between major Precambrian shield units: An interpretation of Western Australian data', *Earth Planet. Sci. Lett.*, **20**, 395–403.

Glikson, A. Y., and Lambert, I. B., 1975. 'Vertical zonation and petrogenesis of the early Precambrian crust in Western Australia', *Tectonophysics* (in press).

Glikson, A. Y., and Sheraton, J. W., 1972. 'Early Precambrian trondjhemitic suites in Western Australia and northwestern Scotland, and the geochemical evolution of shields', *Earth Planet. Sci. Lett.*, **17**, 227–242.

Goodwin, A. M., 1968. 'Archaean protocontinental growth and early crustal history of the Canadian shield', *Int. geol. Cong. 23rd Sess.*, **1**, 69–89.

Goodwin, A. M., 1971. 'Metallogenic patterns and evolution of the Canadian Shield', *Geol. Soc. Aust. spec. publ.* **3**, 157–174.

Green, D. C., and Baadsgaard, H., 1971. 'Temporal evolution and petrogenesis of an Archaean crustal segment at Yellowknife, N.W.T., Canada', *J. Petrol.*, **12**, 177–217.

Green, D. H., 1971. 'Composition of basaltic magmas as indicators of conditions or origin: application to oceanic volcanism', *Phil. Trans. Roy. Soc. Lond.*, **A268**, 707–725.

Green, D. H., 1972. 'Archaean greenstone belts may include equivalents of lunar maria?', *Earth Planet. Sci. Lett.*, **15**, 263–270.

Green, D. H., and Ringwood, A. E., 1967. 'The genesis of basaltic magmas', *Contr. Miner. Petrol.*, **15**, 103–190.

Green, T. H., and Ringwood, A. E., 1968. 'Genesis of the calc-alkaline igneous rock suite', *Contr. Miner. Petrol.*, **18**, 105–162.

Hallberg, J. A., 1972. 'Geochemistry of Archaean volcanic belts in the Eastern Goldfields region of Western Australia', *J. Petrol.*, **13**, 45–56.

Hallberg, J. A., and Williams, D. A. C., 1972. 'Archaean mafic and ultramafic rock associations in the Eastern Goldfields region, Western Australia', *Earth Planet. Sci. Lett.*, **15**, 191–200.

Harrison, N. M., 1968. 'A reassessment of the stratigraphy of the Pre-Cambrian basement complex around Que Que, Gwelo District, Rhodesia', *Trans. geol. Soc. S. Afr.*, **71**, 594–609.

Hart, S. R., Brooks, C., Krogh, T. E., Davis, G. L., and Nava, D. F., 1970. 'Ancient and modern volcanic rocks: a trace element model', *Earth Planet Sci. Lett.*, **10**, 17–28.

Holmes, A., 1948. *Int. geol. Cong. 18th Sess.*, **14**, 254.

Horwitz, R. C., and Sofoulis, J., 1965. 'Igneous activity and sedimentation in the Precambrian between Kalgoorlie and Norseman, Western Australia', *Proc. Aus. Inst. Min. Metall.*, **214**, 45–59.

Hunter, D. R., 1970. 'The Ancient Gneiss Complex in Swaziland', *Trans. geol. Soc. S. Afr.*, **73**, 107–150.

Hunter, D. R., 1974. 'Crustal development in the Kaapvaal craton: the Archaean', *Precambrian Res.*, **1**, 259–294.

Hurley, P. M., Nagy, B., and Pinson, W. H., Jr., 1972. 'Ancient age of the Middle Marker Horizon, Onverwacht Group, Swaziland sequence, South Africa', *Earth Planet. Sci. Lett.*, **14**, 360.

Jahn, B., and Shih, C., 1974. 'On the age of the Onverwacht Group, Swaziland sequence, South Africa', *Geochim. Cosmochim. Acta*, **38**, 873–885.

Jahn, B., Shih, C., and Rama Murthy, V., 1974. 'Trace element geochemistry of Archaean volcanic rocks', *Geochim. Cosmochim. Acta*, **38**, 611–627.

Jakěs, P., and Gill, J. B., 1970. 'Rare earth elements and the island arc tholeiitic series', *Earth Planet. Sci. Lett.*, **9**, 17–28.

Jakeš, P., and Taylor, S. R., 1974. 'Excess europium content in Precambrian sedimentary rocks and continental evolution', *Geochim. Cosmochim. Acta*, **38**, 739–745.

Katz, M. B., 1972. 'Paired metamorphic belts of the Gondwanaland Precambrian and plate tectonics', *Nature*, **239**, 271–273.

Kriewaldt, M. C., 1969. 'Explanatory notes for the Kalgoorlie 1 : 250,000 Sheet', *Geol. Surv. W. Aust.*

McCall, G. J. H., 1969. 'The Archaean succession west of Lake Lefroy', *J. Roy. Soc. W. Aust.*, **52**, 119–128.

MacGregor, A. M., 1932. 'The geology of the country around Que Que, Gwelo District', *Bull. geol. Sur. S. Rhodesia*, **20**, 113.

MacGregor, A. M., 1951. 'Some milestones in the Precambrian of Southern Rhodesia' (anniversary address by President), *Proc. geol. Soc. S. Afr.*, xxvii–lxxiv.

McIver, J. R., 1972. 'Hornblendite from Bon-Accord and a possible komatiite equivalent', *Trans. geol. Soc. S. Afr.*, **75**, 313–316.

McIver, J. R., and Lenthall, D. H., 1974. 'Mafic and ultramafic extrusives of the Barberton Mountain Land in terms of the CMAS system', *Precambrian Res.*, **1**, 327–344.

Mathur, S. P., 1973. 'Crustal structure in southwestern Australia, *Bur. Miner. Resour. Aust. Rec.*, 1973/112.

Milton, D. J., Barlow, B. C., Brett, R., Brown, A. R., Glikson, A. Y., Manwaring, E. A., Moss, F. J., Sedmik, E. C. E., Van Son, J., and Young, G. A., 1972. 'Gosses Bluff impact structure, Australia', *Science*, **175**, 1199–1207.

Miyashiro, A., Shido, F., and Ewing, M., 1969. 'Diversity and origin of abyssal tholeiites from the mid-Atlantic Ridge near 24° and 30° north latitude', *Contr. Miner. Petrol.*, **23**, 38–52.

Naqvi, S. M., and Hussain, S. M., 1972. 'Petrochemistry of early Precambrian metasediments from the central part of the Chitaldrug schist, belt, Mysore, India', *Chem. Geol.*, **10**, 109–135.

Naqvi, S. M., and Hussain, S. M., 1973a. 'Geochemistry of Dharwar metavolcanics and composition of the primeval crust of peninsular India', *Geochim. Cosmochim. Acta*, **37**, 159–164.

Naqvi, S. M., and Hussain, S. M., 1973b 'Relation between trace and major element composition of the Chitaldrug metabasalts, Mysore, India and the Archaean mantle', *Chem. Geol.*, **11**, 17–30.

Naqvi, S. M., Rao, V. D., and Narain, H., 1974. 'The protocontinental growth of the Indian Shield and the antiquity of its rift valleys', *Precambrian Res.*, **1**, 345–398.

Nesbitt, R. W., 1971. 'Skeletal crystal forms in the ultramafic rocks of the Yilgarn block, Western Australia: evidence for an Archaean ultramafic liquid', *Geol. Soc. Aust. spec. Publ. 3*, 331–350.

O'Beirne, W. R., 1968. 'The acid porphyries and porphyroid rocks of the Kalgoorlie area', *Ph. D. thesis*, Univ. W. Aust. (unpubl.).

Oosthuyzen, E. J., 1970. 'The geochronology of a suite of rocks from the granitic terrain surrounding the Barberton Mountainland', *Ph.D. Thesis*, Univ. Witwatersrand (unpubl.).

Osborn, E. F., 1962. 'Reaction series for subalkaline igneous rocks based on different oxygen pressure conditions', *Am. Miner.*, **47**, 211–226.

Pearce, J. A., and Cann, J. R., 1971. 'Ophiolite origin investigated by discriminant analysis using Ti, Zr and Y', *Earth Planet. Sci. Lett.*, **12**, 339.

Pearce, J. A., and Cann, J. R., 1973. 'Tectonic setting of basic volcanic rocks determined using trace element analyses', *Earth Planet. Sci. Lett.*, **19**, 290–300.

Pettijohn, F. J., 1943. 'Archaean sedimentation', *Bull. geol. Soc. Am.*, **54**, 925–972.

Phaup, A. E., 1973. 'Chemical analyses of the rocks, ores and minerals of Rhodesia', *Rhodesia geol. Surv. Bull.*, **71**, 292.

Pichamuthu, C. S., 1967. 'The Precambrian of India', In *The Precambrian*, K. Rankama (Ed.), Interscience, New York, Vol. 3, 1–96.

Pichamuthu, C. S., 1974. 'The Dharwar craton', *J. geol. Soc. India*, **15**, 339–346.

Pichamuthu, U., 1970a. 'The problem of the basement in Mysore State', *Curr. Sci.*, 39.

Pichamuthu, U., 1970b. 'Charnockite-Peninsular gneiss relationships in Mysore State', *Curr. Sci.*, 39.

Pyke, D. R., Naldrett, A. J., and Eckstrand, O. R., 1973. 'Archaean ultramafic flows in Munro township, Ontario', *Bull. geol. Soc. Am.*, **84**, 955–978.

Radhakrishna, B. P., 1967. 'Reconsideration of some problems in the Archaean complex of Mysore', *J. geol. Soc. India*, **8**, 102–109.

Radhakrishna, B. P., 1974. 'Peninsular gneissic complex of the Dharwar craton—a suggested model for its evolution', *J. geol. Soc. India*, **15**, 439–454.

Roddick, J. C., Compston, W., and Durney, D. W., 1973. 'Rb-Sr age of the granite below an Archaean unconformity, Jones Creek, Western Australia', *ANZAAS 45th Congr. Abstracts. Sec.*, **3**, 13.1.

Schilling, J. G., 1971. 'Sea-floor evolution: rare earth evidence', *Phil. Trans. R. Soc. London*, **A268**, 663–706.

Sen, S. K., 1974. 'A review of some of the geochemical characters of the type area charnockites, Pallavaram, India', *J. geol. Soc. India*, **15**, 413–420.

Shido, F., Miyashiro, A., and Ewing, M., 1971. 'Crystallization of abyssal tholeiites', *Contr. Miner. Petrol.*, **31**, 251–266.

Smith, T. E., and Longstaffe, F. J., 1974. 'Archaean rocks of shoshonitic affinities at Bijou Point, Northwestern Ontario', *Can. J. Earth Sci.*, **11**, 1407–1413.

Sreenivas, B. L., and Srinivasan, R., 1974. 'Geochemistry of granite-greenstone terrain of South India', *J. geol. Soc. India*, **15**, 390–406.

Srinivasan, R., and Sreenivas, B. L., 1968. 'Sedimentation and tectonics in Dharwars (Archaean) of Mysore State, India', *Ind. Miner.*, **13**, 42–45.

Srinivasan, R., and Sreenivas, B. L., 1972. 'Dharwar stratigraphy', *J. geol. Soc. India*, **13**, 75–85.

Stowe, C. W., 1968. 'The geology of the country south and west of Selukwe', *Geol. Surv. Rhodesia Bull.*, **59**, 209.

Talbot, C. J., 1968. 'Thermal convection in the Archaean crust', *Nature*, **220**, 552–556.

Talbot, C. J., 1973. 'A plate tectonic model for the Archaean crust', *Phil. Trans. R. Soc. London*, **A273**, 413–428.

Taylor, S. R., Capp, A. C., Graham, A. L., and Blake, D. H., 1969. 'Trace element abundances in andesites. II. Saipan, Bougainville and Fiji', *Contr. Miner. Petrol.*, **23**, 1–26.

Thorpe, R. S., 1972. 'Possible subduction zone for two Precambrian calc-alkaline plutonic complexes from southern Britain', *Bull. Geol. Soc. Am.*, **83**, 3663–3668.

Turek, A., and Compston, W., 1971. 'Rubidium-strontium geochronology in the Kalgoorlie region', *Geol. Soc. Aust. spec. Publ. 3*, 72–73 (Abstract).

Veizer, Jan, 1973. 'Sedimentation in geologic history: Recycling vs. evolution or recycling with evolution', *Contr. Miner. Petrol.*, **38**, 261–278.

Venkatasubramanian, V. S., and Narayanaswamy, R., 1974. 'The age of some gneissic pebbles in Kaldurga conglomerate', *J. geol. Soc. India*, **15**, 318–319.

Viljoen, M. J., and Viljoen, R. P., 1969a, 'The geology and geochemistry of the lower ultramafic unit of the Onverwacht Group and a proposed new class of igneous rocks', *Geol. Soc. S. Afr. spec. Publ. 2*, 55–86.

Viljoen, M. J., and Viljoen, R. P., 1969b. 'Evidence for the existence of a mobile extrusive peridotitic magma from the Komati formation of the Onverwacht Group', *Geol. Soc. S. Afr. spec. Publ. 2*, 87–112.

Viljoen, R. P., and Viljoen, M. J., 1969c. 'The geological and geochemical significance of the upper formations of the Onverwacht Group', *Geol. Soc. S. Afr. Spec. Publ. 2*, 113–152.

Viljoen, M. J., and Viljoen, R. P., 1969d. 'A prop-osed new classification of the granite rocks of the Barberton region', *Geol. Soc. S. Afr. spec. Publ. 2*, 153–188.

Viljoen, M. J., and Viljoen, R. P., 1969e. 'The geochemical evolution of the granitic rocks of the Barberton region', *Geol. Soc. S. Afr. spec. Publ. 2*, 189–220.

Viljoen, R. P., and Viljoen, M. J., 1969f. 'Evidence for the composition of the primitive mantle and its products of partial melting from the study of the rocks of the Barberton Mountainland', *Geol. Soc. S. Afr. spec. Publ. 2*, 275–296.

Viljoen, M. J., and Viljoen, R. P., 1969g. 'A re-appraisal of granite-greenstone terrains of shield areas based on the Barberton model', *Geol. Soc. S. Afr. spec. 2*, 245–274.

Viljoen, R. P., and Viljoen, M. J., 1971. 'The geological and geochemical evolution of the Onverwacht volcanic group of the Barberton Mountainland, South Afr', *Geol. Soc. Aust. spec. Publ. 3*, 133–151.

Visser, D. J. L. (compiler), 1956. 'The geology of the Barberton area', *South Africa Geol. Surv. spec. Publ. 15*, 194 pp.

Viswanathan, S., 1974. 'Contemporary trends in geochemical studies of early Precambrain greenstone-granite complexes', *J. geol. Soc. India*, **15**, 347–379.

Weber, W., 1971. 'The Evolution of the Rice Lake–Gem Lake Greenstone Belt, Southeastern Manitoba', *Geol. Assoc. Canada, spec. Pap. 9*, 97–103.

White, A. J. R., Jakes, P., and Christie, D. M., 1971. 'Composition of greenstones and the hypothesis of seafloor spreading in the Archean', *Geol. Soc. Aust. spec. Publ. 3*, 47–56.

Williams, D. A. C., and Hallberg, J. A., 1973. 'Archaean layered intrusions of the Eastern Goldfields region, Western Australia', *Contr. Miner. Petrol.*, **38**, 45–70.

Williams, I. R., 1970. 'Explanatory notes for the Kurnalpi 1 : 250 000 Sheet', *Geol. Surv. W. Aust.*

Williams, I. R., 1973. Structural subdivision of the Eastern Goldfields province, Yilgarn Block'. *Geol. Surv. W. Aust. ann. Rep.*, **1973**, 53–59.

Williams, I. R., Gower, C. F., and Thom, R., 1971. Explanatory notes on the Edjudina 1 : 250 000 Geological Sheet. *Geol. Surv. W. Aust. Rec.* 1971/26.

Wilson, A. F., 1958. 'Advances in the knowledge of the structure and petrology of the Precambrian rocks of southwestern Australia', *J. Roy. Soc. W. Aust.*, **41**, 57–83.

Wilson, A. F., 1971. 'Some geochemical aspects of the sapphirine-bearing pyroxenites and related highly metamorphosed rocks from the Archaean ultramafic belt of south Quairading, Western Australia', *Geol. Soc. Aust. spec. Publ. 3*, 401–412.

Wilson, J. F., 1973. 'The Rhodesian Archaean craton, an essay in cratonic evolution', *Phil. Trans. R. Soc. Lond.*, **A273**, 389–411.

Wilson, H. D. B., Andrews, P., Moxham, R. L., and Ramlal, K., 1965. 'Archaean volcanism in the Canadian Shield', *Can. J. Earth Sci.*, **2**, 161–175.

Windley, B. F., 1973. 'Crustal development in the Precambrian', *Phil. Trans. Roy. Soc. Lond.*, **A273**, 315–581.

Yoder, H. S., Jr., 1973. 'Contemporaneous basaltic and rhyolitic magmas', *Am. Miner.*, **58**, 153–171.

Volcanism in the Western Superior Province in Manitoba

J. J. M. W. HUBREGTSE

Mineral Resources Division, Department of Mines,
Resources & Environmental Management, Winnipeg.

Abstract

Two periods of volcanism are recognized in the Archean Knee Lake–Oxford Lake greenstone belt. The younger (high-K) calc–alkaline and shoshonitic rocks unconformably overlie the sub-alkaline greenstones, which have a cyclic nature. Their chemical characteristics and some inferences about the generation of these volcanics are presented. An island arc model seems to be incompatible with the evolution of this belt.

Introduction

This paper is an interim report on the geochemistry of the volcanic rocks of an Archean greenstone belt in the Superior Province in Manitoba. The area of study is the Knee Lake–Oxford Lake belt, situated 570 km NNE of Winnipeg. This belt extends for approximately 240 km in an E–W direction. Only the central part (Fig. 1) will be considered here. This has a maximum width of 25 km.

Geological Setting

A simplified stratigraphic column is given in Table 1. The division of the volcano-sedimentary belt into the mainly volcanic Hayes River Group (HRG) and the sedimentary Oxford Lake Group (OLG) was made by Wright (1932), and has been followed in subsequent work. Wright (1932) recorded an unconformity between the two groups. As a result of Barry's mapping (1959, 1960 and 1964) the outlines of the HRG and OLG became known in detail. The cyclic nature of the HRG was first noted by Gilbert and Elbers (1972). Present work has led to the recognition of volcanics in the OLG.

A basement under the HRG has not been recognized. The only possible basement is a tonalitic gneiss (Fig. 1), which structurally underlies the greenstone belt. However, the

Table 1

Younger granites
Intrusive contact—
Oxford Lake Group $\begin{cases} \text{Sedimentary subgroup (conglomerate, greywacke)} \\ \text{Volcanic subgroup (porphyritic flows and pyroclastics)} \end{cases}$
Unconformity—
Older granites (tonalites, trondhjemites, granodiorites, granites)
Intrusive contact—
Hayes River Group (basalts, rhyolites, minor sediments)
?? Tonalitic gneiss ??

280

contact between the greenstones and this tonalitic gneiss is either a tectonic one or is masked by younger intrusions.

The HRG forms a monoclinal, vertically dipping, S-facing structure, which is terminated in central Oxford Lake by an E–W fault that seems to fade out to the east (Fig. 1). A synclinal flexure in the HRG strata underlies the eastern part of the area mapped. The lower portion of the HRG has been intruded by the older and younger granites (Fig. 1).

The mainly WNW-trending OLG rocks unconformably overlie the HRG strata and the older granites. The present outline of the sedimentary OLG basin bifurcates in the central part of the area mapped, as the E–W trending western extremity pinches out along the E–W fault. Volcanics of the OLG outcrop in two areas: near Jackson Bay, and at western Knee Lake (Fig. 1). The OLG rocks comprise a synclinorial basin reaching a maximum width of c. 21 km.

The basal conglomerate of the OLG sedimentary subgroup consists of erosion products of the HRG, the older granites and the OLG volcanic subgroup. Its deposition is fol-

lowed by greywacke sedimentation with intercalations of argillite and arkose. The latest major intrusive event recorded is the emplacement of the younger granites.

Hayes River Group

The HRG is mainly comprised of pillowed and massive basalt, intravolcanic gabbros and minor felsic pyroclastics and flows, intercalated with greywackes, cherts and iron formations. Ultramafic sills are not uncommon, but flows displaying spinifex textures are rare. Campbell, Elbers and Gilbert (1972) estimate that the amount of felsic volcanics, chemical and clastic sediments is less than 10%. Five volcano-sedimentary cycles were recognized in the Knee Lake area from detailed mapping and chemical investigation. In general, each cycle is defined by a thick basaltic lower portion and a much thinner felsic and sedimentary upper portion. As felsic pyroclastic cones developed only locally on basaltic floors, they are not present everywhere along the cycle boundaries, particularly in the Oxford Lake area where delineation of cycle boundaries

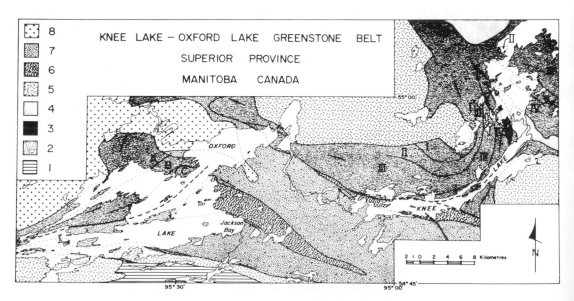

Fig. 1. Geological map of the Knee Lake–Oxford Lake area, Manitoba, Canada. 1: tonalitic gneiss. 2, 3 and 4: Hayes River Group. 2: basalt and subordinate rhyolite and sediment. 3: gabbro. 4: greywacke. 5: older granites. 6: Oxford Lake Group volcanic subgroup. 7: Oxford Lake Group sedimentary subgroup. 8: younger granites. Full lines in the Hayes River Group define cycle boundaries, I–V and A–C in the Knee Lake and Oxford Lake area, respectively. Geology by F. J. Elbers, H. P. Gilbert and J. J. M. W. Hubregtse

becomes more difficult. The chemistry of the basaltic rocks, however, indicates that cyclicity is as persistent as in the Knee Lake area.

Petrochemistry

Based on upper limits of 53·5, 62·0 and 70·0 SiO_2 (wt.%) for basalt, andesite, and dacite, respectively, the HRG rocks analyzed are distributed as follows: basalts 45 (60%), andesites 11 (15%), dacites 13 (17%), and rhyolites 6 (8%). Some chemical characteristics of HRG basalts and andesites are given in Table 2. The average K_2O wt.% of all basalts is 0·27, markedly lower than an average K_2O value of 0·35 for basalts from northwestern Ontario (Goodwin, 1972). Glikson (1971) compiled average K_2O values of 0·26, 0·15, and 0·16–0·26 for the Onverwacht Group and Kalgoorlie System basalts and modern oceanic tholeiites, respectively. Jakeš and White (1972) report 0·44 for Melanesian island arc tholeiites. The Ni and Cr values of the HRG basalts are several times higher than those reported for the island arc association. Although quite variable, Sr and Ba values reported here are lower. The HRG average Ni and Cr values are similar to those of the western Australian Archean basalts (Glikson, 1971). The low-K HRG basalts compare most closely with modern low-K ocean floor basalts. Moreover, quench textures as observed in modern ocean floor basalts are very common, particularly in central Oxford Lake.

According to their normative components the majority of HRG basalts consists of tholeiites and olivine–tholeiites (Table 2). Figs 2a–e show that all five cycles pass from the tholeiitic into the calc–alkaline field. The tholeiitic members show iron enrichment relative to magnesium, although this tendency decreases towards higher cycles as calc–alkaline affinities increase.

The greater amount of data from cycles III and V allow us to have a closer look at the differentiation within a cycle. Green (1973) who initiated chemical studies in the Knee Lake area, recognized for each basalt–rhyolite cycle a lower low-Al tholeiite, a middle high-Al tholeiite, and a thin upper felsic calc–alkaline unit. High-Al and low-Al refer to Al_2O_3/SiO_2 wt.%, higher and lower than 0·300, respectively, as indicated in Table 2. In the field the change from low-Al to high-Al tholeiites is often marked by the presence of glomeroporphyritic plagioclase basalts. The resorbed nature of the plagioclase phenocrysts led Green (1973) to the conclusion that precipitation and subsequent resorption of plagioclase (80–90% An) was the dominant process in the early within-cycle differentiation, as it may account for the generation of high-Al tholeiite from low-Al tholeiite. Subsequently, according to Green (1973), the calc–alkaline members would have been derived from the high-Al tholeiitic magma by further fractionation accompanied by increasing oxygen fugacity. Indeed, fractionation of

Table 2. Hayes River Group basalts and andesites

		42 Basalts				
	Average (wt. %)		Norm composition			
Cycle	K_2O	$Al_2O_3/$ SiO_2	q+hy	hy+ol	ol+ne	Average and range for 6 trace elements in 10 basalts and 2 andesites (ppm)
I	0·62	0·330	2	1	1	Ni 124 (40–191)
II	0·14	0·317	3	—	—	Cr 364 (191–530)
III	0·23 l	0·324	4	3	2	Cu 96 (41–198)
	m	0·281	4	1	—	Zn 100 (61–202)
	u	0·305	4	—	—	Ba 130 (50–450)
IV	0·20	0·288	8	2	—	Sr 180 (76–374)
V	0·34 l	0·272	3	2	—	
	u	0·304	2	—	—	

An-rich plagioclase may have governed the derivation of basalts with repeatedly changing Al_2O_3/SiO_2 values, but it does not seem to bear a relationship to the generation of a mafic–tholeiitic felsic–calc–alkaline association, since the low-Al and high-Al units of cycle III and V are subcycles that range from tholeiitic mafites to calc–alkaline acidic rocks, regardless of their Al_2O_3/SiO_2 ratios (see Fig. 2b and d and Table 2). Moreover, the lowest subcycle of cycle III is high-Al instead of low-Al tholeiite.

A possible mechanism for the generation of a mafic–tholeiite felsic–calc–alkaline cycle may be found in Yoder's (1973) model, which explains near-contemporaneous formation of rhyolitic and quartz normative basaltic liquids at $P_{H_2O} = 20$ kb by partial melting of a quartz normative basaltic parent, as the result of adiabatic decompression without producing intermediate members. The majority of the HRG basalts is quartz normative (Table 2) and there is field evidence that a 'Daly gap' does exist in the HRG. As Yoder points out, hybridization between both liquids can take place during relatively short periods. Gunn and Watkins (1969) gave a detailed account of hybrid rocks formed in a tholeiite–rhyolite

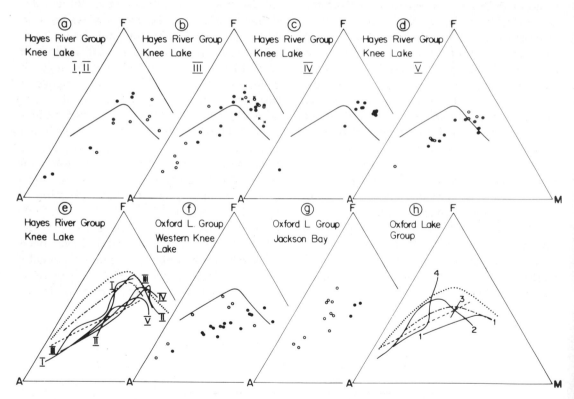

Fig. 2. AFM—plots of chemical analyses of HRG and OLG volcanics. a: HRG cycle I—full circles; HRG cycle II—open circles. b: HRG cycle III; lower subcycle—full circles; middle subcycle—crosses; upper subcycle—open circles. c: HRG cycle IV. d: HRG cycle V; lower subcycle—full circles; upper subcycle—open circles. e: Combination plot for HRG I–V—full lines; Thingmuli, Iceland—dot lines; Talasea, New Britain—dot dash lines; Cascades—dash lines (all taken from Lowder and Carmicheal, 1970). f: OLG volcanics from western Knee Lake; calc–alkaline rocks—full circles; alkaline rocks—open circles. g: OLG volcanics from Jackson Bay, Oxford Lake; high-K calc–alkaline rocks—full circles; shoshonitic rocks—open circles. h: Combination plot for OLG rocks from western Knee Lake (1 and 2), and Oxford Lake (3 and 4); Cascades—dash lines; Hawaiian alkaline trend (Macdonald and Katsura, 1964)—dot lines; Alkaline trend from Sicily channel islands (Barberi, Borsi, Ferrara and Innocenti, 1969)—dot dash lines.
Full line in a, b, c, d and f defines tholeiitic and calc–alkaline fields (Irvine and Baragar, 1968)

composite dike (Stretishorn, eastern Iceland). They found that the hybrid rocks were comparable with calc–alkaline andesite and dacite, and that contamination trends are linear on chemical variation diagrams. On an AFM diagram a curved tholeiitic trend would be straightened to a pseudo-calc–alkaline trend, as the plots of the contaminated rocks are shifted into the calc–alkaline field towards intermediate compositions. Of significance may be the abnormally high $MgO/FeO + Fe_2O_3$ values of the HRG intermediate and felsic rocks in particular, compared with published calc-alkaline and tholeiitic trends (Fig. 2e). We assume that the high-Mg rhyolites were formed by contamination with the mafic magma. The process of iron-enrichment is believed to have operated in the mafic magma after it was separated from its source. Subcycles within one cycle approximately show the same degree of iron enrichment which reflects the physical conditions prevailing during fractionation. The overall decrease in iron enrichment towards higher cycles could reflect a subsidence of the crystal fractionation site rather than the source. It also represents a development towards a mildly calc–alkaline differentiation trend (cycle V), which, however, may still have been influenced by hybridization.

These hypotheses have to be tested, but preliminary evidence suggests that recurring operation of the processes mentioned by Gunn and Watkins (1969) and Yoder (1973) could very well explain the association of the mafic quartz normative tholeiite and the pseudo-calc–alkaline intermediate and felsic rocks of the HRG.

The Volcanic Subgroup of the Oxford Lake Group

The OLG volcanic subgroup at western Knee Lake is constituted essentially by pyroclastic deposits intercalated with volcanogenic sediments and porphyritic flows. Chemical analyses of the volcanics plotted on an AFM diagram (Fig. 2f) define a calc–alkaline trend. Alkaline rocks from the same area plot along a different trend. Both types are closely associated in time and space as they are present in mixed pyroclastic deposits.

Analyses from the OLG volcanic subgroup at Oxford Lake indicate two chemically distinct rock groups (Fig. 2g), a high-K calc–alkaline, and a shoshonitic association, using Mackenzie and Chappell's (1972) classification. The high-K calc–alkaline basalts and andesites are present as rare flows and sills in the OLG sedimentary subgroup in central Oxford Lake. The shoshonitic rocks are restricted to the Jackson Bay area.

Chemical trends from both areas are compared with typical calc–alkaline and alkaline trends in Fig. 2h. The abnormal trend of the Jackson Bay rocks is discussed below.

Jackson Bay Area

The OLG volcanic subgroup at Jackson Bay is comprised of highly porphyritic rocks. The primary mineralogy is generally replaced by a greenschist facies assemblage. However, in the mafic members, pseudomorphs after pyroxene, feldspars, biotite and rarely olivine can be seen. The proportion of feldspars increases with acidity whereas the percentage of mafic minerals diminishes. Biotite is persistant throughout the suite. Quartz phenocrysts are present in the dacites. The mafic and intermediate flows display autobrecciation and the felsic rocks are primarily pyroclastic.

A striking feature of the eleven samples analyzed is the declining K_2O trend with increasing SiO_2 (Fig. 3), indicating that this suite does not develop along the lines of the alkaline basalt series. According to Mackenzie and Chappell's (1972) classification on the basis of K_2O and SiO_2 wt.%, the samples are absarokite (2364), shoshonites (2374, 315), high-K low-Si andesite (289), high-K andesite (319, 310), andesite (322), high-K dacites (539, 541, 791) and dacite (574). This demonstrates that the Jackson Bay volcanics as a whole, range from shoshonitic through high-K calc–alkaline to calc-alkaline types, with increasing SiO_2 content. A flat K_2O–SiO_2 trend is shown to be characteristic for the shoshonitic association (Joplin, 1968).

Declining trends have been found for some shoshonites from Quaternary volcanoes in eastern Papua (Jakeš and Smith, 1970). We therefore believe that the Jackson Bay volcanics were derived from a shoshonitic volcano. Fig. 3 shows a decreasing $FeO + Fe_2O_3/MgO$ ratio with increasing SiO_2, which explains the discordant trend in the AFM-plot (Fig. 2h). The mafic members in particular are Al-rich and Mg-poor. Also notable are the positive correlations between K and Ba, and Na and Sr, respectively.

Although only the mafic rocks (less than 55% SiO_2) meet the chemical properties of the shoshonitic series, the Sr (441–940 ppm) and Ba (430–1010 ppm) contents of the entire suite, fall well within the range reported for the Papuan shoshonites (Jakeš and Smith, 1970); the Ba/Sr ratios (av. 1·06) are also very

similar, except for the absarokite (2364) (1·84). The Ba and Sr contents reported here are lower than those of the Wyoming shoshonites (1360–2140 ppm Ba, 485–1245 ppm Sr) (Nicholls and Carmicheal, 1969).

Considering a possible model for the magma generation, plagioclase fractionation is ruled out as an important factor because of the flat Sr trend (Fig. 3). Moreover, the overall Sr–Ca relationship is positive and the early Sr and Ca enrichment with decreasing Al precludes plagioclase involvement. Jakeš and Smith (1970) suggest that fractional melting of a mica phase is responsible for the flat and declining K_2O–SiO_2 trends of the shoshonitic rocks. Indeed, the high-K, Ba, Al, Al/Si, and Ba/Sr of the absarokite (2364) in particular, are compatible with melting of mica, possibly phlogopite. This mineral is found to be stably

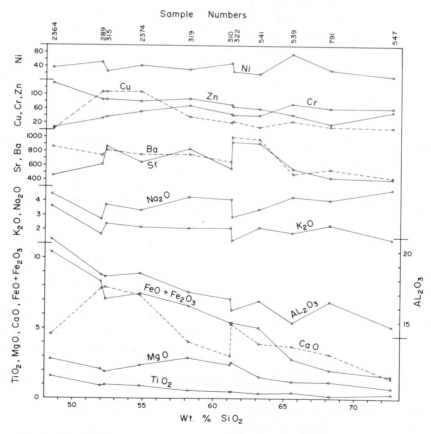

Fig. 3. Graphical representation of volcanic rocks from Jackson Bay, Oxford Lake Group volcanic subgroup; trace elements in ppm, major elements in wt.%

coexistent with mantle phases like pyroxene and olivine. Titan–phlogopite is even more persistent under mantle conditions and its breakdown could explain the concordant K–Ti relationship in Fig. 3 (Forbes and Flower, 1974).

Ni displays a flat trend ranging between 12 and 77 ppm, (Fig. 3). Cr shows a smooth trend of overall enrichment with SiO_2, varying between 7 and 74 ppm. Cu values are between 8 and 108 ppm. These values are considerably lower than those reported for shoshonites from eastern Papua (Mackenzie and Chappell, 1972). It seems most likely that the low Ni and Cr values, for the absarokite in particular, indicate that olivine and pyroxene did not contribute to the magma composition.

Fractionation of phlogopite alone cannot account for the chemical properties of the entire Jackson Bay suite, as the Ba/Sr ratio is close to unity. The almost equal Ba and Sr values are interpreted to be the result of fractionation of hornblende in addition to phlogopite, because the lower Ba/Sr value of hornblende could well compensate for the higher Ba/Sr value of phlogopite. Kesson and Price (1972) stress the importance of kaersutite as an accessory mantle phase and its role in the petrogenesis of alkaline rocks. Indeed the chemical characteristics of kaersutite (Kesson and Price, 1972) are detectable in the Jackson Bay suite, especially in the mafic and intermediate members by a positive correlation between $FeO + Fe_2O_3$ and Zn, similar Al_2O_3/SiO_2 ratios, and the almost invariably greater abundance of Zn over Cr, Cu and Ni. However, melting of phlogopite and kaersutite alone, cannot account for an increase in acidity. A possible explanation could be sought in the results of the melting experiments carried out by Modreski and Boettcher (1973). Melting of the assemblage phlogopite + enstatite + vapour yields SiO_2-rich melts, as enstatite incongruently melts to forsterite and liquid after all phlogopite has entered the melt. The effect of other phases like diopside and spinel on the liquid composition is negligible. Modreski and Boettcher also observed that the composition of the liquid decreases in K and increases in Si with decreasing pressure (from 30 to 10 kb), depending on the amount of enstatite that is involved in the reaction. From these experiments we infer that the Si-undersaturated and K-rich rocks can be derived from a rising, partially melting, mica- and hornblende-bearing mantle source. The latter would rise because the undersaturated and mafic K-rich rocks were extruded prior to the relatively K-poor intermediate and felsic rocks.

The incongruent melting behaviour of enstatite, yielding forsterite could explain the low Ni and Mg values of the mafic members, as these elements are taken up by olivine. Knowledge of the Fe/Mg partition between mica, amphibole and other mantle phases is necessary to interpret the low Mg/Fe ratios of the early derivatives from Jackson Bay. It seems reasonable, however, to assume that the early melts are enriched in iron as magnesium preferentially remains in refractory phases.

Discussion

Considering the chemical data alone, it would be tempting to draw a parallel between an island arc association and this greenstone belt. However, some discrepancies should be mentioned.

The HRG basalts resemble ocean-floor tholeiites. Their tholeiitic trends are disguised by hybridization. As the base and the top of the HRG are not identified, the differentiation towards a mildly calc–alkaline trend in the study area is only part of the overall chemical evolution which remains to be established.

The volcanologic evolution of the area is not continuous as there is an important time break between the deposition of the sub-alkaline HRG and the partially high-K OLG. In this respect there is a similarity to the Kirkland Lake area, Ontario (Cooke and Moorhouse, 1968). In the area studied the interval is marked by granite intrusions, uplift and erosion.

A subduction zone is not a prerequisite for the magma generation models presented here. Moreover, Mackenzie and Chappell (1972) point out that the Papuan shoshonites are not likely related to a Benioff zone.

In summary, the evolution of the Knee Lake–Oxford Lake area is one of cyclic deposition of oceanic tholeiite-like basalts, followed by granite intrusions, contributing to the formation of a (mini)craton. High-K calc–alkaline and shoshonitic volcanism within the newly created (mini)craton is in accord with Joplin's (1968) findings, as it is related to the formation of an intracratonic basin.

Acknowledgements

I am indebted to Dr F. J. Elbers, Mr H. P. Gilbert, Dr W. Weber, Dr S. Singh, Dr W. D. McRitchie and Mr N. L. Green for their advice and constructive criticism.

This contribution is published with permission of the Minister, Manitoba Department of Mines, Resources and Environmental Management, Winnipeg.

References

Barberi, F., Borsi, S., Ferrara, G., and Innocenti, F., 1969. 'Strontium Isotopic Composition of Some Recent Basic Volcanites of the Southern Tyrrhenian Sea and Sicily Channel', *Contr. Miner. Petrol.*, **23**, 157–162.

Barry, G. S., 1959. 'Geology of the Oxford House–Knee Lake Area', *Man. Dep. Mines Publ.*, **58–3**, 1–39.

Barry, G. S., 1960. 'Geology of the Western Oxford Lake–Carghill Island Area', *Man. Dep. Mines Publ.*, **59–2**, 1–37.

Barry, G. S., 1964. 'Geology of the Parker Lake Area', *Man. Dep. Mines Publ.*, **62–1**, 1–26.

Campbell, F. H. A., Elbers, F. J., and Gilbert, H. P., 1972. 'The stratigraphy of the Hayes River Group in Manitoba—A preliminary report', *Man. Mines Br. Geol. Pap. 2/72*, 1–33.

Cooke, D. L., and Moorhouse, W. W., 1968. 'Timiskaming volcanism in the Kirkland Lake area, Ontario, Canada', *Can. J. Earth Sci.*, **6**, 117–132.

Forbes, W. C., and Flower, M. F. J., 1974. 'Phase relations of titan-phlogopite, $K_2 Mg_4 Ti Al_2 Si_6 O_{20} (OH)_4$: a refractory phase in the upper mantle?', *Earth Planet. Sci. Lett.*, **22**, 60–66.

Gilbert, H. P., and Elbers, F. J., 1972. 'Parker Lake–Knee Lake–Oxford Lake Area', *Man. Mines Br. Geol. Pap. 3/72*, 34–37.

Glikson, A. Y., 1971. 'Primitive Archean element distribution patterns: chemical evidence and geotectonic significance', *Earth Planet. Sci. Lett.*, **12**, 309–320.

Goodwin, A. M., 1972. 'The Superior Province. In Variations in tectonic styles in Canada', Raymond A. Price and R. J. W. Douglas (Eds.), *Spec. Pap. Geol. Soc. Can.*, **11**, 527–623.

Green, N. L., 1973. 'The volcanic stratigraphy and petrochemistry of the Gods Lake Subgroup, Knee Lake, Manitoba', *M.Sc. Thesis*, Univ. of Manitoba, 1–132.

Gunn, B. M., and Watkins, N. D., 1969. 'The petrochemical effect of the simultaneous cooling of adjoining basaltic and rhyolitic magmas', *Geochim. Cosmochim. Acta*, **33**, 341–356.

Irvine, T. N., and Baragar, W. R. A., 1971. 'A Guide to the chemical classification of the common volcanic rocks', *Can. J. Earth Sci.*, **8**, 523–548.

Jakeš, P., and Smith, I. E., 1970. 'High Potassium Calc–Alkaline Rocks from Cape Nelson, Eastern Papua', *Contr. Miner. Petrol.*, **35**, 119–124.

Jakeš, P., and White, A. J. R., 1972. 'Major and Trace Element Abundances in Volcanic Rocks of Orogenic Areas', *Bull. Geol. Soc. Am.*, **83**, 29–40.

Joplin, J. A., 1968, 'The shoshonitic association: A review', *J. Geol. Soc. Aust.*, **15**, 275–294.

Kesson, S., and Price, R. C., 1972. 'The Major and Trace Element Chemistry of Kaersutite and Its Bearing on the Petrogenesis of Alkaline Rocks', *Contr. Miner. Petrol.*, **35**, 119–124.

Lowder, G. G., and Carmichael, I. S. E., 1970. 'The volcanoes and Caldera of Talasea, New Britain: Geology and Petrology', *Bull. Geol. Soc. Am.*, **81**, 17–38.

Macdonald, G. A., and Katsura, T., 1964. 'Chemical composition of Hawaiian lavas', *J. Petrol.*, **5**, 82–133.

Mackenzie D. E., and Chappell, B. W., 1972. 'Shoshonitic and Calc–Alkaline Lavas from the Highlands of Papua, New Guinea', *Contr. Miner. Petrol.*, **35**, 50–62.

Modreski, P. J., and Boettcher, A. L., 1973. 'Phase relationships of phlogopite in the system $K_2O–MgO–CaO–Al_2O_3–SiO_2–H_2O$ to 35 kilobars: a better model for mica in the interior of the Earth', *Am. J. Sci.*, **273**, 385–414.

Nicholls, J., and Carmichael, I. S. E., 1969. 'A commentary on the Absarokite–Shoshonite–Banakite Series of Wyoming, U.S.A.', *Schweiz. Miner. Petrogr. Mitt.*, **49**, 47–64.

Wright, J. F., 1932. 'Oxford House Area, Manitoba', *Geol. Surv. Can. Summ. Rep.*, Pt. C., 1–25.

Yoder, Jr., H. S., 1973. 'Contemporaneous Basaltic and Rhyolitic Magmas', *Am. Miner.*, **58**, 153–171.

Physico-chemical Conditions During the Archean as Indicated by Dharwar Geochemistry

S. M. NAQVI

National Geophysical Research Institute, Hyderabad.

Abstract

The Dharwar formations (3·2–2.2 b.y.) of India are mainly composed of various types of metavolcanics and metasediments, and they have been intruded by granites and dykes of varying composition. The geochemistry of these formations from different parts of Karnataka, has shown that the Archean crust of the 'Dharwar Protocontinent' was of the thin, unstable oceanic type and during early Archean times the primary differentiation of the Earth into crust, mantle and core was incomplete. During the late Archean and early Proterozoic, the primary differentiation of the Earth and the consequent thickening, stabilization and granitization of its crust appears to have been rapid, and the processes of cratonization marked by the appearance of alkali olivine basaltic dykes were probably completed by the middle Proterozoic (2.1 b.y.). The chemical differences both in major and trace elements between the Archean–early Proterozoic and middle Proterozoic basalts, granites and sediments on the one hand, and the similarities between all types of rock formations from the middle Proterozoic onward on the other, suggest that the physico-chemical conditions prevailing during the Archean era have never been repeated.

Introduction

As rocks are the equilibrium product of a given set of physico-chemical processes it is expected that these processes will leave their 'fingerprints' in their chemistry. Therefore, if the physico-chemical conditions throughout the geologic past have been changing, the chemistry of the rocks should reflect them. The careful study of the sediments of a given region can reveal the changes which the outer lithosphere has seen and the chemistry of the igneous rocks can be used to decipher the changes in the inner lithosphere. The Dharwar formations, a Super Group representing one billion years of Precambrian era (3·2–2·2 b.y.) are mainly made up of metasediments and metavolcanics. This Super Group can be subdivided into at least two broad divisions namely (i) greenstone belts and (ii) greenstone-like geosynclinal piles (Fig. 1). The greenstone belts like those of Kolar and Holenarsipur start with basic and ultrabasic rocks of komatiitic affinities at their base, whereas the geosynclinal piles like those of Chitaldrug and Shimoga start with an orthoquartzite–carbonate base and develop upward into a thick sequence of eugeosynclinal sediments and volcanics. These eugeosynclinal sequences do not show rhythmic alternating layers of volcanics and sediments, the characteristic of greenstone belts of

290

South Africa and other shield regions (Anhaeusser, 1973). The rocks of the Dharwar Super Group have been folded, cross-folded, metamorphosed from greenschist to granulite facies and intruded by many granites and at least five sets of dykes. All these formations are engulfed by an unclassified crystalline group of rocks termed the Peninsular gneisses, a complex sequence consisting largely of middle rank gneisses. After the Eparchean Interval these gneisses are followed by middle and later Proterozoic geosynclinal and platform deposits of the Cuddapah and Kurnool systems. The complex geology of the Dharwar protocontinent (Fig. 1) therefore provides an excellent opportunity to study the early history of the earth and continental evolution. As the record older

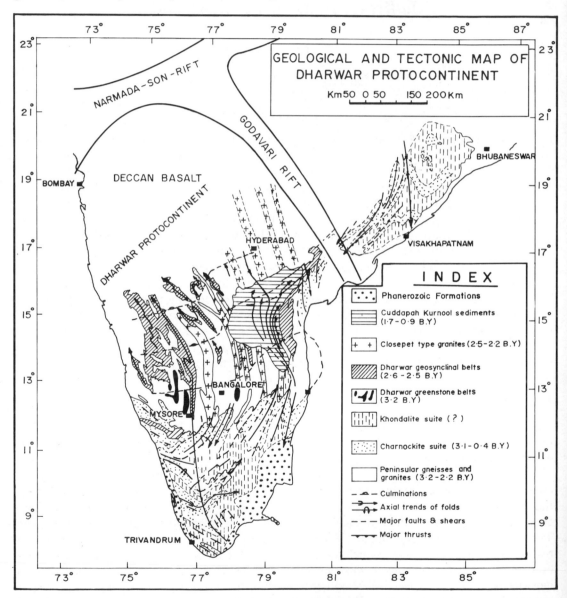

Fig. 1. Geological and Tectonic map (modified after Narayanaswamy 1970) of the Dharwar Protocontinent showing the distribution of Dharwar greenstone belts, Dharwar geosynclinal belts and the middle-late Proterozoic Cuddapah–Kurnool sediments (reproduced with permission of Council of Scientific and Industrial Research, New Delhi)

than 3·2 b.y. has not been identified as yet, and the relation between gneisses and schists is obliterated, the study of the Dharwar Super Group appears to be the only clue for understanding the physico-chemical conditions prevailing during the interval 3·2–2·2 b.y. and the 'missing billion'. The Precambrian rocks found in the Dharwar protocontinent can also be classified into five petrological types, namely (1) the ultrabasics/ultramafics, (2) the metavolcanics, (3) the sediments and metasediments, (4) the gneisses and granites (acidic rocks), and (5) the basic intrusives. These five groups, except for the ultramafics, are found to be well-spread both in time and geological environments, i.e. in the middle Archean greenstone belts, late Archean–early Proterozoic eugeosynclinal belts, and middle-late Proterozoic geosynclines and platforms. A synopsis of the results of the geochemical studies on these rocks is presented below.

The Ultrabasic Rocks

A zone of schistose rocks having an ultrabasic base followed by mostly chemical and fine-grained detrital sediments (often interbedded with thin, fine-grained, ripple marked and current bedded quartzites) metamorphosed to amphibolite facies and surrounded by gneisses and granites, is found as a linear, en echelon belt about 320 km long extending in a NS direction from Sindhuvalli in the south to Shivani in the north (Fig. 1). This zone has been given various local names such as Holenarsipur, and Nuggihalli schist belts, according to the townships near the exposures, and appears to be a true greenstone belt. Its field relationship with the surrounding gneisses and granites (Divakara Rao et al.; 1975), suggest that this is probably the oldest sequence of the Dharwar protocontinent. The ultramafic/ultrabasic rocks include dunite, peridotite, anorthosite, pyroxenites and amphibolites. The schistose rocks include talc–schists, tremolite–actinolite schists, garnetiferous–hornblende schists, thin-layered amphibolites and kyanite–staurolite schist, indicating high P–T metamorphism of the zone. The geochemical data from some of

these ultramafic rocks, show that the MgO content of the peridotites and dunites is higher than those commonly reported, even that of peridotite–komatiite from Barberton (Divakara Rao et al., 1975). However, the bulk chemical composition (Fig. 2) of the ultramafics falls within the field of peridotitic komatiites of Viljoen and Viljoen (1969). They contain high Co, K, Rb and Cs concentrations (Divakara Rao and Satyanarayana 1973). These data indicate that the Archean mantle beneath the Dharwar protocontinent was peridotitic and highly undifferentiated in nature. The high K, Rb and Cs concentrations can be attributed to later hydrothermal enrichment and potash activation.

The Metavolcanics

The basic volcanic rocks are found as scattered patches along the Dharwar schist belts and have been locally named as Holenarsipur, Nuggihalli, Ramagiri and Kolar traps (all found in the true greenstone belts) and Lingadahalli, Jogimardi, Mardihalli, Bellara, Belegatta and Gadag traps (all associated with eugeosynclinal sedimentation of greywackes and greywacke-conglomerates showing graded bedding). The middle and late Proterozoic basalts are found in the Cuddapah basin of Andhra Pradesh. Metabasalts associated with Dharwar greenstones and geosynclinal piles are quartz/olivine tholeiites and resemble the low K^+ oceanic tholeiites in their major element composition. Whereas the Kolar and Holenarsipur metabasalts are komatiitic and closely resemble the basaltic komatiite of Barberton, the metabasalts from the geosynclinal piles, i.e. Chitaldrug schist belt, are found to be soda-rich (spilitized), calc–alkaline basalts of the island arc type like those of Canada and Australia (Fig. 2). However, some of the Kolar metabasalts from the upper horizons also occupy a place in the metatholeiitic field (Fig. 2). These metabasalts, irrespective of their association, show very low K_2O contents compared with younger continental tholeiites, and higher amounts of siderophile elements, as well as

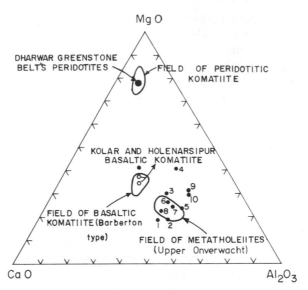

Fig. 2. Shows the compositional characteristics of the metabasalts from the Dharwar greenstone belts, Dharwar geosynclinal belts and later platforms and geosynclines. (1) Dharwar geosynclinal metavolcanics, (2) Alkali–olivine basalt dykes, Chitaldrug (3) Olivine tholeiitic dykes, Chitaldrug (4) Olivine tholeiite dykes, Shimoga (5) Quartz tholeiitic dykes, Chitaldrug (6) Kolar Hornblendic Schists, (7) Kolar amphibolites, (8) Kolar meta-basalts from upper horizons, (9) Basalts from Bhowali Himalaya, (10) Epidiorite from Bhowali—Himalayas

lithophiles like Zr and Y comparable with recent oceanic tholeiites (Naqvi and Hussain, 1973a, 1973b NGRI unpublished data). Basic enclaves, dated as 2·9 b.y. (Venkatasubramaniam et al., 1971) are representatives of an older basalt–sedimentary greenstone cycle, are found all along the Peninsular gneisses, and show low K_2O and high siderophile contents along with Y and Zr (Satyanarayana et al., 1974). The Cuddapah basalts have high K_2O and are in all aspects similar to younger continental basalts (Naqvi, Divakara Rao and Hari Narain, 1974).

The Sediments and Metasediments

Like basalts, the sedimentary–metasedimentary rocks are also found in the three different geological environments which successively prevailed during the Precambrian era. The sediments of the true greenstone belts, namely those of Kolar and Holenarsipur being mainly chemical and volcaniclastic deposits and metamorphosed to amphibolite facies, are difficult to differentiate clearly from the accompanying metabasalts (Fig. 2). The thin-layered paraamphibolites interbedded with thin and current bedded quartzites from the Holenarsipur schist belt, and some amphibolitic xenoliths from the Peninsular gneisses appear to be volcaniclastic sediments. However, on the basis of their chemistry alone their identification is not possible (Satyanarayana et al., 1975). In geosynclinal piles, i.e. Shimoga–Chitaldrug schist belts, mainly clastic sediments of the orthoquartzite–carbonate and geosynclinal facies are found. Assuming isochemical regional metamorphism, the chemical data of these geosynclinal clastic metasediments have been interpreted in terms of their provenance, and the nature and composition of the Pre-Dharwar (2·6 b.y.) crust (Naqvi and Hussain, 1972; Satyanarayana et al., 1973a). A comparison of the composition of the metasediments with that of the average crust has shown that most of these sediments are enriched in elements like Fe, Mg, Ca, Ti, Mn, Co, Cr, Ni and V of basic parentage and depleted in elements like K, Si, Zr and Y of acidic parentage (Naqvi and Hussain, 1972, p. 127). The pebble composition of the conglomerates clearly indicates the presence of acid plutonic, sedimentary and basic volcanic rocks in the source area, and some of these metasediments contain about 600 ppm Cr and 250 ppm Ni indicating the presence of ultramafic rocks in that area. However, the overall composition of these metasediments (Fig. 3) shows the predominance of basic rocks in the area which supplied the debris to the Dharwar geosyncline.

The Gneisses and Granites

The unclassified gneisses and granites of Karnataka State occupy about 50,000 km^2. Their relationship with the schistose rocks is complicated and has led to an age-old con-

troversy of their being older or younger than the Dharwar schists. In fact, both types of gneisses, i.e. older and younger than the Dharwar schist belts, exist in Karnataka (Radhakrishna, 1967, 1974). Honnali gneisses near Shimoga have been dated at 3·2 b.y. (Crawford, 1969), whereas gneisses from Bangalore and some other places are 2·5–2·6 b.y. old (Crawford, 1969; Venkatasubramaniam et al., 1971). More than one period of granitic activity has been recognized in almost all the shield areas of the world (Glikson and Sheraton, 1972). Similarly, in the Indian Shield at least two or three periods of acid igneous activity can be identified from the available geochronological data. However, the later events like potash activation seem to have affected the earlier gneisses and granites considerably. The best representatives of the older gneisses and granites appear to be their pebbles preserved in the different Dharwar metaconglomerates. Although these conglomerates have undergone greenschist metamorphism, the textural, mineralogical and chemical integrity of the pebbles does not

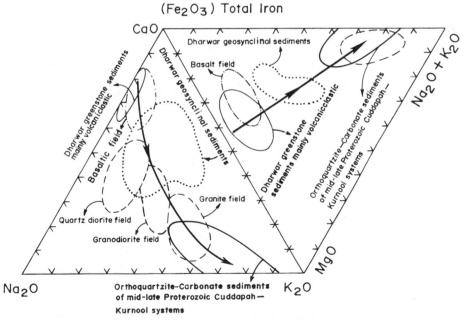

Fig. 3. Shows the compositional change of the sediments from middle and late Archean greenstone belts to mid-late Proterozoic geosynclinal and platform deposits. The early Proterozoic Dharwar geosynclinal sediments are transitional or intermediate between the two

294

appear to have changed due to the blanketing effect of their matrices. These conglomerates are mostly found at the base of the geosynclinal facies, and show graded bedding and disrupted framework; most of the metasediments and metavolcanics overlie them (Naqvi, 1973). Moreover, the pebbles from the Kaldurga conglomerates have been dated at 3.2 b.y. (Venkatasubramaniam and Narayanaswamy, 1974). The geochemistry of 104 pebbles from Kaldurga, Aimangala, Kurmerdikere, Talya, Gadag, Ubrani and other conglomerates which are definitely sedimen-

tary and petromictic show that most of the granite-gneiss pebbles have low K_2O and high Na_2O/K_2O ratios ranging from 0·71 to 38·75 (mode between 4–5), in which more than 95% have Na_2O/K_2O ratios more than 1 (Fig. 4). On the other hand, 75% of the 138 samples of the Peninsular gneisses collected and analyzed from the adjoining localities of the conglomerates are found to have Na_2O/K_2O ratios less than 1 (NGRI unpublished data). The 3·2 b.y. old (Crawford, 1969) Honnali gneisses are also soda-rich and have very high Na_2O/K_2O ratios (Fig. 4). Younger granitic activities have

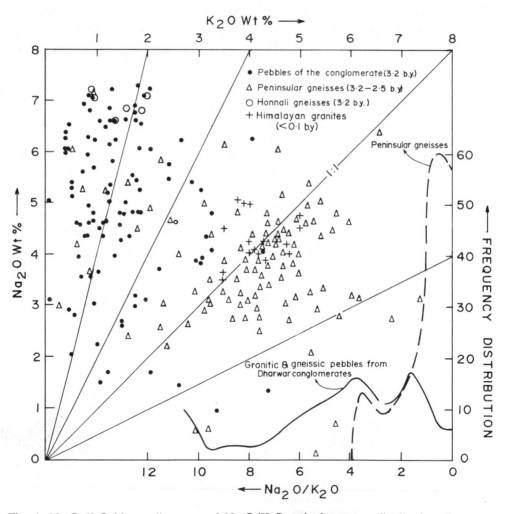

Fig. 4. Na_2O–K_2O binary diagram and Na_2O/K_2O ratio frequency distribution diagram showing that 95% of the granitic and gneissic pebbles from Dharwar conglomerates have a ratio of more than 1, whereas 60% of the Peninsular gneisses and granites have a Na_2O/K_2O ratio of less than 1. The later granites and gneisses even from the Himalayas have Na_2O/K_2O ratios similar to the Peninsular gneisses and granites

been reported from the Cuddapah basin, Aravalli region, Singhbhum and Himalayas ranging in age from 2·1 b.y. to <0·1 b.y. In their chemical composition (e.g. Na_2O/K_2O ratios ≈ 1, high K_2O values) these later granites closely resemble the 2·6–2·4 b.y. old granites and gneisses of the Dharwar region.

The Basic Intrusives

At least five phases of basic intrusive activity are indicated by field and paleomagnetic data of the many sets of dykes recognized in the South Kanara district of Karnataka State. The youngest set, trending ENE and cutting all other trends of dolerites and granite-gneisses, has been dated at 2·1 b.y. (Naqvi, Divakara Rao and Hari Narain, 1974). At Chitaldrug many dykes cut across the metabasalts (2·5 b.y.) and at Nuggihalli these dykes may be divided into an older NE trending set, mostly epidioritic, and a NW trending set of unmetamorphosed dolerites (Naqvi et al., 1974; Satyanarayana et al., 1973b). The geochemistry of the dykes from Shimoga, Chitaldrug and Nuggihalli in Karnataka has shown that there are marked compositional differences between the epidiorites and dolerite dykes. The epidiorites are quartz/olivine

tholeiites and so far no alkali olivine basalts have been reported from them. The post-orogenic dykes show variation in their composition and it is in these that for the first time alkali olivine basalts appear in the Dharwar protocontinent. These alkali olivine dykes are equivalent to or younger than those representing the 2·2 b.y. intrusive igneous activity (Fig. 5). These younger dykes differ considerably in their major, minor and trace element composition from the Archean–early Proterozoic metabasalts belonging to the greenstones and geosynclinal piles, especially by having high K_2O and low MgO and ferromagnesian trace elements, and are identical in composition (Fig. 5) with the dolerites and basalts of the Cuddapah basin and other younger basalts like those of Deccan, Rajmahal and Bhowali (Naqvi et al., 1974; Divakara Rao et al., 1974).

Discussion

The Archean–early Proterozoic boundary is fixed at 2·5 b.y. It marks the change from the 'nuclear stage' of crustal development to the geosynclinal stage, whereas the middle Proterozoic was the era of platforms (Ronov, 1972). This Archean–Proterozoic transition is

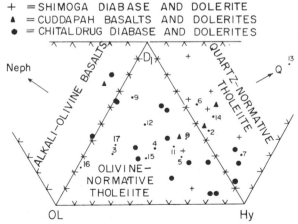

Fig. 5. Normative compositional diagrams showing the Alkali–olivine basaltic nature of some of the younger basaltic dykes (2·1 b.y. onwards)
   1–9: Archean–early Proterozoic metabasalts
  10–15: Average of different Proterozoic dykes
  16: Basalts from Bhowali Himalayas
  17: Epidiorites from Bhowali Himalayas

thought to be the greatest tectonic discontinuity in the Earth's recorded history (3·8 b.y.). These nuclear, geosynclinal and platformal stages are represented by the Dharwar greenstones, Dharwar geosynclines and Cuddapah–Kurnool deposits, respectively. During the past few years the geochemical data from Archean–Proterozoic formations throughout the world have been interpreted in terms of crustal growth and crust–mantle physico-chemical conditions existing during those periods (Glikson, 1972; Condie, 1973; Mueller and Rogers, 1973; Goodwin, 1974; Naqvi, Divakara Rao and Hari Narain, 1974). In the case of the other shields, considerable insight has been made into the 'missing billion' (Anhaeusser, 1973; Windley, 1973), whereas in the case of the Indian Shield, as a result of lack of modern facilities and equipment, data from rocks beyond 3·3 b.y. are not available. However, the available data as summarized above show some interesting trends. A comparison of the geochemical data of the basic, sedimentary and acidic rocks formed during the period 3·2–1·4 b.y. in three different geological environments, namely greenstone belts, geosynclinal piles and middle-late Proterozoic geosynclines and platforms, indicates the evolutionary processes that the crust and mantle went through beneath the Dharwar protocontinent.

The sediments are the products of large-scale chemical and mechanical fractionation processes (Pettijohn et al., 1972, p. 80). The evolution of the sialic continental crust should be reflected in sandstone compositions. Sandstones dating from the time when there was no sialic crust, being derived solely from mafic terrains, might have been largely volcaniclastic. Sandstones derived from thin sialic crust should be intermediate in composition between those expected from the earliest and all later sandstones derived mainly from sialic crust (Pettijohn et al., 1972, p. 578). The geochemistry of Dharwar metasediments from geosynclinal piles clearly shows that the late Archean–early Proterozoic era was a transitional period from simatic to sialic crustal development, and thus sediments showing a mixed but predominantly basic

provenance were laid down. These sediments are transitional between the predominantly chemical and fine-grained volcaniclastic sediments of true greenstone belts, like those of Kolar and Holenarsipur, and those of Cuddapah and Kurnool systems, which are predominantly orthoquartzitic and indicate a completely granitized provenance (Fig. 3).

For the estimation of crustal thicknesses during the Archean–Proterozoic era the geochemical polarity indices and values appear to be useful tools (Naqvi et al., 1974; Condie, 1973). Using θ values as the indices for the depth of generation and emplacement of basaltic magmas through and over a fully cratonized crust ($\theta = <36$) or unstable, thin crust like those of island arcs ($\theta = >36$), it may be inferred that basalts from both the greenstone belts and geosynclinal piles were emplaced through and over a thin unstable (5–15 km) oceanic type of crust (Fig. 6). The partial melting of this oceanic type of thin crust appears to have given rise to sodic granites (3·2 b.y.) as suggested by Glikson (1972). This is evidenced from the almost complete absence of potash granitic pebbles and predominance of sodic granitic pebbles in the Dharwar conglomerates (Fig. 4). The emplacement of potash-rich granites and gneisses appears to be a post-Dharwar event. Probably a widespread potash activation and metasomatism of the Dharwar Protocontinent took place around 2·5 b.y. Or, as suggested by Heier and Adams (1964), Na^+ was released from the mantle at quite high P–T conditions, earlier than K^+. Furthermore, Na^+ can easily enter into amphiboles and replaces Ca^{++} in plagioclase at higher P–T conditions. Therefore, it is expected that at an earlier stage of cooling, Na^+, and at a later stage (probably between 2·6–2·4 b.y.), Na^+ was released in large quantities. And thus, to a large extent, if not totally, the outer lithosphere became rapidly granitized between the period 2·5–2·1 b.y. This rapid granitization and cratonization is also reflected in the chemical data from the 2·1 b.y. old intrusive rocks; their high K_2O, and low 'θ' values and resemblance to younger continental basalts in almost all compositional aspects including trace elements

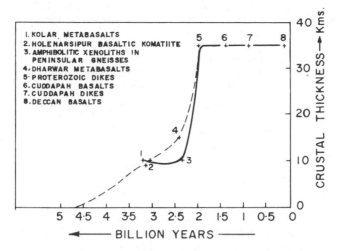

Fig. 6. Shows the rapid thickening of the crust between
2·5–2·1 b.y. Thickness of the crust is estimated from θ
values

indicate their emplacement through and over a fully evolved continental crust (Fig. 6). These dykes appear to mark thc culmination of the cratonization process and primary differentiation. Since the physico-chemical conditions prevailing during the Archean–early Proterozoic periods have never been repeated, the composition of basalts, sediments, granites and dolerites appears to have generally become monotonous after 2·1 b.y. ago.

Conclusions

From the geochemical data so far available from the Dharwar Protocontinent, it can be inferred that during the middle and late Archean the crust was of the unstable, basaltic, oceanic type and the protomantle was highly undifferentiated, suggesting incomplete primary differentiation of the Earth as a whole. The early Proterozoic was a transitional era in which rapid cratonization appears to have taken place. By the middle Proterozoic the primary differentiation and cratonization of the Dharwar region was probably complete through granitization. The culmination of the stabilization of the crust appears to be marked by the appearance of alkali–olivine basalts in the Middle Proterozoic intrusives. The granitic material in the middle-late Archean was sodic, and between 2·6–2·4 b.y. potash activation/metasomatism or potassic granitic activity became very widespread. The non-repeatibility of the greenstone basalt–sedimentary sequence, and the greenstone-like geosynclinal basalt–sedimentary sequence after the middle Proterozoic reflects a definite trend of evolution, which was probably controlled by cooling of the Earth and its consequent primary differentiation.

References

Anhaeusser, C. R., 1973. 'The evolution of the Precambrian crust of Southern Africa', *Phil. Trans. Roy. Soc. Lond.*, **A273**, 359–388.

Condie, K. C., 1973. 'Archean magmatism and crustal thickening', *Geol. Soc.* Am. Bull., **84**, 2981–2992.

Crawford, A. R., 1969. 'Reconnaissance Rb–Sr dating of the Precambrian rocks of the southern peninsular India', *J. geol. Soc. India*, **10**, 117–166.

Divakara Rao, V., and Satyanarayana, K., 1973. 'Anomalously high K, Rb concentration in the serpentinised ultramafics of Mysore State, Peninsular India, and their significance', *Chem. Geo.*, **12**, 51–60.

Divakara Rao, V., Satyanarayana, K., Naqvi, S. M., Hussain, S. M., and Dileep Kulmar, V., 1974. 'Geochemistry and Petrogenesis of basic rocks from Bhowali, U.P.', *Geoph. Res. Bull.*, **12**, No. 1, 63–74.

298

Divakara Rao, V., Satyanarayana, K., Naqvi, S. M., and Hussain, S. M., 1975. 'Geochemistry of Dharwar ultramafics and the Archean mantle', *Lithos*, **8**, 77–91.

Glikson, A. Y., 1972. 'Early Precambrian evidence of a primitive ocean crust and island nuclei of sodic granite', *Geol. Soc. Am. Bull.*, **83**, 3323–3344.

Glikson, A. Y., and Sheraton, J. W., 1972. 'Early Precambrian trondhjemitic suites in Western Australia and North Western Scotland and the geochemical evolution of shields', *Earth Planet. Sci. Lett.*, **17**, 227–242.

Goodwin, A. M., 1974. 'Precambrian belts, plumes and shield development', *Am. Journ. Sci.*, **274**, 987–1028.

Heier, K. S., and Adams, J. A. S., 1964. 'The geochemistry of alkali metals', *Phys. Chem. Earth*, **5**, 253–382.

Mueller, P. M., and Rogers, J. J. W., 1973. 'Secular chemical variation in a series of Precambrian mafic rocks, Beartooth mountains, U.S.A.', *Geol. Soc. Am. Bull.*, **84**, 3645–3652.

Naqvi, S. M., and Hussain, S. M., 1972. 'Petrochemistry of Early Precambrian metasediments from the Central part of the Chitaldrug schist belt, Mysore, India', *Chem. Geo.*, **10**, 109–135.

Naqvi, S. M., 1973. 'Geological structure and aeromagnetic and gravity anomalies in the central part of the Chitaldrug schist belt, Mysore, India', *Geol. Soc. Am. Bull.*, **84**, 1721–1732.

Naqvi, S. M., and Hussain, S. M., 1973a. 'Geochemistry of Dharwar metavolcanics and composition of the primeval crust of peninsular India', *Geochim, Cosmochim. Acta*, **37**, 153–164.

Naqvi, S. M., and Hussain, S. M., 1973b. 'Relation between trace and major element composition of Chitaldrug metabasalts, Mysore, India and the Archean mantle', *Chem. Geo.*, **11**, 17–30.

Naqvi, S. M., Divakara Rao, V., Satyanarayana, K., and Hussain, S. M., 1974. 'Geochemistry of post-Dharwar basic dykes and the Precambrian crustal evolution of peninsular India', *Geol. Magazine*, **111**, 229–236.

Naqvi, S. M., Divakara Rao, V., and Hari Narain, 1974. 'Protocontinental growth of the Indian shield and the antiquity of its rift valleys', *Precambrian Research*, **1**, 345–398.

Narayanaswami, S., 1970. 'Tectonic setting and manifestation of the upper mantle in the Precambrian rocks of South India', *Proc. 2nd U.M.P. Symposium*, Hari Narain (Ed.), Hyderabad, 378–403.

Pettijohn, F. J., Potter, P. E., and Siever, R., 1972. *Sand and Sandstones*, Springer–Verlag, Berlin, 618 pp.

Radhakrishna, B. P., 1967. 'Reconsideration of some problems in the Archean complex of Mysore', *J. geol. Soc. India*, **8**, 102–109.

Radhakrishna, B. P., 1974. 'Peninsular gneissic complex of the Dharwar craton—A suggested model for its evolution', *J. geol. Soc. India*, **15**, 439–456.

Ronov, A. B., 1972. 'Evolution of rock composition and geochemical processes in the sedimentary shell of the earth', *Sedimentology*, **19**, 157–172.

Satyanarayana, K., Divakara Rao, V., Naqvi, S. M., and Hussain, S. M., 1973a. 'On the provenance of Dharwar Schists, Shimoga, Mysore', *Bull. Geoph. Res.*, **11**, 35–41.

Satyanarayana, K., Divakara Rao, V., Naqvi, S. M., and Hussain, S. M., 1973b. 'Geochemistry and petrogenesis of dykes from Nuggihalli schist belt, Mysore', *Bull. Geoph. Res.*, **11**, 273–280.

Satyanarayana, K., Naqvi, S. M., Divakara Rao, V., and Hussain, S. M., 1974, 'Geochemistry of Archean amphibolites from Mysore State, Peninsular India', *Chem. Geo.*, **14**, No. 4, 305–316.

Venkatasubrahmaniam, V. S., Iyer, S. S., and Pal, S., 1971. 'Studies on Rb–Sr. Geochronology of the Precambrian formation of Mysore State, India', *Am. J. Sci.*, **270**, 43–53.

Venkatasubrahmaniam, V. S., and Narayanaswamy, R., 1974. 'The age of some gneissic pebbles in Kaldurga conglomerates, Karnataka, South India', *Jour. Geol. Soc. India*, **15**, 318–319.

Viljoen, M. J., and Viljoen, R. P., 1969. 'Evidence of the existence of a mobile extrusive peridotitic magma from the Komati formation of the Onverwacht group', *Trans. Geol. Soc. South Africa*, Special Publication No. 2, 87–123.

Windley, B. F., 1973. 'Crustal development in the Precambrian', *Phil. Trans. Roy. Soc. Lond.*, **A273**, 321–341.

A Model for the Origin of the Early Precambrian Greenstone–Granite Complex of North-eastern Minnesota

JOSEPH G. ARTH

U.S. Geological Survey, Reston, Virginia 22092

Abstract

The early Precambrian greenstone–granite complex of north-eastern Minnesota formed 2700–2750 m.y. ago during a tectonothermal event which produced new crustal rocks. These include four major igneous rock types:

(1) Tholeiites derived from the shallow mantle.
(2) Tonalites and dacites from the deeper mantle.
(3) Granites (quartz monzonites) from the lower parts of the newly formed crust, and
(4) Syenodiorites and syenites from a deep mantle source.

Introduction

The geology of the early Precambrian terrane of north-eastern Minnesota has been studied for more than 100 years. The early work was summarized by Grout et al. (1951) and Goldich et al. (1961), and the recent work is summarized in *Geology of Minnesota: A Centennial Volume*, edited by Sims and Morey (1972). Arth and Hanson (1975) presented new chemical data that bears on the genesis of the various igneous rock types, and they proposed a model for the origin of each. This paper summarizes their approach and conclusions.

Brief Sketch of the Geologic History

Activity began with the extrusion of basalt and dacite. As bimodal tholeiitic–dacitic volcanism continued, rocks equivalent in composition to the dacitic volcanic rocks were intruded into the volcanic piles as hypabyssal dacite porphyry and diapiric plutonic tonalite. Erosion of the volcanic piles and unroofing of the tonalitic bodies accompanied explosive dacitic volcanism and provided a source for thick sequences of graywacke and argillite. After formation of the graywacke sequences, potassic granites of dominantly quartz monzonite (adamellite) composition intruded the volcanic and sedimentary piles. Emplacement of both the tonalite and adamellite bodies probably played a major role in producing both the tightly folded structures and the generally greenschist facies metamorphic grade of the volcanic and sedimentary rocks. The last early Precambrian igneous event was the intrusion of small postkinematic alkalic stocks of syenodiorite and syenite.

Ages and Initial $^{87}Sr/^{86}Sr$ Ratios

Geochronology in the area by fission track, K–Ar, Rb–Sr, and U–Pb methods shows that all the igneous rocks formed in the period

2750–2700 m.y. ago. No older rocks have been reported. The initial $^{87}Sr/^{86}Sr$ ratios, given by whole-rock isochrons and apatite separates, are listed in Table. 1. Initial ratios

rocks whose genesis has been discussed in the literature. This comparison provided several possible magmatic processes to explain each rock type. Mathematical models for trace-

Table 1. Initial $^{87}Sr/^{86}Sr$ ratios of early Precambrian rocks of north-eastern Minnesota

Rock unit	Initial $^{87}Sr/^{86}Sr$	Reference
Icarus pluton	0·7009 ± 0·0003	1
Linden syenite	0·7009 ± 0·0004	2
Vermilion Granite	0·7005 ± 0·0012	3
Giants Range Granite	0·7002 ± 0·0019	2
Saganaga Tonalite	0·7009 ± 0·0002	1
Keewatin Greenstones	0·7005 ± 0·0009	3
Northern Light Gneiss	0·7006 ± 0·0002	1

References:
1. Hanson et al. (1971).
2. Prince and Hanson (1972).
3. Peterman et al. (1972).

for all rock types are lower than 0·7010 and thus are too low to allow derivation from pre-existing granodioritic or more potassic crustal rocks. The isotopic data do allow origin from a trondhjemitic, basaltic, or ultramafic source, or possibly by rapid recycling (<50 m.y.) of more potassic rocks, such as volcanogenic graywackes, which themselves had no potassic progenitors.

Major- and Trace-element Chemistry of the Igneous Rocks

Further constraints can be placed on the source of the various igneous rocks by considering the major- and trace-element chemistry. Petrography and major-element chemistry allow classification into four major groups: tholeiitic basalt and basaltic andesite; potassium-poor dacites, tonalites, and trondhjemites; moderately potassic granitic rocks of quartz monzonite (adamellite) composition; and alkali-rich rocks of syenodioritic to syenitic composition. Three of these four groups are sufficiently uniform in composition for representation by the averages shown in Table 2.

The major-element composition of each rock was compared with melt compositions which are reported from partial melting and phase equilibria studies, and also with igneous

element behaviour during such processes were then used to see if they could predict the observed concentrations of Rb, Sr, Ba and rare-earth elements in each rock. The results of this procedure yielded models for the origin of each rock type, which, although not unique, satisfy the constraints imposed by our present understanding of field relations, isotopic data, phase equilibria, and trace-element behaviour.

Table 2. Average compositions for three of the four early Precambrian igneous rock groups of north-eastern Minnesota

	Metatholeiites	Dacites, trondhjemites, tonalites	Quartz monzonites
SiO_2	50·2	66·7	73·2
Al_2O_3	13·7	17·1	14·7
TiO_2	1·4	0·3	0·2
Fe_2O_3	2·4	0·9	0·5
FeO	10·2	1·5	1·0
MnO	0·2	0·04	0·03
MgO	6·7	1·4	0·4
CaO	8·3	3·8	1·1
Na_2O	2·8	5·8	4·0
K_2O	0·3	1·2	4·6
Rb (ppm)	5·0	27·0	185·0
Sr (ppm)	150·0	730·0	203·0
Ba (ppm)	70·0	530·0	650·0

Models for Genesis of the Igneous Rock types

The potassium-poor tholeiites fit a model which explains the primitive basalts as the result of about 25% melting of peridotite in the shallow mantle (<60 km). Basalts and basaltic andesites of lower $Mg/(Mg+Fe)$ ratio would result by fractional crystallization of plagioclase and clinopyroxene from the more primitive basalt magmas. Fractional crystallization apparently produced few, if any, andesitic rocks.

The dacitic volcanic rocks cannot be explained as the result of fractional crystallization of potassium-poor tholeiitic magma. The model proposed by Hanson and Goldich (1972) and Arth and Hanson (1972) is that the dacites and compositionally equivalent tonalites and trondhjemites were derived by partial melting of a tholeiitic basalt source at sufficient depth (>45 km) to produce a residue of garnet and clinopyroxene. This eclogite model fits all the observed data but is not unique. An alternate model by Goldschmidt (1922) was utilized by Hietanen (1943) to explain the gabbro–diorite–tonalite–trondhjemite suite of southwest Finland by fractional crystallization of a mildly alkaline hornblende gabbro magma. A recent trace-element and isotopic study by Arth et al. (unpub. data) reaffirms Hietanen's model and indicates that the very distinctive major- and trace-element and isotopic data described in Minnesota could also be explained by the fractional crystallization model. However, this alternate model does not explain the apparent geological absence of suitable parent and intermediate rocks in Minnesota and other Precambrian bimodal terranes. Hopefully, future work will provide clarification of this dilemma.

The model proposed to explain the quartz monzonites suggests that they are the product of 20 to 50% melting of metagraywacke that left a residue composed predominantly of plagioclase, amphibole, garnet, and pyroxene or biotite. This model implies that the quartz monzonites postdate formation of the thick volcanogenic graywacke–argillite sequences, as is observed. The processes of sedimentation were required to enrich the alkali elements in the graywacke compared with their K-poor tholeiitic and dacitic source terrane. Burial and infolding lead to metamorphism and anatexis of these graywackes. The anatectic melts would be expected to approach minimum melt compositions, which agrees with the observed compositions of the quartz monzonites. The graywackes, because of their Rb/Sr ratios, would have to be buried and melted within less than 50 m.y. of their formation in order to generate quartz monzonite melts having low initial $^{87}Sr/^{86}Sr$ ratios. This short time span is consistent with the 50-m.y. history suggested for the entire tectonothermal event.

The postkinematic stocks of syenodiorite and syenite are thought to have formed in the deep mantle by melting of mixtures of garnet peridotite and eclogite. The $Mg/(Mg+Fe)$ ratios of some of the alkalic rocks are sufficiently low to imply that they are differentiates of one of the more primitive magmas.

Implications for Crustal Formation

The most significant conclusion to be drawn from these models is that all the igneous rocks, from early basalt and dacite to postkinematic syenite, originated directly from the mantle or from rapidly recycled mantle-derived material. The event thus represents processes of upper mantle differentiation rather than remobilization of older crust. If other greenstone–granite terranes formed by similar processes, we may conclude that a significant part of our present continental crust formed from the mantle during the world-wide tectonothermal activity of 2·7 b.y. to 3·3 b.y. ago.

References

Arth, J. G., and Hanson, G. N., 1972. 'Quartz diorites derived by partial melting of eclogite or amphibolite at mantle depths', *Contr. Min. and Pet.*, **37**, 161–174.

Arth, J. G., and Hanson, G. N., 1975. 'Geochemistry and origin of the early Precambrian crust of north-eastern Minnesota', *Geochim. Cosmochim. Acta*, **39**, 325–362.

302

Goldich, S. S., Nier, A. O., Baadsgaard, H., Hoffman, J. H., and Krueger, H. W., 1961. 'The Precambrian geology and geochronology of Minnesota', *Minnesota Geol. Survey Bull.*, **41**, 193 pp.

Goldschmidt, V. M., 1922. 'Stammestypen der eruptivegesteine', *Videns-Selsk. Skrift., Kristiania, I. Mat.-Nat. Klasse*, **10**, 1–12.

Grout, F. F., Gruner, J. W., Schwartz, G. M., and Theil, G. A., 1951. 'Precambrian stratigraphy of Minnesota', *Geol. Soc. Am. Bull.*, **62**, 1017–1078.

Hanson, G. N., Goldich, S. S., Arth, J. G., and Yardley, D. H., 1971. 'Age of the early Precambrian rocks of the Saganaga Lake–Northern Light Lake area, Ontario-Minnesota', *Can. J. Earth Sci.*, **8**, 1110–1124.

Hanson, G. N., Goldich, S. S., 1972. 'Early Precambrian rocks in the Saganaga Lake–Northern Light Lake area, Minnesota-Ontario, part II: Petrogenesis', *Geol. Soc. Am. Mem.*, **135**, 179–192.

Hietanen, Anna, 1943. 'Über das Grundgebirge des Kalantigebietes im südwestlichen Finnland', *Finland Comm. Géol., Bull.*, **130**, 105 pp.; Also published as: 'Suomalainen Tiedeakatemia Toimituksia, Ser. A, III', *Geologica-Geographica*, 6.

Peterman, Z. E., Goldich, S. S., Hedge, C. E., and Yardley, D. H., 1972. 'Geochronology of the Rainy Lake region, Minnesota-Ontario', *Geol. Soc. Am. Mem.*, **135**, 193–215.

Prince, L. A., and Hanson, G. N., 1972. 'Rb–Sr isochron ages for the Giants Range Granite, northeastern Minnesota', *Geol. Soc. Am. Mem.*, **135**, 217–224.

Sims, P. K., Morey, G. B. (Eds.), 1972. *Geology of Minnesota: A Centennial Volume*, St. Paul, Minn., Minnesota Geol. Survey, 632 pp.

Metamorphic Patterns and Development of Greenstone Belts in the Eastern Yilgarn Block, Western Australia

R. A. BINNS, R. J. GUNTHORPE, and D. I. GROVES

Department of Geology, University of Western Australia, Nedlands, 6009. W.A. Australia.

Abstract

Throughout greenstone belts in the eastern Yilgarn Block, metamorphic grade varies from prehnite pumpellyite to mid-amphibolite facies in previously folded domains characterized by predominantly static recrystallization, and from mid-amphibolite facies to the amphibolite–granulite facies transition in more intensely deformed dynamic-style domains that are associated with deformed granitic intrusions. Distribution of the different metamorphic domains with respect to regional gravity anomalies, and to exposed levels of fractionated ultramafic dyke-like intrusions, suggests that metamorphic grade was governed by structural setting within broadly synformal greenstone belts. The nature of relic phases surviving in metavolcanic rocks at various metamorphic grades implies a non-progressive metamorphism imposed relatively rapidly on previously unaltered sequences. Structural and limited geochronological data indicate a single, widespread metamorphic episode at 2600–2700 m.y., probably related to remobilization of an older granitic basement and occurring during a marked change in thermal regime in the Archaean crust.

Nickel sulphide deposits of ultramafic affinity in the eastern Yilgarn Block are preferentially located in higher grade metamorphic domains.

Introduction

An understanding of metamorphic phenomena is important to most research on Archaean granite–greenstone terrains. The following discussion for the Eastern Goldfields Province of the Yilgarn Block, Western Australia arises from a broad-scale survey conducted during 1972–74 to provide background for a study of relationships between metamorphism and nickel mineralization. Although over 5000 thin sections from some 1500 data points were examined, the con-siderable area covered ($270,000 \text{ km}^2$) leaves abundant scope for refinement of detail. Results obtained so far nevertheless carry significant geotectonic implications. They indicate the lack of analogy between the evolution of eastern Yilgarn greenstone belts and modern situations, and thus stress the unique character of Archaean cratons.

Variation in Metamorphic Grade and Style

Four types of metamorphic domain have been defined throughout the province (Fig. 1)

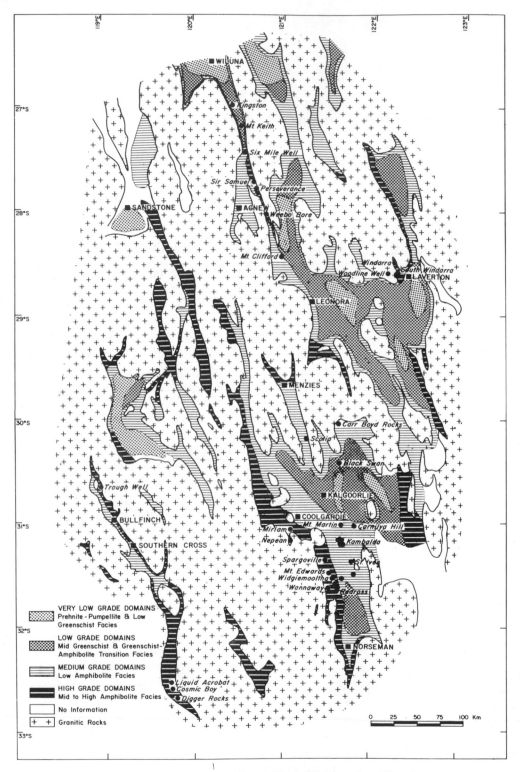

Fig. 1. Metamorphic map of the eastern Yilgarn Block, Western Australia. Contact alteration related to granitic plutons is shown only in closely sampled areas. Nickel deposits (circles) are concentrated in amphibolite facies domains. Nomenclature of domains does not correspond to the classification of Winkler (1974)

based on mineralogical criteria set out in Table 1. At map scale the domain boundaries are necessarily somewhat generalized. Most are isograds, although some are structural discontinuities. *Very low grade domains* embrace assemblages within the prehnite–pumpellyite and lower greenschist facies, commonly with incomplete reconstitution. More thorough recrystallization in *low grade domains* ranges from mid-greenschist facies (biotite zone) to the greenschist–amphibolite facies transition. *Medium grade domains* are of low- to mid-amphibolite facies, whereas in *high grade domains* the assemblages vary from mid- to high-amphibolite facies or locally to the amphibolite–granulite facies transition. The mineralogy of scarce aluminous metasediments denotes low to intermediate pressure metamorphism consistently throughout the province, andalusite being typical of moderate grades and sillimanite of the highest grade domains. Kyanite is exceptionally rare and restricted to probably anomalous structural settings, whereas cordierite is widespread and may be accompanied by almandine at elevated facies. Accessory staurolite also occurs in high grade domains. Although sodic amphiboles occur in metamorphosed iron formations of unusual geochemistry, no suggestions of blueschists, eclogites, or high pressure amphibolite–granulite zones have been detected.

Layered mafic–ultramafic sills within the greenstone succession (Williams and Hallberg, 1973) and large ultramafic dyke-like intrusions (e.g. Mount Keith, Forrestania) are also metamorphosed, although primary phases may survive. Their recrystallized assemblages are compatible with those of the supracrustal rocks they intrude.

Table 1. Metamorphic assemblages in the Eastern Goldfields Province, Yilgarn Block

Domains	Very low grade	Low grade	Medium grade	High grade	Domains	Very low grade	Low grade	Medium grade	High grade
Tholeiites					*Iron formations*				
chlorite					chlorite				
epidote–clinozoisite					minnesotaite				
actinolite					grunerite				
pale hornblende					stilpnomelane				
dark hornblende					actinolite				
stilpnomelane					hedenbergite				
clinopyroxene					eulite				
orthopyroxene					fayalite				
biotite									
Na–plagioclase					*Pelites and Semipelites*				
Ca–plagioclase					white mica				
					chlorite				
Komatiites					biotite				
chlorite					andalusite				
tremolite–actinolite					sillimanite				
olivine					cordierite				
lizardite					almandine				
antigorite					chloritoid				
clinozoisite					staurolite				
Mg–c					Na–plagioclase				
Mg–hornblende					Ca–plagioclase				
anthophyllite									
talc					*Felsic Volcanics*				
clinopyroxene					white mica				
orthopyroxene					chlorite				
spinel					biotite				
					prehnite				
Dunites					pumpellyite				
lizardite					stilpnomelane				
antigorite					epidote–clinozoisite				
talc					actinolite				
forsterite					hornblende				
chlorite					almandine				
brucite					Na–plagioclase				
pyroaurite–stichtite					Ca–plagioclase				
Cr–spinel									
anthophyllite									
enstatite									
spinel									

Quartz occurs in all assemblages except dunites and komatiites. K-feldspar occurs in felsic volcanics and metasediments, including where muscovite is unstable. Carbonate and opaque assemblages not listed. Domains are as shown on Fig. 1, but do not correspond to the nomenclature of Winkler (1974), which was not available at the time of the study.

Two contrasting styles of metamorphism are represented at the regional scale. From the very low to the medium grade domains primary structures (e.g. pillows in metabasalts) are commonly well preserved and delicate micro-textural aspects of parent rocks are retained even where mineralogical reconstitution was complete (Figs. 2A, 2B; also 5B). This has been termed *static-style* metamorphism to emphasize textural preservation and lack of mineral orientation in more competent volcanic rocks, although sequences in such areas are generally folded and steeply dipping. Mild cleavage in pyroclastic units and complex microfolding in banded iron or thin cherty horizons denote heterogeneous strain. In contrast, relatively intense deformation has produced penetrative foliations and lineations throughout high grade areas of *dynamic-style* metamorphism. In the extreme case, original textures were entirely destroyed (Fig. 2C), and former structures are extremely disturbed or no longer discernible. An exception is provided by the distinctive spinifex texture of ultramafic metavolcanic rocks, which is commonly preserved although the platelets may be deformed.

Although a broad separation into static-style or dynamic-style domains is generally possible, there are areas with characteristics that fall between the two extremes, suggesting gradational relationships. Narrow zones of penetrative deformation may occur in essentially static-style areas, and rocks exhibiting relic texture or structure may occur as boudins within dynamic-style domains.

Relationship to Granite Distribution

Most very low and low grade domains in the eastern Yilgarn are located towards the central portions of wide greenstone belts. In detail, however, grade variations within static-style domains do not correlate well with the present outcrop distribution of granitic intrusions. Superimposed contact effects have been recognized around some granitic bodies, ranging from narrow massive hornfels aureoles (hornblende and pyroxene hornfels facies) around discordant plutons to transgressive zones 1 or 2 km wide of foliated amphibolites

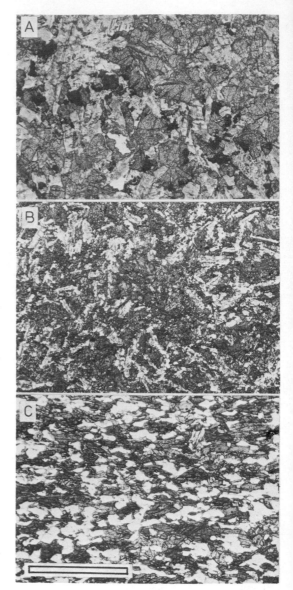

Fig. 2. Photomicrographs of tholeiitic metabasites to illustrate variation in metamorphic grade and style. A; Coarse-grained metabasalt with relic clinopyroxene and altered plagioclase from a very low grade domain. B: Metabasalt from a medium grade static-style domain; although thoroughly recrystallized to aluminous hornblende and andesine, relic texture is preserved and there is no mineral alignment. C: Thoroughly recrystallized amphibolite from a high grade dynamic-style domain, showing aligned hornblendes and granoblastic plagioclase. Scale bar: 1·0 mm

and schists around apparent forcefully emplaced intrusions. At some boundaries contact alteration is lacking or minimal.

The metamorphic pattern within by the Wiluna–Agnew greenstone belt (Fig. 3) demonstrates the difficulty in attributing large-scale grade variation in static domains to the direct thermal influence of granitic rocks. Ignoring the immediate margins of the belt, a pronounced longitudinal increase in grade occurs southwards parallel to regional structure. Significantly, the lowest grade rocks lie close to an unconformity with little-metamorphosed Proterozoic sediments of the Bangemall Basin—a situation repeated in other greenstone belts east of Wiluna (Fig. 1).

High grade dynamic-style metamorphic domains are concentrated along the flanks and at terminations of greenstone belts, or form screens between granitic intrusions (Fig. 1), although they do not occur in all such structural positions. Adjacent granitic rocks typically possess strongly aligned metamorphic fabrics, and smaller similarly deformed plutons commonly lie within the high grade domains. Coarser grained amphibolites and schists, invaded by deformed and metamorphosed aplites, may occur within a few hundred metres of the actual contact with gneissic granites, which generally show structures and textural elements implying their magmatic parentage. Migmatitic structures indicative of anatexis are conspicuously rare.

Structural Relationships between Static and Dynamic Domains

Major folds in high grade dynamic belts are tightly appressed compared with those in static environments, and more complex polyphase structures are present at the smaller scale. In a detailed structural and metamorphic study of one critical area—the southern portion of the Kalgoorlie greenstone belt between Widgiemooltha and Norseman—N. J. Archibald (pers. comm.) has established that the main phase of metamorphic recrystallization occurred synchronously in static-style and dynamic-style terrains. Projects in the Diemals–Mount Jackson greenstone belt 150 km north of Southern Cross (L. S. Andersen and others, pers. comm.), support the contemporaneity of metamorphic recrystallization in static- and dynamic-style areas and provide a key to interpreting structural and petrological observations elsewhere in the Eastern Goldfields Province. In both these areas there is evidence of syn- and post-metamorphism deformation related to the emplacement of granitic domes.

There is abundant evidence favouring the largely post-tectonic nature of the major metamorphic event in static-style terrains. For example, mineral orientation is not commonly found in sheared or cleaved zones within massive metabasalts, and the amphiboles in complexly folded metasedimentary horizons between metabasalts cut indiscriminately across axial surfaces and bedding laminae.

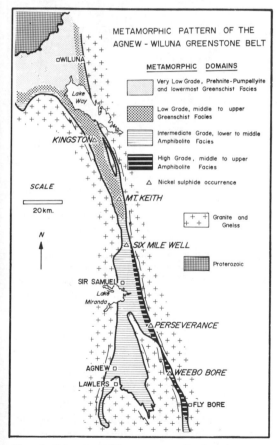

Fig. 3. Metamorphic pattern of the Agnew–Wiluna Greenstone Belt, illustrating in more detail the disposition of domains shown in Fig. 1. The major dyke-like dunitic intrusive referred to in the text extends from Weebo Bore through Perseverance, Six Mile Well, Mt. Keith, and Kingston, and passes approximately 1 km west of Wiluna township

Near boundaries between static and dynamic terrains, localized cleaved zones showing aligned fabrics in the former appear equivalent to the dominant foliation of the latter, but the actual mineral assemblages developed are typically the same as in associated undeformed rocks. Syntectonic fabrics (e.g. rolled garnets) characterize the high grade dynamic belts, but post-tectonic porphyroblast outgrowth is surprisingly common, indicating an extended history of deformation and recrystallization under uniform P–T conditions. The geometry of lineations and foliations in high grade amphibolites and schists is generally compatible with that of associated granitic intrusions. Nowhere has regional-scale superimposition of metamorphism been confirmed. For example, penetratively deformed greenschist belts have not been found cutting amphibolite facies static domains, as might be expected if there were two separate events. Retrograde metamorphism is minor in the eastern Yilgarn, being mostly restricted to narrow zones in especially prone lithologies such as olivine-rich ultramafic rocks.

Gravity Anomalies

A significant correspondence exists between the low and very low grade domains of Fig. 1 and the substantial regional gravity highs relative to background over granitic areas shown in Fig. 4 (from Fraser, 1974). With minor exceptions south of Southern Cross, possibly related to Proterozoic "east–west" dykes and accentuated by a regional gradient increasing westwards, no comparable Bouguer anomalies are associated with the medium and high grade domains. The correlation implies that thick sequences of folded supracrustal rocks underlie lower grade static-style areas, but that such sequences are comparatively thin beneath amphibolite facies domains. The reconnaissance data superficially indicate broadly synformal greenstone belts in which metamorphic grade is related to structural level above a granitic substratum.

Relationships between Metamorphism and Stratigraphy

Regional stratigraphic mapping is difficult in

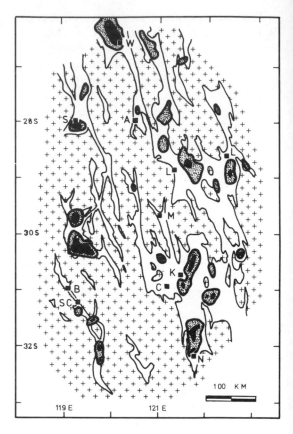

Fig. 4. Bouguer anomalies in the Eastern Goldfields Province, after Fraser (1974). High values generally correspond to lower grade metamorphic domains on Fig. 1. Stippled areas, −30 to −20 mgal; black, above −20 mgal. Bouguer values over granitic and high grade metamorphic areas are typically in the −60 to −40 mgal range (A. R. Fraser (in press) Reconnaissance helicopter gravity survey of Southwest of Western Australia, *Bur. Miner. Resear. Aust. Rep.*)

the eastern Yilgarn due to structural complexity (particularly strike faulting), poor exposure, and disruption by granitic plutons. However, where broad stratigraphic appraisals have been attempted, mainly near Kalgoorlie (Williams, 1970; Gemuts and Theron, 1975) but also in current work on the Diemals–Mount Jackson belt, the sequences recognized appear to transgress domain boundaries and to extend from static into dynamic-style terrains. There are indications that lower grade domains may be at least in part preferentially occupied by higher members of the Yilgarn succession. For example metamorphosed felsic volcanogenic rocks and the comparatively young (R. J. Marston, pers. comm.) conglomerate–coarse clastic sequences with appreciable granitic detritus known at various localities (Glikson, 1971; McCall et al., 1970; Durney, 1972) are rare in high grade dynamic-style belts. Any relationship between metamorphic grade and stratigraphic level, however, must be only a very broad one in view of the general evidence for deformation prior to recrystallization.

A major discontinuous dyke-like ultramafic intrusion extends for about 200 km along the length of the Wiluna–Agnew greenstone belt (Fig. 3). It postdates early deformation phases, but its degree of fractionation, as assessed from relic igneous phases and geochemistry of altered rocks, parallels the metamorphic grade of its environment. In the southern amphibolite facies domains, it was composed exclusively of dunite. Pyroxenitic differentiates were developed marginal to the dunite in the greenschist facies domain, whereas a lateral sequence from dunite through orthopyroxenite to gabbro with minor granophyre occurred in the northern prehnite–pumpellyite facies domain. The preferred interpretation is that successively higher levels of emplacement of the dyke are exposed from south to north, supporting the gravity-based model above of a relationship between metamorphic grade and structural level within a synformal greenstone belt. Erosion after a longitudinal tilting of only a few degrees, perhaps caused by isostatic compensation for Proterozoic sedimentation in the Bangemall Basin, may be sufficient to explain the presently exposed metamorphic pattern in the belt.

Distribution of Relic Phases

In volcanic and subvolcanic peridotitic komatiites from the two lower grade domains, olivine microphenocrysts and blades within spinifex-structured ultramafic rocks have been replaced by lizardite, antigorite or chlorite, but relic quench-textured clinopyroxene crystals are commonly well preserved in a chloritized glass (Fig. 5A). In equivalent rocks from medium grade domains, however, olivine phenocrysts or blades remain unaltered, whereas the groundmass clinopyroxene and glass have reacted to produce amphibole–chlorite aggregates with a distinct palimpsest texture (Fig. 5B). The relic olivines are brownish-coloured and pleochroic, with abundant oriented opaque inclusions evidently expelled during non-hydrous metamorphic reheating (cf. Champness, 1970; Pitt and

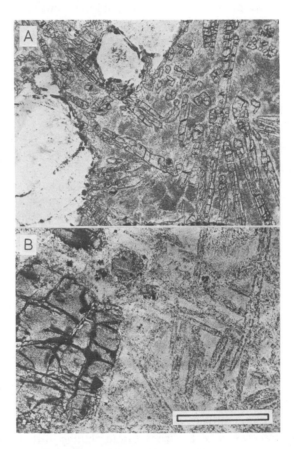

Fig. 5. Photomicrographs of peridotitic komatiites to illustrate varying survival of relic igneous phases at different metamorphic grade. A: Example from a prehnite–pumpellyite facies environment; the olivine microphenocrysts are entirely serpentinized, while fresh quench clinopyroxene is preserved in a chloritized glassy groundmass. B: An equivalent rock from the mid-amphibolite facies; the olivine phenocrysts remain fresh, although brownish-coloured and clouded by minute inclusions, while the groundmass has transformed to an actinolite–chlorite aggregate in which the former quench texture is still preserved. Scale bar: 0·5 mm

Tozer, 1970). They may be distinguished from colourless metamorphic olivines which grew as porphyroblasts transgressing original fabrics (Oliver et al., 1972). Similarly, tholeiitic metabasalts from the lower grade domains contain relic clinopyroxene (Fig. 2A), whereas plagioclase laths are thoroughly altered to pseudomorphs of albite, clinozoisite, chlorite, white mica, hydrogarnet, and stilpnomelane. In analogues from medium grade domains the pyroxene is replaced by hornblende, but relic plagioclase laths may retain labradorite cores that are commonly brownish and clouded by tiny inclusions (cf. MacGregor, 1931). Such relationships are observed too frequently throughout the eastern Yilgarn to be fortuitous, and despite complications of reaction kinetics they imply that volcanic rocks in higher grade static-style domains did not pass through an earlier phase of alteration equivalent to that in the lower grade domains. Thus the regional-scale static metamorphism was not progressive but rather the main phase of recrystallization did not occur until peak metamorphic temperatures had been established within formerly little-altered supracrustal rocks. In medium grade domains, the olivines of komatiites and the calcic plagioclases of tholeiites were inherently stable, and became consumed only to the extent necessary to balance other metamorphic reactions.

The characteristics of ultramafic dyke-like intrusions indicate that similar conditions applied in the high grade domains of dynamic-style. Under low grade conditions dunites have generally been altered on a constant-volume basis to serpentinites with relic cumulate textures. The consequent change in their Mg/Si ratio requires that such previously serpentinized dunites should recrystallize to bladed forsterite–talc rocks by dehydration at higher temperature (cf. Evans and Trommsdorff, 1974). Such products do occur in some medium grade static domains and also in high grade dynamic-style domains, but in the latter there is also an abundance of olivine-rich tectonites in which former igneous olivine has transformed directly into metamorphic olivine without undergoing prior serpentinization.

Metamorphic Evolution of East Yilgarn Greenstone Belts

The widespread static metamorphism ranging from prehnite–pumpellyite to mid-amphibolite facies was imposed throughout the Eastern Goldfields Province on previously deformed, steeply dipping greenstone sequences that were probably disrupted and structurally thickened by strike faults. This, together with the distribution of relic igneous minerals and the lack of close correlation between grade and stratigraphy, precludes comparison with Phanerozoic burial metamorphic environments or modern ocean-floor metamorphism (Miyashiro et al., 1971). The overall metamorphic pattern cannot have developed by the direct thermal influence of granitic intrusions in their present setting. Regional scale metamorphism in higher grade dynamic-style belts appears to have been contemporaneous with that in adjacent static domains.

A coincidence between gravity highs and lower grade domains, and the relationship in the Wiluna–Agnew belt between degree of fractionation in a major ultramafic dyke-like intrusion and its metamorphic characteristics, suggest a broad but definite dependence of grade on structural position within folded yet broadly synformal greenstone belts. Highest temperatures were attained in elongate zones also subjected to the most intense polyphase deformation. These may denote former deeply buried settings now uplifted by diapirism, where metamorphism was intimately related to uprise of granitic intrusions that were also deformed and recrystallized but do not themselves appear the sole source of heat and stress necessary to create those zones. The depth represented by these high grade, dynamic-style domains is restricted, however, by the predominance of andalusite–sillimanite parageneses in metapelites.

Geochronological studies to test the contemporaneity of static and dynamic metamorphism are in progress on the Diemals–Mount Jackson greenstone belt (with M. J. Gorton and J. R. de Laeter). Preliminary results of this study and information obtained elsewhere in the east Yilgarn (Turek, 1966; W. Compston, J. C. Roddick V. M. Oversby, pers. comm.) fail to discriminate effectively between metamorphic and granite ages. The ubiquity of 2600–2700 m.y. radiometric dates with no indication of zonation throughout the Eastern Goldfields Province thus combines with evidence presented above to suggest that the large-scale metamorphic patterns of Fig. 1 arose during the one very widespread event which was linked with the generation or remobilization of granitic bodies.

Gravity evidence alone seems to require a granitic substratum to greenstone belts during the metamorphic episode, and a better understanding of the extensive granite regions between belts is obviously the next step in deciphering the geotectonic significance of eastern Yilgarn metamorphic patterns. Studies carried out so far, particularly by L. F. Bettenay, indicate a complex variation in granitic petrography and structure, with biotite adamellites and granodiorites being predominant. Of particular importance is the occurrence of enclaves of complexly deformed and partly remobilized banded gneisses in zones of deformed granitic rocks marginal to dynamic-style metamorphic belts. These occur to the west of Norseman–Widgiemooltha (N. J. Archibald, pers. comm.), south-east of Southern Cross and to the north of Coolgardie (L. F. Bettenay, pers. comm.) and also in the central and northern parts of the province. Detailed structural, trace element and isotopic studies have yet to be carried out, but the inference is strong that the banded gneisses represent a former sialic basement to the greenstone belts, much remobilized during the 2600–2700 m.y. metamorphic episode.

The suggestions of Hallberg and Williams (1972) and Glikson (1972) that komatiites and pillowed tholeiitic basalts in east Yilgarn greenstone belts were formed at spreading centres in deep Archaean ocean basins is contradicted by field relationships. These lithologies are intimately associated with banded iron formation sequences whose internal characteristics and lateral extent suggest a similar origin to the shallow shelf-iron formations of the Proterozoic Hamersley Range Province (Trendall and Blockley, 1970), and also with considerable thicknesses of felsic volcanics and their pyroclastic or conglomeratic associates (O'Bierne, 1968). The occurrence of spinifex-textured ultramafic fragments in felsic volcanogenic conglomerates and of komatiitic intrusions in banded iron formations appears to rule out the possibility of tectonic justaposition of these associations. The lack of blueschist facies metamorphic rocks throughout the province has already been noted. It appears, therefore, that field data currently available in the east Yilgarn do not support any applications of plate tectonic models, but rather that the greenstone succession commenced forming as an essentially concordant volcanogenic sequence within large shallow marine basins on a granitic basement. It is difficult to estimate the extent of these basins because lack of complete understanding of the metamorphism and deformation have so far prevented meaningful stratigraphic mapping at the regional scale. However, strong suggestions of continuity exist for sequences containing abundant banded iron formations that extend longitudinally for at least 600 km from south of Southern Cross to north of Sandstone. These have a present width (including intervening granite) of at least 200 km in the Mount Jackson area, and this would presumably have been considerably foreshortened by folding. Tectonism evidently commenced before basin infilling ceased, as is indicated by the unconformable conglomeratic units with granitic, greenstone, and banded iron formation clasts, referred to above and which also suffered the regional-scale static metamorphism.

The true age of vulcanism and sedimentation represented by Yilgarn greenstone sequences, and the rapidity or otherwise of their burial and the onset of early tectonism, remain unknown. Despite high temperatures at which many lavas were patently erupted (Pyke et al., 1973; Green 1975) and thicknesses likely to have exceeded 10 km, the important evidence provided by distribution of relic igneous minerals indicates that these deformed sequences remained relatively unaltered, that is in a low heat flow regime, until 2600–2700 m.y. The thermal setting of the Archaean crust then appears to have suddenly reversed, its formerly stable lower sialic layer becoming reactivated and the overlying greenstones deformed and metamorphosed over a huge area. Whether the distinction between coeval dynamic and static metamorphic terrains reflects greater rigidity (at lower temperature) of the latter within a uniform stress field imposed over entire greenstone belts, or localization of both anomalous heat flow and stress along the former, cannot be resolved at present and is a topic deserving further enquiry. However, the striking persistence of dynamic-style metamorphism along a 600 km NNW-trending zone extending from Norseman past Coolgardie to Sandstone (Fig. 1) hints at a relationship with fundamental lineaments in the Archaean lower crust or mantle.

The broad metamorphic patterns described above provide a regional context within which continuing and more detailed studies should yield improved understanding of the Yilgarn Block. The literature referring directly or indirectly to metamorphic and structural aspects of other Archaean cratons suggests that similar overall patterns may apply in granite–greenstone terrains of varying age. Perhaps the most important consequence of our investigation is the strong indication of a relatively rapid reversal in heat flow regimes in the Yilgarn crust at about 2600–2700 m.y., and evidence for comparable events might be sought elsewhere. The predominance of volcanic rocks ranging from ultramafic to felsic in composition within the greenstone succession undoubtedly requires thermal disturbances

deep in the mantle at an early stage, and perhaps movement upwards of these to subcrustal levels explains the sudden change, but further speculation of this nature seems unwarranted until more data have been obtained. For the Yilgarn Block, at least, comparative youth and the repetitive association of komatiites with other rock types appear to render relationships with lunar maria cratering events unlikely (cf. Green, 1972).

Metamorphism and Mineralization

Economic and significant subeconomic iron–nickel sulphide deposits associated with ultramafic rocks in the east Yilgarn are virtually confined to amphibolite facies domains (Fig. 1). Numerically, most lie in the higher grade dynamic-style environments (e.g. Perseverance, Windarra, Nepean, Redross, Cosmic Boy), but the Kambalda orebodies are a notable exception in this respect. Although this relationship may arise from broad-scale stratigraphic controls with the higher grade domains representing deeper ultramafic-dominated sequences, the structural and textural characteristics and geochemistry of ores indicate substantial metamorphic modification of former magmatic sulphides. Sulphurization of ultramafic rocks during serpentinization, upgrading of sulphide content by volume changes, and mechanical segregation or strain-induced diffusion during metamorphism are important factors governing the viability of many deposits.

Genetic studies of Yilgarn gold mineralization within the context of metamorphism represent an important field for further study, and may have valuable exploration implications.

Acknowledgements

This study forms part of a project sponsored by the Australian Research Grants Committee, the Australian Mineral Industries Research Association, and the University of Western Australia Research Committee. We thank our colleagues and students whose advice and work has contributed importantly to the investigation, particularly N. J. Archibald and F. M. Barrett and also L. S. Anderson, L. F. Bettenay, M. J. Gole, M. J. Gorton, R. J. Marston and K. G. McQueen, and are grateful to the many mining companies and their staffs, and to officers of C.S.I.R.O. and the Geological

Survey of Western Australia, who provided much valuable material and field information.

Dr G. A. Chinner and Professor H. G. F. Winkler kindly reviewed the manuscript.

References

Champness, P. E., 1970. 'Nucleation and growth of iron oxide in olivines (Mg, Fe)$_2$SiO$_4$', *Mineralog. Mag.*, **37**, 790–800.

Durney, D. W., 1972. 'A major unconformity in the Archaean, Jones Creek, Western Australia', *J. Geol. Soc. Aust.*, **19**, 251–260.

Evans, B. W., and Trommsdorff, V., 1974. 'On elongate olivine of metamorphic origin', *Geology*, **1**, 131–132.

Fraser, A. R., 1974. 'Reconnaissance helicopter gravity survey of the south-west of Western Australia, 1969', *Bur. Min. Res. Rec.*, 1974/26.

Gemuts, I., and Theron, A. C., 1975. 'The Archaean between Coolgardie and Norseman—stratigraphy and mineralisation', *Geology of Australian Ore Deposits*, **3**, Australasian Institute of Mining and Metallurgy (in press).

Glikson, A. Y., 1971. 'Archaean geosynclinal sedimentation near Kalgoorlie, Western Australia', *Geol. Soc. Aust. Spec. Publ.*, **3**, 443–460.

Glikson, A. Y., 1972. 'Early Precambrian evidence of a primitive ocean crust and island nuclei of sodic granite', *Geol. Soc. Am. Bull.*, **83**, 3323-3344.

Green, D. H., 1972. 'Archaean greenstone belts may include terrestrial equivalents of lunar maria', *Earth Planet. Sci. Lett.*, **15**, 263–270.

Green, D. H., 1975. 'Genesis of Archaean peridotitic magmas and constraints on Archaean geothermal gradients and tectonics', *Geology*, **3**, 15–18.

Hallberg, J. A., and Williams, D. A. C., 1972. 'Archaean mafic and ultramafic rock associations in the Eastern Goldfields region, Western Australia', *Earth Planet. Sci. Lett.*, **15**, 191–200.

MacGregor, A. G., 1931. 'Clouded feldspars and thermal metamorphism', *Mineralog. Mag.*, **22**, 524–538.

McCall, G. J. H., Braybrooke, J. C., and Middleton, D. D., 1970. 'The Merougil Creek sub-area', *J. Proc. R. Soc. West. Aust.*, **53**, 9–32.

Miyashiro, A., Shido, F., and Ewing, M., 1971. 'Metamorphism in the Mid-Atlantic Ridge near 24° and 30°N', *Phil. Trans. Roy. Soc. Lond.*, **A268**, 589–603.

O'Beirne, W. R., 1968. 'The acid porphyries and porphyroid rocks of the Kalgoorlie area', *Ph.D. Thesis*, Univ. of Western Australia (unpublished).

Oliver, R. L., Nesbitt, R. W., Hausen, D. M., and Franzen, N., 1972. 'Metamorphic olivine in ultramafic rocks from Western Australia', *Contr. Mineral. Petrol.*, **36**, 335–342.

Pitt, G. D., and Tozer, D. C., 1970. 'Optical absorption measurements on natural and synthetic ferromagnesian minerals subjected to high pressures', *Phys. Earth Planet. Interiors*, **2**, 179–188.

Pyke, D. R., Naldrett, A. J., and Eckstrand, O. R., 1973. 'Archaean ultramafic flows in Munro Township, Ontario', *Bull. Geol. Soc. Am.*, **84**, 955–978.

Trendall, A. F., and Blockley, J. G., 1970. 'The iron formations of the Precambrian Hamersley Group, Western Australia', *Geol. Surv. W. A. Bull.*, **119**.

Turek, A., 1966. 'Rubidium–strontium isotope studies in the Kalgoorlie-Norseman area, Western Australia', *Ph.D. Thesis*, Aust. Nat. Univ. (unpublished).

Williams, D. A. C., and Hallberg, J. A., 1973. 'Archaean layered intrusions in the Eastern Goldfields Region, Western Australia', *Contr. Mineral. Petrol.*, **38**, 45–70.

Williams, I. R., 1970. 'Explanatory notes, Kurnalpi 1 : 250,000 Geological Series', *Geol. Surv. W. A.*

Winkler, H. G. F., 1974. *Petrogenesis of Metamorphic Rocks*, 3rd Edn., Springer–Verlag, New York, Berlin, 320 pp.

Tectonic Relations between
High- and Low-Grade Regions

Shallow and Deep-level Exposures of Archaean Crust in India and Africa

R. M. SHACKLETON

Department of Earth Sciences, The University, Leeds, U.K.

The relation between low-grade and high-grade Archaean terrains has often been discussed. Many workers have interpreted particular high-grade terrains as older than their surroundings (Hepworth, 1967) and Windley and Bridgwater (1971) concluded that they are generally older. However, the granulites of the Wheat Belt in Western Australia have been regarded as coeval with the Kalgoorlie greenstones (Glikson and Lambert, 1973) and the granulite metamorphism in Southern India has been regarded as younger than the low-grade metamorphism of the Dharwars (Crawford, 1974). No doubt each relationship occurs somewhere and the evidence must be independently assessed in each case.

In the well-exposed Mysore region of Southern India and in the Limpopo belt and adjacent Rhodesian and Kaapvaal cratons in southern Africa, greenschist and granulite facies rocks occur close together and their relationships can be studied.

Three models can be proposed to explain the relationships in these areas. In the first, the granulites are older than any of the associated rocks and are exposed as a result of upfolding in a younger orogenic belt. In a second model, the granulites are the result of metamorphism in an orogenic belt (Limpopo and Eastern Ghats belts) which is younger than the adjacent lower-grade greenstone belts. In accord with this model, a sedimentary series (Messina Formation) in the Limpopo belt is supposed to be confined to the orogenic belt and younger than the nearby greenstone belt successions. In a third model, the granulite terrains represent deeper (sub-Conrad) levels of extensive orogenic domains which also include the greenstone belts. In this case the granulites would continue beneath the Archaean cratons, the structures in the high-grade belts would be continuous with, rather than superimposed on, those on either side and the supracrustal rocks within the high-grade terrains would not be younger than those in the adjacent greenstone belts but either the same age or older. The evidence regarding these age relationships, in Mysore and the Limpopo belt, will be reviewed. It is thought to support the third model.

Between the greenschist and the granulite facies rocks, rocks in amphibolite facies always intervene, both in southern India (Sargur and Kolar greenstone belts) and in the Limpopo region, producing a broad metamorphic zonation (Pichamuthu, 1967). The greenstone belts nearest to the Limpopo granulites (Murchison, Pietersburg and Sutherland ranges to the south, Mweza to the north) are in the amphibolite facies and those in the greenschist facies are farther away (Barberton, Que Que, Bulawayo). Superimposed on this regional zoning is a local zoning: the marginal parts of the greenstone belts, adjacent to the

317

surrounding granitic rocks are in the amphibolite facies. In Mysore a regional northerly plunge is correlated with the southward rise in grade (Pichamuthu, 1961).

The metamorphic sequence appears to be continuous. Metamorphic discontinuities at unconformities occur, for example at Gwelo, Rhodesia (Amm, 1946) and in Mysore (M. Ramakrishnan, pers. comm.) with amphibolite facies rocks below and greenschist facies rocks above; in parts of southern Mysore (Pichamuthu, 1961, 1962) and along the northern side of the Limpopo belt (James, 1975), granulite–facies terrains are bounded by mylonite zones. Elsewhere, however, there are gradations, from greenschist to amphibolite facies in many of the African greenstone belts, from amphibolite facies to granulite facies in the northern Transvaal (Graham, 1973) and in the West Nicholson–Beitbridge area of Rhodesia (Robertson, 1968). In Mysore the southern parts of the Sargur and Kolar schist belts are approaching granulite facies (Pichamuthu, 1962). Usually the granulite facies terrains are surrounded by migmatitic gneisses which separate them from the well-defined greenstone belts. In southern Mysore, swarms of cross-cutting metadolerite dykes show clouding of the feldspars within the granulite–facies terrains and for a short distance into the amphibolite facies terrains. This clouding is attributed to a regional thermal metamorphism of granulite facies (Pichamuthu, 1959) later than the main granulite metamorphism. Its distribution reinforces the argument for continuity from the granulites into the amphibolite facies terrain. Although the simplest interpretation of the apparently continuous metamorphic zonation is that it is all of one age, a similar distribution would be produced if a high-grade metamorphism were locally superimposed on an older lower-grade metamorphism.

Geochronological evidence concerning the ages of the granulite, amphibolite and greenschist facies metamorphisms is inconclusive. In Rhodesia, Bulawayan volcanic rocks were erupted about 2600 m.y. ago (Hawkesworth et al., 1975) and were metamorphosed (in greenschist or amphibolite facies) before 2550 m.y. (the age of the Great Dyke) while in the Limpopo belt, the syntectonic Bulai and Singelele granites, supposedly associated with the granulite–facies metamorphism, are dated at 2690 ± 60 m.y. (Van Breeman and Dodson, 1972). The granulites are cut by the southern satellite intrusions of the Great Dyke (Robertson, 1968). The granulite metamorphism thus apparently occurred about 2690 m.y. ago. the data suggests that the granulite metamorphism is a little older than the greenschist metamorphism of the Bulawayan but it is doubtful whether the difference is significant. In Mysore, greenschist metamorphism of late Dharwar pillow lavas in the Chitradurga area probably occurred about 2450 m.y. ago, this being the age of the late-tectonic Chitradurga granite which intrudes the volcanics (Crawford, 1969). Granulite–facies metamorphism post-dates the Peninsular gneiss (Pichamutha, 1961) which is dated at 2585 ± 35 m.y. (Crawford, 1969). The relation of the granulite–facies metamorphism to the late-tectonic Closepet granite dated at $2380 \pm$ m.y. (Crawford, 1969) is not clear. The Nilgiri charnockites and gneiss were dated at 2615 ± 80 m.y. and the Madras charnockites at $2580 + 95$ m.y. (Crawford, 1969). Crawford (1974) suggests, mainly on eivdence from Ceylon, that the granulite–facies metamorphism in Southern India occurred about 2000 m.y. ago and that the charnockite ages are those of the rocks that were metamorphosed, not the metamorphism. There is no definite evidence from Mysore to support this view: the data, as in the Limpopo belt, do not demonstrate any significant difference between the dates of the greenschist and granulite–facies metamorphisms.

Structural evidence, both in Mysore and in the Limpopo belt appears at first sight to suggest that the granulites occur in belts in which the structures truncate those in the lower grade rocks. South and southeast of Mysore City the north–south Dharwar trend is truncated by east–west trending structures in the charnockites of the Eastern Ghats belt (Pichamuthu, 1962) and this truncation is suggested on a regional scale by maps showing trends (Pichamuthu, 1962, Crawford, 1974;

1 : 2 million tectonic map of India, 1963). But these maps are generalized and when the published data are plotted, and if trends in Ceylon are also taken into account, the truncation is much less convincing. Similarly, in the Limpopo belt, there are tectonic units trending east–west and east–west structures truncate or are superimposed on structures trending more nearly north–south. In the western Transvaal structures in the Archaean Kraaipan Series, which maintain a regular meridional trend for 200 km, are replaced, apparently by a superimposed set of east–west folds, at the southern margin of the Limpopo belt, at the Malopo River in south-east Botswana. But elsewhere, the main planar fabric of the Limpopo belt can be traced north of Bulawayo, far into the Rhodesian craton, and the deformation at least within the southern part of the Rhodesian craton is of similar age to that in the Limpopo belt (Coward and James, 1974). This main deformation is attached to the granulite–facies metamorphism in the Limpopo belt and to the greenschist metamorphism in the Fort Victoria and other greenstone belts. Later deformations are concentrated in the Limpopo belt, producing major shear zones. It is these later deformations which give the impression that the Limpopo belt is younger than the cratons. The deformation in the Limpopo belt is much more intense, and the outcrop patterns due to superimposed phases of folding are much more complex than in the greenstone belts, but the differences seem to be of intensity rather than age.

The relative ages of the supracrustal rocks in the high-grade and low-grade terrains is uncertain, both in Mysore and in the Limpopo belt. In Mysore it can be clearly seen that successive groups in the Dharwar sequence, each including thick tholeiitic basalts whose major and trace element chemistry is attributed to an oceanic type of continental crust (Naqvi and Hussain, 1973), rest uncomformably on older acid gneisses. Moreover, there are shallow-water current-bedded orthoquartzites within the tholeiitic volcanics. It is clear that these greenstone belts accumulated on continental crust and that at least for some time during their accumulation the sea remained shallow. Mixtites near the base of the successions suggest submarine sliding into rapidly subsiding basins. Owing to the general northerly plunge, the younger Dharwar formations outcrop extensively in the north and the older ones mainly in the south. Tholeiitic lavas form a major part of the Dharwar succession, which also includes orthoquartzites, limestones, a manganiferous horizon, banded ironstones, local conglomerates, and pelites and greywackes. The orthoquartzites and limestones occur in the lower part of the sequence, below the manganiferous marker horizon (Pichamuthu, 1974), and the pelites mainly in the upper part. In the high-grade terrains of southern India, recognizable supracrustal rocks include quartzites (often with grunerite), manganiferous granulites (gondites) and garnet–sillimanite schists with graphite, quartz and orthoclase (khondalites). All these have possible equivalents in the Dharwars but the proportions are quite different, with more quartzites and limestones in the high-grade terrains and no recognized equivalent of the thick tholeiitic volcanics; these might be represented by mafic charnockites but it is doubtful whether such rocks occur in proportions comparable with basic volcanics in the Dharwars. A further difference between the low-grade and high-grade terrains is that in the latter there are many layered complexes (Subramanian, 1956; De, 1966) which have no analogues in the Dharwar terrains. It may be concluded from these differences, and from the northerly plunge, that the high-grade supracrustal rocks in southern India are not Dharwars but are older.

In the high-grade domains in the Limpopo belt there is a group of metasediments known as the Messina Formation which consists of thick orthoquartzites (often fuchsitic), marbles, pelites and banded ironstones. As in southern India, there seems to be an absence of basic rocks which could represent the thick basic volcanic sequences of the greenstone belts. Most of the basic rocks in the Limpopo belt are parts of layered igneous complexes, predominantly anorthositic, intruded in the Messina Formation.

Clues to the relationship between the

Messina Formation and the greenstone belt successions may be found in the Murchison Range in the Eastern Transvaal, in the Belingwe greenstone belt of Rhodesia and in the Matsistama and Tati schist belts of Botswana. In the Murchison Range the lowest exposed unit appears to be the Rooiwater layered complex, parts of which closely resemble the layered anorthosite complexes in the Limpopo belt. This complex is overlain by thick southward-facing series of metasediments, including pelites, dolomitic sediments, banded ironstones and quartzites; south of these, as though overlying them, are southward-facing pillow lavas. The Rooiwater layered complex and the overlying sediments appear to correspond to the anorthositic layered complex and Messina Formation in the Limpopo belt.

In the Belingwe greenstone belt in Rhodesia, early Archaean gneissic basement is overlain unconformably by a sedimentary series which includes quartzite, siltstones and limestone; these sediments are overlain by basic volcanic rocks (Bickle et al., in press).

In the Matsitama schist belt in Botswana the stratigraphic sequence is uncertain but there are greenschists which appear to represent basic volcanic rocks and metasediments including limestone, current-bedded ortho-quartzite, a conglomerate and pelites. To the east and south-east of the Matsitama belt, limestones and quartzites of the Messina Formation lie structurally below both the Matsitama belt and the Tati greenstone belt to the north-east. In the Tati belt, pillow shapes and current bedding show that the succession faces away from the structurally underlying Messina Formation. Near Messina there are gneisses whose complex history is shown by the presence of intensely deformed and migmatized basic dykes which cut an earlier foliation. These gneisses may represent the basement of the Messina Formation (Bahnemann, 1972). If so, they could be comparable with the ancient basement gneisses of the Shabani and Belingwe areas (Hickman, 1974; Hawkesworth et al., 1975).

The evidence from the Murchison Range, the Shabani and Belingwe areas and eastern Botswana suggests that the Messina Formation represents the sediments which overlie the gneissic basement and underlie the Bulawayan and equivalent volcanic sequences of the greenstone belts. Thus it would be analogous to the shallow-water Permo–Triassic formations of the Alps, which rest unconformably on pre-Permian basement and are overlain by deeper-water schistes lustrées.

The metamorphic, geochronological, structural and stratigraphic evidence suggests that the high-grade and low-grade terrains in the Limpopo belt and Mysore represent deep and shallow levels of the same tectonic units. There is metamorphic and structural continuity from one to the other. Exposure of the high-grade rocks implies erosion of a great thickness, perhaps 20 km or more, of overlying rocks. This must mean that there were high mountains, underlain by deep roots, which would lead to isostatic uplift as the upper levels were eroded away. The high-grade terrains thus represent zones of drastic crustal thickening as implied also by the very intense deformation which they show.

It is concluded that it is the third of the suggested models which applies; the granulite terrains represent deeper (sub-Conrad) levels of extensive orogenic domains which also include the greenstone belts. The granulites should continue beneath the Archaean cratons. The structures in the high-grade rocks are continuous with those on either side. The supracrustal rocks in the high-grade terrains are not younger and are probably older, than those in the adjacent greenstone belts.

References

Amm, F. L., 1946. 'The Geology of the Lower Gwelo Gold Belt', *Bull. Geol. Surv. Sth. Rhod.*, **37**.

Bahnemann, K. P., 1972. 'A review of the structure, the stratigraphy and the metamorphism of the basement rocks of the Messina District, Northern Transvaal', *D.Sc. Thesis*, Univ. Pretoria.

Bickle, M. J., Martin, A., and Nisbet, E. G., 'Basaltic and peridotitic komatiites and stromatolites above a basal unconformity in the Belingwe greenstone belt, Rhodesia', in press, *Geology*.

Coward, M. P., and James, P. R., 1974. 'The deformation patterns of two Archaean greenstone belts in Rhodesia and Botswana', *Precambrian Research*, **1**, 235–258.

Crawford, A. R., 1969. 'Reconnaissance Rb–Sr dating of the Precambrian rocks of southern Peninsular India', *J. geol. Soc., India*, **10**, 117–166.

Crawford, A. R., 1974. 'Indo-Antarctica, Gondwanaland, and the distortion of a granulite belt', *Tectonophysics*, **22**, 141–157.

De, A., 1966. 'Anorthosites of the Eastern Ghats, India', In Y. W. Isachsen (Ed.), *Origin of Anorthosite and Related Rocks*, Mem. 18, New York State Museum and Science Service. State Educ. Dept. Albany, New York, pp. 425–434.

Glikson, A. Y., and Lambert, I. B., 1973. 'Relations in space and time between major Precambrian shield units: an interpretation of Western Australian data', *Earth Planet. Sci. Lett.*, **20**, 395–403.

Graham, R. H., 1973. 'The southern margin of the Limpopo mobile belt, Transvaal South Africa', *17th Ann. Rep. Res. Inst. Afr. Geol., Univ. Leeds*.

Hawkesworth, C. J., Moorbath, S., O'nions, R. K., and Wilson, J. F., 1975. 'Age relationships between greenstone belts and "granites" in the Rhodesian Archaean craton', *Earth Planet. Sci. Lett.*, **25**, 251–262.

Hepworth, J. V., 1967. 'The photogeological recognition of ancient orogenic belts in Africa', *Q. Jl. Geol. Soc. Lond.*, **123**, 253–292.

Hickman, M., 1974. '3500 m.y. old granite in southern Africa', *Nature*, **251**, 296.

James, P. R., 1975. 'A deformation study across the northern margin of the Limpopo Belt, Rhodesia', *Ph.D. Thesis*, Univ. Leeds.

Naqvi, S. M., and Hussain, S. M., 1973. 'Relation between trace and major element composition of the Chitaldrug metabasalts, Mysore, India and the Archaean mantle', *Chemical Geol.*, **11**, 17–30.

Pichamuthu, C. A., 1959. 'The significance of clouded plagioclase in the basic dykes of Mysore State, India', *J. geol. soc. India*, **1**, 68–79.

Pichamuthu, C. A., 1961. 'Tectonics of Mysore State', *Proc. Indian Acad. Sci.*, **53**, 135–139.

Pichamuthu, C. A., 1962. 'Some observations on the structure, metamorphism and geological evolution of Peninsular India', *J. geol. Soc. India*, **3**, 106–118.

Pichamuthu, C. A., 1967. 'The Precambrian of India', In K. Rankama (Ed.), *The Precambrian*, Vol. 3, Interscience, New York, 1–96.

Pichamuthu, C. A., 1974. 'The Dharwar Craton', *J. geol. Soc. India*, **15**, 339–346.

Robertson, I. D. M., 1968. 'Granulite metamorphism of the basement complex in the Limpopo metamorphic zone', In *Symposium on the Rhodesian Basement*, D. J. L. Vissar (Ed.), *Trans. Geol. Soc. S. Afr.* (Annexure), **71**, 125–133.

Subramanian, A. P., 1956. 'Mineralogy and Petrology of the Sittampundi Complex, Salem district, Madras State, India', *Bull. Geol. Soc. Am.*, **67**, 317–390.

Van Breeman, O., and Dodson, M. H., 1972. 'Metamorphic chronology of the Limpopo Belt, Southern Africa', *Geol. Soc. Am. Bull.*, **83**, 2005–2018.

Windley, B. F., and Bridgwater, D., 1971. 'The Evolution of Archaean low and high-grade terrains', *Geol. Soc. Australia Spec. Pub. No. 3*, 33–46.

The Pre-cleavage Deformation of the Sediments and Gneisses of the Northern Part of the Limpopo Belt

M. P. Coward, B. C. Lintern, and L. I. Wright

Dept. of Earth Sciences, The University, Leeds, England.

Abstract

A strong penetrative fabric can be traced from the gneisses of central Botswana into the granites and greenstones of northern Botswana and south-west Rhodesia, but probably the most important deformation events pre-date this tectonic foliation. The greenstone belts were intruded and deformed into large nappes. The Matsitama schist belt in northern Botswana forms a 15 km thick pile of imbricated sediments.

The Matsitama sediments have been correlated with the metasediments of the Limpopo belt to the south where they show evidence of high-grade metamorphism and are interbanded with granites and gneisses. In this region the sediments were structurally repeated (imbricated?), and intruded by masses of granite and large wedges of metagabbro and anorthosite.

The sediments between Matsitama and Shabani were probably deposited in a basin, in the centre of which basic and ultrabasic rock of ocean floor type was emplaced.

It is suggested that an early spreading centre caused a thinning of the granitic and gneissic crust, with the possible formation of true oceanic crust, only a short time before closure of the basin produced the deformations of the Limpopo belt.

Introduction

Recent geochronological work by Hawksworth et al. (1975) and Hickman (1974) with structural and stratigraphic mapping by Bickle et al. (in press) has shown the presence of two ages of greenstone belts in Rhodesia. The younger belts dated at 2600–2700 m.y. rest on a basement of granite, gneiss and earlier greenstone belt rocks giving ages of 3400–3600 m.y. The older gneisses and greenstone belt rocks are, however, confined to the central part of Rhodesia. The greenstone belts in the western part of Rhodesia and in Botswana probably belong to the younger set (2600–2700 m.y.) and are separated from each other by granite and tonalite intrusions with no evidence of earlier gneiss.

The younger greenstone belts and the granites and tonalites are deformed by the Limpopo belt deformation and intruded by the Great Dyke dated at 2580 m.y. (Allsopp,

1965). Thus the younger Rhodesian greenstones were formed, intruded by granite and tonalite, and deformed in a relatively short period of time.

A strong penetrative fabric can be traced from the gneisses of central Botswana into the granites and Archaean greenstone belts of northern Botswana and south-west Rhodesia. This fabric is defined by a preferred orientation of mineral grains and deformed objects such as pillows, vesicles and conglomerate pebbles. This fabric can be traced across the granite and tonalite diapirs which intrude and separate the greenstone belts, and also across the migmatites and granites which intrude the sediments of the Limpopo belt (Coward et al., in press). It is, however, a finite fabric, and therefore may not be synchronous across the whole area. The intensity of strain varies, from 25% shortening across the granites and greenstone belts in central Rhodesia, to over 60% across the greenstone belts and gneisses in

northern Botswana (Coward, et al., in press).

Prior to this cleavage-producing phase of deformation and before the intrusion of the granites and tonalites important deformation events occurred. The aim of this paper is to describe these early structures and their significance in terms of Archaean tectonics and greenstone-belt evolution.

Pre-Cleavage Deformation

Litherland (1973) has shown that the Tati and Vumba greenstone belts in northern Botswana (Fig. 1) have a similar stratigraphy and were once part of a continuous sheet of supracrustal rock. This sheet was folded, intruded by tonalite diapirs and then intensely deformed to produce the present deformation pattern. From way-up evidence in pillow lavas and graded beds, however, Mason (1968) and Key (1972) concluded that the Tati greenstone

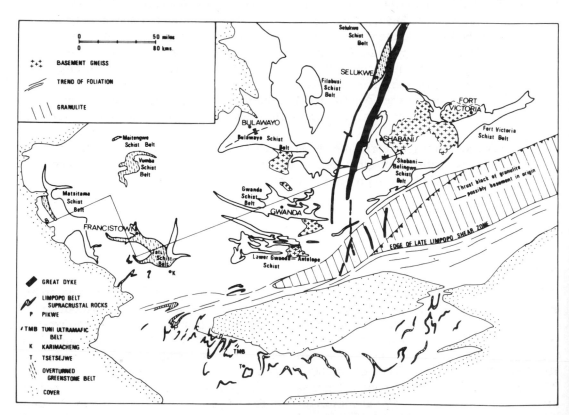

Fig. 1. Map of the southern part of Rhodesia and north-east Botswana showing locations mentioned in the text. A–B is the line of the composite section Fig. 5A.

belt was overturned prior to the deformation. Hence the Tati and Vumba belts form the remains of a large overturned fold limb.

West of Tati, in the Matsitama belt, graded bedding and cross-bedding show most of the sediments to be right way-up, but in the core of a major synform the beds are inverted and the succession is downward facing (Fig. 2). The right way-up sediments are 10 km thick, at least 20 km thick when the rock is unstrained. This is a vast thickness for Archaen or any other sedimentary pile, but the sequence shows several repetitions of stratigraphy and may form part of an imbricate pile (Figs. 2 and 3).

In Rhodesia, the Lower Gwanda and Antelope belts young upwards into gneissic rock which shows several phases of deformation pre-dating the deformation of the greenstone belts. This may represent allochthonous basement gneisses thrust over the greenstone-belt rocks prior to the intrusion of the granite. The greenstone belts, gneisses and granites were intensely folded by the later cleavage-producing deformation. A small allochthonous greenstone belt at Selukwe in Central Rhodesia has been described by Stowe (1974). This belt was emplaced as an overturned nappe prior to being intruded by granite and

deformed by the cleavage-producing phase of deformation.

The greenstone belts of Bulawayo and Shabani are autochthonous and sit on older basement gneiss; the sediments south of Shabani show clear unconformity with shales, conglomerates and basic volcanics resting in an erosional trough on older gneissic granite (Coward et al., in press). Both the Shabani and Bulawayo belts were folded into tight near-upright synforms before being intruded by diapiric granite and deformed by the cleavage-producing phase (Coward et al., in press).

South of the Tati greenstone belt, the sediments of the Limpopo belt are of different facies from those of central Rhodesia, but there is no evidence to suggest that they have suffered any earlier deformation events or that they are significantly different in age. They are similar in rock type and stratigraphy to the Matsitama belt (Fig. 3) and across the northern margin of the Limpopo belt near Rhodesia, there is a gradual change from sediments of Rhodesian greenstone-belt type (the Karrimacheng group) to rocks of the Messina formation (Fig. 3). The sediments are tightly folded and often shown an inverted stratigraphy. They contain large sheets of layered metagabbro–anorthosite near the base of the

Fig. 2. Profile through the Matsitama belt, Northern Botswana, showing major F_2 synform with associated cleavage and earlier (F_1) fold producing downward-facing succession. Solid arrowheads indicate younging direction

sedimentary sequence and within these sheets gravity layering confirms the way-up of the stratigraphy. The layered sheets and the sediments have been deformed into folds which pre-date migmatization and intrusion of large masses of granite (Fig. 4).

Thus the sediments of north-east Botswana and south-west Rhodesia have been involved

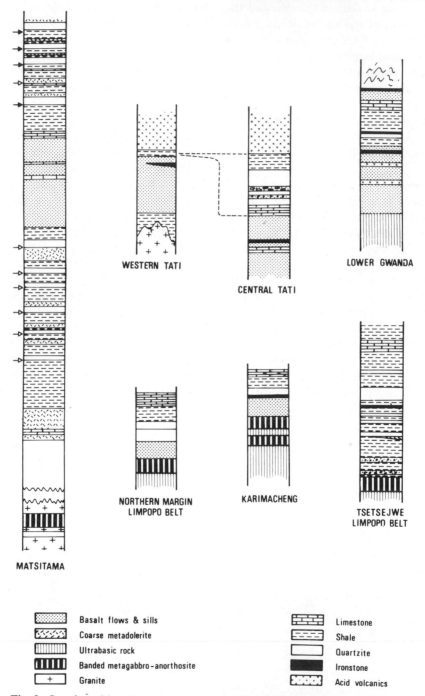

Fig. 3. Stratigraphic columns through greenstone belts and supracrustal rocks of the Limpopo belt. Solid arrows point to the bases of imbricate slices, open arrows to postulated imbricate zones

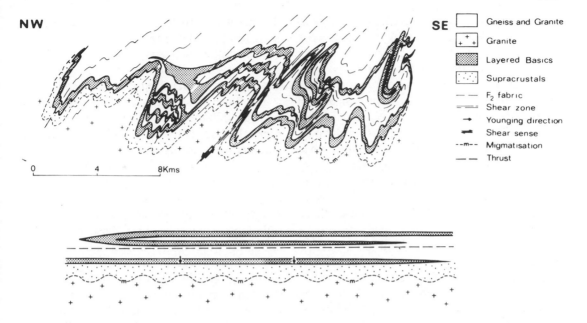

Fig. 4. Profile through the structures in the central part of the Limpopo belt, between Banes Drift and Pikwe. Arrows show younging directions. Lower section shows the form of the structure prior to F$_2$ and F$_3$ Limpopo deformation. The lowermost layer of basic and supracrustals faces downwards into migmatitic granite. The gneisses stratigraphically below the layered basics show deformation events pre-dating the first structures in the supracrustal rocks and may represent an earlier basement

in large scale thrust and nappe tectonics. Probable basement gneisses occur in thrust slices in Rhodesia and there is also gneissic material stratigraphically beneath, but structurally above the inverted sediments in the Limpopo belt (Fig. 4). The greenstone belts in the central part of Rhodesia are folded but apart from the Selukwe belt, they are autochthonous and represent the foreland to the thrust and nappe pile. The Selukwe belt may have travelled a considerable distance as an overturned sheet. The belt consists of an inverted sequence of chrome-bearing ultrabasic rocks structurally underlain by conglomerates with chrome-bearing fragments, and basic volcanics. The nearest ultrabasic rocks of similar type are some 80 km to the south, south of Belingwe, or in the base of the Filabusi belt, some 50 km to the south-west.

The correlation of all these nappe structures into one deformation phase is by no means certain. However, they all pre-date the main metamorphism, migmatization and granitization. In northern Botswana and southern Rhodesia, there is no cleavage associated with

this early deformation, and any early fabric within the Limpopo belt has been obliterated by later migmatization and intense deformation. It is difficult to identify the individual thrust planes, but in the Matsitama belt, the junction between the inverted and right way-up beds is a zone of intense stylolitization. Sulphide mineralization occurs intermittently along this zone. The movement direction of the thrusts and nappes is not definitely known but is presumed to be to the north-north-east near-normal to early fold hinges within the Rhodesian greenstone belts.

Stratigraphic Correlations

The thrust and overturned relics of greenstone belt material imply that the belts were once much larger in extent. Litherland (1973) has suggested that the Matsitaba, Maitengwe, Vumba and Tati belts in Botswana were once continuous; inclusions of greenstone-belt material can be found in the intervening granites and the stratigraphy of the Vumba and

Tati belts is very similar. Similarly, the Shabani and Bulawayan belts in Rhodesia were probably once continuous, as were the Lower Gwanda, Gwanda and Buchwa belts. This continuity of greenstone belts is supported by the gradual regional change in lithological types and sedimentary facies. We suggest that the individual greenstone belts from Shabani in the east to Matsitama in the west, once formed one large sedimentary basin or group of closely connected basins. The sediments in the Limpopo belt to the south are similar to those of Matsitama and may form the southern part of this basin.

Fig. 3 shows detailed stratigraphic sections through the greenstone belts at Lower Gwanda, Tati and Matsitama and the sediments of the Limpopo belt, Fig. 5A gives the composite section assuming the sediments were once continuous. Details of the stratigraphy in the east are taken from Bickle et al. (in press) and reconnaissance work by the authors. In the east the sediments of the Shabani belt rest on a basement gneiss. They are dominantly shelf sediments with ripple-marked siltstones, conglomerates and limestones in which algal structures have been reported (Bickle et al., in press). Within the sediments are pillow lavas and pisolitic tuffs (Oldham, 1968) indicative of shallow-water volcanism.

In the western part of the Shabani belt and in the lower part of the Gwanda belt, the rocks are predominantly basic and ultrabasic and similar rocks dominate the lower parts of the Tati, Karrimacheng and Lower Gwanda belts. Less than 20% of these rocks show structures indicative of extrusive origin. Most of the basic rocks have a coarse doleritic texture and where they are intermixed with sediments they are rimmed by contact metamorphic spots. The basic and ultrabasic rocks are dominantly sheets parallel or sub-parallel to the layering in the volcanics and clastic sediments. The sheets are most common near the base of the succession where the rock is almost wholly ultrabasic.

In the west, the Matsitama belt is made up of dominantly clastic sediments and limestones with less basic volcanic and intrusive material.

Sedimentary structures such as dune bedding and trough bedding in the clastic rocks indicate a shallow-water environment. The sediments of the Limpopo belt seem to be of similar age and facies to the Matsitama rocks.

The basic rocks of the Tati belt are overlain by acid and andesitic volcanics and intrusive rocks (Mason, 1968; Key, 1973). Similar acid igneous rocks occur within the Gwanda and Lower Gwanda belts. Key (1974) considers these volcanic rocks as typical of island arc deposits. The basic rocks in the centre of the basin are overlain and partly mixed with clastic sediments, indicating that these greenstone belts were never too far from a source region.

The basic rocks in the centre of the basin show similar lithologies and structures to the sheeted complex which forms a distinct layer in oceanic ridge deposits (Gass, 1968, and Masson-Smith, 1963). Walker (1975) has described sheets of basic rock intruding the volcanic rocks on the east coast of Iceland and considers these sheets to be equivalents of layer 3 within the Icelandic crust.

The Rhodesian basic rocks interfinger with shelf sediments on the eastern and western edges of the basin and probably cannot be considered as truely oceanic. It is possible, however, that these basic and ultrabasic rocks represent the beginnings of ocean floor development and indicate the presence of an Archaean spreading centre part-way between Shabani and Matsitama.

Discussion

The basic and ultrabasic rocks may mark a phase of crustal separation and ocean floor development at the end of the Archaean, while the sediments at Shabani in the east, and those at Matsitama, and in the Lompopo belt in the west form shallow-water shelf deposits resting on earlier continental crust.

Any crustal spreading indicated by the lower basic rocks was followed by a period of thrust and nappe development, with considerable shortening of any original greenstone basin. The allochthonous rocks of Botswana and south-west Rhodesia have been intruded by far more granitoid material than the

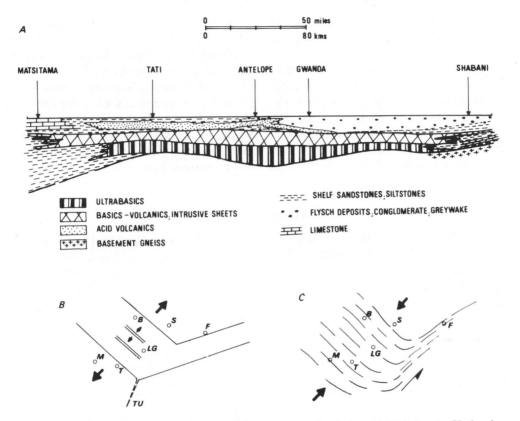

Fig. 5. A. Composite section through the greenstone basin from Matsitama to Shabani. Stratigraphy at Shabani after Bickle et al, and reconnaissance work by the authors. B. Schematic model for greenstone-belt development in south-east Rhodesia, showing triple junction south to south-east of the lower Gwanda belt. F—Fort Victoria, S—Shabani, B—Bulawayo, LG—Lower Gwanda–Antelope, T—Tati and M—Matsitama. TU denotes the failed spreading arm at Tuni. C. Model for the shortening of the greenstone belts during the early phases of the Limpopo deformation (after Coward et al., in press)

autochthonous foreland rocks of central Rhodesia. The shallow-water sediments in the Limpopo belt show more intense metamorphism, the sediments around Pikwe show evidence of partial melting and granulite facies metamorphism (Wakefield, 1974). Thus the nappe pile either lies across a region of high heat flow or represents rocks depressed to lower levels of the crust. A highly schematic model for the evolution of the greenstone belts is given in Fig. 5B where the high-grade rocks of north-eastern Botswana and the granulites of southern Rhodesia (Robertson, 1968) overlie a region of high heat flow at a triple junction. This junction could also explain the curvature of the greenstone outcrop pattern, with

a major spreading arm trending north-northwest through western Rhodesia and a less important spreading arm trending east–west through southern Rhodesia. The location of a third 'failed arm' may be indicated by a belt of ultramafic rocks, the Tuni ultramafic belt (Fig. 1) exposed south of the outcrop of the cover rocks in eastern Botswana (Wright, 1973). This postulated triple junction appears to have controlled the position of the major ductile shear zone during the Limpopo deformation events (Coward, et al., in press).

We suggest that this period of crustal shortening and granite intrusion represents a period of crustal collision, closing of the original supracrustal basin and consumption and

330

subduction of any true ocean floor material which may once have occurred within the basin.

This period of greenstone-belt production and deformation appears to have taken place within a short period of time; any greenstone-belt ocean was shorter lived than the Phanerozoic oceans of the proto-Atlantic and Tethys. The period of nappe formation and granite intrusion was followed by regional penetrative deformation of the whole area, but this deformation had ceased by 2580 m.y. when the Great Dyke and its satellites were intruded along a series of north-north-east trending fractures in Rhodesia. This late phase of basic and ultrabasic rock in the form of the Great Dyke may indicate a late phase crustal spreading, ocean floor development and incipient greenstone belt formation which failed to gain momentum.

Acknowledgements

Part of this work was financed by an NERC research grant and by funds from the Research Institute of African Geology, University of Leeds. One of the authors (B.C.L.) was employed by Bamangwato Concessions Limited, Selebi–Pikwe, Botswana, and all of the authors would like to thank this company for their generous help during fieldwork.

References

Allsopp, H. L., 1965. 'Rb–Sr and K–Sr age measurements in the Great Dyke of southern Rhodesia', *J. Geophys.*, **70**, 977–984.

Bickle, M. J., Martin, A., and Nisbet, E. G., (in press). 'Basaltic peridotitic komatiites and stromatalites above a basal unconformity in the Belingwe greenstone belt, Rhodesia', *Geology*.

Coward, M. P., James, P. R., and Wright, L. I. (in press). 'The movement pattern across the northern margin of the Limpopo mobile belt, southern Africa', *Bull. Geo. Soc. Amer.*

Gass, I. G., 1968. 'Is the Troodos Massif of Cyprus, a fragment of Mesozoic oceanic crust?', *Nature, Lond.*, **220**, 30–42.

Gass, I. G., and Masson-Smith, D., 1963. 'The geology and gravity anomalies of the Troodos Massif, Cyprus', *Phil. Trans. Roy. Soc. Lond.*, **A225**, 417–467.

Hawksworth, C. J., Moorbath, S., O'Nions, R. H., and Wilson, J. F., 1975. 'Age relationships between greenstone belts and "granites" in the Rhodesian Archaen craton', *Earth Planet. Sci. Lett.*, **25**, 251–262.

Hickman, M. H., 1974. '3500-Myr-old granite in southern Africa', *Nature, Lond.*, **251**, No. 5473, pp. 295–296.

Key, R. M., 1973. 'An interim report of the geology of Sheet 2127B, the Tati area', *Rept. Geol. Surv. Botswana* (unpublished), 63 pp.

Key, R. M., 1974. 'Some aspects of the geochemistry of the metavolcanic rocks of the Tati schist belt, N.E. Botswana', *18th ann. Rep. res. Inst. Afr. Geol. Univ. Leeds*.

Litherland, M., 1973. "Uniformation approach to Archaean 'schist relics'", *Nature, Lond.*, **242**, 125–127.

Oldham, J. W., 1968. 'A short note on recent geological mapping in the Shabani area', *Trans. geol. Soc. S. Afr.*, **71**, (Annex), 189–194.

Mason, R., 1968. 'The geology of the southern Tati area', *Rep. Geol. Surv. Botswana* (unpublished).

Robertson, I. D. M., 1968. 'Granulite metamorphism of the basement complex in the Limpopo metamorphic zone', *Trans. geol. Soc. S. Afr.*, **71** (Annex), 125–134.

Stowe, C. W., 1974. 'Alpine-type structures in the Rhodesian basement complex at Selukwe', *Jl. Geol. Soc. Lond.*, **130**, 411–425.

Wakefield, J., 1974. 'The geology of the Pikwe Ni–Cu province, eastern Botswana', Unpubl. Ph.d. Thesis, Univ. Leeds.

Walker, G. P. L., 1975. 'Intrusive sheet swarms and the identity of crustal layer 3 in Iceland', *Jl. geol. Soc. Lond.*, **131**, 143–161.

Wright, L. I., 1973. 'Preliminary report on the structures present in the central zone of the Limpopo belt, Eastern Botswana', *17th ann. Rep. res. Inst. Afr. Geol., Univ. Leeds*, 116–117.

The Pikwitonei Granulites in Relation to the North-western Superior Province of the Canadian Shield

I. F. ERMANOVICS and W. L. DAVISON

Geological Survey of Canada.

Abstract

The Pikwitonei Granulite Belt is an exhumed portion of what is assumed to have been an early Archean craton (> 3000 Ma), a remnant of which now constitutes the Superior Province and at least parts of the south-east half of the Churchill Province. During the Kenoran orogeny, the Superior craton broke and was tilted up along a line that now approximates the Nelson Front and the Uchi Volcanic Belt. The north-western portion of the Superior Province craton that broke along the Nelson Front is now occupied by the Sachigo Volcanic Belt and Berens Batholithic Belt. The Berens Batholithic Belt was produced when this portion of the plate was tilted down to deep crustal levels to produce granitic rocks 2700 Ma ago. The portion that broke along the Uchi Volcanic Belt is overlain by thick accumulations of supracrustal rocks of the southern Superior Province.

Introduction

The Canadian Shield is of comparable age with other old shields of the world (3500 Ma, Goldich et al., 1970), but the Kenoran orogeny with its intense plutonic activity 2700 Ma ago has disrupted and transformed most of the earlier crustal rocks. An appreciation of how early crustal rocks become disrupted during subsequent orogenies may lead to a wider recognition of early crustal terrains, and to an understanding of early crustal orogenies themselves. It is the aim of this paper to describe briefly the Pikwitonei Granulite Belt, and to discuss its relation to orogenic belts of the western Superior Province and to the adjoining Thompson Paragneiss Belt of the Churchill Province.

Discussion of the applicability of terms such as 'province', 'sub-province', or 'belt' to the Pikwitonei complex is beyond the scope of the present paper, and as a matter of convenience, the terminology of Douglas (1973) is followed herein.

Pikwitonei Granulite Belt

The Pikwitonei Belt lies along the north-west flank of the Superior Province in north-central Manitoba (Fig. 1) and consists, for the most part, of a gneissic complex in which plagioclase-rich (enderbitic) hypersthene-bearing gneisses are prominent. The prevailing east–west trend of western Superior Province units continues into the Belt (Fig. 2) but

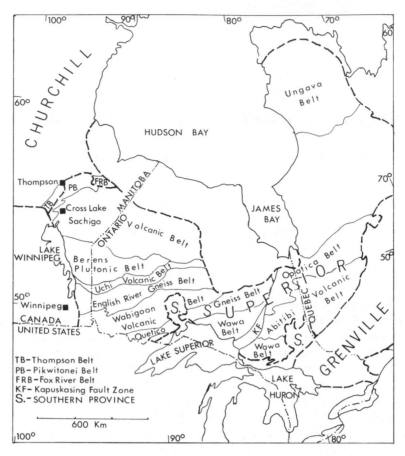

Fig. 1. Index map of tectonic belts of the Superior Province (after Douglas, 1973)

is truncated on the north-west by the Nelson Front of the Church Province. Adjacent to the Nelson Front, retrograded Pikwitonei gneisses form a zone which contains several stocks of late (Hudsonian) biotite granite. The disappearance of orthopyroxene-bearing rocks approximately defines the south-east side of the Belt, but it is not entirely clear whether the boundary represents a metamorphic isograd, as suggested by Rousell (1965), or a major unconformity, as argued by Bell (1966, 1971).

Wilson and Brisbin (1961, 1962) first drew attention to the Belt, which they referred to as the 'Nelson River gneissic belt' (or 'zone'), pointing out that the Nelson River 'gravity high' of Innes (1960) approximately coincides with the axis of the Belt. Drawing on the available geological, magnetic and gravity data, they concluded that the east–west

Superior trends cross the Belt in some places, but not everywhere.

In the Thompson–Moak Lake area, Patterson (1963) found that easterly-trending 'charnockite suite' gneisses abut a narrow zone of highly deformed north- to north-east-striking rocks on the west. Minor components in the granulite–facies complex of the area include recognizable derivatives of quartzose and pelitic sediments, siliceous limestone, and iron-formation together with numerous mafic lenses and sill- or dyke-like bodies of (?)meta-gabbro.

In the Cross Lake area, near the southern end of the belt, Rousell (1965) traced a sequence of progressively metamorphosed and metasomatized Cross Lake Group basalts and sediments through a migmatitic zone into a complex of biotite granodiorite and hornblende tonalite gneisses which, in turn, graded into hypersthene-bearing gneisses to the north-west. On the basis of potassium–argon ages, Rousell (1965, p. 62) concluded that metamorphism of the Cross Lake Group took place during the Hudsonian orogeny, so that 'the boundary between the Superior and Churchill provinces lies somewhere south of Cross Lake'. Rousell also described an unconformity at the base of the Group on Cross Lake, and sill-like bodies of layered gabbro–anorthosite along the base at two other localities (Fig. 2). A K–Ar determination on the underlying granodioritic gneiss gave a Kenoran age (2500 Ma) which he interpreted as a survival value.

During the course of mapping Cross Lake area, Wekusko Lake area on the west, and Sipiwesk area to the north, Bell (1962, 1965, 1966, 1971a) examined the granulite facies complex, together with bordering units, along 195 km of its length within the Upper Nelson River region. Noting that a distinctive pattern on aeromagnetic maps corresponds to the distribution of granulite facies rocks within a belt of high magnetic intensity, Bell (1966, p. 133) traced an eastward extension of the unit along the northern edge of the Superior Province. He proposed that the granulite terrain be named 'Pikwitonei Sub-province' (of the Superior), since the granulites 'differ in age and lithology from rocks to the north and

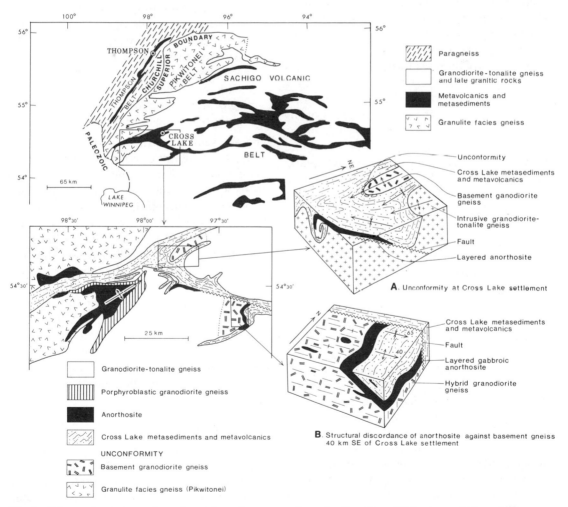

Fig. 2. Schematic geology of portions of north-western Superior Province, Wekusko and Cross Lake map areas

south'. As originally defined, the Pikwitonei encompassed certain amphibolite–facies equivalents of granulites in the north-west, and scattered bodies of anorthosite, but not the retrograde zone to the west, nor a strip of Cross Lake Group rocks, with flanking anorthosite masses, which crosses the Pikwitonei near its southern end (Fig. 2).

Potassium–argon ages of 2680 Ma for biotite from a quartzofeldspathic granulite and 2435 Ma for hornblende from anorthosite of the Wekusko Lake area (Wanless et al., 1966, pp. 55–57) failed to support Rousell's (1965, p. 62) theory that the granulite–facies rocks evolved during the Hudsonian orogeny at the time of granitization of Cross Lake strata. Instead, the ages appeared to lend weight to Bell's (1966, p. 35) contention that the Pikwitonei represents an Archean craton, and the retrograde zone a buffer, against which 'the southerly directed forces of the Hudsonian orogeny were dissipated'. The generally younger ages of Cross Lake biotites (Wanless, 1969) suggest that the effects of the orogeny were not confined to the Churchill side of the craton. A K-Ar age of 2715 Ma for biotite from a sample of pyroxene–bearing quartz monzonite, which Bell had collected 50 km east of his Sipiwesk map-area (Wanless et al., 1968, p. 91), appeared to demonstrate the validity of his eastward projection of the subprovince.

In a series of publications, beginning in 1970 and continuing up to his untimely death in 1973, Bell elaborated on the thesis that the Pikwitonei (which he now regarded as a separate *province*) represents the basement on which younger Archean supracrustal rocks of the Superior Province were deposited. Inasmuch as structural features and potassium–argon ages point to deformation of both basement and overlying units during the Kenoran orogeny, the basement complex must have formed during an earlier orogeny. Considering these relationships, Bell (1971b, p. 33) anticipated that the Pikwitonei gneisses would eventually prove to be older than 3000 Ma.

Instead of consisting almost entirely of granulite–facies areas, Bell's Pikwitonei Province incorporates retrograded granulites, and its boundaries outline additional parts of 'an ancient continental surface that consists, at least locally, of high-grade gneisses and granodioritic plutons', thus allowing for inclusion of basement rocks below the unconformity at Cross Lake and elsewhere (Coats et al., 1972, p. 7). As before, the strip of Cross Lake strata, and synforms of gneisses derived from other Archean units that cross the Pikwitonei from east to west, constitute Superior Province elements that divide the Pikwitonei terrain into several segments.

Following the usage of Turner and Verhoogen (1960) and de Waard (1965), Bell (1971) identified the following pyroxene and hornblende granulite sub-facies:

plagioclase–clinopyroxene–
    orthopyroxene–quartz
plagioclase–clinopyroxene–
    orthopyroxene–garnet
plagioclase–hornblende–
    clinopyroxene–orthopyroxene
plagioclase–clinopyroxene–
    orthopyroxene–hornblende–garnet

The rocks are generally pale brown, medium-grained, granoblastic and may be stratiform or foliated and gneissic. Some of these rocks are probably of volcanic or sedimentary origin. Enderbites are greenish and show evidence of diaphthoresis (hornblende after pyroxene, minor quartz and biotite). Charnockites, locally massive, are generally gneissic and contain augen-shaped microcline metacrysts in a matrix of blue quartz, plagioclase, hornblende, biotite and partially replaced hypersthene. The Pikwitonei rocks are denser (S.G. 2·70 to 3·13 gm/cm^3) than those of the surrounding region and the ubiquitous magnetite content (1 to 5% by volume) produces a typical 'bird's eye maple' aeromagnetic pattern.

Thompson Paragneiss Belt

The Thompson nickel-bearing complex (Figs. 1 and 2) consists of metamorphosed sediments, volcanics and peridotites, and associated metasomatically reconstituted amphibolite facies and hornblende granulite facies gneisses. The complex was considered by Bell (1971) to be sufficiently distinctive to warrant its designation as a new sub-province—the Wabowden Sub-province of the Churchill Province. The south-eastern boundary of the Sub-province was set at the north-western limit of the Pikwitonei Belt where the contact is marked by a zone of retrograded granulite gneiss (Nelson Front). The north-western boundary was initially placed by Bell (1971) at the Setting Lake fault, a zone of mylonite with varying width, ranging from 0·9 to 1·5 km. However, by 1972 Bell (in preparation) had extended the boundary beyond the mylonite zone to include para-gneiss farther west.

Relatively uniform K/Ar ages of 1800 Ma (Wanless, 1969) and a N35°E structural trend distinguish the Thompson paragneisses from Pikwitonei granulites (Kenoran structural trend N70E) at the Nelson Front. Rance (1966) and Cranstone (in Coats et al., 1972) suggested that the migmatites, gneisses and granulites are Pikwitonei rocks remetamorphosed during Hudsonian orogeny. The depositional age of the Thompson rocks is not known although some evidence suggests that they may be Archean. Just south of Thompson, complexly folded orthopyroxene granulites occur conformably within the lower grade paragneisses. Unlike granulites in the Pikwitonei Granulite Belt they are not retrograded and Bell (1971) suggested that these rocks could be either autochthonous remnants or allochthonous slices of rock. The rocks of the Thompson Paragneiss Belt may be of various ages and they may have undergone the effects of two or more major orogenies possibly overlapping in time. It is likely that pervasive cataclasis (culminating in the Setting Lake fault), granulite facies rocks, Thompson belt supracrustal rocks, nickel mineralization, and overlapping orogenies coincided in space to produce the Thompson Paragneiss Belt.

Thompson Nickel Deposits

The section which follows is taken from Coats et al. (1972) and is included here because the Thompson deposits may be Archean in age. Metasediments and metavol-canic rocks occur as discontinuous schistose units, close to the Setting Lake fault (mylonite), within the belt of Thompson migmatitic paragneisses.

Two types of deposits are recognized but probably have a common origin. The first group comprises disseminated pyrrhotite, pendlandite and minor amounts of pyrite and chalcopyrite in ultramafic rocks. The second group comprises massive bodies of similar sulphides in gneisses and metasediments. The massive sulphide bodies contain numerous wall rock inclusions, including ultramafic rock. Because the known volume of ultramafic rock near each deposit is small relative to the amount of massive sulphide it is argued that the associated ultramafic rock cannot have been the source of the sulphides.

At the Thompson mine and elsewhere the ore occurs in biotite schist folded conformably with 'skarn' (diopside–tremolite–clinozoisite–phlogopite–carbonate rock) and 'quartzite' (with variable amounts of feldspar, biotite, muscovite). Elsewhere, serpentinized peridotite, biotite gneiss, schists, calcareous sediments, and amphibolite serve as host rocks to ore. For the Manibridge nickel deposits Coats and Brummer (1971) suggest a magmatic origin inasmuch as the sulphides crystallized prior to serpentinization of the olivine–pyroxene ultramafic host rock. The age of the ultramafic rocks is not known and they may be related to the volcanic rocks (Archean?) or to a later period of ultramafic intrusion.

Sachigo Volcanic Belt (Cross Lake Sub-province)

The belt extends from the Pikwitonei orthopyroxene isograde southward for 190 km to the northern boundary of the Berens Batholithic Belt. Ermanovics (1973a) mapped across the earlier-proposed Sachigo–Berens boundaries (Gibb, 1968; Wilson, 1971 and Bell, 1971) and placed the boundary to coincide with a zone of cataclasis and mylonite.

The Sachigo Belt comprises metavolcanic and metasedimentary rocks of Archean age, typically distributed as thin curvilinear, bifurcated belts (Fig. 2). The remaining terrain constitutes 80% of the Belt in the following proportions:

(a) Agmatite and migmatite, massive and foliated (15%).
(b) Quartz diorite, granodiorite, small quantities of quartz monzonite; massive to weakly foliated (35%).
(c) Quartz diorite (tonalite) and granodiorite (trondhjemite); gneissic augen foliation; flattened skialiths and amphibolite (40%).
(d) Stratiform tonalitic and trondhjemitic orthogneiss; lit-par-lit layers of reconstituted mafic dykes and possibly paragneiss (10%).

Rocks in categories (a) and (b) are essentially post-metamorphic (late Kenoran); category (c) includes synkinematic and inferred pre-kinematic plutonic rocks. Synkinematic and earlier plutonic rocks (category (c)) are penetratively deformed and some show no vestige of an earlier igneous texture. Inclusions in these rocks are similar to existing rocks of the supracrustal belts, but judging from the unconformable relationships at Cross Lake, such inclusions may constitute vestiges of early Archean supracrustal rocks not now exposed or recognized.

Stratiform gneisses (category (d), Fig. 3) are problematical rocks, yet because of their structurally discordant trends with respect to supracrustal belts they provide evidence for an older Archean basement (Ermanovics, 1973a). Such gneisses are granoblastic, medium-grained, and range from leucocratic to melanocratic (quartz, antiperthitic oligoclase, biotite, hornblende, epidote). Garnet and relicts of hypersthene are present but rare. Texturally and compositionally these rocks

Fig. 3. Tonalite gneisses of the Sachigo Volcanic belt. a. Layered hornblendic and quartzo–feldspathic gneiss. 19 km south-east of Poplar River (GSC 202076-I). b. Porphyritic gneissic quartz diorite. The zircon U–Pb age of this rock is 3000 Ma (Kough et al., 1974). Latitude 51°15′ on east shore of Lake Winnipeg (GSC 160266). c. Laminated granoblastic hybrid tonalite gneiss containing biotite, hornblende, quartz and oligoclase. This rock may be 'pre-volcanic' at Island Lake, Manitoba (GSC 163180). d. Highly contorted, layered tonalite gneiss. Hammer lies on pegmatitic dyke. West shore of Playgreen Lake, near north end of Ross Island (GSC 123734)

are identical to those of the retrograde amphibolite (granulite) facies gneisses of the Pikwitonei. Amphibolite and mafic, augen (quartz, oligoclase) quartz diorite interbands are common components of both the Sachigo and Pikwitonei gneisses. Although similar stratiform gneisses are more commonly found adjacent to supracrustal belts they have not been found in contact with them (Ermanovics 1970, 1973a, and Ermanovics et al., 1975). Their preservation near supracrustal belts is probably more a result of Kenoran tectonics than of their original distribution.

Tholeiitic basalt sequences, grading upward to acidic assemblages, are generally interpreted as occupying basal positions in supracrustal belts, although in the absence of 'top' directions this interpretation in the past may have been made more by custom than by fact. However, at Cross Lake (Horwood, 1935, Rousell, 1965, Bell, 1971), Munro Lake maparea (Elbers and Gilbert, 1972), Goose Lake (Elbers, 1974, pers. comm.) and elsewhere, the basal member is a clastic metasediment containing tonalitic boulders and is overlain by basaltic flows. The significance and orogenic timing of quartz-rich, basal metasediments has not been evaluated except at the Cross Lake unconformity. The age of volcanism is not known.

Metamorphism of the Archean volcanic rocks varies across the Sachigo Volcanic Belt and ranges from upper greenschist to upper amphibolite facies. Narrow remnant belts (0·5 to 1·5 km wide) generally show uniform grades of metamorphism. Mafic volcanic rocks associated with high-grade pelitic rocks are schistose and commonly indicate retrograded assemblages comprising epidote, albite, two amphiboles, biotite and chlorite. Wide volcanic belts show increasing grades of metamorphism toward their margins, which also affects adjacent synkinematic tonalitic rocks. Epidotized albite–oligoclase and elongated, poikilitic hornblende overgrown by secondary amphibole, biotite and chlorite is interpreted as protracted or multiple metamorphism in mafic and tonalitic rocks. Three to four phases of deformation locally produce 3 fold styles (Ermanovics et al.,

1975). Judging from K–Ar ages deformation domains become progressively smaller and localized toward the end of the Kenoran orogeny. Because the isograds are folded, the main metamorphism is interpreted as being prekinematic.

Unconformity of the Cross Lake Group of the Sachigo Volcanic Belt

The stratigraphic section at Cross Lake is shown in Table 1. Below the unconformity at Cross Lake lies a hybrid basement granodiorite gneiss containing metasedimentary and metavolcanic skialiths (Fig. 4). The gneiss contains oligoclase, biotite, streaky quartz and secondary muscovite. The unconformity was first described by Rousell (1965) and basement gneiss was interpreted as occupying the flanks of an antiform comprising various tonalitic rocks (Fig. 2). Similar gneiss 40 km south-east of Cross Lake Settlement was described by Horwood (1935) as 'pre-Keewatin tonalite'. The gneissosity of the basement gneiss is vertical and strikes east. Metasediments of the Cross Lake Group are folded and plunge 40 to 65° east (Fig. 2). A layered gabbroic anorthosite sheet lies below Cross Lake metasediments, and was intruded along the unconformity before folding (Bell, 1971). The base of the sheet lies at 90° discordance to the gneissosity of the basement gneiss (Fig. 2). The composition, metamorphic grade and extent of the basement gneiss has yet to be established but it is known that the gneiss is complicated by 'anorthosite outliers, crosscutting granitic material, a mafic stock and late gabbro sills' (Bell, 1971). Horwood (1935) made a few modal measurements on the granodiorite–quartz diorite pebbles in Cross Lake metasediments and 'pre-Keewatin tonalite' basement at Cross Lake and as far removed as 120 km at Molson Lake. He concluded the compositions were identical. Bell (1971) correlated these clasts with the Pikwitonei gneisses. However, since it is possible that these basement gneisses were never raised to granulite grade and since neither unconformities occur within the Pikwitonei

Table 1. Stratigraphic and tectonic elements across the Superior-Churchill boundary (interpreted from Bell, 1971, Coats et al., 1972, Douglas, 1973, and Ermanovics, 1971, 1973)

	CHURCHILL PROVINCE (Hudsonian Orogeny)		NORTHWESTERN SUPERIOR PROVINCE (Kenoran Orogeny)		
NW	THOMPSON PARAGNEISS BELT		PIKWITONEI GRANULITE BELT	SACHIGO VOLCANIC BELT (Wekusko Lake Map Area)	BERENS BATHOLITHIC BELT	SE

PROTEROZOIC — Aphebian

ARCHEAN — Early Archean

MYLONITE (SETTING LAKE FAULT) CONTACT

Late granodioritic rocks
Migmatitic paragneiss
Thompson Nickel Deposits
Ultramafic rocks
Metavolcanics; amphibolite
Metasediments

RETROGRADED GRANULITE CONTACT (NELSON FRONT)

Late granodioritic rocks
Anorthosite

ORTHOPYROXENE ISOGRAD CONTACT

Late granodioritic rocks
Granodiorite–tonalite gneiss
Anorthosite; intruded along unconformity and opx isograd
Cross Lake Group:
Metavolcanics
Metasediments; basal cgl.
UNCONFORMITY
Granodiorite gneiss with metasedimentary skialiths
? OPX ISOGRAD ?

MYLONITE ZONE CONTACT

Late granodioritic rocks

Hybrid granodiorite–tonalite gneiss

Metavolcanic and metasedimentary rocks; amphibolite

Meta quartz diorite and volcanogenic(?) gneiss; granodiorite gneiss

LITHOLOGIC TRANSITION CONTACT

CATACLASTIC CONTACT
Granulite facies gneiss (orthopyroxene)

INTRUSIVE CONTACT
Granulite facies gneiss; enderbite, charnockite and granodiorite gneiss

Belt and because the basal gneiss is compositionally indistinct from regional 'granodiorite-tonalite gneiss' (Table 1), it is suggested that the basement gneiss at Cross Lake should not be included in the Pikwitonei suite of rocks. The Pikwitonei granulites are probably older than the basement gneiss because neither at Cross Lake nor at Goose Lake (Elbers, pers, comm.) can the basement rocks be inferred to have been metamorphosed prior to erosion. It is possible that the basement rocks in the area of the unconformity may have been post-metamorphic intrusions following the main metamorphism to granulite grade in the Pikwitonei Belt. Subsequent Kenoran metamorphism appears to have deformed Cross Lake Group and basement rocks simultaneously.

The age and indeed the stratigraphic position of the unconformity at Cross Lake relative to the supracrustal rocks throughout the Sachigo Belt is unknown. In the absence of age data the problem is one of correlation involving older Hayes River Group metavolcanics and younger Oxford Group metasediments. Various workers (see references in Bell, 1971) have interpreted the basal conglomerate member of the Oxford Group (containing both massive and gneissic tonalite pebbles) as lying disconformably or unconformably upon older Hayes River volcanics. Thus the possibility, of course, arises that two mafic volcanic sequences exist—one below and one above Oxford Group sediments.

Unconformity of the Pikwitonei Granulite Belt

Bell (1971) suggested that the orthopyroxene isograd—the Pikwitonei–Sachigo boundary—may be a metamorphosed unconformity. The boundary is in places a metamorphic discontinuity. Among most of its length the orthopyroxene isograd is separated from Cross Lake Group rocks (variable amphibolite facies) by layered anorthosite, porphyroblastic granodiorite gneiss, and granodiorite tonalite gneiss. Nowhere are

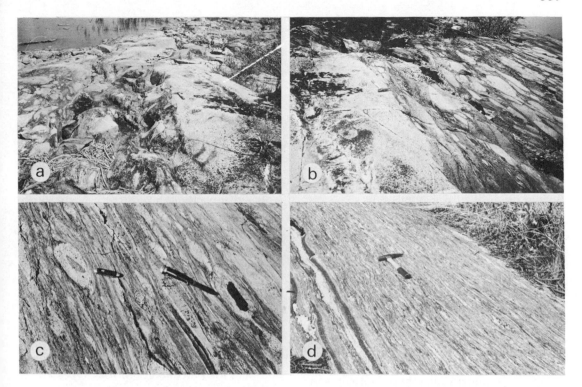

Fig. 4. The Cross Lake unconformity and Cross Lake Group conglomerate. a. Unconformity between basement biotite granodiorite gneiss (right) and conglomerate (GSC 202227-N). b. Unconformity between basement biotite granodiorite gneiss (left) and stretched granodiorite boulders lying in a greywacke matrix (GSC 202227-0). c. Granodiorite boulders with mafic skialiths (GSC 117932). d. Conglomerate with stretched pebbles and remnant beds of metagreywacke with conformable pegmatite segregations. Island 2 km west of where Minago River empties into the main body of Cross Lake

Cross Lake rocks, regionally transecting and truncating granulite gneiss, observed to be in contact with granulite grade gneiss. The same observation was made earlier regarding layered tonalite gneiss and supracrustal rocks in the southern portion of the Sachigo Belt. Pikwitonei rocks were probably at granulite grade at the time of the Kenoran orogeny, whereas basement rocks at the unconformity at Cross Lake were probably not metamorphosed. Considering the possibility that the granulite grade rocks may be older, two unconformities may exist in the area south of the Nelson Front. Bell (1971) considered the Pikwitonei craton to have been static relative to younger rocks during Kenoran and Hudsonian deformation; however, rocks of the Churchill and Superior Provinces were folded parallel to or brought into near-conformity with the margins of the Pikwitonei Belt.

Anorthosite

The origin of the anorthosite with respect to the Pikwitonei and Sachigo orogenic belt is puzzling. Bell (1971) considered the anorthosite to have been injected as a sill or sills before Kenoran deformation, but after granulite metamorphsism. He therefore placed the anorthosites within the Sachigo suite of rocks. Locally, the anorthosite is associated with prophyroblastic granodiorite gneiss, and Bell (1971) compared the 'granulite–anorthosite–porphyroblastic granodiorite gneiss' association with similar suites of rocks in other areas of the world in which 'granitic' rocks may represent localized K-metasomatism as a result of anorthosite intrusion into granulite facies gneisses. Anorthosites, other than those described by Bell, have not been recognized in Manitoba in the Superior province anywhere north of latitude 51. Large areas of the sills have been modified by K-metasomatism; however, locally they are well-preserved at the base of Cross Lake Group sediments at the orthopyroxene isograd and rarely within Pik-witonie granulite gneiss.

Bell (in press) describes the layered sills as having the following composition:

Anorthosite–gabbo anorthosite
Porphyritic hornblende anorthosite
Porphyritic gabbroic anorthosite
Gabbro
Hornblende diorite
Vanadium-bearing titaniferous magnetite.

The petrography of the anorthosites is described by Bell (in press) in great detail. Of the many textural varieties, the porphyritic hornblende anorthosites and football anorthosites, in which the average plagioclase composition is An_{75} and as calcic as An_{84}, are the most fascinating (Fig. 5). In these rocks, bytownite crystals range from $0 \cdot 5$ to 60 cm in length and massive anorthosite occurs with plagioclase approaching 98% of rock volume. Rounded phenocrysts may have concentric poikilitic growth rims marked by hornblende and biotite inclusions. Amphibolite and gneissic metagabbro in deformed portions of the sills are related to consanguineous gabbro, gabbro–porphyry and hornblendite layers.

Fractional crystallization of the sills from a plagioclase-rich gabbroic anorthosite magma occurred after injection (Bell, 1971). It is suggested here that the anorthosites are intrusive sheets which are related to a period of basalt extrusions which overlie the metasediments of the Cross Lake Group.

Berens Batholithic Belt

The Berens Batholithic Belt attains a width of 240 km in Manitoba and is distinguished from other areas by its great abundance of 'granitic' plutonic rock (96%) and dearth of recognizable supracrustal rocks (4%). The characteristics of 46,800 km^2 of the belt have been mapped and interpreted by Ermanovics (1970, 1971, 1973a, 1973b). The northern boundary is placed (Ermanovics 1973a) to coincide with a mylonite zone which separates a variety of homogeneous massive and metamorphosed and foliated granodioritic rocks to the south from the predominantly gneissic and layered tonalitic rocks of the Sachigo Belt to the north. The mylonite zone, which locally attains a width of $2 \cdot 4$ km, shows variable strain and *en echelon* strike progression. The southern contact of the Berens Belt is a similar, but more gradual, lithological change which in places coincides with small zones of cataclasis located on or just south of the Bloodvein River–Aikens Lake drainage systems (Ermanovics, 1973a and 1970).

In 15,600 km^2 of the central portion of the Berens Belt, Kenoran plutons (category (a) and (b) below) attain their greatest abundance and here the percentages of rock types are as follows:

(a) Quartz monzonite; massive (19%)
(b) Quartz diorite–granodiorite; massive (32%)
(c) Granodiorite gneiss (migmatite gneiss) (46%)
(d) Metasedimentary–volcanic rocks and amphibolite (3%).

From these percentages it is estimated that nearly all of the Berens Belt was molten (categories (a) and (b)) or in a process of remobilization (category (c)) during and toward the culmination of the Kenoran orogeny. A further estimate suggests that the lower crust beneath the Berens Belt provided 272,000 km^3 of average granodioritic magma within an area of 34,000 km^2, late in Archean time. Such plutons are estimated, from aeromagnetic data, to have a depth of at least 16 km, and judging from the high positive gravity response must be underlain by denser material below 16 km.

Granodioritic gneiss (category (c)) is believed to be equivalent to transformed tonalite–granodiorite gneiss commonly found in the Sachigo Belt. The transformation is

Fig. 5. Porphyritic and cataclastic hornblende anorthosite (photos by C. K. Bell). a. Football anorthosite. Single plagioclase megaphenocrysts containing hornblende chadacrysts that are concentrated in concentric growth rings. Large subhedral hornblende crystals fill the interstices. On the left, the plagioclase crystals coalesce to form massive anorthosite (GSC 117942). b. Porphyritic hornblende anorthosite. Matrix varies from pure hornblende to 'pea size' hornblende gabbro. Two, nearly euhedral, crystals 10 in. above hammer head show Carlsbad twins. Bay, southeast shore, Nelson River, 5 km below Kiskitto Lake outlet (GSC 117973). c. Porphyritic hornblende anorthosite. The megaphenocrysts coalesce (left-hand side) to form gabbroic anorthosite. The thin leucocratic dykes are aplite. East shore, Nelson River, east of Kiskitto Lake outlet (GSC 123728). d. Block structure in massive, fine-grained anorthosite. Blocks rounded by cataclastic milling. Crushed plagioclase matrix fills cracks in anorthosite blocks. Remnant amphibolite fragments preserved in matrix below hammer. From near centre of the West Channel body (GSC 117962). e. Hammer on block of layered anorthosite that is engulfed by cataclastic anorthosite. Interlayers of amphibolite are plastically deformed adjacent to blocky anorthosite. West side of bay, northcentral shore, Kiskitto Lake (GSC 117960). f. Anorthosite cataclasite with remnants of sheared amphibolite and pegmatite. The large plagioclase fragments still retain albite and Carlsbad twinning although they have been dynamically milled into sub-rounded shapes. Nelson River, 2·6 km below Kisipachewuk Rapids (GSC 117474)

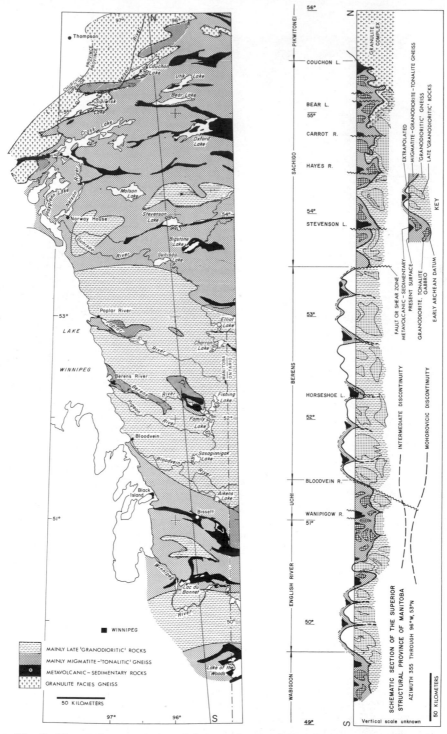

Fig. 6. Map and cross-section of the western Superior Province. Rock types have been reduced to a minimum in order to demonstrate major tectonic elements. Late 'granodioritic' rocks and granitized 'granodioritic' gneiss, produced during the Kenoran orogeny, 'exhume' older rocks during isostatic rebound. The Berens Batholithic Belt has a high positive gravity signature with respect to adjacent Belts, which suggests that at 15 km depth it is underlain by denser rock than that exposed at the surface

predominantly granitization by large scale lit-par-lit injection of granodiorite and quartz monzonite. Supracrustal rocks or relict dyke systems can be inferred from inclusions in north-west-striking mylonite zones. Rock vestiges of once larger areas of supracrustal rocks indicate middle to upper amphibolite facies metamorphism of basaltic, dacitic, and pelitic rocks (Ermanovics, 1970b). Late granitic rocks (categories (a) and (b)) have produced small aureoles of contact hornfels in regional amphibolite grade gneisses.

All evidence indicates that the Berens Batholithic Belt constitutes remobilized and uplifted Sachigo Belt type terrain. Widespread remobilization and magma intrusion (calc–alkaline, quartz diorite to quartz monzonite trend) closed the Kenoran orogeny.

South-western Superior Province

South of the Berens Batholithic Belt the geological pattern is quite different from the rest of the north-western Superior Province (Fig. 7). The following characteristics distinguish the southern and south-western Superior Province from the north-western Superior Province (Berens and Sachigo Belts):

(a) Lineal ENE distribution of supracrustal and gneissic rocks;
(b) Greater abundance of supracrustal rocks;
(c) Greater abundance of granoblastic paragneiss and migmatite in intervolcanic belt areas.

Whether the metavolcanic rocks of the south-western Superior Province are younger than those of the north-western Superior Province or the south-western Superior Province is simply an epi- to meso-zonal section with respect to the 'catazonal' north-western Superior is not known.

In Manitoba the potassic Berens Batholithic Belt is separated from the Uchi Volcanic Belt by a 'quartz diorite line' (Ermanovics, 1973). The transition zone contains diapiric quartz diorites, minor migmatitic paragneiss, and remnants of layered tonalite gneiss folded about NW and NE trending axes. These structures are truncated by ESE-trending Uchi Belt metavolcanics. Locally, close to the contact, they are reoriented parallel to the metavolcanic belt (Ermanovics, 1971; Fig. 3). South of the Uchi metavolcanics in the English River Gneiss Belt for a distance of 35 km metasediments, paragneiss and migmatite are clearly folded (and possibly refolded) along axes trending ESE (McRitchie and Weber, 1971).

The boundary between the Berens Belt and Uchi metavolcanics was found by Hall (1971) to involve a change of crustal thickness (Fig. 6). Using seismic refraction and reflection data, he determined the Mohorovicic discontinuity to be 36 km beneath the Berens Batholithic Belt, and at 30 km beneath the English River Gneiss Belt. The Riel or intermediate discontinuity, 18 km deep in the Berens Belt, drops to 22 km beneath the English River gneisses. Within the 'quartz–diorite line' and trending parallel to the Rice Lake metavolcanics a string of small ultrabasic bodies extends for 110 km. Some bodies are wedged between the metavolcanic–quartz–diorite rock interface, but all are associated with quartz–diorite. The lineal distribution of the ultramafics suggests that their emplacement was controlled by a fault related to the sloping crustal discontinuities (Fig. 6).

Age Determinations

Proceeding from classical geological argument, three stages and depth zones for the north-western Superior orogen are proposed based on an assumed repeated magmatic–orogenic cycle (Fig. 7).

This cycle is as follows:

(1) Deposition of lavas and sediments;
(2) Metamorphism, folding and synkinematic granites;
(3) Postmetamorphic or postkinematic granites.

An important feature of this proposal is that a postkinematic pluton of one stage becomes the prekinematic or synkinematic material for the subsequent stage. Postkinematic plutons of earlier stages thus tend to lose their identity, which imposes severe limitations upon attempts to determine their age.

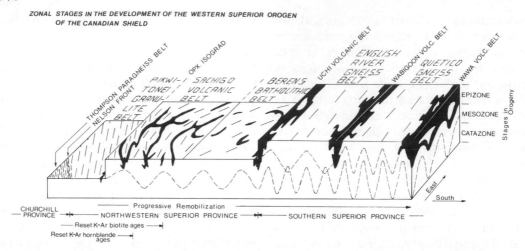

Fig. 7. Cartoon block diagram showing the major tectonic elements of the western Superior Province. In the southern Superior Province supracrustal assemblages (black shading) are thicker, migmatized paragneisses more abundant, and early tonalite gneisses less common than in north-western Superior Province. The Pikwitonei and Sachigo–Berens are interpreted as equivalent to older rocks underlying the English River–Quetico Gneiss Belts. The various belts may also represent various depths of erosion. K–Ar ages are progressively reset from a mean Kenoran value of about 2500 Ma to a mean Hudonian value of about 1800 Ma

The Pikwitonei craton represents one orogenic cycle whose products were raised to granulite–facies grade. The datum for this orogeny is not known. Synkinematic rocks in this cycle are represented by charnockite and enderbite, and postkinematic rocks by granodiorite.

The Sachigo Volcanic Belt records the second stage recognized as the Kenoran orogeny. Synkinematic rocks are tonalitic and may be related to volcanism; the layered anorthosite sheet may have been intruded at this time. The datum for the Kenoran orogeny may be represented by the Cross Lake unconformity and layered tonalitic and granodiorite gneisses throughout the Sachigo Belt. Sedimentary inclusions in basement rocks at the unconformities suggests the existence of supracrustal rocks that are pre-Cross Lake and may belong to the Pikwitonei cratonic stage. Postkinematic plutonic activity in the Sachigo Belt reached a maximum in the Berens Batholithic Belt at the close of the Kenoran orogeny.

The third stage is represented by south-western Superior Province rocks which on the whole show deformational styles which suggest that regional deformation and strain rates kept pace with regional metamorphism toward the end of the Kenoran orogeny. Additionally, volcanism may have occurred later (Krogh and Davis, 1971) or continued longer than in the north-western Superior.

K–Ar ages of biotites and hornblendes in all kinds of rocks vary from 2600 to 1800 Ma in the western Superior Province (Fig. 7). The end of the Kenoran orogeny is defined on the basis of K–Ar ages as approximately 2500 Ma ago (Stockwell, 1964). K–Ar ages of 2200 Ma and younger are confined to the north-western Sachigo Volcanic Belt and are recorded in rocks of the Cross Lake Group and retro-graded Pikwitonei granulites. Because no structural or metamorphic fabric is recognized to coincide with these ages, they are considered to have been reset during an Hudsonian thermal event. For example, 160 km south of the Nelson Front biotite ages are reset, but hornblende ages are not; however, 90 km south of the Nelson Front hornblende is also reset (Fig. 7). Within the 'dry' Pikwitonei granulites K–Ar ages of biotite and hornblende are erratic and range from 1800 to 3000 Ma.

Rb–Sr mineral and whole rock ages (87 Rb, $\lambda = 1 \cdot 39 \times 10^{-11} \, y^{-1}$) in southern Superior Province in Manitoba indicate the peak of metamorphism to have been around 2730 Ma ago and a second pulse at 2530 Ma. (Turek and Peterman, 1971; Penner and Clark, 1971; Farquharson and Clark, 1971). Poskinematic granitic rocks of the Berens indicate an age of 2700 Ma. Rb–Sr data are not available for the Sachigo or Pikwitonei rocks. However, preliminary Rb–Sr data on boulders of the Cross Lake Group and basal Goose Lake metasediments can be bracketed by two reference isochrons yielding ages of 3100 and 2800 Ma (pers. comm., 1974, G. S. Clark, Univ. Manitoba).

U–Pb determinations on zircons have shown the greatest degree of success in penetrating the metamorphic barrier at 2700 Ma. Krogh and Davis (1971) discovered a 2900 Ma year old zircon in trondhkemite gneiss in the Sachigo Belt. Since then, a granoblastic gneiss (possibly dacite) and a cataclastic quartz diorite have yielded zircon ages of 2900 Ma and 3000 Ma in a tectonic slice wedged between the Berens Belt and the Rice River metavolcanics of the Uchi Volcanic Belt (Krogh et al., 1974). Zircon ages of synkinematic quartz diorites and gneisses corroborate the Rb–Sr findings that a climax of orogeny occurred around 2700 Ma ago in the Uchi Volcanic Belt and English River gneisses in Manitoba. A previous plutonic event is recognized in the 3000 Ma year old zircon ages of the granoblastic gneiss and quartz diorite, whose whole rock Rb–Sr age was reset at 2700 Ma (Wanless et al., in prep.). One zircon age of 3100 Ma has been found in the southern Superior Province (pers. comm. 1975, T. E. Krogh, Geophysical Laboratory, Washington, D.C.) in a hornblende–biotite tonalite gneiss. This rock type is recognized commonly in the north-western Superior but less commonly in the southern Superior because of remobilization of early rocks.

Summary and Conclusion

The present paper argues for tectonic activity to have lasted for 700 Ma in post-Pikwitonei time (<3100 Ma) and the evidence indicates that trondhjemitic rocks were exposed at the time of Kenoran volcanism.

Remnants of one or more ancient Archean cratons are assumed to be represented by granulite facies gneiss, enderbites, charnockites and trondhjemitic rocks in northern Quebec, south-eastern Manitoba, and along the Superior-Churchill Province boundary just south of the Nelson Front (The Pikwitonei Belt). In Manitoba the craton broke along a zone parallel to the Front and orogenic activity occurred along this suture until 1700 Ma ago. Tonalitic orthogneiss and paragneiss yielding ages up to 3100 Ma are readily recognized in the Sachigo and Berens Belts as pre-volcanic (pre-Kenoran) in age. Such rocks are more difficult to recognize in the southern and south-western Superior Province because they appear to be intimately associated with thick, deformed supracrustal assemblages. Sparse age determinations and relations of intrusions to regional structures suggest that the axis of plutonic activity in the western Superior Province progressed from north to south between 3100 Ma and 2400 Ma ago. Presently available data fail to demonstrate a parallel age progression for the corresponding supracrustal assemblages.

Orogenic plutonic activity may proceed in stages whereby post-kinematic plutons of one stage become the pre-kinematic or synkinematic materials for the subsequent stage. Post-kinematic plutons of earlier stages thus tend to lose their identity. In the Berens Batholithic Belt, chemically reconstituted and thus isotopically rejuvenated 'younger' rocks underlie remnants of 'older' rocks (Fig. 6).

Acknowledgement

Dr R. F. Emslie critically read the manuscript and made a number of suggestions which greatly improved this paper.

References

Bell, C. K., 1962. 'Cross Lake map-area, Manitoba', *Geol. Surv. Canada*, Paper 61–22.

Bell, C. K., 1965. 'Reconnaissance mapping in Upper Nelson River area (63 NE)', *Geol. Surv. Canada*, Paper 65–1, 94–95.

Bell, C. K., 1966. 'Churchill–Superior Province boundary in north-eastern Manitoba', *Geol. Surv. Canada*, Paper 66–1, 133–136.

Bell, C. K., 1971a. 'Geological Investigations across the boundary between the Churchill and Superior Tectonic provinces', *Geol. Surv. Canada*, Paper 71–1, part A, 120–122.

Bell, C. K., 1971b. 'Boundary geology, Upper Nelson River area, Manitoba and northwestern Ontario', In *Geoscience Studies in Manitoba*, A. C. Turnock (Ed.), Geol. Assoc. Canada, Spec. Paper 9, 11–39.

Bell, C. K., in press. 'The Geology of Wekusko Lake map-area, Manitoba', *Geol. Surv. Canada, Memoir*.

Coats, C. J. A., and Brummer, J. J., 1971. 'Geology of the Manibridge nickel deposits Wabowden, Manitoba', In *Geoscience Studies in Manitoba*, A. C. Turnock (Ed.), Geol. Assoc. Canada, Spec. Paper 9, 155–165.

Coats, C. J. A., Quirke, Jr., T. T., Bell, C. K., Cranstone, D. A., Campbell, F. H. A., 1972. 'Geology and mineral deposits of the Flin Flon, Lynn Lake and Thompson areas, Manitoba, and the Churchill-Superior Front of the Western Precambrian Shield', *Field Excursion Guidebook A31–C31*, 24th Int. Geol. Cong. Canada, 1972.

De Waard, D., 1965. 'A proposed subdivision of the granulite facies', *Am. J. Sci.*, **263**, 455–461.

Douglas, R. J. W., 1973a. 'Geological Provinces', In *The National Atlas Canada*, Geol. Geol. Surv. Canada, 27–28.

Douglas, R. J. W., 1973b. 'Tectonics', In *The National Atlas Canada*, Geol. Surv. Canada, 29–30.

Elbers, F. J., and Gilbert, H. P., 1972. 'Munroe Lake Area', In *Summary of Geological Field Work*, Mines Br., Manitoba Publ. 72–3, 39–40.

Ermanovics, I. F., 1970. 'Geology of Berens River–Deer Lake map-area, Manitoba–Ontario—preliminary analysis of tectonic variation in the area', *Geol. Surv. Canada*, Paper 70–29.

Ermanovics, I. F., 1971. '"Granites". "granite-gneiss" and tectonic variation of the Superior Province in southeastern Manitoba', In *Geoscience. Studies in Manitoba*, A. C. Turnock (Ed.), Geol. Assoc. Canada, Spec. Paper 9, 77–82.

Ermanovics, I. F., 1973a. 'Precambrian geology of the Norway House and Grand Rapids map-area', *Geol. Surv. Canada*, Paper 72–29.

Ermanovics, I. F., 1973b. 'Evidence for early Precambrian Mg-rich crust and pre-Kenoran sialic crust in northwestern Superior Province of the Canadian Shield', In *Volcanism and Volcanic Rocks*, Geol. Surv. Canada, Open File 164.

Ermanovics, I. F., Park, R. G., Hill, J., and Geotz, P., 1975. 'Geology of Island Lake map-area Manitoba and Ontario', In *Report of Activities*, Geol. Surv. Canada, Paper 75–1A, 311–316.

Farquharson, R. B., and Clark, G. S., 1971. 'Rb–Sr geochronology of some granitic rocks in southeastern Manitoba', In *Geoscience Studies in Manitoba*, A. C. Turnock (Ed.), Geol. Assoc. Canada, Spec. Paper 9, 111–118.

Gibb, R. A., 1968. 'A geological interpretation of the Bouguer anomalies adjacent to the Churchill-Superior boundary in Northern Manitoba', *Can. J. Earth Sci.*, **5**, 439–453.

Goldich, S. S., Hedge, C. E., and Stern, T. W., 1970. 'Age of the Morton and Montevideo Gneisses and related rocks, Southwestern Minnesota', *Bull. Geol. Soc. Am.*, **81**, 3671–3696.

Hall, D. H., 1971. 'Geophysical determinations of deep crustal structure in Manitoba', In *Geoscience Studies in Manitoba*, A. C. Turnock (Ed.), Geol. Assoc. Canada, Spec. Paper 9, 83–88.

Horwood, H. C., 1935. 'A pre-Keewatin (?) tonalite', *Roy. Soc., Canada, Trans.*, **29**, sec. 4, 139–147.

Innes, M. J. S., 1960. 'Gravity and isostasy in Northern Ontario and Manitoba', *Publ. Dom. Obs.*, **21**, 6.

Krogh, T. E., and Davis, G. L., 1971. 'Zircon U–Pb ages of Archean metavolcanic rocks in the Canadian Shield', *Geophys. Lab. Yr. Bk.*, **70**, 241–242.

Krogh, T. E., Ermanovics, I. F., and Davis, G. L., 1974. 'Two episodes of metamorphism and deformation in the Archean rocks of the Canadian Shield', *Geophys. Lab. Yr. Bk.*, **73**, 573–575.

McRitchie, W. D., and Weber, W., 1971. 'Metamorphism and deformation of the Manigotagan gneissic belt, Southeastern Manitoba', In *Mines Br.*, Publ. 71–1, W. D. McRitchie and W. Weber (Eds.), 235–284.

Patterson, J. M., 1963. 'Geology of the Thompson–Moak Lake area', *Manitoba Mines Br.*, Publ. 60–4.

Penner, A. P., and Clark, G. S., 1971. 'Rb–Sr age determinations from the Bird River area, southeastern Manitoba', In *Geosc. Studies in Manitoba*, A. C. Turnock (Ed.), Geol. Assoc. Canada, Spec. Paper 9, 105–110.

Rance, H., 1966. 'Superior–Churchill structural boundary, Wabowden, Manitoba', *Ph.D. Thesis*, Univ. of Western Ontario.

Rousell, D. H., 1965. 'Geology of the Cross Lake

area', *Mines Br., Manitoba*, Publ. 62–4

Stockwell, C. H., 1964. 'Age determinations and geological studies', *Geol. Surv. Canada*, Paper 64–17. Pt. 2, 1–21.

Stockwell, C. H., 1969. *Tectonic Map of Canada*, Geol. Surv. Canada, Map 1251A.

Turek, A., and Peterman, Z. E., 1971. 'Advances in the geochronology of the Rice Lake–Beresford Lake area, Southeastern Manitoba', *Can. J. Earth Sci.*, **8**, 572–579.

Turner, F. J., and Verhoogen, J., 1960. *Igneous and Metamorphic Petrology*, 2nd Edn., McGraw-Hill, New York.

Wanless, R. K., Stevens, R. D., Lachance, G. R., and Rimsaite, J. Y. H., 1966. 'Age determinations and geological studies: K–Ar isotopic ages', *Report 6*, Geol. Surv. Canada, Paper 65–17.

Wanless, R. K., Stevens, R. D., Lachance, G. R., and Edmonds, C. M., 1968. 'Age determinations and geological studies. K–Ar isotopic ages', *Report 8*, Geol. Surv. Canada, Paper 67–2, Part A.

Wanless, R. K., 1969. *Isotopic Age Map of Canada*, Geol. Surv. Canada, Map 1256A.

Wilson, H. D. B., and Brisbin, W. C., 1961. 'Regional structure of the Thompson–Moak Lake nickel Belt', *Bull. Can. Inst., Mining Met.*, **64**, 815–822.

Wilson, H. D. B., 1962. 'Tectonics of Canadian Shield in northern Manitoba; *in* The Tectonics of the Canadian Shield', *Roy. Roc. Can., Spec. Publ.*, No. 4, 60–75.

Wilson, H. D. B., 1971. 'The Superior Province in the Precambrian of Manitoba', In *Geoscience Studies in Manitoba*, A. C. Turnock (Ed.), Geol. Assoc. Canada, Spec. Paper 9, 41–49.

Geochronology

Age and Isotope Constraints for the Evolution of Archaean Crust

STEPHEN MOORBATH

Department of Geology and Mineralogy, Oxford University, Oxford, England.

1. Age and Nature of the Oldest Commonly Occurring Rocks

Rocks yielding isotopic dates in the range *c.* 2500–2800 m.y. are found on all continents. A considerable proportion of the continental, sialic crust may already have existed at that time. Detailed Rb–Sr and U–Pb age work shows, for example, that about 50% of the area of the North American continent may have been in existence at about 2500 m.y. ago (Muehlberger et al., 1967). On the basis of exposed stratigraphic thicknesses in Archaean terrains, high pressure and temperature mineral assemblages in Archaean granulite terrains, trace element contents and ratios in Archaean volcanic rocks, etc., it appears that large portions of the Archaean crust were 25 km or more in thickness (Condie, 1973; McIver and Lenthall, 1974; Hunter, 1974).

Geological events within the 2500–2800 m.y. are of a very wide variety and represent widely differing crustal levels. The most diverse igneous, sedimentary and metamorphic rocks were being produced by a range of geological processes hardly less wide than those which occurred in later geological times, except that the biosphere probably only played a minor part in rock-forming processes. The characteristic bimodal Archaean 'granite-greenstone' association appears to have attained its maximum development at around this time on several continents (e.g. Green and Baadsgaard, 1971; Hanson et al., 1971; Arth and Hanson, 1975; Hawkesworth et al., 1975).

Recent age work shows that rocks exceeding 3000 m.y. in age are more prevalent than previously thought. The best documented examples come from the Godthaab and Isua areas of West Greenland, where the Amîtsoq gneisses have yielded concordant Rb–Sr and Pb/Pb whole rock ages, as well as zircon U–Pb ages, which average about 3750 m.y. (Black et al., 1971; Moorbath et al., 1972; Baadsgaard, 1973; Moorbath et al., 1975.) At Isua, a supracrustal succession of greenstone belt facies containing volcanics as well as clastic and chemical sediments (banded ironstones) has yielded similar ages (Moorbath et al., 1973; Moorbath et al., 1975). The stratigraphic relationship between the Isua supracrustals and surrounding Amîtsoq gneisses is not yet clear (Bridgwater and McGregor, 1974; James, 1975), but it is evident that the characteristic Archaean 'granite–greenstone' association was already in existence at about 3800 m.y. ago, only some 800 m.y. after the Earth's formation. The Isua area is so far unique in that it has escaped major tectonothermal and rock-forming events since about 3700 m.y. ago. In this, it provides a strong contrast with the Godthaab area, some 150 km to the southwest, where major plutonism and tectonism occurred during the interval of about 3000–2600 m.y. ago (Pankhurst et al., 1973a; 1973b), post-dating the deposition the Malene supracrustals and the emplacement of major anorthosite bodies (McGregor, 1973; Bridgwater et al., 1974).

Other 'old' dates have recently been reported on granite gneisses from the

Minnesota River Valley (Rb–Sr whole rock isochron age *c.* 3800 m.y., Goldich and Hedge, 1974) and on granulite gneisses from West Langøy, Vesteraalen, North Norway (Pb/Pb whole rock isochron age 3460 ± 70 m.y., Taylork 1975). Rhodesian basement gneisses in the Mushandike and Mashaba areas have yielded Rb–Sr whole rock isochron ages of about 3500–3600 m.y. (Hickman, 1974; Hawkesworth et al., 1975), confirming beyond doubt the field evidence (Wilson, 1973; Bickle et al., 1975) that the main Bulawayan greenstone belts, which yield Rb–Sr whole rock ages of about 2600–2700 m.y. (Hawkesworth et al., 1975), were deposited on much older granitic crust. No reliable dates have yet been reported from the Ancient Gneiss Complex of Swaziland, whose stratigraphical relationship to the Barberton greenstone belt supracrustal rocks has been much debated (Anhaeusser, 1973; Hunter, 1974), but it appears that the latter were perhaps deposited at around 3300–3500 m.y. ago (Hurley et al., 1972; Sinha, 1972; Jahn and Shih, 1974).

2. Constraints for Evolution of Precambrian Crust from Isotopic Evidence

Many geologists believe that Archaean (and post-Archaean) gneisses, whose ultimate origin is not recognizable from field evidence, are somehow derived by large-scale reworking of 'much older' sialic crust. (The term 'reworking' is used here in the sense of partial or complete melting leading to mobilization and reconstitution as an essentially new rock, and not in the frequently used sense of deformation and/or isochemical recrystallization.) On the whole, these views are not compatible with the results of some recently published Sr and Pb isotopic data on Archaean gneisses, or on greenstone-belt volcanics and associated cross-cutting granitic plutons, as will be discussed below.

(a) *Strontium isotopes*

Table 1 gives a summary of some recently reported Rb–Sr whole rock ages and initial $^{87}SR/^{86}Sr$ ratios of Archaean gneisses from Greenland, Scotland, Rhodesia and North America. Data for two suites of Proterozoic gneisses are also shown, one from south-west Greenland, the other from Colorado, U.S.A. Also tabulated are measurements on basaltic-to-andesitic greenstone-belt volcanics and associated (cross-cutting or presumed later) calc–alkaline plutons from Rhodesia and North America. All initial $^{87}Sr/^{86}Sr$ ratios are in close agreement in the general range 0·700–0·702, and fall rather close to the value presumed to be characteristic for the upper mantle at the corresponding age.

Fig. 1(a) and 1(b) shows averaged $^{87}Sr/^{86}Sr$ growth lines, calculated from the mean of the measured Rb/Sr values for each rock unit in the cited publications. The slopes of the lines are proportional to the mean Rb/Sr ratio for a given rock unit which are mostly within the range 0·2–0·4 for the tabulated gneisses and granites, although each rock unit exhibits considerable internal dispersion in Rb/Sr values. The data for Greenland, Scotland and Rhodesia are plotted in relation to a (hypothetically) linear upper mantle growth curve extending from 0·699 to 0·703 in 4600 m.y., corresponding to a Rb/Sr ratio of 0·02. The positions of all initial ratios fall almost on, or slightly above, the linear upper mantle growth line, and demonstrate that later rock units cannot possibly have been derived by the reworking of earlier ones of the type exposed in Greenland or Rhodesia. Limitations of space preclude plotting of the North American data. Arth and Hanson (1975) have suggested, for example, that there may be no older continental crust in the north-eastern Minnesota area because (i) there is no stratigraphic evidence for it and the area has yielded no isotopic ages greater than 2750 m.y., (ii) the initial $^{87}Sr/^{86}Sr$ ratios for all the igneous rocks are below 0·701 and preclude derivation of the magmas by remelting of older granitic crust and (iii) trace element models for the origin of any of the igneous rocks can be constructed without considering derivation from, or assimilation with, pre-existing crust. If there was an older granitic crust in north-eastern Minnesota (such as the possibly 3800 m.y. old Minnesota River valley gneisses, some 500 km distant) it behaved passively

Table 1. Rb–Sr whole rock ages and initial $^{87}Sr/^{86}Sr$ ratios of Precambrian rock units

Rock Unit and Area	Age (m.y.)	Initial $^{87}Sr/^{86}Sr$	Reference
(I) *Greenland and Scotland*			
Amîtsoq gneiss, Godthaab and Isua areas,			
West Greenland	3750 ± 50	0·7010 ± 0·0005	Moorbath et al., 1972
Nûk Gneiss, Godthaab,			
West Greenland	3040 ± 50	0·7026 ± 0·0004	Pankhurst et al., 1973
Ketilidian granite			
gneiss, Southwest	1890 ± 90	0·7022 ± 0·0010 ⎫	Van Breemen et al., 1974
Greenland	1780 ± 20	0·7032 ± 0·0005 ⎭	
Grey gneiss, Outer			
Hebrides, Scotland	2690 ± 140	0·7014 ± 0·0007	Moorbath et al., 1975a
(II) *Rhodesia*			
Basement granite			Hickman 1974; Hawkesworth
gneiss	3600 ± 200	0·701 ± 0·001	et al., 1975
Gwenoro gneiss	2780 ± 30	0·7011 ± 0·0001	Hawkesworth et al., 1975
Maliyami formation			
(Main greenstone			
belt)	2720 ± 70	0·7010 ± 0·0002	ibid
Sesombi tonalite			
pluton	2690 ± 70	0·7008 ± 0·0004	ibid
(III) *North America*			
(a) *Greenstone belt volcanics*			
Michipicoten,			
Ontario	2600 − 2700	0·7018 ± 0·0002	Hart and Brooks, 1969
Coutchiching and			
Keewatin, Minnesota			Hart and Davis, 1969;
and Ontario	2600 − 2700	0·7010 ± 0·0004	Paterman et al., 1972
Yellowknife, N.W.T.	2625 ± 160	0·7022 ± 0·0023	Green and Baadsgaard, 1971
(b) *Gneisses and granites*			
Minnesota River			
Valley Gneiss	approx. 3800 (?)	approx. 0·700 (?)	Goldich and Hedge, 1974
Northern Light,			
Saganaga, Icarus,			
Grants Range,			Hanson et al., 1971;
Algoman, Vermilion,			Peterman et al., 1972;
Minnesota-			Prince and Hanson,
Ontario border	2600 − 2700	0·7010 ± 0·0005	1972
Yellowknife,			Green and Baadsgaard,
N.W.T.	2640 ± 40	0·7011 ± 0·0008	1971
Chibougamou,	2740 ± 230	0·7009 ± 0·0005 ⎫	
Quebec	2680 ± 80	0·7005 ± 0·0002 ⎭	Jones et al., 1974
Wind River Range,			
Wyoming	2630 ± 20	0·702 ± 0·001	Naylor et al., 1970
Twilight Gneiss,			
Colorado	1805 ± 35	0·7015 ± 0·0004	Barker et al., 1969

* ^{87}Rb decay constant $1·39 \times 10^{-11}$ y^{-1}

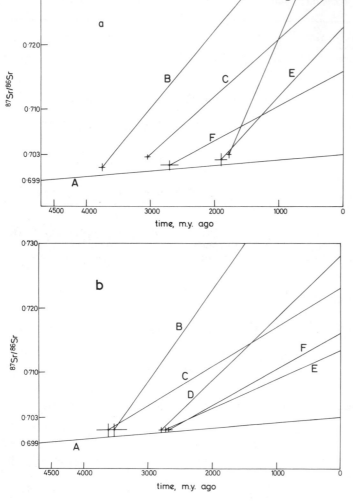

Fig. 1. Average ^{87}Sr/^{86}Sr growth lines for: (a) Greenland and Scotland. A, hypothetical upper mantle growth line: B, Amîtsoq gneiss, Godthaab and Isua area; C, Nûk gneiss, Godthaab area; D, E, Ketilidian gneisses, southwest Greenland; F, Grey gneiss, Outer Hebrides, Scotland. (b) Rhodesia. A, hypothetical upper mantle growth line; B, Mushandike granite gneiss; C, Mashaba granite gneiss; D, Gwenoro migmatitic gneiss; E, Maliyami formation greenstone belt; F, Sesombi tonalite pluton (cross-cutting the Maliyami formation). For references see Table 1

and was not the source for any of the later igneous rocks. In Rhodesia, the ancient granitic basement to the main greenstone belts has, indeed, behaved passively, as is evident from Fig. 1(b).

The measured ages and initial ^{87}Sr/^{86}Sr ratios for unaltered greenstone belt volcanics and later plutons almost certainly relate to the time of crystallization. In the case of a gneiss complex, they *could* relate to the time of regional metamorphism. In that case linear extrapolation of a given ^{87}Sr/^{86}Sr growth line back to the upper mantle growth line yields a maximum age limit for the gneiss precursor. This usually does not exceed about 100 m.y., and in some cases it is less than 50 m.y. Since

high-grade metamorphism frequently causes a decrease in the Rb/Sr ratio, the pre-metamorphic part of the ^{87}Sr/^{86}Sr growth line could have been significantly steeper, so that the age difference between formation of the precursor and its metamorphism would be correspondingly reduced. A non-linear upper mantle growth line (convex upwards, cf. Faure and Powell, 1972) would, of course, have a similar effect.

(b) Lead isotopes

It is well known that the Pb isotope compositions of several major lead ore deposits of different ages and from different localities approximate to a single-stage growth curve, although a re-examination of ore lead isotope systematics in the light of more precise data shows that the single-stage development model fails in detail (Doe, 1970; Richards, 1971; Doe and Stacey, 1974).

It has been widely held that close conformity of ore leads to single stage evolution ('conformable leads') indicates possible derivation from upper mantle source regions, whilst an increasing degree of departure from single-stage evolution for many ore leads seems best explained by the mixing of several isotopically heterogeneous source materials, with increasing involvement of crustal rocks. It appears, furthermore, from a study of Pb isotopic variation in modern oceanic basalts, that upper mantle source regions have been heterogeneous with respect to U/Pb over a long time (Doe, 1970). The situation is complicated by the problem of choice of parameters for construction of the primary growth curve, although this does not significantly affect the following discussion of Precambrian rock leads. The primary growth curve in Fig. 2 is based on the recommendations of Oversby (1974) using recently revised decay constants for ^{235}U and ^{238}U, and the latest values for the Pb isotope composition of troilite. A geologically and isotopically plausible value of 4570 m.y. is chosen for the Earth's age. (The effect of using a slightly different value for the latter is discussed by Doc and Stacey, 1974.) In the Oversby model, Archaean ore leads are grouped around a single-stage growth curve with a primary ^{238}U/^{204}Pb (μ_1) value close to 7·5.

With the relatively recent development of techniques for the extraction and precise isotopic analysis of rock leads, several sets of analyses have been reported on ancient gneisses. It appears to be a method of great promise. Good Pb/Pb whole rock or K-feldspar isochron ages may be obtained which are in excellent agreement with corresponding Rb–Sr whole rock isochron ages. Furthermore, in some cases, the Pb/Pb slope age agrees closely with the age obtained from the intersection of the isochron with the primary growth curve. This implies derivation of the rock unit from a source region of constant μ_1 from the time of formation of the Earth to the intersection age. (Note that the μ_1 value is conventionally calculated to the present time, with due allowance for decay of U to Pb.) However, the agreement between slope and intersection age is not always perfect. In addition, different areas within a single major rock unit may show slight relative displacements, i.e. the same slope, but slightly different intercepts of the Pb/Pb isochrons on the primary growth curve, corresponding to differences in calculated μ_1 values of up to 10%. This is observed, for example, in the Lewisian of north-west Scotland (Moorbath and Park, 1971; Moorbath et al., 1975a; also unpublished data). This demonstrates again that the single-stage evolutionary model fails in detail and that the calculated μ_1 value for a given major rock unit either represents derivation from an already heterogeneous U/Pb source region, or represents the sum total of all upper mantle and/or crustal processes that have produced the analyzed rock unit. In that case the rock leads are best described by a multi-stage evolutionary model, which may be too complex to be solved in detail (Gale and Mussett, 1973).

Despite all this, the astonishing and fundamental fact remains that Pb/Pb systematics in the few ancient gneisses that have been analyzed appear to be rather closely approximated by derivation from a single stage U/Pb system at a time which does not differ significantly from the age at which the rock unit acquired its present geological characteristics. Using the present model parameters, the calculated μ_1 values usually fall in the approximate range 7·3 to 8·0. Modern oceanic volcanics mostly yield μ_1 values of around 7·8–8·0 (Oversby, 1974). It appears that Pb in modern volcanic rocks and in ancient gneiss complexes may be derived from comparable source regions (for further discussion, see below).

Fig. 2 shows Pb/Pb whole rock isochrons for several Archaean rock suites, plotted in relation to the primary growth curve. These are the 3700 m.y. old Amîtsoq gneisses from West Greenland (Black et al., 1971; Moorbath et al., 1975b), the c. 2800 m.y. old granulite

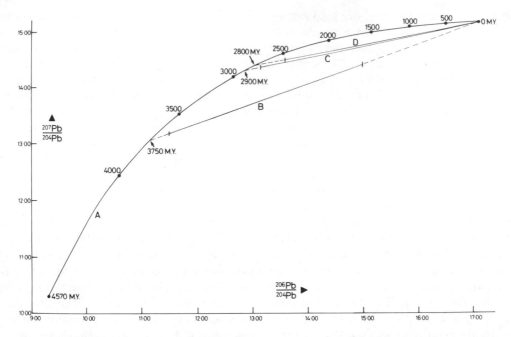

Fig. 2. Pb/Pb whole rock isochrons for Archaean gneisses. (A) Primary growth curve, (B) Composite for Amîtsoq gneisses, West Greenland (Black et al., 1971: Moorbath et al., 1975b; also unpublished work), (C) Fiskenaesset and Nordland granulite gneisses and associated anorthositic rocks, West Greenland (Black et al., 1973), (D) Composite for Lewisian gneisses, north-west Scotland (Moorbath and Park, 1971; Moorbath et al., 1975a; also unpublished work)

The continuous parts of the isochrons indicate the range of rock lead isotopic compositions actually observed. Most of the samples in each case fall towards the unradiogenic end. For all three isochrons, the ages calculated from the slopes are in close agreement with the indicated intersection ages.

Parameters for primary growth curve: age of Earth = 4750 m.y., decay constants, $^{238}U = 0.1551 \times 10^{-9}$ yr^{-1}, $^{235}U = 0.9849 \times 10^{-9}$ yr^{-1}, meteorite Pb isotope composition $^{206}Pb/^{204}Pb = 9.307$, $^{207}Pb/^{204}Pb = 10.294$, $^{238}U/^{235}U = 137.8$, primary $^{238}U/^{204}Pb$ $\mu_1 = 7.6$.

(Oversby, 1974)

gneisses and associated anorthositic rocks from the Fiskenaesset and Nordland regions of West Greenland (Black et al., 1973), and the c. 2700 m.y. old Lewisian gneisses from north-west Scotland (Moorbath and Park, 1971; Moorbath et al., 1975a; also unpublished data). The slope and intersection ages are in reasonable agreement in each case, whilst the μ_1 value for all the rock units lies close to 7·6.

The dispersion of points along a Pb/Pb isochron owes its origin to severe U/Pb fractionation at the measured slope age. This wide range of observed $^{238}U/^{204}Pb$ values in gneisses (and other crustal rocks) contrasts very strongly with the homogeneous μ_1 value calculated for the ultimate source region of the

rocks. Severe U depletion is common in some gneisses, yielding present-day unradiogenic Pb isotope ratios falling close to the primary growth curve. Many samples of Amîtsoq gneiss, for example, contain the most unradiogenic present-day Pb so far measured· on the Earth. This depletion in U (and several other incompatible elements) may be a metamorphic phenomenon (Heier, 1973a, 1973b), or it may have occurred during the formation of the immediate precursors of the gneisses from the relatively homogeneous μ_1 source region which is presumed to be the upper mantle.

A corollary of the above discussion, and of the data plotted in Fig. 2, is that the 2700–2800 m.y. old gneisses from West Greenland

and north-west Scotland cannot possibly be derived by reworking of rocks similar to the 3700 m.y. old Amîtsoq gneisses. Any rock derived by reworking of Amîtsoq gneisses *must* contain the same type of unradiogenic Pb. There is no geologically plausible, or even imaginable, way in which later rock units could approximate so closely to a single-stage evolution model of the type outlined above and yet be produced by the reworking of a much older rock unit with an isotopically very different (in this case much more ancient and highly unradiogenic) type of Pb.

Very recent, preliminary Pb isotope data on the Icarus syenodiorite pluton of north-eastern Minnesota has similarly led Arth and Hanson (1975) to conclude that the measured points represent primitive Pb derived directly from the upper mantle at about 2700 m.y. ago, the source region having a μ_1 value close to 7·4.

Thus the conclusions to be drawn from the Pb isotope evidence closely parallel those from the Sr isotopes, although not nearly as much Pb isotope data are as yet available. The hypothesis of major reworking of much more ancient continental, sialic crust can be rigorously tested and constrained for any given case by means of Sr and Pb isotope measurements.

3. Discussion

The following discussion is partly based on an earlier paper by Moorbath (1975).

The major conclusion from the type of data discussed above is that several Archaean and Proterozoic gneiss complexes, as well as greenstone-belt volcanics and associated later plutons, are predominantly juvenile additions to the continental crust at, or close to, the measured age, implying continental growth on a major scale during the early Precambrian. Also, the interval between time of extraction of juvenile igneous material from upper mantle and/or subducted oceanic lithosphere (on a hypothetical plate tectonic model) and time of production of the gneiss complex may be less than 50–150 m.y., thus falling within the analytical uncertainty of most age measurements on Archaean rocks. (The present author

confidently predicts that much of this discussion will eventually also be found to be valid for comparable Proterozoic rocks, as indicated by the limited data in Table 1.) It is possible that during such an 'accretion' event, major amounts of juvenile igneous material compositionally broadly equivalent to the island-arc tholeiite and calc-alkaline series were produced at depth in an ancient proto-island arc regime (Ringwood, 1974). Some of this juvenile material came directly to the surface of pre-existing continental or oceanic crust where it gave rise to the volcanic assemblage characteristic of greenstone belts. The remainder crystallized at depth as the plutonic equivalent of the high level rocks. Major element geochemistry of many ancient high-grade gneiss terrains, such as the *c.* 2700 m.y. old Scourian of north-west Scotland (Sheraton et al., 1973) suggests close affinity with calc–alkaline igneous rocks. Indeed, some authors favour an igneous origin for Precambrian pyroxene–granulite gneisses, involving transfer into the crust of a mantle-derived magma of approximately andesitic composition. Thus, Spooner and Fairbairn (1970) suggest that the part that reaches the upper crust will crystallize as normal extrusive and intrusive igneous rocks, whilst any part that is trapped in the lower crust at about 10 kb pressure will cool slowly to a minimum of about 500°C and crystallize directly into plagioclase–pyroxene assemblages characteristic of the granulite facies.

At any rate, at depth the juvenile plutonic material slowly crystallizes with igneous or metamorphic mineral assemblages, but at the same time already undergoes profound geochemical differentiation to produce a compositionally layered lower crust of the type visualized by Fyfe (1970, 1973), Brown and Fyfe (1970) and Heier (1973a, 1973b), in which the major and minor components of granite (*sensu lato*) as well as the incompatible elements migrate upwards to form calc–alkaline plutonic rocks, whilst leaving behind a thick residue of depleted granulite facies rocks at greater depth. Highly complex, gradational crustal layering results, whilst the conventional distinction between what is 'igneous' and 'metamorphic' is not always clear-cut. Intermediate zones of variably depleted calc–alkaline amphibolite facies gneisses must also be quantitatively important.

Plate tectonic models for continental growth and crustal evolution may well be adequate to explain the development of major Precambrian crustal units. The isotopic data cannot by themselves prove a plate tectonic model, but they are not inconsistent with it. On such a model, the isotopic data would suggest that the entire sequence of events including plate formation—plate separation (ocean floor spreading)—plate convergence—subduction—partial melting of basic oceanic lithosphere and/or upper mantle—extraction and emplacement of basaltic and calc–alkaline volcanic and plutonic rock types—accretion of new continental crust—profound geochemical differentiation of newly accreted continental crust—can be regarded as one continuous or semi-continuous 'superevent'. The major constraint on all this, for any given crustal segment, is a close grouping of rock-forming ages, low initial $^{87}Sr/^{86}Sr$ ratios and close approximation to single stage Pb/Pb evolution of the ultimate precursors. Such an 'accretion superevent' probably does not exceed 50–150 m.y. in duration, a value in general agreement with that found between the most recently formed complementary ocean ridges, island arcs and continental margins.

For the examples cited in this paper, the isotopic evidence suggests that the process of repeated reworking of significantly older sialic crust must be rejected. It seems, in any case, rather unlikely that relatively low-density, continental crust can be buried or subducted sufficiently deeply for really massive reworking of the type envisaged by many geologists. True reworking of older sialic crust *may*, of course, happen on a more limited scale above ancient and modern subduction zones, whilst it is probable that in certain tectonic regimes, such as orogenic belts formed in sites of continental collision with major crustal thickening, sialic crust can be truly reworked on a moderate scale. Localized melting of ancient, sialic crust may also occur in the heating zone above major bodies of basic and ultrabasic magma introduced into high crustal levels, as for example in the Tertiary igneous province of north-west Scotland (Moorbath and Bell, 1965). The numerous complicating factors involving partial melting of, and assimilation with, ancient sialic crust are clearly shown up by the complexities of initial Sr and Pb isotope ratios in many relatively high-level granitic rocks, as well as volcanics, of late Precambrian and Phanerozoic age (Doe, 1970; Faure and Powell, 1972).

With this evidence for relatively short periods of accelerated crustal growth (here termed 'accretion superevents'), the suspicion arises that they may have occurred episodically, rather than randomly or semi-continuously, during the Earth's history. They may well correspond with the most commonly observed groupings of radiometric dates relating to rock-forming events, which are *very approximately* 3800–3500, 2800–2500, 1900–1600, 1200–900, 500–0 m.y. ago. This problem, as well as the closely related one of the actual rate of continental growth throughout geological time, can only be satisfactorily solved by a large number of additional age and isotope measurements on major sialic rock units of Archaean and post-Archaean age.

References

Anhaeusser, C. R., 1973, 'The evolution of the early Precambrian crust of southern Africa', *Phil. Trans. Roy. Soc. Lond.*, **A273**, 359–388.

Arth, J. G., and Hanson, G. N., 1975. 'Geochemistry and origin of the early Precambrian crust of northeastern Minnesota', *Geochim. Cosmochim. Acta*, **39**, 325–362.

Baadsgaard, H., 1973. 'U–Th–Pb dates on zircons from the early Precambrian Amîtsoq gneisses, Godthaab District, West Greenland', *Earth Planet. Sci. Lett.*, **19**, 22–28.

Barker, F., Peterman, Z. E., and Hildreth, R. A., 1969. 'A rubidium-strontium study of the Twilight Gneiss, West Needle Mountains, Colorado', *Contr. Mineral. and Petrol.*, **23**, 271–282.

Bickle, M. J., Martin, A., and Nisbet, E. G., 1975. 'Basaltic and peridotitic komatiites and stromatolites above a basal unconformity in the Belingwe greenstone belt, Rhodesia', *Geology*, in press.

Black, L. P., Gale, N. H., Moorbath, S., Pankhurst, R. J., and McGregor, V. R., 1971. 'Isotopic dating of very early Precambrian amphibolite facies gneisses from the Godthaab District, West Greenland', *Earth Planet. Sci. Lett.*, **12**, 245–259.

Black, L. P., Moorbath, S., Pankhurst, R. J., and Windley, B. F., 1973. '$^{207}Pb/^{206}Pb$ whole rock age of the Archaean granulite facies metamorphic event in West Greenland', *Nature Phys. Sci.*, **244**, 50–53.

Bridgwater, D., and McGregor, V. R., 1974. 'Field work on the very early Precambrian rocks of the Isua area, southern West Greenland', *Geol. Surv. Greenland Rep.*, No. 65, 49–53.

Bridgwater, D., McGregor, V. R., and Myers, J. S., 1974. 'A horizontal tectonic regime in the Archaean of Greenland and its implications for early crustal thickening', *Precambrian Research*, **1**, 179–198.

Brown, G. C., and Fyfe, W. S., 1970. 'The production of granitic melts during ultrametamorphism', *Contr. Mineral. and Petrol.*, **28**, 310–318.

Condie, K. C., 1973. 'Archaean magmatism and crustal thickening', *Bull. geol. Soc. Am.*, **84**, 2981–2991.

Doe, B. R., 1970. *Lead Isotopes*, Springer–Verlag, Berlin.

Doe, B. R., and Stacey, J. S., 1974. 'The application of lead isotopes to the problems of ore genesis and ore prospects evaluation: a review', *Econ. Geol.*, **69**, 757–776.

Faure, G., and Powell, J. L., 1972. *Strontium Isotope Geology*, Springer–Verlag, Berlin.

Fyfe, W. S., 1970. 'Some thoughts on granitic magmas', In *Mechanisms of Igneous Intrusion*, G. Newall and N. Rast (Eds.), Geol. J. Special Issue, No. 2, 201–216.

Fyfe, W. S., 1973. 'The granulite facies, partial melting and the Archaean crust', *Phil. Trans. Roy. Soc. Lond.*, **A273**, 457–462.

Gale, N. H., and Mussett, A. E., 1973. 'Episodic uranium-lead models and the interpretation of variations in the isotopic composition of lead in rocks', *Rev. Geophys. and Space Phys.*, **11**, 37–86.

Goldich, S. S., and Hedge, C. E., 1974. '3800-Myr granitic gneiss in south-western Minnesota', *Nature*, **252**, 467–468.

Green, D. C., and Baadsgaard, H., 1971. 'Temporal evolution and petrogenesis of an Archaean crustal segment at Yellowknife, N.W.T., Canada', *J. Pet.*, **12**, 177–217.

Hanson, G. N., Goldich, S. S., Arth, J. G., and Yardley, D. H., 1971. 'Age of the early Precambrian rocks of the Saganaga Lake—Northern Light Lake Area, Ontario—Minnesota', *Can. J. Earth Sci.*, **8**, 1110–1124.

Hart, S. R., and Brooks, C., 1969. 'Rb–Sr mantle evolution models', *Carnegie Inst. Year Book 68 for 1968–1969*, 426–429.

Hart, S. R., and Davis, G. L., 1969. 'Zircon U–Pb and whole rock Rb–Sr ages and early crustal development near Rainy Lake, Ontario', *Bull. Geol. Soc. Am.*, **80**, 595–616.

Hawkesworth, C. J., Moorbath, S., O'Nions, R. K., and Wilson, J. F., 1975. 'Age relationships between greenstone belts and "granites" in the Rhodesian Archaean Craton', *Earth Planet. Sci. Lett.*, **25**, 251–262.

Heier, K. S., 1973a. 'Geochemistry of granulite facies rocks and problems of their origin', *Phil. Trans. Roy. Soc. Lond.*, **A273**, 429–442.

Heier, K. S., 1973b. 'A model for the composition of the deep continental crust', *Fortschr. Mineral.*, **50**, 174–187.

Hickman, M. H., 1974. '3500-Myr-old granite in southern Africa', *Nature*, **251**, 295–296.

Hunter, D. R., 1974. 'Crustal development in the Kaapvaal Craton, 1. The Archaean', *Precambrian Research*, **1**, 259–294.

Hurley, P. M., Pinson, W. H., Nagy, B., and Teska, T. M., 1972. 'Ancient age of the Middle Marker Horizon, Onverwacht Group, Swaziland Sequence, South Africa', *Earth Planet. Sci. Lett.*, **14**, 360–366.

Jahn, B. M., and Shih, C. Y., 1974. 'On the age of the Onverwacht Group, Swaziland Sequence, South Africa', *Geochim. Cosmochim. Acta*, **38**, 611–627.

James, P. R., 1975. 'Deformation of the Isua Block, West Greenland: a remnant of the earliest stable continental crust'. In press.

Jones, L. M., Walker, R. L., and Allard, G. O., 1974. 'The rubidium–strontium whole rock age of major units of the Chibougamou greenstone belt, Quebec', *Can. J. Earth Sci.*, **11**, 1550–1561.

McGregor, V. R., 1973. 'The early Precambrian gneisses of the Godthaab district, West Greenland', *Phil. Trans. Roy. Soc. Lond.*, **A273**, 343–358.

McIver, J. R., and Lenthall, D. H., 1974. 'Mafic and ultramafic extrusives of the Barberton Mountain Land in terms of the CMAS system', *Precambrian Research*, **1**, 327–343.

Moorbath, S., 1975. 'Constraints for the evolution of Precambrian crust from strontium isotopic evidence', *Nature*, **254**, 395–398.

Moorbath, S., and Bell, J. D., 1965. 'Strontium isotope abundance studies and rubidium-strontium age determinations on Tertiary igneous rocks from the Isle of Skye, northwest Scotland', *J. Pet.*, **6**, 37–66.

360

Moorbath, S., O'Nions, R. K., Pankhurst, R. J., Gale, N. H., McGregor, V. R., 1972. 'Further rubidium-strontium age determinations on the very early Precambrian rocks of the Godthaab district, West Greenland', *Nature Phys. Sci.*, **240**, 78–82.

Moorbath, S., O'Nions, R. K., and Pankhurst, R. J., 1973. 'Early Archaean age for the Isua iron-formation, West Greenland', *Nature*, **245**, 138–139.

Moorbath, S., and Park, R. G., 1971. 'The Lewisian geochronology of the southern region of the Scottish mainland', *Scott. J. Geol.*, **8**, 51–74.

Moorbath, S., Powell, J. L., and Taylor, P. N., 1975a. 'Isotopic evidence for the age and origin of the "grey gneiss" complex of the southern Outer Hebrides, Scotland', *J. geol. Soc. Lond.*, **131**, 213–222.

Moorbath, S.,, O'Nions, R. K., and Pankhurst, R. J., 1975b. 'The evolution of early Precambrian crustal rocks at Isua, West Greenland—geochemical and isotopic evidence', Submitted to *Earth Planet. Sci. Lett.* In press.

Muehlberger, W. R., Denison, R. E., and Lidiak, E. G., 1967. 'Basement rocks in continental interior of United States', *Bull. Am. Assoc. Petrol. Geol.*, **51**, 2351–2380.

Naylor, R. S., Steiger, R. H., and Wasserburg, G. J., 1970. 'U–Th–Pb and Rb–Sr systematics in 2700×10^6-year old plutons from the southern Wind River Range, Wyoming', *Geochim. Cosmochim. Acta*, **34**, 1133–1159.

Oversby, V. M., 1974. 'New look at the lead isotope growth curve', *Nature*, **248**, 132–133.

Pankhurst, R. J., Moorbath, S., and McGregor, V. R., 1973a. 'Late event in the geological evolution of the Godthaab District, West Greenland', *Nature Phys. Sci.*, **243**, 24–26.

Pankhurst, R. J., Moorbath, S., Rex, D. C., and Turner, G., 1973b. 'Mineral age patterns in ca. 3700 m.y.-old rocks from West Greenland', *Earth Planet. Sci. Lett.*, **20**, 157–170.

Peterman, Z. E., Goldich, S. S., Hedge, C. E., and Yardley, D. H., 1972. 'Geochronology of the Rainy Lake region, Minnesota–Ontario', *Geol. Soc. Am. Memoir* 135, 193–215.

Prince, L. A., and Hanson, G. N., 1972. 'Rb–Sr isochron ages for the Giants Range Granite, northeastern Minnesota', *Geol. Soc. Am. Memoir* 135, 217–224.

Richards, J. R., 1971. 'Major lead ore bodies—mantle origin?', *Econ. Geol.*, **66**, 425–434.

Ringwood, A. E., 1974. 'The petrological evolution of island arc systems', *J. geol. Soc. Lond.*, **130**, 183–204.

Sheraton, J. W., Skinner, A. C., and Tarney, J., 1973. 'The geochemistry of the Scourian gneisses of the Assynt district', In *The Early Precambrian of Scotland and Related Rocks of Greenland*, R. G. Park and J. Tarney (Eds.), Keele University, England, pp. 13–30.

Sinha, A. K., 1972. 'U–Th–Pb systematics and the age of the Onverwacht Series, South Africa', *Earth Planet. Sci. Lett.*, **16**, 219–227.

Spooner, C. M., and Fairbairn, H. W., 1970. 'Strontium 87/Strontium 86 initial ratios in pyroxene granulite terrances', *J. geophys. Res.*, **75**, 6706–6713.

Taylor, P. N., 1975. 'An early Precambrian age for granulite gneisses from Vikan i Bo, Vesteraalen, North Norway', *Earth Planet. Sci. Lett.*, **27**, 35–42.

Van Breemen, O., Aftalion, M., and Allaart, J. H., 1974. 'Isotopic and geochronologic studies on granites from the Ketilidian mobile belt of South Greenland', *Bull. Geol. Soc. Am.*, **85**, 403–412.

Wilson, J. F., 1973. 'The Rhodesian craton—an essay in cratonic evolution', *Phil. Trans. Roy. Soc. Lond.*, **A273**.

Thermal Regimes

Archean Thermal Regimes, Crustal and Upper Mantle Temperatures, and a Progressive Evolutionary Model for the Earth

Richard St. J. Lambert

Department of Geology, University of Alberta, Edmonton, Alberta, Canada.

Abstract

The problem of temperature gradients in the outer part of the Earth in the past is considered from petrological and geophysical angles. Archean thermal gradients can be guessed at from studies of regional metamorphism. In Africa, parts of Canada and the N. Atlantic Craton, high T/P gradients are commonly recorded but there is some evidence that this may be due to local igneous activity. Elsewhere in the N. Atlantic Craton, Canada and the U.S.S.R. lower gradients are known. Crustal gradients reaching 700 to 1000°C at 30 km appear to be normal maxima, while high P/T gradients are entirely absent. Heat generation and the attainment of equilibrium situations in the upper mantle and crust can be calculated for various simple models in the range 3600–2600 Ma. Preferred models all have a thick basaltic crust overlying K-rich mantle in turn overlying K-poor mantle. Other models fail to produce reasonable gradients under the high heat generation conditions of the Archean (two to three times greater than at present). The preferred model gives a very thin lithosphere which will have been too thin to be subducted. The generation of K-rich upper crust by differentiation produces thicker lithosphere, lower thermal gradients at depth and continental stability, presumably regionally effective by 2600 Ma. These considerations, combined with lunar evolution data lead to the following evolutionary pattern for the Earth:

4500–3900 (3800?) Ma	chaotic phase
3900 (3800)–3400 Ma	anarchic phase
3400–2600 Ma	scum tectonics: the Archean proper
2600–?600 Ma	skin or single-plate tectonics
?600 Ma–today	multi-plate tectonics

Introduction

This paper is concerned with the problem of the thermal regimes near the Earth's surface in the Archean, which is taken as the time prior to 2600 Ma ago. Three approaches may be used to define such regimes: consideration of the nature and origin of igneous complexes; regional metamorphism; and estimates of heat flow allied with calculations of thermal gradients. A wide variety of opinions have been expressed on all three subjects, but so far as one is aware, no author has attempted to combine information from all sources into a unified theory. In the following, the metamorphic and geophysical approaches will be considered, but the igneous problem will be ignored.

Regional Metamorphism

The commonly expressed view of Archean metamorphic regimes is that they are typically of the low-pressure type, but the conclusion which will be reached in the following will be rather less clear-cut, namely that gradients of the high-pressure type can be excluded, but that all others are possible.

The reviews by Windley and Bridgwater (1971), Saggerson and Owen (1971) and Saggerson and Turner (1972) emphasize the low-pressure character of Archean metamorphic complexes. Saggerson and Owen, and Saggerson and Turner concentrate on African Precambrian complexes of all ages, but particularly draw attention to the widespread occurrence of cordierite, both in amphibolite– and granulite–facies terrains. The review of Windley and Bridgwater, in which attention is drawn to the contrast between gneissic and granite–greenstone Archean Complexes, also summarizes Archean metamorphism by reference to high-temperature, medium- to low-pressure environments. Reference to development of a low-pressure facies in the Kalgoorlie greenstone belt has also been made by Glikson (1971), and in the Kaapvaal craton by Hunter (1974).

That such metamorphism was not the only type known in the Archean, however, is clear from the account of the Siberian shields by Kratz and Glebovitsky (1972) who emphasize the occurrence of early high- or intermediate-pressure granulites, or kyanite–sillimanite series, followed by later Archean higher thermal gradient metamorphism in the 'almandine–amphibolite' facies: also, Late Archean or Lower Proterozoic regional metamorphisms were of andalusite–sillimanite type. Specific examples of series which probably developed in a higher pressure environment are the Lewisian Scourie assemblage

garnet–pyroxene–granulites (8–11 kb at about 800°C, Muecke 1969) and perhaps the Yellowknife greenstone belt (Boyle 1961) which passes from the greenschist to epidote–amphibolite facies where remote from later intrusions. The metagreywackes adjacent to the Yellowknife greenstones contain garnet–staurolite–cordierite assemblages, as yet unstudied in detail, associated laterally with the well-known cordierite–andalusite assemblages surrounding later granites (Boyle, 1961; Ramsay, 1973). A broadly similar situation in a lower grade environment has been described by Jolly (1974) in the Abitibi belt, where prehnite–pumpellyite assemblages are succeeded by greenschists, in turn passing into a series of 'hornfelsic' facies adjacent, and only adjacent, to plutons. This association of the lower-pressure facies series with igneous intrusions is a significant feature of the Archean (as of later terrains) and, because many Archean complexes are extensively cut by or almost entirely made of igneous complexes, the suspicion must prevail that much of the metamorphism is associated with the igneous process.

Altogether a more important problem, however, is how to translate the information gained from these broad generalizations into actual PT values. Unfortunately there are as yet virtually no detailed studies of the mineralogy of truly regionally metamorphosed Archean rocks, and none which yield closely defined PT conditions to a degree comparable with studies such as those on the Alps and Caledonides. Reference to Fig. 1 (based on Lambert, 1972) shows that basic assemblages only yield highly generalized geothermal gradients, and that even kyanite-free andalusite-sillimanite series could develop along a geotherm such as (2), which is by no means extreme (36°C/km). Admittedly, use of the Holdaway (1971) phase diagram rather than that of Richardson, Bell and Gilbert (1969) would alter this figure to 40–45°C/km, but many Archean complexes contain kyanite-sillimanite sequences (references cited above) for which it makes less difference which experimental data is used. The significance of cordierite is also by no means clear, as it is

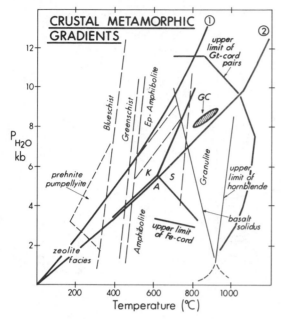

Fig. 1. Approximate PT distribution of basaltic facies, with aluminosilicate stability fields, outer limits of the garnet–cordierite stability field and upper limit of iron–cordierite stability. The shaded area indicates the PT region of occurrence of most of the granulite–facies garnet–cordierite pairs discussed by Hensen and Green (1973). The two geotherms represent a calculated Archean greenstone geotherm (1) and a higher geotherm (2) associated with basic magma intrusion at depth

stable in common compositions right up to the potential limits shown on Fig. 1, following the data of Hensen and Green (1973) on garnet–cordierite pairs. The latter authors give some data which may be relevant to the present question, though they themselves offer caution over the use of their data (p. 158). Data from eight granulite facies garnet–cordierite pairs are centred around the shaded area of Fig. 1 except an assemblage at 970°C/10 kb from Enderby Land. Gradients averaging 33–34°C/km would produce such assemblages: however, the age of most of these assemblages is not clear and only half of the examples are likely to be Archean. Currie (1971) gave data on garnet–cordierite pairs which differ significantly from that of Hensen and Green in absolute P and T, but which would indicate only slightly higher P/T gradients than the assemblages studied by Hensen and Green. There is an indication from these studies on

garnet and cordierite that granulite gradients may be higher pressure than is sometimes imagined.

I conclude that we have a long way to go before we can assert that Archean metamorphic gradients were truly of any particular character. Published evidence seems to give different conclusions in different areas, experimental data has hardly yet been applied, and detailed phase studies of natural systems are quite inadequate. The only certain conclusion is that blueschists and eclogites are absent, and therefore, presumably, so are the modern tectonic conditions which give rise to them.

Heat Generation and Heat Flow

The alternative method of finding temperature gradients is to calculate them directly from known and estimated thermal parameters. In the following, it is assumed that there is a thin lithosphere in which heat transport is generally by conduction and that the whole process is at equilibrium. It is equally assumed that below the depths considered heat transfer will be by convection (and possibly radiative transfer), both of which will increase the rate of movement of thermal energy above that due to conduction. Consequently, temperature gradients at depth will be less than those nearer the surface, and at the same time, since convection will probably be associated with the presence of a partial melt near the Earth's surface, there will be a steady chemical differentiation process. The thermal equilibrium assumption is much more questionable, but the critical areas of the Earth's surface for tectonic considerations are not continental, but are oceanic, particularly in the Archean and Proterozoic. Current tectonic models for the early stages of Earth's history concentrate on events in proto-continents, thereby missing out on discussion of the great majority, perhaps 80–90%, of the surface. The Pacific Plate, the unique oceanic plate, has characteristics which make one feel that investigation of the possible nature of oceanic regions in earlier times is essential to our understanding of the early history of the Earth. Ancestral Pacific or oceanic plates or

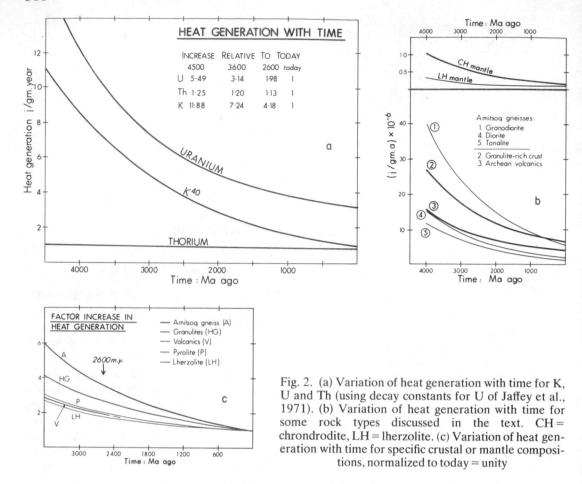

Fig. 2. (a) Variation of heat generation with time for K, U and Th (using decay constants for U of Jaffey et al., 1971). (b) Variation of heat generation with time for some rock types discussed in the text. CH = chrondrodite, LH = lherzolite. (c) Variation of heat generation with time for specific crustal or mantle compositions, normalized to today = unity

regions of the Earth should, if anything like the Pacific region, have reached a thermal equilibrium in their older areas, just as the present Pacific tends towards a uniform heat flow in the west. In the following, therefore, oceanic thermal models are investigated, to see what effects the higher heat production in the past might have on possible stable states.

For simplicity, uniform compositional models are adopted, allowing application of the basic heat flow equation for one uniform layer, $T_s - T_z = qz/K - Az^2/2K$, which can be generalized for two or more layers as required (Jaeger, 1965, p. 9). In this relationship T_s is temperature at the upper surface of a layer of thickness z, conductivity K and heat generation A. The heat flow at the upper surface is q and the temperature at depth z is T_z. Other authors have assumed variations of A with depth, either exponentially or linearly

decreasing downwards, but in view of the uncertainties in all of q, A and K, such variations were felt to be unnecessary for the present purpose.

Relative heat production in the Archean is the easiest to evaluate of the required parameters, although actual distribution and quantity of heat-producing elements K, U and Th is only poorly understood. Fig. 2 shows heat production of K, U and Th with time and relative heat-production rates for some of the rock compositions used in this study. It can be seen that heat production by Th, at any reasonable relative abundance, is much less significant in the Archean than that of K and U; also that the most important basic and ultrabasic rock types have values of A at 2600 Ma and 3600 Ma of very close to two and three times present-day levels, respectively. Rocks which have relatively high K/U have even higher heat productions in the remote past.

Abundances of K, U and Th are not well understood and a very wide range of estimates are available for both crust and mantle. The problem has been extensively discussed (Clark and Ringwood, 1964; Wasserburg et al., 1964; MacDonald, 1965; Shaw, 1967; Hanks and Anderson, 1969; Lubimova, 1969; Gast, 1972; Herrin, 1972 and Heier, 1973 and references therein). Generally speaking, each successive investigation has tended to reduce previous estimates of abundances, and it has become clear that K and U, especially the latter, must be concentrated in the outer Earth, and that the mantle is laterally inhomogeneous with respect to K, U and Th. In the course of this attempt to construct Archean thermal models, it was found that both low overall abundances and concentration of K, U and Th near the surface are absolutely necessary to give a stable solution. For the calculations given below, data as in Table 1 were adopted.

The other parameters, q and K, are more speculative. Because they directly control thermal gradients, it became essential to adopt conventional values based on reasonable assumptions. q_0, heat flow today, reaches low values of 48 mW/m^2 (mean) in the Cretaceous areas of the Pacific and 38 hfu in Precambrian shields (Sclater and Francheteau, 1970). The world average is close to 63 hfu (Lee and Uyeda, 1965), this high value being due to convective systems (e.g. oceanic ridges) which may be regarded as transient. Taking a value of 45 hfu as a compromise for the long term, leads to the conclusion that values of 90 and 135 hfu could be expected in 'stable' Archean areas at 2600 Ma and 3600 Ma, respectively, if a simple extrapolation is reasonable and thermal equilibrium always maintained. As will be seen, however, such values yield temperatures at shallow depth which are higher than is reasonable.

Conductivity (K) values for the ensuing calculations were chosen from the array used in recent papers. Birch (1955) showed that common crustal rocks had conductivities which varied from 0·021 j/cm sec deg to double that value, the higher conductivities being in ultrabasic rocks. Values decline with increasing temperature, towards a common level in the range 0·021–0·025. Minerals with particularly high K include hornblende, pyroxene and quartz (Beck, 1965) as well as olivine

Table 1. Element abundances and heat generation

Rock-type	K ppm	U ppm	Th ppm	Heat generation 3600 Ma 10^{-6} j/gm.a.
Archean volcanics (Eade and Fahrig, 1971)	5900	0·6	2·0	12·5
Granulite crust (based on Heier, 1973)	11000	0·25	1·10	12·3
Archean amphibolite facies gneisses (Eade and Fahrig, 1971)	24000	1·3	9·9	42
Low-K average mantle (Model III, Hanks and Anderson, 1970)	136	0·0232	0·074	0·41
Lherzolite mantle	59	0·019	0·05	0·28
Pyrolite I ($\cong$ K-rich mantle)	2350	0·189	0·79	4·5
Pyrolite II ($\cong$ K-poor mantle)	760	0·059	0·25	1·4

Note: the curves of Figs. 4 and 5 were calculated using some of these abundances adjusted in order to give the stated values of q_{3600}.

(Fukao, 1969), a value of 0·042 j/cm sec deg for the latter being suggested for upper mantle conditions. Accordingly, in the following, values of 0·021–0·029 have been chosen for crusts of various compositions, of 0·029 for the uppermost mantle and of 0·042–0·046 for the deeper mantle layer in the range studied. The mantle has been considered opaque to radiative heat transfer in the range of temperature under consideration, other than as provided for in Fukao's calculation.

The problem was approached by extrapolating back from modern thermal gradients and by considering possible limiting cases for the Archean at 2600 Ma and 3600 Ma. Results from the first approach will be discussed elsewhere in relation to Proterozoic tectonics, but are in close accord with the results outlined here. One-layer models with a uniform mantle composition and no crust give impossibly high gradients, even with the very low K, U and Th abundances of the lherzolite (LH) model (Fig. 3). An automatic

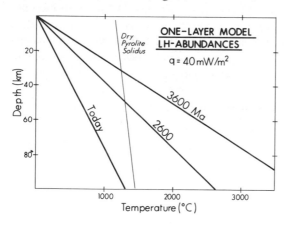

Fig. 3. Thermal gradients in a hypothetical refractory mantle with LH heat generation, one-layer model

redistribution of K, U and Th must occur under such gradients, concentrating them upwards in a sialic crust. Accordingly, two-layer models were examined, with similar results. For any reasonable combination of basaltic crustal thickness and composition overlying a uniform mantle, with total K, U and Th equivalent to a homogeneous LH mantle, temperatures at 2600 Ma and 3600 Ma at

100 km depth are impossible. This is essentially because a basalt crust, 30 km thick over the Earth, only contains 33% of the total heat production, even given the very low LH model concentrations of K, U and Th (Table 1) which are below the levels required to give present-day q. Where heat sources are primarily at greater depths than those being considered, gradients tend to be linear.

It is therefore concluded that because all reasonable one-layer and two-layer models have built-in suicidal tendencies through convection and accompanying fractionation, three- or four-layer models are always required, with a K-rich mantle overlying K-poor mantle. Examples of gradients in a three-layer model with K-rich mantle 70 km thick underlying 20 to 30 km basalt are shown on Fig. 4, compared with Green's 1975 Model III Archean geotherm. The thickness of basalt is seen to be unimportant, unlike variations of either K or the thickness of K-rich mantle. Also, the curve for $q = 69$ (two- or three-layer model) corresponds closely to Green's Model III to a depth of 40 km (the geotherms of Fig. 4 were calculated prior to publication of Green's 1975 article). Below 47 km, Green's model implies some convective transfer of heat and/or a much thicker 'K-rich' mantle layer with lower average K-content: it cannot be simulated easily, however, in terms of a simple three-layer model without introducing very high values of conductivity.

For higher values of q_{3600}, such as 88, shown on Fig. 4, melting must develop at <40 km and partial melting and fractionation will occur continuously below that depth until all heat-producing elements are transferred to high levels. Eventually, therefore, q will fall, leading to the lower gradients of Fig. 4. I would argue that large-scale continent production had not occurred by 3600 Ma and that, therefore, the $q = 88$ gradient of Fig. 4 is invalid. The problem created by the lower q_{3600} values is, of course, that they are too low by comparison with $q_0 = 40$ to 60, assuming thermal equilibrium. The equivalent heat flow today, 23, represents heat production from an Earth with an average mantle composition of only 50–60 ppm K, with all crustal K in the mantle.

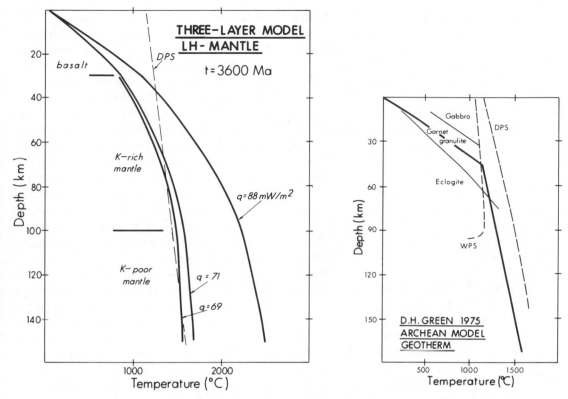

Fig. 4. (a) Preferred three-layer models for 3600 Ma age with low q, indicating a shallow asthenosphere (80 km maximum). (b) The Model III geotherm of Green (1975) for comparison with (a). Note that the $q - 71$ geotherm is close to Green's geotherm to about 40 km depth ((b) reproduced with permission of The Geological Society of America)

While this is possible, this ratio of observed/predicted heat flow is too high to be satisfactory, even assuming the present to be non-representative: it can hardly be assumed that heat flow through today's continents is too high on a transient basis by a factor of 50%.

The alternative solution is to assume that the Earth was warming up 3600 Ma ago, leading to lower surface heat flow than in a thermal equilibrium situation. MacDonald (1965) gave thermal histories showing such a possibility, but Hanks and Anderson (1969) criticized these models for ignoring heat production (1600 j/gm) during core formation (Birch, 1955). This is approximately equivalent to the heat produced in $4 \cdot 10^8$ a by a low-K lherzolite mantle 3600 Ma ago, and, if the core was produced quickly would necessarily raise the temperature of the Earth substantially: Birch estimated 1570°C. Hanks and Anderson (1969) calculated that core formation could

occur ~3600 Ma ago if the Earth accreted in $2 \cdot 10^5$ a, or even earlier if accretion was more rapid. They conclude: 'Core formation and large-scale differentiation of the Earth upon accretion seem most likely' (p. 28). The geochemical Earth accretion models of Clark, Turekian and Grossman (1972) and Grossman (1975) imply very short accretion times, in line with Hanks and Anderson's reasoning. However, Grossman's calculations require that U is preferentially incorporated in early condensing fractions, and thus it is possible to envisage a model in which a core forms effectively instantly on accretion, overlain by a stratified mantle with U below, K in the middle reaches and volatiles in the outer portions. Not until melting or convective overturn developed extensively in the mantle would U be released upwards, so it is just possible that the outer Earth was comparatively cool 3600 Ma ago and was still warming up to the point of extensive fractionation. The very high K/U and very low Th/U abundances of the Amîtsoq gneisses (Lambert and Holland, this volume, and their low U/Pb, Baadsgaard, pers. comm.) are pointers in this direction.

The preferred oceanic model of Fig. 4 ($q_{3600} = 69$) gives a partial melt in a dry mantle at 75 km, or in the event of a hydrous mantle at even lower depth. Because convective transfer of heat will tend to reduce and not increase this depth, it may be regarded as a *maximum* depth to an Archean low-velocity zone. Also, if q_{3600} was any higher, the most likely variant on the above calculation, the depth would also be less. Green (1975) has argued that such a gradient would not permit eclogite formation or thence the triggering of subduction. Perhaps, however, the question of whether a thin hot lithosphere can be subducted at all is more significant. Griggs (1972) shows a density profile for an 80 km thick inclined lithospheric slab, in which 60% of the density excess in the depth range 100–300 km and only 8% in the top 100 km: thus (even allowing for extensive variation of parameters) it is probable that the early rate of downward movement must be slow, until olivine in the

slab begins to invert to a spinel structure at 300 to 400 km depth. I would argue, qualitatively, that in the Archean, with heat production three times as great as at present, that the chances of a slab developing are negligible, for any downwarped lithosphere will warm up rapidly compared with today's situation. The time required to heat a slab is also proportional to the square of the thickness of the slab, so any subducted thin Archean lithosphere would equilibrate relatively quickly with the mantle. Yet another argument, based on the assumptions above, is that the upper mantle may have been K-rich relative to today, perhaps producing heat at an order of magnitude greater rate.

The subsequent evolution of this model Archean oceanic lithosphere and shallow asthenosphere can be seen on Fig. 5, where the same parameters are used to calculate the standard 2600 Ma geotherm. the maximum depth of the low-velocity zone is 90 km and, as argued above, it was probably less than that rather than more. The factors inhibiting subduction are still all present: little likelihood of eclogite formation; high heat generation; thin slabs; and a K-rich environment in the upper

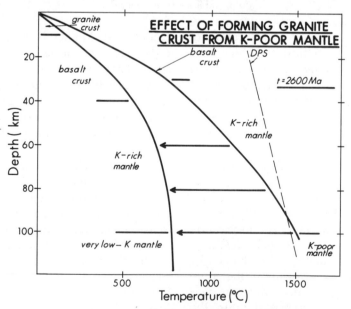

Fig. 5. Evolution of a K-rich upper mantle (three-layer model) to give a continental crust (four-layer model) with a low geotherm and a mesosphere almost devoid of potassium, uranium and thorium

mantle. However, progressive fractionation will have occurred and the granodiorite–greenstone association produced, known to be in the age range 3300 Ma to 2600 Ma at least. Each successive fractionation will tend to produce a K-richer crust, the effect of which is shown in Fig. 5. The K, U and Th concentrations in the 'granite' crust are those of the average Canadian shield amphibolite facies rocks (Eade and Fahrig, 1971) and the K, U and Th are presumed to have been extracted from the K-poor lower mantle, leaving it almost devoid of them. The result is quite dramatic, producing a geotherm of even lower temperature than the modern shield geotherm, because the model assumes a 60-km K-rich upper mantle, absent below the modern shields. Its accuracy is therefore a little questionable, but it vividly illustrates the kind of change which must occur following continent creation, and thereby the reason for the fundamental stability of continental structures. Extracting the 'granite' from the K-rich layer produces a similar but less dramatic change. Therefore, in those regions of Earth where appreciable continental scum had accumulated, continental stability slowly developed, the original thermal energy having been used by transporting melts to the surface and cooling them quickly there.

Discussion: A Model for Earth's Early History

The inferred crustal metamorphic gradients and the calculated gradients for 2600 Ma ago (Figs. 1 and 5) are in very good agreement, curve 1 on Fig. 1 being very close to the pre-granite crust curve on Fig. 5. The standard higher T/P geotherm (curve 2) on Fig. 1 corresponds therefore to the kind of gradient which would exist in the late Archean or early Proterozoic where some basic magma had penetrated to high levels, convecting additional heat into the system. Both lines of reasoning support the concept of a thin, non-subducting lithosphere, and therefore a non-plate-tectonic situation. It will be reasoned elsewhere that a similar situation was maintained until late in the Proterozoic, but that the latter time was marked by relatively stable continental and thick oceanic crust, which overlay a thin but continuous lithosphere, behaving essentially as one plate, though occasionally fracturing or being intruded by diapiric magma from ancient analogues of hot spots. This plate, or skin, could slowly drift as one unit over the thick K-rich asthenosphere, giving rise to the peculiar polar wandering paths discovered from the Proterozoic continental areas. Eventually, the oceanic lithosphere cooled enough to commence the critical stage of eclogite formation, producing local subduction which eventually developed into a world-wide system. It may be relevant to this discussion that andesites seem scarce in the Proterozoic; perhaps because there was no quartz–eclogite to be partially melted in subduction zones.

The earlier history of Earth, particularly pre-3600 Ma, is less easily discerned. Lacking evidence from actual examples, the nature of the outer Earth is obscure. However, it is now known that Mars, Moon and Mercury all have an outwardly broadly similar appearance, indicative of intense meteorite bombardment and cratering, in the case of the Moon until 3900 to 3800 Ma ago (Turner and others, 1973). It seems most natural to assume that Earth suffered a similar bombardment and that the outer skin of the Earth was repeatedly reworked in this phase. Rocks from this phase are at present unknown, but it may be noted that each of the oldest complexes (Godthaab area, Morton Gneiss, Ancient Gneiss complex of Kaapvaal) contains abundant quartz–diorite or tonalite, the 'wet' or 'hornblendic' differentiation equivalent of the 'dry' anorthositic lunar highland differentiate. From 3800 Ma to 3400 Ma the only convincingly dated sequences are those of Godthaab, Isua, the Minnesota River valley and the Ancient Gneiss Complex (accepting the Hunter (1974) interpretation). The rock types in these areas (except Minnesota) are many and varied, but typically include tonalites, iron-rich units (iron formation) and amphibolites. The Kaapvaal craton includes abundant gneisses containing assemblages which indicate fairly high thermal gradients (Hunter, 1974, p. 265). To fit such

372

complexes into the lowish thermal gradient situation outlined above requires that heat be convected into these complexes, a possible source being the intrusives now occurring as pyroxene-granulites which are known or inferred in at least two of these areas (Himmelberg and Phinney, 1967; Hunter, 1974). Although the date of gneissification is not known unequivocally for any of these complexes, it is reasonable to associate it with their general age of origin, implying strongly deformational histories at an early age of Earth. These gneissic complexes pre-date the oldest known greenstone belts, which are frequently comparatively undeformed internally and little metamorphosed, and indicate a more static type of surface.

On the basis, therefore, of analogy with lunar history, of a generalized and simple series of solutions to early heat-flow problems, and utilization of the present-day and very fragmentary knowledge of the metamorphic and geochronologic histories of early Archean rocks, the following Earth evolution is proposed:

1. Chaotic phase; 4500–3900 (3800?) Ma.
2. Anarchic phase; 3900 (3800)–3400 Ma.
3. Scum tectonics (greenstone phase); 3400–2600 Ma.
4. Skin or single-plate tectonics (Proterozoic); 2600–1000 (600?) Ma.
5. Plate tectonics, 1000 or 600? to present.

'Chaos' is chosen to indicate a totally disordered state of affairs in the earliest phase; 'anarchy' is a state bordering on chaos; 'scum' is chosen to designate small proto-continental units drifting on a steadily but irregularly convective upper mantle; and 'skin' indicates a uniform lithosphere, stable but too thin to suffer prolonged and stable subduction processes. The reasons for adopting the post-2600 Ma sequence will be discussed in a parallel paper, in which the present-day thermal structure will be extrapolated backwards in time.

Acknowledgements

The writer wishes to acknowledge indebtedness to E. R. Oxburgh and S. W. Richardson for interesting him in heat-flow studies, and to his colleagues in the University of Alberta, in both Physics and Geology, for discussions of this work. The project was supported by the National Research Council of Canada grant A-7489.

References

Beck, A. E., 1965. 'Techniques of measuring heat flow on land', *Amer. Geophys. U. Geophys.*, Monograph 8, 24–57.

Birch, F., 1955. 'Physics of the crust', *Geol. Soc. Amer. Spec. Paper 62*, A. Poldervaart (Ed.), 101–117.

Boyle, R. W., 1961. 'The geology, geochemistry and origin of the gold deposits of the Yellowknife district', *Mem. Geol. Surv. Can.*, **310**.

Clark, S. P. Jr., and Ringwood, A. E., 1964. 'Density distribution and constitution of the mantle', *Rev. Geophys.*, **2**, 35–88.

Clark, S. P. Jr., Turekian, K. K., and Grossman, L., 1972. 'Model for the early history of the Earth', In *The Nature of the Solid Earth*, E. C. Robertson (Ed.), McGraw-Hill, New York, 3–18.

Currie, K. L., 1971. 'The reaction 3 cordierite = 2 garnet + 4 sillimanite + 5 quartz as a geological thermometer in the Opinicon Lake region, Ontario', *Contr. Mineral. Petrol.*, **33**, 215–226.

Eade, K. E., and Fahrig, W. F., 1971. 'Geochemical evolutionary trends of continental plates—a preliminary study of the Canadian Shield', *Geol. Surv. Can. Bull.*, **179**, 51.

Fukao, Y., 1969. 'On the radiative heat transfer and the thermal conductivity in the upper mantle', *Bull. Earth Res. Inst.*, **47**, 549–569.

Gast, P., 1972. 'The chemical composition of the earth, the moon, and chondritic meteorites', In *The Nature of the Solid Earth*, E. C. Robertson (Ed.), McGraw-Hill, New York, 19–40.

Glikson, A. Y., 1971. 'Structure and metamorphism of the Kalgoorlie system southwest of Kalgoorlie, Western Australia', *Geol. Soc. Austr. Sp. Publ. 3*, 121–132.

Green, D. H., 1975. 'Genesis of Archean peridotitic magmas and constraints on Archean geothermal gradients and tectonics', *Geology*, **3**, 15–18.

Griggs, D. T., 1972. 'The sinking lithosphere and the focal mechanism of deep Earthquakes', In *The Nature of the Solid Earth*, E. C. Robertson (Ed.), McGraw-Hill, New York, 361–384.

Grossman, L., 1975. 'The most primitive objects in

the Solar system', *Sci. Amer.*, **232**, No. 2, 30–38.

Hanks, T. C., and Anderson, D. L., 1969. 'The early thermal history of the Earth', *Phys. Earth Planet. Interiors*, **2**, 19–29.

Heier, K. S., 1973. 'Geochemistry of granulite facies rocks and problems of their origin', *Phil. Trans. Roy. Soc. Lond.*, **A273**, 429–442.

Hensen, B. J., and Green, D. H., 1973. 'Experimental study of the stability of cordierite and garnet in pelitic compositions at high pressures and temperatures', *Contr. Mineral. Petrol.*, **38**, 151–166.

Herrin, E., 1972. 'A comparative study of upper mantle models: Canadian Shield and Basin and Range provinces', In *The Nature of the Solid Earth*, E. C. Robertson (Ed.), McGraw-Hill, New York, 216–231.

Himmelberg, G. R., and Phinney, Wm. C., 1967. 'Granulite-facies metamorphism, Granite-Falls—Montevideo area, Minnesota', *J. Petrology*, **8**, 325–348.

Holdaway, M. J., 1971. 'Stability of andalusite and the aluminium silicate phase diagram', *Amer. J. Sci.*, **271**, 97–131.

Hunter, D. R., 1974. 'Crustal development in the Kaapvaal craton, I. The Archean', *Precambrian Res.*, **1**, 259–294.

Jaeger, J. C., 1965. 'Application of the theory of heat conduction to geothermal measurements', *Amer. Geophys. U. Geophys. Monograph 8*, W. H. K. Lee (Ed.), 7–23.

Jaffey, A. H., Flynn, K. F., Glendenin, L. E., Bentley, W. C., and Essling, A. M., 1971. 'Precision measurements of half lives and specific activities of U^{235} and U^{238}', *Phys. Rev. C.*, **4**, 1889–1906.

Jolly, W. T., 1974. 'Regional metamorphic zonation as an aid in study of Archean terrains: Abitibi region, Ontario', *Can. Mineralogist*, **12**, 499–508.

Kratz, K. O., and Glebovitsky, V. A., 1972. 'Metamorphic belts of the U.S.S.R.', *Proc. 24th Intl. Geol. Congr. Montreal*, **1**, 348–353.

Lambert, R. St J., 1972. 'The metamorphic facies concept—continued', *Proc. 24th Intl. Geol. Congr. Montreal*, **2**, 100–108.

Lee, W. H. K., and Uyeda, S., 1965. 'Review of heat flow data', *Amer. Geophys. U. Geophys. Monograph 8*, 87–191.

Lubimova, E. A., 1969. 'Thermal history of the Earth', *Amer. Geophys. U. Geophys. Monograph 13*, P. J. Hart (Ed.), 63–77.

MacDonald, G. J. F., 1965. 'Geophysical deductions from observations of heat flow', *Amer. Geophys. U. Geophys. Monograph 8*, W. H. K. Lee (Ed.), 191–211.

Muecke, G. K., 1969. 'The petrogenesis of the granulite-facies rocks of the Lewisian of Sutherland, Scotland', Unpubl. D. Phil. Thesis, Oxford University.

Ramsay, C. R., 1973. 'Controls of biotite zone mineral chemistry in Archean metasediments near Yellowknife, Northwest Territories, Canada', *J. Petrology*, **14**, 467–488.

Richardson, S. W., Bell, P. M., and Gilbert, M. C., 1969. 'Experimental determination of kyanite-andalusite and andalusite-sillimanite equilibria, aluminium silicate triple point', *Amer. J. Sci.*, **267**, 259–272.

Saggerson, E. P., and Owen, L. M., 1971. 'Metamorphism as a guide to depth of the top of the mantle in southern Africa', *Geol. Soc. S. Africa Sp. Publ. 2*, 335–349.

Saggerson, E. P., and Turner, L. M., 1972. 'Some evidence for the evolution of regional metamorphism in Africa', *Proc. 24th Intl. Geol. Congr. Montreal*, **1**, 153–161.

Sclater, J. G., and Francheteau, J., 1970. 'The implications of terrestrial heat flow observations on current tectonic and geochemical models of the crust and upper mantle of the Earth', *Geophys. J. Roy. Astr. Soc.*, **20**, 509–542.

Shaw, D. M., 1967. 'U, Th and K in the Canadian Precambrian shield and possible mantle compositions', *Geochim. Cosmochim. Acta*, **31**, 1111–1113.

Turner, G., Cadogan, P. H., and Yonge, C. J., 1973. 'Argon selenochronology. pp. 1889–1914', In *Proc. Fourth Lunar Sci. Conf.*, vol. 2, W. A. Gose (Ed.), Pergamon, New York.

Wasserburg, G. J., MacDonald, G. J. F., Hoyle, F., and Fowler, W. A., 1964. 'Relative contributions of uranium, thorium and potassium to heat production in the Earth', *Science*, **148**, 465–467.

Windley, B. F., and Bridgwater, D., 1971. 'The evolution of Archean low- and high-grade terrains', *Geol. Soc. Austr. Sp. Publ. 3*, 33–46.

General Geochemistry

The Geochemistry of Archean Rocks

Richard St. J. Lambert

Department of Geology, University of Alberta, Edmonton, Alberta, Canada.

V. E. Chamberlain

University Museum, Oxford, England.

J. G. Holland

Department of Geological Sciences, Science Laboratories, South Road, Durham, England.

Abstract

Major and trace element abundances in Archean rocks are compared on the Alkali–FeO–MgO, K_2O–Na_2O–CaO and normative O–Ab–Or plots, and for K v. Rb, Sr, Ba, and U, and U v. Th. U.S.S.R. shields appear to be more fractionated than the Canadian Shield, and igneous complexes are in general more fractionated than their associated basement. Fe/Mg ratios of average compositions from individual geological units vary widely, greenstone averages covering a wide range. Granulite facies rocks and the Kaapvaal tonalites are more magnesian than the average, whereas the Amîtsoq gneiss and the Kaapvaal younger granitic intrusions are distinctly iron-rich. In both major and trace elements the Ancient Gneiss Complex and its included tonalites is intermediate in composition between N. Atlantic Craton granulite and amphibolite facies averages, raising the question of whether it should be considered to be a high-level equivalent or fraction of the N. Atlantic granulites. K/U ratios are almost invariably >10,000, with the Scourian and Amîtsoq gneisses being extreme. Th/U ratios vary widely, but are generally higher than in assumed crustal averages. Some aspects of the trace element chemistry of the greenstone belts and perhaps the gneissic complexes can be tentatively correlated with the existence of high thermal gradients and a hydrous potassic uppermost mantle containing a hornblende-bearing layer to about 40 km depth.

Introduction

This paper summarizes Archean geochemical data for major elements and some trace elements in the form of selected diagrams and brief discussion. The format is selective, by means of restricting the range of elements to be discussed, and to a lesser extent by neglecting some specific rock types or regions, gener-ally where data quantities are limited, or restricted to unusual rock compositions. To a certain extent, the paper is aimed at compar-ing gneissic and greenstone terrains. Equally, little or no attempt is made to seek new com-parisons between Archean and later rocks, this subject having been well reviewed in

377

recent years by Ronov and Migdisov (1970); Eade and Fahrig (1971, 1973); Glikson (1970, 1971); Glikson and Sheraton (1972); Anhaeusser (1973); Engel et al. (1974) and Hunter (1974b). Detailed studies of Archean complexes, utilized in preparing this paper, have been given by Baragar (1966, 1968); Condie (1967); Baragar and Goodwin (1969); Viljoen and Viljoen (1969a,b and c, 1971); Anhaeusser (1971); Bowes et al. (1971); Condie and Lo (1971); Goodwin (1971); Hallberg (1972); Tarney et al. (1972); Green and Baadsgaard (1971); Windley et al. (1973); Sheraton et al. (1973); Holland and Lambert (1973); Wildeman and Condie (1973); Hunter (1974a,b); Naqvi et al. (1974); Condie and Baragar (1974) and Arth and Hanson (1975).

The principal conclusions regarding the geochemistry of the Archean, summarizing the above papers, are that Archean complexes (*as sampled at the present surface*) are low in K relative to Na; may have higher Fe relative to Mg; are generally low in Al and Ti; that most rock types appear to be higher in Cr and Ni; that basic rocks also have more Co, Rb and Mn (and Ba?) than their modern counterparts and there is higher K/Rb and Sr/Rb in silicic rocks; that in some areas, K, Rb, Pb and Th in silicic rocks increase in absolute abundance with time; that the heavy rare-earth elements and Y are usually but not always depleted relative to chondritic abundances, and that U may be relatively scarce. In addition, many other inter-element ratios can be shown to differ from those in younger rocks. Some of these conclusions are, however, more open to discussion than others. The chief and largely unresolved problems in interpretation of these data lie in a lack of knowledge of the effect of relative erosion level, particularly in gneissic terrains, and in lack of knowledge of the effects of high-grade metamorphism on element abundances, except in the case of granulite grade metamorphism (Heier, 1973 and earlier references). While the latter problem is believed to be of little significance, the former is important, because shield regions (and parts thereof, such as greenstone belts) undoubtedly basify downwards.

The main unresolved question at the present time regarding these conclusions is not their validity on the basis of the data presented, either by virtue of inadequate sampling or analytical inadequacy, but whether any individual characteristics can be related to a purely temporal cause such as overall evolution of the mantle, or whether they are due to local processes which operated solely in the Archean. Included under the latter heading would be geochemical effects controlled by f_{H_2O}, f_{O_2} or PT conditions in the uppermost mantle, which may change from time to time. For instance, Green (1975) has argued that eclogite formation could not occur in the Archean and that subduction was unlikely, a view endorsed by Lambert (this volume) who also argues that thermal gradients must have been significantly steeper then. Strong and Stevens (1974) also discussed the consequence of high Archean thermal gradients and reasoned that Archean upper mantle differentiation would occur under lower pressure (but higher P_{H_2O}) than in later times. All such variables would control magma compositions, particularly trace element abundances.

Major Element Abundances

In order to enquire further into some of the problems raised above, the relative major element chemistry of greenstone sequences, their related granitic complexes and the greywacke suites, and the major gneissic terrains were compared.

Considering first alkalies and CaO, some regional averages are given in Fig. 1 and some averages and individual analyses from specific areas in Fig. 2. The general distinctiveness of K_2O/Na_2O between the Proterozoic and Archean is clearly seen. Likewise, there is a clear distinction between the data for the igneous and metamorphic segments of the U.S.S.R. Archean shields (data from Ronov and Migdisov, 1970). Their average Archean igneous rock has higher K_2O/CaO than the metamorphics of the same area, presumably reflecting broad density differences. Na_2O/CaO is likewise higher in the igneous complexes. However, more puzzling is the

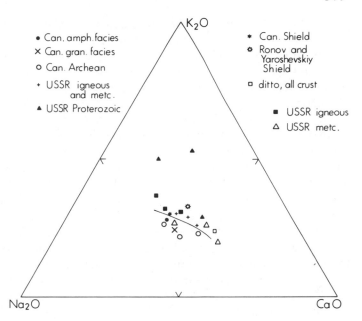

Fig. 1. K₂O–Na₂O–CaO by weight for Archean regional average compositions. The line separates all Canadian Archean averages from all U.S.S.R. shield averages

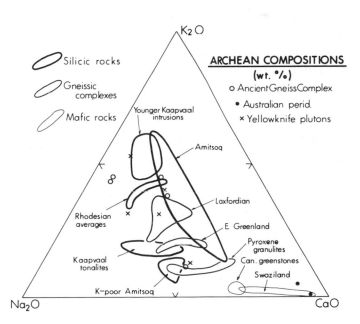

Fig. 2. K₂O–Na₂O–CaO by weight for examples of Archean complexes

contrast between the U.S.S.R. and Canadian shield averages: the line of Fig. 1 separates Na₂O-richer Canadian Archean from K₂O-richer U.S.S.R. Archean.

The more detailed diagram, Fig. 2, shows a wide range of compositions. The main trend comprises successively ultramafic and mafic groups (the averages for all greenstones lying close to the Canadian examples), granulites of the N. Atlantic Craton, E. Greenland gneisses and Kaapvaal tonalites, the Lewisian Laxfordian complex, Rhodesian granites (mostly from Robertson, 1973) and the younger Kaapvaal intrusions. Individual metamorphic rocks from the Ancient Gneiss Complex and plutons from Yellowknife scatter fairly widely but most of the latter are similar to their Rhodesian analogues. The Amîtsoq gneiss (Lambert and Holland, this volume) is clearly exceptional. The main feature of this Figure is that it clearly portrays the intermediate character of the N. Atlantic Craton and Kaapvaal Craton, compared with greenstone belts and their related intrusives (Rhodesia, Yellowknife) which show a marked bimodal character.

The relationship between quartz and feldspars in many of the same complexes as in Figs. 1 and 2 is shown in Fig. 3. The sodic character of Archean rocks is well displayed, and the U.S.S.R. Proterozoic is markedly potassic. The U.S.S.R. Archean igneous and metamorphic are again separate, with the igneous rocks more siliceous, perhaps again indicating a higher degree of differentiation than their associated basement. The other groups show their anticipated relationships, the Canadian shield rocks being more sodic and less siliceous than their Soviet counterparts, and the Kaapvaal craton rocks being divided again into two clear groups by K content, but interestingly, not by normative quartz. The Kaapvaal tonalites again overlap the range of the N. Atlantic Craton gneissic complexes.

Finally, the iron–magnesium relationships are examined in Fig. 4. The most exceptional complexes are the Amîtsoq gneiss and the other N. Atlantic craton gneisses, the former being exceptionally Mg-poor, but the latter rather Mg-rich. Shield averages, the averages of various sets of Archean greywackes, Canadian shield gneissic complexes and the Yellowknife plutonics define a sequence which trends away from the Noranda belt (Baragar, 1968: note, however, that a few Noranda rocks lie in the 'Yellowknife' area in Fig. 4) and thence via the Swaziland greenstones to peridotitic compositions. The basis for identifying high Fe in Archean greenstones is clearly seen: the W. Australian and Yellowknife analyses (in the case of the former, plotted as the range of the most commonly formed compositions) are clearly separate from the Noranda–Swaziland types. Evidently upper mantle processes did not uniquely provide high Fe/Mg in Archean basalts. Another problem is seen in the Kaapvaal younger intrusions and the tonalites in the Kaapvaal Craton: they are clearly geochemically distinct. Finally, as before the gneissic complexes are intermediate in composition between the greenstones and their associated batholiths. Discussion is deferred until after consideration of some aspects of trace element chemistry.

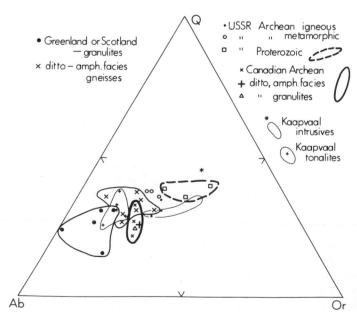

Fig. 3. Normative Q–Ab–Or for some silicic Archean complexes

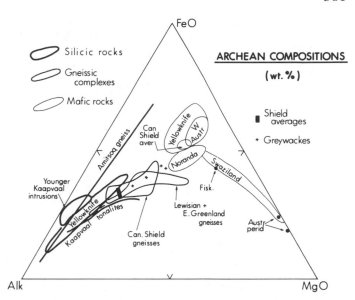

Fig. 4. Alkalies–FeO–MgO by weight for examples of Archean complexes

Trace Element Geochemistry

Discussion of trace elements in the Archean has ranged across all the elements which are reasonably abundant and easily and comparatively accurately analyzed. However, the transition metals, although some conclusions concerning them have been formulated by previous authors (references above) will not be discussed because: (1) we are not entirely convinced that analytical accuracy is always adequate; (2) many facets of their geochemistry depend on crystal field effects which are not yet fully documented and (3) some conclusions may be based on the chances of data comparisons. Included in the latter may be Cr, Ni abundances, reputed to be higher in Archean basalts: however, Sigvaldason (1974) gives Cr, Ni analyses for Iceland 'plume' basalts with low K_2O (<0·16%) which correspond closely with the high values for Archean basalts. Presumably whatever process produced one composition also produced the other, so that it is not necessarily age-dependent. This does not, however, alter the conclusion that the presently available average of Cr, Ni in all analyzed Archean basalts is higher than that in averages of modern basalts. Another problem, though, is that Archean greenstone belts are invariably seen in cross-section, often exposing the lower, olivine-rich portions of the

sequence, whereas we rarely see the corresponding levels of modern sequences. The same kind of argument will affect the conclusions for any transition metal in basic rocks.

The large-ion lithophile elements are also probably 'stratified' in the crust, but we do not have the problem of seeing only certain types of cross-section in terrains of particular age. The relationships of K with Rb, Sr and Ba are therefore examined in Figs. 5–7, respectively. Taking the greenstone averages and the NE Minnesota differentiated volcanics as a guide, we find that K/Ba = 20–30 and K/Sr near 20 appear to be characteristic, but that the affinity breaks down in K/Rb, which in greenstone belts is generally lower (c. 200) than in the NE Minnesota volcanics (c. 400). The greenstones are known to be significantly higher in Rb and probably higher in Ba and Sr than modern counterparts. The difference between the NE Minnesota greenstones and the others is enigmatic.

The relationship between the other main rock groups for each of the element ratios in Figs. 5–7 is different. The Scourian pyroxene–granulites and Greenland equivalents have very high K/Rb, similar K/Sr and much lower K/Ba than the greenstone trend. The associated amphibolite facies gneisses have only slightly higher K/Rb, much higher K/Sr and similar K/Ba compared with the green-

382

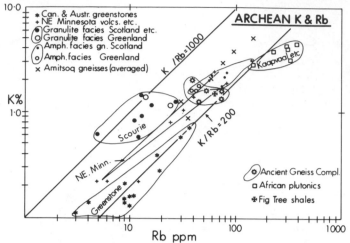

Fig. 5. K and Rb in Archean complexes

stones. The members of the Ancient Gneiss Complex, except the younger K-rich intrusions, lie between the N. Atlantic granulite and amphibolite groups in each case. These patterns confirm a greater affinity between the Ancient Gneiss Complex and the N. Atlantic Craton gneisses than between the Ancient gneisses and any of the members of the greenstone regions, as was generally seen in the major element chemistry.

In each case, however, the Amîtsoq gneisses are unique, their K/Rb remaining high in the K-rich members, and the trends in K/Ba and K/Sr being quite different from the other groups. The quartz–monzonite group of Arth and Hanson (1975) appears to be chemically related (in LIL elements) to the much older Amîtsoq gneiss. The younger intrusives in the Kaapvaal are very much like granites of later ages in K, Ba, Sr and Rb, utilizing the average

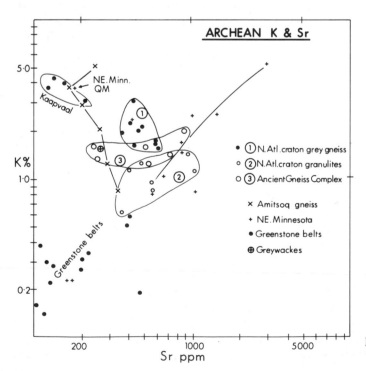

Fig. 6. K and Sr in Archean complexes

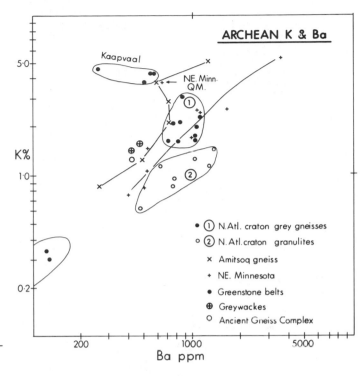

Fig. 7. K and Ba in Archean complexes

granites, A, B, C of Glikson and Sheraton (1972) for comparison. There are enough similarities between the Kaapvaal granites and the Amîtsoq gneiss to hazard a guess that they were generated by similar processes. This comparison extends to the REE (Hunter, 1974b; O'Nions and Pankhurst, 1974), where both types show heavy REE depletion patterns. However, Rb and Th (and perhaps U, by analogy) are lower in the Amîtsoq gneiss.

Finally, members of the greywacke suite and the Fig Tree shales do not show any peculiarities in their K, Sr and Rb geochemistry, but are low in Ba, like the Ancient Gneiss Complex, in so far as data are available.

The much more limited quantity of data on Th and U is shown in Figs. 8 and 9. Unpublished data on pure pyroxenic Scourian pyroxene–granulites (Lambert, Burwash and Muecke) are included in both Figures and

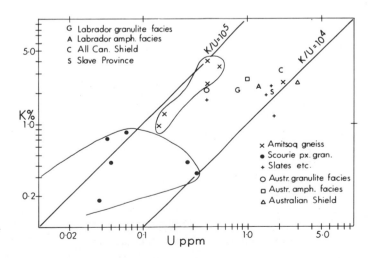

Fig. 8. K and U in Archean complexes

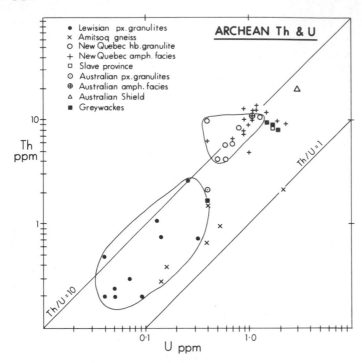

Fig. 9. U and Th in Archean complexes

show that these rocks are exceptionally depleted in U, but have average Th/U = 5·6, although Th/U ranges widely. The average K/U is close to 40,000, including data plotting off-scale on Fig. 8. The Amîtsoq gneiss (Lambert and Holland, this volume) is also exceedingly low in U, except the intrusive granite sheet (110869) which has 2·19 ppm U. The averages for the Australian Shield and Musgrave Ranges (Lambert, 1971), the Canadian Shield (Eade and Fahrig, 1971) and Wyoming and Minnesota (Rogers, Condie and Mahan, 1970) are shown for comparison. We conclude that the Amîtsoq gneiss and Scourie pyroxene–granulites are exceptional, but for different reasons (Lambert and Holland, this volume; Holland and Lambert, 1975, respectively). The main conclusion is, however, that all analyzed Archean rocks are depleted in U relative to K, if a K/U ratio of 10,000 is taken as the norm. Thorium shows a more complicated pattern. The Amîtsoq gneiss is again extremely exceptional, with Th/U close to 2, but otherwise most Archean rocks seem to have Th/U in the range 8 to 15, significantly higher than the value of 4 commonly assumed in heat production calculations. Some investigation of the significance of this fact appears to be required, because Th is not normally

regarded as an exceptional element. At first sight, Th appears to be of normal abundance compared with K in the Archean (that is, it is U which is exceptionally low) but there must be some reason for the breaking of a well-known geochemical association in opposite directions in the typical shield rocks and the Amîtsoq gneiss.

Discussion and Conclusion

In terms of major element chemistry, the conclusions from the sub-sets of elements considered are:

(a) That there may be significant differences between bulk compositions of the surface of shields in terms of quartz and feldspar components;

(b) That average igneous rocks (intrusives and volcanics together) are more differentiated than the associated gneisses;

(c) That iron–magnesium ratios vary widely; and

(d) That the gneissic complexes differ significantly from the greenstone–granitic pluton association, and that the Amîtsoq gneisses differ from all other series.

These differences are reflected in trace element chemistry, which varies greatly from area to area. It is argued that sampling problems may be a cause of the differences in abundances of trace transition metals from area to area, but that LIL elements may be more readily compared. These show again the characteristic differences between the greenstone–granitic pluton association and the gneissic cratons observed in the major elements. Also, the Kaapvaal complex appears to lie in an intermediate geochemical position between the K, Rb and U poor granulites of the N. Atlantic Craton and their associated amphibolite facies gneisses. We conclude that the Kaapvaal Ancient Gneiss Complex may be a higher level equivalent of the N. Atlantic granulites, being gradational towards the broadly 'supracrustal' chemistry of the grey gneisses of the N. Atlantic craton, presumably as the result of igneous processes rather than sedimentary, because the majority of the rocks analyzed from the Ancient Gneiss Complex and included in these comparisons appear to be igneous or meta-igneous.

The geochemical distinctions between Archean and later rocks which seem to be conclusively demonstrated relate to abundances of most of the large-ion lithophile elements (K, Rb, Sr, Y, Ba, REE, Pb, Th and U). This immediately suggests an incompletely fractionated upper mantle, which ties itself in well with the reasoning of Lambert (this volume) that reasonable thermal gradients in the Archean require a K-rich uppermost mantle to a depth of 100 km in the model given.

Considering the highly hydrous nature of the greenstone sequences, and the model of Glikson (1971, Fig. 4, stages 1–4) as one which broadly fits the suggested thermal pattern, the process of igneous recycling would likely maintain a highly hydrous upper mantle. To 40 km and 100°C on the preferred geotherm, hornblende would be stable, and by extrapolation, phlogopite to at least 60 km. A mineralogically layered outer Earth, characterized by about 20 (or 30) km oceanic crust, 20 km hornblende-bearing ultramafics, and 29 or more km of phlogopite-bearing ultramafics, would act as a filter or trap for most of the above-listed elements, particularly if garnet occurred in association with hornblende. The partitioning data of Griffin and Rama Murthy (1969) and Philpotts and Schnetzler (1970) for amphibole show that partial melts which equilibrated in the hornblende layer with residual hornblende (with high K/Rb, K/Sr and K/Ba) would be relatively enriched in Rb, Sr and Ba compared with their modern oceanic counterparts, which are believed to originate in a relatively H_2O-free environment. Such hornblendes would be rich in Al and Ti, two elements believed to be relatively low in Archean basalts, but the Si/Al balance would have to be very finely adjusted in order to give the observed fractionation sequences. It is therefore concluded that the combination of relatively high thermal gradients, a hydrous uppermost mantle and a K-rich zone extending to about 100 km affords a framework within which many of the observed Archean geochemical patterns could develop.

Acknowledgements

The first author wishes to acknowledge discussions with colleagues at the University of Alberta on the general subject matter of this paper. We wish to thank R. A. Burwash for the analyses of U and Th in the Amîtsoq and Scourian gneisses, Prof. E. A. Vincent for use of facilities in the Department of Geology and Mineralogy, Oxford, and the National Research Council of Canada for financial support through grant A-7489.

References

Anhaeusser, C. R., 1971. 'Cyclic volcanicity and sedimentation in the evolutionary development of Archean greenstone belts of shield areas', *Geol. Soc. Austr. Sp. Publ. 3*, 57–70.

Anhaeusser, C. R., 1973. 'The evolution of the early Precambrian crust of Southern Africa', *Phil. Trans. R. Soc. Lond.*, **A273**, 359–388.

Arth, J. G., and Hanson, G. N., 1975. 'Geochemis-

386

try and origin of the early Precambrian crust of Northeastern Minnesota', *Geochim. Cosmochim. Acta*, **39**, 325–362.

Baragar, W. R. A., 1966. 'Geochemistry of the Yellowknife volvanic rocks', *Can. J. Earth Sci.*, **3**, 9–30.

Baragar, W. R. A., 1968. 'Major-element geochemistry of the Noranda volcanic belt, Quebec–Ontario', *Can. J. Earth Sci.*, **5**, 773–790.

Baragar, W. R. A., and Goodwin, A. M., 1969. 'Andesites and Archaean volcanism in the Canadian Shield', *Andesite symp.*, A. R. McBirney (Ed.), Oregon Dept. Geol. Min. Indust., **65**, 121–142.

Bowes, D. R., Barooah, B. C., and Khoury, S. G., 1971. 'Original nature of Archaean rocks of North-West Scotland', *Geol. Soc. Austr. Sp. Publ. 3*, 77–92.

Condie, K. C., 1967. 'Geochemistry of early Precambrian greywackes from Wyoming', *Geochim. Cosmochim. Acta*, **31**, 2135–2149.

Condie, K. C., and Baragar, W. R. A., 1974. 'Rare-earth element distributions in volcanic rocks from Archean greenstone belts', *Contr. Mineral. Petrol.*, **45**, 237–246.

Condie, K. C., and Lo, H. H., 1971. 'Trace element geochemistry of the Louis Lake batholith of early Precambrian age, Wyoming', *Geochim. Cosmochim. Acta*, **35**, 1099–1119.

Eade, K. E., and Fahrig, W. F., 1971. 'Geochemical evolutionary trends of continental plates—a preliminary study of the Canadian Shield', *Geol. Surv. Can. Bull.*, **179**, 51 pp.

Eade, K. E., and Fahrig, W. E., 1973. 'Regional, lithological and temporal variation in the abundances of some trace elements in the Canadian Shield', *Geol. Surv. Can.*, Paper 72–46, 46 pp.

Engel, A. E. J., Itson, S. P., Engel, C. G., and Stickney, D. M., 1974. 'Crustal evolution and global tectonics: a petrogenic view', *Bull. Geol. Soc. Amer.*, **85**, 843–858.

Glikson, A. Y., 1970. Geosynclinal evolution and geochemical affinities of early Precambrian systems, *Tectonophysics*, **9**, 397.

Glikson, A. Y., 1971. 'Primitive Archaean element distribution patterns: chemical evidence and geotectonic significance', *Earth Planet. Sci. Lett.*, **12**, 309–320.

Glikson, A. Y., and Sheraton, J. W.; 1972. 'Early Precambrian trondhjemitic suites in Western Australia and Northwestern Scotland and the geochemical evolution of shields', *Earth Planet. Sci. Lett.*, **17**, 227–242.

Goodwin, A. M., 1971. 'Metallogenic patterns and evolution of the Canadian Shield', *Geol. Soc. Austr. Sp. Publ. 3*, 157–174.

Green, D. H., 1975. 'Genesis of Archean peridotitic magmas and constraints on Archean geother-

mal gradients and tectonics', *Geology*, **3**, 15–18.

Green, D. C., and Baadsgaard, H., 1971. 'Temporal evolution and petrogenesis of an Archaean crustal segment at Yellowknife, N.W.T., Canada', *J. Petrology*, **12**, 177–217.

Griffin, W. L., and Rama Murthy, V., 1969. 'Distribution of K, Rb, Sr and Ba in some minerals relevant to basalt genesis', *Geochim. Cosmochim. Acta*, **33**, 1389–1414.

Hallberg, J. A., 1972. 'Geochemistry of Archaean volcanic belts in the Eastern Goldfields region of Western Australia', *J. Petrology*, **13**, 45–56.

Heier, K. S., 1973. 'Geochemistry of granulite facies rocks and problems of their origin', *Phil. Trans. Roy. Soc. Lond.*, **A273**, 429–442.

Holland, J. G., and Lambert, R. St J., 1973. 'Comparative major element geochemistry of the Lewisian of the mainland of Scotland', *The early Precambrian of Scotland and related rocks of Greenland*, R. G. Park and J. Tarney (Eds.), Univ. of Keele Press, pp. 51–62.

Holland, J. G., and Lambert, R. St J., 1975. 'The chemistry and origin of the Lewisian gneisses of the Scottish mainland: the Scourie and Inver assemblages and sub-crustal accretion', *Precambrian Res.* (in press).

Hunter, D. R., 1974a. 'Crustal development in the Kaapvaal Craton, I. The Archaean', *Precambrian Res.*, **1**, 259–294.

Hunter, D. R., 1974b. 'Crustal development in the Kaapvaal Craton, II. The Proterozoic', *Precambrian Res.*, **1**, 295–326.

Lambert, I. B., 1971. 'The composition and evolution of the deep continental crust', *Geol. Soc. Austr. Sp. Publ. 3*, 419–428.

Naqvi, S. M., Rao, V. D., and Narain, H. 1974. 'The protocontinental growth of the Indian Shield and the antiquity of its rift valleys', *Precambrian Res.*, **1**, 345–398.

O'Nions, R. K., and Pankhurst, R. J., 1974. 'Rare element distribution in Archaean gneisses and anorthosites, Godthåb area, West Greenland', *Earth Planet. Sci. Lett.*, **22**, 328–338.

Philpotts, J. A., and Schnetzler, C. C., 1970. 'Phenocryst-matrix partition coefficients for K, Rb, Sr and Ba, with applications to anorthosite and basalt genesis', *Geochim. Cosmochim. Acta*, **34**, 307–322.

Robertson, I. D. M., 1973. 'Potash granites of the southern edge of the Rhodesian Craton and the northern granulite zone of the Limpopo belt', *Geol. Soc. S. Africa Sp. Publ. 3*, 265–276.

Rogers, J. J. W., Condie, K. C., and Mahan, S., 1970. 'Significance of thorium, uranium and potassium in some early Precambrian greywackes from Wyoming and Minnesota', *Chem. Geol.*, **5**, 207–213.

Ronov, A. B., and Migdisov, A. A., 1970. 'Evolution of the chemical composition of the rocks in the shields and sediment cover of the Russian and

North American platforms', *Geochem. International*, **7**, 294–325.

Sheraton, J. W., Skinner, A. C., and Tarney, J., 1973. 'The geochemistry of the Scourian gneisses of the Assynt district', In *The early Precambrian of Scotland and related rocks of Greenland*, R. G. Park and J. Tarney (Eds.), Univ. of Keele, Newcastle, Staffs., 13–30.

Sigvaldason, G. E., 1974. 'Basalts from the centre of the assumed Icelandic mantle plume', *J. Petrology*, **15**, 497–524.

Strong, D. F., and Stevens, R. K., 1974. 'Possible thermal explanation of contrasting Archean and Proterozoic geological regimes', *Nature*, **249**, 545–546.

Tarney, J., Skinner, A. C., and Sheraton, J. W., 1972. 'A geochemical comparison of major Archean gneiss units from NW Scotland and E Greenland', *Proc. 24th Intl. Geol. Congr. Montreal*, **1**, 162–174.

Viljoen, M. J., and Viljoen, R. P., 1969a. 'An introduction to the geology of the Barberton granite-greenstone terrain', *Geol. Soc. S. Africa Sp. Publ. 2*, 9–28.

Viljoen, M. J., and Viljoen, R. P., 1969b. 'The geology and geochemistry of the lower ultramafic unit of the Onverwacht Group and a proposed new class of igneous rocks', *Geol. Soc. S. Africa Sp. Publ. 2*, 55–85.

Viljoen, M. J., and Viljoen, R. P., 1969c. 'The geochemical evolution of the granitic rocks of the Barberton region', *Geol. Soc. S. Africa Sp. Publ. 2*, 189–219.

Viljoen, R. P., and Viljoen, M. J., 1971. 'The geological and geochemical evolution of the Onverwacht Volcanic Group of the Barberton Mountain Land, South Africa', *Geol. Soc. Austr. Sp. Publ. 3*, 133–150.

Wildeman, T. R., and Condie, K. C., 1973. 'Rare earths in Archean greywackes from Wyoming and from the Fig Tree Group, South Africa', *Geochim. Cosmochim. Acta*, **37**, 439–453.

Windley, B. F., Herd, R. K., and Bowden, A. A., 1973. 'The Fiskenaesset complex, West Greenland, part 1', *Medd. om Grønland*, **196**, 80 pp. (also, *Grønlands geol. Unders., Bull.*, 106).

A Comparison of Modern and Archaean Oceanic Crust

and Island-arc Petrochemistry

BERNARD M. GUNN

Université de Montreal.

Introduction

The thick, folded sequences of Archaean meta-volcanic rocks have been the subject of many petrochemical investigations, not only on the Canadian Shield but also the shield regions of Australia and South Africa. The definitive results of these investigations have been, it must be admitted, of limited value. Variation diagrams of a typical Archaean sequence (Fig. 1) show such a broad scatter of data points as to cover the entire span of known igneous and sedimentary rocks so that either it must be admitted that the Archaean

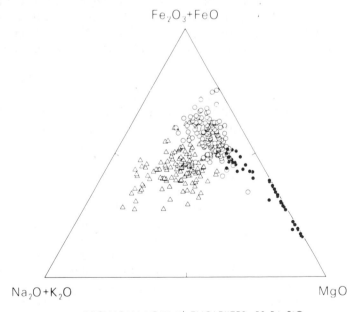

Fe$_2$O$_3$+FeO

Na$_2$O+K$_2$O   MgO

○ ARCHAEAN LOW-Al THOLEIITES, 50-54 SiO$_2$

• ARCHAEAN METAPICRITES AND PERIDOTITES

△ ARCHAEAN ANDESITES, DACITES, HIGH-Al

Fig. 1. Ternary diagram of ΣFe, MgO, alkalies of some Archaean metabasalt flows and gabbroic sills.

389

volcanoes were amazingly variable chemically, or that elemental migration within the rocks during halmyrolysis, diagenesis, metamorphism, recrystallization and orogenesis has been so extreme as to mask the original compositions.

As a typical example, the writer made, some years ago, a geochemical traverse of 45,000 ft of Archaean metavolcanic rocks in the Chibougamau district of Quebec. This section comprised a lower section of 8000 ft of pillow-basalt grading upwards into volcaniclastic breccia, tuff, volcanic metagreywacke and shale. This was overlain by a second formation of 12,000 ft of pillow basalt and a third formation of 5000 ft of breccia, tuff and greywacke intercalated by sills of gabbro, norite, granophyre and peridotite. Intelligent interpretation of the resulting data proved virtually impossible. The massive greenstone, pillowed flows had the general characteristics of mid-oceanic ridge basalts and had the expected high Ni and Cr contents, but variation diagrams showed a wide scatter of points unrelated to any likely fractionation of olivine, pyroxene, etc., or to any likely primary melt series. Although some primary clinopyroxene remained in some thick sills, the majority of rocks were completely reconstituted to albite, epidote, chlorite, tremolite and calcite. High-iron rocks contained abundant pyrite, possibly of secondary origin and the high calcium rocks much calcite. Recalculation to a pyrite- or calcite-free composition often gave unrealistically low compositions. High-Ti rocks contained veins of secondary ilmenite, and secondary silicification was common. Even if only those analyses with similar Si, Al, Fe, Mg, Ca to average modern sea-floor basalt were used, the K, Na, Sr, Ba, though on the average low enough to be in the range of such modern equivalents, were erratic in the extreme with K_2O ranging from 20 ppm to 1%, Rb from 0·5 ppm to 100 ppm, Sr from 100 to 400 ppm. Other writers evidently encountered similar difficulty, Wilson et al. (1965) published an average of '20 Archaean andesites' but these rocks have the same composition as their 'average of 53 basalts' except for elevated soda and are obviously spilitized basalts.

The andesitic tuffs of the Chibougamau regions were even more variable, there being a negative correlation between grainsize and K_2O content suggesting the presence of large amounts of illite. Even massive flows such as the Waite andesite in the Noranda district have such variable alumina and alkalies that no comparison can be made between one region or even one flow and the next, i.e. local and regional variations are masked by the within-flow noise level. Fine, black cherty 'Archaean potassic rhyolites' have such high K_2O coupled with impossibly low soda as to suggest that these are metamorphosed siliceous clays.

Even 10 years ago little information was available on the possible range of compositions either within the calc–alkaline suite or within the sea-floor basalts or what might be the significance of any such variations were they detected. Further, although some classic papers have been published on the keratophyre and spilite suites, as recently as 1966 Donnelly described keratophyres and spilites of the Virgin Islands as being produced by a sodic magma 'typical of the early stages of volcanism in a geosynclinal environment'.

The chemical 'noise level' was undoubtedly so high in the Archaean series that no significant changes in composition with time could be detected, with the exception of the work of Baragar and Goodwin (1969) who showed a progressive increase in alumina in a thick section near Noranda. Yet a study of modern basaltic lava piles invariably shows compositional periodicity and general shift, often towards a more alkaline magma type with time (e.g. Gunn and Watkins, 1970). That Archaean lavas could show little or no apparent change in average bulk composition between pillow basalts at the top and bottom of a lava-pile 12 km thick, seemed impossible. The lack of simple trends in the Archaean calc–alkaline sequences seemed to conflict with the work of Nockolds and Allen (1953). Their work on the igneous chemistry of many regions suggested that in any one volcanic province, a simple series of related chemistry, yielding simple, often almost linear fractionation trends, should be found. However, differ-

Table 1. Average composition of apparent primary magma types from recent mid-Atlantic oceanic crust, together with the means of plagioclase and olivine cumulate rocks. Note the regular increase in 'Alkalinity' shown by the regular increase in Ti, K, Na, Sr, Ba, Rb relative to silica in the first four low-Al tholeiites. Note also similarity between the average of 30 Archaean andesites of 58–62% silica and the modern Mont Pelee average from the Antillean island arc

NUMBER	AV 20	AV 16	AV 11	AV 15	AV 46	AV 23	AV 34	AV 33	AV 7	AV 17
		ARCHAEAN AND RECENT AVERAGES						GEOKEM DATA FILE.		
SIO2	50.54	50.54	50.34	49.66	49.82	48.69	49.19	50.14	46.14	48.83
AL2O3	15.15	15.26	15.62	15.98	15.80	17.89	16.49	16.42	17.02	15.09
TIO2	.746	.907	1.017	1.219	2.000	.437	.453	1.019	.834	.775
FE2O3	9.910	10.374	10.670	11.624	10.850	6.660	10.160	9.680	11.140	10.430
MNO	.165	.177	.194	.174	.170	.112	.148	.150	.174	.148
MGO	8.31	7.177	7.030	7.028	6.70	7.24	7.69	7.27	4.95	10.37
CAO	13.030	12.720	12.420	12.246	10.530	15.250	12.440	12.940	16.540	12.090
NA2O	1.840	2.080	2.240	2.225	3.430	1.560	2.300	2.060	2.610	1.980
K2O	.191	.236	.273	.307	.616	.101	.208	.215	.462	.198
P2O5	.080	.100	.120	.420	.420	.050	.090	.120	.130	.090
TOTAL	100.002	99.990	100.010	99.927	100.144	99.990	99.949	100.014	100.000	100.001
CR	145.0	112.95	112.80	87.5	124.45	144.9	119.0	94.4	141.2	274.2
NI										
RB	3.5		4.5	5.5	77.6	1.8	4.0		2.5	3.3
SR	77.5	97.3	167.5	117.3	159.7	76.2	93.0	115.0	129.5	107.3
BA	65.7	74.3	74.4	76.8	39.9	45.2	60.0	72.4	57.7	55.0

C.I.P.W. NORMS

NUMBER	AV 20	AV 16	AV 11	AV 15	AV 46	AV 23	AV 34	AV 33	AV 7	AV 17
QZ	1.64	1.45	1.07	1.68	3.00	0.00	0.00	1.33	0.00	1.09
OR	1.13	1.39	1.61	1.83	3.62	0.17	0.35	1.27	2.73	1.17
PLAG	48.09	49.63	49.01	48.92	54.92	60.17	54.55	52.36	40.57	48.46
NE										
DI	29.86	25.63	24.93	24.03	15.00	22.78	21.12	23.10	39.44	22.30
HY	17.66	16.56	16.50	14.60	15.52	10.26	12.56	15.53	0.00	14.02
OL	0.00	0.00	0.00	0.00	0.00	2.26	3.16	0.00	2.46	0.00
MT	3.59	3.76	3.87	4.24	3.93	2.43	3.62	3.51	4.04	3.47
IL	1.42	1.72	1.94	2.31	3.42	.83	.94	1.94	1.58	1.47
AP	.17			.37	.42	.11	.20	.26	.28	.20
TOTAL	99.26	99.21	99.21	99.12	99.33	99.49	99.23	99.29	99.16	99.22
AN-PERCENT	66.3	62.4	61.7	64.2	45.7	77.0	63.0	65.4	81.4	64.1

AV 20 = BASALT, THOLEIITIC, LT 0.85 PC TIO2, LEG 37, GUNN AND ROOBOL(1975)
AV 16 = BASALT, THOLEIITIC, 0.85-1.0 PC TIO2, LEG 37, GUNN AND ROOBOL(1975)
AV 11 = BASALT, THOLEIITIC, 1.0-1.2 PC TIO2, LEG 37, GUNN AND ROOBOL(1975)
AV 13 = BASALT, THOLEIITIC, GT 1.2 PC TIO2, LEG 37, GUNN AND ROOBOL(1975)
AV 46 = TOLEIITIC BASALTS, MORE THAN 1.5 PC TIO2, LEG 37, DREDGED SAMPLES, GEOKEM FILE.
AV 23 = BASALT, PLAGIOCLASE, LT 0.7 PC TIO2, LEG 37, GUNN AND ROOBOL(1975)
AV 34 = BASALT, PLAGIOCLASE, 0.7-0.75 PC TIO2, LEG 37, GUNN AND ROOBOL(1975)
AV 33 = BASALT, PLAGIOCLASE, GT 0.75 PC TIO2, LEG 37, GUNN AND ROOBOL(1975)
AV 7 = BASALT, PLAG, LOW MG, ALTERED, LEG 37, GUNN AND ROOBOL(1975)
AV 17 = BASALT, PICRITIC, HIGH TI SERIES, LEG 37, GUNN AND ROOBOL(1975)

Table 1 continued on page 392

Table 1—*continued*

ARCHAEAN AND RECENT AVERAGES.   GEOKEM DATA FILE.

NUMBER	AV 30	AV 62	AV 67	AV 17	AV 98	AV 42	AV 89	AV 3	AV 5	AV 23
SI02	57.38	56.15	46.26	48.42	49.33	49.77	49.76		47.24	45.26
AL203	17.38	17.26	10.65	15.46	17.80	15.07	15.17		12.73	15.27
TI02	.854	.947	2.610	1.654	1.430	14.180	.600			
FE203	7.590	8.970	12.070	16.860	11.430	14.180	13.275			
MNO	.114	.146	.202	.237	.185	.224	.205			
MGO	4.44	5.01	21.38	2.32	6.75	6.52	7.36			
CAO	4.950	6.910	8.100	11.230	12.830	9.500	10.845			
NA20	4.410	3.710	1.480	2.810	2.400	3.265	2.065			
K20	.910	.841	.186	2.230	.360					
P205	.110	.150	.150	.460	.080	.240				
TOTAL	100.096	100.094	100.488	100.083	100.185	100.062	100.058	100.022	100.007	99.998
CR	14.8	0.0		47.9	280.3	275.8	261.			
CO		0.0			69.2	69.2	64.			
NI	97.6	105.4	6500.0	16.2	101.2	98.4	167.			
CU	74.0	77.2		48.4	113.2	105.7	112.			
ZN		86.1		107.4	95.6		94.			
GA	40.0									
RB		12.9		42.5	13.4	16.1				
SR	231.4	136.7	58.3	222.5	24.0	145.0	135.			
BA	25.3	0.0	52.0	88.5	151.0	80.0	145.			
PB		0.0		1.5	0.0	81.9				
TH	1.6	0.0	.8							

C.I.P.W. NORMS

	AV 30	AV 62	AV 67	AV 17	AV 98	AV 42	AV 89	AV 3	AV 5	AV 23
QZ	10.24	7.00	0.00	2.92	0.00	0.00	1.74			
OR	5.43	4.97	1.10	1.37	0.89	0.57	1.94			
PLAG	61.23	59.35	32.75	53.76	54.32	52.54	44.94			
DI	17.00	4.44	15.16	10.67	13.52	16.72	17.93			
HY	17.70	18.28	11.86	10.85	19.44	19.57	21.93			
OL		0.00	32.85	0.00	0.00	0.00	0.00			
MT	2.95	3.25	4.38	6.11	4.14	5.14	4.81			
IL	2.75	1.80	1.16	3.00	1.04	2.57				
AP	1.62	.33	.33	1.00	.17	.31	.25			
COR	.32	0.00	0.00	0.00	0.00	0.00	0.00			
TOTAL	99.53	99.42	99.58	98.82	99.33	99.00	99.00	99.23	99.23	99.25
AN PERCENT	37.7	45.6	60.4	54.3	62.6	49.1	62.5	96.0	83.9	66.4

AV 23 BASALT, PICRITIC, LOW TI SERIES, LEG 37, GUNN AND ROOBOL (1975)
AV 5   PICRITE, LEG 37, GUNN AND ROOBOL (1975)
AV 3   PERIDOTITE, LEG 37, MID-ATLANTIC RIDGE   GUNN AND ROOBOL (1975)
AV 89 META-BASALTS, THOLEIITIC, CANADIAN ARCHAEAN, LT 2.5 PC SODA, GEOKEM FILE.
AV 42 META-BASALTS, THOLEIITIC, ARCHAEAN, LT 16.5 PC AL203, GEOKEM FILE.
AV 98 META-BASALTS, HIGH-AL, ARCHAEAN, GEOKEM FILE.
AV 17 META-BASALTS, LOW-MG, ARCHAEAN, GEOKEM FILE.
AV 67 META-PICRITES AND PERIDOTITES, ARCHAEAN, GEOCHEM FILE
AV 62 ANDESITE, ARCHAEAN, 54-58 PC SI, HIGH AL, GEOCHEM FILE.
AV 30 ANDESITE, ARCHAEAN, 58-62 PC SI, HIGH AL, GEOKEM FILE.

ARCHAEAN AND RECENT AVERAGES.   GEOKEM DATA FILE.

NUMBER	AV-23	AV-36
SIO2	60.24	64.23
AL2O3	17.58	16.61
TIO2		
FE2O3	7.310	5.732
MNO	.186	.089
MGO	2.23	3.08
CAO	7.060	3.746
NA2O	3.730	4.050
K2O	1.013	1.557
P2O5	.150	.216
TOTAL	100.012	100.158
CR	21.4	26.7
CO	16.3	14.1
NI	34.7	45.3
CU	61.7	8.6
ZN	25.6	
GA		
RB	306.0	398.0
SR		508.7
BA	145.3	
PB	2.8	2.1
TH		

C.I.P.W. NORMS

	AV-23	AV-36
QZ	14.68	26.36
OR	5.99	9.19
PLAG	57.33	57.65
DI	14.14	
HY	2.07	
MT		
IL		
AP	.33	
COR		1.44
TOTAL	99.16	99.70
AN PERCENT	45.7	36.3

AV 23 ANDESITES, MONT PELEE, MARTINIQUE. GUNN,ROOBOL,SMITH,1974,BGSA,85,1027.
AV 36 DACITES, ARCHALAN, 62-68PC SI, GEOCHEM FILE.

ent modern calc–alkaline regions, described by different authors, seemed to vary quite widely in factors such as Fe/Mg ratio and in alkali content and although one could sample a small area geographically in an Archaean calc–alkaline sequence, inevitably a great span of time might be represented. Little was known of the compositional changes likely in a given region over time periods which might exceed that of the whole of the Phanerozoic.

To fill in gaps in the knowledge then existing of the chemical characteristics of the ocean floors, oceanic islands and island arcs and continental andesite series, an extensive programme was begun, both to acquire a large volume of new data, free of interlaboratory uncertainty, and to assemble a data-file from the literature, of chemical data from comparable regions. For example, we have carried out an extensive programme on the characteristics of island arcs, based entirely on the Lesser Antilles, but published data has been assembled from the Scotia Arc, the Tongan Arc, the New Hebrides, the Aleutians and Kuriles arc to check the validity of our findings.

Laboratory Methods

About 800 to 1000 new chemical analyses have been made per year by X-ray fluorescence methods for the major elements plus the trace elements Cr, Co, Ni, Cu, Zn, Ga, Rb, Sr, Ba and in special cases also S, Cl, Y, Pb, Th, Ce, La, Zr. Determination of the light elements is made on a fused glass bead diluted in a tetraborate flux and an iterative matrix correction and data reduction is made by PDP-11 on-line minicomputer, or by a central CYBER 74 computer. A data file (GEOKEM) is stored on both cards and tape and includes, as well as our own analyses, an assemblage of data from the literature totalling about 15,000 analyses. Regretably, few of the data other than our own include accurate trace elements and the accuracy of much of the older data, especially for minor elements is suspect. A set of about 40 computer routines is stored on disc and available through remote terminal for statistical, presentation, and plotting treatment, calculation of meso- epi- and katanorms, etc. For rapid sorting and retrieval, whole sections of the GEOKEM file may be transferred from magnetic tape to disc.

Sample Section

As the Archaean rocks appear to represent ancient oceanic crust and superposed island-arc volcanics, an effort was made to sample recent sea-floor and island-arc areas especially. Thanks to the great help and cooperation of Dr N. D. Watkins, extensive collections were also analyzed from a wide range of tholeiitic through alkali–basalt to basanitic and limburgitic oceanic islands. These include the plagioclase tholeiite islands of Amsterdam and St. Paul on the mid-Indian Ocean Rise (Gunn et al., 1971, Gunn et al., 1975), the ankaramitic suites of the Crozet group (Gunn et al., 1970, 1972), and the tholeiitic to basanitic suites of Kerguelen (Watkins et al., 1974).

In the Atlantic ocean only Iceland (Gunn, Watkins and Coy-Yll, 1970), the Azores (Self and Gunn, in prep.), the Cape Verde and Fernando de Noronha (Gunn and Watkins, 1975) have been sampled in detail.

To obtain data from island arcs a number of islands in the Lesser Antilles were sampled including Martinique (Gunn, Roobol and Smith, 1974) Antigua (Gunn and Roobol, 1974), Guadaloupe (Gunn, Roobol and Smith, in prep.), Tobago, Desirade, and Carriacou (Lewis and Gunn, 1972). Continental andesites are represented by Central Mexico (Gunn and Mooser, 1970; Watkins et al., 1971), North Chile (Roobol et al., 1974), South Central Chile (unpublished data) and Mt Shasta in the Cascade Province (Gunn and Payette, unpublished data).

Fresh sea-floor basalts were analyzed from Legs 2 and 3 of the Deep Sea Drilling Project and from the Leg 37 cores of the mid-Atlantic Ridge. Dredged samples were obtained from the research vessel *Hudson* from the mid-Atlantic Ridge, and others were supplied by Dr F. Aumento whose help is gratefully acknowledged. Samples of the Juan de Fuca Rise were provided by Dr S. Barr.

Many other studies have been made of other regions as well but these are of less direct relevance to the Archaean problem. As all of this data have been acquired in the same laboratory, by the same methods and monitored against the same standards, the usual interlaboratory descrepancies have been avoided. The number of concrete facts which have emerged from amassing this large amount of data are remarkably few in number, and many of the seemingly consistent patterns not previously recognized still await explanation.

Applications and Limitations of Petrochemistry to Archaean Problems

The mere analysis of ancient igneous rocks does not necessarily increase our knowledge of even the original composition of these rocks. An analysis of a greenschist metabasalt may reveal no more than information as to the stable bulk chemistry of an assemblage of chlorite, tremolite, albite and hematite. All elements not finding accommodation in the stable mineral assemblage may, and often have been, entirely lost from the rock. Even lava samples collected within minutes of extrusion may no longer be representative of the magma from which it formed, at least for some elements. For example, repeated analyses of samples of the recent Kirkufell lavas of Iceland gave such erratic values of elements such as F, Cl, S that no reliable information as to the composition of the magma for these elements could be gained. Other elements may be reduced in amount after the first rainfall and although apparently reliable and consistent data may be had from massive sills and flows as old as Jurassic (Gunn, 1966), it is uncommon to be able to analyze volcanic rocks older than Miocene without finding such profound diagenetic and metasomatic changes that the chemical data tell us little more than can be gained in the field from a hand lens, i.e. information of the type 'that this is an andesitic rock of general calc–alkaline parentage'. Massive, unmetamorphosed granites of Archaean age may retain their essential original character including mineralogy, but these plutons are invariably concentrically zoned so that only slopes of trends can be compared with other granitic series, individual analyses having little value.

A petrochemical study of the freshest Quaternary basalt or andesite series may be expected to yield the following information:

(a) The original chemical composition of the magma (except for F, Cl, S, Se, H_2O, CO_2, etc.) from which may be inferred the normative mineral composition, the ratios of alkalis, or alkali feldspar to plagioclase and mafic minerals, the plagioclase composition, the amounts of titania, phosphate, strontium, potash, alumina and the iron–magnesium ratio. From consideration of all these, the classification of the rock may be made. It must be stressed that in the absence of any one of these criteria, classification may become difficult.

(b) From the slope of the covariant element-pairs, the nature and extent of the crystal fractionation processes can be inferred, and in simple cases, e.g. of olivine, ankaramitic, or plagioclase fractionation the composition of the phenocryst phases may be written down.

(c) The relationship of the lava to specific petrographic provinces or cones. Thus the Cascade province is characterized by its unusually high Sr and low K. Of two adjacent cones on the island of Martinique, one, Morne Jacob, has 50% more K, Rb, Ba than the nearby Mont Pelée, at the same silica content. Flows occurring in the median valley can be quite confidently assigned to one or the other on their chemical composition.

(d) Periodic and progressive magma changes with time. Thus a study of 70 successive high-alumina basalt flows in Oregon showed both periodic composition changes overlain by a trend towards greater alkalinity (Gunn and Watkins, 1970). The Leg 37 drill cores of oceanic ridge basalt show both gravity stratification and periodicity.

(e) The relationship of the series being studied to a specific volcanic association, e.g. to tholeiitic flood basalt, continental orogenic andesite or to a nephelinite–basanite–phonolite association. This usually involves assignment to some classification of alkalinity which must lean heavily on the accurate determination of Ti, P, Na, K, Rb, Ba, Th, Sr especially, relative to silica.

(f) The probable tectonic province of the lava suite. Thus the low-K, low-Sr andesite suite now occurs only in orogenically active belts in regions of thin, oceanic crust, and high-K, high-Sr andesites only in regions of thick continental crust.

Unless at least some of the information listed above is forthcoming, chemical analysis cannot be justified. In the case of many Archaean metavolcanic rocks whose composition has been modified by diagenesis, halmyrolysis and metamorphism, the only unequivocal information may be:

(g) The present composition of a metamorphic rock.

(h) In the presence of different stages of alteration, the nature and type of the alteration, and the probable types of secondary minerals involved in alteration, such as smectite, illite, heulandite, laumontite, adularia, opal, prehnite, calcite, etc. If the alteration is not complex, by extrapolation some indication of the original composition may be gained, e.g. in the Bermuda drill core, basaltic flows could be shown to be both spilitized and altered in K, Rb, Ba content with the formation of illite. As all stages of alteration were present, an original composition close to Oceanic Ridge Basalt could be inferred (Aumento and Gunn, in press). As all the alteration processes to which Archaean metavolcanics have been subject have not yet been defined, we cannot yet place fine limits on their original composition.

The Chemical Characteristics of Oceanic Ridge Basalt

As is well known, a typical oceanic ridge basalt is hypersthene normative, and characterized by low K_2O (<0·5%), low Rb (<5 ppm), low Sr (<200 ppm) and TiO_2 (<1·5%). High-alumina and high-iron variants are known but no crystal fraction series have been previously described.

Much of our new knowledge of the possible range of composition of modern oceanic ridges come from 300 analyses of the Leg 37 cores, and although we have 500 other analyses of dredged and shallow-core material, their interrelationships are often totally obscure. For example, two successive dredge hauls on the mid-Atlantic ridge may bring up both material from recent, *in situ* flows and from ancient fault slices. While many more variants will undoubtedly be discovered with further drilling, the following types may be distinguished at this time:

(a) A series of undifferentiated primary tholeiitic basalts ranging from 9% to 7% MgO. The amount of incompatible elements (TiO_2, P_2O_5, Sr, K, Ba, Rb) increases sharply with decrease in MgO, by a factor of two. As this is not compatible with any form of crystal fractionation, it is inferred that these basalts represent decreasing percentages of partial melt of a mantle parent. These compositional differences are at the low end of the tholeiitic basalt-transitional-alkali basalt-basanite series. Thus while oceanic ridge lavas have 0·5–1·25% TiO_2, 0·07–0·5% P_2O_5, 100–150 ppm Sr, Hawaiian tholeiites have 2·5%, 0·22% and 330 ppm compared with 2·3%, 1·3% and 1400 ppm for typical basanites.

In the undifferentiated sea-floor basalts a real range of 0·2–0·5% K_2O is present, i.e. over this range K is covariant with Ti, P, Rb, Ba, etc. Higher values usually appear to be associated with alteration or with alkaline rocks associated with seamounts. Magnesian gabbros, etc., may have as low as 0·05% K_2O and peridotites even less.

There is a broad spectrum of alumina content from about 14·5 to 16·5% in apparently undifferentiated rocks. A similar range in alumina has been noted in other basalt series, thus the alkali-basalt, hawaiite, mugearite series of the Azores all have high and low alumina variants.

(b) Olivine accumulation leads to the formation of picrite series with those that stem from the more magnesian, low Ti parents being the better developed. The Mg/Cr, or Mg/Ni covariant trends (Figs. 2, 3) are exactly the same as those found in the Hawaiian picrites where the parental magma also with 8–10% MgO has four times the TiO_2, P_2O_5 content of the sea-floor equivalent (Gunn, 1971). The older Mauna Loa picrites of Hawaii are in turn less alkaline than the younger Kilauea picrites (Wright, 1971).

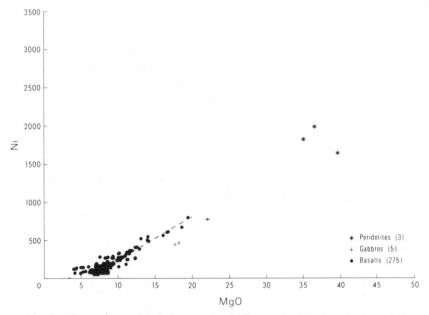

Fig. 2. Magnesium–nickel diagram for 300 oceanic ridge basalts and picrites, together with associated peridotite and eucrite, Leg 37 drill cores

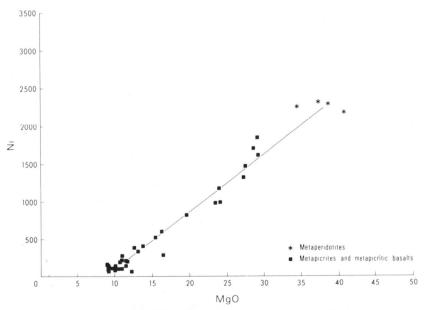

Fig. 3. Magnesium–nickel diagram for Archaean metabasalts, metapicrites, peridotites and metagabbros. Trend is similar to that of Fig. 2 and also to that of 1959 Kilauea Iki picrites, indicating similar olivine compositions and similar Mg–Ni ratio in parental material

(c) A plagioclase accumulation series stems from a high alumina parent of about 16·5% alumina. Accumulation of calcic bytownite gives rocks of up to 24% Al_2O_3, the series being distinguished by constant low soda (1·5%), steeply increasing CaO and low, constant K and Sr (the partition coefficients for Na, K, Sr being close to 1 for calcic bytownite). Again, several plagioclase series occur stemming from parents of differing Ti, P, Sr levels as for the picrites, but the picrite and plagioclase series parents differ in their Fe/Mg ratio. Mg-depleted rocks occur, very strongly depleted also in Ni, Cr, but no Al-depleted rocks have been found.

(d) Peridotite lenses occur in conjunction with eucritic gabbros of the same Fe/Mg ratio as the peridotites, these apparently being part of a rhythmically banded complex.

High Fe tholeiites with 16% Fe and 2·00% TiO_2 occur among dredge hauls but these do not have the correspondingly high P, K, Sr, etc., suggesting that they do not represent a more alkaline basalt, but may be the products of protracted closed-system fractionation.

The Archaean Basaltic Group

Having defined limits for the freshest of the sea-floor basalts, i.e. into high and low Ti undifferentiated tholeiites, picrite and plagioclase series, high Mg gabbros and peridotites, a series of computer searches were made through 1400 Archaean analyses to retrieve similar types.

Spilitized Basalts

No fresh sea-floor basalt, which we have ourselves analyzed, has more than 2·5% Na_2O, but 180 Canadian metabasalts had more, with an average of 3·2% with individual analyses having up to 5%. By contrast, 293 metabasalts of all types had less, so that roughly 40% of Archaean lavas may have been partially spilitized. This is a lower percentage than was found in the spilitized tholeiites of Bermuda (Aumento and Gunn, in press) but elevated temperatures during later dike injection in Bermuda may have accelerated alteration. When we made the same retrieval on all recorded sea-floor dredged rocks, however, about the same proportion of high-soda values were found, i.e. 40%, possibly suggesting that soda metasomatism affects about the same number of sea-floor basalts now as 3 b.y. ago.

Archaean Metapicrites

Prior to the Leg 37 drill cores, only rare picritic rocks had been recovered from the sea-floor, but more than 10% of all Archaean rocks analyzed have 9–50% MgO. This may be in part due to preferential sampling of

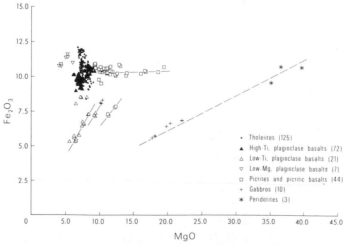

Fig. 4. Iron–Magnesium diagram for modern oceanic ridge rocks as in Fig. 2

obviously magnesian rocks, but may also be due to the superficial sampling of the sea-floor to date and the apparent gravity stratification present, the denser, more magnesian lavas being found at depth. The Archaean rocks include metapicrites, as well as metagabbro, meta-pyroxenite, meta-anorthosite and meta-norite. The common occurrence of clinopyroxenite sills and possible peridotitic flows in Archaean sequences cannot be matched as yet by any pyroxenites from the oceanic crust, but otherwise the distribution of rock types is similar. The Cr/Mg and Ni/Mg ratios in olivine-controlled series is identical both to modern oceanic and Hawaiian picrites. No anorthosite–pyroxenite complexes similar to the Archaean Doré Lake complex (Allard, 1970) have yet been found in the oceanic crust, but the anorthosite is a bytownitic, low Sr type, in every way comparable with plagio-clase cumulates already described.

One of the few features of difference between modern and Archaean oceanic crust is the apparent absence in the latter of the alkali basalt islands and ultra-alkaline nephelinite or bermudite dike complexes.

Both Leg 37 and other oceanic picrite–peridotite rocks show a gap between magnesium contents of about 25–35%, the more

basic rocks which include eucrite cumulates having a lower, constant Fe/Mg ratio while the picrites have a ratio decreasing with increasing olivine. The same break in Fe/Mg can be detected in the Archaean rocks (Fig. 5), but a more complete span of Mg is present, perhaps due to the inclusion of many differentiated, basic and ultra-mafic sill complexes in the data.

Archaean High-alumina Series

High-alumina plagioclase basalts appear to be under-represented in the Archaean rocks compared with modern sea-floor samples. We have only forty analyses of Archaean basalts with more than 16·5% alumina. Few have over 20% Al_2O_3 and little trace element data are available. Like the sea-floor basalts, they tend to have low Ti and P, but the K is un-usually variable even for Archaean rocks, ranging from 0·1 to 1·0%, possibly due to the ready alteration of the plagioclase. The pres-ence of bytownitic anorthosite in the Doré Lake complex shows that the Archaean mag-mas did, under closed conditions, fractionate calcic plagioclase extensively, and that this was of a low Sr type. Tholeiitic plagioclase basalts are similar to the high-alumina basalts of island arcs but the latter, at least those of the

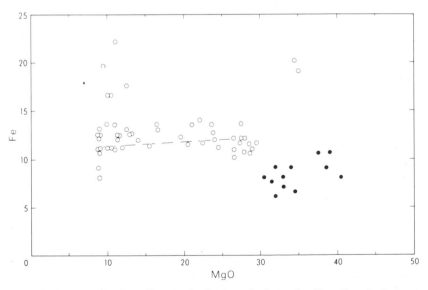

Fig. 5. Iron–magnesium diagram for basic rocks from the Canadian Archaean showing presence of metapicrites and also more magnesian rocks intermediate to peridotite not yet found in the modern sea-floor

Lesser Antilles, have only 10–20 ppm Ni, while plagioclase tholeiites of oceanic ridges and islands have 60–100 ppm Ni. The Archaean high-alumina basalts discussed here are of the latter type with 60–120 ppm Ni.

Modern Orogenic Andesite Series

The wide possible chemical range of the high-alumina basalt, basaltic andesite, andesite, dacite, rhyolite series and the possible tectonic affiliations of the different types is only now beginning to be realized. Based on a least-mean-squares average composition at 60% SiO_2 (Gunn, 1975), the andesite may be classed into low-K andesite (0·75–1·25% K_2O), 'normal' andesite (1·25–2·25%) and high-K andesite (2·25–3·5%). The low-K andesites are almost entirely confined to the island arcs though there is some overlap with 'normal' andesite in the Aleutian and Kurile arcs. Normal andesites are found in the South-Central Chile province, the Central American and Mexican provinces, the Cascade province (which has the lowest K of all the continental suites) while high-K rocks make up the North Chilean and San Juan provinces. In general, the high-K suites have also the highest Sr, Rb, Ba, Th but unlike the tholeiitic–alkali basalt group, there is no parallel increase in P and Ti. However, within a province or volcanic island Sr varies inversely with K and in several instances it has been shown that the oldest volcanoes or flows in a group have the highest K, Rb, Ba and lowest K/Rb ratio, but the lowest Sr abundances. The youngest flows or volcano have the highest Sr, Mg and soda. Different vents or even parasitic cones produce andesites quite unrelated by crystal fractionation. For example, in the Cascade province where the inverse Sr–K effect is extreme (Fig. 6), a young Mt Shastina flow has over 1000 ppm Sr but 1·2% K, 19 ppm Rb, while the older Sand-Flat flow from Mt Shasta has only 500 ppm Sr but 1·7% K and 42 ppm Rb at similar silica contents (63%). Thus, andesite flows may vary extremely erratically in the same region, but a low-K, low-Sr andesite will almost certainly occur only in a region of thin oceanic crust, and a high-K andesite with

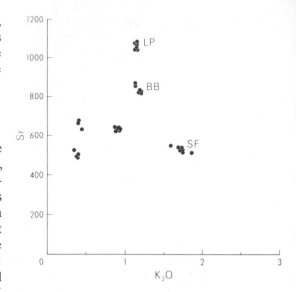

Fig. 6. Strontium–potassium diagrams for Mt Shasta and parasitic cones, Cascades province, U.S.A. LP = Lava Park andesite flow, Mt Shastina, BB = Black Butte dacite, SF = Sandflat andesite of Mt Shasta, together with basaltic andesites of Everitt Hill parasite cones. Note lack of any coherent relationship between the flow units. Lava Park and Sandflat flows both have 61% silica

usually 600 ppm or more Sr will occur only in a region of more than 50 km of crust. Two-feldspar andesites appear to be confined to regions of thick continental crust.

Alteration mainly has the effect of intensifying these differences, thus low-K andesites of the prehnite–pumpellyite grade of metamorphism from the Limestone Caribbees have altered to spilitic basalts and keratophyres with bulk loss of Ca, K, Rb, Ba (Gunn and Roobol, 1974), but high K-altered andesites of the Greater Antilles appear to have enhanced K contents, often over 4% (Jolly, 1970).

High iron is not, contrary to popular opinion, a characteristic of island-arc andesites, those of Central America and even the extremely potassic San Juan province being extremely iron-enriched. The well-known Cascade province is characterized by unusually low Fe/Mg ratio, and it is probably from this that the myth originated.

In addition to the wide spectrum of orogenic andesites, low alumina icelandites and similar rocks occur in the tholeiitic suites of Iceland and in the sodic Deception Island lavas as well. The hawaiites of the alkali basalt series are often superficially similar to the orogenic andesite, and might be especially difficult to distinguish in altered rocks.

Archàean Andesite

The marked, and erratic levels of alumina and alkalies in the Archaean andesite series together with the erratic behaviour of the andesite suite in general as described above, make interpretation almost impossible. However, the average of large numbers of Archaean andesites is a low-K, low-Sr andesite of unmistakable island-arc type. It is suspected that the potassic Archaean rhyolites are largely due to the secondary formation of illite, but this is dependent on climate, the low-K rhyolites of Guadaloupe altering to kaolinite with loss of K (Gunn, Roobol and Smith, in preparation). Archaean sodic rhyolites do match, quite closely, the soda-rhyolites of Guadaloupe and St Lucia.

Conclusions

Study of both recent and Archaean complexes can be used to complement each other. The extreme, meta-picrite, meta-pyroxenite and peridotite trends found in Archaean rocks can be accepted as being probably genuine, and the similarity between ancient and modern oceanic crust is sufficiently close that we can predict that layered complexes of the Doré Lake type will be found by drilling in modern sea-floors. However, the minor variants of the undifferentiated tholeiites cannot be defined in Archaean rocks because of alteration but once recognized in unaltered modern rocks, they can be tentatively identified as also existing in the older series if samples can be found in which the TiO_2, P_2O_5, Sr, K_2O, Rb, Ba and probably REE vary in sympathy. It must be strongly urged that future studies of Archaean rocks should include more careful determination of the minor elements as at least half of the currently available data do not include phosphate.

As low values of Ni, Cr and high Co appear to distinguish high alumina basalts of island-arc andesites from the plagioclase basalts of the oceanic ridges, these elements are also important. The low average K/Rb ratio of about 300 for the Archaean rocks (compared with about 470 for MORB) has probably no meaning as Rb is rapidly adsorbed by illite clays.

Many Archaean basalts have low alumina, less than 14%. Alteration of calcic plagioclase to albite involves either relocation of the

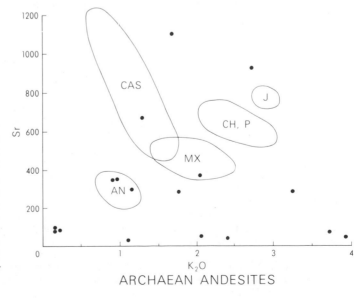

Fig. 7. Sr–K diagram for some Canadian Archaean andesites and dacites of 54–68% SiO_2. The fields shown are for the calculated averages at 60% SiO_2 of modern andesites, An = Lesser Antilles island arc, MX = Mexican volcanic province, CAS = Cascades province, CH, P = North Chile–Peru province, J = Jamaican intrusives. The South Central Chilean, Central American and Aleutian Provinces largely coincide with the Mexican province

ARCHAEAN ANDESITES

402

alumina or gain in soda and the alumina content of the amphibole forming at the expense of pyroxene is dependent to some degree on metamorphic grade. Rocks of the Island of Desirade in the Lesser Antilles are almost completely reconstituted to prehnite, albite, chlorite, epidote and quartz and have characteristically lower alumina than less-altered lavas from Antigua. We believe this is possibly a feature of alteration. We also suspect that the very constant Fe/Ti covariance often seen in altered rocks including those of the Archaean may be due to the stability of one of

the oxidation states of titanomagnetite. The general similarity between Archaean and modern oceanic crust is sufficiently close that it appears that the very processes which created the primordial crust of the Earth are those still operating at the mid-oceanic ridges today. It seems that the upper mantle must still be essentially of the same composition now as 3 b.y. ago, possibly suggesting that a state of dynamic equilibrium exists in the upper mantle indicating renewal, either by convective movement or replacement of basaltic material, by addition from the lower mantle.

References

Allard, G. O., 1970. 'The Doré Lake Complex, Chibougamau Quebec. A metamorphosed Bushveldt-type layered intrusion', In *Symposium on the Bushveld Igneous Complex*. Geol. Soc. South Africa Spec. Publ. No. 1, pp. 477–491.

Aumento, F., and Gunn, B. M. 'Deep Drill–1972. Petrology and Geochemistry of the Bermuda seamount', (in press).

Baragar, W. R. A., and Goodwin, A. M., 1969. 'Andesites and Archaean Volcanism of the Canadian Shield', *Proc. And. Conf. State of Oregon, Dept. of Geology and Mineral. Industries, Bull.*, 65.

Donnelly, T. W., 1966. 'Geology of St. Thomas and St. John, U.S. Virgin Islands', *Geol. Soc. America Mem.* 98, pp. 85–176.

Gunn, B. M., 1966. 'Modal and element variation in Antarctic tholeiites', *Geochim. Cosmochim. Acta*, 30, 881–920.

Gunn, B. M., 1971. 'Trace element partition during olivine fractionation of Hawaiian basalts', *Chem. Geol.*, 8, 1–13.

Gunn, B. M., 1975. 'The systematic petrochemical differences in andesite suites', *Bull. Volcanol.*, 28, 2.

Gunn, B. M., Coy-Yll, R., Watkins, N. D., Abranson, C. E., and Nougier, J., 1970. 'Geochemistry of an oceanite-ankaramite-basalt suite from East Island, Crozet Archipelago', *Contr. Mineral. Petrol.*, 28, 319–339.

Gunn, B. M., and Mooser, F., 1970. 'Geochemistry of the volcanoes of Central Mexico', *Bull. Volcano.*, 34, (2), pp. 577–616.

Gunn, B. M., and Watkins, N. D., 1970. 'Geochemistry of the Steens Mountain Basalts, Oregon', *Bull. Geol. Soc. Amer.*, 81, (5), 1497–1516.

Gunn, B. M., Abranson, E. C., Nougier, J., Watkins, N. D., and Hajash, A., 1971. 'Amsterdam

Id., an isolated volcano in the southern Indian Ocean', *Contr. Mineral Petrol.*, 32, 79–92.

Gunn, B. M., Abranson, E. C., Watkins, N. D., and Nougier, J., 1972. 'Petrology and Geochemistry of the Crozet Islands: A summary', pp. 825–829 in Adie, R. J. (Ed.) *Proc. 2nd Conference in Antarctic Geology and Geophysics* (Oslo, 1970).

Gunn, B. M., and Roobol, M. J., 1974. 'The geochemistry of the Limestone Caribbees', Preprint, 18 pp., *VII Caribbean Geol. Conf.*, Held Guadaloupe, in press.

Gunn, B. M., Roobol, M. J., and Smith, A. L., 1974. 'Petrochemistry of the Pelean-type volcanoes of Martinique', *Bull. Geol. Soc. Am.*, 85, pp. 1023–1030.

Gunn, B. M., Watkins, N. D., Nougier, J., and Baksi, A. K., 1974. 'Kergulen: Continental fragment or oceanic island?', *Geol. Soc. Amer. Bull.*, 85, pp. 201–212.

Gunn, B. M., Watkins, N. D., Trzcienski, W. E., and Nougier, J., 1975. 'The Amsterdam St-Paul volcanic province and the formation of low Al-tholeiitic andesites', *Lithos*, 13, pp. 62–73.

Gunn, B. M., and Watkins, N. D., (in press). 'The Geochemistry of the Cape Verdes and Fernando de Noronha'.

Gunn, B. M., Roobol, M. J., and Smith, A. L. (in preparation). 'Geochemistry of Guadaloupe'.

Jolly, W. T., 1970. 'Zeolite and prehnite-pumpellyite facies in South Central Puerto Rico', *Contr. Mineral. Petrol.*, 27, pp. 204–224.

Lewis, J. F., and Gunn, B. M., 1972. 'Aspects of Island arc evolution and magmatism in the Caribbean: Geochemistry of some West Indian plutonic and volcanic rocks', *Trans. VI Carib. Conf.*, pp. 171–177.

Nockolds, S. R., and Allen, R., 1953. The geochemistry of some igneous rock series. I. Calc-alkalic rocks', *Geochim. Cosmochim. Acta*, 4, pp. 105–142.

Roobol, M. J., Francis, P. W., Ridley, W. I., Rhodes, M., and Walker, G. P. L., 1974. 'Physico-chemical characters of the Andean volcanic chain between latitudes 21° and 22° south', Preprint. 11 pp. *IAVCEI Symp. Internat. de Geologia*, Santiago, Chile.

Self, S., and Gunn, B. M. (in preparation). Volume and age relations in alkaline volcanics from Terceira, Azores.

Watkins, N. D., Gunn, B. M., Baksi, A. K., York, D., and Ade-Hall, J., 1971. 'Palaeomagnetism, geochemistry and K/Ar ages of the Rio Grande de Santiago Volcanics, Central Mexico', *Bull. Geol. Soc. Am.*, **82**, pp. 1955–1968.

Watkins, N. D., Gunn, B. M., and Coy-Yll, R., 1970. 'Major and trace element variations during the initial cooling of an Icelandic lava', *Amer. J. Sci.*, **268**, 24–49.

Wilson, H. D. B., Andrews, P., Moxham, R. L., and Ramlal, K., 1965. 'Archaean volcanism in the Canadian Shield', *Can. J. Earth Sci.*, **2**, pp. 161–175.

Wright, T. L., 1971. 'Chemistry of Kilauea and Mauna Loa Lavas in space and time', *U.S. Geol. Surv. Prof.*, Paper, 735, 40 pp.

Geochemistry of Archaean High-grade gneisses, with Implications as to the Origin and Evolution of the Precambrian Crust

JOHN TARNEY

Department of Geological Sciences, The University of Birmingham, England.

Introduction

The purpose of this paper is to review the geochemical characteristics of the high-grade gneiss terrain of the North Atlantic craton (Bridgwater et al., 1973) with reference to four areas: mainland Scotland, the Outer Hebrides, East Greenland and to a lesser extent the Fiskenæsset area of West Greenland. A number of authors in recent years have discussed various aspects of the geochemistry of the gneisses with respect to local areas and local problems (e.g. Sheraton, 1970; Bowes et al., 1971; Tarney et al., 1972; Sheraton et al., 1973; Drury, 1973; Holland and Lambert, 1973). The emphasis here will be placed on overall geochemical characteristics, in particular those bearing upon the nature, origin and evolution of the gneiss terrain during the Archaean.

The geology of the region is complex, but is adequately reviewed in Park and Tarney (1973), and by Windley (1972) and Kalsbeek and Myers (1973) for the Fiskenæsset region. Briefly, in mainland Scotland a central zone of granulites and retrogressed granulites is flanked to the north and south by amphibolite facies grey gneiss, originally thought to be reworked gneisses of the central zone, but now regarded as juxtaposed different crustal levels (Sheraton et al., 1973) with considerable deformation and migmatization at the tectonic junctions. The relationships on the Outer Hebrides are basically similar, although different in detail (Graham and Coward, 1973). The

Angmagssalik region of East Greenland, just north of the Nagssuqtoqidian boundary, has suffered mild Proterozoic reworking (Wright et al., 1973) but the main rock types can be traced without difficulty into the undeformed Pre-Ketilidian zone (Andrews et al., 1973). The dominant gneiss in the whole East Greenland region is a granodioritic grey gneiss. This is repeatedly interleaved tectonically with high-grade metasedimentary supracrustals. Near Angmagssalik itself large slices of basic to acid granulites occur, invaded by anorthosite. The geological relationships between the three areas have been discussed by Tarney et al. (1972) and Wright et al. (1973). Essentially similar rock types occur at Fiskenæsset in West Greenland (Windley, 1972), but the grey gneisses locally develop pyroxenic assemblages.

In most of the areas considered there are three main Archaean gneiss groups: granodioritic grey gneiss, high-grade metasedimentary supracrustals, and pyroxene granulites. In all three groups the range of composition varies between ultramafic and acid. Acid grey gneiss dominates the amphibolite facies terrains, but in granulite facies terrains the composition is distinctly more basic (cf. Sheraton et al., 1973). Barker and Peterman (1974) noted that in many Archaean gneisses the compositional variation is bimodal: a banded granodiorite/tonalite gneiss with basic amphibolite layers, lenses or

405

schlieren. This holds for the amphibolite facies terrains discussed here, the basic component probably representing variably fragmented sills or dykes, possibly of more than one generation.

The character of the granulite facies terrains is more heterogeneous. Bowes et al. (1974) noted large numbers of ultramafic lenses scattered throughout the central zone of the Lewisian. In fact there are numerous others (e.g. Davies, 1974), associated with perhaps ten times as much mafic gneiss, and often severely disrupted and interbanded with the acid gneisses. There are also gneisses of various intermediate compositions, so that there is almost a complete range of gneisses between ultramafic and acid. AFM plots for the granulite facies gneisses from the Drumbeg area of the Lewisian (Fig. 1) show a rather smooth calc-alkaline trend with little evidence of bimodality. Such a compositional variation may be partly original, partly tectonic and even partly metasomatic (cf. Sheraton et al., 1973).

Grey Gneiss Composition

The dominant gneiss type making up the Greenland–Scotland Archaean craton is a granodioritic plagioclase–quartz–biotite–hornblende ± minor microcline gneiss ('Grey Gneiss'). It is stressed that this is also a dominant gneiss composition in the granulite facies Central Zone of Scotland, although microcline is notably absent here. Mean analyses of grey gneisses representing six different areas of the North Atlantic craton are shown in Table 1. Considering the different sampling techniques employed by the various workers, the apparent uniformity of the grey gneiss composition with regard to all major and trace elements is remarkable. Notable features are the high levels of Ba, Sr, Zr and Ce, and the moderately high Ba/Rb, Ba/Sr and Ce/Y ratios. If yttrium reflects the behaviour of the heavy rare-earth elements (Yb), then the rare-earth patterns for these grey gneisses would appear to be relatively highly fractionated, with Y levels 2–9 times chondrites and Ce and La

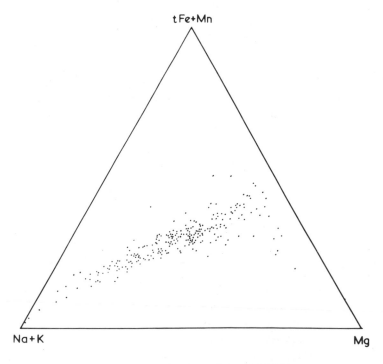

Fig. 1. AFM plot of Lewisian gneisses from Assynt, showing smooth calcalkaline trend (from Sheraton et al., 1973)

Table 1. Mean compositions of grey gneisses from the North Atlantic craton

| | Grey gneisses | | | | | | | | | |
	A	B	C	D	E	F	1	2	3	4
SiO_2	69·2	66·7	67·6	67·17	68·15	67·62	64·4	63·2	65·3	51·21
TiO_2	0·34	0·34	0·33	0·31	0·39	0·38	0·62	0·50	0·25	0·85
Al_2O_3	15·9	16·04	15·19	15·79	16·32	15·61	15·5	15·3	18·3	20·1
Fe_2O_3	3·4	1·94	1·75	1·68	3·60	1·89	6·54	0·99	2·34	3·25
FeO	—	1·47	1·28	1·41	—	1·05	—	4·26	—	3·9
MnO	0·06	0·04	0·05	0·04	0·03	0·04	—	0·07	0·03	0·107
MgO	1·5	1·44	1·28	1·09	1·00	1·46	3·12	3·42	1·42	4·40
CaO	3·2	3·18	3·22	3·54	3·98	3·42	2·22	2·76	4·38	7·56
Na_2O	4·2	4·90	4·15	4·74	4·45	4·97	3·74	3·51	6·08	3·87
K_2O	2·3	2·09	2·58	2·10	2·07	1·13	2·44	2·09	0·99	1·38
P_2O_5	—	0·14	0·11	0·12	0·20	0·15	—	0·11	0·13	0·327
S	—	140	280	230	178	460	—	—	—	170
Cl	—	170	130	100	—	130	—	—	—	260
Cr	—	32	30	23	21	26	—	—	—	64
Ni	—	20	18	13	13	40	—	—	—	26
Rb	—	74	77	44	60	15	88	58	14	25
Sr	—	580	475	580	460	615	424	457	1064	721
Y	—	7	6	8	17	3	—	—	—	18
Zr	—	193	182	207	215	278	—	—	—	164
Nb	—	6	5	6	—	6	—	—	—	8
Ba	—	713	840	1040	673	984	—	566	505	690
Ce	—	71	43	68	—	49	—	33	13	62
La	—	32	—	—	28	25	—	—	—	—
Pb	—	22	16	16	15	14	—	—	—	18
Th	—	11	8	10	4	1·5	—	—	—	<5
K/Rb		234	278	396	286	625	230	299	586	458
Rb/Sr		0·127	0·162	0·076	0·130	0·024	0·207	0·127	0·013	0·035
Ba/Rb		9·6	10·9	23·6	11·2	65·6	—	9·8	36·1	27·6
Ba/Sr		1·2	1·8	1·8	1·5	1·6	—	1·2	0·5	0·9
Ce/Y		10	7	9	—	16	—	—	—	3

A. Average of 268 Lewisian grey gneisses, Laxford Assemblage, Holland and Lambert (1973).
B. Average of 39 Lewisian grey gneisses, north of Laxford Bridge (Sheraton et al., 1973).
C. Average of 22 Lewisian grey gneisses, Harris and Lewis, Outer Hebrides (Sheraton et al., 1973).
D. Average of 41 grey gneisses, Angmagssalik, E. Greenland (Sheraton et al., 1973).
E. Average of 21 grey gneisses, Fiskenæsset, W. Greenland (Tarney, Cheesman and Windley, unpubl. data).
F. Average of 36 (retrogressed granulite-facies grey gneisses, Assynt (Sheraton et al., 1973).
1. Average of 23 Early Precambrian greywackes, Wyoming (Condie, 1967).
2. Early Precambrian greywacke, Minnesota (Arth and Hanson, 1975).
3. Saganaga Tonalite, Minnesota (Arth and Hanson, 1975).
4. Average of 9 tonalites, S. Harris, Outer Hebrides (Skinner, 1970).

levels 50–90 times chondrites. Note that the grey gneiss component of the granulite facies Lewisian ((F), Table 1) is similar to the rest apart from K_2O and Rb and Th which are low by factors of 2 and 5, respectively. These three elements, together with U, are characteristically depleted in granulite facies gneisses (Lambert and Heier, 1968; Tarney et al., 1972) and it would seem that they have been removed from these Lewisian grey gneisses too.

Suitable parent rock types for the grey gneisses might be granodiorite/tonalite or greywacke. Archaean tonalites (e.g. Saganaga tonalite, Table 1 (3) and S. Harris tonalite, Table 1 (4)) have comparable abundance levels of Sr, Ba, Zr and Ce but lower Ba/Sr ratios. However, their major element composition is divergent in more than one respect: few gneisses in the areas studied have such high alumina contents for example.

The comparison with greywacke is much closer. Archaean greywackes from Wyoming (Table 1 (1)) and the Vermilion district of Minnesota (Table 1 (2)) are close to the grey gneiss composition apart from rather high levels of Ti, Mg and Fe. This similarity has been noted previously with regard to the Laxford grey gneisses of the Scottish mainland by Holland and Lambert (1973) who suggested that they represented a supracrustal sequence overlying the granulites. Certainly obvious metasedimentary gneisses occur within the grey gneisses throughout a large part of the North Atlantic craton as in other Archaean gneiss cratons, but have been regarded as sedimentary intercalations in a dominantly igneous lava sequence (Sheraton, 1970;

Barker and Peterman, 1974), or as tectonic intercalations of supracrustals into a gneissic basement (Wright et al., 1973) rather than sediments of more extreme composition in a sequence of monotonous greywackes, subsequently gneissified (e.g. Holland and Lambert, 1973).

This problem can be resolved by reference to TiO_2–SiO_2 abundance levels in the whole range of gneisses from the areas studied (Fig. 2). There is a well-defined negative correlation between TiO_2 and SiO_2 in Archaean gneisses (Fig. 2A). There is also a well-marked correlation between TiO_2 and SiO_2 in sediments, metasediments and migmatitic metasediments associated with the Caledonian orogen in Britain (Torridonian, Moinian, Monian, Dalradian, Ordovician formations) but this is approximately 0·6% TiO_2 higher at equivalent SiO_2 levels than the Archaean gneisses (Fig. 2D). Sediments falling below the approx-

imate dividing line tend to be calcareous, and would tend to fall above the line if calculated $CaCO_3$ and water-free. Archaean sediments and metasediments from Australia, N. America and the North Atlantic craton (Fig. 2C) have a TiO_2–SiO_2 distribution comparable with the more recent sediments, although there is a greater low TiO_2 scatter, possibly implying more general low TiO_2 source rocks. Nevertheless the distinction is sufficiently good to indicate that the Archaean gneisses were not dominantly of sedimentary derivation.

The large volume occupied by grey gneiss in the craton as a whole would also make a sedimentary source difficult, unless a widespread primary pseudo-sedimentary breccia comparable with the upper lunar crust is postulated. On the other hand, the similarity of the Archaean gneiss TiO_2–SiO_2 distribution to that of the British calc–alkaline igneous trend

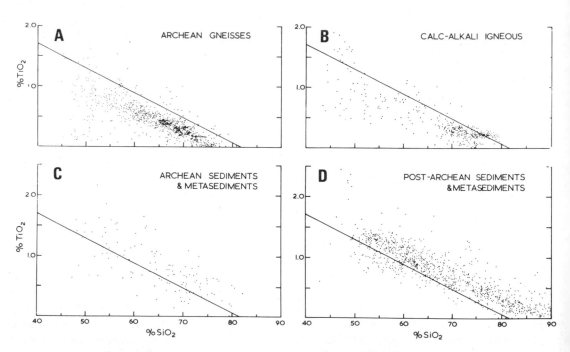

Fig. 2. TiO_2–SiO_2 distribution in Archaean gneisses. A. Archaean gneisses from Scotland, Outer Hebrides, East Greenland and West Greenland. B. Continental margin calcalkaline igneous rocks of Late Precambrian to Lower Palaeozoic age, England and Wales. C. Archaean clastic sediments and metasediments from Scotland, Greenland, North America and Australia. D. Late Precambrian to Palaeozoic sediments and metasediments (including migmatitic metasediments) from Britain. Diagonal line for reference. Archaean gneisses show a well marked Ti–Si negative correlation similar to calcalkaline igneous rocks, and most fall below the line. Clastic sediments and metasediments mostly plot above this line, although calcareous sediments may fall below it. Sources of data: numerous

(Fig. 2B) suggests that comparisons with calc–alkaline igneous rocks might be more appropriate. The overall gneiss composition is distinctly calc–alkaline (Fig. 1). It would appear that the grey gneiss composition in Table 1 represents a widespread Archaean granodioritic rock type in high-grade terrains. Rb–Sr isotopic ages (e.g. Lyon et al., 1973; Moorbath et al., 1975) suggest comparison with the Nûk plutonic phase in West Greenland (McGregor, 1973).

Geochemistry of High-grade Gneisses and the Origin of the Continental Crust

The geochemical comparisons made in Table 1 could suggest that granodioritic grey gneiss may represent some sort of crustal protolith in the high-grade terrains (Wynne-Edwards and Hasan, 1972) although how and when it formed is uncertain.

Evidence from nearby planets suggests a major episode of differentiation at an early stage in their history. Whether accreted homogeneously (Ringwood, 1972) or inhomogeneously (Anderson, 1975) the outer regions of the Moon seem to have undergone complete melting and differentiation to form a 60 km thick outer brecciated crust some 4·6 b.y. ago with effective cessation of igneous activity 3·2 b.y. ago (Toksöz et al., 1975; Taylor and Jakeš, 1975). The evolution of Mercury and Mars may have been not too dissimilar (Toksöz et al., 1975).

Although Fanale (1971) has argued for a catastrophic early differentiation and degassing of the Earth rather than a progressive core growth (e.g. Runcorn, 1962), the Earth's crust records little of this early history. Rb–Sr ages from Archaean shield areas fall between 3·8 b.y. (Black et al., 1971; Goldich and Hedge, 1974) and ~2·8 b.y. (e.g. Moorbath et al., 1975), but initial $^{87}Sr/^{86}Sr$ ratios preclude derivation of the gneisses from a pre-existing high Rb/Sr crustal source.

The general similarity of the average major element composition of the crust to andesite led Taylor (1967) to suggest lateral and progressive growth of continents through addition of mantle-derived calc–alkaline andesite.

With the advent of plate tectonics Ringwood (1969, 1974) amplified this into a two-stage model, (a) mantle melting at ocean ridges to give oceanic crust and (b) re-melting of the oceanic crust during subduction to give lithophile element-enriched andesite, dacite, etc. More recently Jakeš (1973) suggested that island arcs, built up from a base of island arc tholeiite and capped by calc–alkaline andesitic dacite lavas, produce a primitive vertically zoned crust which, if accreted at continental margins and further modified by Andean-type plutonism, could contribute to continental growth. The geochemical similarity between island arc volcanics and Archaean greenstones (Jahn et al., 1974; Arth and Hanson, 1975) might appear to lend credence to this model.

With regard to Archaean crustal development, Glikson (1971) argued that greenstone belts represent primitive oceanic crust subsequently enveloped in granitic crustal material generated from the mantle. Bowes et al. (1971) speculated that the Lewisian high-grade granulite terrain could be a deformed greenstone belt, although Sheraton et al. (1973) expressed reservations. Whether the more basic components of the high-grade gneiss terrains have any affinity with greenstone belts can, however, be tested geochemically.

Representative analyses of basic to acid gneisses from the Lewisian and from East Greenland are listed in Table 2 and compared with representative volcanics from Archaean greenstone belts, modern island arcs and calc–alkaline andesite. While major element comparisons could be made, the minor element abundance levels of the gneisses and volcanics are very different. Notable are the high concentrations of Ba, Sr, Zr, Ce and La in high-grade gneisses, acid through basic. Not only are the Ba and Sr values much higher than in Archaean or island arc volcanics but Ba/Sr ratios are generally higher than unity except in the melanocratic gneisses. This is illustrated in Fig. 3. Ocean basalts, island arc tholeiites and andesites have low Ba contents and low Ba/Sr ratios (Fig. 3F). Taylor's (1969) average andesites also have Ba/Sr ratios less than unity. Only shoshonites have comparable Ba and Sr levels with Archaean gneisses (Jakeš and White, 1972), but these form an insignificant proportion of calc–alkaline provinces. Similar remarks apply to Archaean volcanics.

If continental growth by lateral accretion of *oceanic* island arc or calc–alkaline volcanics took place in the Archaean, and if geochemical fractionation of lithophile elements from the lower to the upper crust occured in order to account for the trace element levels in the upper crust as envisaged by Jakes (1973), then granulite facies gneisses ought to be depleted in lithophile elements. While lithophile elements such as K, Rb, Th and U conform to this pattern, others such as Ba, Sr and Zr do not; in fact they are notably enriched in granulites.

Confirmatory evidence comes from rare-earth abundances in Archaean gneisses. Although no complete rare-earth measurements are yet available, Ce and La values are high, especially in some mafic granulites where Ce levels exceeding 100 ppm are not uncommon. On the other hand, Y levels are low, only 2 to 10 times chondrite values, hence, if heavy rare-earth behaviour parallels that of Y, rare-earth patterns must be highly fractioned. The rare-earth patterns of island arc and Archaean greenstone belt volcanics, however, are characteristically flat (Jakeš and White, 1972; Jahn et al., 1974).

It would appear from the trace element data that suggestions that granulite facies terrains like the Lewisian are gneissified greenstone belt or oceanic volcanics are incorrect. Jamieson and Clarke (1970) noted consistent differences in K, Rb, Ba, Sr, Zr and P between oceanic and continental basalts. The mafic gneisses in high-grade terrains are distinctly non-oceanic: to state the obvious, the earliest continental rocks are continental and calc–

Table 2. Comparisons of high-grade gneisses with Archaean and oceanic volcanics

	A	B	C	D	E	F	G
SiO_2	54·89	45·78	52·22	56·28	63·56	51·54	51·01
TiO_2	0·69	1·16	0·78	0·72	0·46	0·60	1·25
Al_2O_3	16·27	13·27	13·85	16·18	15·83	7·10	12·45
Fe_2O_3	3·37	5·32	5·55	3·94	2·69	1·98	3·39
FeO	4·73	9·34	4·97	3·22	1·83	6·45	6·57
MnO	0·12	0·21	0·16	0·09	0·06	0·13	0·13
MgO	5·31	8·58	7·26	4·23	2·57	17·22	10·02
CaO	8·83	12·55	8·76	6·72	4·92	11·93	9·13
Na_2O	3·41	1·47	3·09	4·50	4·69	1·18	2·15
K_2O	0·89	0·54	0·61	0·87	0·91	0·77	1·35
P_2O_5	0·13	0·09	0·13	0·32	0·17	0·20	0·43
Cr	169	237	263	106	52	1150	368
Ni	88	129	134	67	44	387	181
Rb	7	8	6	9	9	17	35
Sr	363	151	392	719	631	376	1040
Y	13	19	18	15	5	13	13
Zr	154	52	110	262	191	90	176
Nb	5		5	5	5		
Ba	520	236	455	668	728	407	1210
Ce	39	27	37	76	44	73	117
La	14		13	31	20	34	50
Pb	11	4	11	15	12	7	11
Th	1	<5	2	1	1	<5	<5
K/Rb	1026	603	830	793	888	353	300
Ba/Rb	72	34	75	73	86	24	35
Ba/Sr	1·4	1·8	1·2	0·9	1·2	1·1	1·2
Rb/Sr	0·020	0·051	0·016	0·013	0·013	0·044	0·033

	1	2	3	4	5	6	7
SiO_2	50·2	57·6	49·9	51·57	57·40	79·20	59·52
TiO_2	1·38	1·00	1·51	0·80	1·25	0·23	0·70
Al_2O_3	13·7	15·4	17·2	15·91	15·60	11·10	17·2
Fe_2O_3	13·7	1·82	2·0	2·74	3·48	0·52	6·8
FeO		5·96	6·9	7·04	5·01	0·90	
MnO	0·22	0·15	0·17	0·17			0·16
MgO	6·65	3·75	7·2	6·73	3·38	0·36	3·42
Cao	8·33	5·64	11·8	11·74	6·14	2·06	7·03
Na_2O	2·76	3·70	2·7	2·41	4·20	3·40	3·68
K_2O	0·29	0·84	0·16	0·44	0·43	1·58	1·60
P_2O_5	0·14	0·22	0·12	0·11	0·44		0·28
Cr		103	297	50	15	4	56
Ni		63	97	30	20	1	18
Rb	5		1·1	5	6	15	31
Sr	150	263	136	200	220	90	385
Y		14	43			23	21
Zr		160	95	70	70	125	110
Nb							
Ba	70	248	10·5	75	100	175	270
Ce	15			2·6		15	24
La		7		1·1	2·4	5·5	7
Pb			2				2
Th			0·14	0·5	0·3	1·6	0·7
K/Rb	481		1050	1000	890	870	430
Ba/Rb	14		10	15	17	11	8·7
Ba/Sr	0·45	0·9	0·08	0·38	0·45	1·9	0·7
Rb/Sr	0·033		0·008	0·025	0·026	0·17	0·08

A–E Means of Lewisian granulites and retrogressed granulites. A. 39 basic pyroxene gneisses. B. 3 garnet pyroxene gneisses. C. 11 basic hornblende gneisses. D. 40 basic hornblende–biotite gneisses. E. 46 hornblende-biotite gneisses (from Sheraton et al., 1973).

F–G Means of E. Greenland granulites. F. 11 ultramafic granulites. G. 25 basic granulites (Wright et al., 1973).

1.   Average of 5 Archaean tholeiites, Vermilion district, Minnesota (Arth and Hanson, 1975).

2.   Average of 232 Archaean andesites, Canadian Shield (Baragar and Goodwin, 1969).

3.   Ocean ridge basalt.

4–6. Island Arc tholeite, andesite, dacite (values from Jakeš and White, 1972).

7.   Mean circum-Pacific andesite (Taylor, 1969).

alkaline in character. It is interesting to note further that younger calc–alkaline provinces of Andean-type bordering continental margins (Fig. 3D) and intracontinental calc–alkaline complexes (Fig. 3E) are much richer in Ba, Sr, Zr, Ce and La and have much higher Ba/Sr ratios than their island arc counterparts (cf. Jakeš and White, 1972).

Strontium isotope studies on the Lewisian gneisses indicate that a large proportion of the gneiss terrain in Scotland could not have been in existence much before 3·0 b.y. (Lyon et al., 1973; Moorbath et al., 1975), at least with its present Rb/Sr ratio, hence indicating deriva-tion from a low Rb/Sr source such as the mantle at this time. Similar constraints apply to other Archaean areas (Moorbath et al., 1972; Arth and Hanson, 1975), although not necessarily conforming with accepted field evidence (Chadwick et al., 1974). While Kushiro (1972) has argued that very wet melt-ing of mantle peridotite can yield small amounts of acid magma, more recent experi-mental evidence (Nicholls and Ringwood, 1973) gives this suggestion considerably less support. Indeed the proportion of acid vol-canics in the truly oceanic regions of the Earth is relatively small. Since the earliest Archaean

412

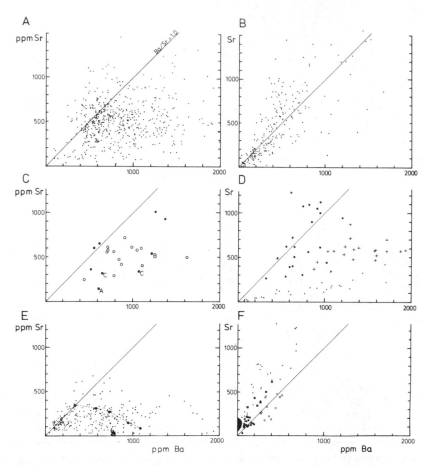

Fig. 3. Ba–Sr distributions in Archaean gneisses from North Atlantic craton.
A. Acid gneisses (>60% SiO_2) from Sheraton et al. (1973). B. Basic and
ultrabasic gneisses (<60 SiO_2) from Sheraton et al. (1973). C. mean gneiss
groups from Tarney et al. (1972) and Sheraton et al. (1973). Filled circles,
basic gneisses; open circles, acid gneisses. Squares, mean Shield gneiss averages from Australia (A), Canada (C) and Brazil (B). D. Intracontinental
calc–alkaline complex, Angmagssalik, East Greenland. Filled circles, picrites
and gabbros; crosses, diorites; dots, granites (Tarney, unpub. data). E. Continental margin calc–alkaline igneous rocks of late Precambrian age from
England and Wales (Thorpe, 1970, and others). Large circles: Ordovician
(Borrowdale volcanics) mean basalts–andesites–dacites–rhyolites from Fitton
(1972). F. Shaded area, ocean ridge basalts; squares, mean Circum-Pacific
basalts, andesites from Taylor (1969); dots, island arc volcanics; filled circles,
Archaean andesites and basalts; open circles, Archaean salic volcanics
(Baragar and Goodwin, 1969, and others)

crustal gneisses (basic through acid) in the
high-grade terrains are relatively highly fractionated with regard to trace elements, this
implies only very *small* degrees of partial
melting (in one or more stages) from typical
mantle compositions, during their production.
How then in such a short time are exceptionally large volumes of fractionated

crustal material produced? Extreme non-
uniformitarian solutions would appear to be
ruled out by the fact that contemporary
greenstone-belt volcanics are relatively
unfractionated with regard to trace elements,
and have low values similar to modern ocean-
floor basalts or island arc tholeiites, suggesting, as pointed out by Jahn et al. (1974), that

the composition of the Earth's upper mantle may not have changed appreciably with time.

There would appear to be two alternative solutions to this problem. It is clear that many of the geochemical characteristics of younger continental margin calc–alkaline igneous suites, including high Ba/Sr ratios (Fig. 3D, E), are already present in the underlying Archaean basement, and no doubt could be transferred directly through high degrees of melting. Lower crustal Archaean granulite terrains (such as the Lewisian) have a similar trace element geochemistry, but notably low Rb–Sr ratios and $^{87}Sr/^{86}Sr$ ratios, which could account for the observed low initial strontium ratios of younger calc–alkaline plutons or their grey gneiss equivalents. Such plutons might, however, be expected to retain the un-radiogenic lead isotope ratios of the granulite source (cf. Moorbath et al., 1969).

While the geochemical characteristics of some younger crustal segments could be explained in this way, it is difficult on present isotopic evidence to envisage the mechanism being applicable on a world-wide scale. The reverse argument would then be that many Archaean gneisses have been generated from the mantle by processes similar to those now operating at Andean-type continental margins which produce the cordilleran batholiths and associated volcanics. The required trace element levels could be attained by appealing to a three-stage process of mantle melting (cf. Ringwood, 1974): generation of ocean crust, remelting of subducting ocean crust at continental margins, and igneous fractionation at the margin itself, perhaps producing a primitively zoned crust (see below).

Depleted Granulites and the Nature of the Lower Crust

The geochemical nature of granulite terrains and their distribution has been reviewed by Lambert and Heier (1968), Tarney et al. (1972), Sheraton et al. (1973) and Heier (1973), and will not be enlarged upon here. Sufficient to state that granulite terrains are invariably more mafic in composition than the amphibolite facies grey gneiss areas. There is a duality in their lithophile element abundances: K, Rb, Cs, Th and U are moderately to strongly depleted, while Ba, Sr, Zr, Ce and La are normal or even enriched compared with amphibolite facies gneisses. Granulites are therefore characterized by high K/Rb and Ba/Rb ratios, but low K/Sr, K/Ba and Rb/Sr ratios (Tarney et al., 1972), while $^{87}Sr/^{86}Sr$ ratios tend to be low (Evans, 1965; Spooner and Fairbairn, 1970).

Hypotheses accounting for the depleted characteristics include:

(a) That the low abundance of K, Rb, Th, U is a primary feature (Holland and Lambert, 1973).
(b) That granulites represent the residuum after removal of a granitophile melt (e.g. Fyfe, 1973).
(c) That development of high pressure mineral assemblages leaves minerals with sites unable to take the large lithophile elements, which are removed along with the hydrous component (Lambert and Heier, 1968; Heier, 1973).
(d) That depletion is specifically a Precambrian process operative mainly on diopside-normative granulites under open system conditions, with removal of K and Rb, etc., being assisted by mantle degassing (Tarney et al., 1972; Sheraton et al., 1973).

Disproving that the depleted characteristics of granulites is a primary feature is difficult in specific instances, but it is equally difficult to find an acceptable source rock. Rocks of the anorthosite series have some similar trace element characteristics (e.g. low K and Rb, and high K/Rb ratios), but the more differentiated members tend to be potash-rich. Lewisian granulites, on the other hand, show no increase in K or Rb on going from basic to acid gneisses (Fig. 4), an unusual feature with few parallels among igneous rock series. None of the granulites from Scotland or Greenland has the high alumina contents characteristic of anorthosites. Later anorthosites are associated with the granulites of East Greenland, however (Moorlock et al., 1972).

414

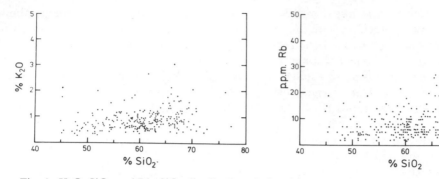

Fig. 4. K_2O–SiO_2 and Rb–SiO_2 distributions in Lewisian granulites and retrogressed granulites. Note low values of K and Rb in both basic and acid gneisses, and lack of any obvious enrichment in acid gneisses contrary to usual igneous fractionation trends

The presence of abundant leucocratic acid gneisses, even potassic pegmatites, with depleted characteristics in the granulite terrain of mainland Scotland (Sheraton et al., 1973) argues against removal of K- and Rb-rich granitophile melts as being a major cause of depletion. It might be expected that Ba would be removed also, but Ba levels remain high in the granulite terrains.

Lambert and Heier's (1968) suggestion that high-pressure granulites should show the most depleted geochemical characteristics is logical, but not necessarily borne out by fact. The Lewisian granulites are amongst the most depleted so far described, yet are intermediate pressure granulites and hornblende–granulites, considerably retrogressed to amphibolite facies assemblages. Furthermore, high-pressure eclogite–granulites from the Bohemian massif of Central Europe have relatively high levels of K and Rb, and fairly low K/Rb ratios (Tarney unpubl. data), yet have very dry, anhydrous mineral assemblages.

Clearly granulite facies conditions can be attained under closed system conditions. It is only when open-system conditions occur simultaneously with granulite facies metamorphism that loss of K, Rb, Th and U takes place. It is difficult to conceive transport of these elements without the agency of a fluid phase, presumably hydrous. Removal could in fact take place at hornblende–granulite facies (hornblendes have notably high K/Rb ratios—Hart and Aldrich, 1967) with assem-blages reverting to granulite facies as pH_2O falls, perhaps assisted by CO_2 flushing as noted by Heier (1973). The source of such fluids might well be the mantle, suggesting that granulites with depleted characteristics may be more indicative of an original situation near the base of the lower continental crust. The fact that granulite facies terrains with depleted characteristics tend to be more basic than the average crust would tend to support this suggestion.

Seismological evidence supports a more mafic lower crust (e.g. Woollard, 1970) with either a sharp velocity gradient near the Moho, or perhaps more commonly, a gradual increase to sub-Moho values. Basic granulite is compatible with deeper crustal velocities (Christensen and Fountain, 1975), although an eclogite or peridotite component appears necessary in some cases (cf. Heier, 1973). The development of such a compositionally zoned continental crust may have taken place by several mechanisms.

(a) A primary zonation by differentiation at the time of crustal formation, with noritic gabbros, and perhaps minor anorthosite at the base, grading upwards through tonalite to granodiorite ('grey gneiss') and more potash-rich compositions at high crustal levels. The oldest crustal rocks so far dated (3·8 b.y. gneisses from W. Greenland and Minnesota) are relatively potash-rich, and thought to be upper crustal gneisses (cf. Black et al., 1971).

(b) A granodioritic crust with the lower crust representing original and perhaps thickened oceanic crust which may have underplated the granodioritic crust whilst the latter was still in a relatively plastic condition. The high proportion of ultramafic gneisses and associated basic gneisses in the granulite facies areas of the Lewisian might then represent detached fragments of the oceanic lithosphere, boudinaged and gneissified. Geochemically these gneisses have the lowest lithophile element concentrations of any basic gneisses in the complex, but the levels are higher than in typical ocean basalts (Table 2).

(c) Emplacement of basic magmas into the lower crust (see Sheraton et al., 1973). Prior to the 2·2 b.y. dyke swarm there was surprisingly little basic magma emplacement in the grey gneiss areas, although there is some on the Outer Hebrides (Watson and Lisle, 1973), rather more in East Greenland (Wright et al., 1973) and (the Ameralik dyke swarm) in West Greenland (McGregor, 1973). With the higher geothermal gradients in the Archaean, more basic magmatism might have been expected. Did denser basic magma infill the plastic lower crust rather than rise in fractures to higher crustal levels under gas pressure as it did in later Proterozoic times? Continental basalts tend to be more lithophile-element enriched than oceanic basalts (Jamieson and Clarke, 1970), agreeing with the geochemical data on basic gneisses in granulite facies terrains.

The relative merits of such models need to be assessed by detailed geochemical and isotopic studies on other granulite facies terrains, supplemented by detailed seismic studies in continental shield areas.

A point worthy of note is that development of a more mafic lower crust, with removal of the low melting and heat-producing constituents K, Rb, Th and U during the widespread granulite facies event $\sim 2 \cdot 6^+$ b.y. ago would leave the lower crust more refractory, and presumably more firmly welded to the underlying lithospheric mantle. This may have substantially increased effective plate thickness. Did this coincide with the change from Archaean tectonics to the proto-plate tectonics of the Proterozoic?

Acknowledgements

I should like to thank J. W. Sheraton, A. C. Skinner, S. A. Drury, R. S. Thorpe, B. S. P. Moorlock, A. H. Hickman, P. J. F. Jeans, A. H. Alseyegh, G. L. Hendry and B. H. Lazell who supplied much of the data used in the text Figures. B. H. Lazell and N. Donellan kindly assisted in the plotting routines.

References

Anderson, D. L., 1975. 'Formation and composition of the Moon', *Proc. Soviet-American Conf. Geochemistry of Moon and Planets, Moscow* (in press).

Andrews, J. R., Bridgwater, D., Gormsen, K., Gulson, B., Keto, L., and Watterson, J., 1973. 'The Precambrian of South-East Greenland', In Park, R. G., and Tarney, J. (Eds.), *The Early Precambrian of Scotland and related rocks of Greenland*, Univ. Keele, 143–156.

Arth, J. G., and Hanson, G. N., 1975. 'Geochemistry and origin of the early Precambrian crust of northeastern Minnesota', *Geochim. Cosmochim. Acta*, **30**, 325–362.

Baragar, W. R. A., and Goodwin, A. H., 1969. 'Andesites and Archean volcanism of the Canadian shield', *Bull. Dept. Geol. Min. Resources, State of Oregon*, **65**, 121–142.

Barker, F., and Peterman, Z. E., 1974. 'Bimodal tholeiitic-dacitic magmatism and the Early Precambrian crust', *Precam. Res.*, **1**, 1–12.

Black, L. P., Gale, N. H., Moorbath, S., Pankhurst, R. J., and McGregor, V. R., 1971. 'Isotopic dating of very early Precambrian amphibolite facies gneisses from the Godthaab district, West Greenland', *Earth Planet. Sci. Lett.*, **12**, 245–259.

Bowes, D. R., Barooah, B. C., and Khoury, S. G., 1971. 'Original nature of the Archaean rocks of Northwest Scotland', *Spec. Publs. geol. Soc. Austral.*, **3**, 77–92.

Bowes, D. R., Wright, A. E., and Park, R. G., 1964. 'Layered intrusive rocks in the Lewisian of the North-West Highlands of Scotland', *Quart. J. geol. Soc. Lond.*, **120**, 153–191.

Bridgwater, D., Watson, J., and Windley, B. F., 1973. 'The Archaean craton of the North Atlantic region', *Phil. Trans. R. Soc. Lond.*, **A273**, 493–512.

Chadwick, B., Coe, K., Gibbs, A. D., Sharpe, M. R., and Wells, P. R. A., 1974. 'Field evidence relating to the origin of 3000 Myr gneisses in southern West Greenland', *Nature, Lond.*, **249**, 136–137.

Christensen, N. I., and Fountain, D. M., 1975. 'Constitution of the lower continental crust based on experimental studies of seismic velocities in granulite', *Bull. Geol. Soc. Am.*, **86**, 227–236.

Condie, K. C., 1967. 'Geochemistry of Early Precambrian greywackes from Wyoming', *Geochim. Cosmochim. Acta*, **31**, 2135–2149.

Davies, F. B., 1974. 'A layered basic complex in the Lewisian, south of Loch Laxford, Sutherland', *J. geol. Soc. Lond.*, **130**, 279–284.

Drury, S. A., 1973. 'The geochemistry of Precambrian granulite facies rocks from the Lewisian Complex of Tiree, Inner Hebrides, Scotland', *Chem. Geol.*, **11**, 167–188.

Evans, C. R., 1965. 'Geochronology of the Lewisian basement near Lochinver, Sutherland', *Nature, Lond.*, **207**, 54–56.

Fanale, F. D., 1971. 'A case for catastrophic early degassing of the Earth', *Chem. Geol.*, **8**, 79–105.

Fitton, J. G., 1972. 'The genetic significance of almandine-pyrope phenocrysts in the calc-alkaline Borrowdale Volcanic Group, Northern England', *Contr. Mineral. Petrol.*, **36**, 231–248.

Fyfe, W. S., 1973. 'The granulite facies, partial melting and the Archaean crust', *Phil. Trans. R. Soc. Lond.*, **A273**, 457–461.

Glikson, A. Y., 1971. 'Primitive Archaean element distribution patterns: chemical evidence and geotectonic significance', *Earth Planet. Sci. Lett.*, **12**, 309–320.

Goldich, S. S., and Hedge, C. E., 1974. '3800-Myr granitic gneiss in south-western Minnesota', *Nature, Lond.*, **252**, 467–468.

Graham, R. H., and Coward, M. P., 1973. 'The Laxfordian of the Outer Hebrides', In Park, R. G., and Tarney, J. (Eds.), *The Early Precambrian of Scotland and Related Rocks of Greenland*, Univ. Keele, 85–93.

Hart, S. R., and Aldrich, L. T., 1967. 'Fractionation of potassium/rubidium in amphibole: implications regarding mantle composition', *Science*, **155**, 325–327.

Heier, K. S., 1973. 'Geochemistry of granulite facies rocks and problems of their origin', *Phil. Trans. R. Soc. Lond.*, **A273**, 429–442.

Holland, J. G., and Lambert, R. St. J., 1973. 'Comparative major element geochemistry of the Lewisian of the mainland of Scotland', In Park, R. G., and Tarney, J. (Eds.), *The Early Precambrian of Scotland and Related Rocks of Greenland*, Univ. Keele, 51–62.

Jahn, B., Shih, C., and Murthy, V. R., 1974. 'Trace element geochemistry of Archaean volcanic rocks', *Geochim. Cosmochim. Acta*, **38**, 611–627.

Jakeš, P., 1973. 'Geochemistry of continental growth', In Tarling, D. H., and Runcorn, S. K. (Eds.), *Implications of Continental Drift to the Earth Sciences*, Academic Press, New York, Vol. 2, 999–1009.

Jakeš, P., and White, A. J. R., 1972. 'Major and trace element abundances in volcanic rocks of orogenic areas', *Bull. Geol. Soc. Am.*, **83**, 29–40.

Jamieson, B. G., and Clarke, D. B., 1970. 'Potassium and associated elements in tholeiitic basalts', *J. Petrol.*, **11**, 183–204.

Kalsbeek, F., and Myers, J. S., 1973. 'The geology of the Fiskenæsset region', In *Grønlands geol. Unders. Rapp.*, **51**, 5–18.

Kushiro, I., 1972. 'Effect of water on the composition of magmas formed at high pressures', *J. Petrol.*, **13**, 311–334.

Lambert, I. B., and Heier, K. S., 1968. 'Geochemical investigation of deep-seated rocks in the Australian shield', *Lithos*, **1**, 30–53.

Lyon, T. D. B., Pidgeon, R. T., Bowes, D. R., and Hopgood, A. M., 1973. 'Geochronological investigation of the quartzofeldspathic rocks of the Lewisian of Rona, Inner Hebrides', *J. geol. Soc. Lond.*, **129**, 389–404.

McGregor, V. R., 1973. 'The early Precambrian gneisses of the Godthaab district, West Greenland', *Phil. Trans. R. Soc. Lond.*, **A273**, 343–358.

Moorbath, S., O'Nions, R. K., Pankhurst, R. J., Gale, N. H., and McGregor, V. R., 1972. 'Further rubidium–strontium age determinations on the very early Precambrian rocks of the Godthaab District, West Greenland', *Nature. Phys. Sci.*, **240**, 78–82.

Moorbath, S., Powell, J. L., and Taylor, P. N., 1975. 'Isotope evidence for the age and origin of the "grey gneiss" complex of the southern Outer Hebrides, Scotland', *J. geol. Soc. Lond.*, **131**, 213–222.

Moorbath, S., Welke, H., and Gale, N. H., 1969. 'The significance of lead isotopes studied in ancient, high grade metamorphic basement complexes, as exemplified by the Lewisian rocks of North-West Scotland', *Earth Planet. Sci. Lett.*, **6**, 245–256.

Moorlock, B. S. P., Tarney, J., and Wright, A. E., 1972. 'K–Rb ratios of intrusive anorthosite veins from Angmagssalik, East Greenland', *Earth Planet. Sci. Lett.*, **14**, 39–46.

Nicholls, I. A., and Ringwood, A. E., 1973. 'Effect of water on olivine stability in tholeiites and the production of silicasaturated magmas in the island arc environment', *J. Geol.*, **81**, 285–300.

Park, R. G., and Tarney, J., 1973. *The Early Precambrian of Scotland and related rocks of Greenland*, Univ. Keele, 194 pp.

Ringwood, A. E., 1969. 'Composition and evolution of the upper mantle', *Am. Geophys. U. Monograph.*, **13**, 1–17.

Ringwood, A. E., 1972. 'Some comparative aspects of lunar origin', *Phys. Earth Planet. Interiors*, **6**, 366–376.

Ringwood, A. E., 1974. 'The petrological evolution of island arc systems', *J. geol. Soc. Lond.*, **130**, 183–204.

Runcorn, S. K., 1965. 'Changes in the convection pattern in the Earth's mantle and continental drift: evidence for a cold origin of the Earth', In Blackett, P. M. S. et al. (Eds.), 'Symposium on Continental Drift', *Phil. Trans. Roy. Soc. Lond.*, **258A**, 228–251.

Sheraton, J. W., 1970. 'The origin of the Lewisian gneisses of Northwest Scotland, with particular reference to the Drumbeg area, Sutherland', *Earth Planet. Sci. Lett.*, **8**, 301–310.

Sheraton, J. W., Skinner, A. C., and Tarney, J., 1973. 'The geochemistry of the Scourian gneisses of the Assynt district', In Park, R. G., and Tarney, J. (Eds.), *The Early Precambrian of Scotland and Related Rocks of Greenland*, Univ. Keele, 13–30.

Skinner, A. C., 1970. Unpublished Ph.D. Thesis, University of Birmingham.

Spooner, C. M., and Fairbairn, H. W., 1970. 'Strontium 87/strontium 86 initial ratios in pyroxene granulite terrains', *J. Geophys. Res.*, **75**, 6706–6713.

Tarney, J., Skinner, A. C., and Sheraton, J. W., 1972. 'A geochemical comparison of major Archaean gneiss units from North-West Scotland and East Greenland', *Rept. 24th Int. geol. Congress, Montreal*, **1**, 162–174.

Taylor, S. R., 1967. 'The origin and growth of continents', *Tectonophysics*, **4**, 17–34.

Taylor, S. R., 1969. 'Trace element chemistry of andesites and associated calc alkaline rocks', *Bull. Dept. Geol. Min. resources, State of Oregon*, **65**, 43–63.

Taylor, S. R., and Jakeš, P., 1975. 'Chemical zoning and early differentiation in the Moon', *Proc. Soviet-American Conf. Cosmochemistry of Moon and Planets, Moscow* (in press).

Thorpe, R. S., 1970. Unpublished Ph.D. Thesis, University of Birmingham.

Toksöz, N. N., and Johnston, D. H., 1975. 'The evolution of the moon and the terrestrial planets', *Proc. Soviet-American Conf. Cosmochemistry of Moon and Planets, Moscow* (in press).

Watson, J., and Lisle, R. J., 1973. 'The pre-Laxfordian complex of the Outer Hebrides', In Park, R. G., and Tarney, J. (Eds.), *The Early Precambrian of Scotland and Related Rocks of Greenland*, pp. 45–50. University of Keele.

Windley, B. F., 1972. 'Regional geology of early Precambrian high-grade metamorphic rocks in West Greenland', *Rapp. Grønlands geol. Unders.*, **46**, 1–46.

Woollard, G. P., 1970. 'Evaluation of the isostatic mechanism and role of mineralogical transformations from gravity and seismic data', *Phys. Earth Planet. Interiors*, **3**, 484–498.

Wright, A. E., Tarney, J., Palmer, K. F., Moorlock, B. S. P., and Skinner, A. C., 1973. 'The geology of the Angmagssalik area, East Greenland and possible relationships with the Lewisian of Scotland', In Park, R. G., and Tarney, J. (Eds.), *The Early Precambrian of Scotland and related rocks of Greenland*, Univ. Keele, 157–177.

Wynne-Edwards, M. R., and Hasan, Z., 1972. 'Grey gneiss complexes and the evolution of the continental crust', *Rept. 24th. Int. geol. Congress, Montreal*, **1**, 175.

Trace-element Models for the Origin of Archean Volcanic Rocks

KENT C. CONDIE

Department of Geoscience, New Mexico Institute of Mining and Technology, Socorro, New Mexico 87801.

Abstract

Progressive melting and fractional crystallization models based on trace-element distributions indicate that Archean volcanic magmas are produced both by varying degrees of melting in upper mantle sources and by progressive fractional crystallization. Results also suggest one or more pre-greenstone magmatic epochs, continuously replenished mantle source rocks for calc–alkaline magmas, and lateral variability in upper mantle composition.

Trace-element evidence for the existence of undepleted Archean upper mantle is not unequivocal. Together with initial Sr^{87}/Sr^{86} ratios, existing data for alkali and related elements suggest that both undepleted and depleted mantle source areas existed in the Archean. The relatively large amounts of transition trace elements in Archean volcanic rocks when compared with modern volcanic rocks of the same bulk composition may reflect secular changes in the proportions of upper mantle minerals that concentrate these elements.

Introduction

Trace-element distributions in Archean volcanic rocks are valuable in formulating models for the origin of greenstone magmas and in documenting the existence of undepleted upper mantle. Before it is possible to study these problems, however, it is first necessary to evaluate the effects of alteration and low-grade metamorphism on trace-element concentrations. Studies of the effects of these processes on young volcanic rocks are valuable in this respect. The results of such studies are summarized in Table 1. It is clear that the transition elements and rare earths are least susceptible to changes during alteration and metamorphism and hence should be given most weight in defining geochemical properties of original greenstone volcanics.

Table 1. Effects of deep-sea alteration and low-grade metamorphism on element mobilities in young volcanic rocks

Enrichment	Little or no change	Depletion
Fe^{+3}, K, Cs, Rb, H_2O, total Fe	Ti, Y, Zr, Zn, Cr, V, Sr, Sc, Hf, Co, Nb, REE, Ni	Si, Ca, Al, Mg

Major references: Frey et al. (1968); Hart (1969), Hattori et al. (1972); Christensen et al. (1973); Hermann et al. (1974); Hart et al. (1974).

A description of trace-element distributions in Archean volcanic rocks and a comparison with distributions in young volcanic rocks is given in Condie (1975). Tables summarizing average compositions of various rock types are also given in that reference and will not be repeated here.

419

420

Magma Models

It is well known that trace elements partition themselves between solid and liquid phases in such a manner that the distribution coefficient (equal to the concentration of a particular element in a solid phase divided by that in a coexisting liquid) is a constant at a given temperature, pressure, and bulk composition. With this information it is possible to use the Rayleigh fractionation law and related equations for partial melting (Shaw, 1970) to evaluate the roles of fractional crystallization and progressive melting in the formation of greenstone magmas.

of producing the average andesite. The degree of crystallization illustrated is that which most closely matches the andesite compositions yet is consistent with major element data. It is clear from the results that in neither greenstone belt can the andesites be produced by such a mechanism. This conclusion remains unchanged even when the most favourable distribution coefficients are selected from the published ranges. Varying the ratios of clinopyroxene and plagioclase being removed or allowing removal also of small amounts of olivine, orthopyroxene, and magnetite does not greatly change the positions of the Model 1 curves. Fractional crystallization models were

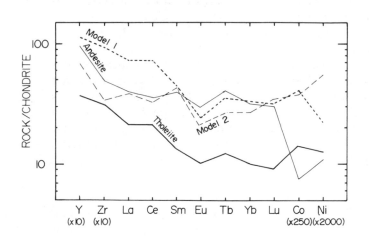

Fig. 1. Chondrite-normalized trace-element distributions in average tholeiite and andesite from the Abitibi greenstone belt (after Condie and Baragar, 1974) and model magma distribution patterns. Model 1: 75% fractional crystallization of tholeiite, with plagioclase and clinopyroxene removed in the ratio of 4 : 1. Model 2: 10% equilibrium melting of plagioclase peridotite with melting ratios of plagioclase : olivine : clinopyroxene : orthopyroxene = 5 : 0 : 2 : 2 and modal ratios = 1 : 4 : 3 : 3. Assumed concentrations in plagioclase peridotite (in ppm): $Y = 10$, $Ni = 1000$, $Co = 100$, $Zr = 60$, $La = 1$, $Ce = 3$, $Sm = 1$, $Eu = 0.3$, $Tb = 0.3$, $Yb = 0.75$, $Lu = 0.25$

Two examples of model calculations are given in Figs. 1 and 2. In these figures chondrite-normalized trace-element concentrations are shown for average tholeiite from the lower part of the stratigraphic succession of the Abitibi and South Pass greenstone belts (after Condie and Baragar, 1974). Also shown are average andesite compositions from higher stratigraphic levels. Several models, two of which are illustrated in each of the figures, have been tested for the origin of the andesite. Distribution coefficients employed in the calculations are given in Condie and Harrison (1975). Model one in each case evaluates shallow fractional crystallization ($\leqslant 35$ km) of the average tholeiite as a means

also tested for greater depths where amphibole and garnet are important liquidus phases. Calculations indicate that with exception of Zr, Y, and perhaps Co and Ni, it is possible with 75–85% fractional crystallization at 35–100 km depth to produce the South Pass-type andesites by removing clinopyroxene, amphibole, and garnet in the ratios of approximately 5 : 4 : 1.

Model 2 in each Figure represents eclogite or plagioclase peridotite which are partially melted to produce the average andesites of each belt. The degree of melting in each case was arrived at by successive approximations and is consistent with bulk compositions of the given parent rocks. The Model 2 curve in the

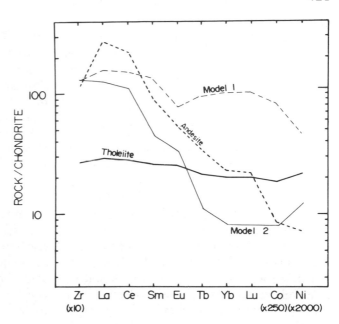

Fig. 2. Chondrite-normalized trace-element distributions in average tholeiite and andesite from the South Pass greenstone belt (after Condie and Baragar, 1974) and model magma distribution patterns. Model 1: 85% fractional crystallization of tholeiite with plagioclase and clinopyroxene removed in the ratio of 4 : 1. Model 2: 20% equilibrium melting of eclogite with melting ratios of garnet : clinopyroxene = 6 : 4 and modal ratios = 4 : 6. Assumed concentrations in eclogite (in ppm): $Y = 10$, $Ni = 225$, $Co = 60$, $Zr = 55$, $La = 3$, $Ce = 10$, $Sm = 2$, $Eu = 0.8$, $Tb = 0.5$, $Yb = 1.8$, $Lu = 0.3$

Abitibi belt closely approximates the average andesite curve. The magnitude of the negative Eu anomaly in the model curve is critically dependent upon the Eu^{+3}/Eu^{+2} ratio which varies with oxygen fugacity. The most serious disagreement is with Co and Ni. Removal of an acceptable amount of olivine or/and magnetite during shallow fractional crystallization could partially, although not completely, alleviate this problem. In the case of the South Pass belt, the overall slope of the melting model curve is similar to the average andesite curve although displaced downwards for most elements. About 10% of shallow fractional crystallization removing a combination of clinopyroxene, plagioclase, and perhaps olivine could raise the Model 2 curve such that it approximately coincides with the andesite curve.

One of the most striking geochemical differences between modern and Archean volcanic rocks is the extremely low Y and heavy REE contents in Archean rocks of rhyolitic composition (Condie, 1975). Since garnet is the only major mineral in the crust and upper mantle that selectively accepts large amounts of these elements, it would appear that garnet may have played a major role in the production of Archean rhyolites but not in the production of Phanerozoic rhyolites. The fact that

transition trace elements are enriched in Archean rhyolites (and dacites) strongly suggests that they have not been produced by fractional crystallization since the large distribution coefficients of transition elements in fractionating magmas would result in their rapid removal.

Existing trace-element data indicate Archean volcanic successions that range from mafic to intermediate in composition as a function of stratigraphic height cannot be produced by simple progressive melting or progressive fractional crystallization (Arth and Hanson, 1972; Condie and Baragar, 1974; Condie and Harrison, 1975). As with young calc–alkaline volcanic assemblages (Gill, 1974; DeLong, 1974; Condie and Hayslip, 1975), it would appear that Archean volcanic successions are produced by complex multistage processes involving both progressive melting and fractional crystallization. Furthermore, trace-element model studies seem to demand one or more pre-greenstone magmatic episodes and a continuously replenished mantle source for many greenstone successions (Condie and Baragar, 1974). Continuously replenished magma source rocks could be accomplished in Archean subduction zones. Data also suggest that the composition of Archean upper mantle source areas varied geographically. For example, plagioclase peridotite is the most satisfactory parent rock for Abitibi-type andesites while eclogite is more satisfactory for South Pass-type andesites.

Detection of Undepleted Archean Mantle

There is a good theoretical basis for an Archean upper mantle less depleted in many trace elements than the present upper mantle. Episodes of melting in the Earth throughout geologic time, for instance, would progressively deplete the upper mantle in those elements whose bulk distribution coefficients are less than one. Most alkali and related elements fall into this category, and hence should be relatively undepleted in the Archean upper mantle. Unambiguous evidence of undepleted mantle, however, is difficult to find. It is possible, at least in principle, to detect the existence of undepleted mantle indirectly by measuring the concentrations of trace elements in volcanics that were at one time in equilibrium with such mantle. In practice, however, one encounters the problems of alteration of the volcanics and the effects of varying degrees of fractional crystallization and melting.

Hart et al. (1970) analyzed many individual Archean tholeiite samples from various greenstone belts in the Canadian Shield for their alkali-element contents. By comparing Archean averages with modern rise and arc tholeiites they concluded that Cs and Rb were higher in the Archean samples. They interpreted such results to reflect an *undepleted* mantle source for Archean tholeiites from the Canadian Shield. However, subsequent studies of the effects of deep-sea alteration and low-grade metamorphism of modern rise tholeiites indicate that alkali elements are substantially enriched during these processes. Hence, one must use extreme caution in interpreting such results in terms of mantle source composition. Jahn et al. (1974) have recently reported alkali and related element concentrations from Archean volcanic rocks in northern Minnesota. Because of the similarities in trace-element concentrations between the Archean volcanics and modern

counterparts, they interpret the results to reflect *depleted* Archean mantle. Their data show, however, that K, Ba, and Rb vary by an order of magnitude in tholeiites of the Ely Formation strongly suggesting mobilization of these elements during alteration or low-grade metamorphism.

Recent geochemical studies of the Midlands–Bulawayo greenstone belt in Rhodesia (Condie and Harrison, 1975) also bear on this problem. The textures, mineralogy, and major element composition of tholeiites from the Mafic Formation in this belt indicate that they represent quenched liquids and that they were produced by about the same degree of partial melting of mantle source rocks as modern rise tholeiites. The consistently low K_2O and Rb contents of these tholeiites were interpreted by the authors to reflect a depleted mantle source (at least as depleted as that source from which rise tholeiites are derived today). The alternative interpretation that the low alkalies reflect intense and uniform spilitization was eliminated on the basis of normal Na_2O contents and the absence of petrographic evidence of albitization.

Initial Sr^{87}/Sr^{86} ratios may also be useful in detecting undepleted mantle. Initial Sr^{87}/Sr^{86} ratios of Precambrian diabase dykes and Tertiary basalts from north-west Wyoming and south-west Montana are shown in Fig. 3 as a function of whole-rock Rb–Sr age (after Mueller and Rogers, 1973). The initial ratios define a growth curve for parent mantle with a Rb/Sr ratio of 0·04 as compared with the growth curve for parent mantle of modern rise tholeiites of 0·025 (Faure and Powell, 1972). One interpretation of these data is that the region in northern Wyoming and adjacent Montana is underlain by undepleted mantle (having a higher Rb/Sr ratio) which has served as a source for tholeiitic magmas for the last 2·9 b.y. This interpretation would necessitate producing basaltic magmas at different times from mantle that had not been previously tapped in order to preserve the constant Rb/Sr ratio of the source.

Most Archean volcanic rocks are enriched in transition metals when compared with pos-

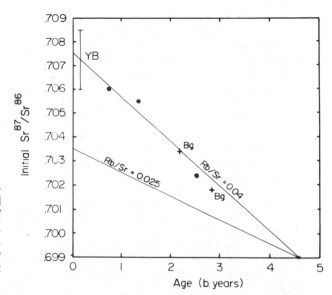

Fig. 3. Initial Sr^{87}/Sr^{86} ratios for Precambrian diabase dykes from Wyoming and south-west Montana plotted as a function of age (after Mueller and Rogers, 1973). Bighorn dyke ratios (Bg) from Stueber et al. (1974). Yb = range of ratios for Tertiary basalts in Yellowstone Park area (Doe et al., 1970)

sible modern counterparts. Such enrichments persist from tholeiite to rhyolite compositions and increase in magnitude in this direction. Calculations indicate that the enrichments can be explained by crystal-melt equilibria in a source that contains greater amounts of these elements than the present-day upper mantle. The depletions in transition metals in young rhyolites require in addition shallow fractional crystallization, while the high transition metal content of Archean rhyolites does now allow such crystallization. An Archean upper mantle depleted in transition metals is difficult to explain in that most of these elements have bulk solid/liquid distribution coefficients greater than one and hence would be refractory during melting episodes. One possible cause for such relative enrichments in Archean volcanics is a progressive change as a function of time in the relative proportions of source minerals that contain these elements. If for instance spinel, which has very large distribution coefficients for most transition metals yet a high melting point, were more abundant in modern than in Archean upper-mantle ultramafic source rocks, modern mantle-derived tholeiites would be expected to be lower in their transition metal contents than Archean ones. Furthermore, if low-K tholeiites inverted to eclogite were the principal sources of calc–alkline magmas in the Archean, the relatively high transition metal contents would be inherited by derivative andesites, dacites, and rhyolites.

Existing data, although sparse and not entirely unequivocal, suggest that the Archean upper mantle was quite heterogeneous in trace-element composition (especially in alkali and related elements). Some regions may have been undepleted while others were as depleted as modern oceanic upper mantle. The perhaps more interesting question of whether *average* Archean upper mantle was undepleted or not awaits a greater abundance of unambiguous trace-element and isotopic data.

Acknowledgements

The author's research on Archean greenstone belts is sponsored in part by National Science Foundation Grant GA-36724.

References

Arth, J. G., and Hanson, G. N., 1972. 'Quartz diorites derived by partial melting of eclogite or amphibolite at mantle depths', *Contr. Mineral. and Petrol.*, **37**, 161–174.

424

Christensen, N. I., Frey, F. A., MacDougall, D., Melson, W. G., Peterson, M. N. A., Thompson, G., and Watkins, N., 1973. 'Deep sea drilling project: properties of igneous and metamorphic rocks of the oceanic crust', *Trans. Am. Geophys. Union*, **54**, 972–1035.

Condie, K. C., 1975. 'Trace element distributions in Archean greenstone belts', *Earth Sci. Revs.*, (in press).

Condie, K. C., and Baragar, W. R. A., 1974. 'Rare-earth element distributions in volcanic rocks from Archean greenstone belts', *Contr. Mineral. and Petrol.*, **45**, 237–246.

Condie, K. C., and Harrison, N. M., 1975. 'Geochemistry of the Archean Bulawayan Group, Midlands greenstone belt, Rhodesia', *Precamb. Res.*, (in press).

Condie, K. C., and Hayslip, D. L., 1975. 'Young bimodal volcanism at Medicine Lake Volcanic Center, northern California', *Geochim. Cosmochim. Acta*, (in press).

DeLong, S. E., 1974. 'Distribution of Rb, Sr, and Ni in igneous rocks, central and western Aleutian Islands, Alaska', *Geochim. Cosmochim. Acta*, **38**, 245–266.

Doe, B. R., Christiansen, R. L., and Hedge, C. E., 1970. 'Radiogenic tracers and the basalt-rhyolite association, Yellowstone National Park and vicinity', *Geol. Soc. America Abs.*, with Programs, **2**, 538.

Faure, G., and Powell, J. L., 1972. *Strontium Isotope Geology*, Springer–Verlag, New York, 188 pp.

Frey, F. A., Haskin, M. A., Poetz, J. A., and Haskin, L. A., 1968. 'Rare earth abundances in some basic rocks', *J. Geophys. Res.*, **73**, 6085–6098.

Gill, J. B., 1974. 'Role of underthrust oceanic crust in the genesis of a Fijian calc-alkaline suite', *Contr. Mineral. and Petrol.*, **43**, 29–45.

Hart, S. R., 1969. 'K, Rb, Cs contents and K/Rb and K/Cs ratios of fresh and altered submarine basalts', *Earth Planet. Sci. Lett.*, **6**, 295–303.

Hart, S. R., Brooks, C., Krogh, T. E., Davis, G. L., and Nava, D., 1970. 'Ancient and modern volcanic rocks: a trace element model', *Earth Planet. Sci. Lett.*, **10**, 17–28.

Hart, S. R., Erlank, A. J., and Kable, E. J. D., 1974. 'Sea floor basalt alteration: Some chemical and Sr isotopic effects', *Contr. Mineral and Petrol.*, **44**, 219–230.

Hattori, H., Sugisaki, R., and Tanaka, T., 1972. 'Nature of hydration in Japanese Paleozoic geosynclinal basalt', *Earth Planet. Sci. Lett.*, **15**, 271–285.

Hermann, A. G., Potts, M. J., and Knake, D., 1974. 'Geochemistry of the rare earth elements in spilites from the oceanic and continental crust', *Contr. Mineral. and Petrol.*, **44**, 1–16.

Jahn, B-M., Shih, C-Y., and Murthy, V. R., 1974. 'Trace element geochemistry of Archean volcanic rocks', *Geochim. Cosmochim. Acta*, **38**, 611–627.

Mueller, P. A., and Rogers, J. J. W., 1973. 'Secular chemical variation in a series of Precambrian mafic rocks, Beartooth Mountains, Montana and Wyoming', *Geol. Soc. Amer. Bull.*, **84**, 3645–3652.

Shaw, D. M., 1970. 'Trace element fractionation during anatexis', *Geochim. Cosmochim. Acta*, **34**, 237–243.

Stueber, A. M., Heimlich, R., and Ikrammudin, M., 1974. 'Rb–Sr ages of Precambrian mafic dikes from the Bighorn Mountains, Wyoming', *Trans. Am. Geophys. Union*, **56**, 1198.

Paleomagnetism

Late Archean–Early Proterozoic Paleomagnetic Pole Positions from West Greenland

W. F. Fahrig

Geological Survey of Canada, Ottawa.

D. Bridgwater

Greenland Geological Survey, Copenhagen

Abstract

Archean gneisses of amphibolite facies from the Godthaabsfjord region (Rb–Sr whole-rock ages 3700 to about 2600 m.y.) are stably magnetized in a direction $D = 225°$, $I = 60°$, $\alpha_{95} = 6·6°$. Late Archean–Early Proterozoic Kangamiut dykes in a 75 km coastal section at right-angles to the Nagssugtoqidian Front (between Kangamiut and Itivdleq) are stably magnetized in a direction $D = 220°$, $I = 56°$, $\alpha_{95} = 2·1°$. This direction is indistinguishable from that of the Godthaabsfjord rocks. Amphibolite and foliated garnetiferous gneiss forms the centres of some Kangamiut dykes and these are thought to be primary features (Windley, 1970). The amphibolitic rocks are magnetically less stable than the doleritic dyke margins, but appear to be magnetized in the same direction as the margins. The trend of the Kangamiut dykes in the coastal section varies systematically from S. to N. but magnetization directions are constant. This suggests that either the magnetization and the trend are both primary features, or the magnetization was regionally impressed after tectonic rotation of the dykes. Contact tests suggest that the granulite facies country rocks (Pb/Pb whole-rock ages 2800 m.y.) preserve a pre-dyke magnetic direction. Amphibolite boudins to the north of the Nagssugtoqidian boundary, which were derived from Kangamiut dykes as a result of intense deformation and metamorphism, have a magnetization direction of $D = 277°$, $I = 69°$, $\alpha_{95} = 14°$.

These data show that the Godthaab gneisses and the Kangamiut dykes were magnetized at the same time, but at some time after the magnetization of the granulite country rock between Kangamiut and Itivdleq. This magnetization pre-dated the magnetization of the metamorphosed dykes to the north of the Nagssugtoqidian 'Front'.

Introduction

In this paper we describe paleomagnetic results from two areas in the Archean gneiss complex of West Greenland (Fig. 1), the major geological features of which are described in

428

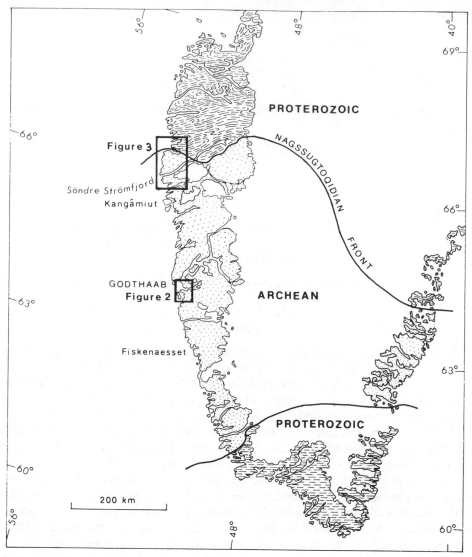

Fig. 1. Index map showing the two areas in West Greenland from which samples were obtained for paleomagnetic work

Bridgwater et al. (1973a) and Bridgwater et al. (in press). Most of the material (279 drill cores from 34 sites) was collected during a joint programme sponsored by the Greenland Geological Survey and the Geological Survey of Canada in 1972. The measurements have all been carried out in Ottawa. The research programme was designed to study two main problems: (1) The magnetic properties of the early Archean gneisses from the Godthaabs-fjord area (McGregor, 1973); and (2) Possible changes in the magnetization of late Archean or early Proterozoic basic dykes and their

country rocks at a series of sites across the boundary of the Nagssugtoqidian mobile belt in which the northern extension of the Archean gneiss complex is reworked by Proterozoic deformation and metamorphism.

The Amphibolite Facies Rocks of the Godthaab Area

The major rock units described by McGregor (1973) from the Archean gneiss complex near Godthaab were sampled (Fig. 2). These include several varieties of Amîtsoq

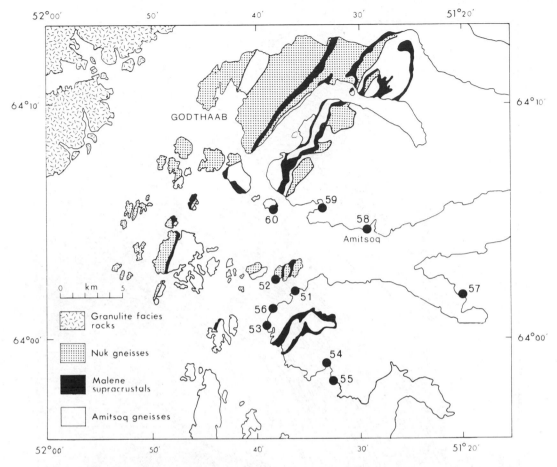

Fig. 2. Geological sketch map of the Godthaab area (Black et al., 1971) showing paleomagnetic sampling sites (reproduced with permission of Elsevier Scientific Publishing Company, Amsterdam)

gneiss; inclusions of basic and iron-rich material in the Amîtsoq gneisses, some of which may represent the highly metamorphosed equivalents of the Isua supracrustal rocks; deformed Ameralik dykes, and a variety of supracrustal rocks and younger granitic gneisses. The rock units sampled range in Rb/Sr whole-rock age from approximately 3700 m.y. to 2600 m.y. All except the youngest intrusive granitic rocks have been severely deformed, metamorphosed, and rotated from their original attitude by major tectonic events in the period 2600 to 3100 m.y. ago. The youngest tectonic event recognized from the area is the imposition of a strong fabric on parts of the gneiss complex accompanying and locally outlasting the regional formation of granulite facies assemblages in the gneiss complex to the

north and south of Godthaab. This event is thought to have occurred about 2800 m.y. ago (Black et al., 1973). The Godthaabsfjord gneisses were affected by a mild regional metamorphism about 1600–1700 m.y. ago which has reset the Rb/Sr mineral ages, but did not completely reset the K/Ar in hornblende and biotite (Pankhurst et al., 1973). Basic dykes older than 1700 m.y. in the area show no obvious metamorphic effect of this late event.

The Border Zone of the Nagssugtoqidian Mobile Belt

The Archean gneiss complex between Sukkertoppen and Itivdleq (Fig. 3) consists of granulite facies quartzofeldspathic gneisses interlayered with high-grade supracrustal

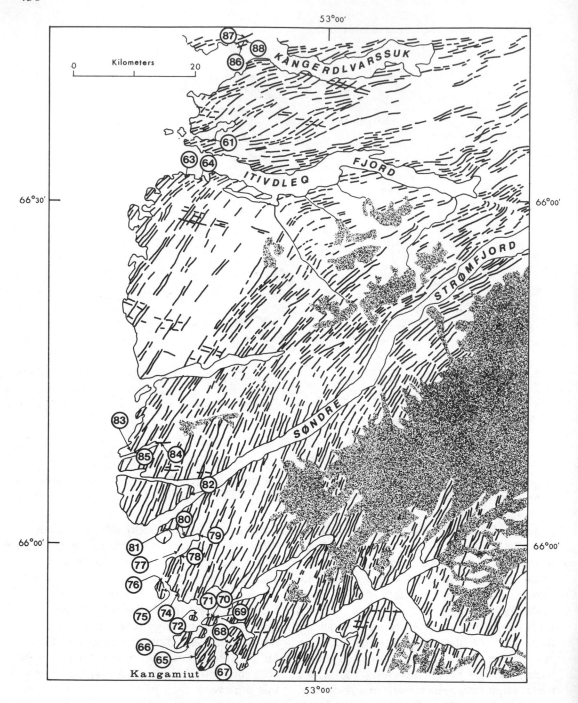

Fig. 3. Basic dyke trends in the Søndre Strømfjord region (Escher et al., 1975) and location of paleomagnetic sampling sites (reproduced by permission of the National Research Council of Canada from the *Canadian Journal of Earth Sciences* **12**, 1975, pp. 158–173)

units. These were last affected by high-grade regional metamorphism about 2800 m.y. ago. The gneisses are intruded by several genera- tions of basic dykes (Berthelsen and Bridgwater, 1960; Windley, 1970; Bridgwater et al., in press) which occur in major regional swarms and which can be traced for many tens of kilometres (Fig. 3). To the north of Itivdleq the gneisses and the dykes cutting them are severely affected by one or more periods of major deformation and both structures in the gneisses and the dykes become parallel to the dominantly E–W trend of the Nagssugtoqidian. The effect of the various phases of deformation found within the Nagssugtoqidian mobile belt is very inhomogeneously distributed, so that within the belt comparatively large 'eyes' of gneiss which preserve the earlier Archean structures are found surrounded by rocks in which the earlier structures have been completely rotated into parallelism with the new structures.

Detailed work, largely by A. Escher, J. Escher, J. Watterson and students from Liverpool University, has established that there is a complex sequence of events in the border-zone of the mobile belt (Bridgwater et al., 1973b, 1973c; Escher et al., 1975; Watterson, 1974). These are summarized in Table 1 (modified from Watterson, 1974).

Dyke Chronology and Petrology

There are at least five major suites of Precambrian basic dykes intruded into the Archean gneiss complex between Godthaabsfjord and Itivdleq (Berthelsen and Bridgwater, 1960) separable by intersections between dykes of different swarms and by the relation between the dykes and major faults. They are subdivided into two main groups, two generations of early dykes trending, respectively, EW and SE, characterized by a high content of mafic minerals and commonly showing evidence of accumulation of olivine and orthopryoxene within the dyke fissure, and three generations of iron-rich tholeiitic dolerites, the majority of which trend approximately NNE or SW. In the northern part of the Archean block, dykes of the second group represented by the NNE-trending Kangamiut dyke swarm are extremely abundant. The Kangamiut dykes appear to have crystallized from iron-rich hydrous tholeiitic magmas; however, they show considerable range in composition and internal structure apparently related to the tectonic conditions under which they were emplaced. In the Sukkertoppen area, Windley (1970) describes primary quartz ferro–dolerite–garnet amphibolite dykes in which the margins are microporphyritic quartz-bearing hornblende dolerites while the

Table 1. Major events in the border zone of the Nagssugtoqidian (Modified from Watterson, 1974)

Age	Event
1650–1800[1]	Uplift.
	Later Nagssugtoqidian deformation } Resetting of magnetic directions in dykes north of Itivdleq.
c. 2400–2700[2]	Emplacement of Kangamiut dykes } In areas not affected by earlier Nagssugtoqidian movements striking NNE, elsewhere parallel to ENE Nagssugtoqidian fabric.
	Emplacement of early N–S and E–W dykes[3].
	Brittle deformation parallel to early Nagssugtoqidian fabric with production of pseudotachylite.
	Pegmatites.
c. 2500–2800[4]	Ductile simple shear strain. Imposition of intense tectonic fabrics on previously isotropic rocks. Displacements mainly transcurrent dextral. Probably contemporaneous with last phases of folding affecting Archean complex further south.
2700–2900[5]	Regional high-grade metamorphism.

[1] K/Ar biotite ages, Larsen and Møller, 1968.
[2] K/Ar ages on whole-rock samples of hornblende dolerites 2560 and 2790 m.y. (Bridgwater, 1970 and 1971) interpreted as close to age of intrusion, but suspect due to possible presence of excess argon.
[3] Berthelsen and Bridgwater, 1960.
[4] Chessex, quoted in Bridgwater et al., 1973c.
[5] Black et al., 1973.

432

centres show a wide variety of rock types ranging from ophitic hornblende dolerites to foliated garnet amphibolites arranged in layers parallel to the margins. Intersections between different dykes belonging to the same swarm show that the internal structures are primary.

In the area between Kangamiut and Itivdleq (sampled in this study) the NNE Kangamiut dykes range from normal tholeiitic dolerites or hornblende dolerites to composite bodies with doleritic margins and centres of foliated quartz–albite–biotite–garnet rock (Fig. 4). These are separated in several dykes from the marginal dolerites by layered hornblendic gabbro corresponding to the rocks described by Windley. The central zones may be over 20 m wide and make up over half the dyke and can be traced for several kilometres in individual dykes. The central zones commonly show a marked fissility parallel to the dyke contacts and in places appear almost mylonitic in the field. Contacts between the outer more basic rocks and the leucocratic central zones are generally abrupt, and apophyses of the quartz–albite rocks transgress layering in the border zones.

The authors differ in their interpretation of the genesis of the peculiar primary zonation of these dykes. One of us (W.F.F.) suggests that the original dyke magma originated in the mantle as a tholeiitic basalt. The initial dyke injection was rapid and normal. In passing upward, however, the dyke material passed through, or possibly formed magma chambers within, a volatile-rich, probably amphibolite substratum. The interaction of the basalt magma with material of this zone would produce high pressures, so that during the waning stage of dyke intrusion a volatile-rich

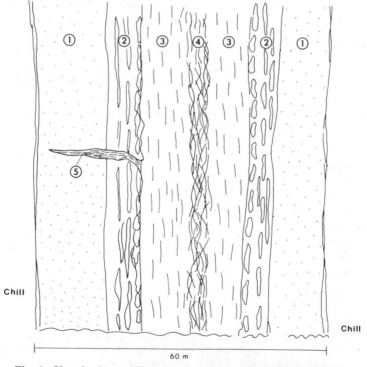

Fig. 4. Sketch of zoned Kangamuit dyke at site 81, Fig. 3. Numbered zones are as follows: (1) Massive medium-grained hornblende dolerite with less than 1% garnet. (2) Coarse-grained inhomogeneous hornblende-rich dolerite, with rhythmic layers parallel to the margins. (3) Quartz–albite–biotite–garnet rock with marked schistosity parallel to the margins. (4) Mylonitic zone in quartz–albite rock. (5) Vein of quartz–albite garnet rock cutting outer zones

amphibolitic liquid would be forcibly injected and would consolidate as massive amphibolite. In some cases later surges up the central fissure, under high pressure, would produce the cataclastic gneissic structures found in the central zone of some dykes. The anomalous feature of such a system is the fact that the country rocks are dry rocks of largely granulite facies. Since formation of zoned, primary amphibolitic dykes such as these are a rarity in geological history, they may reflect peculiar conditions of formation. Possibly, rocks of granulite facies at an early stage of, or even prior to, Nagssugtoqidian deformation, were thrust over a slab of hydrous oceanic crust. The oceanic crust would become the hot amphibolite substratum alluded to above.

The second author (D. B.) explains the zoned nature as largely the result of the filter pressing of the late crystallizing fractions from the dyke magma during active shearing along the length of the dyke during emplacement.

He envisages a primary hydrous iron-rich tholeiitic magma which crystallized as a hornblende dolerite at the margins of the fissures when the rate of shearing parallel to the dyke walls was relatively small compared with the speed of crystallization. As crystallization slowed down, the effects of (intermittent) movement in the dyke fissure became the dominant control of the type of rocks formed, producing first the rhythmically layered structures described by Windley (1970) and last, the leucocratic centres seen in the dykes described here. The highly foliated centres are ascribed to a relatively high rate of shearing compared with the speed of crystallization as the last fraction of the dyke crystallized.

Fig. 5 shows an AFM plot from samples taken across a typically zoned dyke plotted on the same diagram as results from Windley (1970) and analyses from other dykes emplaced into the Archean gneiss complex between 2600 and 1800 m.y. ago.

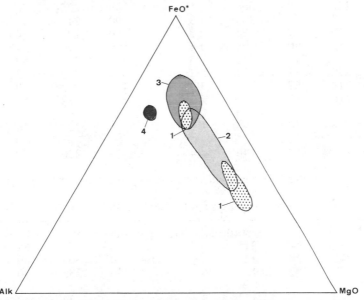

Fig. 5. A.M.F. plot of analyzed samples from late Archaean–early Proterozoic dykes from West Greenland. (1) E–W and N–S dykes with local mafic cumulates (Berthelsen and Bridgwater, 1960). Earlier than main Kangamiut swarm ('MD 1'). (2) Main swarms of Proterozoic basic dykes ('MD 2' and 'MD 3' of several authors). (3) Zoned Kangamiut dykes, margins and rhythmically layered hornblende gabbros. (4) Leucocratic centre of dyke illustrated in Fig. 4. Data from Henriksen, 1969; Rivalenti and Sighinolfi, 1971; Williams, 1973; Windley, 1970; and the present study

Sampling and Measurement

Six to eight cores 2·5 cm in diameter were drilled, and oriented with a solar compass at each of the sites whose locations are shown on Figs. 2 and 3. Two specimens were then cut from each core. The mean value obtained from the two specimens representing the original cores was used as basic datum for those cores that proved to be homogeneously

Table 2. Kangamiut dyke directions and pole positions

Site	N	D	I	k	α	A.F.	D	I	k	α
					Sites south of the Nagssugtoqidian Front					
61	7	219	58	139	5·1	500	216	44	119	5·5
63	6	222	79	84	7·3	400	205	64	290	3·9
64	8	215	59	779	2·0	500	220	62	623	2·2
65	8	238	68	162	4·4	500	231	59	426	2·7
66	6	224	74	1065	2·1	500	214	56	369	3·5
67	8	228	71	505	2·5	500	217	59	1202	1·6
68	8	229	72	384	2·8	500	215	60	792	2·0
69	8	231	63	422	2·7	500	226	52	533	2·4
70	7	346	64	346	3·3	500	227	55	401	3·0
71	8	227	74	1288	1·5	500	219	60	1212	1·6
72	7	251	70	77	7·0	500	239	59	1152	1·8
73	8	235	76	69	6·7	500	221	54	296	3·2
74	7	224	66	462	2·8	500	215	53	1811	1·4
75	6	229	66	483	3·0	500	221	57	1132	2·0
76	6	222	66	169	5·2	500	220	59	1448	1·8
77	6	222	78	368	3·5	500	225	56	278	4·0
78	7	227	73	333	3·3	500	217	55	1067	1·9
79	7	230	67	784	2·2	500	222	54	845	2·1
80	6	252	80	194	4·8	500	216	54	166	5·2
81	3	240	72	285	7·3	500	221	59	960	4·0
82	6	239	71	174	5·1	500	229	52	296	3·9
83	8	218	67	309	3·2	400	213	50	588	2·3
85	6	216	72	87	7·3	500	212	52	547	2·9
All cores	157	228	70	93	1·2		220	56	143	0·9
All sites	23	228	70	146	2·4		220	56	190	2·1

Mean north pole position from 23 cleaned sites listed above, 86·2°W17·1°N dp = 3·0°, dm = 2·2° k = 190 α = 2·1

Mean north pole position assuming Bullard's fit of Greenland and North America 109·1°21·5°N

Site	N	D	I	k	α	A.F.	D	I	k	α
					Sites north of the Nagssugtoqidian Front					
86	6	282	74	316	3·8	500	278	68	57	9·1
87	5	242	79	125	6·9	100–150	303	60	15	20
88	6	267	77	136	5·8	400	235	70	49	9·6
All cores	17	268	77	122	3·1		274	69	20	7·5
All sites	3	267	77	236	8·1		277	69	33	21·0

Mean north pole position from 3 cleaned sites above 121·8°W48·8°N dp = 24° dm = 20° k = 33 α = 14°
Mean north pole position assuming Bullard's fit of Greenland and North America 140·3W57·2N

Notes: N = number of cores collected at the site; D = declination, I = inclination, k = Fisher's estimate of precision, α = half angle of the cone of confidence. A.F. = peak alternating cleaning field in oersteds, dp, dm are the semi axes of the elipse of confidence.

magnetized after optimum alternating field demagnetization.

Most of the samples from Fig. 3 are lithologically similar, i.e. Kangamiut dolerite dykes. A small proportion of the samples were obtained from the garnetiferous amphibolite centres of some dykes and a few samples are country rock obtained for a contact test. Half a dozen very different rock types are represented by the Godthaab cores and this diversity is within-site (see Table 3). Since most of the Kangamiut samples are very stably magnetized, and the Godthaab samples show great variation in stability, samples from the two areas were treated quite differently in the laboratory.

Optimum A.F. cleaning fields for the Kangamiut sites were readily obtained by step-cleaning two specimens in peak fields from 50 to 800 oersteds. Test specimens directions changed little in the range 350 to 800 oe except to become somewhat dispersed as the 800 oe field was approached. The samples from most sites were therefore bulk-cleaned in peak fields of 500 oe (Table 2). All of the specimens from the central amphibolitic zones were cleaned in steps because these specimens exhibited less stability during A.F. test demagnetization.

All of the Godthaab samples were demagnetized in 25 oe steps to 800 oe. During this procedure, many of the specimens showed no evidence of an end-point and retained or developed pronounced within-sample inhomogeneity. These samples could not be used in the site means and in the cases of sites 58 and 60 no specimens showed any indication of magnetic stability. The optimum field for the remaining cores was chosen on the basis of the end-point for each core. The range of A.F. cleaning used for each site is given in Table 3.

Paleomagnetic Results—Kangamiut Dykes

The doleritic parts of the Kangamiut dykes, from site 83 (Fig. 3) southward, are stably magnetized and their N.R.M. directions differ systematically from the dipole field direction. Optimum alternating field demagnetization decreased the inclination of their magnetization by 15° and generally decreased their within-site, and between-site, dispersion (Fig. 6, Table 2).

Dykes 61, 63 and 64, about 50 km north of the main collecting area (Fig. 3) have an orientation about 30° clockwise from those to the south. These dykes are garnetiferous throughout, though not gneissic, and lack ophitic texture. The declination of their magnetization, however, does not differ from that present in the dykes to the south. The only noticeable difference is a lower inclination of magnetization exhibited by dyke 61. These data suggest either that the difference of trend from site 61 south is primary, or that the magnetization was

Table 3. Directions and pole positions of Godthaab rocks (chiefly Amîtsoq gneisses)

Site	N	n	D	I	k	α	A.F.	D	I	k	α	Rock Type
51	7	5	217	79	219	5·2	375	223	61	299	4·4	4 Amîtsoq, 1 Ameralik
52	8	2	231	67	—	—	100 to 300	218	52	—	—	1 Malene, 1 Nûk
53	6	3	104	83	232	8·1	150 to 350	189	78	262	7·6	3 Amîtsoq
54	8	5	253	74	16	20	75 to 300	221	61	23	16	4 Amîtsoq, 1 ultrabasic
55	8	4	234	67	21	21	100 to 650	241	55	29	17	4 Amîtsoq
56	12	10	242	81	48	6·9	150 to 800	234	60	30	9·0	7 Diabase, 3 Amîtsoq
57	7	5	227	54	35	13	175 to 550	221	48	200	5·4	3 Amîtsoq, 2 Ameralik
59	8	7	279	77	19	14	75 to 375	229	58	39	9·7	4 Ameralik?, 3 Amîtsoq
All cores		41	239	76	21	4·7		227	59	33	3·8	
All sites		8	235	75	39	8·9		225	60	57	7·5	

Mean north pole position from 8 cleaned sites listed above 86°W20°N dp = 9·9 dm = 7·5 k = 57 α = 7·5
Mean north pole assuming Bullard's fit of Greenland and North America 110°W25°N

Notes: N = number of cores drilled at site, n = number of cores after excluding those that are inhomogeneously magnetized (θ > 25°) after cleaning at optimum field. A.F. = optimum cleaning field range used for each site. See explanation in text. Remaning symbols the same as in Table 2.

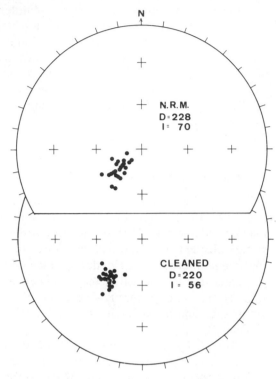

Fig. 6. Equal area plot of mean site magnetization direction of 23 Kangamiut dykes; all directions north-seeking down

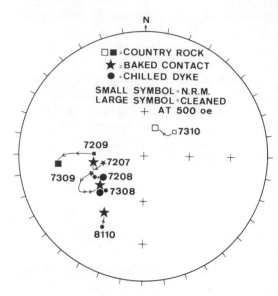

Fig. 7. Results of contact test of Kangamiut dykes

imposed on the dykes after tectonic rotation—presumably of dykes 61 to 64.

Sites 86 to 88 are on large amphibolite boudins derived from the metamorphosed Kangamiut dykes. Samples from site 87 are rather unstable, showing large random changes in magnetization direction in cleaning fields above 150 oe. Samples from the other two sites are stably magnetized and after cleaning in fields of 400 to 500 oe, exhibit a steep westerly direction (Table 2, Fig. 8). This suggests that the magnetization of the Kangamiut dykes to the south pre-dated the magnetization (presumably related to metamorphism and deformation) of these altered Kangamiut dykes.

An attempt was also made to establish whether the present magnetization of the Kangamiut dykes occurred at the time of their intrusion. At two sites (72 and 73) cores were drilled from chilled dyke margin, baked country rock and from country rock at least 30 m distant from the dyke. In both dykes the

magnetization of chilled margins and baked contacts are in good agreement, while the two country rock samples (7310 and 7209) are distinctly different (Fig. 7). A baked country rock specimen from site 81 (8110) also shows good agreement with the cleaned Kangamiut direction. These results are interpreted to indicate that the magnetization of the dykes post-dates the predominantly granulite facies

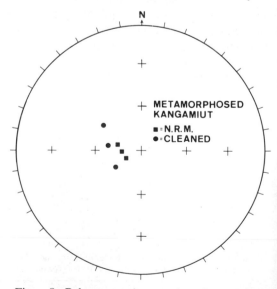

Fig. 8. Paleomagnetic results from three metamorphosed Kangamiut dykes from north of the Nagssugtoqidian Front

metamorphism of the country rocks, and may have formed at the time of dyke intrusion.

The samples of amphibolite gneiss and massive amphibolite from the centres of some of the Kangamiut dykes were A.F. cleaned in 50 oe steps to 800 oe. A series of 7 cores were collected from garnet gneiss through increasingly massive amphibolite to the doleritic border zone of site 81. The garnet gneiss samples exhibited random directional changes above 100 or 200 oe and the series of amphibolite samples as the dolerite border is approached showed progressively increased stability in response to A.F. cleaning, with end-points in the direction of the remainder of the Kangamiut dykes. This suggests that magnetization of the amphibolitic central zones took place at about the same time as did that of the dolerite borders; there is certainly no evidence that it was magnetized significantly later than the doleritic zones.

Paleomagnetic Results—Godthaab Rocks

Site locations in the Godthaab area are shown in Fig. 2. Samples from a number of these sites yielded Rb/Sr whole-rock results that fit the ~3·7 b.y. isochron (Black et al., 1971). The various lithologic types within the sites showed great variation in stability. The number of stably magnetized cores and their rock type are given in Table 3. The Amîtsoq gneisses, the diabase, and Ameralik dykes proved most stable.

The stably magnetized cores, exclusive of those from the Amîtsoq gneisses, have a magnetization direction after optimum cleaning of $D = 230°$, $I = 56°$, $\alpha_{95} = 5·2°$, whereas the Amîtsoq samples alone have a magnetization direction $D = 228°$, $I = 62°$, $\alpha_{95} = 5·2°$. There is no significant difference between these two directions, so the data have been combined in Table 3 (Fig. 9). It is concluded that all of the various Godthaab rock types acquired their magnetization at the same time.

The mean magnetization direction of the 8 (of 10) Godthaab sites that contain a stable remanent magnetization is very similar to that obtained from Kangamiut dykes south of the Nagssugtoqidian Front. This is reasonably

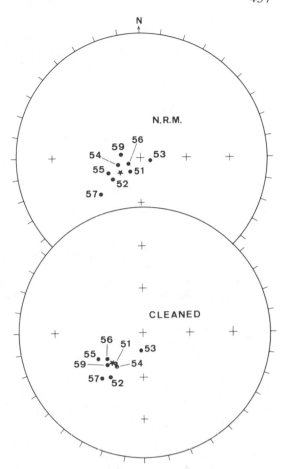

Fig. 9. Equal area plot of site mean directions of 8 sites from the Godthaab area; all directions north-seeking down

good evidence that the Godthaab rocks and the Kangamiut dykes were magnetized at the same time. This is a bit surprising as the magnetization of the granulite gneisses north of Godthaab appears to be pre-dyke in age.

Discussion of Results

(1) Coincidence between the paleomagnetic pole of the Godthaab rocks and that of the Kangamiut dykes suggests that they received their magnetic imprint at the same time. The magnetization of the dykes is probably primary, i.e. formed at or near the time of intrusion. The Godthaab rocks were probably magnetized during the cooling history of the complex following injection of the Qorqut

438

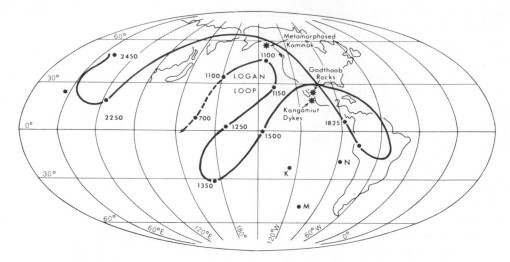

Fig. 10. Pole positions from West Greenland (assuming Bullard's fit) plotted on a polar wandering curve (Christie et al., in press) for the Superior Orogen of the Canadian Shield

granite about 2500 m.y. ago. The age of Godthaab magnetization, and hence the age of Kangamiut magnetization, is probably about 2500 m.y. We do not consider that the mild thermal event that reset some Rb/Sr mineral ages at 1600–1800 m.y. was strong enough to remagnetize the gneisses.

(2) It appears that although the granulite gneiss which forms the country rock for the Kangamiut dykes gives younger Rb/Sr ages than the Godthaab rocks, it retains a magnetization which is older than the magnetization of the Godthaab rocks. This is probably a matter of the differences in opaque mineralogy between the two metamorphic facies.

(3) The metamorphosed Kangamiut dykes north of Itivdleq (sites 86 to 88) have a magnetization direction that differs from that of the dykes to the south. This magnetization was probably imposed no later than, and probably before, the 1700–1900 age indicated by K–Ar determinations.

(4) It would be of special interest to compare the paleomagnetic result from this very ancient crust with results from adjacent parts

of the Canadian Shield which have had a similar history (Bridgwater et al., 1975). Unfortunately, there are no published data from the coast of Labrador so we can compare them only with results from elsewhere in the Canadian Shield. The Superior Orogen is the logical reference plate for N. American paleomagnetism and the poles for Greenland have been plotted on a revised polar wandering curve for the Superior (Christie et al., in press) assuming Bullard's fit for Greenland and North America (Fig. 10). The Godthaab and Kangamiut pole positions, relative to this curve, suggest an age of magnetization in the range 1800–2000 m.y. Such a suggestion should, however, be viewed with great caution as the Labrador fold belt lies between the Superior Orogen and the Nain Orogen on the coast of Labrador. There is every reason to suppose that during the development of the fold belt, movement took place between the Superior and the Nain–Greenland Archean block. The extent of this movement can only be determined by carefully integrated radiometric and paleomagnetic studies.

Acknowledgements

We wish to thank the Director of the Greenland Geological Survey for permission to publish this paper, Dr V. McGregor for help and advice in sampling in the Godthaab area, and G. Freda for carrying out the laboratory measurements.

References

Berthelsen, A., and Bridgwater, D., 1960. 'On the field occurrence and petrogrpahy of some basic dykes of supposed pre-Cambrian age from the southern Sukkertoppen district, western Greenland', *Bull. Grønlands geol. Unders.*, **24** (also *Meddr. Grønland*, **123**, 3) 43 pp.

Black, L. P., Gale, N. H., Moorbath, S., Pankhurst, R. J., and McGregor, V. R., 1971. 'Isotopic dating of very early Precambrian amphibolite facies gneisses from the Godthaab district, West Greenland', *Earth Planet. Sci. Lett.*, **12**, 245–259.

Black, L. P., Moorbath, S., Pankhurst, R. J., and Windley, B. F., 1973. '$^{207}Pb/^{206}Pb$ whole-rock age of the Archaean granulite facies metamorphic event in West Greenland', *Nature phy. Sci.*, **244**, 50–53.

Bridgwater, D., 1970. 'A compilation of K/Ar age determinations on rocks from Greenland carried out in 1969', *Rapp. Grønlands geol. Unders.*, **28**, 47–55.

Bridgwater, D., 1971. 'Routine K/Ar age determinations on rocks from Greenland carried out for GGU in 1970', *Rapp. Grønlands geol. Unders.*, **35**, 52–60.

Bridgwater, D., Watson, J., and Windley, B. F., 1973a. 'The Archaean craton of the North Atlantic region', *Phil. Trans. R. Soc. Lond.*, **A273**, 493–512.

Bridgwater, D., Escher, A., and Watterson, J., 1973b. 'Tectonic displacements and thermal activity in two contrasting Proterozoic mobile belts from Greenland', *Phil. Trans. R. Soc. Lond.*, **A273**, 513–533.

Bridgwater, D., Escher, A., Nash, D., and Watterson, J., 1973c. 'Investigations on the Nagssugtoqidian boundary between Holsteinsborg and Kangamiut, West Greenland', *Rapp. Grønlands geol. Unders.*, **55**, 22–25.

Bridgwater, D., Collerson, K. D., Hurst, R. W., and Jesseau, C. W., 1975. 'Field characters of the very early Precambrian rocks from Saglek, Coast of Labrador', *Geol. Surv. Can.*, Paper 75–1, part 1, 287–296.

Bridgwater, D., Keto, L., McGregor, V. R., and Myers, J. S., in press. 'Archaean gneiss complex of Greenland', In Escher, A. and Watt, W. S., *The Geology of Greenland*, Geol. Surv. Greenland publications, Copenhagen.

Christie, K. W., Davidson, A., and Fahrig, W. F., in press. 'The Paleomagnetism of Kaminak dikes— no evidence of significant Hudsonian plate motion'.

Escher, A., Escher, J., and Watterson, J., 1975. 'The reorientation of the Kangamiut dyke swarm, West Greenland', *Can. J. Earth Sci.*

Henriksen, N., 1969. 'Chemical relations between metabasaltic lavas and metadolerites in the Ivigtut area, South-West Greenland', *Meddr dansk geol. Foren.*, **19**, 27–50.

Larsen, O., and Møller, J., 1968. 'K/Ar age determinations from western Greenland. I. Reconnaissance programme', In *Report of Activities for 1967, Rapp. Grønlands geol. Unders.*, **15**, 82–86.

McGregor, V. R., 1973. 'The early Precambrian gneisses of Godthaab district, West Greenland', *Phil. Trans. R. Soc. Lond.*, **A273**, 343–358.

Pankhurst, R. J., Moorbath, S., Rex, D. C., and Turner, G., 1973. 'Mineral age patterns in ca. 3700 m.y. old rocks from West Greenland', *Earth Planet. Sci. Lett.*, **20**, 157–170.

Rivalenti, G., and Sighinolfi, G. P., 1971. 'Geochemistry and differentiation phenomena in basic dikes of the Frederikshaab district, South-West Greenland', *Atti Memorie Soc. Tosc. Sci. nat.*, **A77**, 358–380.

Watterson, J., 1974. 'Investigations on the Nagssugtoqidian boundary in the Holsteinsborg district, central West Greenland', *Rapp. Grønlands geol. Unders.*, **65**, 33–37.

Williams, H. R., 1973. 'The geology of an area to the northeast of Bjørnesund, near Fiskenaesset, West Greenland', Unpublished Ph.D. thesis, Univ. Exeter, 290 pp.

Windley, B. F., 1970. 'Primary quartz ferrodolerite/garnet amphibolite dykes in the Sukkertoppen region of West Greenland', In Newall, G., and Rast, N. (Eds.), *Mechanism of igneous intrusion, Geol. J. Spec. Issue* 2, 79–92.

Metallogeny

Mineralization in Archaean Provinces

JANET WATSON

Department of Geology, Imperial College, London, England

Abstract

The mineral deposits of Archaean provinces are considered in relation to two stages in the tectonic evolution of the provinces. The first stage was one of crustal mobility during which the fundamental structure of each province was established. The broad characteristics of deposits formed during this stage are now well-established but the significance of regional variations in the range and relative importance of the metals represented remains to be established. Variations in the style of mineralization in greenstone belts reflect both primary variations in volcanism and the effects of metamorphism and granite-formation. The possibilities of regional geochemical variations in the mantle or of progressive changes in volcanic regime during the Archaean era also require consideration. The second stage in the history of Archaean provinces was that dominated by cratonic regimes. Mineral deposits in cratonic cover-sequences are largely concentrated in early Proterozoic formations; the very limited interchange of sediment between Archaean cratons and adjacent terrains restricted the development of younger deposits. Basic, alkaline and kimberlitic intrusions with associated deposits invaded the Archaean cratons intermittently over very long periods and old lineaments in or at the margins of the cratons acted as channels for mineralizing solutions even in Tertiary times.

1. The Geological Context

Mineral deposits formed within the time-span ~3800-2400 Ma occur widely both in tectonic provinces which have remained essentially stable since the end of Archaean times and in polycyclic Archaean complexes incorporated in Proterozoic or Phanerozoic provinces. Their original relationships tend to be best preserved in the Archaean provinces and it is therefore from these that the bulk of our information about the processes of mineralization which were active during the early stages of Earth history is drawn. The constitution and architecture of such

Archaean provinces are discussed extensively elsewhere in this volume and for my present purposes it is sufficient to draw distinctions between four principal geological assemblages:

(a) The oldest remnants of continental crust (>3500 Ma): granitic gneisses, associated in some localities with metamorphosed supracrustal rocks, which occupy areas of no more than a few hundred square kilometres.

(b) Granite/greenstone-belt provinces, usually incorporating areas of banded gneiss but characterized by the low metamorphic grade of rocks in the greenstone belts.

(c) Granite and gneiss provinces incorporating more highly metamorphosed remnants of greenstone belts.

(d) Gneiss and granulite provinces characterized by high metamorphic grades and strong penetrative deformation, often incorporating remnants of supracrustal rocks and layered anorthosites.

The oldest assemblages (a) survive only as remnants in Archaean terrains of other types and have so far yielded only one deposit of economic significance. The relationships in time and space of the three principal Archaean assemblages (b)–(d) remain controversial. In general terms, the granite/greenstone-belt assemblages (b) which are economically the most productive appear to provide samples of higher crustal levels than the assemblages dominated by gneisses and granulites (c) and (d). A vertical passage has been observed in certain provinces from granites and volcanics downward into gneisses or granulites which may represent their more highly altered equivalents but may also incorporate older material. Lateral variations may also be connected with regional variations in tectonic regime or in the character and thickness of the crust. Despite these variations, Archaean provinces share common features—notably the relative abundance of dominantly basic igneous suites on the one hand and of tonalitic rocks on the other, and the scarcity of (Archaean) indicators of cratonic crustal regimes—which distinguish them from terrains of Proterozoic or Phanerozoic age.

In addition to mineral deposits formed during the evolution of the assemblages mentioned above, which clearly originated during periods of crustal mobility, Archaean tectonic provinces may carry Proterozoic or Phanerozoic deposits formed under relatively stable crustal regimes. These deposits, located in cratonic cover-formations or in cratonic igneous intrusions, are scarcely less distinctive than, although very different from, those in the underlying basement and reflect the exceptional inertia of the Archaean units in the continental cratons.·

2. Archaean Mineralization

The general features of Archaean mineral deposits have become familiar through many regional and general surveys (e.g. Pereira and Dixon, 1965, Goodwin, 1971, Warren, 1972, Laznicka, 1973, Anhaeusser, 1974). Table 1 which is drawn from these and other sources illustrates the consistency in the styles of mineralization which has been emphasized by most authors. The great majority of deposits appear (Watson, 1973) to have been concentrated directly or at only one remove by processes connected with the rise of magma from sources in the mantle. They can be grouped according to setting as follows:

1. In the volcanic successions of greenstone belts

(a) Sulphide deposits of Cu, Zn (Pb, Sb) mainly related to intermediate to acid volcanic centres.

(b) Jaspilitic iron-formations (Algoman type) and manganiferous sediments interstratified with volcanic sequences.

(c) Gold–quartz and gold–telluride deposits segregated after deformation of greenstone belts, often during periods of granite emplacement.

2. In ultrabasic–basic intrusive bodies

(a) Chromite deposits in stratiform intrusives ranging from peridotites to gabbroic anorthosites.

(b) Ni(Cu) deposits in ultrabasic–basic bodies located in greenstone belts.

3. In pegmatites

Li, Be, W, Sn, Mo, Ta minerals located in late-tectonic pegmatites.

(i) *Mineralization in Volcanic Assemblages*

The most striking feature of the deposits associated with Archaean volcanic sequences is their general uniformity both in time and in space which is, of course, directly related to the wide distribution in time and space of the enclosing greenstone belts. The oldest known supracrustal rocks, those of the Isua area in West Greenland (>3700 Ma), include bedded iron-formations not unlike the Algoman type which was common in greenstone belts formed

Table 1

PROVINCE Name and type	ARCHAEAN							PROTEROZOIC PHANEROZOIC	
	Cr/Ni	Cu	AuAg	Zn(Pb)	Fe/Mn	Pegma-tite	Others	Cratonic cover	Cratonic intrusives
AUSTRALIA									
Yilgarn block b (c)	Ni	Cu	Au		(Fe)	LiBe W	(SbAs)		
Pilbara block b c	(Cr)	Cu	Au	(Zn)	Fe	LiBe W Sn	Sb	BIF (Pb) (U)	
N. Australia	(CrPtNi)	(Cu)	Au					U Fe	
AFRICA									
Kaapval craton b c	(Cr)		Au		Fe	BeLiTa Sn	Sb(Hg)	Au U BIF	Pt Cr FeTi Sn diamond Cu
Limpopo belt d	Ni	Cu							
Rhodesian craton b (c)	CrNi	Cu	Au		(Fe)	LiBe W Sn			Cr
Tanganyika b c d			Au		(Fe)				diamonds
West Africa b c			Au		(Mn)	Sn Ta		Au	diamonds
SOUTH AMERICA									
Guyana shield c d			Au		Fe				diamonds
Brazilian shield c								Au (U)	
INDIA									
Dharwar province b c d	(Cr)		Au		FeMn				
Singbhum area c					Fe				
NORTH AMERICA AND GREENLAND									
Superior province b c	Ni	Cu	Au	Zn(Pb)	(Fe)	LiBeMo		U AgCo BIF	AgCo Ni
Slave province b			Au			(LiBeW Sn)		PbZn	
Wyoming massif c								(Au)	Cr FeTi
Pre-Ketilidian-Nain massif d (a)	(Cr)				(Fe)				
EUROPE									
Kola-Karelia area c (a)	NiPt	Cu			(Fe)				Apatite REE
Ukraine c					(Fe)			BIF	
ASIA									
Anabar massif d									diamonds
Aldan massif c d			Au		Fe				diamonds

down to >2700 Ma (Allaart, this volume). The oldest dated gold deposits—those of the Barberton belt in South Africa, dated at >3400 Ma (Anhaeusser, 1974 quoting Saager and Köppel)—are well over a thousand million years older than those of the Birrimian of West Africa. Such comparisons point to a considerable degree of consistency in the style of mineralization during the Archaean stages of crustal evolution and are particularly interesting in the light of evidence now emerging that igneous activity in Archaean provinces was episodic rather than continuous.

Within the characteristic pattern, however, there are significant variations which throw light on the controls governing both the formation and preservation of ore deposits. The distribution of sulphide deposits dominated by CuZn, with minor amounts of Pb, Au and Ag, is restricted. These deposits are of major importance in the Abitibi belt of the Superior province and are fairly widespread in adjacent greenstone belts of the Superior and western Churchill province, but apart from some occurrences in the Slave province of Canada and Pilbara block of Western Australia they are otherwise of little importance. They appear to have originated as exhalative deposits at volcanic centres emitting felsic volcanics and pyroclastics during the late stage of evolution of the parent greenstone belts and are confined to belts in which andesitic, dacitic and rhyolitic lavas are relatively abundant and ultramafic rocks relatively scarce (e.g. Goodwin, 1971). Goodwin (1968, 1971) concludes that the Abitibi belt and the other greenstone belts which show a calc–alkaline trend and carry substantial CuZn mineralization were developed as island arcs flanking a Superior 'protocontinent' made of granites, gneisses and older greenstone belts. The possibility that the Abitibi belt—the largest and most richly mineralized of the Canadian shield—is relatively young and could mark a new stage in Archaean volcanism has been raised by Goodwin. U–Pb dating of intrusives show that the belt is >2750 Ma in age (Krogh and Davis, 1970), not significantly different from at least some of the belts less rich in CuZn within the 'protocontinent' (the Uchi belt, for example, gives U–Pb dates between 2690 and 2761 Ma, Krogh et al., 1974) and certainly no younger than greenstone belts in Western Australia which lack CuZn mineralization. Although tectonic and magmatic factors therefore appear to be of more importance than age in controlling this style of mineralization, it may be significant that no deposits seem to have been found in belts older than 3000 Ma.

Gold mineralization, which has both a very wide geographical distribution and a wide age-range, shows restrictions which appear to relate to the maximum temperatures attained in the crust. Although the source of the gold is now usually thought to have been the greenstone belts themselves, the workable gold deposits were concentrated after the period of eruption in response to changes induced by the rise of granites into and around the greenstone belts. The optimal thermal conditions for mineralization are thought to have been ~450–550°C, roughly those of the greenschist metamorphic facies (Anhaeusser, 1974). Gold is virtually absent from most provinces in categories c and d (Table 1). In the Superior province, economic mineral deposits are almost confined to regions where greenstone-belt volcanics are in greenschist facies (cf. Goodwin, 1971, 1972) although similar volcanics at higher metamorphic grades are present. In the Kaapvaal and Rhodesian cratons, the quantity of gold decreases outward from the centre in harmony with increases in metamorphic grade towards the high-grade Limpopo, Zambezi and Mozambique belts. These relationships suggest that gold deposited under optimum thermal conditions may have been mobilized and dispersed where subsequent events led to the development of higher grades of metamorphism. Such an inference is consistent with the fact that many polycyclic Proterozoic provinces incorporating Archaean greenstone belts are barren of gold. The western Churchill province, in which greenstone belts very like those of the adjacent Superior province occur, carries gold only in the areas (Flin Flon, Lynn Lake, the Kaminak belt) where these have remained in greenschist facies (Davidson, 1972).

In so far as crustal temperature is related to depth, the presence or absence of gold may be determined by the depth of erosion. This factor is considered by Laznicka (1973) to be responsible for many apparent contrasts between the mineral deposits characteristic of provinces of different ages. As I have argued elsewhere, however (Watson, in press), there is little correlation between age and depth of erosion. The low grade and low-pressure style of metamorphism in many Archaean greenstone belts suggest that provinces of type (b) reveal high-level crustal sections from which only a few kilometres of overburden have been removed. Gay (1975) has inferred from other data that the cover on the Kaapvaal craton may have been no more than 1·5 km, an inference which is supported by the survival of a major Archaean antimony deposit (Murchison Range) and a number of minor deposits of both antimony and mercury. The depth of section revealed in such provinces is unlikely to be greater than that revealed in many Proterozoic and Palaeozoic provinces.

In provinces of types (c) and (d), Archaean deposits of minerals which are sensitive to high temperatures are lacking. Chromite deposits in anorthositic complexes and certain NiCu sulphide deposits, on the other hand, are seen in these provinces and the characteristic metals appear to have undergone only limited migrations in response to metamorphism of amphibolite or granulite facies. Chromite layers in the Fiskenaesset anorthosite of south-west Greenland can be identified at consistent horizons through multiple folds in rocks of granulite or amphibolite facies (Windley, Herd and Bowden, 1973). Nickel sulphide ores in the Pikwe–Selebi deposit of the Limpopo belt remain within the presumed host intrusion, although they are concentrated at fold hinges and have undergone amphibolite–facies metamorphism (Gordon, 1973). The inertia of these deposits in relation to high-grade metamorphic processes is in accord with their distribution through provinces representing a wide range of metamorphic environments (Table 1).

(ii) *Mineralization Related to Pegmatites*

The granites, tonalites and quartzofeldspathic gneisses which form some 50% of most Archaean provinces are often regarded as barren and have in consequence received little attention. Suites of pegmatites carrying a wide range of accessory elements—principally Li, Be, W, Sn—are, however, of economic importance (Table 1). These pegmatite suites have a number of common features. They are generally located in granite/greenstone belt terrains (b)—intrusive pegmatites in amphibolite—or granulite–facies Archaean terrains (d) usually have simple mineralogical compositions—and are generally emplaced in or near greenstone belts. Most have been classified as late-tectonic or post-tectonic and appear to be geochemically distinct from the older granitic rocks in each province. A number of pegmatites from southern and western Africa have yielded dates in the region of 2600 Ma, a sample from the Pilbara block of Western Australia a date of 2900 Ma.

The mineralized pegmatites are commonly potassic and in this respect resemble the relatively small late-tectonic granites of provinces such as that of Rhodesia rather than the more widely distributed tonalitic or granodioritic batholiths. They carry the oldest known concentrations of tin, tantalum and wolfram, elements which were not concentrated on a large scale until about 1000 Ma. They seem to have much the same suite of minor elements wherever they occur—lithium especially is almost ubiquitous—and this suite is not one universally present in late-tectonic pegmatites of younger provinces. Their appearance, therefore, suggests the attainment of some definite stage in the processes leading to differentiation of granitic materials. It would be interesting to know whether similar pegmatites occur in association with the most ancient potassic granites such as the 3700 Ma Amîtsoq gneiss of West Greenland.

3. The Termination of Archaean Styles of Mineralization

Isotopic studies (e.g. Moorbath, 1975) are now tending to suggest that plutonic processes involving granite formation were episodic in Archaean times and that early and later phases were separated by periods of relative quiescence. The last very widespread phase of greenstone-belt vulcanicity, granite formation and high-grade metamorphism seems to have taken place at 2700–2600 Ma, although essentially similar activity continued locally in

West Africa down to the emplacement of the Eburnian granites at *c.* 2000 Ma.

Several of the characteristic styles of Archaean mineralization are now known to have had a very long time-range. Iron-formations (some of which carry gold) are associated with the oldest (3700 Ma) supra-crustal assemblages of West Greenland and with volcanic sequences formed at intervals over at least a billion years. Ores of Cr and CuNi and gold–quartz vein deposits range over an equally long period from about 3400 Ma onward. No later billion years of geological time has been characterized by so uniform a suite of ore deposits of comparable importance. There are, however, some indications of progressive changes which may reflect an evolution of the processes of magma-generation or differentiation. The formation of sulphide ores dominated by CuZn does not appear to have taken place on a significant scale prior to 3000 Ma, and throughout Archaean times these ores remained relatively poor in Pb. The appearance of volcanic sequences associated with PbZn deposits in Broken Hill and Mount Isa marked a further change in early Proterozoic times leading on to a style of mineralization characteristic of Proterozoic and Phanerozoic periods. Mineralized pegmatites—almost the only Archaean deposits not connected directly or at one remove with mantle-derived igneous suites—appear in the record after 3000 Ma carrying concentrations of lithophile elements. The apparent link between these pegmatites and late-tectonic granites of a type very sparsely represented before 2900 Ma suggests that their arrival may be connected with the generation of magmas from new sources or under new conditions.

More extensive changes in the style of mineralization were connected with the decline in volcanicity of greenstone-belt type. The youngest greenstone belts are probably those of West Africa invaded by the Eburnian granites at ~2000 Ma, but in most Archaean provinces volcanicity of this type ended some five hundred million years earlier. The termination of the associated mineralization was therefore diachronous with respect to the Archaean/Proterozoic boundary (if this is to be regarded as a definable time-plane), but was linked with the more general changes in crustal behaviour which characterized the transitional period round about the end of the Archaean era.

4. Proterozoic–Phanerozoic Mineralization in Archaean Tectonic Provinces

Mineral deposits formed in Archaean tectonic provinces during Proterozoic and Phanerozoic times are commonly associated either with units of the cratonic cover or with intrusive igneous bodies. The styles of mineralization are again somewhat monotonous (Table 1) and contrast in several ways with those formed under cratonic regimes in younger tectonic provinces. The principal deposits can be grouped as follows:

1. *Associated with sedimentary formations*
   (a) Gold and/or uranium in clastic formations.
   (b) Banded iron formations.

2. *Associated with intrusive igneous rocks*
   (a) Deposits including Cr, Pt, FeTi or Ni in differentiated basic bodies such as the Bushveld complex.
   (b) Rare-earth elements, apatite, etc., in carbonatites and alkaline complexes.
   (c) Diamonds in kimberlites.

The principal deposits of the cratonic cover are the very distinctive early Proterozoic banded iron formations characterized by alternations of cherty and iron-rich layers and the equally distinctive deposits of the Witwatersrand, Huronian, Jacobina and similar successions which carry uranium and/or gold derived ultimately from the erosion of older mineralized terrains. The great bulk of both these types of deposit accumulated within a few hundred million years of the cessation of mobility in the underlying basement, that is, before about 2000 Ma (Fig. 1). Although later Proterozoic and Phanerozoic cover-formations are present in many Archaean tectonic provinces, these are seldom mineralized.

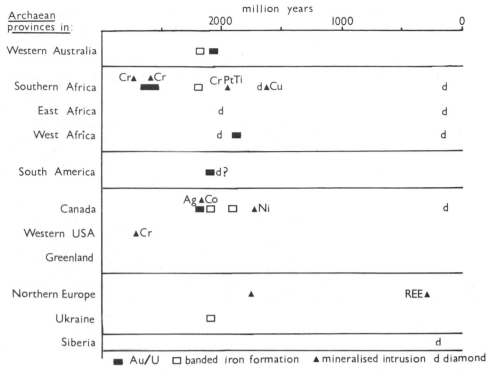

Fig. 1. Distribution in time of ore deposits in the cratonic cover on Archaean provinces (rectangles) and in intrusive bodies emplaced in these provinces under cratonic conditions

The restricted time range of the banded iron formations is generally attributed to factors connected with organic evolution and has little significance in the present context. The restriction of rich gold–uranium deposits to a short time-span, on the other hand, may be directly related to the tectonic behaviour of the Archaean provinces. In many of these provinces, the evidence suggests that unroofing of the basement took place soon after the cessation of Archaean activity and that very little basement material was removed by erosion after this initial period (Watson, in press). Material from such mineralized complexes therefore made only very small contributions to the later Proterozoic and Phanerozoic sediments accumulating on the basement. It is interesting to note that these younger formations seldom seem to carry copper or lead–zinc concentrations of the types found elsewhere in cratonic successions of late Proterozoic and Phanerozoic age.

Ore deposits associated with cratonic igneous intrusions in Archaean provinces fall into two categories which differ not only in magmatic affinity but also in time-range (Fig. 1). Most of the great differentiated basic complexes which carry concentrations of chromium, nickel, platinum, iron–titanium or other metals— the Great Dyke, the Bushveld complex, the Nipissing diabase, etc.—were emplaced in late Archaean or early Proterozoic times. Younger igneous complexes of broadly similar type occur in provinces stabilized in Proterozoic times but are scarce in Archaean provinces. Once again, the tectonic inertia of many Archaean crustal units through later Proterozoic and Phanerozoic times is reflected in the reduction of activities leading to ore formation. On the other hand, most of the mineralized alkaline complexes and carbonatites and the diamondiferous kimberlites which characterize a number of Archaean provinces date from the later periods of geological time. The occurrence of a few alkaline and kimberlitic intrusions yielding mid-Proterozoic ages and the presence of detrital diamonds in early Proterozoic formations in

West Africa and probably also in Guyana show that similar magma-forming processes continued intermittently over a least two thousand million years. The persistence of kimberlite-forming processes at depth beneath Archaean provinces and the spatial relationship between the central parts of these provinces and the concentrations of diamondiferous kimberlites (e.g. Dawson, 1970) may indicate that the mantle portion of the lithosphere was not recycled during later Proterozoic and Phanerozoic times.

The possibility that other persistent processes of mineralization relate to heterogeneities in the mantle has been referred to by Anhaeusser with reference to the chrome deposits of southern Africa. Anhaeusser (1974) points out that approximately half the world's production of chromite comes from a well-defined area, elongated north–south, which embraces parts of the granite/greenstone-belt terrains of the Kaapvaal and Rhodesian cratons and part of the intervening Limpopo belt. Chromite occurs in a variety of mantle-derived differentiated intrusive complexes some of which (Selukwe, Rhodesia) are associated with greenstone-belt volcanics and predate the local tectonism and metamorphism, while others (Great Dyke, Bushveld complex) represent cratonic bodies emplaced after the termination of Archaean activity. The basic complexes themselves, with the possible exception of the Bushveld complex, have analogues in other Archaean terrains but are distinguished by their high concentrations of chromite. Collectively they scan a period of more than 700 Ma and it seems possible that during the period of their emplacement the underlying part of the mantle had an above-average content of chromium.

5. Mineralization Along Lineaments and at Province Boundaries

The economic importance of Archaean/Proterozoic boundaries, which was stressed particularly by Derry (1961), is attributable to two principal factors. Some deposits are concentrated near the margins of Archaean tectonic provinces because their accumulation was controlled by regional geological contrasts between these provinces and the adjacent provinces. Early Proterozoic deposits located in the cratonic cover commonly illustrates this control, since such cover rocks tend to be absent from the interior parts of Archaean provinces and to thicken towards their junction with adjacent mobile belts. The banded iron formations of Labrador and the Great Lakes in North America and of the Hamersley Range in Western Australia provide examples of this type of distribution.

The operation of a different set of controls is seen where province boundaries are defined by deep lineaments which represent fundamental discontinuities in the crustal structure. Many such lineaments originated in late Archaean or Proterozoic times as transcurrent or transform dislocations (cf. McConnell, 1974; Sutton and Watson, 1974) and have persisted through the rest of geological time, filling a variety of roles in response to changes in crustal regime (Watterson, 1975). Lineaments of this type at the boundaries of Archaean provinces may carry mineral deposits dating from roughly the time of their initial formation, the most notable example being the nickeliferous ultramafic bodies of the Thompson lineament at the western border of the Superior province. I regard this structure, which has been interpreted as a plate suture by Gibb and Walcott (1971), as one of a set of north-easterly lineaments which traverses the North American and Greenland cratons (cf. Sutton and Watson, 1974). Other lineaments of the same set are characterized by igneous intrusives or by zones of mineralization dating from subsequent periods which distinguish them from the adjacent terrains.

Kimberlites of Mesozoic age, not so far known to be diamondiferous, are located along the Nagssugtoqidian front in West Greenland. Small alkaline intrusives appear along the Kapuskasing lineament in the Superior province which is interpreted by some authors as an early rift (Parsons, 1961, Tuttle and Gittins, 1966). Furthermore, the Tertiary gold mineralization of western North America tends in some areas to concentrate in north-easterly zones parallel to the basement

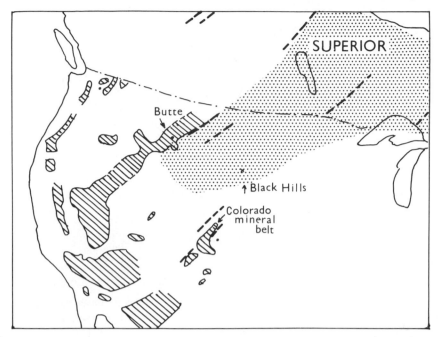

Fig. 2. Precambrian lineaments and Archaean basement complexes (stippled) in relation to domains of gold mineralization (lined, after Noble, 1970) in Western U.S.A. (reproduced with permission of The Geological Society of America)

lineaments. Fig. 2, on which the outlines of mineralized areas defined by Noble (1970) are superimposed on a map of basement lineaments, suggests that gold is concentrated in north-easterly zones close to the borders of the small Archaean province of Wyoming and, although the present expressions of these zones reflect later Proterozoic or Phanerozoic events, it seems probable that they occupy the site of still older marginal lineaments.

The distribution of Pb–Zn deposits of Mississippi Valley type in the Palaeozoic cover of the North American craton appears to be influenced by the location of basement 'highs' which represent areas of intermittent uplift. Fig. 3, based on the U.S.G.S. basement map of North America (1967) shows that some of the highs associated with mineral deposits have a north-easterly alignment parallel to the early Proterozoic set of basement lineaments. The Pine Point Pb–Zn deposit is located on the projection of the lineament at the southern margin of the Slave Archaean province which is associated in the cover with such sedimen-

tary features as the Presqu'ile reef complex of Middle Devonian age. There appears, therefore, once more to be a connection between the localization of a widespread mineralization of Phanerozoic age and the presence of very ancient deep lineaments in the basement.

Putting together the evidence mentioned in the last few pages we obtain a picture of the distribution of mineralization in time and space with respect to the small Archaean tectonic provinces which have survived to the present day. The long Archaean period of mineralization in association with mobile tectonic regimes was followed by an early Proterozoic period in which major concentrations of various metals were formed under cratonic regimes (Fig. 1). From mid-Proterozoic times onwards, the contributions of rock material received by these provinces from adjacent regions in the form of sedimentary or magmatic accessions became very small. The contributions from Archaean provinces to adjacent areas by the transfer of erosional debris or material in solution were also small

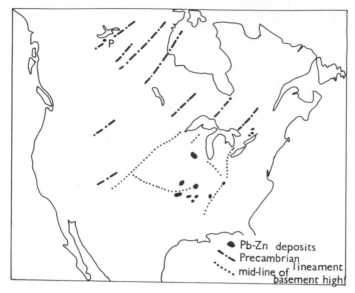

Fig. 3. Precambrian lineaments in North America (based on Sutton and Watson, 1974) in relation to the position of basement 'highs' beneath the Phanerozoic cratonic cover and to the distribution of lead–zinc deposits of Missisippi Valley type. P = Pine Point

and, consequently, these provinces have played little part in the recycling of metals from mantle or crustal sources during the last 1800 Ma of geological time. Activity continued on a larger scale in marginal features and in deep lineaments which occasionally tapped sources of mantle material and channelled fluids of crustal derivation.

References

Anhaeusser, C. R., 1974. 'Archaean metallogeny in southern Africa', *Univ. Witwatersrand, Inf.*, Circ 91.

Davidson, A., 1972. 'The Churchill province', In *Variations in Tectonic Styles in Canada*, R. A. Price and R. J. W. Douglas (Eds.), *Geol. Assoc. Canada, Sp. Pub.* 11, 381–434.

Dawson, J. B., 1970. 'The structural setting of African kimberlite magmatism', In African magmatism and tectonics, T. N. Clifford and I. G. Gass (Eds.), Oliver and Boyd, 321–336.

Derry, D. R., 1961. 'Economic aspects of Archaean-Proterozoic boundaries', *Econ. Geol.*, **56**, 635–647.

Gay, N. H., 1975 (in press), *Tectonophysics*.

Gibb. R. A., and Walcott, R. I., 1971. 'A Precambrian suture in the Canadian shield', *Earth Planet. Sci. Lett.*, **10**, 417–422.

Goodwin, A. M., 1968. 'Archaean protocontinental growth and early crustal history of the Canadian shield', *23rd Int. Geol. Congr. Prague*, **1**, 69–89.

Goodwin, A. M., 1971. 'Metallogenic patterns and evolution of the Canadian shield', *Geol. Soc. Aust. Sp. Pub.* 3, 157–174.

Goodwin, A. M., 1972. 'The Superior province', In *Variations in Tectonic Styles in Canada*, R. A. Price and R. J. W. Douglas (Eds.), *Geol. Assoc. Canada, Sp. Pub. 11*, 528–623.

Gordon, P. S. L., 1973. 'The Selebi-Pikwe nickel–copper deposits, Botswana', *Symposium on Granites, Gneisses and Related Rocks*, R. A. Lister (Ed.), *Geol. Soc. S. Africa, Sp. Pub.*, *3*, 167–188.

Krogh, T. E. and Davis, G. L., 1971. 'Zircon U-Pb ages of Archaean metavolcanic rocks in the Canadian shield', *Carnegie Inst. Yr. Bk.*, **70**, 241–242.

Krogh, T. E., Ermanovics, I. F., and Davis, G. L., 1974. 'Two episodes of metamorphism and deformation in the Canadian shield', *Carnegie Inst. Yr. Bk.*, **74**, 573–575.

Laznicka, P., 1973. 'Development of non-ferrous metal deposits in geological time', *Can. J. Earth Sci.*, **10**, 18–25.

McConnell, R. B., 1974. 'Evolution of taphrogenic lineaments in continental platforms', *Geol. Rund.*, **63**, 389–429.

Moorbath, S., 1975. 'Evolution of Precambrian crust from strontium isotopic evidence', *Nature, Lond.*, **254**, 395–398.

Noble, J. A., 1960, 'Metal provinces of the Western United States', *Geol. Soc. Amer. Bull.*, **81**, 1607–1624.

Parsons, G. E., 1961. 'Niobium-bearing complexes east of Lake Superior', *Ont. Dept. Mines, Geol. Rep. 3*.

Pereira, J., and Dixon, C. J., 1965. 'Evolutionary trends in ore deposition', *Trans. Instn. Mining Metal*, **74**, 505–527.

Sutton, J., and Watson, J., 1974. 'Tectonic evolution of continents in early Proterozoic times', *Nature, Lond.*, **247**, 433–435.

Tuttle, O. F., and Gittins, J., 1966, *Carbonatites*, Interscience, London, 515–528.

Warren, R. G., 1972. 'A commentary on the metallogenic map of Australia and Papua, New Guinea', *Bur. Min. Res., Bull.*, 145.

Watson, J., 1973. 'Influence of crustal evolution on ore deposition', *Trans. Instn. Mining Metal.*, **82**, B107–114.

Watson, J., 1973, in press. 'Vertical movements in Proterozoic structural provinces', *Phil. Trans. Roy. Soc. Lond.*

Watterson, J., 1975. 'Mechanism for the persistence of tectonic lineaments', *Nature, Lond.*, **253**, 520–522.

Windley, B. F., Herd, R. K., and Bowden, A. A., 1973. 'The Fiskenaesset complex, West Greenland, Part 1', *Grønlands geol. Unders.*, Bull-106.

Gold Metallogeny in the Archaean of Rhodesia

R. E. P. FRIPP,

Department of Geology, University of the Witwatersrand, Jan Smuts Avenue, Johannesburg, 2001, South Africa.

Introduction

Early workers involved with gold deposits in the Archaean proposed that there is a genetic relationship between vein gold deposits and greenstone-belt metamorphism. Recent experimental data pertaining to gold solubility in hydrothermal systems has revived interest in this thesis (e.g. Fyfe and Henley, 1973). Furthermore, recent field studies have indicated that significant amounts of gold occur as stratiform ore bodies in mineralized banded iron formations, and that these may be derived from exhalative hydrothermal systems (Fripp, 1975).

A synthesis and analysis of the local and regional controls to the distribution of gold in the Archaean of Rhodesia reveals that the more important ore bodies can be genetically linked to volcanic and hydrothermal activity, and that the localization of mineralization is related to particular stratigraphic levels and lithologic sequences. These levels may reflect Archaean geothermal gradients and related, localized hydrothermal systems.

This synthesis also reveals that the gold deposits are mostly generated at an early stage in the evolution of the greenstone sequences. Consequently geological parameters such as the extent of metamorphism, stratigraphic level and subsequent deformation can provide useful guides to exploration.

Regional Geology

The Archaean rocks of Rhodesia, excluding the Great Dyke, comprise a typical granite–greenstone terrain. The greenstones occur as tightly folded belts surrounded by various granitoids, including significant amounts of migmatite and banded, schistose tonalitic gneisses which are regarded as representing the basement on, or adjacent to which the greenstone volcano-sedimentary sequences were deposited. Pertinent data have been reviewed by Bliss and Stidolph (1969) and Wilson (1973).

The stratigraphy of the greenstone belts has been divided into three lithologically distinct groups (MacGregor, 1932 and 1947). The lowermost Sebakwian group is comprised dominantly of magnesium-rich ultramafic and mafic volcanic rocks with thin interlayered felsic volcaniclastic rocks, argillite, limestone, banded chert and iron formation. This is overlain by the Bulawayan group, a thick sequence of mafic (basaltic) and felsic volcanic rocks with interlayered iron formation, chert, limestone, argillite and felsic pyroclastics. Mafic pillow lavas and andesitic tuffs and agglomerates are well represented. The Shamvaian group is comprised largely of volcaniclastic and epiclastic coarse and fine-grained sediments, including conglomerate, grit, greywacke, arkose, and quartzite, with lesser

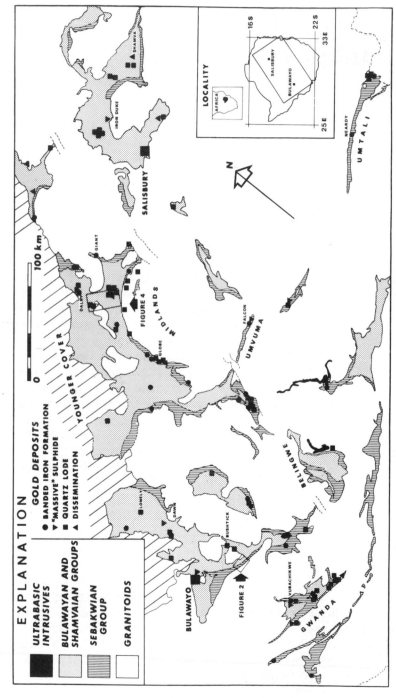

Fig. 1. Regional Archaean geology of Rhodesia, showing the distribution of the 100 larger producers (>600 kg gold up to 1964). The Great Dyke is omitted

amounts of interlayered iron formation, chert and argillite (Phaup, 1968a). Intrusive (penecontemporaneous?) ultramafic layered and undifferentiated bodies occur in places and represent a minor part of the sequence. The distribution of the Sebakwian, Bulawayan and Shamvaian groups, as interpreted by the author, are shown in Fig. 1.

Metamorphism

In general the greenstone-belt rocks are metamorphosed to greenschist facies with chlorite and tremolite present in the mafic rocks. However, relatively unmetamorphosed and even fresh unaltered basic rocks have been noted to occur over wide areas such as in the Midlands belt (Harrison, 1970; Bliss, 1970) and the Belingwe belt (A. Martin, pers. comm.). Metamorphic grade tends to be higher at the margins of most belts, where it commonly reaches amphibolite grade, and at a few localities kyanite and andalusite are developed in quartz–alumina schists (e.g. Swift, 1956; Fripp, 1974). However, the margins of some belts are diffuse. Such is the case at the south-east margin of the Bulawayo belt where large 'horses' of greenstone belt have been introduced lit-par-lit by tonalitic gneiss over a distance of about 5 km, while the greenstones at the 'contact' contain fresh quenched pyroxenes (Smith and Fripp, 1973). It is evident that greenstone-belt metamorphism is certainly not uniform and that the distribution and behaviour of heated waters and granitic rocks (basement, partly reactivated basement and intrusive high-level stocks) may impose important controls on local hydrothermal and metamorphic environments. These in turn can influence the distribution of gold mineralization within a specific greenstone belt.

Distribution and Geology of the Gold Deposits

The distribution of 100 larger gold mines, all of which produced more than 600 kg of gold up to the end of 1964, is shown in Fig. 1. The Neardy and Iron Duke mines are exceptions and have produced little gold. Ninety-two of these occur within the greenstone belts,

the remainder in the surrounding granitoids. Fifty deposits occur within rocks of the Sebakwian group and 42 within rocks of the Bulawayan and Shamvaian groups (Table 1). Notably the Sebakwian group is volumetrically much smaller than the Bulawayan and Shamvaian groups, yet it contains the majority of the mines.

Table 1. Lithostratigraphic distribution of different types of gold deposit shown in Fig. 1

Type	Sebakwian group	Bulawayan and Shamvaian groups	Granitoids	Total
Stratiform	27	3	0	30
'Massive'	3	4	0	7
Quartz lode	18	30	8	56
Dissemination	2	5	0	7
TOTAL	50	42	8	100

These deposits have been classified into 4 categories, based upon the geological nature of the ore bodies. These are: (1) stratiform deposits of mineralized banded iron formation, (2) stratabound 'massive' sulphides, (3) quartz lodes, veins, stockworks and siliceous shear zones and (4) stratabound disseminated mineralization in clastic rocks.

The deposits have been classified on the basis of existing descriptions of the mines (particularly dormant ones), communications with mine geologists and personal knowledge.

1. Stratiform deposits

About 25% of Rhodesia's gold has been derived from mineralized banded iron formation. The gold is derived preferentially from the carbonate and sulphide facies iron-formation, within which the sulphides occur as distinctive centimetre and millimetre-thick layers, alternating with iron carbonate-rich and chert-rich layers. The sulphides comprise arsenopyrite and/or pyrite and/or pyrrhotite and detailed studies indicate that the gold occurs dominantly as minute grains of the order of 50 μm in diameter occluded within the arsenopyrite grains. In pyrite-rich deposits the gold is generally not microscopically visible (Fripp, 1975).

Individual beds of gold-bearing iron formation are generally not more than 5 m thick, and are locally interlayered with ferruginous limestones, sulphidic argillite and mafic and felsic aquagene tuffs. In most cases these units are interbedded with a thicker sequence of magnesium-rich mafic and ultramafic rocks characteristic of the Sebakwian group. Most of these stratiform deposits therefore occur within this group (Fig. 1 and Table 1). In addition, these deposits tend to be locally grouped so that one or several parallel closely spaced beds of iron formation produce gold from several mines distributed along a strike length of up to 10 or more kilometres (Fig. 2).

Variation in the silver content of the sulphidic iron formation tends to be limited and the gold fineness tends to be high. In the Vubachikwe group (Gwandá), for example, gold fineness varies only from 900 to 970‰ (Eales, 1961).

In addition to gold, stratiform beds of iron formation in places contain economic concentrations of other metals. The Giant mine (Midlands) and the Falcon mine (Umvuma) have produced copper as a by-product. In the Umtali belt (Fig. 1) copper and lead occur in a group of stratiform deposits along a strike length of about 10 km. In some of these gold and silver are also present in significant amounts, and the Neardy mine has produced Cu–Pb–Ag from millimetre and centimetre thick sulphide layers in an argillaceous variety of iron formation (Fig. 3). Here a pyrite–rich bedded 'massive' sulphide horizon lies below the metalliferous iron formation and grades into it (Fripp, 1974).

2. Stratabound 'Massive' Sulphides

Massive sulphide deposits which have produced gold are uncommon in Rhodesia. The Iron Duke mine (Salisbury) was initiated by gold extraction from the supergene enriched capping of a massive sulphide, and is now merely a pyrite producer. However, there are a number of stratabound gold deposits com-

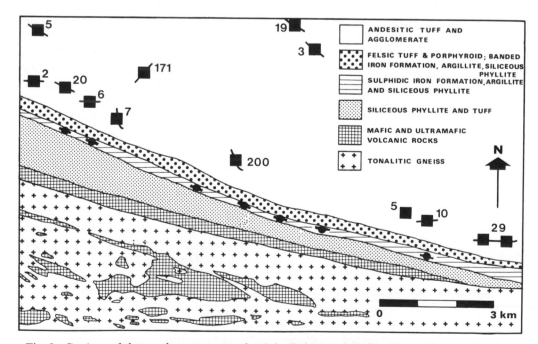

Fig. 2. Geology of the south-eastern margin of the Bulawayo belt (locality on Fig. 1). Illustrates the strong stratigraphic control of gold distribution, both in veins (squares) and iron formation (circles)

Fig. 3. Photograph of hand specimen of stratiform chalcopyrite–galena–pyrrhotite–pyrite ore in argillaceous iron formation; Neardy mine. Also contains silver, and further along strike gold

prised of intermixed cherty, sulphidic and chloritic material, up to 5 m in thickness, and containing evidence of soft-sediment deformation. These are interlayered with extrusive volcanic rocks and are present in the Sebakwian and Bulawayan groups (Fig. 1, Table 1).

3. Lode Deposits

More than half of Rhodesia's gold has been extracted from vein systems which transect greenstone-belt stratigraphy. The veins vary in thickness from less than 1 m up to about 5 m and extend along strike for up to 2 km. At some mines only one vein is exploited, and at many others several veins provide ore. Many veins have slickensided walls indicating thrust or reverse faulting while others are extension features. The veins may be arranged en échelon in parallel or as a 'splay' system. Some siliceous shear zones and stockworks also contain economic gold mineralization, as at the Bushtick mine, Bulawayo (Amm, 1950, p. 99).

At most lode deposits the veins are comprised dominantly of quartz, accompanied by lesser amounts of carbonate and a few percent of sulphides. These sulphides consist mainly of pyrite and, in places, lesser amounts of pyrrhotite, sphalerite, galena, arsenopyrite, stibnite, chalcopyrite and scheelite are present. The gold is typically coarse-grained and in places visible. In a few deposits gold tellurides and aurostibite are also present. There is a tendency for larger gold deposits to be associated with ores containing pyrite together with chalcopyrite, stibnite and galena, rather than pyrrhotite, sphalerite and arsenopyrite (Pretorius and Hempkins, 1968), although exceptions do exist.

Gold distribution within some individual lodes is very variable, whereas in others values are consistent. At many mines the quartz veins have been found to continue, without significant physical changes, beyond payability limits with either a sharp or gradational cut-off. Mineral zoning has also been described. For example, chalcopyrite, sphalerite and galena increase with depth at the Lonely mine (MacGregor, 1928; Greenberg, 1945). The large Globe and Phoenix mine is zoned vertically with lead decreasing and antimony and copper increasing with depth (Phaup, 1964, p. 3). Changes in the silver content (i.e. gold fineness) within veins have been established at some deposits. At the Lonely mine the fineness decreases regularly with depth from about 9400 to $870\permil$ over 700 m (MacGregor, 1928; Greenberg, 1945).

The vein wall rocks exhibit distinctive wall rock alteration (propylitization), which can extend from less than a metre to several metres away from the vein. In mafic wall rocks the most common alteration assemblage is chlorite, epidote, tremolite and albite, with lesser amounts of disseminated sulphide. Mehliss (1964) has described extensive alteration and carbonation of the pillowed sequence transected by the vein at the Dawn mine. Here quartz and sericite replace the feldspar, and iddingsite the olivine. In most deposits fragments of bleached and propylitized wall rock occur in the quartz veins.

Vein gold deposits are grouped in some areas, suggesting the presence of locally rich metallogenic subprovinces of the order of

100 km² in extent. Within any group the individual lodes tend to have a similar mineralogy and a base metal association, distinct from groups elsewhere (Phaup, 1964 and 1968b). Gold fineness variation is known to be large within some groups (Eales, 1961). Furthermore, in some areas it has been found that the veins indicate a common genetic stress field (Mehliss, 1963; Goldberg, 1964; Stowe, 1968).

There are distinct regional and local stratigraphic controls to the distribution of the lode deposits. Most of the larger ones occur within the thick mafic–felsic volcanic pile of the Bulawayan group (Pretorius and Hempkins, 1968; Table 1 and Fig. 1). Locally, it can be demonstrated that groups of deposits occur at particular stratigraphic levels within the Bulawayan. For example, the Masterpiece–Golden Valley–Glasgow group of deposits in the Midlands belt are distributed in an arc, sub-parallel to an F_1 axial trace deformed by the F_2 Gatooma anticline (Fig. 4). They occur preferentially within a portion of the volcanic pile over a stratigraphic interval of about 1·5 km and probably pre-date much of the greenstone-belt deformation. Furthermore, detailed local evidence for vein deformation is widespread in the descriptions of individual deposits (Rhodesia Geological Survey bulletins) and many veins are described as being folded and boudinaged.

4. *Stratabound Disseminated Mineralization*

Some mines have exploited mineralization which is disseminated through various classic rocks. For example, the Motor and Centre lodes at the Cam and Motor mine in the Midlands belt are comprised of a few percent of pyrite and arsenopyrite disseminated through calcareous volcaniclastic arkose, greywacke and carbonaceous shale (Hartman, 1953; Collender, 1964; Fig. 5). The ore bodies of the Shamva mine near Salisbury comprise bedded and disseminated pyrite within calcareous volcanic arkose and greywacke, over a thickness of up to 70 m (Mehliss, 1963).

These deposits therefore are stratabound and relatively thick. Furthermore, they have gradational margins and occur mostly in the Bulawayan and Shamvaian groups (Fig. 1 and Table 1).

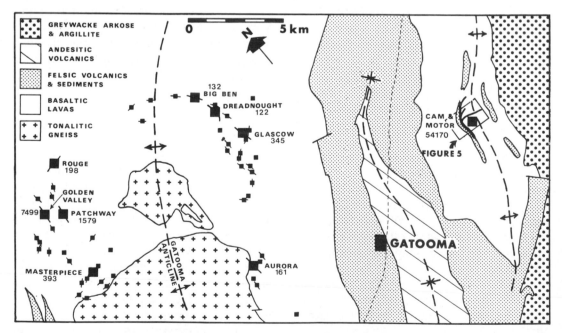

Fig. 4. Geology of the area around Gatooma, Midlands belt (locality on Fig. 1). Gold-bearing quartz veins are distributed through a narrow stratigraphic interval, deformed by the Gatooma anticline (after Bliss, 1970)

Relationships between different types of deposit. Apart from the distinctive regional stratigraphic controls which these deposits exhibit (Table 1), it is possible in places to demonstrate local relationships between the various types of deposit. A detailed map of the south-western part of the Bulawayo belt (Fig. 2) shows a change from Sebakwian-type stratigraphy at the base to Bulawayan-type volcanic rocks above. Near the top of the lower ultramafic–mafic unit, in part intruded by tonalitic gneiss and with possible unrecognizable 'granitized' felsic volcaniclastics (Smith and Fripp, 1973), a unit of iron formation and argillite contains stratiform gold deposits over a strike length greater than 10 km. Within the overlying 4 km thick succession of mafic and felsic volcaniclastic rocks there are many small, and a few large lode deposits, about 2 km above the stratiform deposits. Since the rocks have undergone considerable finite strain the true stratigraphic interval is possibly 50% greater. There is evidently a close spatial association between lode deposits and stratiform deposits here, and this relationship is demonstrable elsewhere (e.g. the Lonely mine (Greenberg, 1945), and around the Dalny mine in the Midlands belt (Bliss, 1970)).

At some better developed mines it is possible to demonstrate a close spatial relationship between gold-bearing disseminated mineralization and lode gold deposits. The Cam and Motor mine is situated on the west limb of an anticline comprised dominantly of basaltic pillowed and massive lavas. Interbedded with these is a large lentil of felsic volcanic rocks which contains the Motor and Centre lodes described above (Fig. 5). Stratigraphically beneath these mineralized sediments are 3 large mineralized quartz lodes which strike approximately perpendicular to the strata; these terminate at the base of the sedimentary lentil, and do not cut through it. These are the Petrol, Cam and Cam Hanging Wall lodes; the Cam lode has been extensively mined over a strike length of 1300 m (Fig. 5). The quartz lodes characteristically contain more arsenopyrite than pyrite, some massive stibnite and carbonates, and the wall rocks are strongly impregnated with carbonate and sulphides. In contrast, the stratabound mineralization contains more pyrite than arsenopyrite, up to 2% carbon and abundant carbonate, and

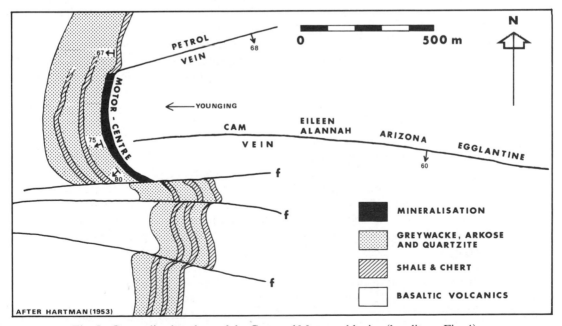

Fig. 5. Generalized geology of the Cam and Motor gold mine (locality on Fig. 4)

some copper and zinc sulphides (Hartman, 1953).

Discussion

The physical, mineralogical and geological characteristics of the deposits reviewed above suggest that distinct genetic relationships exist both between the different types of deposit and between each type and its geological environment. All of the deposits are considered to be genetically related to volcanic hydrothermal activity and, in the broadest sense, to be volcanogenic (cf. Ridler, 1970; Hutchinson et al., 1971). This is supported by the field evidence and new work on the role of hydrothermal fluids in greenstone-belt metamorphism and gold transport.

Experimental data on gold solubility has been reviewed elsewhere (Fripp, 1974 and 1975) and only the salient points will be summarised here. Gold solubility has been investigated in two principal hydrothermal systems, both of which are pertinent to the genetic model presented below.

Gold solubility in hydrothermal alkaline sulphide solutions has been investigated by Weissberg (1970). He found gold to be soluble in the range 100 to 200 ppm at temperatures from 150° to 280°C, which he suggested was due to the formation of a stable gold complex involving H_2S and HS.

Gold solubility in aqueous sulphide solutions for the pH range 4 to 9·5 was investigated by Seward (1973). In the temperature range 160° to 300°C he found that gold is soluble from 40 to 230 ppm at a pH of 6; in more alkaline solutions the solubility decreased by a factor of 20. Seward concluded that three gold–thio complexes dominate in these solutions: $Au (HS)_2^-$ in neutral solution, $Au_2 (HS)_2S^{2-}$ in alkaline solution and the molecular species $AuHS^0$ in acid solution. Furthermore, from field and thermodynamic data he suggested that arseno–thio and antimino–thio complexes such as $Au(AsS_3)^{2-}$ may be important gold transporting compounds in these relatively low-temperature solutions.

The solubility of gold within the temperature range 300°C to 500°C has been investigated by Henley (1973), using chloride solutions. He found that gold solubility increased steadily from 10 ppm at 300°C to 200 ppm at 450°C and at 480°C increased markedly to about 1000 ppm at 510°C. Henley suggested that molecular chloride species such as $Au_2Cl_6(HCL)_2$ were involved. This temperature range was favoured since it covers the range of greenschist facies metamorphism from 380°C to 480°C (Winkler, 1967, pp. 117 and 177) and that of the albite–epidote hornfels facies from 400°C to 500°C (Winkler, 1967, p. 70). These results led Fyfe and Henley (1973) to propose that most gold transported in hydrothermal solutions during metamorphic events would be precipitated out between 300°C and 400°C.

Metallogenic Model

The Sebakwian stratiform mineral deposits in banded iron formations, including those containing gold and the Neardy copper–lead–silver–gold group, are considered to be derived from subaqueous volcanic exhalations of thermal brines (Hutchinson et al., 1971; Fripp, 1975). This is supported by their intimate relationship with volcanic rocks, including aquagene tuffs, and their detailed stratiform petrology which closely resembles exhalative sulphide deposits forming today (e.g. Puchelt, 1973; Honnorez et al., 1973; Ferguson et al., 1974). Furthermore, exhalations are known to be precipitating gold, silver and other metals in modern fumarolic areas (White and Waring, 1963; Weissberg, 1969).

At depths of 200 m or more, the brines could reach the sea floor at temperatures of up to 220°C without boiling (Haas, 1971). Presuming that they contained gold as soluble thio-complexes and possibly arseno–thio complexes, they would precipitate their metals upon contact with cooler marine waters of different pH. The 'massive' sulphide deposits are presumably generated in a similar manner, the intermixed tuffaceous material and the soft-sediment deformation features reflecting more active volcanic environments and relatively rapid formation.

The lode deposits which occur mostly in the Bulawayan reflect a higher temperature environment of deposition. The generally low sulphide content of these deposits suggests that the circulating metamorphic hydrothermal brines contained the gold as dissolved chloride and molecular species (Henley, 1973), with the gold precipitating in the 300°C to 400°C temperature range (Fyfe and Henley, 1973). However, limited fluid inclusion data from Archaean and Palaeozoic lode gold deposits indicate very high partial pressures of carbon dioxide in the fluid (F. J. Sawkins, pers. comm.), and this may have played a significant role in gold transport and deposition. Measurements of the carbon dioxide content of modern geothermal fluids (White and Waring, 1963) and the presence of carbonates in both the stratiform and lode gold deposits supports this view.

The disseminated gold deposits in volcaniclastic Bulawayan rocks may be diagenetic and/or syngenetic in origin, precipitated from gold-bearing brines percolating through the rocks and fed through underlying fractures. The disposition of the quartz lodes in relation to the disseminated lodes at Cam and Motor strongly supports this view (Fig. 5) and the underlying quartz lodes quite possibly represent the original feeder fractures. The lode deposits, from the structural evidence, must have formed within the volcanic piles prior to most of the regional deformation and therefore were generated soon after the burial of the pile. In this regard, New Zealand geothermal brines in buried fractures are known to have temperatures of 250°C at depths up to 1500 m over wide areas (Verhoogen et al., 1970, p. 575).

The different temperature ranges required to generate the contrasting types of deposit at different regional-scale stratigraphic levels reflect greater geothermal gradients during the Sebakwian compared with the Bulawayan. The generalized, diagrammatic mineralization model (Fig. 6) is based in part upon the ideas of Henley (1971), who proposed that downward percolating brines are heated and leach the gold and other metals, including iron and silica, from the greenstone volcanic pile. This is supported by the work of Mantei and Brownlow (1967), which indicates that gold occurs as the native metal lodged meta-stably within crystal-lattice imperfections, preferentially in biotite and hornblende in mafic and ultramafic rocks. The dissolved chloride species are transported upward and on cooling to 300°C to 400°C precipitate the gold, together with large amounts of silica and some sulphides. At these temperatures the through-flowing Na-rich fluids also cause metamorphism of the host rock to the albite–epidote facies.

As mentioned previously, carbon dioxide may also have been important, and the factors which locally controlled the deposition of gold were probably very complex. In addition, the presence of significant lode deposits in the lower ultramafic unit (Table 1), and the few stratiform deposits in the Bulawayan and Shamvaian groups indicates the danger of generalization. Nevertheless, the overall (regional) suggestion is one of a higher geothermal gradient in the older Sebakwian.

No detailed diagnostic evidence is currently available to suggest the exact age relationships between regional greenstone-belt metamorphism and the emplacement of quartz lodes. However, the local spatial association of groups of deposits with more metamorphosed greenstones and the converse lack of mineralization (of any type) within largely unmetamorphosed areas suggests that they are intimately associated in time. Progressively increasing metamorphism will obliterate original relationships, and this is probably true of those areas where remobilized tonaltic basement indicates successively higher grades of metamorphism and partial anatexis. The lode deposits in granitic terrains (Table 1 and Fig. 1) are manifestations of this.

Further studies relating to gold solubility, in particular in carbonate-rich and arsenic-rich solutions, and of fluid inclusions, in particular of their hydrogen and oxygen isotopic compositions and chemistry, may help resolve

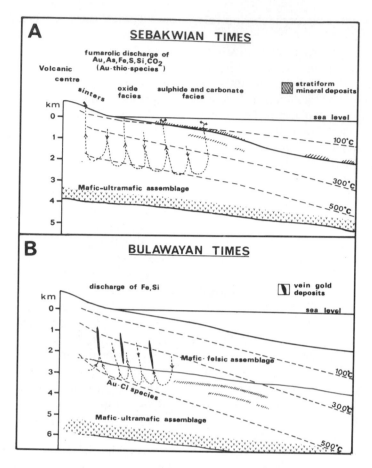

Fig. 6. Generalized model of the evolution of gold deposits in the Archaean, reflected by changing gold-transport mechanism and changing geothermal gradient with time

these problems and lead to a better understanding of metamorphism and gold mineralization in the Archaean. In addition, the interrelationships of the various types of deposits, their local and regional lithostratigraphic controls and the proposed genetic models described above, should provide guides to new ore.

References

Amm, F. L., 1940. 'The geology of the country around Bulawayo', *Bull. geol. Surv. Sth. Rhod.*, No. 35, 307 pp.

Bliss, N. W., 1970. 'The geology of the country around Gatooma', *Bull. geol. Surv. Sth. Rhod.*, No. 64, 240 pp.

Bliss, N. W., and Stidolph, P. A., 1969. 'The Rhodesian Basement Complex: A Review', *Geol. Soc. S. Afr.*, Spec. Pub. No. 2, 305–331.

Collender, F. D., 1964. 'The geology of the Cam and Motor mine Southern Rhodesia', In S. H. Haughton (Ed.), *The Geology of Some Ore Deposits in Southern Africa*, 2, 15–27, Geol. Soc. S. Afr., Johannesburg.

Eales, H. V., 1961. 'Fineness of gold in some Southern Rhodesian mines' (Reply to discussion), *Trans. Instn. Min. Metall.*, 7–, 688–695.

Ferguson, J., Lambert, I. B., and Jones, H. E., 1974. 'Iron sulphide formation in an exhalative-sedimentary environment, Talasea, New Britain, P. N.G.', *Mineral. Deposita*, **9**, 33–47.

Fripp, R. E. P., 1974. 'The distribution and significance of stratabound mineral deposits in Archaean banded iron formation in Rhodesia with particular reference to the Vubachikwe gold mine', Unpub. M. Phil. Thesis, Univ. Rhodesia, Salisbury, Rhodesia, 147 pp.

Fripp, R. E. P., 1975. 'Stratabound gold deposits in Archaean banded iron formation, Rhodesia', *Econ. Geol.* (in press).

Fyfe, W. S., and Henley, R. W., 1973. 'Some thoughts on chemical transport processes, with particular reference to gold', *Miner. Sci. Engng.*, **5**, 295–303.

Goldberg, I., 1964. 'Notes on the relationship between gold deposits and structure in Southern Rhodesia', In S. H. Haughton (Ed.), *The Geology of Some Ore Deposits in Southern Africa*, **2**, 9–13, Geol. Soc. S. Afr., Johannesburg.

Greenberg, R., 1945. 'Applied geology at the Lonely Reef gold mine, Bubi district, Southern Rhodesia', Unpub. M.Sc. Thesis, Univ. Witwatersrand, Johannesburg.

Haas, J. L. Jr., 1971. 'The effect of salinity on the maximum thermal gradient of a hydrothermal system at hydrostatic pressure', *Econ. Geol.*, **66**, 940–946.

Harrison, N. M., 1970. 'The geology of the country around Que Que', *Bull. geol. Surv. Sth. Rhod.*, No. 67, 125 pp.

Hartman, L. W., 1953, 'The geology of the Cam and Motor mine', Unpub. Ph.D. Thesis, Univ. Witwatersrand, Johannesburg, 200 pp.

Henley, R. W., 1971. 'Geochemistry and genesis of Precambrian gold deposits', Unpub. Ph.D. Thesis, Univ. Manchester, 177 pp.

Henley, R. W., 1973. 'Solubility of gold in hydrothermal chloride solutions', *Chem. Geol.*, **11**, 73–87.

Honnorez, J., Honnorez-Guerstein, B., Valette, J., and Wauschkuhn, A., 1973. 'Present day formation of an exhalative sulphide deposit at Vulcano (Tyrrhenian Sea), Part 2: Active crystallisation of fumarolic sulphides in the volcanic sediments of the Baia di Levante', In G. C. Amstutz and A. J. Bernard (Eds.), *Ores in Sediments*, **8**, 139–166.

Hutchinson, R. W., Ridler, R. H., and Suffel, G. G., 1971. 'Metallogenic relationships in the Abitibi belt, Canada: A model for Archean metallogeny', *Trans. Can. Instn. Metall.*, **74**, 106–115.

MacGregor, A. M., 1928. 'The geology of the country around the Lonely mine, Bubi district', *Bull. geol. Surv. Sth. Rhod. No. 11*, 96 pp.

MacGregor, A. M., 1932. 'The Geology of the country around Que Que, Gwelo district', *Bull. geol. Surv. Sth. Rhod.*, No. 20, 113 pp.

MacGregor, A. M., 1947. 'An outline of the geological history of Southern Rhodesia', *Bull. geol. Surv. Sth. Rhod.*, No. 38, 73 pp.

Mantei, E. J., and Brownlow, A. H., 1967. 'Variation in gold content of minerals of the Marysville quartz diorite stock, Montana', *Geochim. Cosmochim. Acta*, **31**, 225–235.

Mehliss, A. T. M., 1963. 'Studies in gold epigenesis', Unpub. Ph.D. Thesis, Univ. Witwatersrand, Johannesburg, 166 pp.

Mehliss, A. T. M., 1964, 'The Dawn gold deposit, Queens area, Bulawayo, Southern Rhodesia', In S. H. Haughton (Ed.), *The Geology of Some Ore Deposits in Southern Africa*, **2**, 41–58. Geol. Soc. S. Afr., Johannesburg.

Phaup, A. E., 1964. 'Gold mines of Southern Rhodesia—Introduction', In S. H. Haughton (Ed.), *The Geology of Some Ore Deposits in Southern Africa*, **2**, 1–7, Geol. soc. S. Afr., Johannesburg.

Phaup, A. E., 1968a. 'Introduction', In D. J. Visser (Ed.), *Symposium on the Rhodesian Basement Complex*, *Annex. Trans. geol. Soc. S. Afr.*, **71**, 3–7.

Phaup, A. E., 1968b. 'Gold in Rhodesia', *Chamb. Mines J.*, **10**, 30–32.

Pretorius, D. A., and Hempkins, W. B., 1968. 'Statistical moments of the frequency distribution of Rhodesian gold mineralization in time and space: a preliminary analysis', *Annex. Trans. geol. Soc. S. Afr.*, **71**, 9–20.

Puchelt, H., 1973. 'Recent iron sediment formation at the Kameni Islands, Santorini (Greece)', In G. C. Amstutz and A. J. Bernard (Eds.), *Ores in Sediments*, **8**, 227–245.

Ridler, R. H., 1970. 'Relationship of mineralisation to volcanic stratigraphy in the Kirkland–Larder Lakes area, Ontario', *Proc. geol. Assoc. Canada*, **21**, 33–42.

Seward, T. M., 1973. 'Thio complexes of gold in hydrothermal ore solutions', *Geochim. Cosmochim. Acta*, **37**, 379–400.

Smith, C. C., and Fripp, R. E. P., 1973. 'The northern contact facies of the Matopos granite, Rhodesia', In. L. A. Lister (Ed.), *Symposium on Granites, Gneisses and Related Rocks, Geol. Soc. S. Afr.*, Spec. Pub., No. 3, 447–453.

Stowe, C. W., 1968. 'The fissure pattern and mineralisation in the Mont d'Or area, Selukwe', In D. J. L. Visser (Ed.), *Symposium on the Rhodesian Basement Complex. Annex. Trans. geol. Soc. S. Afr.*, **71**, 91–102.

Swift, W. H., 1956. 'The geology of the Odzi gold belt', *Bull. geol. Surv. Sth. Rhod.*, No. 45, 44 pp.

Verhoogen, J., Turner, F. J., Weiss, L. E., Wahrhaftig, C., and Fyfe, W. S., 1970. *The Earth*, Holt, Rinehart and Winston, New York, 748 pp.

Weissberg, B. C., 1969. 'Gold–silver ore grade precipitates from New Zealand thermal waters', *Econ. Geol.*, **64**, 95–108.

466

Weissberg, B. C., 1970. 'Solubility of gold in hydrothermal alkaline sulphide solutions', *Econ. Geol.*, **65**, 551–556.

White, D. E., and Waring, G. A., 1963. 'Volcanic emanations', *Prof. Pap. U.S. geol. Surv.*, 440–K, 27 pp.

Wilson, J. F., 1973. 'The Rhodesian Archaean craton—an essay in cratonic evolution', *Phil. Trans. R. Soc. Lond.*, **A273**, 389–411.

Winkler, H. G. F., 1967. *Petrogenesis of Metamorphic Rocks*, Springer–Verlag, Berlin, 237 pp.

Regional Reviews

Archaean Crustal History in North-western Britain

D. R. BOWES

Department of Geology, University of Glasgow, Scotland.

Introduction

Precambrian crystalline rocks, referred to as constituting the Lewisian complex, make up the foreland of the Caledonian orogenic belt in Scotland. They were largely formed in late Archaean times and despite tectonic overprinting and igneous activity in the c.2200 m.y. Inverian episode and the c. 1950–1700 m.y. Laxfordian episode, the rock fabrics mainly reflect the effects of polyphase deformation and metamorphism during the Scourian episode c. 2800–2600 m.y. ago (Bowes and Hopgood, 1973, 1975a). The North-west Highlands (Peach et al., 1907) and the Outer and Inner Hebrides (Jehu and Craig, 1923–34) of Scotland are the main areas of outcrop (Fig. 1(a)). Tectonic slices also occur in the Caledonian belt (Johnstone et al., 1969, Fig. 1) and in parts these show profound effects of tectonic overprinting and metamorphic recrystallization in the Caledonian episode (Ramsay, 1953). Rocks correlated with those of the Lewisian complex occur in Eire, on the island of Inishtrahull (Bowes and Hopgood, 1975b) and in north-western Mayo (Sutton and Max, 1969). Quartzofeldspathic and related gneisses generally predominate except in the part of the North-west Highlands, around Scourie and to the south, where pyroxene granulites and their retrogressed equivalents make up a major portion of the complex (Fig. 1(b); Bowes, 1969).

Geochronology

The age of the dominant metamorphic fabric is 2770 ± 10 m.y. (U–Pb zircon) for gneisses of the northern Outer Hebrides (Pidgeon and Aftalion, 1972), 2690 ± 140 m.y. (Rb–Sr whole rock)—2640 ± 120 m.y. (Pb–Pb whole rock) for the gneisses of central-southern Outer Hebrides (Moorbath et al., 1975), 2710 ± 20 m.y. (U–Pb zircon)—2790 ± 210 m.y. (Rb–Sr whole rock) for gneisses of the southern North-west Highlands (Lyon et al., 1973), 2850 ± 50 m.y. (U–Pb zircon) for gneisses and at least c. 2730 m.y. (U–Pb zircon) for xenocrysts in a granite, both from the northern North-west Highlands (Lyon and Bowes, in press), and 2700 ± 20 m.y. (U–Pb zircon) for pyroxene granulites from the central North-west Highlands (Pidgeon and Bowes, 1972). Initial $^{87}Sr/^{86}Sr$ considerations indicate a crustal history for gneisses in the Loch Maree district going back to $2 \cdot 7$–$2 \cdot 8$ b.y. (Bikerman et al., 1975) while gneiss formation on Inishtrahull was before $2 \cdot 4$ b.y. (Macintyre et al., 1975). The range of ages determined from the U–Pb zircon data, with its small uncertainties, indicates the diachronous nature of this metamorphic event (Fig. 2).

Evidence that both the gneisses and granulites separated from a mantle source only shortly before the metamorphism of the Scourian episode is provided by common lead measurements and initial $^{87}Sr/^{86}Sr$ ratios (Moorbath et al., 1969, 1975; Lyon et al.,

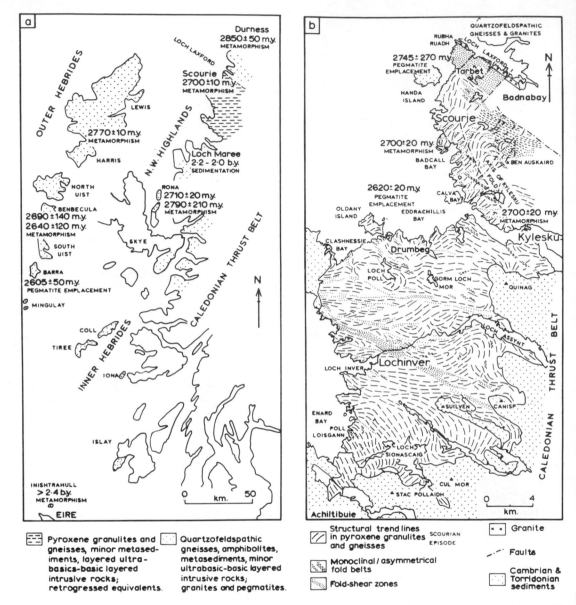

Fig. 1. (a) Distribution and ages of rocks formed in Archaean times in North-western Britain. (b) Structural pattern of deep-level granulite facies rocks and superimposed zones of basement deformation in part of the North-west Highlands of Scotland (after Bowes, 1969 and Sheraton et al., 1973)

1973). There is no isotopic evidence that supracrustal rocks older than *c.* 2850–2900 m.y. formed part of the Lewisian complex and were 'reworked' during the Scourian episode (Sutton, 1973). Rather the rocks predominantly represent juvenile additions to continental crust at, or close to, the measured age, implying continental growth on a considerable scale (Moorbath, 1975). There are

marked similarities in pattern with parts of the Canadian Shield dominated by metavolcanic and associated plutonic rocks where there was extensive new crustal addition, diachronously, between 2750 and 2950 m.y. (Krogh and Davis, 1971). However, such similarities are not evidence of contiguous development.

Studies of the zircon population of a metasedimentary quartzite lens amongst the

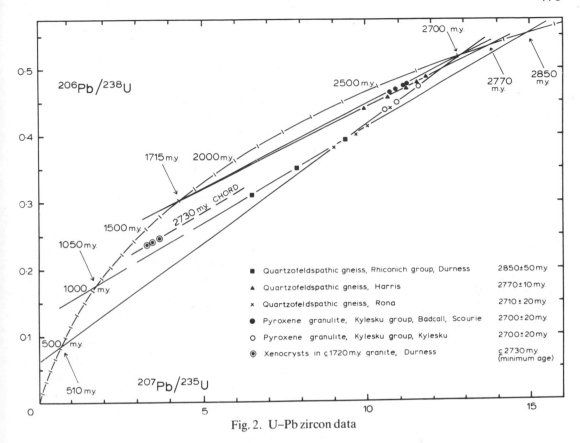

Fig. 2. U–Pb zircon data

c. 2700 m.y. old quartzofeldspathic gneisses of Rona (Fig. 1(a)) indicate apparent inhomogeneity in both its original composition and its response to isotopic disturbances. At least some of the zircons are significantly older than the gneiss-forming event and could have histories going back to 3200 m.y., or earlier. This implies that at least some of the source rocks of the sedimentary unit were in existence before the formation of the Lewisian complex. These could have been comparable with the *c.* 3·5 b.y. old gneisses of north-western Norway (Taylor, 1975), a region with a continuous crystalline basement connection with north-western Britain (Bowes, this volume, Fig. 5). The likelihood of the zircon population being derived both from a mass of >3·2 b.y. crystalline basement and from the *c.* 2·8 b.y. new crustal addition suggests the operation of vertical tectonics co-eval with sedimentation as well as pointing to the new crust being the result of igneous activity at the surface rather than at depth.

Pegmatite emplacement within the Scourian episode, soon after the phase of intense metamorphism, is indicated by ages of 2605 ± 50 m.y. (Barra—Francis et al., 1971; [87]Rb half-life of 5·0 × 10[10] yr used throughout), 2745 ± 270 m.y. (Tarbet—Lyon et al., 1975) and 2620 ± 20 m.y. (Scourie to Lochinver—Evans and Lambert, 1974). Corresponding pegmatites are located in the hinge zones of F_3 folds of the Scourian episode (Fig. 4(d); Khoury, 1968a).

The effects of events during episodes younger than the Scourian episode are reflected by isotopic disturbances apparent from the U–Pb data for *c.* 2700 m.y. old zircons. In one case, a lower intercept of *c.* 1050 m.y. (Fig. 2) suggests the possibility of a mild event, as yet unrecognized by geological criteria, although an event of generally corresponding age is strongly expressed in southern Norway (Brueckner, 1972). The *c.* 500 m.y. disturbance, related to the Caledonian orogeny, is indicative of lead loss, probably due to hydrous solutions circulating through zircon lattices much damaged by radiation. In the areas where the U–Pb data for zircons give evidence of disturbance at this time, there was little or no emplacement of granites and pegmatites in the latter part of the Laxfordian episode. Where the zircon data show a c. 1715 m.y. lower intersection, the isotopic disturbance is related to reheating associated with more common emplacement of these quartzofeldspathic igneous masses. Corresponding effects are shown by Rb–Sr isotopic systems of minerals (Giletti et al., 1961) and where the igneous masses are abundant, the effects are shown by Rb–Sr whole rock systems (Bikerman et al., 1975). This has been demonstrated for the area between Durness and Loch Laxford (Fig. 1; Lyon and Bowes, in press) which includes the type locality for the Laxfordian orogeny (Sutton and Watson, 1951). There the zircons retained c. 2850 m.y. isotopic systems while the Rb–Sr whole rock systems indicate homogenization c. 1750 m.y. ago, the time of abundant granite and pegmatite emplacement. The K–Ar mineral systems record times at which the rocks cooled to blocking temperatures (hornblende—1700 m.y.) and/or were subjected to crustal uplift (biotite—1585 m.y.). As is the case for the Lewisian complex elsewhere, the isotopic data place severe restraints on the extent to which the concept of 'reworking' or 'regeneration' of crystalline rocks can be applied (Watson, 1973). There is no support for the activity of migmatitic fronts in the Laxfordian episode associated with the production of 'virtually new rocks' on a massive scale leaving only relict 'kernels' of the products of Archaean times extant (Sutton and Watson, 1969; Read and Watson, 1975). Rather the major portion of the Lewisian complex has characters consistent with being the products of the Scourian episode with the dominant rock fabric imposed c. 2800–2700 m.y. ago. There has been reheating and tectonic overprinting with new fabrics developed in particular structural situations, such as in elongate fold-shear zones (Fig. 1(b)). For example, in the wide belt south of Loch Laxford, muscovite in a pegmatite dated at 2745 ± 270 m.y. (Rb–Sr whole rock)

recrystallized c. 1750 m.y. ago (Rb–Sr mineral—whole rock isochrons) axial planar to folds deforming the c. 2700 m.y. old metamorphic fabric of the regionally developed granulites (Lyon et al., 1975). Here, as elsewhere, the U–Pb zircon, Rb–Sr whole rock and mineral and K–Ar mineral isotopic systems combine as effective indicators of the nature and intensity of the successive events variably expressed from place to place.

The main suites of younger rocks developed either unconformably on the products of Archaean times, or emplaced into them, are (1) the Loch Maree Group of metasediments and related rocks which were deposited c. 2·2–2·0 b.y. ago, metamorphosed early in the Laxfordian orogenic episode (c. 1950 m.y.) and uplifted c. 1500 m.y. ago in the epeirogenic stage of the Laxfordian cycle (Bikerman et al., 1975), and (2) various igneous suites including a succession of basic minor intrusions, the granitic rocks of c. 1750–1700 m.y. ago and possibly the anorthositic–tonalitic complex of Harris (Dearnley, 1963; Bowes, 1969; van Breeman et al., 1971; Bowes and Hopgood, 1975a).

Lithological Units

The amphibolite facies gneisses that predominate throughout much of the complex are at the same crustal levels as some of the originally deeper crustal level pyroxene granulites. This is the result of tectonic activity in the latter stages of the Laxfordian episode (Lyon et al., 1975). However a c. 2800–2700 m.y. disposition of amphibolite facies metamorphosed acidic lavas, volcanoclastic rocks and derived sediments overlying granulite facies metamorphosed andesitic volcanic rocks, together with basic and acidic material and an apparently minor proportion of sediments, corresponds with the upper parts of dominantly volcanogenic crust-building piles of other Archaean regions, apart from metamorphic grade. This lithologically layered assemblage is referred to as the Gaelic Supergroup. Rocks of corresponding density ranges, which grade down to possible metamorphosed basalts, form 'a Precambrian basement of Lewisian type rocks' under part of the Caledonian belt of Southern Scotland (Powell, 1971). There the 10 km thick lower crustal layer of a 30 km thick crust is interpreted as being anorthositic in nature. However, under the Caledonian foreland of northern Scotland, the Moho is at a remarkably uniform estimated depth of 26 ± 2 km (Smith and Bott, 1975). No comparable anorthositic layer is recorded, but a refractor making up much of the crust, and probably continuous beneath a substantial part of the Caledonian foreland, is tentatively identified with granulite facies rocks of the Lewisian complex.

The exposed granulite facies rocks of the Kylesku group south of Scourie (Fig. 1(b); Fig. 3(a), (b); Khoury, 1968a) consist largely of plagioclase (An_{33-44}), clinopyroxene, orthopyroxene and variable proportions of quartz. Some primary hornblende occurs in parts, as does garnet, scapolite, biotite and opaque ores. For the Lochinver region the data of Evans and Lambert (1974) show the garnet to be Gr_{15} Sp_2 Alm_{50} Pyr_{33}, the brown hornblende to be rich in Ti, Al^{iv} and $Na + K$ and both augite and hypersthene to be highly aluminous with Al_2O_3 in the clinopyroxene up to 6·7 wt.%. The average composition of the granulites between Scourie and Kylesku, with major elements expressed as standard cell and trace elements as parts per million is:

$K_{0·9}$ $Na_{6·7}$ $Ca_{6·6}$ $Mg_{4·8}$ $Mn_{0·1}$ $Fe_{4·6}$ $Al_{16·7}$ $Ti_{0·4}$ $Si_{55·0}$ $(0_{152·9}[OH]_{7·1})_{160}$
Ba 440, Ce 50, Co 48, Cr 150, Cu 48, Ga 18, La 19, Nb 5·5, Ni 88, Rb 3, Sr 460, Zn 83, Zr 115.

This andesitic composition (Bowes et al., 1971) corresponds with that for granulites around Drumbeg (Sheraton, 1970). Generally the rocks show conspicuous depletion in K, Rb, Th and U, very high K/Rb, Ba/Rb and very low Rb/Sr, K/Ba and K/Sr ratios (Sheraton et al., 1973), characters related to purging of 'incompatible' elements during amphibolite facies followed by granulite facies metamorphism deep in the crust during the Scourian episode, probably aided by mantle degassing. These geochemical characteristics

have largely survived subsequent tectonic activity and retrogression, even in the localized zones in which new rock fabrics were formed in Proterozoic times (Tarney et al., 1972). While the average composition is andesitic, there is a wide compositional range, amongst rocks with igneous geochemical characters, with SiO_2 47–76%, MgO 9–0·5% and CaO 13–3%. Within this volcanogenic pile are small masses of distinctive metasediments, including calc–silicate rocks (Barooah, 1970).

The amphibolite facies quartzofeldspathic and related gneisses of the Rhiconich group (Dash, 1969) occur, on geophysical evidence, above pyroxene granulites (Bott et al., 1972). Their average composition (Bowes, 1972) is:

$K_{2·1}$ $Na_{7·5}$ $Ca_{2·7}$ $Mg_{2·0}$ $Mn_{0·03}$ $Fe_{1·9}$ $Al_{15·0}$ $Si_{60·9}$ $Ti_{0·3}$ $P_{0·1}$ $(0_{153·6}[OH]_{6·4})_{160}$
Ba 592, Ce 71, Co 39, Cr 32, Cu 26, Ga 17·5, La 58, Nb 6, Ni 30, Rb 101, Sr 588, Zn 57, Zr 144.

There is considerable variation as shown by a 88–64% SiO_2 range, with the more siliceous varieties probably of metasedimentary derivation, at least in part, together with the small units of zircon-bearing quartzite and calc–silicate rock (Chowdhary et al., 1971; Chowdhary, 1971). Amphibolites of tholeiitic composition (Bowes, 1972), which show a community of structural relations with the c. 2850 m.y. old gneisses also occur (Chowdhary and Bowes, 1972). There are corresponding assemblages in other parts of the North-west Highlands and Islands (Bowes, 1967; Hopgood and Bowes, 1972; Holland and Lambert, 1973; Sheraton et al., 1973), with remnants of metasedimentary units common in parts of the Outer Hebrides (Coward et al., 1969).

Layered ultrabasic–basic igneous complexes (Fig. 1(e)), interpreted as derived from stratiform masses by tectonic disruption, occur throughout the North-west Highlands (Bowes et al., 1964). Amongst the pyroxene granulites they generally occur as small lenses of metaperidotite, metapyroxenite, garnet–pyroxene rock of gabbroic composition (Bowes et al., 1971) and meta-anorthosite

474

Fig. 3. (a), (b) Pyroxene granulite showing early foliation (S_1) deformed by an intrafolial fold (F_2); ordinary light photomicrograph also shows axial planar S_2 foliation and weak S_4 cleavage; 6·5 km SSE of Scourie. (c) Quartzofeldspathic gneiss showing S_1 deformed by F_2 folds with axial planar S_2 foliation: Rona. (d) Quartzite showing F_2 fold and L_2 lineation affected by almost coaxial folds formed c. 1000 m.y. later during the Laxfordian episode; Rona. (e) Layered metaperidotite with wedge bedding; Drumbeg. (f) Metaperidotite (ordinary light) with a band of highly serpentinized olivine with abundant iron ore between serpentinized olivine–hypersthene–hornblende assemblages; Gruinard Bay, north of Loch Maree.

with larger masses at Achiltibuie, Drumbeg and near Scourie. Their layering and structures resembling current and wedge bedding pre-date the granulite facies metamorphism and related tectonics (Fig. 4(b)). Amongst the quartzofeldspathic gneisses they are represented completely, or in part, by talc–tremolite–carbonate–serpentine rocks, but with some original forsteritic olivine–chrome spinel bands still extant (Fig. 3(f)). Peridotitic to anorthositic masses also affected by deformation and metamorphism during the Scourian episode also occur in various parts of the Hebrides (Davidson, 1943; Watson, 1969).

Structural Sequence

Tectonic overprinting during the Laxfordian episode hinders the elucidation of a complete structural sequence of the Scourian episode in the quartzofeldspathic gneiss of much of the North-west Highlands and Hebrides. The early phases of foliation (S_1, S_2) and lineation development and intrafolial fold (F_2) formation (Fig. 3(c), (d)) responsible for the dominant c. 2800–2700 m.y. rock fabric over such large areas (Hopgood and Bowes, 1972; Bowes and Hopgood, 1973, 1975a) are consistently prominent. Locally igneous masses emplaced early in the recognized sequence (Hopgood, 1971) record events of Archaean times (Francis et al., 1971). The successively formed open folds recognized in the deeper level granulite facies terrain as representative of the latter phases of the Scourian episode have yet to be uniquely identified. However, as the most prominent open folds of the banding of the granulites between Kylesku and Scourie have NW-trending hinge zones (Kylesku fold—Fig. 1(b)) and the most prominent folds formed in the quartzofeldspathic gneisses during the Laxfordian episode have NW-trending hinge zones in many regions (e.g. Dash, 1969), at least some of the folds now seen are probably earlier-formed folds that have been tightened. Likewise, some linear features are composite, representing metamorphic mineral growth of the Scourian

episode and coincident microfolds of the Laxfordian episode (Fig. 3(d)).

In the more competent rocks of the deeper levels, a metamorphic banding (S_1; Fig. 3(a), (b)) is evidence of the first deformational phase during which the rocks were at amphibolite facies. The bands are generally thin and distinct from the gross basaltic–andesitic–dacitic compositional layers. Hornblende-rich pods, resulting from metamorphic segregation, and a prominent mineral lineation were also developed but no associated folds have been recognized over large areas. This could be related to the penetrative nature of subsequent deformation or to direct crystallization to foliated metamorphic mineral assemblages of igneous masses which moved by viscous flow. Sheraton et al. (1973) have suggested that the repeated imposition of successively formed flat-lying foliation could be related to progressive flattening of rather plastic deep crust against the sub-horizontal crust-mantle interface, or the lower crust–upper crust interface. Development of intrafolial folds affecting the banding of the layered igneous masses as well as the lithological layering and foliation of the main lithological pile (F_2; Fig. 3(a), (b); Fig. 4(a), (b)), and tectonic fragmentation was partly co-eval with and partly followed by the development of the dominant metamorphic fabric. In places there is a weak axial planar foliation (S_2; Fig. 3(b)). The development of localized asymmetrical to monoclinal folds which deform the dominantly flat-lying metamorphic foliation and intrafolial folds (Fig. 4(c); Khoury, 1968a), with limited metamorphic mineral growth in some hinge zones, marks the initiation of the structural control of uprising gas and magma, with the c. 2600 m.y. pegmatites situated in fold hinge zones (F_3; Fig. 4(d)).

Crustal uplift towards the close of Archaean times, and associated with the latter stages of the Scourian orogenic episode, is indicated by the successive development of upright, open folds with NW–SE (F_4; Fig. 4(e)), E–W (F_5) and approximately N–S (F_6) axial trends and little or no related mineral growth. There was only very limited structurally controlled emplacement of granitic intrusions, due to the

476

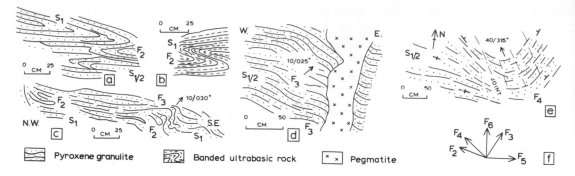

Fig. 4. Nature and relations of structures developed during the Scourian episode in the Scourie–Kylesku region (partly after Khoury, 1968a)

refractory nature of the granulites, and no evidence of faulting to suggest that there was any marked epeirogenic uplift stage associated with the Scourian cycle. However, the abundant vertical NW–SE joints axial planar to the F_4 folds (Fig. 4(e)), including the Kylesku fold (Fig. 3(b)), exerted marked control on basement deformation in the next cycle of crustal development in the region which began early in Proterozoic times. In this, the uprise of geoisotherms associated with localized gneiss formation in the Inverian episode was strongly influenced by the channelling of uprising mantle degassing products in NW–SE-trending narrow zones through the granulites (Khoury, 1968b). No corresponding effects in the higher level amphibolite facies quartzofeldspathic gneisses have been recognized by either geological or isotopic methods (Lyon et al., 1973; Lyon and Bowes, in press). However, the varying expression of basement deformation during the Laxfordian episode has been recognized with the fold-shear zones affecting the deeper level granulites corresponding to hinge zones of folds deforming the quartzofeldspathic gneiss (Bowes 1969).

Crustal Considerations

The correspondence both in sequence of development and orientations of successively formed structural elements resulting from polyphase deformation in the generally coeval Scourian episode of north-western Britain and the Pre-Svecokarelian episode of Fennoscandia (Bowes, 1975) suggest spatial

connection in late Archaean times. This matching of complex structural frameworks implies that the early Precambrian rocks of Britain, and possibly of the continental shelf of the eastern North Atlantic Ocean, are part of the Baltic Shield (Bowes, in press). Whether at this time there was connection with Greenland, which shows such an extensive early Archaean crustal history (Bridgwater et al., 1973) not shown by the Lewisian complex, remains an open question.

Many of the features of Archaean crustal history in north-western Britain are consistent with the operation of hot line tectonics associated with horizontal rolls formed normal to mantle convection that extended deeper than at the present time, associated with the steeper thermal gradients (Sun and Hanson, 1975). In such a regime there would be the uprise of very large volumes of volcanic material along sub-parallel fractures formed diachronously over the c. 2800–2900 m.y. period (to c. 2900–3000 m.y. in Finland—Bowes, this volume), with possible upwelling of older basement between major fracture zones. This dominantly volcanogenic activity would be followed by the formation of major compressional-type fractures and metamorphic activity as known for younger orogenic belts, with both lateral and vertical variation in metamorphic grade comparable with that shown in the granulite–greenstone terrain of Western Australia (Binns, and Gunthorpe, this volume). Such a model would explain the development of the thick and mainly volcanogenic Gaelic Supergroup and its metamorphism and deformation in the Scourian episode only a short interval after the material separated from the mantle. The dominance of flat-lying foliation over such large areas would appear to indicate that any greenstone-belt type tectonics was above the present level of erosion of the rocks now exposed in North-western Britain.

References

Barooah, B. C., 1970. 'Significance of calc-silicate rocks and meta-arkose in the Lewisian complex south-east of Scourie', *Scott. J. Geol.*, **6**, 221–225.

Bikerman, M., Bowes, D. R., and Van Breemen, O., 1975. 'Rb–Sr whole rock isotopic studies of Lewisian metasediments and gneisses in the Loch Maree region, Ross-shire', *J. geol. Soc. Lond.*, **131**, 237–254.

Bott, M. H. P., Holland, J. G., Storry, P. G., and Watts, A. B., 1972. 'Geophysical evidence concerning the structure of the Lewisian of Sutherland, N.W. Scotland', *Jl geol. Soc. Lond.*, **128**, 599–612.

Bowes, D. R., 1967. 'Petrochemistry of some Lewisian granitic rocks', *Mineral. Mag.*, **36**, 342–363.

Bowes, D. R., 1969. 'The Lewisian of Northwest Highlands of Scotland', In M. Kay (Ed.), *North Atlantic—Geology and Continental Drift. Mem. Amer. Assoc. Petrol. Geol.*, **12**, 575–594.

Bowes, D. R., 1972. 'Geochemistry of Precambrian crystalline basement rocks, North-West Highlands of Scotland', *24th Int. geol. Congr.*, **1**, 97–103.

Bowes, D. R., 1975. 'Scotland—Finland Precambrian correlations', *Bull. geol. Soc. Finl.*, **47** (in press).

Bowes, D. R., Barooah, B. C., and Khoury, S. G., 1971. 'Original nature of Archaean rócks of North-West Scotland', *Spec. Publ. Geol. Soc. Aust.*, **3**, 77–92.

Bowes, D. R., and Hopgood, A. M., 1973. 'Framework of the Precambrian crystalline complex of Northwestern Scotland', In R. T. Pidgeon et al. (Eds.), *Geochronology—isotope geology of Scotland*. 3rd European Congress of Geochronologists, East Kilbride, A1–14.

Bowes, D. R., and Hopgood, A. M., 1975a. 'Framework of the Precambrian crystalline complex of the Outer Hebrides, Scotland', *Krystalinikum*, **11**, 7–23.

Bowes, D. R., and Hopgood, A. M., 1975b. 'Structure of the gneiss complex of Inishtrahull, Co. Donegal', *Proc. R. Irish Acad.*, **75B**, 369–390.

Bowes, D. R., Wright, A. E., and Park, R. G., 1964. 'Layered intrusive rocks in the Lewisian of the North-West Highlands of Scotland', *Q. Jl geol. Soc. Lond.*, **120**, 153–192.

Bridgwater, D., Watson, J., and Windley, B. F., 1973. 'The Archaean craton of the North Atlantic region', *Phil. Trans. R. Soc. Lond.*, **A273**, 493–512.

Brueckner, H. K., 1972. 'Interpretation of Rb–Sr ages from the Precambrian and Paleozoic rocks of southern Norway', *Am. J. Sci.*, **272**, 334–358.

Chowdhary, P. K., 1971. 'Zircon populations in Lewisian quartzite, gneiss and granite north of Loch Laxford, Sutherland', *Geol. Mag.*, **108**, 255–262.

Chowdhary, P. K., and Bowes, D. R., 1972. 'Structure of Lewisian rocks between Loch Inchard and Loch Laxford, Sutherland, Scotland', *Krystalinikum*, **9**, 21–51.

Chowdhary, P. K., Dash, B., and Findlay, D., 1971. 'Metasediments in the Rhiconich group of the Lewisian between Loch Laxford and Durness, Sutherland', *Scott. J. Geol.*, **7**, 1–9.

Coward, M. P., Francis, P. W., Graham, R. H., Myers, J. S., and Watson, J., 1969. 'Remnants of an early metasedimentary assemblage in the Lewisian complex of the Outer Hebrides', *Proc. geol. Assoc.*, **80**, 387–408.

Dash, B., 1969. 'Structure of the Lewisian rocks between Strath Dionard and Rhiconich, Sutherland, Scotland', *Scott. J. Geol.*, **5**, 347–374.

Davidson, C. F., 1943. 'The Archaean rocks of the Rodil district, South Harris, Outer Hebrides', *Trans. R. Soc. Edinb.*, **61**, 71–112.

Dearnley, R., 1963. 'The Lewisian complex of South Harris', *Q. Jl geol. Soc. Lond.*, **119**, 243–307.

Evans, C. R., and Lambert, R. St J., 1974. 'The Lewisian of Lochinver, Sutherland; the type area for the Inverian metamorphism', *Jl geol. Soc. Lond.*, **130**, 125–150.

Francis, P. W., Moorbath, S., and Welke, H. J., 1971. 'Isotopic data from Scourian intrusive rocks on the Isle of Barra, Outer Hebrides, northwest Scotland', *Geol. Mag.*, **108**, 13–22.

Giletti, B. J., Moorbath, S., and Lambert, R. St J., 1961. 'A geochronological study of the metamorphic complexes of the Scottish Highlands', *Q. Jl geol. Soc. Lond.*, **117**, 233–264.

Holland, J. G., and Lambert, R. St J., 1973. 'Comparative major element geochemistry of the Lewisian of the mainland of Scotland', In R. G. Park and J. Tarney (Eds.), *The Early Precambrian of Scotland and Related Rocks of Greenland*, Univ. of Keele, 51–62.

Hopgood, A. M., 1971. 'Structure and tectonic history of Lewisian Gneiss, Isle of Barra, Scotland', *Krystalinikum*, **7**, 27–60.

478

Hopgood, A. M., and Bowes, D. R., 1972. 'Application of structural sequence to the correlation of Precambrian gneisses, Outer Hebrides, Scotland', *Bull. Geol. Soc. Am.*, **83**, 107–128.

Jehu, T. J., and Craig, R. M., 1923–34. 'Geology of the Outer Hebrides, Parts I-V', *Trans. R. Soc. Edinb.*, **53**, 419–441, 615–641; **54**, 46–89; **55**, 457–488; **57**, 839–874.

Johnstone, G. S., Smith, D. I., and Harris, A. L., 1969. 'Moinian Assemblage of Scotland', In M. Kay (Ed.), *North Atlantic—Geology and Continental Drift. Mem. Amer. Assoc. Petrol. Geol.*, **12**, 159–180.

Khoury, S. G., 1968a. 'The structural geometry and geological history of the Lewisian rocks between Kylesku and Geisgeil, Sutherland, Scotland', *Krystalinikum*, **6**, 41–78.

Khoury, S. G., 1968b. 'Structural analysis of complex fold belts in the Lewisian north of Kylesku, Sutherland, Scotland', *Scott. J. Geol.*, **4**, 109–120.

Krogh, T. E., and Davis, G. L., 1971. 'Zircon U–Pb ages of Archean metavolcanic rocks in the Canadian Shield', *Carnegie Institution Year Book*, **70**, 241–224.

Lyon, T. D. B., and Bowes, D. R., In press. 'Rb–Sr, U–Pb and K–Ar isotopic study of the Lewisian complex between Durness and Loch Laxford, Scotland', *Krystalinikum*, **13**.

Lyon, T. D. B., Gillen, C., and Bowes, D. R., 1975. 'Rb–Sr isotopic studies near the major Precambrian junction, between Scourie and Loch Laxford, Northwest Scotland', *Scott. J. Geol.*, **11** (3) (in press).

Lyon, T. D. B., Pidgeon, R. T., Bowes, D. R., and Hopgood, A. M., 1973. 'Geochronological investigation of the quartzofeldspathic rocks of Rona, Inner Hebrides', *Jl geol. Soc. Lond.*, **129**, 389–404.

Macintyre, R. M., Van Breemen, O., Bowes, D. R., and Hopgood, A. M., 1975. 'Isotopic study of the gneiss complex, Inishtrahull, Co. Donegal', *Sci. Proc. R. Dub. Soc. A.*, **5**, 301–309.

Moorbath, S., 1975. 'Evolution of Precambrian crust from strontium isotopic evidence', *Nature Lond.*, **254**, 395–398.

Moorbath, S., Powell, J. L., and Taylor, P. N., 1975. 'Isotopic evidence for the age and origin of the "grey gneiss" complex of the southern Outer Hebrides, Scotland', *Jl geol. Soc. Lond.*, **131**, 213–222.

Moorbath, S., Welke, H., and Gale, N. H., 1969. 'The significance of lead isotope studies in ancient high-grade metamorphic basement complexes, as exemplified by the Lewisian rocks of Northwest Scotland', *Earth Planet. Sci. Lett.*, **6**, 245–256.

Peach, B. N., Horne, J., et al., 1907. 'The Geological Structure of the North-West Highlands of Scotland', *Mem. geol. Surv. Scotland*.

Pidgeon, R. T., and Aftalion, M., 1972. 'The geochronological significance of discordant U–Pb ages of oval-shaped zircons from a Lewisian gneiss from Harris, Outer Hebrides', *Earth Planet. Sci. Lett.*, **17**, 269–274.

Pidgeon, R. T., and Bowes, D. R., 1972. 'Zircon U–Pb ages of granulites from the Central Region of the Lewisian of northwestern Scotland', *Geol. Mag.*, **109**, 247–258.

Powell, D. W., 1971. 'A model for the Lower Palaeozoic evolution of the southern margin of the early Caledonides of Scotland and Ireland', *Scott. J. Geol.*, **7**, 369–372.

Ramsay, J. G., 1963. 'Structure and metamorphism of the Moine and Lewisian rocks in the northwestern Caledonides', In M. R. W. Johnson and F. D. Stewart (Eds.), *The British Caledonides*, Oliver and Boyd, Edinburgh, 143–175.

Read, H. H., and Watson, J., 1975. *Introduction to Geology*, Vol. 2, 'Earth History—Part I', Macmillan, London.

Sheraton, J. W., 1970. 'The origin of the Lewisian gneisses of Northwest Scotland with particular reference to the Drumbeg area, Sutherland', *Earth Planet. Sci. Lett.*, **8**, 301–310.

Sheraton, J. W., Skinner, A. C., and Tarney, J., 1973. 'The geochemistry of the Scourian gneisses of the Assynt district', In R. G. Park and J. Tarney (Eds.), *The Early Precambrian of Scotland and Related Rocks of Greenland*, Univ. of Keele, 13–30.

Sheraton, J. W., Tarney, J., Wheatley, T. J., and Wright, A. E., 1973. 'The structural history of the Assynt district', In R. G. Park and J. Tarney (Eds.), *The Early Precambrian of Scotland and Related Rocks of Greenland*, Univ. of Keele, 31–43.

Smith, P. J., and Bott, M. H. P., 1975. 'Structure of the crust beneath the Caledonian foreland and Caledonian belt of the north Scottish shelf region', *Geophys. J. R. Astr. Soc.*, **40**, 187–205.

Sun, S. S., and Hanson, G. N., 1975. 'Evolution of the mantle: geochemical evidence from alkali basalts', *Geology*, **2**, 297–302.

Sutton, J., 1973. 'The first three-quarters of the geological record in Britain', In R. G. Park and J. Tarney (Eds.), *The Early Precambrian of Scotland and Related Rocks of Greenland*, Univ. of Keele, 9–12.

Sutton, J., and Watson, J., 1951. 'The pre-Torridonian metamorphic history of the Loch Torridon and Scourie areas in the North-West Highlands and its bearing on the chronological classification of the Lewisian', *Q. Jl geol. Soc. Lond.*, **106**, 241–295.

Sutton, J., and Watson, J., 1969. 'Scourian–Laxfordian relationships in the Lewisian of northwest Scotland', *Geol. Assoc. Canada Spec. Pap.*, **5**, 119–128.

Sutton, J. S., and Max, M. D., 1969. 'Gneisses in the northwestern part of County Mayo, Ireland', *Geol. Mag.*, **106**, 284–290.

Tarney, J., Skinner, A. C., and Sheraton, J. W., 1972. 'A geochemical comparison of major Archaean gneiss units from Northwest Scotland and East Greenland', *24th Int. geol. Congr.*, **1**, 162–174.

Taylor, P. N., 1975. 'An early Precambrian age for migmatitic gneisses from Vikan i Bø, Vesterålen, North Norway', *Earth Planet. Sci. Lett.*, **27**, 35–42.

Van Breemen, O., Aftalion, M., and Pidgeon, R. T., 1971. 'The age of the granitic injection complex of Harris, Outer Hebrides', *Scott. J. Geol.*, **7**, 139–152.

Watson, J., 1969. 'The Precambrian gneiss complex of Ness, Lewis, in relation to the effects of Laxfordian regeneration', *Scott. J. Geol.*, **5**, 269–285.

Watson, J., 1973. 'Effects of reworking on high-grade gneiss complexes', *Phil. Trans. R. Soc. Lond.*, **A273**, 443–455.

Archaean Crustal History in the Baltic Shield

D. R. BOWES

Department of Geology, University of Glasgow, Scotland.

Geochronology

The main occurrence of Archaean rocks in the Baltic Shield is in the north-western U.S.S.R. and the adjacent parts of eastern and northern Finland (Fig. 1). The geological evolution of the region given by Lobach-Zhuchenko et al. (1972), in which K–Ar mineral dates are much used together with dates derived by other methods, separates three developmental stages during Archaean times with age ranges of 3300–3000 m.y., 3000–2800 m.y. and 2800–2600 m.y. The products of the earliest stage include relatively high level oligoclase granites and deeper level supracrustal rocks (Kola Series), some now at granulite facies from which pyroxene has yielded an age of 3300 m.y. by K–Ar methods. The region, which is characterized by intense tectonic activity in mobile belts, has the representatives of different crustal levels because of major tectonic movements in Proterozoic times which followed pre-existing lines. The pre-3000 m.y. old rocks are intruded by c. 2900–2800 m.y. igneous rocks.

Basic magmatism is recorded as separating the first and second developmental stages with the 3000–2800 m.y. time span being characterized by further development of supracrustal rocks, the 'reactivation' of previously formed masses by tectonic activity and plutonism and rapid stabilization associated with 2800 ± 100 m.y. tectonic and metamorphic activity. The third stage is recorded as resulting in a regional activation of the whole Kola Peninsula–White Sea–Karelia region with widespread magmatism and metamorphism, although granulites of the Kola Series remain unaffected. There was much granite emplacement 2500 ± 100 m.y. ago, particularly in pre-existing linear tectonic zones. The rocks show penetrative fabrics like those of earlier-formed granitic gneisses and the c. 2500 m.y. age is interpreted as the time of remobilization.

The oldest dated material in Finland is detrital zircon from a quartzite that is 3000 m.y. old. The rock forms part of the Kuhmo schist belt in which the rocks were metamorphosed to amphibolite facies during the Pre-Svecokarelian episode 2800 m.y. ago and again metamorphosed, with disturbance of some isotopic systems, during the Svecokarelian episode c. 1900 m.y. ago. However, despite these metamorphic events, the old age is preserved, as is a c. 2900 m.y. age for detrital zircons from the Ilomantsi schist belt itself affected by the 2800 m.y. metamorphic event.

The age of the quartzofeldspathic gneisses and related rocks that cover large areas of eastern and northern Finland is generally c. 2800 m.y. (Figs. 2, 3) but with some 2700 m.y. These ages are based on U–Pb data (Wetherill et al., 1962; Kouvo, 1964–75; Kouvo and Tilton, 1966; Kouvo and Sakka 1974; Kouvo, pers. comm.) and record the intense penetrative metamorphic activity of the first two deformational phases of the Pre-Svecokarelian episode (Pl. Fig. 4(a)–(c)).

481

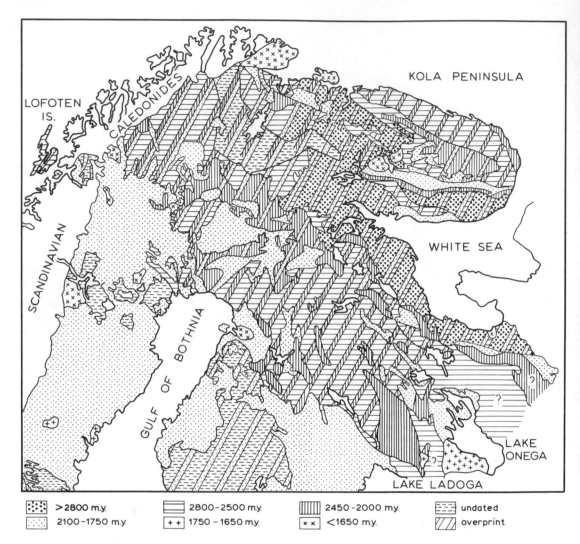

Fig. 1. Geochronological map of part of the Baltic Shield (after Lobach-Zhuchenko, 1972, Fig. 58)

U–Pb zircon ages of the basement complex associated with the granulite belt in Lapland are also 2800 m.y. to 2500 m.y., analogous to the whole rock Rb–Sr isochron age, while the 1900 m.y. age shown by mineral isochrons and by sphene point to disturbance during the Svecokarelian episode. Disturbance is also shown by zircons in gneisses of the c. 2800 m.y. old basement which occur in the cores of fold interference domes formed during the Svecokarelian episode, as at Kuopio and Sotkuma (Figs. 2, 3). However, in some of the basement masses involved in the later orogenic episode, zircons retain their 2800 m.y. ago, as at Kotalahti.

A pegmatite in northern Finland contains 2700 m.y. old zircons and a Re/Os age for a disseminated molybdenum deposit in quartzofeldspathic gneiss SE of Nurmes (Fig. 2) is also 2700 m.y. The time of termination of the Pre-Svecokarelian episode is probably c. 2600 m.y. The age of the unconformity separating the products of this episode from the overlying sedimentary sequence is given by Kahma (1973) as 2500 m.y., with the mean of lead model ages of sulphides in the cover sequence at Outokumpu being 2300 m.y.

The Precambrian crystalline rocks of Norway form part of the Baltic Shield being connected to its main outcrops beneath much, or

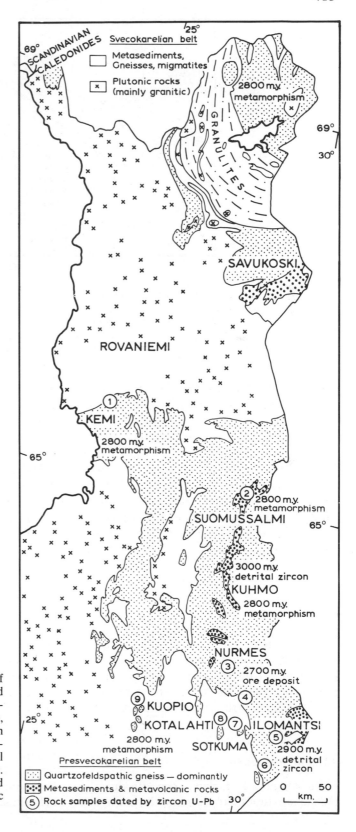

Fig. 2. Distribution and age of Archaean rocks in Finland (based largely on Kahma 1973, Map 1 and Pre-Quaternary Rocks of Finland map, 1971); samples 1–9 correspond with those on Fig. 3. The granulites of Lapland yield Rb–Sr whole rock and mineral isochron and zircon ages of *c.* 2100 m.y. (Meriläinen, 1975), possibly associated with the emplacement of anorthositic complexes

484

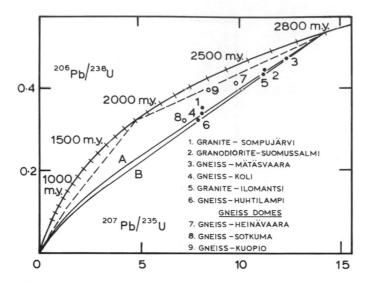

Fig. 3. Concordia diagram showing uranium–lead ratios for zircons from rocks formed in the Pre-Svecokarelian episode in eastern and northern Finland (locations on Fig. 2); curves A and B give ratios expected if samples have lost lead by continuous diffusion (after Kouvo and Tilton, 1966, Fig. 2)

possibly all the Scandinavian Caledonides through which it is seen in various 'windows' (Fig. 6). In the north-western part, in the Vesterålen Isles, migmatitic gneisses of overall intermediate composition have been dated (Pb/Pb whole rock isochron) at 3460 ± 70 m.y. (Taylor, 1975). Previously Heier and Compston (1969) interpreted Rb–Sr isotopic data for high-grade metamorphic rocks from both the Vesterålen and Lofoten Isles as indicative of the effects of a $c.$ 2800 m.y. event followed by other events at 1800 m.y. and 1550 m.y. Events at $c.$ 2850–2700 m.y., associated with the Scourian episode and at $c.$ 1950–1700 m.y., associated with the Laxfordian episode, are recorded in northwestern Britain which is connected to northwestern Norway by a continuity of crystalline basement (Talwani and Eldholm, 1972). Geochronological details for this part of the Baltic Shield are given by Bowes (this volume).

Lithological Units

In much of Finland the products of Archaean times are dominantly gneisses of granodioritic–quartz dioritic affinities (Fig. 4(a)–(c), (e), (f); Harme, 1949; Eskola, 1963). Most are considered to be of igneous derivation (Simonen, 1971), but some are interpreted as metamorphosed sediments (Preston, 1954). Amphibolites (Fig. 4(d)) are interpreted as metamorphosed basaltic lavas and tuffs with agglomeratic, porphyritic and amygdaloidal features remaining in parts. Leptitic rocks are attributed to the metamorphism of the weathering products of basic masses as well as of acidic volcanic and pyroclastic rocks (Nykänen, 1971). Ultrabasic, basic and intermediate minor intrusions are of limited occurrence (Fig. 5(b)). In marked contrast to the Svecokarelian belt is the limited occurrence of large unfoliated plutonic granitic intrusions (Fig. 2) although there are many quartzofeldspathic veins whose emplacement was controlled by structures developed in the latter part of the Pre-Svecokarelian episode (Fig. 5). The long narrow synformal schist belts consist of a supracrustal assemblage of basaltic volcanic rocks, including pillow lavas, banded ironstones, serpentinites and clastic sediments including greywackes, arkoses and subordinate conglomerates.

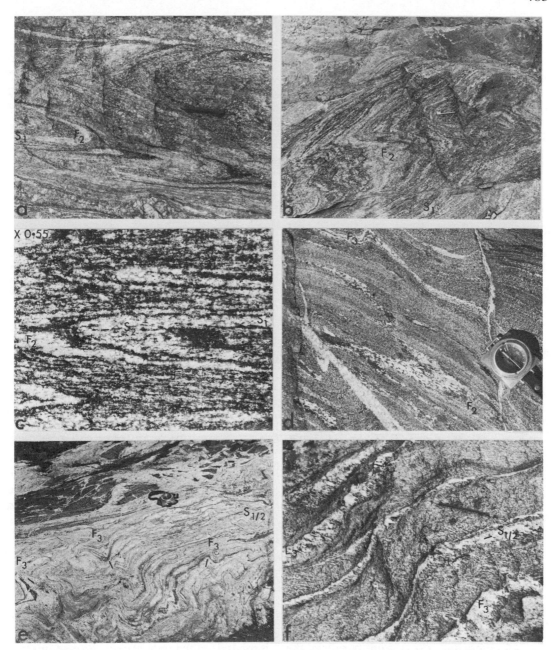

Fig. 4. (a) Quartzofeldspathic gneiss of migmatitic aspect with early foliation (S_1) deformed by intrafolial folds (F_2) now much dissected; 35 km WSW of Ilomantsi. (b) F_2 folds with limited axial planar S_2 foliation deforming S_1 foliation in quartzofeldspathic gneiss; between Kotalahti and Kuopio. (c) Photomicrograph (ordinary light) of intrafolial F_2 folds, with some S_2 foliation, deforming S_1 foliation in quartz–feldspar–hornblende–biotite gneiss; Kuopio. (d) Intrafolial F_2 fold with S_2 foliation and $S_{1/2}$ foliation deformed about F_3 asymmetrical folds and cut by a N–S-trending granite vein; 9 km SW of Kuopio. (e) NW-, N- and NE-trending F_3 folds deforming composite foliation $S_{1/2}$; Kotalahti. (f) Open F_3 folds with prominent L_3 lineation deforming composite foliation $S_{1/2}$; 50 km NW of Sotkuma

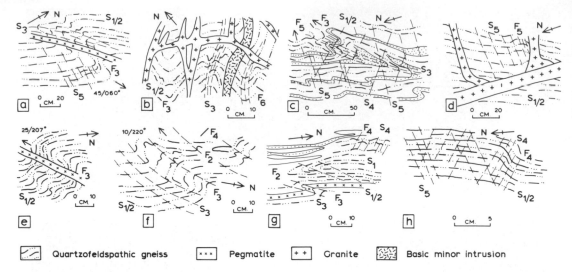

| Quartzofeldspathic gneiss | Pegmatite | Granite | Basic minor intrusion |

Fig. 5. Structural elements and igneous masses formed in the Pre-Svecokarelian episode in Finland; (a)–(d)—Kuopio, (e)–(h)—Sotkuma

Structural Sequence

There is correspondence in the structural sequence of the Pre-Svecokarelian episode in the main outcrop of North Karelia and in both the fingers of basement amongst younger cover rocks affected by the Svecokarelian episode and the basement which forms the cores of fold interference domes. This is shown by the structural sequence in the rocks west of Ilomantsi, north-west of Sotkuma and at Sotkuma and Kuopio (Fig. 2). In each region early phases of foliation (S_1, S_2), formed under amphibolite facies conditions and associated with intrafolial folds (Fig. 4(a)–(d)), are responsible for the dominant rock fabric. In the hinge zones of the intraolial folds deforming S_1, S_2 is axial planar (Fig. 4(c), (d)), but not dominant (Fig. 4(b), (c)). Generally the foliation is composite ($S_{1/2}$) and parallels the lithological layering. Together these fabric elements have been deformed in successive later deformational phases with the formation of asymmetrical to symmetrical and generally upright, open folds (Fig. 5). In parts, folds of the third deformational phase show NE-, N- and NW-trending hinge zones (Fig. 4(e)) but regionally the NNE to NE trend predominates (Fig. 4(d); Fig. 5(a)–(c), (e), (f)) and in parts is reinforced by a prominent lineation (L_3; Fig.

4(f)). Both acidic and basic minor intrusions are structurally controlled by F_3 fold hinge zones (Fig. 5(a), (b), (e)). Locally developed S_3 cleavage, as well as the L_3 lineation are deformed about open, upright F_4 folds with NW-trending hinge zones (Figs. 5(f)) with the development of a new axial planar cleavage (S_4; Fig. 5(g), (H)). Folds of both the fifth and sixth deformational phases are upright and generally very open and contribute to complex fold interference patterns (Fig. 5(c)). Commonly they have controlled (Fig. 5(d)) or partially controlled (Fig. 5(b)) the emplacement of granitic veins. The association of F_6 folds with N–S granitic veins is clear in parts and the common occurrence of N–S veins without obvious structural control (Fig. 4(d); Fig. 5(d)) is related to the regional expression of this fold set.

While the structural sequence and orientations of successively formed fold structures has only been determined for part of the Archaean basement (Fig. 6), their combined use, associated with isotopic data indicative of comparability of age, suggest a correlation between the Pre-Svecokarelian episode of Finland and the Scourian episode of north-western Britain (Bowes, 1975).

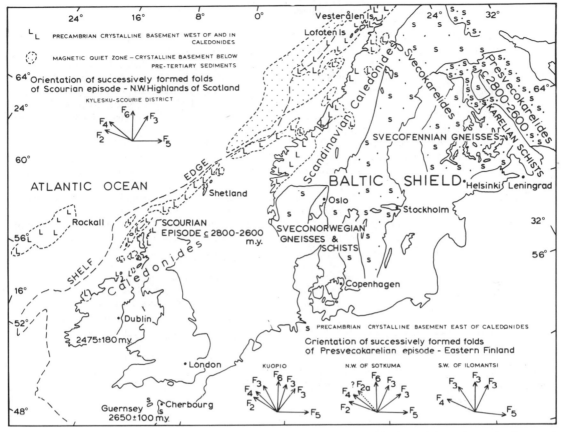

Fig. 6. Distribution of Precambriam crystalline rocks of the Baltic Shield, ages from isotopic data and orientations of successively formed folds of the Pre-Svecokarelian episode, Finland, and of the Scourian episode, Scotland (based on Bowes, 1975, Fig. 6)

Crustal Considerations

Evidence for the existence of continental crust in early Archacan timcs is provided by the geological record of north-western Norway and north-western U.S.S.R. Whether these were parts of nucleii or of a continuous or near-continuous mass under the region of the Baltic Shield has yet to be established. However, the existence of a moderately stable crust by 3000 m.y. ago, followed by extensive igneous activity, including considerable volcanism, c. 3000–2900 m.y. ago, in association with hot line tectonics related to mantle convection (Sun and Hanson, 1975), might account for the occurrence of fragments of pre-3000 m.y. old basement, the existence of 3000–2900 m.y. old zircons now found in metasediments, and the intense metamorphic and tectonic activity of the Pre-Svecokarelian

episode c. 2800–2600 m.y. ago (cf. Bowes, this volume). The existence of long, narrow synformal belts of metasediments and volcanogenic rocks could be indicative of the present level of erosion being near the base or mainly below any greenstone belt-type assemblages and any associated tectonic regime of the Pre-Svecokarelian belt.

Large areas of the Baltic Shield have remained stable since the end of Archaean times with the c. 2800 m.y. old fabrics developed in the Pre-Svecokarelian episode remaining essentially intact. The disturbance of some isotopic systems has resulted in overprinting (Fig. 2). However, this has been dominantly related to mineral rather than whole rock systems and associated with tectonic overprinting rather than 'reworking' or regeneration' of older crystalline material.

488

References

Bowes, D. R., 1975. 'Scotland–Finland Precambrian correlations', *Bull. geol. Soc. Finl.*, **47** (in press).

Eskola, P., 1963. 'The Precambrian of Finland', In K. Rankama (Ed.), *The Precambrian*, Wiley, New York, Vol. 1, 145–263.

Härme, M., 1949. 'On the stratigraphical and structural geology of the Kemi area, northern Finland', *Bull. de la Comm. géol. de Finl.*, **147**.

Heier, K., and Compston, W., 1969. 'Interpretation of Rb–Sr age patterns in high grade metamorphic rocks, north Norway', *Norsk geol. Tidsskr.*, **49**, 257–283.

Kahma, A., 1973. 'The main metallogenic features of Finland', *Geol. Surv. Finl. bull.*, **265**.

Kouvo, O., 1965–75. 'Geological Survey of Finland', *Annual Reports on Activities for the Years 1964–1974*, Otaniemi.

Kouvo, O., and Sakko, M., 1974. 'Examples of disturbances in U–Pb systematics', *Internat. Meeting for Geochron., Cosmochron. & Isotope Geol.*, Paris, Abstr.

Kouvo, O., and Tilton, G. R., 1966. 'Mineral ages from the Finnish Precambrian', *J. Geol.*, **74**, 421–442.

Lobach-Zhuchenko, S. B., and Kratz, K. O., et al., 1972. *Geochronological Boundaries and the Geological Evolution of the Baltic Shield* (In Russian), Precambrian Research Inst., U.S.S.R. Acad. Sci., Leningrad.

Meriläinen, K., 1975. 'The granulite complex and adjacent rocks in Lapland, Northern Finland', *Geol. Surv. Finl. Bull.* In press.

Nykänen, O., 1971. *Geological Map of Finland*, Sheet 4241—Kiihtelysvaara; Explanation to the map of rocks, Geological Survey of Finland.

Preston, J., 1954. 'The geology of the Precambrian rocks of the Kuopio district', *Ann. Acad. Sci. Fennicae*, Ser. A, III, **40**.

Simonen, A., 1971. 'Das finnische Grundgebirge', *Geol. Rundsh.*, **60**, 1406–1421.

Sun, S. S., and Hanson, G. N., 1975. 'Evolution of the mantle: geochemical evidence from alkali basalts', *Geology*, **3**, 297–302.

Talwani, M., and Eldholm, O., 1972. 'Continental margin off Norway: a geophysical study', *Bull. Geol. Soc. Amer.*, **83**, 3575–3606.

Taylor, P. N., 1975. 'An early Precambrian age for migmatitic gneisses from Vikan i Bø, Vesterålen, North Norway', *Earth Planet. Sci. Lett.*, **27**, 35–42.

Wetherill, G. W., Kouvo, O., Tilton, G. R., and Gast, P. W., 1962. 'Age measurements on rocks from the Finnish Precambrian', *J. Geol.*, **70**, 74–88.

The Archaean of Equatorial Africa: A Review*

L. Cahen and J. Delhal

Musée royal de l'Afrique centrale, Tervuren, Belgium.

J. Lavreau

Bureau de Recherches géologiques et minières, Paris, France.

1. Introduction

In most countries of Equatorial Africa (Fig. 1) the term 'Archaean' has never been used or is no longer in use. A number of well-defined ages at *c.* 2·6 b.y. date the massive emplacement of various forms of granitoid rocks over most of the area under review following which characteristic mobile belts make their appearance. The earliest of these belts, the Luizian of Kasai (S. Zaïre) was folded and metamorphosed at 2·47 b.y.; the majority of the early Proterozoic belts are, however, dated at 2·0–2·2 b.y. The age of *c.* 2·6 b.y. is thus a milestone in the early geological history of Equatorial Africa. All terranes earlier than 2·55 b.y. are here included in the Archaean.

Although vast areas of Archaean rocks are recognizable in both N.E. and S.W. Equatorial Africa, only restricted portions of these areas are truly cratonic: a great deal of the southwestern Archaean has been reworked during early Proterozoic tectonometamorphic events 2·47 to *c.* 2·0 b.y. old; the north-eastern Archaean has subsequently undergone two main periods of reworking during early and late Proterozoic times.

---

* Except when otherwise mentioned, the geochronological data quoted are derived from Rb–Sr isochrons with λ 87Rb: 1·39 10⁻¹¹ y⁻¹ and from U–Pb data using the tables published by Stieff et al. (1959).

2. The Archaean of S.W. Equatorial Africa

Large areas of Archaean rocks stretch with some interruptions around the southern and western borders of the Central Zaïre (Congo) Basin from E. Kasai to S. Cameroun (Fig. 1). Important portions of this Archaean are overlain by Phanerozoic cover rocks. Proterozoic fold-belts separate the Archaean of Kasai–Angola from that of Cameroun–Gabon, etc.

The Archaean of Kasai and of N. Angola

Of this vast area, the portion between 21°30'–24°E and 6°–8°S (see Fig. 2) has been studied in detail (Delhal and Ledent, 1971, 1975a; Delhal, 1975). Parts of this area have been cratonized immediately after their formation in late Archaean times but others have been affected by early Proterozoic tectonometamorphic events (2·47 b.y. and *c.* 2·0 b.y.) (Delhal and Ledent, 1973a); later events have had but little influence.

The Archaean of Kasai has a general NE to ENE trend and comprises two major units: the Dibaya granite and migmatite complex and the Kasai–Lomami gabbro–norite and charnockite complex, and a less obvious possibly composite unit which is the basement to the tectonometamorphic events characterizing the former complexes.

489

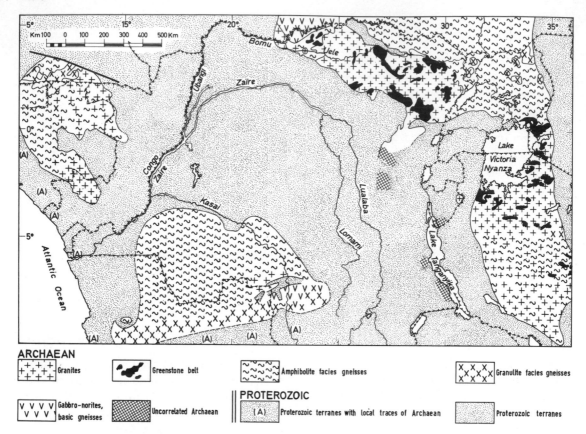

Fig. 1. Geological sketch map of the Archaean of Equatorial Africa (Phanerozoic cover removed)

The Dibava Complex consists almost exclusively of granitic to tonalitic migmatites (2·7 b.y.) and of related calc–alkaline granites (2·65 b.y.); the granites are considered to be the end-products of the migmatization. Isotopic data indicate that the migmatites have had a longer crustal history than the granites and that they are in fact products of the reworking of formations distinctly older than 2·7 b.y. There is no evidence of important juvenile granitic material; the average granodioritic–tonalitic composition of the migmatites is therefore considered to approximate that of the basement formation. A few septa of amphibole–pyroxene rocks are metamorphosed relics of basic bodies in this basement. The Dibayan reworking consists chiefly of cataclasis and recrystallization which occurred in approximately the same bathymetric conditions as those in which the gneisses of the basement were formed, so that

it is not always possible to distinguish the mineral components and textural features of the migmatites which belong to the original gneissification from those produced by Dibayan recrystallization. To the south, the migmatites grade into the anatectic Malafudi granite (2·65 b.y.). The cataclasis characteristic of the migmatites is only slightly marked in the northern portion of the granite massif and not at all further south.

The Kasai–Lomami gabbro–norite and charnockite Complex. As its name suggests, this complex situated to the south of the Dibaya Complex (see Fig. 2) consists of two main components. The northern part contains various types of gabbros, generally noritic, with or without hornblende, sometimes with garnet. The gabbros grade from anorthositic to quartz-bearing types. The granoblastic texture common to this heterogeneous collection of basic rocks and the frequency of metamorphic minerals such as garnet, point to a high-

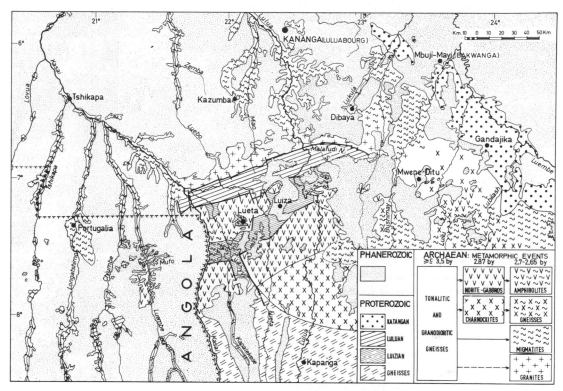

Fig. 2. Geological map of Central Kasai

grade plutonic metamorphism common to all rock types of a basic complex rather than to the discrete intrusion of various basic rock types. Originally this basic complex may have been intrusive or volcanic. The southern acidic part of the Kasai–Lomami Complex contains a great variety of facies, the more common being: enderbites (true charnockites are exceptional); metasedimentary granulites with garnet, sillimanite, and very exceptionally cordierite; pink hololeucocratic alaskitic rocks consisting of mesoperthite and quartz; metadolerites frequently garnetiferous, and locally, quartz veins. All these rocks are locally gold-bearing.

As is the case with the more northerly basic portion, all these rocks have undergone a high-grade metamorphism in the granulite facies which has been dated at 2·87 b.y. A generalized cataclasis locally accompanied by retrograde metamorphism occurred before 2·47 b.y. and is considered to be of 'Dibayan' age.

The metadolerites which cut the gneissic texture of the enderbites and granulites have to a great extent preserved, despite high-grade metamorphism and granoblastic recrystallization, their primary doleritic texture as well as chilled margins. Preservation of doleritic texture and of intrusive relationship of the dolerites implies that metamorphism affected a rigid body poor in water content and hardly subject to folding. These characteristics strongly suggest that the dolerite dykes were hypabyssal intrusions in a pre-charnockitization crystalline basement. The hololeucocratic rocks which define a 2·87 b.y. isochron may be the only products mobilized during the charnockitization process. The $^{87}Sr/^{86}Sr$ ratios of the other rocks (enderbites and garnet-bearing granulites) plot above the isochron which points to their having had a longer crustal history which appears to be traceable to 3·5–3·6 b.y.

The relations between the Dibaya Complex and the Kasai–Lomami Complex cannot be ascertained in the area where these units were first studied as they are separated by Proterozoic belts. To the east, in the region of Mwene Ditu (7°S; 23°30′–24°E), their mutual relations can be defined. Two types of rocks are associated in this transition zone: biotite–tonalitic gneisses and hololeucocratic rocks with mesoperthites. The tonalitic gneisses have the same composition as the gneisses of the Dibaya Complex but are not affected by the Dibayan cataclasis. The mesoperthitic rocks are identical to those of the charnockitic portion of the Kasai–Lomami Complex to the west and are the same age as the latter (c. 2·85 b.y.). This association suggests that the mesoperthitic rocks are either high-temperature granitic intrusions or more likely that they represent the first *in situ* remobilization during a regional metamorphic transformation which either was interrupted or occurred at the bathymetric frontier between the non-charnockitic gneisses of the north and the charnockitic rocks outcropping in the south. This implies that the tonalitic gneisses are an important remnant of the basement to the charnockitization on the one hand and to the Dibayan migmatization on the other.

The existence of a crystalline *Basement* which underwent high grade metamorphism is therefore indicated by (1) the presence of metadolerite intrusions in rocks which reached crystalline state before the charnockitization of 2·87 b.y.; (2) the fact that the granulite facies paragneisses and enderbites have relatively high $^{87}Sr/^{86}Sr$ ratios, which may indicate derivation from rocks about 3·6 b.y. old; (3) the presence of tonalitic gneisses as probable basement both in the Dibaya Complex and in the transition zone described above; (4) the existence, in a more recent graben-like structure (7°45′S; 22°10′E) of the Upper Luanyi gneiss in which pegmatites are imprecisely dated as $\geqslant c$. 3·5 b.y. (Delhal and Ledent, 1973b).

The Dibayan event (2·7–2·65 b.y.) is in fact characterized by the generalized cataclasis and recrystallization; the amphibolite facies of the Dibaya Complex is not characteristic of the Dibayan event alone, and in some cases is an inheritance of an earlier event; only the

southern Malafudi Granite (2·65 b.y.) is free from this generalized cataclasis. The charnockitic event (2·87 b.y.) is characterized by its granulite facies metamorphism without contemporary deformation.

These characteristics (and geochronological data that support them) can be extended to the West into the Kwango region of S. Zaïre and in N. Angola (Delhal and Ledent, 1971; Delhal, 1973; Mendes and Vialette, 1972). In these regions the Dibaya Complex and the charnockitic Complex are known with reasonable continuity, the latter disappearing under the Mesozoic coastal sediments near Salazar (9°30′S; 14°30′E) (Delhal and Fieremans, 1964; Delhal and Torquato, 1975).

Between this latitude and Gabon, there are indications that, underneath the early Proterozoic terranes older than 2·15 b.y., remnants of the Archaean are discerned either for structural reasons (Bas Zaïre) or for similar reasons accompanied by imprecise Pb–α apparent ages (Congo). In the North of Congo, in Gabon, in Equatorial Guinea and in South Cameroun, the influence of the Proterozoic belts is much weaker and little deformed Archaean outcrops widely. Two major units are recognized. The 'Complexe calco-magnesien' of South Cameroun and its continuation in Equatorial Guinea and Gabon (Fig. 1) contains medium-grained enderbites and charnockites with on the one hand rocks containing two pyroxenes, amphibole and biotite and on the other hololeucocratic rocks. Feldspar is antiperthitic, perthitic and occasionally mesoperthitic; K-feldspar is orthoclase with appearance of microline twinning probably in relation to the cataclastic texture. Metadolerites, with pyroxene, amphibole and garnet have preserved much of their original doleritic texture. All in all, these rocks are very comparable with those of the Kasai–Lomami Complex. The age of the South Cameroun rocks is 2·9 b.y. or older (Delhal and Ledent, 1975b). The charnockitic massif is associated with heterogeneous granitoid rocks which extend into the Congo Republic under the name Duchaillu Massif. Two main granite generations are recognized: an older (or grey) granitoid is considered to be the result of granitization of the pyroxene gneiss complex and the other rocks; it is associated with gneisses and micaschists, and consists of biotite ankeritic granite, granodiorite and quartzdiorite. This generation has been dated at *c.* 2·7 b.y. (Bonhomme, 1969). Pegmatites accompany the younger (or pink) potassic granite; they are 2·01 b.y. old (Vachette, 1964; Bonhomme, 1969).

The analogy with the Kasai region is striking in facies, apparent relations of rocks units, and age. This Archaean massif is to the north limited by a series of faults which separate it from the Precambrian terranes which make up most of the rest of Cameroun.

3. The Archaean of N.E. Equatorial Africa

In this area, the Archaean stretches almost continuously from 22°E to the Uganda–Kenya border and from S. Sudan to 8°S in S. Tanzania. It is described under four headings: (Λ) The Uele–Nyanza granite–greenstone belt; (B) the Ganguan schist belt and the Bomu Complex; (C) the West Nile gneissic Complex; (D) the central and southern portions of the Tanzania shield and the Dodoman schist belt.

A. *The Uele–Nyanza granite–greenstone belt*

This important Archaean feature is approximately 1400 km by 350 to 550 km (Fig. 1). It has a general WNW direction and stretches from 4°N–24°E in the Lower Uele region of N. Zaïre to 5°S–35°E in Tanzania, south and east of Lake Victoria–Nyanza. This belt has been separated in two portions by several successive Proterozoic mobile belts in the vicinity of the Western Rift. The ratio of schist and greenstone belts to granitoid rocks is about 1 in 5 or 6.

Several generations of *granitoid* rocks are known:

(a) 3·1 b.y.—granodiorites and tonalites, near Kilo, and near Moto (Fig. 3) (Lavreau and Ledent, 1975); 3·15 b.y.—biotite from granite, E. of Nandi Escarpment (Cahen and Snelling, 1966).

(b) 2·8 b.y.—equigranular granites to inequigranular quartz monzonites (Dodson et al., 1975).

(c) 2·5–2·55 b.y.—medium to coarse-grained inequigranular quartz monzonites, Mumias, Buteba, Masaba (Dodson et al., 1975), Maragoli (Cahen and Snelling, 1966); near Kilo, near Moto, Bondo (Lavreau and Ledent, 1975).

(d) *c.* 2·07 b.y.—potassic granites and pegmatites (Lavreau and Ledent, 1975).

Furthermore, gneissic rocks cover large areas of the granitoid massif in N. Zaïre. Their relationship to the granitoid generations of 3·1 b.y. and 2·5–2·55 b.y. are at present unknown. The three generations in N. Zaïre (a), (c) and (d) were more or less clearly recognized by the first explorers and are now confirmed by age determinations. In the East African part of the belt, generations (b) and (c) are distinguished by their relative relations to the schist and greenstone belts.

In N. Zaïre, the *schist and greenstone* belts have, with the exception of the Ganguan (see paragraph 3B), all been described under the name Kibalian. It is now clear that all outcrops of 'Kibalian' rocks do not form a single unit. Three facies are at present distinguished (Fig. 3) (Lavreau and Ledent, 1975).

(1) The *Eastern Kibalian* of Kilo and Moto which from bottom to top comprises a Basic Group: massive and fine-grained amphibolites at Kilo, spilitic and volcanodetrital rocks with pillow structures at Moto; a Basic to Intermediate volcanic Group: one or two levels of volcanic rocks (albite-, sericite-, chlorite-, talc-schists) with some quartzite members; a Metasedimentary and Metavolcanic Group: quartzites, quartzphyllites more or less enriched in ferro-magnesian minerals (chlorite, carbonates) and iron oxides grading into banded ironstones; these metasedimentary formations alternate with metavolcanic rocks. The Eastern Kibalian is cut by granitoid rocks 3·1 b.y. and 2·5 b.y. old.

(2) The *Western Kibalian* of the Upper Tele and between the Uele river and Isiro (Raucq, 1974; Sekirsky, 1954) (Fig. 3) has

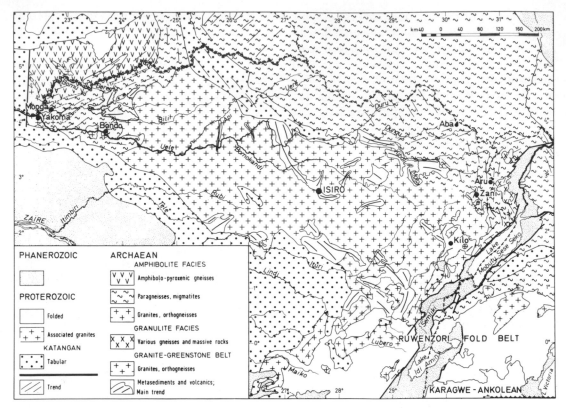

Fig. 3. Geological sketch of N.E. Zaïre

another composition: from bottom to top: clastics, sericite-schists, ferruginous at the bottom, locally amphibolite and banded ironstones; banded ironstones and ferruginous grits; ferruginous grits, schists, quartzites, locally tuff. This Western Kibalian is certainly earlier than granitoid generation (d) c. 2·07 b.y., and is thought to rest upon an undated basement consisting of gneiss with some micaschists.

(3) The *Kibalian of West-Nile (Uganda) and Zani (Zaïre)* consists, respectively, of hornblende-schists, possibly of calc–silicate and graphite-schists, resting uncomformably upon the Western Grey (= Aruan) gneisses (Hepworth, 1964), and of sericite-schists, ferruginous schists and quartzites overlain by chlorite-schists, resting unconformably upon the Aru–Zani gneisses (Lavreau, 1975).

The Aruan gneisses have been correlated with similar rocks from the Southern Base-ment Complex of Uganda dated 2·6± 0·05 b.y. (see 3C). The age of the Aru–Zani gneisses is unknown.

Whereas the banded ironstones of the Eastern Kibalian are in relatively thin bands and are generally insignificant from an economic point of view, those of the Western Kibalian may constitute massive iron ore in economic quantities.

The East African part of the granite–greenstone belt is the northern portion of the Tanzania Shield. The *Nyanzian* is the most important schist and greenstone component. It is very similar to the Eastern Kibalian. From bottom to top it comprises (Pulfrey, 1938; Shackleton, 1946) a Basic volcanic Group: chiefly pillow lavas, locally a banded ironstone member; Intermediate to Acid volcanic Group: rhyolites and sub-acid lavas with intercalated tuffs and agglomerates; Greywacke Group: greywackes and gritty andesitic tuffs near the top, banded ironstone near the base;

Slaty and Andesitic group: andesitic rocks near the top, banded ironstone and tuffaceous, silty and ferruginous slates at a lower level. The banded ironstones have the same characteristics as those of the Eastern Kibalian.

The Nyanzian is cut by the Migori granite of 2·8 b.y. (Dodson et al., 1975). It is probably older than c. 3·15 b.y. (biotite K:Ar age of a probably post-Nyanzian granite: Cahen and Snelling, 1966; Sanders, 1965).

No accurate thickness has been measured either for the Eastern Kibalian or for the Nyanzian. An estimate of 7500 m has been proposed for the latter (Grantham et al., 1945).

The *Kavirondian* of East Africa is an essentially metasedimentary unit younger than the Migori granites of 2·8 b.y. and older than the Buteba and Maragoli granites of 2·55 b.y. It comprises from bottom to top: a Lower division of conglomerates and arkoses with siltstones; a Middle division of mainly pelitic rocks (slates, mudstones and phyllites) with some fine-grained grits (occasionally this division consists of volcanic rocks); and an Upper division of arkoses with pebble bands and conglomerates. An estimate of 1500–3000 m has been given for the thickness of the Kavirondian (Hitchen, 1936; Shackleton, 1946).

B. *The Ganguan Schist Belt and the Bomu Complex*

To the west and north-west of the Uele–Nyanza belt, a vast expanse of high-grade terranes which in Zaïre are known as the Bomu Complex extend into the Central African Republic (Lavreau, 1975). In Zaïre, the Bomu Complex is unconformably overlain by the *Ganguan schist belt* which is folded on EW passing into NS axes. The lithostratigraphy is poorly known; the following rock types are encountered in ascending order: sericitic quartzites and quartz phyllites, quartz-poor talcschists, sericite schists, chlorite schists, black schists and phyllites. Greenstones of uncertain position are often observed. Gold-bearing quartz reefs cutting a predominantly phyllitic formation of the Ganguan near Kule–Matundu (4°15′W–23°45′E) contain galenas whose lead is of practically identical isotopic composition to that of the Barberton leads in the Swaziland Sequence (Cahen, 1967).

The earliest and most characteristic component of the *Bomu Complex* is a vast mass of amphibole–pyroxene rocks on both sides of the river Bomu. These are medium-grained, equigranular homogenous or layered rocks composed of plagioclase and hornblende often associated with one or two pyroxenes and garnet. They have undergone various types of retrograde metamorphism. The Bereme gneisses, mica quartzites, biotite–quartzofeldspathic gneisses and amphibole gneisses form a NW-trending synclinorial structure superposed on the earlier basic unit. To the N and to the S, these gneisses link up with, or are deflected by, NE-trending rocks. To the south, these are composed of quartz-rich gneisses, amphibole–garnet gneisses, granitic gneisses in alternation. These, the Monga gneisses, appear to be associated with mica schists and subordinate quartzites. They show signs of several deformations and have undergone retrograde metamorphism to a variable degree.

Rather important granitoid masses, the 'granite hétérogène concordant' intrude the earlier basic unit in the Central African Republic. They are generally elongated in a NW direction which may well be the characteristic trend of this earlier unit. Near Yakoma (4°05′N, 22°25′E) this NW trend passes into a NE trend and digits of refolded 'granite hétérogène concordant' alternate with gneisses similar to the Monga gneisses of Zaïre. In the latter, several granite intrusions are known, e.g. the coarse-grained porphyritic Bondo granite (2·55 b.y.) which is to be considered as an 'outlier' of the granitoid portion of the Uele–Nyanza belt (Lavreau and Ledent, 1975). Various rock units of uncertain affinities are also known.

C. *The West Nile Gneiss Complex*

This name is used in a wide sense for all the gneisses which outcrop to the north of the Uele–Nyanza belt in N. Zaïre, Central African Republic, Uganda, and Southern Sudan. The eastern part of this complex is known as the

'Basement complex of Uganda' and extends slightly into N.E. Zaïre. The succession of Archaean events originally recognized in the West Nile district, has been found to apply to the rest of Uganda (Hepworth, 1964; Hepworth and MacDonald, 1966; Hepworth, 1967). From oldest to youngest: (1) Deposition of sedimentary and volcanic rocks affected by the following events; (2) First Watian tectonic phase (E–W trends, flat-lying folds) in granulite facies (2·88 b.y., Leggo, 1974); (3) Second Watian tectonic phase (E–W trends with steep axial planes) in granulite facies conditions; (4) intrusion of dolerite dykes; (5?) possible deposition of sedimentary and volcanic rocks; (6) Aruan grey-gneiss amphibolite facies metamorphism and retrograde metamorphism of Watian (2·55 b.y., Leggo, 1974); tight folding on NNE-plunging axes. Near the northern border of the granite–greenstone belt, the Archaean trends tend towards E–W and gneisses with such a foliation, just north of Lake Victoria–Nyanza are 2·55–2·58 b.y. old and are often considered to represent granitized Nyanzian (Leggo, 1974).

The western part of the West Nile Gneiss Complex outcrops in N. Zaïre, Sudan and Central African Republic; it is not well known but appears to differ from the eastern part in several respects. No trace of the Watian tectonic group has been recognized. Two Archaean gneissic formations are known; the oldest, the *Garamba Gneisses*, are biotite gneisses (more rarely muscovite gneisses) generally garnet-bearing with some two-mica gneisses with kyanite and garnet-bearing amphibolite. Between the Sudan–Zaïre border (4°25′N, 28°30′E) and the Garamba (3°50′N–29°E) the foliation trends vary from N10W to N60W from NW to SE. Dips are steep 60°–90° to the NE. These biotite and garnet gneisses are overlain by biotitic gneisses which outcrop eastwards and to the south of the Garamba river (Lavreau, 1975).

Near the Zaïre–Uganda border, in the Upper Kibali region, underlying the E–W trending Eastern Kibalian synform of Zani, the *Aru–Zani–Gneisses* are composed of a lower unit of homogeneous biotitic or granitic gneisses sometimes coarse-grained, and an upper one consisting of muscovite-gneisses, biotite–garnet-schists alternating with quartzites, and amphibole schists. The gneisses are cut by the Mt Aleza (2°30′N; 30°30′E) porphyritic biotite-granite (2·07 b.y.); they have a foliation varying from NW to NE with medium dips and are structurally parallel to the Aruan gneisses.

The correlation of these different gneisses with the Aruan was hitherto considered most likely (Lepersonne, 1974). The existence of several gneissic units opens up other possibilities. The Kibalian of West Nile, though considered by Hepworth (1964) to rest upon the Aruan, might in fact rest upon the undated Aru–Zani gneisses. The relationship of the Garamba and Aru–Zani gneisses to the gneissic facies of the granitoid mass of the Uele–Nyanza belt and to the Bereme and Monga gneisses of the Bomu complex (see 3B) are at present unknown. Two more recent schist and gneiss units have been detected; they are of Proterozoic age (Lavreau, 1975).

D. *The Central and Southern Portions of the Tanzania Shield and the Dodoman Schist Belt*

The central portion consists mainly of migmatites, whereas the southern portion contains migmatites, granites and the Dodoman schist belt. On the south-eastern and eastern border of the 'shield' the *Western Granulites* like the Dodoman schists are intruded and replaced by granites. Near Itiso (5°45′S–36°15′E) two concordant K–Ar mineral ages yield 3·08 b.y. for ultrabasic rocks of the granulite grade (Snelling, 1966). The *Dodoman schist belt* contains metasediments and metavolcanic rocks intruded by granites and pegmatites. It includes quartzites, sometimes ferruginous and banded, sericite-schists, quartz-schists and talc-chlorite- and corundum-bearing rocks, amphibolites and hornblende gneisses. These rocks are accompanied by altered ultrabasic rocks; the banded granitic gneisses pass into unfoliated granite (Hepworth, 1972).

The Dodoman granites and gneisses are 2·6 b.y. (Wendt et al., 1972), the pegmatites,

2·55 b.y. (Webster et al., 1957). The migmatic rocks of the central portion of the shield are 2·50–2·56 b.y. old (Dodson et al., 1973).

To the south of Dodoma (6°10'S–35°45'E) the influence of the early Proterozoic Usagaran belt is manifest and post-Archaean intrusives of various types, among which potassic granites are 2·0 b.y. old (Wendt et al., 1972).

4. Conclusions

As Fig. 1 shows, the Congo (Zaïre) basin is for more than 75% surrounded by Archaean terranes. To the SE the gap, almost entirely occupied by several Proterozoic belts, is only about 500 km wide when account is taken of various Archaean terranes which outcrop under the early Proterozoic Rusizian belt (2·0–2·1 b.y.). The NW gap is possibly more apparent than real: the existence of a large number of granulite facies massifs points to the possible existence of widespread Archaean outcrops. However, the very sparse geochronological data yield practically only pan-African mineral ages. In the Congo (Zaïre) basin, seismic surveys point to the existence of a gneissic basement beneath much of the flat-lying late Proterozoic beds which underly the Phanerozoic cover. It can thus be presumed that the whole area under review was one mass in late Archaean times.

The history of the two portions distinguished in the Archaean of Equatorial Africa since c. 2·85 is certainly very similar, as the granulite facies conditions of 2·85 (±0·05) b.y. are everywhere followed by massive granitoid or migmatite formation in the amphibolite facies at 2·65 (±0·1); the greenstone facies is also the seat of granitic intrusion at the latter time.

In Kasai, the granite-forming processes of 2·65 b.y. are, however, immediately followed by the cratonization of most of the Dibayan Complex (biotites give a 2·65 b.y. age, as do the rocks themselves). The Luizian which developed to the south of the Complex was metamorphosed at 2·47 b.y. without granitization processes; it can be considered as the first Proterozoic fold belt in this region. Numerous indications of an event at 2·1 b.y. are known in the same area, probably corresponding to a later Proterozoic belt.

No equivalent of the Luizian fold belt has yet been recognized nor geochronologically detected in the north-eastern Archaean region, where granite emplacement proceeds until 2·55 b.y. after the deposition of the Kavirondian sediments, and where the first Proterozoic belt is dated at 2·1 b.y. Nevertheless, if the Kibalian rocks of Zani and West-Nile are actually discordant upon 2·55 b.y. old Aruan gneisses, they could represent a counterpart of the Luizian.

Acknowledgements

We thank D. Ledent for numerous unpublished age determinations and J. Lepersonne for a certain number of photogeological interpretations.

References

Bonhomme, M., 1969. 'Compléments à la géochronologie du Bassin de France-ville et de son environnement', Résumé commun., 5e Coll. Geol. Afric., Clermont-Ferrand, 1969, *Ann. Fac. Sci. Univ. Clermont*, **41**, 85–88.

Cahen, L., 1967. 'Least radiogenic terrestrial leads: a comment', *Earth Planet. Sci. Lett.*, **3**, 171–172.

Cahen, L., and Snelling, N. J., 1966. *The Geochronology of Equatorial Africa*, Amsterdam.

Delhal, J., 1973. 'Contribution à la connaissance géologique du Nord-Est Lunda (Angola)', *Mus. roy. Afr. centr., Rapp. ann. 1972 Dépt. Géol. Min.*, 53–62.

Delhal, J., 1975. 'The early Precambrian of Kasai' (in preparation).

Delhal, J., and Fieremans, C., 1964. 'Extension d'un grand complexe charnockitique en Afrique centrale', *C.R. Acad. Sci. Paris*, **259**, 2665–2668.

498

Delhal, J., and Ledent, D., 1971. 'Ages U/Pb et Rb/Sr et rapports initiaux du strontium du Complexe gabbro-noritique et charnockitique du bouclier du Kasai (République démocratique du Congo et Angola)', *Ann. Soc. géol. Belg.*, **94**, 211–221.

Delhal, J., and Ledent, D., 1973a. 'L'âge du Complexe métasédimentaire de Luiza, Région du Kasai, Zaïre', *Ann. Soc. géol. Belg.*, **96**, 289–300.

Delhal, J., and Ledent, D., 1973b. 'Résultats de quelques mesures d'âges radiométriques par la méthode Rb/Sr dans les pegmatites et les gneiss de la Haute-Luanyi, région du Kasai (Zaïre)', *Mus. roy. Afr. centr., Rapp. ann. 1972 Dépt. Géol. Min.*, 102–103.

Delhal, J., and Ledent, D., 1975a. 'L'âge du Complexe granitique et migmatitique de Dibaya (région du Kasai, Zaïre) par les méthodes Rb/Sr et U/Pb', *Ann. Soc. géol. Belg.*, in press.

Delhal, J., and Ledent, D., 1975b. 'Données géochronologiques sur le Complexe calco-magnésien du Sud-Cameroun', *Mus. roy. Afr. Centr., Rapp. ann., 1974,* Dépt. Géol. Min., pp. 71–76.

Delhal, J., and Torquato, J., 1975. 'Extension vers l'ouest jusqu'à la région de Salazar (Angola) du Complexe charnockitique du Kasai (Zaïre)', *Etude géochronologique*, in press.

Dodson, M. H., Bell, K., Gledhill, A. T., and Shackleton, R. M., 1975. 'Age differences between Archaean cratons of Eastern and Southern Africa', in press.

Dodson, M. H., Bell, K., and Shackleton, R. M., 1973. 'Archaean geochronology of East Africa', *Fortschr. Miner.*, **50**, Beiheft 3, I–II, 67 (Abstract).

Grantham, D. R., Temperley, B. P., and McConnell, R. B., 1945. 'Explanation of the geology of degree sheet n° 17 (Kahama)', *Tanganyika Terr., Bull. Dept. Lands & Mines, Geol. Div.*, **15**.

Hepworth, J. V., 1964. 'Explanation of the geology of sheets 19, 20, 28 and 29 (Southern West Nile)', *Geol. Surv. Uganda*, Rept. 10.

Hepworth, J. V., 1967. 'The photogeological recognition of ancient orogenic belts in Africa', *Quart. J. Geol. Soc. London*, **123**, 253–292.

Hepworth, J. V., 1972. 'The Mozambique orogenic belt and its foreland in northeast Tanzania', *J. Geol. Soc.*, **128**, 5, 461–500.

Hepworth, J. V., and MacDonald, R., 1966. 'Orogenic belts in northern Uganda basement', *Nature*, **210**, 726–727.

Hitchen, C. S., 1936. 'Geologic survey of n°2 Mining Area, Kavirondo, *Interim report and map of NW Quadrant'*, *Mining and Geol. Dept., Kenya, Rept. 6.*

Lavreau, J., 1975. 'Etat des connaissances sur les séries gneissiques du Haut-Zaïre septentrional', *Mus. roy. Afr. centr., Rapp. ann.*, 1974 Dépt. Géol. Min., pp. 77–88.

Lavreau, J., and Ledent, D., 1975. 'Etablissement du cadre chronologique du Kibalien (Zaïre)', To be published in *Ann. Soc. Geol. Belg.*

Leggo, J. P., 1974. 'A geochronological study of the basement complex of Uganda', *J. Geol. Soc.*, **130**, 263–277.

Lepersonne, J., 1974. 'Notice explicative de la Carte géologique du Zaïre au 1/2.000.000', *Rép. Zaïre, Dept. Mines, Dir. Géol.*

Mendes, F., and Vialette, Y., 1972. 'Le Precambrien de l'Angola', 24th *Int. Geol. Congr., Montreal*, Sect. 1, 213–220.

Pulfrey, W., 1938. 'Geological survey of n°2 Mining Area. Interim report and map of the SW Quadrant', *Mining and Geol. Dept., Kenya, Rept. 7.*

Raucq, P., 1974. Relations et signification des minerais hématitiques et couches itabiritiques dans une série précambrienne métamorphique. *Ac. roy. Sc. O.–M., Bull. séances*, **3**, 408–411.

Sanders, L. D., 1965. 'Geology of the contact between the Nyanza shield and the Mozambique belt in W. Kenya. (Min. Nat. Res.)', *Geol. Surv. Kenya, Bull. 7.*

Sekirsky, B., 1954. 'Contribution à l'étude de la constitution géologique de l'Uele, *Ann. Soc. géol. Belg.*, **77**, B. 189–199.

Shackleton, R. M., 1946. 'Geology of the Migori gold belt and adjoining areas', *Geol. Surv. Kenya, Rept. 10.*

Snelling, N. J., 1966. Unpublished results.

Stieff, L. R., Stern, T. W., Seili Oshoro, and Senftle, E. E., 1959. 'Tables for the calculation of lead isotope ages', *Geol. Surv. Prof. Paper*, **334A**.

Vachette, M., 1964. 'Ages radiométriques des formations cristallines d'Afrique Equatoriale', *Ann. Fac. Sci. Clermont–Ferrand*, **25**, 31–38.

Webster, R. K., Morgan, J. W., and Smales, A. A., 1957. 'Some recent Harwell analytical work on geochronology', *Trans. Amer. Geoph. Union*, **38**, 543–546.

Wendit, I., Besang, C., Harre, W., Kreuzer, H., Lenz, H., and Müller, P., 1972. 'Age determinations of granitic intrusions and metamorphic events in the Early Precambrian of Tanzania', *Int. Geol. Congress, 24th Montreal 1972*, **1**, 295–314.

The Wyoming Archean Province in the Western United States

KENT C. CONDIE

Department of Geoscience, New Mexico Institute of Mining and Technology, Socorro, New Mexico 87801.

Abstract

The Wyoming Province is composed principally of gneiss–migmatite terranes (~60%) and granitic batholiths (~30%). The structure varies from areas in which only one period of open folding is found to more typically, areas which have undergone two or more periods of folding. Regional metamorphic grade ranges from upper greenschist to locally, lower granulite facies with amphibolite–facies terranes dominating. The andalusite–sillimanite facies series is most widespread with the kyanite–sillimanite series found only along the north-western and southern borders. A complex shear zone separates the Wyoming Province from younger Precambrian terranes to the south-east.

Gneiss–migmatite terranes appear to have formed chiefly from clastic sedimentary parent rocks. Granitic rocks reflect four different tectonic settings and probably several different origins. Greenstone belts may have developed in plate–tectonic environments although the nature of the basement rocks on which they were erupted remains obscure. Quartzite–mica schist terranes along the north-western and southern borders appear to reflect stable tectonic settings and probably a lower geothermal gradient than that characterizing most of the province.

Radiometric ages indicate that some sialic crust existed in the Wyoming Province by ⩾3·2 b.y. ago. The first orogenic event is recorded in the Bighorn, Teton, Sierra Madre, and Granite Mountains at 2·9 b.y. while the major orogenic event (including most granite emplacement) at 2·6–2·7 b.y. is recorded throughout other parts of the province. Diabase dyke swarms were intruded intermittently between 2·9 and 0·74 b.y. A heating event probably associated with the Hudsonian Orogeny (1·6–1·8 b.y.) in surrounding terranes reset mineral ages to ⩽1·8 b.y. in the outer parts of the province.

Introduction

Archean rocks in the western United States occur in the cores of young mountain ranges in Wyoming and adjacent states (Fig. 1) and have been called the Wyoming Province (Condie, 1969a). Because of inadequate exposure, the outer boundaries of the province are poorly known. Relict Archean ages (2·0–2·5 b.y.) from basement rocks in North Dakota and Manitoba (Burwash, et al., 1962; Goldich, et al., 1966), however, indicate that the Wyoming Province may be an arm of the Superior Province in Canada. Of the two known types

499

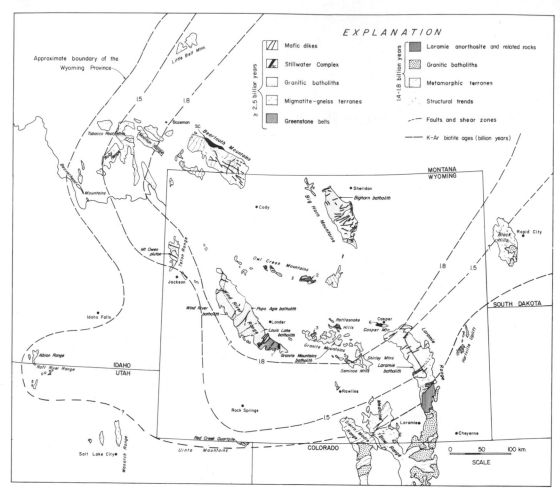

Fig. 1. Geologic map of the Wyoming Province. Greenstone belts: 1—South Pass, 2—Owl Creek, 3—Western Granite Mountains, 4—Rattlesnake Hills, 5—Seminoe Mountains, 6—Casper Mountain

of Archean crust, i.e. high-grade metamorphic and granite–greenstone terranes, the Wyoming Province is an example of the latter.

Rock Types

Gneisses and Migmatites

Gneisses and migmatites compose more than 60% of the Wyoming Province. The principal occurrences are in the Beartooth, Bighorn, northern Wind River, and Laramie Mountains (Fig. 1). Granitic gneiss is most common with varying amounts of biotite schist, migmatite, agmatite, and amphibolite (and hornblende gneiss) occurring in most areas. Remnants of quartzite, iron formation, ultramafic rocks, hornfels, and marble occur in some gneiss terranes. The major mineral assemblages in the province including the Red Creek Quartzite in north-western Utah are summarized in Table 1. The most common assemblage, accounting for 75% of the gneisses and migmatites is the two-feldspar–quartz–biotite assemblage. In the Cherry Creek and Pony Series of south-western Montana amphibolite, biotite schist, anthophyllite schist, quartzite, and marble become much more abundant relative to gneiss and migmatite (Reid, 1957; Heinrich and Rabbitt, 1960). Because field relations are not well known and presently are a subject of disagreement, it is

Table 1. Metamorphic mineral assemblages in the Wyoming Province

---

Majority of the Province

Biotite–quartz–plagioclase–K–Feldspar
Biotite–quartz–plagioclase
Biotite–hornblende–quartz–plagioclase
Biotite–hornblende–quartz–plagioclase–K–feldspar
Biotite–quartz–plagioclase ± garnet ± cordierite
Biotite–quartz–plagioclase–staurolite ± garnet ± cordierite ± anthophyllite
Biotite–quartz–plagioclase–K–feldspar–sillimanite ± cordierite ± garnet
Biotite–quartz–sillimanite
Biotite–quartz–plagioclase ± andalusite ± muscovite

South-west Montana Only

Hornblende–plagioclase–quartz–garnet
Biotite–quartz–plagioclase–garnet ± kyanite(± sillimanite*)
Biotite–quartz–plagioclase–K–feldspar–staurolite ± garnet(± sillimanite*)
Biotite–quartz–plagioclase–K–feldspar ± garnet(± sillimanite*)
Quartz–plagioclase–garnet–anthophyllite ± actinolite

Red Creek Quartzite Only

Quartz–muscovite
Quartz–muscovite–garnet ± kyanite
Quartz–muscovite–garnet–staurolite ± kyanite

---

* Replaces kyanite and biotite.

not clear if all the Precambrian rocks exposed in this area are Archean in age.

Most evidence from the gneiss–migmatite terrances indicates a dominantly sedimentary parentage (Casella, 1969; Butler, 1969; Heimlich, 1969). Zircon morphology studies from gneisses in the Bighorn (Malcuit and Heimlich, 1972) and Beartooth Mountains (Eckelmann and Poldervaart, 1957) indicate two generations of zircon. Small rounded zircons appear to be of detrital origin while larger, light-coloured, often nearly euhedral zircons (sometimes occurring as overgrowths on the detrital zircons) are thought to have developed during metamorphism. The uniform layering in the gneisses; the presence of remnants of quartzite (metachert?), marble, and iron formation; and the gross similarity of the composition of some of the gneisses to average Archean graywackes are also consistent with a dominantly clastic sedimentary parent-rock terrane.

Granitic Rocks

Approximately 30% of the exposed Wyoming Province is composed of granitic plutons and batholiths. Some of the features of the major granitic bodies which are located on Fig. 1 are summarized in Tables 2 and 3. In terms of their tectonic setting and bulk compositions (Heimlich, 1969; Condie, 1969b; Condie and Lo, 1971; Reed and Zartman, 1973), the Wyoming Archean granites can be classified according to the scheme proposed for Archean granites in South Africa (Anhaeusser, et al., 1969; Hunter, 1971). The Wind River and Bighorn batholithic complexes are similar to the ancient gneiss complexes; the Louis Lake and Popo Agie batholiths may be classified as diapiric domes (representing in part or entirely reactivated sialic crust which rises diapirically); the Laramie and Granite Mountains batholiths (which may be part of the same body) as shield-forming batholiths (widespread K-rich granites often erupted over gneiss–greenstone basement); and the Mount Owen and Owl Creek granites as anorogenic plutons (i.e. small K-rich, sharply discordant bodies emplaced at shallow depths). As shown by the radiometric ages in Table 2 and Fig. 2, however, there does not seem to be a relationship between tectonic type and age of emplacement such as that found in South Africa

Table 2. Characteristics of the major granitic batholiths and plutons in the Wyoming Province

Granite	Approximate areal extent (km^2)[4]	Texture[1]	Contacts	Inclusions	Composition ratio[2]	Initial Sr^{87}/Sr^{86} ratio[3]	Age (b.y.)	Inferred depth of emplacement (km)
1. Louis Lake	800	F, H	Concordant, faulted	Common	9	0·702	2·7	15–20
2. Popo Agie	700	P, H	Gradational	?	?		2·6	≥15
3. Laramie	≥2000	H, F	Gradational, locally sharp	Minor	0·2	0·701	2·6	15–20
4. Granite Mtns	≥2500	H, F	Not exposed	Minor	0·2	0·705	2·6	15–20
5. Mt Owen	200	H	Discordant	Uncommon except near contacts	≤0·1	0·732	2·5	≤15
6. Bighorn	≃3000	H, F	Gradational	Common	2		2·9	≥20(?)
7. Wind River	≥1000	F, H	Gradational (?)	Common	≥5(?)			≥20(?)
8. Owl Creek	300	H	Discordant	Common	≤0·1		≥2·7	≤15

[1] F = Foliated. H = Homogeneous. P = Porphyritic.

[2] Composition ratio = areal abundances of $\dfrac{\text{Quartz Diorite} + \text{Granodiorite}}{\text{Quartz Monzonite} + \text{Granite}}$

[3] References given in footnote of Fig. 2.

[4] Including areas of shallow Phanerozoic cover.

(Hunter, 1971). Most of the granites in the Wyoming Province were emplaced over a relatively short period of time between 2·6 and 2·7 b.y.

Geochemical studies of the Laramie batholith are consistent with an origin either by fractional crystallization of a mafic or intermediate magma or by a large degree of melting of the surrounding gneiss–migmatite complex (Condie, 1969b). Major and trace-element data from the Louis Lake batholith favour a model involving partial melting of eclogite in the upper mantle producing a parent granodiorite magma which undergoes a small amount of fractional crystallization at shallow depths (Condie and Lo, 1971). Studies of the Bighorn batholith indicate a progressive northward gradation from well-layered gneisses in the southern Bighorns into migmatites and agmatites in the central Bighorns and finally into homogeneous quartz diorite and quartz monzonite in the northern Bighorns (Heimlich, 1969). This factor and the preservation of relict foliation in the granitic rocks are consistent with origin by alkali metasomatism or/and anatexis of the surrounding gneiss

Table 3. Estimates of granitic-rock abundances in batholiths and plutons from the Wyoming Province

	Quartz Diorite		Granodiorite	Quartz Monzonite	Granite
Louis Lake[1]	5		85	9	1
Laramie and Granite Mountains[2]	<1		14	74	12
Bighorn[3]		70		~25	≤5
Owl Creek[4]	—		—	80	20

[1] Condie and Lo (1971) and unpublished data.
[2] Condie (1969b) and unpublished data.
[3] Heimlich (1969).
[4] Condie, unpublished data.

terrane. Initial Sr^{87}/Sr^{86} ratios (Table 2) allow variable contributions of older sialic crust in the formation of granitic rocks in the province.

Greenstone Belts

Greenstone belts are not well preserved in the Wyoming Province. Only one greenstone belt, the South Pass belt, exhibits readily recognizable volcanic and sedimentary rocks. This greenstone belt is exposed over an area of about 500 km^2 in the southern Wind River Range (Fig. 1). Small fragments of greenstone belts metamorphosed to the amphibolite facies occur in the Owl Creek, Seminoe, and Granite Mountains, in the Rattlesnake Hills, and in Casper Mountain (Fig. 1). The South Pass belt has been the subject of a considerable number of geological and geochemical studies (Condie, 1967; Condie, 1972; Bayley, et al., 1973). The base of the greenstone succession is an intrusive contact with the Louis Lake batholith. The lowest exposed unit is composed of quartzites, tholeiites, graywackes, siliceous iron formation, and a small intrusive serpentinite. This unit is concordant with overlying pillowed, low-K tholeiites and diabase sills. A shear zone separates this part of the section from a small thickness of calc–alkaline volcanic rocks (principally andesites) which are overlain by a minimum of 2500 m of isoclinally folded graywackes. The low-K tholeiites are grossly similar in composition to modern altered rise tholeiites or arc tholeiites and the andesites and dacites to their modern counterparts in the calc–alkaline series (Condie, 1972). The nature of the basement rocks upon which the South Pass or any other greenstone belt in the Wyoming Province formed remains obscure. A plate–tectonic model for the origin of the South Pass belt has been presented by Condie (1972). The model involves an arc–arc collision in which the northern part of the belt collides with and is welded to the southern part.

Amphibolites

Amphibolites compose 10–15% of the gneiss–migmatite terrances where they occur as small lenses ($\geqslant 5$ m long) and as tabular sheets from tens to thousands of metres long. They also occur as major rock types in the greenstone belts and as minor inclusions in granites. Field relationships and chemical compositions suggest that most of the amphibolites in the gneiss–migmatite terranes are metamorphosed diabase dykes and sills. In some areas complete gradations occur between diabase dykes with well-preserved diabasic textures and amphibolites.

Ultramafic and Related Rocks

Ultramafic and related rocks have two major occurrences in the Wyoming Province: (1) as differentiates in stratiform sheets, and (2) as lenses and tectonic fragments in gneiss–migmatite terranes. They also rarely occur as inclusions in granites and one small intrusive serpentinite has been described from the South Pass greenstone belt (Bayley, et al., 1973). The most extensive occurrence is in the layered rocks of the Stillwater Complex in the northern Beartooth Mountains (Fig. 1). This complex is well known from field and petrologic studies (Hess, 1960; Jackson, 1961). It is characterized from the base upwards by ultramafic, norite, lower gabbro, anorthosite, and upper gabbro zones. Features characterizing differentiated sheets such as cumulus textures and rhythmic and cryptic layering are well preserved throughout the complex. Fractional crystallization models indicate that the exposed portion (5–7 km) represents only about 60% of the entire body (Hess, 1960). A second stratiform sheet, the Preacher Creek body, in the east-central Laramie Range has recently been studied in detail (Potts and Condie, 1971; Potts, 1972). This body, which is lense-shaped and about 4 km long and 1·5 km wide, is composed chiefly of layered peridotite and is characterized by cryptic layering and cumulus textures. Unlike the Stillwater Complex, it shows no evidence of intrusion and probably represents a fragment of a larger stratiform complex tectonically emplaced prior to or during regional metamorphism.

Field and petrologic studies of bodies of serpentinized ultramafic rock contained in gneiss–migmatite terranes of the Beartooth Mountains indicate that they represent tectonically rotated fragments of ultramafic plutons or dykes intruded into the dominantly sedimentary precursors of the gneisses (Skinner, 1969).

Diabase Dykes

Mafic dykes ranging in composition from tholeiite to picrite occur in most Archean exposures in the Wyoming Province. Unlike the extensive swarms in the Canadian Shield, however, they do not seem to exhibit a consistent trend between or in some cases within mountain ranges. Some of the larger dykes are shown in Fig. 1. They range in thickness from a few metres to over 100 m and a few individual dykes can be traced for over 30 km. The most extensive studies of the dykes have been in the Beartooth and Bighorn Mountains (Prinz, 1964; Condie, et al., 1969a,b; Heimlich, et al., 1974). The dykes are grossly similar in composition to modern continental tholeiites. Mineralogical and chemical variations across dykes in the northern Bighorns are consistent with an origin by fractional crystallization from the outsides inward (Ross and Heimlich, 1972). Trace-element studies of dykes from various locations in the province indicate a depletion in Sr, a feature which is most satisfactorily explained by removal of plagioclase during fractional crystallization at shallow (≤15 km) depths (Condie, et al., 1969a; Mueller and Rogers, 1973).

Quartzites and Related Rocks

In the Pony and Cherry Creek Series of south-west Montana and in the Red Creek Quartzite of northern Utah, quartzites and mica schists are important rock types (Reid, 1957; Heinrich and Rabbitt, 1960). In south-west Montana the metamorphic section is composed of up to 10% quartzite and the remainder chiefly of paragneisses, amphibolites and various types of mica schists; marbles locally compose up to 30% of the section. In the Red Creek Quartzite, quartzites and quartz–mica schists are the dominant rock types. The relationships of the south-west Montana and north-east Utah areas to the rest of the Wyoming Province are not clear due to lack of exposures.

Structure

The structure of the Wyoming Province varies from area to area. Although structural trends are quite variable, the overall directions of gneissic foliation and fold axes are diagrammatically illustrated in Fig. 1 from detailed published field studies. The structure in the Black Hills is complex showing an overall north to north-west trend. Because this area has been significantly effected by later Precambrian orogeny (at ~1·7 b.y.), however, the structural trends may not reflect the Archean fabric. The simplest structural setting recorded in the Wyoming Province occurs in the Beartooth Mountains which is characterized by passive, open folds with north to north-eastern strikes (Casella, 1969). Individual folds are traceable for distances up to 10 km. Field and petrographic studies in this area indicate that folding was contemporary with regional metamorphism and that with minor exceptions, only one period of folding is recorded. In the Teton and Wind River Ranges an early period of isoclinal folding is followed by later open folding (Reed and Zartman, 1973; Bayley, et al., 1973). Regional metamorphism appears to coincide with the open folding. The Bighorn and Laramie Range gneiss complexes may record more than two periods of folding (Heimlich, 1969; Condie, 1969b).

Metamorphism and Metasomatism

The metamorphic grade in the Wyoming Province ranges from upper greenschist facies to lower granulite facies. Amphibolite–facies terranes, however, are by far the most widespread. Upper greenschist–facies terranes are represented by the South Pass greenstone belt, parts of south-west Montana, and the western Beartooths. Small occurrences of hypersthene granulite associated with sillimanite gneisses, amphibolite, and quartzite in the east-central Wind River Range are indicative of the lower granulite facies. Most of the province is characterized by the andalusite–sillimanite facies series and only in south-west Montana and north-east Utah (the Red Creek Quartzite) is the kyanite–sillimanite series found.

The distribution of facies series indicates a rather steep geothermal gradient beneath most of the Wyoming Province ($\geq$60 deg/km) decreasing around parts of the north-western and southern boundaries ($\leq$40 deg/km). The origin of the higher pressure metamorphism around the edges of the province is poorly understood but is probably related to differences in tectonic environment. A lower geothermal gradient probably reflects a cooler upper mantle near the edges of the province.

As an example of the complex metamorphic history of parts of the province, the sequence of events in the South Pass greenstone belt is summarized in Table 4 (in part after Bayley, et al., 1973). In this area three distinct periods of metamorphism are recognized. An early upper-greenschist–facies metamorphism was widespread and accompanied by isoclinal folding. The northern perimeter of the belt was metamorphosed to the amphibolite (or hornblende–hornfels) facies during emplacement of the Louis Lake batholith at about 2·7 b.y. Lastly, a period (or periods) of retrograde metamorphism is recorded by chlorite

and chloritoid which cross-cut earlier metamorphic fabrics

Geological studies suggest that wide-scale alkali metasomatism was important in the gneiss–migmatite terranes of the Beartooth, Medicine Bow, and Bighorn Ranges (Eckelmann and Poldervaart, 1957; Butler, 1969; Houston, et al., 1968; Heimlich, 1969). The most important evidences are as follows: (1) the replacement of early plagioclase by later sodic plagioclase or microcline; (2) gradation of obvious paragneisses into gneissic granite both parallel and normal to strike without disruption of planar elements; and (3) abundant layering in granitic terranes.

Geochronology

Radiometric ages from the Wyoming Province are summarized in Figs. 2 and 3. Although most Rb–Sr whole rock isochron and zircon concordia dates fall in the range of 2·5–2·7 b.y., several occur at about 2·9 b.y. and appear to record the earliest distinct metamorphic–igneous event or events in the

Table 4. Summary of the metamorphic history of the South Pass greenstone belt

Age (b.y.)	$\geq$2·9	2·7	$\leq$2·0
Metamorphic mineral paragenesis (quartz, plagioclase, and K-feldspar ubiquitous)	Biotite Almandite Actinolite ___ Blue-green hornblende Chlorite ___ Muscovite _____ Epidote __	Andalusite ___ Cordierite Sillimanite	Chlorite ___ Sericite Chloritoid
Type and extent of metamorphism	Regional (widespread)	Contact (localized around contact of Louis Lake batholith)	Retrograde (widespread)
Igneous activity		Emplacement of Louis Lake batholith	Diabase dyke intrusion
Deformation	Isoclinal Folding	Open folding, faulting, shear zone development	

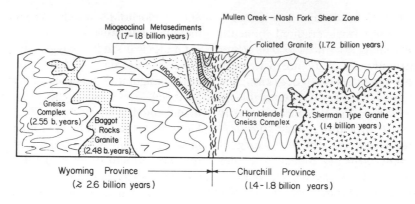

Fig. 2. Summary of Precambrian geochronology of the Wyoming Province (modified after Reed and Zartman, 1973) (reproduced with permission of The Geological Society of America). Lengths of boxes estimate uncertainties. Principal references: South-west Montana—Giletti (1966), P. A. Mueller (1974, pers. comm.); Little Belt Mountains—Catanzaro and Kulp (1964); Beartooth Mountains—Catanzaro and Kulp (1964), Powell, et al. (1969), Brookins (1968), Nunes and Tilton (1971), Baadsgaard and Mueller (1973); Teton Range—Reed and Zartman (1973); Albion Range—Armstrong and Hills (1967); Wind River Range—Naylor et al. (1970), Bassett and Giletti (1963), Condie et al. (1969b); Bighorn Mountains—Heimlich and Banks (1968), Heimlich and Armstrong (1972), Stueber et al. (1974); Granite Mountains—Peterman et al. (1971), Rosholt et al. (1973); Black Hills—Zartman and Stern (1967), Ratte and Zartman (1970); Laramie Range—Hills and Armstrong (1974), Johnson and Hills (1974), Subbarayudu and Hills (1974); Medicine Bow Mountains—Hills et al., (1968)

province. These dates occur in the Teton (Reed and Zartman, 1973), Granite (Peterman, et al., 1971), Bighorn (Heimlich and Banks, 1968; Stueber, et al., 1974), and Sierra Madre Mountains (Divis, 1974). Detrital zircon dates from gneisses in the Beartooth Mountains indicate that they were derived from a probable granitic or gneissic source rock >3·1 b.y. in age (Catanzaro and Kulp, 1964; Nunes and Tilton, 1971).

The most widespread orogenic event in the Wyoming Province coinciding with the Kenoran Orogeny in the Superior Province is the deformation, granitic plutonism, and regional metamorphism recorded by both whole-rock Rb–Sr ages (Fig. 3) and U–Pb zircon ages at 2·6–2·7 b.y. It is recorded in all mountain ranges including the peripheral ones. During this interval of time, the major granites of the province were emplaced and a period of diabase dyke intrusion occurred in the Beartooth Mountains. Although the age of the Stillwater

Complex is still not completely agreed upon, the zircon ages of Nunes and Tilton (1971) suggest an age of about 2·75 b.y.

Initial Sr^{87}/Sr^{86} ratios from rocks in the Wyoming Province are tabulated in histogram form in Fig. 3 and range from about 0·700 to 0·730. The general high values in most of the

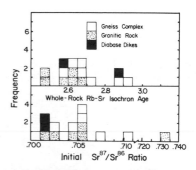

Fig. 3. Histogram plot of whole-rock Rb-Sr isochron ages and initial Sr^{87}/Sr^{86} ratios from the Wyoming Province (references in Fig. 2)

granites (chiefly 0·702–0·706) indicate that they cannot have been derived directly from the upper mantle which would have had a Sr^{87}/Sr^{86} ratio of 0·700–0·701 at this time. The Mount Owen pluton which has an initial ratio of 0·732 appears to have been selectively enriched in Sr^{87} (Reed and Zartman, 1973). It is clear also that the 2·9 b.y. paragneisses of the Granite Mountains which have an initial ratio of 0·7048 (Peterman, et al., 1971) must represent metasedimentary rocks derived from a still older sialic source.

K–Ar whole-rock dates suggest that diabase dykes were intruded in the southern Wind River and Western Owl Creek Ranges at about 2 b.y. (Condie, et al., 1969b). Whole-rock Rb–Sr isochron ages indicate that diabase dykes were intruded in the Beartooth gneisses at 2·6, 1·3, and 0·74 b.y. (Baadsgaard and Mueller, 1973) and at 2·2 and 2·9 b.y. in the Bighorn Mountains (Stueber, et al., 1974). The high initial Sr^{87}/Sr^{86} ratios for the Archean diabase dykes (0·701–0·702) probably reflect minor contamination of the original magmas with sialic crust or contamination during later metamorphism.

The last Precambrian event recorded in the Wyoming Province is a heating event coincident with the Hudsonian Orogeny (1·6–1·8 b.y.) in neighbouring terranes.

South-eastern Boundary of the Wyoming Province

The contact of the Wyoming Province with surrounding Precambrian terranes is exposed in the northern Medicine Bow and Sierra Madre Mountains in south-eastern Wyoming (Fig. 1). In the Medicine Bow exposure, gneisses of the Wyoming Province are in fault contact with gneisses of unknown age (although probably <1·8 b.y.) and younger granites (Fig. 4) (Hills, et al., 1968). The contact zone is complicated by the presence of a miogeoclinal sequence of metasediments unconformably overlying the Archean terrane and also displaced by the shear zone. The shear zone extends into the Sierra Madre Mountains to the west and also probably into the Laramie Range to the east (Hills and Arm-

strong, 1974) where it was intruded by the Laramie anorthosite and associated syenites at about 1·7 b.y. (Subbarayudu and Hills, 1974).

It is clear from the numerous mineral ages and fewer number of Rb–Sr and U–Pb mineral isochron ages from the Wyoming Province that daughter isotopes were significantly redistributed between 1·6 and 1·8 b.y. (and somewhat earlier in some areas; see Fig. 2). This is generally thought to have resulted from heating of the province during granite emplacement and orogeny in surrounding terranes at 1·6–1·8 b.y. (Reed and Zartman, 1973). The effect of this heating appears to diminish inward from the boundaries of the province as indicated on Fig. 1 by the suggested locations of contours of equal K–Ar biotite ages. It would appear that as the Hudsonian Orogeny subsided, the temperature of the Wyoming Province decreased outwards from the centre and the K–Ar biotite ages represent the time when Ar began to be retained in biotite.

Geologic History

The Precambrian geologic history of the Wyoming Province as deduced from available field and radiometric-age data is summarized in Table 5. The earliest crust in the Wyoming Province was ≥3·2 b.y. in age and from the detrital zircons it supplied, it must have been composed in large part of granite–gneiss terranes. Between about 2·9 and 3·2 b.y. dominantly 'eugeoclinal-type' sedimentation and volcanism occurred. The province appears to have evolved into a stable 'miogeoclinal setting' along parts of the north-western and southern borders in response perhaps to a decreasing geothermal gradient. The first orogenic event is recorded in parts of the province at about 2·9 b.y. The major deformation, metamorphism and plutonism, however, occurred between 2·6 and 2·7 b.y. This event is coincident and perhaps continuous geographically with the Kenoran Orogeny in the Superior Province. It is noteworthy that the 2·6–2·7 b.y. event does not appear to be recorded in the Teton, Granite, Sierra Madre, and Bighorn Mountains where the earlier 2·9 b.y. event occurred.

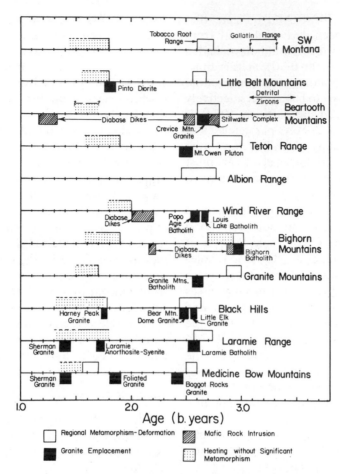

Fig. 4. Diagrammatic cross-section of the Wyoming Province boundary in the Medicine Bow Mountains (after Hills et al., 1968) (reproduced with permission of The Geological Society of America)

Table 5. Precambrian geologic history of the Wyoming Province

Age (b.y.)	Events
0·74, 1·3	Intrusion of diabase dykes in the Beartooth Mountains.
1·6–1·8	Heating of outer parts of the province in response to adjacent orogeny.
2·0–2·4	Intrusion of two or more diabase dyke swarms in central part of province.
2·6–2·7	Widespread deformation, regional metamorphism, and plutonism; emplacement of Stillwater Complex and one or more diabase dyke swarms.
2·9	Deformation, granite emplacement, regional metamorphism and diabase dyke intrusion in parts of the province.
2·9–3·2 +	Dominantly eugeoclinal sedimentation and volcanism; minor miogeoclinal sedimentation.
≥3·2	Formation of sialic crust.

Between 2 and about 2·4 b.y., two or more diabase dyke swarms were intruded in the northern and central parts of the province. Radiogenic daughter isotopes in mineral systems were partially redistributed especially in the outer parts of the province between 1·6–1·8 b.y. The final Precambrian events recorded in the province are the intrusion of diabase dykes at about 1·3 and 0·74 b.y. in the Beartooth Mountains. Dyke swarms in other parts of the province may also have been intruded between 1 and 2 b.y. ago.

References

Anhaeusser, C. R., Mason, R., Viljoen, M. J., and Viljoen, R., 1969. 'A Reappraisal of some aspects of Precambrian shield geology', *Geol. Soc. Amer. Bull.*, **80**, 2175–2200.

Armstrong, R. L., and Hills, F. A., 1967. 'Rb–Sr and K–Ar geochronologic studies of mantled gneiss domes, Albion Range, southern Idaho U.S.A.', *Earth Planet. Sci. Lett.*, **3**, 114–124.

Baadsgaard, H., and Mueller, P. A., 1973. 'K–Ar and Rb–Sr ages of intrusive Precambrian mafic rocks, southern Beartooth Mountains, Montana and Wyoming', *Geol. Soc. Amer. Bull.*, **74**, 209–212.

Bassett, W. A., and Giletti, B. R., 1963. 'Precambrian ages in the Wind River Mountains, Wyoming', *Geol. Soc. Amer. Bull.*, **74**, 209–212.

Bayley, R. W., Proctor, P. D., and Condie, K. C., 1973. 'Geology of the South Pass area, Fremont County, Wyoming', *U.S. Geol. Survey Prof. Paper*, **793**, 39 pp.

Brookins, D. G., 1968. 'Rb–Sr and K–Ar age determinations from the Precambrian rocks of the Jardine-Crevice Mountain area, southwestern Montana', *Earth Sci. Bull.*, **1**, 5–9.

Burwash, R. A., Baadsgaard, H., and Peterman, Z. E., 1962. 'Precambrian K–Ar dates from the western Canada sedimentary basin', *J. Geophys. Res.*, **67**, 1617–1625.

Butler, J. R., 1969. 'Origin of Precambrian granitic gneiss in the Beartooth Mountains, Montana and Wyoming', *Geol. Soc. Amer. Mem.*, **115**, 73–101.

Casella, C. J., 1969. 'A review of the Precambrian geology of the eastern Beartooth Mountains, Montana and Wyoming', *Geol. Soc. Amer. Mem.*, **115**, 53–71.

Catanzaro, E. J., and Kulp, J. L., 1964. 'Discordant zircons from the Little Belt (Montana), Beartooth (Montana) and Santa Catalina (Arizona) Mountains', *Geochim. Cosmochim. Acta*, **28**, 87–124.

Condie, K. C., 1967. 'Geochemistry of early Precambrian graywackes from Wyoming', *Geochim. Cosmochim. Acta*, **31**, 2135–2149.

Condie, K. C., 1969a. 'Geologic evolution of the Precambrian rocks in northern Utah and adjacent areas', *Utah Geol. Surv. Bull*, **82**, 71–95.

Condie, K. C., 1969b. 'Petrology and geochemistry of the Laramie batholith and related metamorphic rocks of Precambrian age, eastern Wyoming', *Geol. Soc. Amer. Bull.*, **80**, 57–82.

Condie, K. C., 1972. 'A plate tectonics evolutionary model of the South Pass Archean greenstone belt, southwestern Wyoming', *24th Intern. Geol. Congress, Sect.*, **1**, 104–112.

Condie, K. C., Barsky, C. K., and Mueller, P. A., 1969a. 'Geochemistry of Precambrian diabase dikes from Wyoming', *Geochim. Cosmochim. Acta*, **33**, 1371–1388.

Condie, K. C., Leech, A. P., and Baadsgaard, H., 1969b. 'Potassium-argon ages of Precambrian mafic dikes in Wyoming', *Geol. Soc. Amer. Bull.*, **80**, 899–906.

Condie, K. C., and Lo, H. H., 1971. 'Trace element geochemistry of the Louis Lake batholith of early Precambrian age, Wyoming', *Geochim. Cosmochim. Acta*, **35**, 1099–1119.

Divis, A., 1974. 'The geochemistry of Sr isotope chronology of a Precambrian polycyclic gneiss', *American Geophys. Union Trans.*, **56**, 1199.

Eckelmann, F. D., and Poldervaart, A., 1957. 'Geologic evolution of the Beartooth Mountains, Montana and Wyoming', *Geol. Soc. Amer. Bull.*, **68**, 1225–1262.

Giletti, B. R., 1966. 'Isotopic ages from southwestern Montana', *J. Geophys. Res.*, **71**, 4029–4036.

Goldich, S. S., Lidiak, E. G., Hedge, C. E., and Walthall, F. G., 1966. 'Geochronology of the Midcontinent region, United States, 2. Northern Area', *J. Geophys. Res.*, **71**, 5389–5408.

Heimlich, R. A., 1969. 'Reconnaissance petrology of Precambrian rocks in the Bighorn Mountains, Wyoming', *Contr. Geol.*, **8**, 47–61.

Heimlich, R. A., and Armstrong, R. L., 1972. 'Variance of Precambrain K–Ar biotite dates, Bighorn Mountains, Wyoming', *Earth and Planet. Sci. Lett.*, **14**, 75–78.

Heimlich, R. A., and Banks, P. O., 1968. 'Radiometric age determinations, Bighorn Mountains, Wyoming', *Amer. Jour. Sci.*, **266**, 180–192.

Heimlich, R. A., Gallagher, G. L., and Shotwell, L. B., 1974. 'Quantitative petrography of mafic dikes from the central Bighorn Mountains, Wyoming', *Geol. Mag.*, **111**, 97–192.

Heinrich, E. Wm., and Rabbitt, J. C., 1960. 'Pre-Beltian geology of the Cherry Creek and Ruby Mountains Areas, southwestern Montana', *Montana Bur. Mines Geol. Mem.*, **38**, Part 1, 1–15.

510

Hess, H. H., 1960. 'Stillwater Igneous Complex, Montana,' *Geol. Soc. Amer. Mem.*, **80**, 230 pp.

Hills, F. A., and Armstrong, R. E., 1974. 'Geochronology of Precambrian rocks in the Laramie Range and implications for the tectonic framework of Precambrian southern Wyoming', *Precamb. Res.*, **1**, 213–225.

Hills, F. A., Gast, P. W., Houston, R. S., and Swainbank, I. G., 1968. 'Precambrian geochronology of the Medicine Bow Mountains, southeastern Wyoming', *Geol. Soc. Amer. Bull.*, **79**, 1757–1784.

Houston, R. S., and others, 1968. 'A regional study of rocks of Precambrian age in that part of the Medicine Bow Mountains lying in southeastern Wyoming', *Geol. Surv. Wyoming Mem.*, **1**, 167 pp.

Hunter, D. R., 1971. 'The granitic rocks of the Precambrian in Swaziland', *Econ. Geol. Res. Unit Infor. Circ. No. 63*, 25 pp.

Jackson, E. D., 1961. 'Primary textures and mineral associations in the ultramafic zone of the Stillwater Complex, Montana', *U.S. Geol. Surv. Prof. Paper 358*, 106 pp.

Johnson, R. C., and Hills, F. A., 1974. 'Rb–Sr geochronology and geology of the Precambrian basement, Box Elder Canyon Area, northern Laramie Range, Wyoming', *Amer. Geophys. Union Trans.*, **55**, 465.

Malcuit, R. J., and Heimlich, R. A., 1972. 'Zircons from Precambrian gneiss, southern Bighorn Mountains, Wyoming', *Amer. Mineral.*, **57**, 1190–1209.

Mueller, P. A., and Rogers, J. J. W., 1973. 'Secular chemical variation in a series of Precambrian mafic rocks. Beartooth Mountains, Montana and Wyoming', *Geol. soc. Amer. Bull.*, **84**, 3645–3652.

Naylor, R. S., Steiger, R. H., and Wasserburg, G. J., 1970. 'U–Th–Pb and Rb–Sr systematics in 2700×10^9-year old plutons from the southern Wind River Range, Wyoming', *Geochim. Cosmochim. Acta*, **34**, 1133–1159.

Nunes, P. D., and Tilton, G. R., 1971. 'Uranium-Lead ages of minerals from the Stillwater Igneous Complex and associated rocks, Montana', *Geol. Soc. Amer.. Bull.*, **82**, 2231–2250.

Peterman, Z. E., Hildreth, R. A., and Nkomo, I., 1971. 'Precambrian geology and geochronology of the Granite Mountains, Central Wyoming', *Geol. Soc. Amer. Abst. for 1971*, **3**, 403–404.

Potts, M. J., 1972. 'Origin of a Precambrian ultramafic intrusion in southeastern Wyoming, U.S.A.', *Contr. Mineral. and Petrol.*, **36**, 249–264.

Potts, M. J., and Condie, K. C., 1971. 'Rare earth element distributions in a proto-stratiform ultramafic intrusion', *Contr. Mineral. and Petrol.*, **33**, 245–258.

Powell, J. L., Skinner, W. R., and Walker, D., 1969. 'Whole-rock Rb–Sr age of metasedimentary rocks below the Stillwater Complex, Montana', *Geol. Soc. Amer. Bull.*, **80**, 1605–1612.

Prinz, M., 1964. 'Geologic evolution of the Beartooth Mountains, Montana and Wyoming, Part 5, Mafic dike swarms of the southern Beartooth Mountains', *Geol. Soc. Amer. Bull.*, **75**, 1217–1248.

Ratte, J. C., and Zartman, R. E., 1970. 'Bear Mountain gneiss dome, Black Hills, South Dakota-age and structure', *Geol. Soc. Amer. Abstr. for 1970*, **2**, 345.

Reed, J. C. Jr., and Zartman, R. E., 1973. 'Geochronology of Precambrian rocks of the Teton Range, Wyoming', *Geol. Soc. Amer. Bull.*, **84**, 561–582.

Reid, R. R., 1957. 'Bedrock geology of the north end of the Tobacco Root Mountains, Madison County, Montana', *Montana Bur. Mines and Geology Mem.*, **36**, 25 pp.

Rosholt, J. N., Zartman, R. E., and N. Komo, I. T., 1973. 'Lead isotope systematics and uranium depletion in the Granite Mountains, Wyoming', *Geol. Soc. Amer. Bull.*, **84**, 989–1002.

Ross, M. E., and Heimlich, R. A., 1972. 'Petrology of Precambrian mafic dikes from the Bald Mountain Area, Bighorn Mountains, Wyoming', *Geol. Soc. Amer. Bull.*, **83**, 1117–1124.

Skinner, W. R., 1969. 'Geologic evolution of the Beartooth Mountains, Montana and Wyoming; Part 8, Ultramafic rocks in the Highline Trail Lakes Area, Wyoming', *Geol. Soc. Amer. Mem.*, **115**, 19–52.

Stueber, A., Heimlich, R. A., and Ikramuddin, M., 1974. 'Rb–Sr ages of Precambrian mafic dikes from the Bighorn Mountains, Wyoming', *Amer. Geophys. Union Trans.*, **56**, 1198.

Subbarayudu, G. V., and Hills, F. A., 1974. 'On the age and origin of syenite in the Laramie anorthosite complex, Wyoming', *Amer. Geophys. Union Trans.*, **55**, 476.

Zartman, R. E., and Stern, T. W., 1967. 'Isotopic age and geologic relationships of the Little Elk Granite, northern Black Hills, South Dakota', *U.S. Geol. Survey Prof. Paper 575-D*, 7 pp.

Progress Report on Early Archean Rocks in Liberia, Sierra Leone and Guayana, and their General Stratigraphic Setting

P. M. HURLEY, H. W. FAIRBAIRN, H. E. GAUDETTE

54-1122, M.I.T., Cambridge, Mass., U.S.A.

Introduction

This report is the outcome of a collaborative effort between the Geochronology Laboratory of the Massachusetts Institute of Technology, the U.S. Geological Survey cooperative programme with the Liberian Geological Survey (G. W. Leo and R. W. White); E. M. Laing, the Director of the Geological Survey Division, Sierra Leone; R. R. Thompson, Chief Geologist of the Sierra Leone Development Company; S. E. Luchsinger and G. Ascanio of the Orinoco Mining Company; V. Mendoza and A. Espejo of the Ministry of Mines and Hydrocarbons, Venezuela; J. Kalliokoski of the Michigan School of Mines; and others who have been cited in references in the text.

Liberia and Sierra Leone

The distribution of age provinces and of some major geologic units in Liberia and Sierra Leone is shown in Fig. 1. Geologic mapping is at a relatively early stage in both countries. Because of this, and also because of generally poor exposures caused by prolonged lateritic weathering, detailed field relationships are not known for some of the dated samples. Nonetheless, some basic features of the geology of the respective age provinces are known and have been summarized by Andrews-Jones (1966), Allen and others (1967) for Sierra Leone, by Leo and White (1967), and White and Leo (1971) for Liberia, and by several earlier workers cited in those papers. The brief summary that follows has been abstracted from these sources.

The Atlantic edge of the West African shield may be divided into three major age provinces (Fig. 1). The Liberian age province, greater than 2500 m.y. (Hurley et al., 1971), extends from eastern Liberia north-westward through western Liberia, Sierra Leone, Senegal, and Mauritania, with the exception of a coastal belt of Pan-African age (c. 600 m.y.). The Eburnean (c. 2000 m.y.) age province (Bonhomme, 1962, Vachette, 1964) extends over Ghana, Ivory Cost, and a small part of eastern Liberia, and is bounded by the Liberian province on the north-west and the Pan-African province of Dahomey, Togo, and Nigeria on the east.

The Liberian age province in Liberia is dominated by granitic rocks (compositional range from granite to quartz diorite) which are mostly gneissic except for extensive exposures of massive rocks in the extreme north (Fig. 1). Similar rocks, together with migmatite and granulite, characterize the province in Sierra Leone. Scattered belts of infolded metasedimentary rocks including iron-formation and metavolcanic rocks trend north to north-east within this age province in both countries. The metamorphic grade of most rocks within the province is amphibolite facies, although relict masses of granulite are

511

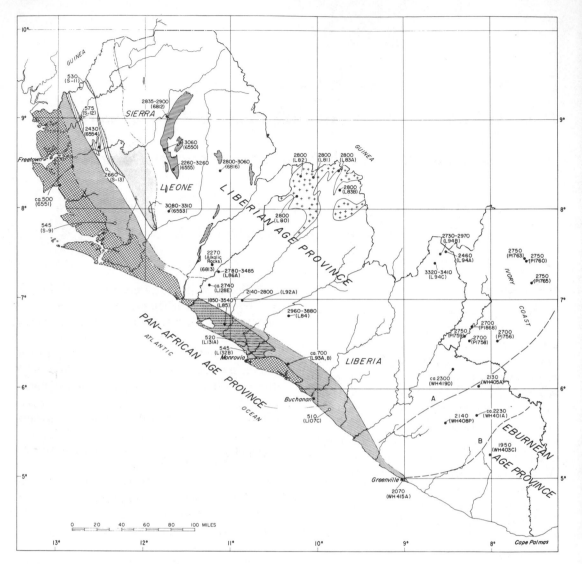

Fig. 1. Isotopic analyses were performed on a 30·5-cm radius, 60°-sector mass spectrometer. All Sr^{87}/Sr^{86} ratios are relative to a value of 0·7080 for the Eimer and Amend standard strontium carbonate and normalized to Sr^{86}/Sr^{88} equal to 0·1194. A value of $Rb^{87}\lambda = 1·39 \times 10^{-11}$ yr^{-1} was used for calculations of age. Unless otherwise indicated, the errors of analysis are estimated to be 2% for Rb/Sr ratio and 0·0015 for Sr^{87}/Sr^{86}, within 95% confidence limits

common within the gneissic terrane in south-western Sierra Leone. Mineral assemblages in infolded metasedimentary rocks in Liberia indicate a relatively low-pressure, high-temperature metamorphic facies series; where the metamorphic grade is highest, these metasedimentary rocks are cut by pegmatites, and incipient anatexis may be recognized. Intrusions within the province are largely of the synkinematic type.

The Eburnean (c. 2000 m.y.) age province includes rocks of the Birrimian System as defined originally in Ghana (Junner, 1940), now mostly slates and volcanics and their gneissic equivalents. These are geosynclinal sequences folded, metamorphosed, and intruded by granites during the time of the Eburnean thermotectonic episode at 2000–1800 m.y. The Birrimian is cut by synorogenic concordant, strongly foliated, batholiths of granite and

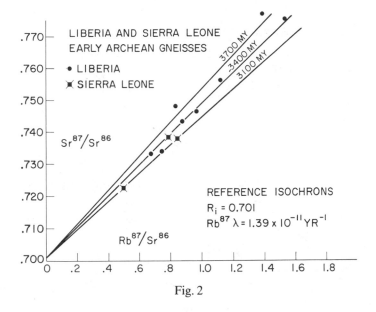

Fig. 2

granodiorite, and smaller more alkaline masses of unfoliated and cross-cutting post-orogenic granites. The system is divided into the Lower Birrimian slates, phyllites, graywackes, and volcanics and highgrade equivalents, and the Upper Birrimian metamorphosed volcanic materials with intercalated phyllite and graywacke and a noteworthy manganiferous horizon.

In the Ivory Coast, Tagini (1965) recognizes a geosynclinal province in the central part of the country and a 'semiplatform' to the west which overlies the older basement in the area of Man (Papon, et al., 1968). It is now clear that the basement underlying this Birrimian geosyncline is represented by the 2700 m.y. Liberian age province to the west.

Age Measurements

Reconnaissance age dating in large regions which have not been well mapped may be carried out either by the 'scatter-gun' approach, or else by focusing the analyses on individual rock units at the outset. We believe in the first approach, and that time and effort can be saved if a preliminary overview of the spread of age values is obtained in order to indicate points for more concentrated attack later on. In this progress report most of the results represent the reconnaissance stage, so

that data arrays do not represent isochrons in the normal sense. Instead we observe groupings of age values on samples from regionally scattered rock units, which first indicate the principal period of thermotectonic activity; and secondly permit the field geologist to piece together the stratigraphy of the region using local field relationships to fill in the gaps.

Almost all of the analyses to date are covered in Figs. 2, 3 and 4. There are clearly three groupings: (a) an early Archean cluster scattered between 3100 and 3700 m.y.; (b) a late Archean (Liberian) group rather closely spaced about a 2800 m.y. reference isochron; and (c) an Eburnean group close to 2050 m.y. A fourth period of activity, at about 500 m.y., has been found in the Pan-African belt along the coast, but is not relevant to this report. The early Archean rocks cannot yet be clearly differentiated in the field, except that they are found to be the basement whenever a local stratigraphy is observable. Perhaps the best example is a coarse granitic gneiss in the Mt Detton area at the north-west end of the Nimba Range in Liberia, which underlies the sedimentary section carrying the iron formation. Three samples from drill cores, with considerably different Rb/Sr ratios, yielded age values of 3320, 3340 and 3410 m.y.

Thus in general it appears that the Liberian thermotectonic event at about 2800 m.y.

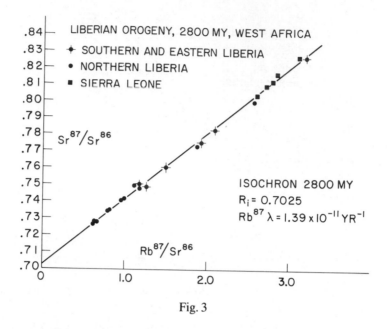

Fig. 3

developed an extensive continental block, incorporating Early Archean relics. This basement was overlain by the Birrimian supracrustals which became later folded and intruded in the extensive Eburnean event of about 2000 m.y. This sequence is so similar to that in the Guayana Shield of South America that the pre-drift reconstruction of a single West African–Guayana craton persisting (and growing in size) from about 3400 m.y. to the opening of the South Atlantic Ocean, is given some credence. This sequence is also found

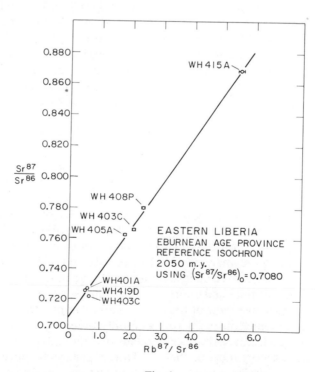

Fig. 4

elsewhere in Africa and the rest of the world, except that the older rocks (>3000 m.y.) are not usually present.

In the remainder of the West African Shield, the Liberian and Eburnean age groupings have been reported by various principal investigators (see references for Bonhomme, Vachette, Lassére, Choubert, Caby, Bertrand, Papon, Charlot, Ferrara, Allégre, Grant, Hurley, Sougy, Cahen, Clifford, Bassot) and their co-investigators. Rocks as old as 3400 m.y. are rare enough so that special efforts are being made to locate them and carry out intensive studies. The cases in Liberia and Sierra Leone are continuing to receive attention, using total-rock Pb as well as Rb–Sr measurements in conjunction with planned field work. The only other actual age measurement known to us on Early Archean rocks is the 3300 m.y. granulite in the In Ouzzal block within the Pan-African belt in the Hoggar (Allègre and Caby, 1972).

Guayana Shield

The Imataca age province in the northern part of the Guayan Shield (Fig. 5) is composed of a generally east–north-east striking belt of complexly folded metasediments and gneisses, and younger granitic intrusions. In this report we shall refer to the oldest unit in excess of 3200 m.y. in age, as the Imataca Series. This has been refolded, injected, and partly assimilated during a thermotectonic event at about 2800 m.y., and then later affected by the Trans-Amazonian orogeny, at about 2100 m.y., during which time granite plutons were introduced.

A number of investigators have examined the high-grade gneisses and granulites of the Imataca Series, with reference to their origin (for bibliography see Kalliokoski (1965) and Bellizzia, 1974). The evidence of apparent stratification, low content of CaO, rounded zircons, quartz-rich horizons and iron formation, and major and minor element chemistry

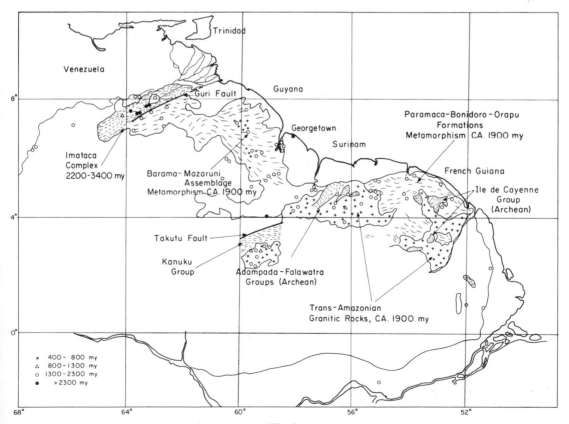

Fig. 5

suggests that these rocks are the metamorphic product of original sediments of the greywacke association, for the most part, and that there has been relatively little metasomatism or migration of components throughout the section. The rocks are now dominantly feldspathic granulites, with or without biotite, K-feldspar, and quartz, and with orthopyroxene or amphibole. Apparently the earlier granulite facies metamorphism left these refractory blocks of crustal material immune to subsequent metamorphic and tectonic processes to a sufficient extent so that their antiquity is still discernible.

The site of the great Guri dam, on the Caroni river, has afforded a good collection of deeply excavated samples of different rock types from within the Imataca Series. A plot of the analyses on these samples to date is given in Fig. 6, showing age values between 3200 and 3700 my. Small chips and separated minerals fall along secondary isochrons indicating that the system was subjected to continued migration of components as late as 1200 m.y., and that the diffusion distances were of the order of centimetres.

We are pursuing a detailed study of diffusion within a single block to determine whether the movement of components was a simple intra-block equilibration, or involved a loss of Rb, for example, from the block as a whole. The block is about 30 cm on the cube-edge and gives an age value of 3500 m.y. Total-rock Pb measurements are being carried out in addition to Rb–Sr. It is not yet possible to differentiate the gneisses in the field, and measurements elsewhere, such as at the El Pao mine, show a mixture of 3700 m.y. bands with 2800 m.y. granitoid phases presumably injected between them, somewhat reminiscent of the Minnesota River Valley and Greenland cases. The question is whether these are segregation or injection effects on a large scale.

This Early Archean event has been called the Gurian by Bellizzia (1974) who refers to an age range of 3000–3400 m.y.

The next discernible event in the Imataca province is at 2800 m.y. (Fig. 7). The rocks developed during this event are injection gneisses, and migmatites, together with an equal abundance of fairly homogeneous granitic rocks. One of the principal regions of this age identified in mapping to date is devoid of iron formations, amphibolite, and evident sedimentary structures of the Imataca Series, and has been referred to as the Cerro la Ceiba quartz monzonite migmatite. It is situated between Ciudad Bolivar and Cerro Bolivar centring on Cerro la Ceiba. It has two principal components, a fine to medium grain, pink, light or dark grey (hornblende)–biotite–quartz–feldspar gneiss intimately penetrated

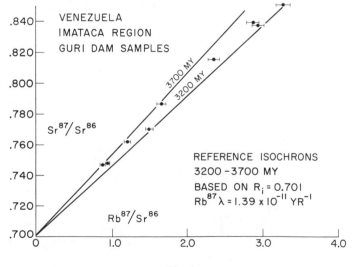

Fig. 6

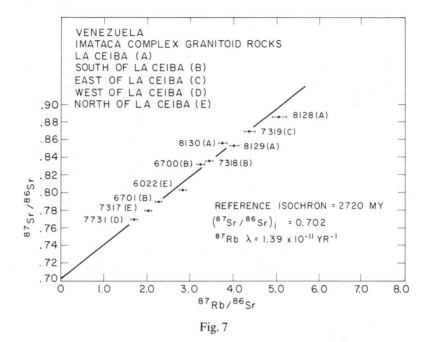

Fig. 7

by a medium- to coarse-grained, pink granitic phase, including pegmatites.

Finally, there are about ten granitic bodies in the Imataca Province large enough to be mapped. These show intrusive relationships, and from the results of this investigation are clearly related to a younger event with ages close to 2100 m.y. (Fig. 8). They are fairly uniform in composition; generally quartz monzonites.

The Pastora Province in Venezuela to the south of the Imataca is separated from it by the extensive Guri Fault System, a zone of multiple faulting, shearing and mylonitization. The sequence of events in the Pastora Province is still in debate. Geologists working in

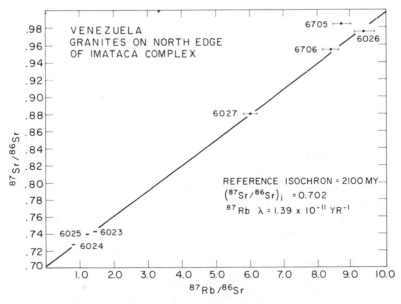

Fig. 8

the region visualize an older peneplaned basement (The Supamo Complex) overlain by the Pastora–Carichapo sedimentary-volcanic sequences (Bellizzia, 1974), and subsequent remobilization, folding and later granite intrusion of both the basement and overlying rocks. This would imply that the Supamo basement antedated the Pastora–Carichapo. The Pastora–Carichapo has been considered by most investigators to be the equivalent of the Barama–Mazaruni of Guyana, the Paramaca–Rosebel–Armina of Surinam, the Paramaca–Bonidoro–Orapu of French Guiana, the Amapa Series of Brazil and the upper and lower Birrimian of Liberia and Ivory Coast, all supracrustals on a pre-existing basement.

Pastora–Carichapo assemblage of Venezuela is bounded on the South by a great fault and a range of mountains 60 km across underlain by rocks of the Kanuku group (Berrangé, 1974) consisting chiefly of biotite and biotite–garnet gneisses in the amphibolite and granulite facies, hypersthene gneisses, and charnockites. The great fault can be traced eastward 400 km where it brings up pyroxene granulites and gneisses in the Bakhuys Mountains of Surinam, correlated with the Kanuku group. These rocks in Surinam have been called the Coeroeni and Falawatra groups and are described by Dahlberg (1974).

To the south and east of this great horst there is another basin in French Guiana in

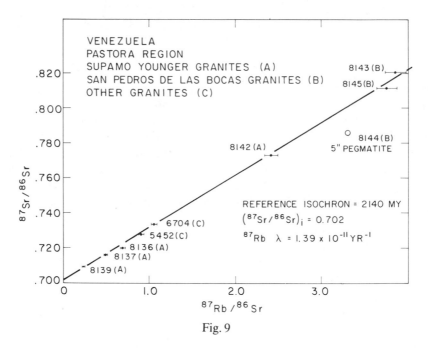

Fig. 9

The granites which cut both the Supamo basement and the Pastora–Carichapo are found to be similar in age to the late granites in the Imataca region, namely 2100 m.y. (Fig. 9). They are therefore correlated with an early expression of the Trans-Amazonian thermotectonic event in South America and the Eburnean in West Africa.

The wide Guyana–Venezuela basin of sediments comprising the area of eugeosynclinal facies defined in the Barama–Mazaruni assemblage of northern Guyana and the

which the stratigraphy appears to correspond with that in the Guyana–Venezuela basin. The basement underlying this basin, the Ile de Cayenne, appears to be Archean (Allègre, pers. comm.; B. Choubert, 1964).

If these supracrustal sequences are equivalent to the Birrimian in west Africa, the Supamo Complex in Venezuela, the Kanuku in Guyana, and the Falawatra in the Bakhuys Mountains of Surinam should be equivalent to the 2700 m.y. Liberian basement. Rb–Sr whole-rock age values on these rocks are plot-

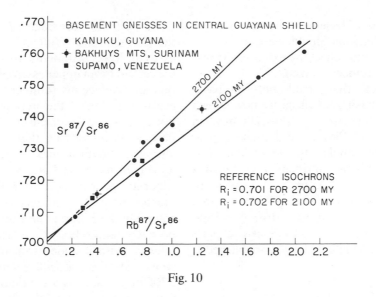

Fig. 10

ted in Fig. 10. Some of the age values are close to 2700 m.y. and the others are clearly pre-Trans Amazonian, but appear to have been variously affcctcd by it, showing a downward resetting from an upper limit of about 2700 m.y. In a further test on the Supamo trondhjemitic gneiss from the Pastora province zircon separations indicated a Concordia-plot agc of 2600 m.y. (Fig. 11). These preliminary data support the 2700 m.y.

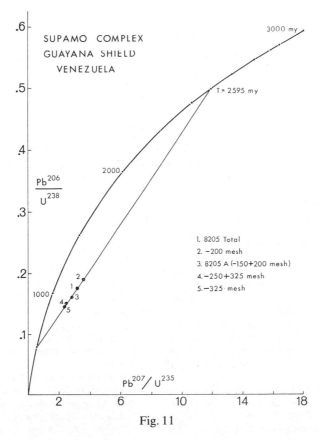

Fig. 11

age of a pervasive basement beneath a major part of the Guayana Shield, despite the fact that most age data reflect the extensive Trans-Amazonian thermotectonism.

Granitic rocks that intrude both the basement and the basin sequences are now known to range in age from 2100–2200 m.y. in Amapa, and the Pastora and Imataca provinces in Venezuela, to 1800–2100 m.y. in French Guiana, Surinam and Guyana (Snelling and McConnell, 1969; B. Choubert, 1964; Priem et al., 1966 and 1968, Hurley et al., 1973 and this work), to <1800 in regions in the west and north-west parts of the Guayana Shield (Hurley, unpublished).

Summary and Reconstruction of the Guiana–West African Shield

The pre-2000-m.y. geosynclinal basins of southern Venezuela, Guyana, Surinam, and French Guiana have a close resemblance to those of the Birrimian of West Africa (McConnell and Williams, 1969). In both cases the lower sedimentary–volcanic complex is overlain by an upper assemblage of arenaceous and rudaceous sediments. Both regions are also marked by the presence of manganese horizons in the lower part of the system. The basements underlying these basins, forming also the Imataca and Liberian forelands, appear to have the Late Archean age of the Liberian orogeny, 2700–2800 m.y. In the north-west part of the reconstructed cratonic assembly the Liberian and Imataca provinces both contain relics of metasedimentary sequences with included iron formation which show scattered age values between 3100 and 3700 m.y. There is a continuity of structural trends between the two regions if South America is rotated to fit into its former position. In both shield areas the geosynclinal basins have been folded, metamorphosed, and invaded by granites in a major event at 2100–1800 m.y.

Acknowledgements

The laboratory work was supported by National Science Foundation grants GA-31872 and A-40242.

References

Allègre, C. J., and Caby, R., 1972. 'Chronologie absolue du Précambrien de l'Ahaggar occidental', *C.R. Aca. Sci. Paris*, D, **275**, 2095–2098.

Allen, P. M., Snelling, N. J., and Rex, D. C., 1967, 'Age determinations from Sierra Leone', In *M.I.T. Fifteenth Annual Progress Report No. 1381-15 to the U.S. Atomic Energy Commission*, P. M. Hurley (Ed.).

Andrews-Jones, D. A., 1966. 'The geology and mineral resources of the northern Kambui schist belt and the adjacent granulites', *Sierra Leone Geol. Survey Bull.*, **6**, 100.

Bassot, J. P., Bonhomme, M., Roques, M., and Vachette, M., 1962. 'Mesures d'ages absolus sur les series précambriennes et paleozoiques du Senegal Oriental', *Bull. Geol. Soc. France, 7th Ser.*, 401–405.

Bellizzia, C. M., 1974. Paleotectonica del escudo de Guayana, *Ninth Inter-Guayana Geological Conference, Spec. Publ. No. 6, Ministerio de Minas e Hidrocarburos, Venezuela*, pp. 251–305.

Berrangé, J. P., 1974. The Tectonic/Geological Map of Southern Guyana, *Ninth Inter-Guayana Geological Conference, Spec. Publ. No. 6, Ministerio de Minas e Hidrocarburos, Venezuela*, pp. 159–178.

Bertrand, J. M. L., 1970. 'Remarques et hypothèses à propos de l'Ahaggar central et oriental', *Notes et Mem. Serv. Geol. Maroc*, **236**, 87–89.

Bertrand, J. M. L., 1974. 'Evolution polycyclique des gneiss precambriéns de l'Aleksod (Hoggar Central, Sahara Algerien)', *C.M.R.S., Centre Rech. Zones Arides, Séries Geol. No. 19*, p. 307.

Bonhomme, M., 1962, 'Comparison a l'etude géochronologique de la Plate-forme de l'ouest Africain', *Ann. Fac. Sci. Clermont–Ferrand*, **25**, Ser. Geol. 8, Etudes Geochronologiques I, p. 31–38.

Caby, R., 1972. 'Evolution préorogeñique site et agencement de la chaine pharusienne dans le NW de l'Ahaggar (Sahara algerién), sa place dans l'orogenèse pan-africaine en Afrique occidentale', *Notes et Mem. Serv. Geol. Maroc.*, **236**, 65–80.

Cahen, L., and Snelling, N. J., 1966. *The Geochronology of Equatorial Africa*, North–Holland Publ. Co., Amsterdam, 195 pp.

Charlot, R., Choubert, G., Faure-Muret, A., and Tisserant, D., 1970. 'Etude geochronologique de Precambrian de l'Anti-Atlas (Maroc)', *Maroc Service Géol. Notes*, **30**, No. 225, 99–134.

Choubert, B., 1964. 'Ages absolus du Précambrian guyanais', *Compt. Rend. Acad. Sci. Paris*, **258**, 631–634.

Choubert, G., and Faure-Muret, A. (Eds.), 1971a. 'Tectonique de l'Afrique', *UNESCO Earth Sciences*, **6**, p. 602.

Clifford, T. N., 1970. 'The structural framework of Africa', in Clifford, T. N., and Gass, I. G. (Eds.), *African Magmatism and Tectonics*, Oliver and Boyd, Edinburgh, pp. 1–26.

Dahlberg, E. H., 1974, Granulites of Sedimentary Origin Associated with Rocks of the Charnockite Suite in the Bakhuys Mountains, N. W. Surinam, *Ninth Inter-Guayana Geological Conference, Spec. Publ. No. 6, Ministerio de Minas e Hidrocarburos, Venezuela*, pp. 415–423.

Ferra, G., and Gravelle, M., 1966. 'Radiometric ages from western Ahaggar (Sahara) suggesting an eastern limit for the West African craton', *Earth Planet. Sci. Lett.*, **1**, 319–324.

Grant, N. K., 1970. 'Geochronology of Precambrian basement rocks from Ibadan, Southwestern Nigeria', *Earth Planet. Sci., Lett.*, **10**, 29–39.

Grant, N. K., 1973. 'Orogeny and reactivation to the west and southeast of the West African Craton', In Nairn, A. E. M., and Stehli, F. G. (Eds.), *The Ocean Basins and Margins*, vol. 1, Plenum Press, New York, pp. 447–492.

Hurley, P. M., Leo, G. W., White, R. H., and Fairbairn, H. W., 1971. 'Liberian age province (about 2700 m.y.) and adjacent provinces in Liberia and Sierra Leone', Geol. Soc. Am. Bull., **82**, 3483–3490.

Hurley, P. M., and Rand, J. R., 1973. 'Outline of Precambrian chronology in lands bordering the South Atlantic, exclusive of Brazil', In Nairn, A. E. M., and Stehli, F. G. (Eds.), *The Ocean Basins and Margins*, vol. 1, Plenum Press, New York, pp. 391–410.

Junner, N. R., 1940. 'Geology of the Gold Coast and Western Togoland', *Bull. Geol. Soc. Am.*, **11**, 40.

Kalliokoski, J., 1965. 'Geology of north-central Guayana Shield, Venezuela', *Bull. Geol. Soc. Am.*, **76**, 1027–1050.

Lassere, M., Lameyre, J., and Buffiere, M.-M., 1970. 'Données geochronologiques sur l'axe precambrian Yetti-Eglab en Algerie et en Mauritanie du Nord' (France), *Bur. Recherches Géol. et Minieres Bull.*, ser. 2, sec. 4, no. 2, pp. 5–13.

Leo, G. W., and White, R. W., 1967. 'Geologic reconnaissance in western Liberia', U.S. Geol. Survey open-file rept., 29 pp., *Symp. on continental drift*, Montevideo, Uruguay, October (in press).

McConnell, R. B., and Williams, E., 1969. 'Distribution and provisional correlation of the Precambrian of the Guiana Shield', *Proc. Eighth Guiana Geological Conference, Georgetown*, Geological Survey of Guyana.

Papon, A., Roques, M., and Vachette, M., 1968. 'Age de 2700 millions d'années déterminé par la méthode au strontium, pour la série charnockitique de Man, en Cote-d'Ivoire', *Acad. Sci. Comptes Rendus*, **266**, 2046–2048.

Posadas, V. G., and Kalliokoski, J., 1967. 'Rb–Sr ages of the Encrucijada granite intrusive in the Imataca Complex, Venezuela', *Earth Planet. Sci. Lett.*, **2**, 210–214.

Priem, H. N. A., Boelrijk, N. A. I. M., Hebeda, E. H., Verschure, R. H., and Verdurmen, E. A. Th., 1968. 'Isotopic age determinations on Surinam rocks, 4. Ages of basement rocks in north-western Surinam and the Roraima tuff at Tafelberg', *Geol en. Mijn.*, **47**, 191–196.

Priem, H. N. A., Boelrijk, N. A. I. M., Verschure, R. H., and Hebeda, E. H., 1966. 'Isotopic age determinations on Surinam rocks, 1', *Geol. en Mijn.*, **45**, 16–19.

Snelling, N. J., and McConnell, R. B., 1969. 'The geochronology of Guyana', *Geologie en Mijnbouw*, **48**, 201–213.

Sougy, J., 1962. 'West African fold belt', *Bull. Geol. Soc. Am.*, **72**, 871.

Spooner, C. M., Berrangé, J. P., Fairnbairn, H. W., 1971. 'Rb–Sr whole-rock age of the Kanuku Complex, Guyana', *Geol. Soc. Am. Bull.*, **82**, 207–210.

Tagini, B., 1975. *Esquisse géotectonique de la Cote d'Ivoire*, Soc. pour le Developpement Minier de la Cote d'Ivoire, Abidjan, unpublished report, p. 100.

Vachette, M., 1964. 'Ages radiometriques de formations cristallines d'Afrique Equatoriale', *Ann. Fac. Sci. Clermont–Ferrand*, **25**, Ser. Geol. 8, Etudes Geochronolgiques I, pp. 31–38.

Vachette, M., 1964. 'Essai de synthèse des déterminations d'ages radiometriques de formations crystallines de l'Ouest Africain (Cote d'Ivoire, Mauritanie, Niger)', *Univ. Clermont Fac. Sci. Ann.*, no. 25, ser. geol. 8, pp. 7–30.

White, R. W., and Leo, G. W., 1971, 'Geologic reconnaissance in western Liberia', *Liberian Geol. Survey Spec. Paper no. 1*, p. 18.

The Atmosphere

Archaean Atmosphere and Evolution of the Terrestrial Oxygen Budget

Manfred Schidlowski

Max-Planck-Institut für Chemie (Otto-Hahn-Institut), Saarstr. 23, D-65 MAINZ, W-Germany.

Abstract

A review is attempted of our present knowledge pertaining to the development of the terrestrial atmosphere during the early history of the Earth. Based on a geochemical ^{13}C mass balance and current concepts for the increase of the stationary sedimentary mass as a function of time, a quantitative model is presented for the evolution of the terrestrial oxygen budget of photosynthetic origin. The implications of this model, as well as probable interdependencies between atmospheric and organic evolution, are discussed in the light of the available geological and paleontological evidence.

According to our present knowledge of planetary atmospheres the gaseous shell of the Earth is quite unique within our solar system. There are, however, good reasons to believe that the Earth's primordial atmosphere did not differ very much in composition from the atmospheres of our neighbouring planets and that, therefore, the present atmosphere must be the result of an evolution which has started with the formation of the planet some $4 \cdot 5 \times 10^9$ yr ago and continued since over the major part of its geological history. It is generally accepted by now that this evolution was largely dependent on the development of the terrestrial biosphere which may be expected to have exerted a profound impact on its inorganic environment. The following paper aims at briefly reviewing the outlines of this evolution, with special reference to the build-up of the terrestrial oxygen budget.

The Reducing Primordial Atmosphere

With the noble gases depleted on Earth by factors between 10^{-7} and 10^{-11} as compared with their cosmic abundances (Brown, 1949; Suess, 1949), we may infer that the parent planetesimals giving rise to the protoplanet did not carry with them gaseous shells of their own. Therefore, the primordial atmosphere must have been a secondary degassing product released from the mantle of the newly formed planet. After the latter had attained modern surface temperatures of about 300°K, the liberated gases were—with few exceptions—trapped within its gravity field, this consequently resulting in the build-up of a gaseous layer around the solid surface. If regarding the average compositions of contemporary volcanic exhalations as representative of the primary degassing products of the Earth, water vapour and CO_2 should have made up

Fig. 1. Assemblage of well-rounded pyrite pebbles, isolated from foot-wall portions of a Witwatersrand conglomerate (Elsburg A3 Reef, Orange Free State) (from Schidlowski, 1970; reproduced by permission of Bundesanstalt für Bodenforschung, Hannover)

the bulk of these volatiles, followed by H_2S, CO, H_2, N_2, CH_4, NH_3, HF, HCl, Ar, and others. As is attested, inter alia, by the Fe^{2+}/Fe^{3+} ratio of primary rocks (cf. von Engelhardt, 1959; Holland, 1962), the material of the Earth's mantle and crust is undersaturated with respect to oxygen, this accounting for a negligible equilibrium concentration of molecular oxygen in these emanations. Accordingly, the primordial atmosphere must have been a reducing one, the reducing state having, in all probability, prevailed during the first $2 \cdot 5 \times 10^9$ yr of the Earth's history during which minerals like pyrite and uraninite (which are thermodynamically unstable in oxygenated environments) were concentrated in ancient detritals (Figs. 1 and 2). The concept of an anoxygenic atmosphere during the Early and Middle Precambrian is further substantiated by the ubiquitous presence in the oldest sedimentary record of banded iron formations for the origin of which a weathering transport of iron in the bivalent state seems to be a necessary prerequisite (Cloud, 1973).

In contrast to the above degassing model based on present-day volcanic emanations, it cannot be excluded that methane and ammonia prevailed among the components discharged from the Earth's interior at the very beginning of atmospheric evolution (Urey, 1959). With hydrogen being the most abundant element in the cosmos, carbon can be actually expected to have existed as CH_4 rather than as CO_2, provided the primitive Earth was not substantially depleted in this element. It has been shown that Jupiter and Saturn, for instance, possess reducing atmospheres of this latter type. As can be inferred from the occurrence of carbonates within the oldest sediments, however, a primary major methane content of the primitive atmosphere should have been definitely replaced by carbon dioxide as from about $3 \cdot 5 \times 10^9$ yr ago.

Regarding the main constituents of the present atmosphere we may assume that N_2 and the noble gases have principally accumulated through the ages, with only He escaping at a certain rate into outer space. As a result of early ocean formation (made possible by surface temperatures favouring a condensation of water vapour), the CO_2 content should have always been close to present levels, the salient buffering mechanism being most probably provided by marine silicate equilibria (Sillén, 1963). The principal problem of atmospheric evolution is, therefore, the build-up of modern

Fig. 2. Detrital grains of uraninite with dust-like inclusions of galena (PbS, white), the latter also forming overgrowths on individual grains. Note typical muffin-shape of grain in lower left corner (polished section, oil immersion) (from Schidlowski, 1966; reproduced by permission of E. Schweizerbart'sche Verlagsbuchhandlung, Stuttgart)

partial pressures of molecular oxygen, which will be dealt with in the following section.

The Origin of the Oxygenic Atmosphere

Because of the pronounced oxygen deficiency in the material of the lithosphere, degassing of the Earth's crust and mantle cannot have supplied molecular oxygen to the atmosphere–ocean system. Thus, we have to resort to a non-geological oxygen source, releasing this component by splitting gaseous oxides such as water vapour or carbon dioxide in an endergonic or 'uphill' reaction. *With solar radiation providing the most convenient energy source for such process, the presence of O_2 in the atmosphere must be ultimately due to a photochemical effect.* Here, we may distinguish between an *inorganic* and an *organic* photochemical effect.

The Inorganic Photochemical Effect

Such effect is provided by photolysis of water vapour in the upper atmosphere, followed by a gravity separation of the reaction products. The photochemically active 'dissociative' part of the solar spectrum comprises the short wavelengths between 1500 and 2100 Å. The net result of this photodissociation is determined by the preferential escape from the gravity field of the hydrogen released. Once hydrogen has reached heights of about 100 km, recombination processes become more and more unlikely (because of low atmospheric density) and its escape chances are increased proportionately.

It has been argued that this process of inorganic O_2 production will come to an end at a certain equilibrium concentration of oxygen, as the latter will absorb (and thereby shield the water vapour from) the photochemically active radiation. According to Berkner and Marshall (1967) self-regulation is likely to occur already at a very low oxygen pressure of 10^{-3} PAL (=Present Atmospheric Level), whereas Brinkmann (1969) and Hesstvedt et al. (1964) assume considerably higher equilibrium levels (the Hesstvedt model even indicating that photolytic splitting of water might, in principle, account for our present atmospheric oxygen pressure, provided the hydrogen released had been continuously removed from the system). These models do not, however, take into account the enormous amount of O_2 which has been transferred, as a result of oxidation processes, to the 'bound' oxygen reservoir of the crust where it is stored as either sulphate or oxide of trivalent iron.

At present, the escape rate of hydrogen atoms due to photolytic water splitting ($2 \times 10^8 \, cm^{-2} \, sec^{-1}$, with a probable error of less than a factor of 2) does not provide an important oxygen source and probably never did so in the geologic past. Part of this hydrogen loss is, furthermore, balanced by the incoming solar proton flux. As can be inferred from the terrestrial ^3He budget, however, this proton influx is smaller by about 3 orders of magnitude than the escape rate of hydrogen (cf. Walker, 1974; 1976).

It should be noted in this context that the stationary oxygen concentrations recorded for the atmospheres of several of our lifeless neighbour planets ($<1\%$) do not reach by far the equilibrium levels of either the Brinkmann or Hesstvedt model. As will be detailed below, the geochemical carbon–oxygen balance sheet provides very strong evidence of an ultimately biogenic derivation of the terrestrial oxygen budget.

The Organic (Biologic) Photochemical Effect

It is generally accepted by now that the build-up of living systems from inorganic matter ('*generatio spontanea*') could be achieved only within a reducing substrate. Accordingly, a primordial atmosphere devoid of oxygen was a necessary prerequisite for the origin of life on Earth. We can, therefore, assume that the oldest self-replicating protein assemblages ('eobionts') existed in an O_2-deficient environment, i.e. they must have utilized the anaerobic process of fermentation to meet their energy requirements (cf. Fig. 5). Browsing on the abiotically produced amino acids of the surrounding substrate, the so-called 'prebiotic soup', these eobionts were actually heterotrophs (or more specifically 'amino-heterotrophs'), utilizing organic food for their livelihood.

However, during subsequent organic evolution mutants made their appearance which were independent of this mode of nutrition, having acquired instead the capability to directly synthesize organic substances from inorganic compounds of their environment. A convenient contrivance serving this purpose was the reduction of CO_2 to carbohydrates which, from the standpoint of energetics, is effected most economically by photosynthetic activity of green plants and certain blue–green algae (Broda, 1975). In this reaction, water acts as hydrogen donor as indicated in the following simplified version of the photosynthesis reaction:

$$CO_2 + H_2O \xrightarrow{h\nu} CH_2O + O_2$$

From carbon dioxide and water these 'autotrophic' life forms were able to synthesize, in a thermodynamic 'uphill' reaction powered by low-energy (visible) solar radiation, organic matter (carbohydrates), with molecular oxygen released as a metabolic by-product. *It is obvious from the stoichiometry of this reaction that for each carbon atom fixed in organic substance one O_2-molecule is being released to the environment.*

Quantitatively, the process of photosynthesis is the most important biochemical reaction taking place on Earth. It is, therefore, reasonable to assume that the oxygen content of the terrestrial atmosphere is ultimately due to photosynthetic O_2 production. Recent model calculations based on the isotopic composition of carbon in Precambrian sediments (Junge et al., 1975) indicate that the build-up of a sedimentary reservoir of organic carbon started prior to $3 \cdot 7 \times 10^9$ yr and that, accordingly, photosynthesis had been 'invented' by then. The oldest paleontological evidence of biogenic oxygen production is provided by algal bioherms from Rhodesia with a minimum age of some $2 \cdot 9 \times [10^9]$ yr (Bond et al., 1973). Further, degradation products of chlorophyll (phytane, pristane, porphyrins) have been reported from sediments some 3×10^9 yr old (Barghoorn and Schopf, 1966; Oró and Nooner, 1967; Kvenvolden and Hodgson, 1969).

Evolution of the Terrestrial Oxygen Budget

It is well known from geochemical investigations (Ronov, 1968) that the sediments of the Earth's crust contain the largest reservoir of organic carbon (about 10^{22} g), the amount fixed in the living biomass being smaller by 4 orders of magnitude. If all molecular oxygen ever released were produced by photosynthetic activity of plants, the sedimentary C_{org} reservoir should be coupled—via photosynthesis reaction—with the overall terrestrial 'free' oxygen budget. If we could trace, therefore, through geological history the size of the sedimentary reservoir of organic carbon, we would get, at the same time, a quantitative picture of the evolution of photosynthetic oxygen.

According to a recent mass balance (Li, 1972), the 10×10^{21} g of organic carbon in the crust are matched by $29{\cdot}8 \times \pm 8{\cdot}9 \times 10^{21}$ g of oxygen of photosynthetic origin, the bulk of

the latter being stored as SO_4^{2-} and Fe_2O_3 in sediments (atmospheric O_2 accounts for only $1{\cdot}3 \times 10^{21}$ g of this total amount). With the stoichiometric equivalent of the sedimentary C_{org} reservoir being 27×10^{21} g of oxygen, the above estimate comes very close to the theoretically expected figure, thus lending decisive support to the concept of a photosynthetic origin of the terrestrial oxygen budget.

As has been fully set out elsewhere (Schidlowski et al., 1975), a reasonable approximation for the quantitative evolution of the terrestrial oxygen budget may be actually obtained by combining sedimentary carbon isotope data with a plausible model for the accumulation of the sedimentary mass as a whole. As a result of the constraints imposed by the global ^{13}C mass balance, the $\delta^{13}C$ values of organic and carbonate carbon are a measure of the relative proportions of both carbon species within the total reservoir of sedimentary carbon (Fig. 3). With the $\delta^{13}C$ values of marine carbonates lying mostly around zero permil (PDB) as from about the

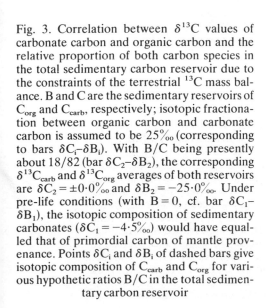

Fig. 3. Correlation between $\delta^{13}C$ values of carbonate carbon and organic carbon and the relative proportion of both carbon species in the total sedimentary carbon reservoir due to the constraints of the terrestrial ^{13}C mass balance. B and C are the sedimentary reservoirs of C_{org} and C_{carb}, respectively; isotopic fractionation between organic carbon and carbonate carbon is assumed to be $25\%_0$ (corresponding to bars $\delta C_i{-}\delta B_i$). With B/C being presently about 18/82 (bar $\delta C_2{-}\delta B_2$), the corresponding $\delta^{13}C_{carb}$ and $\delta^{13}C_{org}$ averages of both reservoirs are $\delta C_2 = \pm 0{\cdot}0\%_0$ and $\delta B_2 = -25{\cdot}0\%_0$. Under pre-life conditions (with $B = 0$, cf. bar $\delta C_1{-}\delta B_1$), the isotopic composition of sedimentary carbonates ($\delta C_1 = -4{\cdot}5\%_0$) would have equalled that of primordial carbon of mantle provenance. Points δC_i and δB_i of dashed bars give isotopic composition of C_{carb} and C_{org} for various hypothetic ratios B/C in the total sedimentary carbon reservoir

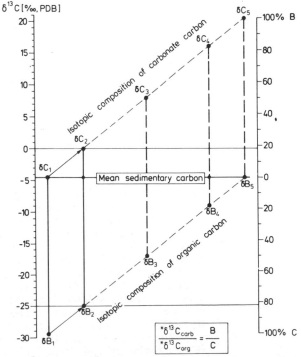

530

start of the sedimentary record, this would imply that the percentage of C_{org} in the sedimentary carbon reservoir was always close to the 'Ronov ratio' $R = C_{org}/(C_{org}+C_{carb}) \cong$ 0·18 (cf. Ronov, 1968; Broecker, 1970). In other words, *organic carbon should have always made up roughly one-fifth of total sedimentary carbon as from at least $3\cdot3\times 10^9$ yr ago.*

Introduction of this ratio into currently accepted models for the accumulation of the sedimentary mass as a function of time would give us absolute figures for the evolution of the sedimentary C_{org} reservoir. According to the principle of 'chemical uniformitarianism' (Garrels and Mackenzie, 1971) based on the uniformity through geologic time of the over-all geochemical mass balance between igneous rocks and acid volatiles on the one hand and

sedimentary rocks on the other, the sedimentary shell can be expected to have always contained its present burden of *total* carbon of about 3%. Assuming that the stationary sedimentary mass has asymptotically approached the present one as a function of an exponentially decreasing degassing rate, and knowing from the available isotope data that C_{org} has always accounted for about one-fifth of total sedimentary carbon, we may calculate the increase of both the sedimentary C_{org} reservoir and its oxygen equivalent through geologic time. A graphic synopsis of the resulting evolution of the stationary oxygen reservoir of photosynthetic pedigree is given in Fig. 4.

In detail, the relationship presented graphically in Fig. 4 is derived as follows. Assuming with Li (1972) an overall terrestrial degassing constant $\lambda = 1\cdot16\times10^{-9}$ y^{-1}, we would get a

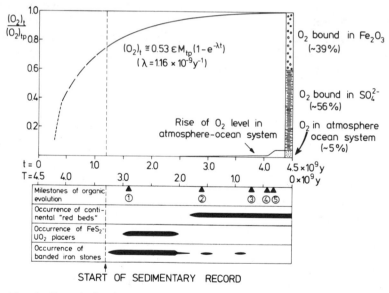

Fig. 4. Cumulative increase of the stationary reservoir of photosynthetically produced oxygen as a function of time, with reservoirs existing at times t expressed as fractions of the present reservoir $[(O_2)_t/(O_2)_{tp}$; $tp = 4\cdot5\times 10^9$ yr]. Column on right side shows partitioning of oxygen between the 'bound' and the 'free' reservoir in the present budget (note that molecular O_2 makes up 5% only of the present oxygen reservoir). Small curve shows tentative rise of free oxygen level in atmosphere–ocean-system as inferred from geological and paleontological evidence represented below. Milestones of organic evolution indicated are: (1) appearance of oldest algal bioherms (photoautotrophic blue–green algae); (2) appearance of eukaryotic biota; (3) appearance of oldest eumetazoan faunas; (4) life conquers continents (Upper Silurian); (5) appearance of exuberant continental floras (Upper Carboniferous) (from Schidlowski et al., 1975; reproduced by permission of Elsevier Publishing Company, Amsterdam)

quantitative picture of mantle degassing as a function of time. Since sedimentary rocks can be visualized as products of a global acid titration process between primary igneous rocks and acid volatiles released from the Earth's interior, the cumulative amount of volatiles degassed at any time t would give us a convenient approximation for the contemporary sedimentary mass M_t, namely

$$M_t \cong M_{tp} (1 - e^{-\lambda t}), \qquad (1)$$

with t_0 being the time of the Earth's creation $4 \cdot 5 \times 10^9$ yr ago and $M_{tp} = 2 \cdot 4 \times 10^{24}$ g the presently existing sedimentary mass. With about 3% of M_t consisting always of carbon and the $\delta^{13}C$ values of ancient carbonates indicating that roughly one-fifth of this latter amount was C_{org}, we would get, for any time t covered by the sedimentary record, both the stationary reservoir of sedimentary organic carbon $(C_{org})_t$ and the corresponding oxygen equivalent $(O_2)_t$:

$$(C_{org})_t \cong 0 \cdot 2 \varepsilon M_{tp} (1 - e^{-\lambda t}) \qquad (2)$$

$$(O_2)_t \cong 0 \cdot 53 \varepsilon M_{tp} (1 - e^{-\lambda t}) \qquad (3)$$

($\varepsilon = 0 \cdot 03$ is the average carbon content of sedimentary rocks, $0 \cdot 53$ is a stoichiometric conversion factor). Naturally, Eq. (3) rests on the premise that the reduced carbon in sedimentary rocks owes its origin to the most advanced form of photosynthesis which releases free oxygen—an assumption supported, inter alia, by the occurrence of breakdown products of chlorophyll within the oldest sediments.

According to this model (Fig. 4), a total oxygen reservoir close to 80% of the present one should have existed already near the start of the sedimentary record some 3×10^9 yr ago. If the partitioning of this oxygen between the 'free' and the 'bound' reservoir had been the same as today (with about 5% of the total budget stored as molecular oxygen in the atmosphere, cf. budget column on Fig. 4), the partial pressure of O_2 in the Early Precambrian atmosphere should have likewise amounted to some 80% of this present value. However, as we can infer from the available geological and paleontological evidence (Fig.

4, lower part), the ancient atmosphere was largely devoid of oxygen prior to about 2×10^9 yr ago. Therefore, we must necessarily conclude that, during the early history of the Earth, a fairly modern total oxygen reservoir ($\geq 80\%$ of the present one) was coupled with negligible O_2 concentrations in the atmosphere.

Such disproportionation between the 'bound' and the 'free' oxygen reservoirs would imply that the ancient oxygen cycle must have been practically 'short-circuited', with all photosynthetic oxygen produced being instantaneously sequestered by the crust as a result of very effective O_2 consuming reactions. In principle, there would be no difficulties to store the 5% of the total oxygen budget presently contained in the atmosphere–ocean system (cf. Fig. 4) within the 'bound' reservoir as well, since the reducing capacity of the crust–mantle system is by no means exhausted at the present level of sedimentary O_2 fixation. Presently, less than half of the sulphur contained in the ocean–sediment system exists as sulphate sulphur, the rest being sulphide sulphur constituting a potential reductant. Further, the Fe^{2+} content of primary rocks exposed at the Earth's surface is likely to provide an almost inexhaustible reducing reserve.

We have tentatively proposed (1975) that the process of banded ironstone formation characteristic of Early and Middle Precambrian times could have provided an oxygen-absorbing reaction of high efficiency. As had already been previously assumed (e.g. MacGregor, 1927; Urey, 1959; Cloud, 1973) the ancient seas were likely to have acted as accumulator systems for dissolved bivalent iron ions, washed away from the continents as a result of an anoxygenic weathering cycle (presently, Fe^{2+} is being oxidized to Fe^{3+} and largely retained in continental weathering crusts because of the low solubility of Fe^{3+} compounds; accordingly, the iron content of present-day sea water is in the order of ppb only). The hydrated ferrous ions within the ancient seas constituted ideal oxygen acceptors when the contemporary autotrophs released molecular oxygen to their environment. With the biological cycle submerged below ocean levels during the Precambrian, the oxygen-scavenging activity of the ferrous ions should have been extremely effective. Only after the ancient oceans had been swept of dissolved bivalent iron (which came to be precipitated as ferric iron in banded ironstone deposits), free oxygen could have accumulated in the seas and, consequently, in the atmosphere.

The incipient build-up of an atmospheric oxygen pressure is heralded by the appearance of the oldest continental red beds between $1·8$–$2·0 \times 10^9$ yr ago (Cloud, 1973; see also Fig. 4, lower part). As a result of the gradually increasing redox potential of the ancient atmosphere, oxidation of bivalent iron now was shifted to the continents. With the supply of Fe^{2+} to the oceans thus being dried up, the latter consequently lost their oxygen-absorbing capacity and more and more O_2 could escape into the atmosphere. Because of the sluggishness of oxidation weathering on the continents (where rocks are being slowly decomposed and their reducing constituents thus made available to oxygen attack), a new stationary state was finally established at a markedly increased atmospheric P_{O_2} level at which the new sink function ultimately matched the rate of oxygen production. The oxygen burden of the present atmosphere must, therefore, be looked upon as being due to the failure of thermodynamics to exercise strict control over the redox balance of the atmosphere–hydrosphere–crust system during operation of the present weathering cycle, with dynamic equilibrium being attained only at a considerable P_{O_2} in the 'free' reservoir.

Atmospheric Evolution and Diversification of Terrestrial Life

With the total terrestrial 'free' oxygen budget having almost certainly resulted from biologic activity, the biosphere has exerted a profound impact on atmospheric evolution. Moreover, grave infringements of the rules of equilibrium chemistry as attested by the co-existence in the present atmosphere of O_2, CH_4, N_2, etc., and by isotopic anomalies dis-

played by individual components (^{18}O enrichment in atmospheric O_2, 'DOLE effect') are likely to indicate that also the contemporary atmosphere is largely controlled by the totality of life on Earth, maintaining these disequilibria as a result of some 'biostationary' state. An interpretation of the atmosphere as an integral part of the biosphere would thus appear legitimate, at least for some of its components. A control of the terrestrial atmosphere by the biosphere has probably also accounted for obvious homeostatic tendencies of the Earth's gaseous shell during the geologic past (Lovelock and Margulis, 1974).

On the other hand, there is reason to believe that the atmosphere, and notably free oxygen, has triggered major biological innovations which consequently enabled life to advance to new levels of organizational complexity (for a recent review, see Tappan, 1974). Students of the paleontological record have long been puzzled by the enormous length of the time span (about 3×10^9 yr according to latest results) required by organic evolution before the appearance of the oldest eumetazoan faunas, whereas all further diversification of life was achieved during the last 700 m.y. of Earth's history. To account for this lag, several workers have entertained the view that the advent of multicellular life had awaited the attainment of an environmental oxygen pressure favourable for oxidative metabolism.

Primarily, photosynthetic O_2-production should have constituted a severe hazard for primitive life forms. Pending the acquisition of oxygen-mediating enzymes, free oxygen is actually toxic to living cells. As has been stressed by Cloud (1973), the process of banded ironstone formation in the Precambrian seas was of paramount importance for decontaminating the ancient environment. By keeping the O_2-content of the substrate at trivial levels for an extended period of time, life was given the opportunity to adapt gradually to its own waste product, finally learning to harness the process of oxidative degradation of organic matter to meet its own energy requirements.

From a synoptic scheme of the two main pathways of catabolism (Fig. 5) it is apparent

Fig. 5. Synopsis of the energy-releasing processes of fermentation and respiration. Oxidative breakdown of the six-carbon sugar glucose to carbon dioxide and water ('respiration') releases about 14 times more energy than anaerobic fermentation resulting in the formation of two molecules of lactic acid (other fermentative processes may yield different end-products, such as ethanol). $\Delta G_0'$ values given are the standard free energies at pH = 7. The energy released is used for 'phosphorylation' of adenosine diphosphate (ADP) to adenosine triphosphate (ATP), the latter constituting the universal energy carrier in all living systems. According to the mode of ATP synthesis we may distinguish, therefore, between *fermentative* and *respiratory* (oxidative) phosphorylation. Note that divergence of the two catabolic pathways occurs after formation of pyruvic acid in the anaerobic phase of glycolysis (adapted from Schidlowski, 1971; reproduced by permission of Geol. Vereinigung, Stuttgart)

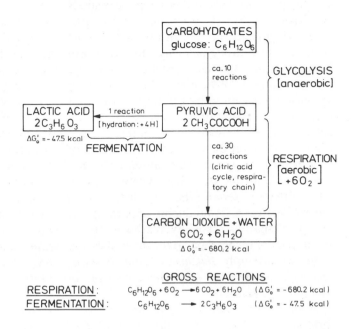

that fermentation and respiration do not differ principally from each other. During an initial anaerobic phase ('glycolysis') they follow the same lines, with pyruvic acid being formed as an intermediate product. Only as from this stage the anaerobic and aerobic processes are clearly separated. Formation of lactic acid as an end-product of anaerobic fermentation is achieved from here by one single reaction, whereas the aerobic process ('respiration') requires a whole series of additional steps (reactions of the citric acid or 'Krebs' cycle and the respiratory chain). Accordingly, *the process of fermentation (in the form of anaerobic glycolysis) still underlies aerobic metabolism as an 'archaic' element.* In the evolution of metabolism the aerobic pathway has obviously been imposed upon anaerobic fermentation, thus constituting a relatively late achievement of life. *The process of respiration is, therefore, no primary attribute of life, but a highly advanced form of catabolism using molecular oxygen as the ultimate hydrogen (or electron) acceptor instead of the end-product of the anaerobic pathway.*

Since facultative anaerobes have been found to switch over from fermentation to

respiration at about 10^{-2} of the present oxygen pressure ('Pasteur effect', cf. Rutten, 1971), the appearance of respiration should have awaited corresponding environmental oxygen levels. With all living eukaryotes contingent on oxidative metabolism and the oldest eukaryotic biota showing up on the fossil record $1·5 \pm 0·3 \times 10^9$ yr ago (Cloud and Gibor, 1970; Schopf, 1974), the oxygen pressure of the environment must have reached the above 'Pasteur' level of 10^{-2} PAL by that time. The energy requirements of the oldest eumetazoans evolving from primitive eukaryotes some $0·7 \times 10^9$ yr ago were probably such as to necessitate considerably higher O_2 concentrations. Moreover, it has been pointed out that biosynthesis of collagen, which must have preceded the acquisition of 'hard parts' by various metazoans and metaphytes towards the close of the Precambrian, is dependent on the availability of free oxygen (Towe, 1970).

It seems appropriate to discuss briefly in this context the coincidence during the Late Precambrian of the appearance of eukaryotic (notably multicellular) life with the incipient build-up of an environmental oxygen pressure. It would not appear unlikely that rising P_{O_2} in the environment has ultimately triggered the 'invention' of the most effective respiratory pathway of metabolism which, in turn, was the energetic prerequisite for the evolution of all 'higher' life. Taking the increased net gain in metabolic energies as a measure of evolutionary success, respiring cells are more 'advanced' (or differentiated) than their anaerobic ancestors. Evolution of multicellular biota and their subsequent functional and morphologic diversification did, in all probability, require the introduction of a most effective ATP-producing apparatus. As was particularly stressed by Warburg (1966), the impact of respiration on the diversification of life would imply *that this diversification was clearly dependent on an increased supply of metabolic energies.*

Hence, from the standpoint of thermodynamics, biological differentiation and organic evolution as a whole must be looked upon as highly improbable phenomena. According to the well-known Boltzmann relationship, the entropy S of a system is directly proportional to its probability W (S = k ln W). Since, from all possible states, the most probable ones will show up in reality, these latter are always characterized by maximum entropy. Any probable state is, furthermore, the result of a spontaneous process, i.e. one proceeding of its own accord *without requiring an energy supply from outside sources.*

Biological diversification being obviously dependent on increased utilization of metabolic energies, the evolutionary trend towards higher (or differentiated) life actually constitutes a 'crime against entropy'. Both maintenance of life as a non-equilibrium steady state with a marked negentropy and its evolutionary diversification are energy-absorbing (endergonic) processes, driving 'upstream' within the universal entropy flux (cf. Blum, 1968; Sylvester-Bradley, 1967). Once the energy requirements of life were served by oxidative metabolism, the resulting increase in available metabolic energies should have enabled major evolutionary innovations. It is reasonable to assume, therefore, that utilization of the oxidative pathway of metabolism has played the role of the 'MAX-WELLian demon' which caused evolution to go into the direction of more order (or diversification) and less entropy.

Acknowledgements

The evolutionary model of the terrestrial oxygen budget presented here has evolved from the current work of our group towards which R. Eichmann and C. E. Junge have extensively contributed. Financial support by the Deutsche Forschungsgemeinschaft (SFB 73) is gratefully acknowledged.

References

Barghoorn, E. S., and Schopf, J. W., 1966. 'Microorganisms three billion years old from the Precambrian of South Africa', *Science*, **152**, 758–763.

Berkner, L. V., and Marshall, L. C., 1967. 'The rise of oxygen in the Earth's atmosphere with notes on the Martian atmosphere', *Adv. Geophys.*, **12**, 309–331.

Blum, H. F., 1968. *Time's Arrow and Evolution*, Princeton University Press, Princeton, 232 pp.

Bond, G., Wilson, J. F., and Winnall, N. J., 1973. 'Age of the Huntsman limestone (Bulawayan) stromatolites', *Nature*, **244**, 275–276.

Brinkmann, R. T., 1969. 'Dissociation of water vapor and evolution of oxygen in the terrestrial atmosphere', *J. Geophys. Res.*, **74**, 5355–5368.

Broda, E., 1975. *The Evolution of the Bioenergetic Processes*, Pergamon, Oxford, 220 pp.

Broecker, W. S., 1970. 'A boundary condition on the evolution of atmospheric oxygen', *J. Geophys. Res.*, **75**, 3553–3557.

Brown, H., 1949. 'Rare gases and the formation of the Earth's atmosphere', In *The Atmospheres of the Earth and Planets*, G. P. Kuiper (Ed.), University of Chicago Press, Chicago.

Cloud, P. E., 1973. 'Paleoecological significances of the banded iron formation', *Econ. Geol.*, **68**, 1135–1143.

Cloud, P. E., and Gibor, A., 1970. 'The oxygen cycle', *Sci. Amer.*, **1970**, 111–123.

Engelhardt, W. von, 1959. 'Kreislauf und Entwicklung in der Geschichte der Erdrinde', *Nova Acta Leopold.*, *N.F.*, **21**, 85–99.

Garrels, R. M., and Mackenzie, F. T., 1971. *Evolution of Sedimentary Rocks*, Norton, New York, 397 pp.

Hesstvedt, E., Henriksen, S., and Hjartarson, H., 1974. 'On the development of an aerobic atmosphere. A model experiment', *Geophys. Norvegica*, **31**, 1–8.

Holland, H. D., 1962. 'A model for the evolution of the Earth's atmosphere', In *Buddington Volume*, A. E. J. Engel, H. L. James and B. F. Leonard (Eds.), pp. 447–477.

Junge, C. E., Schidlowski, M., Eichmann, R. and Pietrek, H., 1975. 'Model calculations for the terrestrial carbon cycle: carbon isotope geochemistry and evolution of photosynthetic oxygen', *J. Geophys. Res.*, **80**, 4542–4552.

Kvenvolden, K. A., and Hodgson, G. W., 1969. 'Evidence for porphyrins in Early Precambrian Swaziland System sediments', *Geochim. Cosmochim. Acta*, **33**, 1195–1202.

Li, Y. H., 1972. 'Geochemical mass balance among lithosphere, hydrosphere and atmosphere', *Amer. J. Sci.*, **272**, 119–137.

Lovelock, J. E., and Margulis, L., 1974. 'Homeostatic tendencies of the Earth's atmosphere', In *Cosmochemical Evolution and the Origins of Life*, J. Oró et al (Eds.), Reidel, Dordrecht, pp. 93–103.

MacGregor, A. M., 1927. 'The problem of the Precambrian atmosphere', *S. Afr. J. Sci.*, **24**, 155–172.

Oró, J., and Nooner, D. W., 1967. 'Aliphatic hydrocarbons in Precambrian rocks', *Nature*, **213**, 1082–1085.

Ronov, A. B., 1968. 'Probable changes in the composition of sea water during the course of geological time', *Sedimentology*, **10**, 25–43.

Rutten, M. G., 1971. *The Origin of Life by Natural Causes*, Elsevier, Amsterdam, 420 pp.

Schidlowski, M., 1966. 'Beiträge zur Kenntnis der radioaktiven Bestandteile der Witwatersrand-Konglomerate I–III', *N. Jb. Miner., Abh.*, **105/106**, 183–202; 310–324; 55–71.

Schidlowski, M., 1970. 'Untersuchungen zur Metallogenese im südwestlichen Witwatersrand-Becken (Oranje-Freistaat-Goldfeld, Südafrika)', *Beih. Geol. Jahrb.*, **85**, 80+46 pp.

Schidlowski, M., 1971. 'Probleme der atmosphärischen Evolution im Präkambrium', *Geol. Rdsch.*, **60**, 1351–1384.

Schidlowski, M., Eichmann, R., and Junge, C. E., 1975. 'Precambrian sedimentary carbonates: carbon and oxygen isotope geochemistry and implications for the terrestrial oxygen budget', *Precambrian Res.*, **2**, 1–69.

Schopf, J. W., 1974, 'The development and diversification of Precambrian life', In *Cosmochemical Evolution and the Origins of Life*, J. Oró et al. (Eds.), Reidel, Dordrecht, p. 119–135.

Sillén, L. G., 1963. 'How has sea water got its present composition?', *Svensk. Kem. Tidskr.*, **75**, 161–177.

Suess, H. E., 1949. 'Die Häufigkeit der Edelgase auf der Erde und im Kosmos', *J. Geol.*, **57**, 600–607.

Sylvester-Bradley, P., 1967. 'Evolution versus entropy', *Proc. Geol. Assoc.*, **78**, 137–148.

Tappan, H., 1974. 'Molecular oxygen and evolution', In *Molecular Oxygen in Biology*, O. Hayaishi (Ed.), North–Holland, Amsterdam, pp. 81–135.

Towe, K. M., 1970. 'Oxygen-collagen priority and the early metazoan record', *Proc. Natl. Acad. Sci., Wash.*, **65**, 781–788.

Urey, H. C., 1959. 'The atmospheres of the planets'. In *Handbuch d. Physik*, 5. Flügge (Ed.), Springer, Berlin, pp. 363–418.

Walker, J. C. G., 1974. 'Stability of atmospheric oxygen', *Amer. J. Sci.*, **274**, 193–214.

Walker, J. C. G., 1976. *Evolution of the Atmosphere*, Hafner, New York, in press.

Warburg, O., 1966. 'Über die Ursache des Krebses', In *Molekulare Biologie des malignen Wachstums*, H. Holzer and A. W. Holldorf (Eds.), Springer, Berlin, pp. 1–16.

Implications for Atmospheric Evolution of the Inhomogeneous Accretion Model of the Origin of the Earth

JAMES C. G. WALKER

National Astronomy and Ionosphere Center, Arecibo Observatory, P.O. Box 995, Arecibo, Puerto Rico 00612, U.S.A.

Abstract

The most important parameter in models of the primitive atmosphere is the oxidation state of the atmospheric source gas, which depends on the oxidation state of the upper mantle. If the interior of the Earth was fairly homogeneous at the time the atmosphere originated, with free iron present in the upper mantle, then hydrogen would have been abundant in the atmospheric source gas, and the primitive atmosphere would have contained substantial concentrations of highly reduced species such as methane. According to the inhomogeneous accretion model of Earth formation, however, the proto-Earth was formed of material low in volatiles that condensed from the solar nebula at high temperatures. The proto-Earth melted and differentiated into core and mantle before it acquired a veneer of volatile-rich, low-temperature condensate. This late-forming veneer provided most of the material of the crust and upper mantle as well as of the atmosphere and ocean. Its overall oxidation state was approximately the same as that of the upper layers of the Earth today. According to this model, the atmospheric source gas would have been no more reduced than modern volcanic gases that have achieved equilibrium with basaltic magma. The primitive atmosphere might have resembled the modern atmosphere with the addition of a few per cent of hydrogen and the removal of all oxygen.

Introduction

It is frequently assumed that life originated on Earth in the presence of an atmosphere containing abundant hydrogen, methane, and ammonia (cf. Gabel and Ponnamperuma, 1972). The assumption of a primitive, highly reducing atmosphere derives from the predominance of hydrogen in the solar system and in the nebula from which the Earth and its atmosphere evolved. It is appealing because of the apparent ease of inorganic synthesis of organic molecules in highly reducing atmospheres. I doubt, nevertheless, whether the assumption is correct. In this paper I shall present some reasons for this doubt.

Primordial Atmosphere

An early reducing atmosphere might possibly have been a remnant of the nebula from which the solar system condensed. I shall call such an atmosphere primordial. Here I shall argue that the mass of nebular gas retained by the Earth was negligibly small (Abelson, 1965).

Compared with the Sun, the inert gases are depleted on Earth relative to non-volatile constituents by amounts so large as to suggest that gases were not retained by the Earth when it accreted (Moulton, 1905; Brown, 1952; Fisher, this volume). The present atmosphere has been formed by the release from the body of the Earth of material that was originally incorporated in the form of non-gaseous chemical compounds. If there was ever a primordial atmosphere it must have been almost completely dissipated before the secondary atmosphere accumulated. The primordial atmosphere may have been swept away by the enormously enhanced solar wind during the T Tauri phase of evolution of the primitive Sun (Cameron, 1973).

We may use the terrestrial abundance of neon to set an upper limit on the amount of primordial atmosphere that could have survived the dissipation. The mass of neon in the atmosphere is 6.48×10^{16} gm (Verniani, 1966) and the neon content of rocks appears to be negligible (Canalas et al., 1968; Fanale and Cannon, 1971; Phinney, 1972). Using the table of solar system abundances compiled by Cameron (1968) we calculate corresponding upper limits on the amounts of other elements that could have been retained in a primordial atmosphere of solar composition. The results in Table 1 show that the primordial atmosphere must have been reduced at one time to a mass less than 1% of the present mass of the atmosphere (5.1×10^{21} gm).

Table 1. Upper limits on the masses of the most abundant elements in a primordial atmosphere

	H	He	C	N	O	S
10^{16} gm	3500	1100	22	4.6	51	2.2

It does not seem very likely that the dissipation of the primordial atmosphere would have stopped when such a small amount of gas remained. Even Jeans escape, unaided by a T Tauri solar wind, could have dissipated 3.6×10^{19} gm of hydrogen in about 10^4 years (Hunten, 1973a). There is, moreover, evidence that the inert gases on the Earth today are not the remnants of a primordial atmosphere. Studies of meteorites have revealed two components of their light inert gas complements (Pepin and Signer, 1965). One, called 'solar', exhibits relative concentrations that are characteristic of the Sun; it is thought to consist of trapped solar wind particles. The other, called 'planetary', exhibits a much less rapid decrease of concentration with increasing atomic mass than does the solar component. The planetary component is believed to be gas adsorbed from the solar nebula by a process that favours the heavier species. The terrestrial complement of inert gases is planetary rather than solar (when allowance is made for xenon incorporated into shales), indicating that inert gases were incorporated into the Earth adsorbed on to solids and not in a primordial atmosphere of solar composition.

In summary, we find that if the Earth ever had a primordial reducing atmosphere (a remnant of the solar nebula), this atmosphere was dissipated, probably completely, before the accumulation of the present, secondary atmosphere. Any life that originated in the primordial atmosphere could hardly have survived the disappearance of this atmosphere, so it is on the secondary atmosphere that we must concentrate in exploring the conditions under which terrestrial life originated. In the next section I shall argue that a highly reducing secondary atmosphere could have developed only if the crust and upper mantle were much less oxidized than they are today.

Highly-reducing Secondary Atmosphere

A key question with regard to the secondary atmosphere is the oxidation state of the gases released from the solid Earth. With hydrogen escaping from the top of the atmosphere into space a highly reducing atmosphere could hardly have arisen unless the atmospheric source gases themselves were highly reducing.

Present-day volcanic gases are not highly reducing. Water vapour and carbon dioxide are much more abundant than hydrogen and carbon monoxide. Methane and ammonia are almost never found (White and Waring, 1963). In fact, the oxidation state of modern volcanic gases is in approximate agreement with theoretical predictions based on equilibrium with basalt melts (Holland, 1962, 1964; Fanale, 1971a; Nordlie, 1972). There is therefore every reason to believe that the average oxidation state of gases released from the solid Earth depends on the oxidation state of the material of the crust and upper mantle. We may conclude that a highly reducing secondary atmosphere could have originated only if the upper layers of the Earth were much more reduced than they are today.

In practice, this means that free iron would have to have been present in the upper mantle. Holland (1962) has shown that a primitive upper mantle containing free iron could have produced a highly reducing secondary atmosphere. Such an atmosphere therefore implies that the Earth accreted cold and more or less homogeneously, with the free iron that is now in the core initially distributed throughout the mantle. Presumably, the reducing atmosphere could have lasted until radioactive heating had softened the interior enough to permit the free iron to settle to the core.

There are several geochemical arguments against such a model of Earth formation, however (Ringwood, 1959, 1966; Turekian and Clark, 1969; Clark et al., 1972). The most compelling concerns the high observed abundance of nickel in ultramafic rocks and basalts. Nickel would have been almost entirely removed from these rocks if they had ever been in contact with the iron–nickel alloy that now forms the core. A second argument concerns the oxidation state of the upper mantle as reflected in the Fe_2O_3/FeO ratio in ultramafic rocks and basalts. This ratio is large enough to suggest that the rocks of the upper mantle were never intimately mixed with metallic iron.

Both of these lines of evidence suggest that free iron was never present in the upper mantle and this suggests in turn that there was never a highly reducing secondary atmosphere. Suppose, however, that the geochemical evidence has been misinterpreted and that a secondary atmosphere did arise and survive for a long time before the segregation of the core. Gravitational energy released by core formation would probably have been sufficient to cause widespread melting of the upper mantle and complete destruction of the crust (Hanks and Anderson, 1969; Clark et al., 1972). It does not seem likely that any terrestrial life could have survived this upheaval. The abundant volcanism would have caused any pre-existing atmosphere to come to equilibrium with the new oxidation state of the material of the upper mantle. The result would have been a lifeless Earth without a highly reducing atmosphere.

In a speculative subject like the early history of the Earth and its atmosphere all deductions must be regarded as tentative. The conclusion that I am about to draw may not be the only one or even the correct one. Nevertheless, I feel that any terrestrial life that originated in a possible primordial atmosphere or in a possible highly reducing secondary atmosphere would not have survived the dissipation of the primordial atmosphere or the tectonic upheaval associated with core formation. Therefore, I feel, modern terrestrial life did not originate in a highly reducing terrestrial atmosphere. It is not at all clear that a highly reducing atmosphere is necessary, anyway (Abelson, 1966; Rutten, 1971; Hubbard et al., 1971). It may be sufficient that the atmosphere lacked appreciable free oxygen (Eck et al., 1966). Later in this paper I will try to deduce the properties of a prebiological atmosphere that is consistent with the inhomogeneous accretion model of Earth formation.

Inhomogeneous Accretion of the Earth

In this section I shall present the main features of a theory of the origin of the solar system and the Earth that is based largely on papers by Turekian and Clark (1969), Clark et al. (1972), Grossman (1972), Grossman and

540

Larimer (1974), Cameron and Pine (1973), Cameron (1973), and Lewis (1972, 1974).

The solar system started as a cloud of gas, dust, and ice of a few solar masses at a temperature of about 10°K collapsing under its own gravitational attraction. Compression caused the temperature to rise to several thousand degrees, vapourizing all but the most refractory compounds, while conservation of angular momentum flattened the cloud into a disc. The properties of this disc have been calculated by Cameron and Pine (1973). Their results are summarized in Fig. 1. The thickness of the disc was governed by its own gravitational attraction and therefore was smaller near the centre, where matter was most abundant, than near the edges. Much of the disc was optically opaque and in a state of turbulent convection.

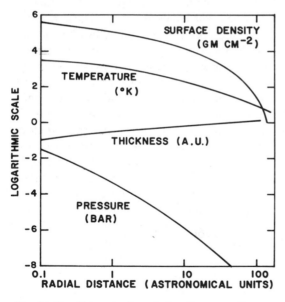

Fig. 1. Conditions in the nebular disc according to Cameron and Pine (1973). Pressure and temperature are shown for the central plane of the disc (reproduced with permission of Academic Press, Inc., New York)

The gravitational field of the disc caused the refractory solids to settle to the mid-plane; there they began to accumulate into larger bodies. Accretion was facilitated by the turbulence and also by viscous drag, which caused particles of different size to move at different speeds relative to the gas. At this time temperatures in the inner regions of the nebula were sufficiently high to prevent the condensation and accretion of any but the most refractory materials. The nuclei of the inner planets therefore consisted largely of corundum (Al_2O_3), perovskite ($CaTiO_3$), gehlenite ($Ca_2Al_2SiO_7$), and metallic iron (Grossman and Larimer, 1974).

The turbulence in the nebula transported angular momentum outward and matter inward, so the Sun grew while gas pressures decreased. Simultaneously, the nebula cooled by emission of radiation to space, allowing less refractory minerals to condense and accrete.

The rate of growth of the planetary nuclei increased rapidly with their masses and the strengths of their gravitational fields. The largest nuclei grew at the expense of their smaller neighbours. In the intermediate stages of planetary growth it is likely that the rate of accretion was large enough to melt the planets (Clark et al., 1972) or possibly even to volatilize some of the less refractory silicates (Cameron, 1973). The molten planets segregated efficiently into nickel–iron cores and silicate mantles. Volatiles that had been incorporated in the earlier stages were largely driven out of the planets at this time.

Meanwhile, the Sun continued to grow at the expense of the nebula, while nebular temperatures continued to fall, permitting the condensation of progressively less refractory compounds. The rates of growth of the planetary nuclei decreased as they gathered up more and more of the available condensed material and as the nebular density decreased. The planets therefore began to cool and solidify, driving to their surfaces the radioactive elements and other trace elements that are not readily incorporated into the silicate lattice. By this time the inner solar system contained hot, differentiated, volatile-free planets, solid debris not yet gathered up by the planets, and a rapidly cooling and dissipating nebular gas.

When temperatures fell below 750°K any metallic iron still exposed to the nebular gas began to be oxidized, first to the ferrous state and then, below 400°K, to the ferric state

(Grossman and Larimer, 1974). In the same range of temperatures the stable high-temperature compound of carbon, carbon monoxide, began to react with hydrogen to form methane and other less volatile hydrocarbons (Lewis, 1972). Below 350°K magnesium silicates reacted with water vapour to form hydrated silicates. Lower temperatures were probably not achieved in the inner solar system before the nebula was dispersed.

The cooling of the nebula was interrupted when the Sun had accumulated enough mass to become a star. During its approach to the main sequence it passed through a T Tauri phase during which it ejected into space several tenths of a solar mass of material in the form of a very strong solar wind (Hayashi, 1961; Ezer and Cameron, 1963, 1971). This wind carried the remaining nebular gas out of the inner solar system and also dissipated any primordial atmospheres that had accumulated around the inner planets (Cameron, 1973).

The protoplanets were left to gather up the solid debris left in their regions of the solar system from the condensation that had occurred during cooling of the nebula. This debris included refractory material that had condensed early but had not been accumulated right away. The refractory component provided the source of the nickel and iron in the upper mantle. The iron had been oxidized as a result of exposure to cold nebular gas. The debris also included volatile-rich material, resembling Type 1 carbonaceous chondrites, that had condensed when the nebula was quite cool. This volatile-rich material was the source of the atmosphere and ocean.

At the time of its dissipation the nebula was, of course, cooler in its outer parts than in its inner parts. Planets closer to the Sun are therefore composed, on average, of more refractory materials than planets further from the Sun. Lewis (1974) has shown how this theory can explain the observed differences in the mean densities of the inner planets and some of the satellites of the outer planets. Additional considerations involved in the formation of the outer planets themselves have been discussed by Cameron (1973).

According to the theory of inhomogeneous accretion, therefore, the Earth differentiated into core and mantle before it ever acquired the material that now comprises the upper mantle, crust, ocean, and atmosphere. The iron in the material that formed the surface layers was oxidized by the time it arrived on Earth. Volatiles were incorporated in the form of low-temperature condensates resembling carbonaceous chondrites, carbon dioxide in the form of the carbon in hydrocarbons, and water as the hydrogen in hydrocarbons and possibly also in hydrated silicate minerals. The inert gases were added largely as adsorbed material in the late forming condensate. Their abundance ratios, therefore, are 'planetary'. [For evidence that primordial inert gases are still being released from the upper mantle see Dymond and Hogan (1973) and Fisher (1974; this volume).]

I have said nothing yet about the time scale for the evolution of the solar nebula and accretion of the inner planets. Theoretical arguments indicate that the duration of the whole process I have described is geologically very short, less than 10^7 years, possibly less by several orders of magnitude (Clark et al., 1972; Cameron, 1973; Hills, 1973). The time that is of most importance for the atmosphere, however, is the time for the release of volatiles from the late-forming and late-accreting veneer. Degassing of the volatile-rich veneer may have resulted from impact heating, and occurred as fast as material was accreted (Fanale, 1971a,b). This would have led to an essentially instantaneous origin of the atmosphere. The volatile-rich material may, however, have arrived in small enough pieces to escape degassing on impact. Degassing in this case could have resulted from radioactive heating and more normal tectonic processes. Instantaneous and gradual degassing models have been compared for Venus by Walker (1975) and for Earth by Walker (1976). None of these models yield highly reducing secondary atmospheres. In order to illustrate how the composition of the primitive atmosphere may be estimated I shall arbitrarily choose, here, the instantaneous degassing model.

Primitive Steam Atmosphere

I assume that infalling material was heated to temperatures high enough to cause the release of any water of hydration and the conversion of hydrocarbons largely to water and carbon dioxide by reactions of the sort

$$CH_2 + 3Fe_3O_4 \rightarrow CO_2 + H_2O + 9FeO$$

(Walker et al., 1970). The high temperatures may have been local, corresponding to widely separated impacts, or the whole surface of the globe may have been hot. Either way, we can start with an atmosphere having approximately the composition and mass of the excess volatiles (Rubey, 1951; Ronov, 1968; Walker, 1976). Water vapour would have predominated (300 bar), with carbon dioxide the next most abundant gas (50 bar), and other constituents present only in trace amounts. I shall now discuss what would have happened to this steam atmosphere and how long it might have lasted.

The properties of a steam atmosphere have been examined by Ingersoll (1969). Two features are important to us. First, latent heat of condensation in a convecting steam atmosphere causes atmospheric temperature to decrease very slowly with height. There is therefore no cold trap, such as the tropopause in the modern atmosphere, which concentrates terrestrial water at the bottom of the atmosphere. An atmosphere that is predominantly water vapour near the ground remains predominantly water vapour to very great heights. At very great heights the water vapour is exposed to photolysis by solar ultraviolet radiation, and the hydrogen that results from photolysis escapes into space. The rate of loss of hydrogen is limited by the flux of ultraviolet photons with enough energy to dissociate water. Studies by Hunten (1973) and Walker (1975) suggest that an oceanic mass of water vapour can be converted to oxygen in this way in a time of about 35 m.y.

The second important feature of Ingersoll's investigation concerns the surface temperature of a planet with an atmosphere of steam and the question of whether the steam will condense or whether the greenhouse effect

due to the steam will maintain temperatures too high for condensation. The determining factor here is the flux of energy that must be transported upwards through the atmosphere, and this depends on the planetary albedo and the planet's distance from the Sun. Several studies of this question (the runaway greenhouse effect) indicate that a steam atmosphere on Earth would condense, although Venus might be too close to the Sun to permit condensation (Sagan, 1960; Gold, 1964; Rasool and De Bergh, 1970; Pollack, 1971).

These studies have considered only solar heating, however, and have not allowed for additional energy provided by accretion of the surface layers of the Earth. Let us therefore estimate how long a steam atmosphere could have been sustained on Earth by accretional energy. We can then estimate how much terrestrial water could have been converted to oxygen during the period of the steam atmosphere.

From Ingersoll (1969) we find that an additional energy source comparable with the solar source (about $10^6 \, \text{erg cm}^{-2} \, \text{sec}^{-1}$) would be sufficient to maintain a steam atmosphere. This we equate to the rate of release of gravitational energy by infalling material, $Rg \, dm/dt$, where R cm is the radius of the Earth, g cm sec^{-2} is the gravitational acceleration, and dm/dt gm cm^{-2} sec^{-1} is the rate of accretion per unit area. We find that $dm/dt = 1 \cdot 6 \times 10^{-6}$ gm cm^{-2} sec^{-1} would be required. At this rate it would take 10^6 years to accumulate a layer 200 km thick, which provides an indication of how long the steam atmosphere could have survived.

In 10^6 years only 3% of the present mass of the ocean would have been converted to oxygen by photolysis of water vapour and escape of hydrogen. We may assume that weathering rates were rapid at this time due to the high surface temperature and perturbations caused by meteorite impacts. The oxygen should therefore have been consumed as fast as it was produced. Oxidation of the ferrous iron, carbon, and sulphide in a layer of basalt about 3 km thick would be sufficient to consume all of the oxygen produced during the lifetime of the steam atmosphere (Walker, 1974).

I conclude, therefore, that the primitive steam atmosphere, if it existed at all, did not last long or result in the loss of a large amount of water. Once the water had condensed, Earth was left with an atmosphere composed principally of carbon dioxide (about 50 bar). The carbon dioxide would have started to dissolve in the water and react with the rocks to form carbonates. At modern rates of erosion and weathering it would have taken about 300 m.y. to produce the volume of carbonate rock now in the crust (Garrels and Mackenzie, 1969, 1971), but I suspect that weathering would have been much more rapid in the presence of high carbon dioxide pressures and possibly enhanced tectonic activity.

With water gone into the oceans and carbon dioxide gone into the rocks, Earth was left with an atmosphere that was more or less modern, except for the absence of oxygen. It was in the presence of this atmosphere, I feel, that terrestrial life evolved.

Prebiological Atmosphere

Most of Earth's nitrogen is in the atmosphere today (Eugster, 1972), and it presumably would have been in the prebiological atmosphere also. Although nitrogen may originally have been incorporated into the Earth in the form of ammonium ions in silicate lattices (Eugster and Munoz, 1966), the equilibrium form in near-surface terrestrial environments is N_2. Molecular nitrogen is much more abundant than ammonia in volcanic emanations (White and Waring, 1963). So I assume that the prebiological atmosphere contained about 1 bar of molecular nitrogen.

The carbon dioxide partial pressure should not have been greatly affected by the absence of oxygen from the atmosphere or by the absence of life, so I assume that it was more or less the same then as it is now, say within a factor of 10. Temperatures in the troposphere are not directly affected by oxygen or life, so I assume that surface temperatures were more or less the same then as now. The Sun may have been less luminous than it is now, but

very small traces of ammonia in the anoxygenic atmosphere could have provided enough of a greenhouse effect to keep the surface temperature above the freezing point of water (Sagan and Mullen, 1972). Surface temperatures control the average water vapour content of the troposphere, so this too would have been about the same. There are, of course, all sorts of reasons why the prebiological atmosphere could not have been exactly like the modern atmosphere minus oxygen, but there do not seem to be any compelling reasons why it should have been grossly different.

The key question is how reducing was the prebiological atmosphere; that is, what was the partial pressure of hydrogen? I shall assume that this partial pressure was controlled by a balance between two processes—the release of hydrogen from the interior of the Earth in volcanism, and the escape of hydrogen from the top of the atmosphere to space. The hydrogen content of volcanic gases is determined, approximately, by equilibrium with the molten rocks from which the gases originate. Although the ratio of hydrogen to water vapour is small (Holland, 1962, 1964), tectonic activity and volcanism may have been much more rapid in the early days, and I cannot think of any other source of hydrogen that is likely to have been significant.

Since I am discussing an instantaneous degassing model of the prebiological atmosphere I am, of course, assuming that the volcanic gases were derived from recycled surface material and were not being released to the atmosphere for the first time. Volcanic hydrogen, in particular, arose from the reaction of reduced rocks with water carried down into the Earth as wet sediments or water of hydration. When hydrogen escaped from the atmosphere, therefore, it left behind it the oxygen with which it had at one time been combined. This permits us to set an upper limit on the amount of hydrogen lost from the primitive Earth by considering the fate of this oxygen. The oxygen produced by dissociation of one-tenth of an oceanic mass of water would be sufficient to produce all of the Fe_2O_3 in the top 50 km of the mantle by oxidation of

FeO (Clark et al., 1972). Dissociation of one-sixth of an ocean would produce all the oxygen now combined with carbon in crustal carbonate rocks. Let us therefore take one-third of the hydrogen content of the present ocean as an upper limit on the hydrogen loss integrated over the age of the Earth (this equals 5×10^{22} gm H_2). Brinkmann (1969) used an upper limit that was smaller than this by a factor of 10.

We must now consider the factors that control the rate of escape of hydrogen from an atmosphere like that of the Earth. Provided the temperature at the level of escape is more than about 500°K hydrogen escapes almost as soon as it reaches the escape level, and the rate of escape is governed not by the temperature but by the processes that transport hydrogen into the upper atmosphere (Hunten, 1973a, 1973b; Hunten and Strobel, 1974; Walker, 1976). These transport processes can be treated quite generally to show that the escape flux, under a wide range of conditions, is given approximately by $2 \times 10^{13} f/(1+f)$ atoms $cm^{-2} sec^{-1}$ where f is the ratio of hydrogen atoms (including combined hydrogen) to all other molecules in the lower atmosphere. If, for example, we assume that hydrogen constituted one molecule in ten in the prebiological atmosphere, then f would have been $0 \cdot 2$, the escape flux would have been $3 \cdot 6 \times 10^{12}$ atoms $cm^{-2} sec^{-1}$, and all of the hydrogen in one-third of an ocean of water would have been dissipated in 45 m.y.

We can now derive the hydrogen abundance in the prebiological atmosphere for various assumed rates of volcanic hydrogen production by equating the production rate to the hydrogen escape rate. We can set an upper limit on the duration of a given rate of production by equating the total production to the hydrogen content of one-third of an ocean of water. Illustrative results are shown in Table 2.

We see that an atmosphere containing a per cent or so of hydrogen could have survived for a time approaching 10^9 years, provided there was sufficient tectonic activity. It can be shown that the oxygen content of such an atmosphere would have been vanishingly small (Walker, 1976). This is the atmosphere under which terrestrial life may have originated. I do not see how more reducing atmospheres could have been sustained for long enough times.

Table 2. Degassing rates and lifetimes of prebiological atmospheres with different hydrogen contents

Hydrogen abundance (per cent H_2)	Degassing rate		Lifetime (10^6 yrs)
	(molecules $cm^{-2} sec^{-1}$)	(gm yr^{-1})	
10	$1 \cdot 8 \times 10^{12}$	10^{15}	53
1	$2 \cdot 0 \times 10^{11}$	10^{14}	480
$0 \cdot 1$	$2 \cdot 0 \times 10^{10}$	10^{13}	4800

Conclusion

I have presented a very tentative account of the origin of the Earth and its atmosphere, calling attention to circumstances that could have produced the atmosphere of hydrogen, methane, and ammonia favoured for the origin of terrestrial life. Evidence involving inert gas abundances and the composition of the upper mantle suggests that a highly reducing atmosphere, if it ever existed, had a short life and a violent end. Attention should therefore, I feel, be directed to weakly reducing atmospheres containing a few per cent of hydrogen.

Conclusions in this field, however, depend very much on personal taste and inclination. There is little firm knowledge and much room for speculation. It is tiresome to write and to read a paper that is larded with expressions of doubt and uncertainty. Such expressions should be understood in this paper wherever I have failed to state them explicitly.

Acknowledgements

My ideas on this subject have benefited from discussions with a number of friends and colleagues, particularly K. K. Turekian and D. M. Hunten. I am grateful to all of them.

The National Astronomy and Ionosphere Center is operated by Cornell University under contract with the National Science Foundation.

References

Abelson, P. H., 1965. 'Abiotic synthesis in the Martian environment', *Proc. U.S. Natl. Acad. Sci.*, **54**, 1490–1497.

Abelson, P. H., 1966. 'Chemical events on the primitive Earth', *Proc. U.S. Natl. Acad. Sci.*, **55**, 1365–1372.

Brinkmann, R. T., 1969. 'Dissociation of water vapor and evolution of oxygen in the terrestrial atmosphere', *J. Geophys. Res.*, **74**, 5355–5368.

Brown, H., 1952. 'Rare gases and the formation of the Earth's atmosphere', In *The Atmospheres of the Earth and Planets*, G. P. Kuiper (Ed.), University of Chicago Press, Chicago, 2nd Edn., 258–266.

Cameron, A. G. W., 1968. 'A new table of abundances of the elements in the solar system', In *Origin and Distribution of the Elements*. L. H. Ahrens (Ed.), Pergamon, New York, 125–143.

Cameron, A. G. W., 1973. 'Accumulation processes in the primitive solar nebula', *Icarus*, **18**, 407–450.

Cameron, A. G. W., and Pine, M. R., 1973. 'Numerical models of the primitive solar nebula', *Icarus*, **18**, 377–406.

Canalas, R. A., Alexander, E. C., and Manuel, O. K., 1968. 'Terrestrial abundance of noble gases', *J. Geophys. Res.*, **73**, 3331–3334.

Clark, S. P., Turekian, K. K., and Grossman, L., 1972. 'Model for the early history of the Earth', In *The Nature of the Solid Earth*, E. C. Robertson (Ed.), McGraw-Hill, New York, 3–18.

Dymond, J., and Hogan, L., 1973. 'Noble gas abundance patterns in deep sea basalts—primordial gases from the mantle', *Earth Planet. Sci. Lett.*, **20**, 131.

Eugster, H. P., 1972. 'Ammonia, in minerals and early atmosphere', In *The Encyclopedia of Geochemistry and Environmental Sciences*, R. W. Fairbridge (Ed.), Van Nostrand, New York, 29–33.

Eugster, H. P., and Munoz, J., 1966. 'Ammonium micas: Possible sources of atmospheric ammonia and nitrogen', *Science*, **151**, 683–686.

Ezer, D., and Cameron, A. G. W., 1963. 'The early evolution of the Sun', *Icarus*, **1**, 422–441.

Ezer, D., and Cameron, A. G. W., 1971. 'Pre-main sequence stellar evolution with mass loss', *Astrophys. Space Sci.*, **10**, 52–70.

Fanale, F. P., 1971a. 'History of Martian volatiles: Implications for organic synthesis', *Icarus*, **15**, 279–303.

Fanale, F. P., 1971b. 'A case for catastrophic early degassing of the Earth', *Chemical Geology*, **8**, 79–105.

Fanale, F. P., and Cannon, W. A., 1971. 'Physical adsorption of rare gas on terrigenous sediments', *Earth Planet. Sci. Lett.*, **11**, 362–368.

Fisher, D. E., 1974. 'The planetary primordial component of rare gases in the deep Earth', *Geophys. Res. Letters*, **1**, 161–164.

Gabel, N. W., and Ponnamperuma, C., 1972. 'Primordial organic chemistry', In *Exobiology, Frontiers in Biology*, Vol. 23, C. Ponnamperuma (Ed.), North–Holland, Amsterdam, 95–135.

Garrels, R. M., and Mackenzie, F. T., 1969. 'Sedimentary rock types: Relative proportions as a function of geological time', *Science*, **163**, 570–571.

Garrels, R. M., and Mackenzie, F. T., 1971. *Evolution of Sedimentary Rocks*, Norton, New York.

Gold, T., 1964. 'Outgassing processes on the moon and Venus', In *The Origin and Evolution of Atmospheres and Oceans*, P. J. Brancazio and A. G. W. Cameron (Eds.), John Wiley, New York, 249–256.

Grossman, L., 1972. 'Equilibrium condensation in the primitive solar nebula', *Geochim. Cosmochim. Acta*, **36**, 597–619.

Grossman, I., and Larimer, J. W., 1974. 'Early chemical history of the solar system', *Rev. Geophys. Space Phys.*, **12**, 71–101.

Hanks, T. C., and Anderson, D. L., 1969. 'The early thermal history of the Earth', *Phys. Earth Planet. Interiors*, **2**, 19–29.

Hayashi, C., 1961. 'Stellar evolution in early phases of gravitational contraction', *Publ. Astronom. Soc. Japan*, **13**, 450–452.

Hills, J. G., 1973. 'On the process of accretion in the formation of the planets and comets', *Icarus*, **18**, 505–522.

Holland, H. D., 1962. 'Model for the evolution of the Earth's atmosphere', In *Petrologic Studies: A Volume in Honor of A. F. Buddington*, A. E. J. Engel, H. I. James, and B. F. Leonard (Eds.), Geological Society of America, New York, 447–477.

Holland, H. D., 1964. 'On the chemical evolution of the terrestrial and cytherean atmospheres', In *The Origin and Evolution of Atmospheres and Oceans*, P. J. Brancazio and A. G. W. Cameron (Eds.), John Wiley, New York, 86–101.

Hubbard, J. S., Hardy, J. P., and Horowitz, N. H., 1971. 'Photocatalytic production of organic compounds from CO and H_2O in a stimulated Martian atmosphere', *Proc. U.S. Natl. Acad. Sci.*, **68**, 574–578.

Hunten, D. M., 1973a. 'The escape of light gases from planetary atmospheres', *J. Atmos. Sci.*, **30**, 1481–1494.

Hunten, D. M., 1973b. 'The escape of H_2 from Titan', *J. Atmos. Sci.*, **30**, 726–732.

546

Hunten, D. M., and Strobel, D. F., 1974. 'Production and escape of terrestrial hydrogen', *J. Atmos. Sci.*, **31**, 305–317.

Ingersoll, A. P., 1969. 'The runaway greenhouse: A history of water on Venus', *J. Atmos. Sci.*, **26**, 1191–1198.

Lewis, J. S., 1972. 'Low temperature condensation from the solar nebula', *Icarus*, **16**, 241–252.

Lewis, J. S., 1974. 'The temperature gradient in the solar nebula', *Science*, **186**, 440–443.

Moulton, F. R., 1905. 'On the evolution of the solar system', *Astrophys. J.*, **22**, 165–181.

Nordlie, B. E., 1972. 'Gases—Volcanic', In *The Encyclopedia of Geochemistry and Environmental Sciences*, R. W. Fairbridge (Ed.), Van Nostrand, New York, 387–391.

Pepin, R. O., and Signer, P., 1965. 'Primordial rare gases in meteorites', *Science*, **149**, 253–265.

Phinney, D., 1972. '^{36}Ar, Kr, and Xe in terrestrial materials', *Earth Planet. Sci. Lett.*, **16**, 413–420.

Pollack, J. B., 1971. 'A nongrey calculation of the runaway greenhouse: Implications for Venus past and present', *Icarus*, **14**, 295–306.

Rasool, S. I., and DeBergh, C., 1970. 'The runaway greenhouse and the accumulation of CO_2 in the Venus atmosphere', *Nature*, **226**, 1037–1039.

Ringwood, A. E., 1959. 'On the chemical evolution and densities of the planets', *Geochim. Cosmochim. Acta*, **15**, 257–283.

Ringwood, A. E., 1966. 'The chemical composition and origin of the Earth', In *Advances in Earth Science*, P. M. Hurley (Ed.), MIT Press, Boston, 287–356.

Ronov, A. B., 1968. 'Probable changes in the composition of sea water during the course of geological time', *Sedimentology*, **10**, 25–43.

Rubey, W. W., 1951. 'Geologic history of sea water: An attempt to state the problem', *Bull. Geol. Soc. Am.*, **62**, 1111–1147.

Rutten, M. G., 1971. *The Origin of Life by Natural Causes*, Elsevier Publishing Co., Amsterdam.

Sagan, C., 1960. *The Radiation Balance of Venus*, California Institute of Technology, Jet Propulsion Lab., *Tech. Rept. No. 32–34*, 23 pp.

Sagan, C., and Mullen, G., 1972. 'Earth and Mars: Evolution of atmospheres and surface temperatures', *Science*, **177**, 52–56.

Turekian, K. K., and Clark, S. P., 1969. 'Inhomogeneous accumulation of the Earth from the primitive solar nebula', *Earth Planet. Sci. Lett.*, **6**, 346–348.

Verniani, F., 1966. 'The total mass of the Earth's atmosphere', *J. Geophys. Res.*, **71**, 385–391.

Walker, J. C. G., 1974. 'Stability of atmospheric oxygen', *Am. J. Sci.*, **274**, 193–214.

Walker, J. C. G., 1975. 'Evolution of the atmosphere of Venus', *J. Atmos. Sci.*, in press (June, 1975).

Walker, J. C. G., 1976. *Evolution of the Atmosphere*, Hafner, New York.

Walker, J. C. G., Turekian, K. K., and Hunten, D. M., 1970. 'An estimate of the present-day deepmantle degassing rate from data on the atmosphere of Venus', *J. Geophys. Res.*, **75**, 3558–3561.

White, D. E., and Waring, G. A., 1963. 'Volcanic emanations', *U.S. Geol. Surv. Prof. Paper 440-K*.

Rare Gas Clues to the Origin of the Terrestrial Atmosphere

DAVID E. FISHER

Rosenstiel School of Marine and Atmospheric Science, University of Miami, Florida 33149

Abstract

The bulk of the atmospheric ^{4}He and ^{40}Ar is radiogenic, and as such represents a volatile component decoupled from a non-volatile parent in the Earth's lithosphere. Studies of the evolving ratios of a radiogenic/non-radiogenic pair, such as ^{40}Ar/^{36}Ar or of the radiogenic pair ^{4}He/^{40}Ar will provide a sensitive record of the terrestrial degassing process, and are therefore pertinent to ideas bearing on the thermal and chemical evolution of the Earth and its atmosphere. The non-radiogenic rare gases (Ne, Ar, Kr, and Xe) show varying elemental and/or isotopic abundances in the terrestrial atmosphere, in different types of meteorites, and on the Moon. Aside from extinct radioactivity and cosmic-ray effects, these variations are not well understood. Observed correlations, however, can suffice to put boundary conditions on the evolution of the terrestrial atmosphere.

Historical Introduction

The first application of measured terrestrial abundances of the rare gases to models of formation of the Earth's atmosphere was in the realization that the rare gases are indeed very rare compared with their nuclear neighbours. Fig. 1 shows estimated crustal and atmospheric abundances of the rare gases and their neighbours. It was recognized early in the game that the extremely low gas abundances (a depletion factor of roughly 10^6 for Xe) were an indication that the accreting Earth was essentially incapable of retaining an atmosphere and that therefore the Earth's atmosphere must have resulted from a later degassing of the solid Earth (Holland, 1963; Brown, 1952).

This conclusion is extended by noting that the rare gases are also grossly underabundant when compared with the molecular species which are the main constituents of our present-day atmosphere, as shown in Fig. 2. If these constituents had always been in a volatile phase, it is hard to understand the great depletion for the rare gases of interspersed mass. Rather, the present-day atmosphere must have been locked in a non-volatile phase during the Earth's accretion; for example, water in hydrated minerals and CO_2 in carbonates. This limits the initial temperature of the Earth to that at which such compounds are stable, roughly a few hundred degrees, and leads to the question: How and when did the Earth degas its interior to form our atmosphere and oceans?

This question is related to ideas on the process of core formation in the Earth. A model that is frequently discussed at the present time involves the melting of the Earth at depths of ~500 km within the first billion years, due to the accumulation of heat of

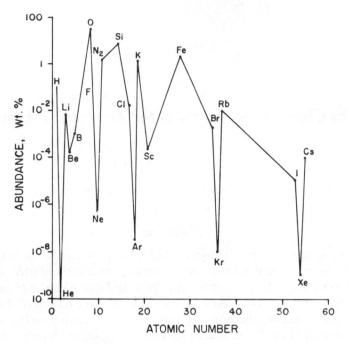

Fig. 1. Crustal and atmospheric abundances of the elements in
the region of the rare gases (Mason, 1966)

accretion and/or radioactive heat and the subsequent release of gravitational energy as molten iron sinks to form the core. This event might have served as a world-wide big burp, degassing the internal Earth to form the atmosphere at that time. Such 'catastrophic' degassing models have been discussed by Schwartzman (1973a), Fanale (1972), Chase and Perry (1972), and Damon and Kulp (1958). Alternatively, the atmosphere may have been released in a continuous degassing process taking place throughout geologic time (Rubey, 1951; Turekian, 1959, 1963; Ozima and Kudo, 1972). Actually, some combination of these processes probably took place. There have also been suggestions that at least part of

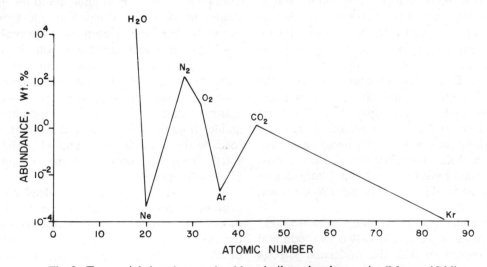

Fig. 2. Terrestrial abundances of stable volatile molecular species (Mason, 1966)

the rare-gas component of the atmosphere may have been added externally by solar-wind bombardment of the Earth, either directly or through carriers (Cameron, 1963; Tilles, 1965).

Measurements of the various rare gases in both the modern and ancient atmosphere and in deep-Earth rocks can put some stringent boundary conditions on these models of atmospheric evolution.

The Radiogenic Rare Gases

In Figs. 1 and 2 the isotopes of He and Ar used were the rare ones ^{3}He and ^{36}Ar and ^{38}Ar; the common isotopes ^{4}He and ^{40}Ar are radiogenic, and essentially all of them now present on Earth were provided subsequent to Earth accretion through the decay of their parents U, Th, and K (von Weizäcker, 1937). They are thus examples of volatile daughters having decoupled from non-volatile parents, and the evolution of their atmospheric and deep-Earth abundances must be indicative of the processes of terrestrial degassing.

(a) *Atmospheric* $^{40}Ar/^{36}Ar$

Since ^{36}Ar is non-radiogenic (has always existed as a gas), the various degassing models predict distinct evolutionary histories for the ^{40}Ar/^{36}Ar ratio in the atmosphere, depending on the mode of injection of K into the Earth's crust, and on the degassing of Ar from the crust and/or mantle. These models and histories have been reviewed by Alexander (1975); the growth of the ^{40}Ar/^{36}Ar in the atmosphere according to the various models is summarized in Fig. 3. Thus a series of ^{40}Ar/^{36}Ar measurements in rocks of different ages, each of which has trapped a sample of the ambient atmosphere at the time of its formation, might distinguish between the models. However, even if one could be confident of having found samples which trapped the ancient environment, the decay of K in the sample subsequent to this formation will contribute more ^{40}Ar, altering the original ratio. And it is not clear at the present time how much of the 'atmospheric' Ar found in old rocks was trapped at the time of rock formation and how much was incorporated subsequently. If one can find the proper sample, i.e.

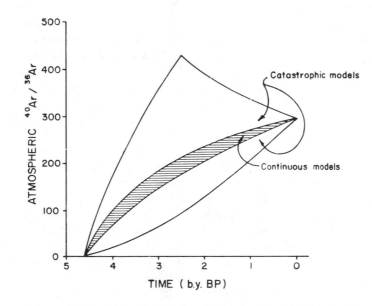

Fig. 3. The evolution of the atmospheric ^{40}Ar/^{36}Ar ratio, as determined by catastrophic and continuous degassing models (after Alexander, 1975)

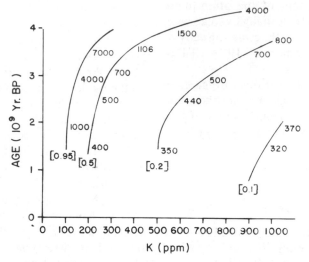

Fig. 4. The evolution of the internal-earth $^{40}Ar/^{36}Ar$ ratio, in catastrophic degassing models (after Ozima, 1973) (reproduced with permission of *Nature*)

one which trapped the ambient atmosphere at time of its formation and has remained a closed system with respect to K and Ar since then, and which has a low K content and a high trapped Ar content so that subsequent K-decay does not overwhelm the original ratio, application of the 'isochron' method should reveal both the age and the original $^{40}Ar/^{36}Ar$ ratio, as discussed by Alexander (1975). The key, as he notes, is in finding these proper samples, and so far they have not been found.

(b) Deep-Earth $^{40}Ar/^{36}Ar$

Ozima (1973) has pointed out that the present-day $^{40}Ar/^{36}Ar$ ratio *within the mantle* will be a sensitive indicator of the reality of the catastrophic model. In such models, the degassing is defined by any two of these four quantities: (1) The time of degassing; (2) The fraction of Ar degassed; (3) The K content of the Earth; (4) The present-day value of the $^{40}Ar/^{36}Ar$ ratio within the Earth. He presents a graph similar to Fig. 4, showing sets of solutions for these quantities which satisfy the total Ar abundance and the present-day atmospheric ratio of 295·5.

As noted by several authors (Funkhouser et al., 1968; Dalrymple and Moore, 1968; Noble and Naughton, 1968) the rare gases found in deep-sea basaltic glasses are clearly those trapped within the rock upon its eruption, due to the high hydrostatic pressures and rapid chilling, and as such represent mantle-ambient gases. A characteristic of these trapped gases is the presence of 'excess' radiogenic ^4He and ^{40}Ar, i.e. in amounts greater than can be accounted for by the decay of U, Th, and K subsequent to their eruption. Similar 'excess' ^{40}Ar contents have been measured in ultramafic intrusives (Kirsten and Gentner, 1966), but their origin is not so clear.

Ozima (1973) therefore accepted the $^{40}Ar/^{36}Ar$ trapped in submarine basalts as representing the isotopic ratio in the upper mantle; the published value available to him was ~1000. He assumed that the time of a catastrophic degassing, if it existed, must predate the oldest known rocks (~3·9 b.y.). According to his graphical solution, then, to reconcile any such catastrophic degassing with an internal-Earth isotopic ratio of ~1000 necessitates an unrealistically high K content for the whole Earth of ≳1500 ppm (Fig. 4). On this basis he ruled out the catastrophic model, noting however, following Schwartzman (1973b), that if the ultramafic values for the ratio (~10,000) were indicative of mantle gases, then his conclusion was invalid.

In a recent paper (Fisher, 1975a) I have presented further data on He and Ar isotopes measured in the glassy margin of deep-sea basalts, from both the Atlantic and Pacific Oceans, on both Ridge and non-Ridge samples. Several of the data provide an $^{40}Ar/^{36}Ar$ ratio even greater than 10,000 and thus Ozima's argument against the catastrophe must be relinquished. Ozima has, however, recently completed further calculations bearing on this problem (Ozima, 1975). These have not yet been evaluated with the present data. In the meantime, his $^{40}Ar/^{36}Ar$ argument can be replaced with one based on $^{4}He/^{40}Ar$ measurements in these same rocks.

(c) Deep-Earth $^{4}He/^{40}Ar$

Schwartzman (1973a) predicts a present-day mantle $^{4}He/^{40}Ar$ ratio of between 1·36 and 2·23, depending on Th/U ratio, for a catastrophic degassing model with very early degassing. He notes that this is in agreement with a ratio estimated for the former environment of lherzolite nodules from Hawaiian nepheline mellilite basalt (Grämlich and Naughton, 1972) and points out that higher values (2–10) found by Wasserburg et al. (1963) in volcanic gases might well be due to

preferential leaking of ^{4}He from wall rocks. But ratios measured in the above-quoted paper (Fisher, 1975a) reach values of 15–20, much too high for Schwartzman's model, and the preferential leaking argument is much weaker for sub-oceanic mantle sources than for the terrestrial volcanic emanations of Wasserburg et al. Since the gas trapping mechanism cannot be 100% efficient, and since in any gas loss ^{4}He will be lost preferentially to ^{40}Ar, the highest measured $^{4}He/^{40}Ar$ ratios are minimum values for the magma-ambient ratios. These high values cannot be accounted for at all if the magmatic K/U ratio is similar to the crustal value of 10^{4}, even less if the ratio is chondritic (3–6×10^{4}), as seen in Fig. 5. Thus the data rule out all models which ascribe the terrestrial atmosphere to *any* type of degassing of a mantle which is chondritic or crustal with respect to the K/U ratio. K/U ratios measured on ultramafic rocks lead to estimated mantle values of $\sim 3 \times 10^{3}$ (Fisher, 1975b; Wakita et al., 1967; Green et al., 1968; Fisher, 1970). Assuming this value does give maximum $^{4}He/^{40}Ar$ values of 15–20 (Fig. 5), but only for an age $\lesssim 10^{9}$ years; for a

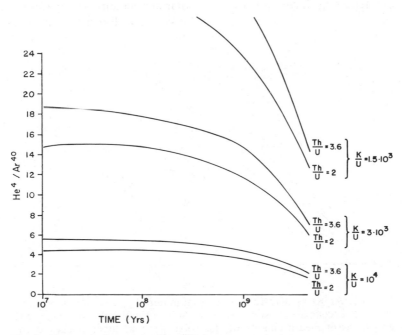

Fig. 5. The evolution of the $^{4}He/^{40}Ar$ ratio in a closed system, in terms of the Th/U and K/U ratios

552

catastrophic degassing model this is the age of the catastrophe, and is much too young. Since the catastrophe is normally identified with formation of the Earth's core, it must have occurred at least within the first billion years of Earth history; i.e. the age must be $\gtrsim 3 \cdot 5 \times 10^9$. This necessitates a K/U ratio in the source region $\lesssim 1 \cdot 5 \times 10^3$. Such a value is barely within the widest limits set by K/U measurements on ultramafic rocks: $1–10 \times 10^3$ (Fisher, 1970); $1–5 \times 10^3$ (Green et al., 1968). A K/U ratio even lower would be necessary if core formation (and an associated big burp) is to have generated the atmosphere within the first 100 m.y. of Earth history, as suggested by Oversby and Ringwood (1971).

Please recognize that these data do *not* argue that catastrophic core formation, with or without an associated burp, could not have occurred. Rather, the argument is that the mantle has not been a closed system with respect to Ar and He since any such postulated event, and therefore that significant later or continuous degassing from the interior Earth *must* have occurred. The relative contributions to our present-day atmosphere from a *possible* early catastrophic event and the *necessary* later degassing have yet to be worked out.

In summary, the ^{4}He and ^{40}Ar data present an effective argument against an early catastrophic degassing event being the sole source of the Earth's atmosphere if the mantle K/U ratio is $\gtrsim 1 \cdot 5 \times 10^3$, as it probably is, and if the ^{4}He/^{40}Ar data measured in deep-sea glasses are indicative of world-wide mantle abundances. Since the very high ratios are found in both Atlantic and Pacific basalts, in both Ridge and non-Ridge samples, and since the lower values can always be explained by incomplete trapping of the mantle gases (with preferential loss of He), this last assumption may be valid. Further data on ^{4}He/^{40}Ar ratios in more submarine glasses of wider geographic location and various petrographic types, and on K/U ratios in a wider variety of possible mantle materials, should settle the question conclusively.

The Non-radiogenic Rare Gases

Meteorites contain two distinct components of 'primordial' gases: a *solar* component picked up on the surface of grains directly from the solar wind, and a *planetary* component somehow trapped from the primitive solar nebula during formation of the first solid bodies. The Earth may have gained its own

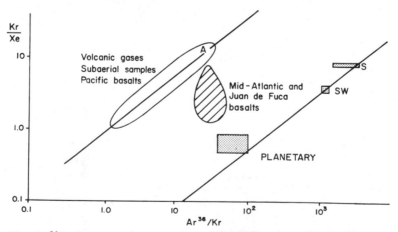

Fig. 6. ^{36}Ar/Kr vs. Kr/Xe in terrestrial materials. *A* is the atmospheric point, *S* encompasses estimates of the solar primordial component, *SW* encompasses measurements of the solar wind in lunar fines and the meteorite Pesyanoe, and *Planetary* encompasses the planetary primordial component as measured in meteorites. See Fisher (1974) for references

rare gases by either or both of these same processes, although the elemental composition of terrestrial atmospheric rare gases is identical to neither the solar nor the planetary component. The differences from the solar component cannot be accounted for by gravitational fractionation, nor by injection of ions from the solar wind, either directly or through carriers (Cameron, 1963; Tilles, 1965). The difference from the planetary component is an underabundance of Xe relative to either Kr or Ar. Gravitational fractionation should produce the opposite effect. Fanale and Cannon (1971) suggest that the atmospheric Xe depletion is the result of physical sorption by terrestrial shales (Canalas et al., 1968).

Further data pertinent to the problem would be the elemental composition of gases ambient in the interior of the Earth, since these should be at a different evolutionary stage than those in the atmosphere, depending upon their mode of terrestrial incorporation and subsequent evolution. They might be expected to be *non-existent* if the total atmospheric rare gases were provided by direct capture from the solar wind after accretion of the Earth, *solar* in relative abundances if the atmospheric gases were captured from the solar wind during the Earth's accretion and fractionated during a subsequent degassing, *planetary* in relative abundance if they were captured by the same process responsible for the planetary gases in meteorites and subsequently fractionated during degassing, or *atmospheric* in relative abundances if no process operated to fractionate them subsequent to the Earth's degassing or if the interior is simply a recirculated atmospheric mix. If the interior reservoir is not very large, it will also be fractionated by the degassing process; in this case it is the sum of the interior and atmospheric gases that will be planetary or solar in abundance.

The problem of determining the presence of primordial gases is not an easy one; one method that shows promise is that of plotting Ar/Kr vs. Kr/Xe, as shown in Fig. 6 (Fisher, 1974). Gases from igneous subaerial rocks, water, terrestrial dust, and some deep-sea basalts all lie in the upper envelope, which includes the atmospheric point. I have discussed the significance of this distribution previously (Fisher 1970, 1974). Rocks which contain a mixture of primordial gases and atmospheric contamination would be expected to lie in the region between the atmospheric envelope and either of the two types of primordial gases, and indeed several suites of rocks from the Juan de Fuca and Mid-Atlantic Ridges show such a distribution (Fisher, 1973, 1974; Dymond and Hogan, 1973). I interpret these data as indicating the presence of a mantle-ambient component of planetary primordial gases.

This interpretation suggests that the atmospheric rare gases were not captured directly from the solar wind, but instead were trapped with the accreting Earth by a process similar to that which trapped the planetary gases within meteorites. As the Earth subsequently degassed, these gases were fractionated to form the present atmospheric component, depleting Xe relative to the other gases. Such a process may have been either incomplete removal of Xe from the interior, or selective sorption of degassed Xe. If incomplete degassing was the responsible process, one would expect to find Xe/Kr ratios in the deep-Earth gases even higher (Kr/Xe ratios even lower) than the planetary ratio. Such is not the case in the present work. My data therefore support the suggestion of Fanale and Cannon (1971) that selective sorption of Xe from the atmosphere is responsible for this fractionation.

Canalas et al. (1968) calculated that if all the noble gases sorbed on to terrestrial shales were readmitted to the atmosphere, the total atmospheric Kr/Xe ratio would be lowered from 13 to ~4·5. Later data presented by Phinney (1972) have Kr/Xe ratios an order of magnitude lower, combined with even greater absolute concentrations; if these values are to be taken as the average, the effect would be to lower the atmosphere-plus-shale Kr/Xe ratio to the planetary value, or even below it. Precise calculations are hampered by order-of-magnitude variations in estimates of total terrestrial shale, and of the variation in rare-gas

554

sorption by shale (and perhaps other sedimentary sinks) which might have taken place under various surface conditions and at different times in the Earth's history. The data do show, however, that preferential sorption of Xe from the atmosphere by sedimentary rocks such as shale is a perfectly adequate process for changing the atmospheric Kr/Xe ratio from its original value (suggested here to be in the planetary primordial region) to its present-day value.

Final confirmation of this interpretation awaits isotopic measurements. Unfortunately the Xe isotopic ratios were difficult to determine in these samples because of the small absolute abundances; the best I can do at the present time is to say that they are identical with atmospheric to within $\pm 5\%$ for the isotopes 129–136. The planetary values are also identical with atmospheric within this accuracy for these isotopes (Bogard et al., 1973). A larger distinction has been shown to exist between planetary and atmospheric for the rare isotopes 124–126; further studies will attempt to measure these isotopes in these rocks.

It should be pointed out that the observed Xe light-isotope differences are a distinct embarrassment to this model, since Phinney's data show the shale-sorbed Xe to have atmospheric isotopic ratios. Further speculation as to the origin of these isotopic differences is useless at the present time; perhaps further insights will come with the successful determination of these ratios in the trapped component of the deep-sea basalts. It would also be interesting to determine the elemental ratios of the non-radiogenic gases present in the terrestrial atmosphere as a function of time, in a manner analogous to that discussed for the atmospheric $^{40}Ar/^{36}Ar$ ratio. According to the model discussed here, these ratios will vary from original planetary to present-day atmospheric with a time-dependence different depending upon the relative con-

tributions of an early degassing event and a subsequent continuous degassing. But such measurements are subject to the same uncertainties as discussed for the $^{40}Ar/^{36}Ar$ data, in particular to the problem of finding suitable samples.

Summary

1. The great depletion of terrestrial rare gases, compared both with non-volatile nuclear species of interspersed mass and to volatile molecular species in the terrestrial atmosphere, indicates that the Earth's atmosphere formed by degassing subsequent to solid-state accretion, and that the Earth accreted 'cold'.

2. The secular variation of the atmospheric $^{40}Ar/^{36}Ar$ ratio and the elemental Ar, Kr and Xe abundances should be sensitive indicators of the degassing process. However, experimental problems, in particular the identification of the proper geologic material, have prevented any definite information being obtain at the present time.

3. The present-day $^{40}Ar/^{36}Ar$ ratio within the Earth cannot distinguish between catastrophic and continuous degassing models.

4. The present-day $^{4}He/^{40}Ar$ ratio within the Earth provides strong evidence against an early catastrophe being the sole source of the terrestrial atmosphere, unless rather strong limits on the K/U ratio in the mantle are exceeded. Any sort of degassing of a chondritic mantle is ruled out as the source of the atmosphere.

5. The abundances of the non-radiogenic rare gases within the Earth indicate that the terrestrial rare gases did not originate in a post-accretion solar wind bombardment, that their original abundance ratios were similar to those of the planetary primordial component observed in meteorites, and that they were fractionated subsequent to degassing, probably by selective geological sorbtion of Xe.

Acknowledgements

Research supported by the Oceanography Section, National Science Foundation, NSF Grant DES74-22092.

References

Alexander, E. C., 1975. 'On the time evolution of the atmospheric $^{40}Ar/^{36}Ar$ ratio', preprint.

Bogard, D. D., Reynolds, M. A., and Simms, L. A., 1973. 'Noble gas concentrations and cosmic ray exposure ages of eight recently fallen chondrites', *Geochim. Cosmochim. Acta*, **37**, (11), 2417–2431.

Brown, H., 1952. 'Rare gases and the formation of the Earth's atmosphere', In *Atmospheres of the Earth and Planets*, 2nd Edn., G. Kuiper (Ed.), Univ. Chicago Press.

Cameron, A. G. W., 1963, 'Interpretation of Xe measurements', In *The Origin and Evolution of Atmospheres and Oceans*, 235 pp.

Canalas, R. A., Alexander, E. C., and Manuel, O. K., 1968. 'Terrestrial abundance of noble gases', *J. Geophys. Res.*, **73**, 3331.

Chase, C. G., and Perry, E. C., 1972. 'The oceans: growth and oxygen isotope evolution', *Science*, **177**, 192.

Dalrymple, G. B., and Moore, J. G., 1968 'Argon-40: Excess in submarine pillow basalts from Kilauea Volcano, Hawaii', *Science*, **161**, 1132.

Damon, P. E., and Kulp, J. L. (1958). 'Inert gases and the evolution of the atmosphere', *Geochim. Cosmochim. Acta*, **13**, 280–292.

Dymond, J., and Hogan, L., 1973. 'Noble gas abundance patters in deep sea basalts', *Earth Planet Sci. Lett.*, **20**, 131.

Fanale, F. P., and Cannon, W. A., 1971. 'Physical adsorption of rare gas on terrigenous sediments', *Earth Planet. Sci. Lett.*, **11**, 362.

Fanale, F. P., 1972. 'A case for catastrophic early degassing of the Earth', *Chem. Geol.*, **8**, 79–105.

Fisher, D. E., 1970a. 'Heavy rare gases in a Pacific Seamount', *Earth Planet Sci. Lett.*, **9**, 331.

Fisher, D. E., 1970b. 'Homogenized fission track analysis of uranium in some ultramafic rocks of known potassium content', *Geochim. Cosmochim. Acta*, **34**, 630.

Fisher, D. E., 1972. 'Uranium content and radiogenic ages of hypersthene, bronzite, amphoterite and carbonaceous chondrites', *Geochim. Cosmochim. Acta*, **36**, 15.

Fisher, D. E., 1973. 'Primordial rare gases on the deep Earth', *Nature*, **244**, 344.

Fisher, D. E., 1974. 'The planetary primordial component of rare gases in the deep Earth', *Geophys. Res. Lett.*, **1**, 161.

Fisher, D. E., 1975a. 'Trapped He and Ar and the formation of the atmosphere by degassing', *Nature*, **256**, 113.

Fisher, D. E., 1975b. 'Geoanalytic applications of particle tracks', *Earth Sci. Revs*, in press.

Funkhouser, J. G., Fisher, D. E., and Bonatti, E., 1968. 'Excess argon in deep-sea rocks', *Earth Planet. Sci. Lett.*, **5**, 95–100.

Grämlich, J. W., and Naughton, J. J., 1972. 'Nature of source material for ultramafic minerals from Salt Lake Crater, Hawaii, from measurement of helium and argon diffusion', *J. Geophys. Res.*, **77**, 3032–3042.

Green, D. H., Morgan, J. W., and Heier, K. S., 1968. 'Thorium, uranium, and potassium abundances in peridotite inclusions and their hosts', *Earth Planet. Sci. Lett.*, **4**, 155.

Holland, H. D., 1963. 'On the chemical evolution of the terrestrial and cytherean atmospheres', In *The Origin and Evolution of Atmospheres and Oceans*, P. J. Brancazio and A. G. W. Cameron (Eds.), Wiley, New York, p. 86.

Kirsten, T., and Gentner, W., 1966. *Z. Naturforsch*, **21a**, 119.

Mason, B., 1966. *Principles of Geochemistry*, Wiley, New York.

Noble, C. S., and Naughton, J. J., 1968. 'Deep ocean basalts: Inert gas content and uncertainties in age dating', *Science*, **162**, 265.

Oversby, V. M., and Ringwood, A. E., 1971. *Nature*, **234**, 463–465.

Ozima, M., 1973. 'Was the evolution of the atmosphere continuous or catastrophic?' *Nature Phys. Sci.*, **246**, 41.

Ozima, M., 1975. *Geochim. Cosmochim, Acta.*, in press.

Ozima, M., and Kudo, K., 1972. 'Excess argon in submarine basalts and an earth-atmosphere evolution model', *Nature Phys. Sci.*, **239**, 23–24.

Phinney, D., 1972. '^{36}Ar, Kr and Xe in terrestrial materials', *Earth Planet. Sci. Lett.*, **16**, 413.

Rubey, W. W., 1951. 'Geologic history of sea water', *Bull. Geol. Soc. Am.*, **62**, 1111.

Schwartzman, D. W., 1973a. 'Ar degassing and the origin of the sialic crust', *Geochim. Cosmochim. Acta*, **37**, 2479.

Schwartzman, D. W., 1973b. 'On argon degassing models of the Earth', *Nature Phys. Sci.*, **245**, 20.

Tilles, D., 1965. 'Atmospheric noble gases: Solar-wind bombardment of extraterrestrial dust as a possible source mechanism', *Science*, **148**, 1085.

Turekian, K. K., 1959. 'The terrestrial economy of helium and argon', *Geochim. Cosmochim. Acta*, **17**, 37–43.

Turekian, K. K., 1963. 'Degassing of argon and helium from the Earth', In *The Origin and Evolution of Atmospheres and Oceans*, P. J. Brancazio and A. G. W. Cameron (Eds.), Wiley, New York, pp. 74–82.

Wakita, H., Nagasawa, H., Uyeda, S., and Kuno, H., 1967. 'Uranium, thorium, and potassium contents of possible mantle materials', *Geochem. J.*, **1**, 183.

556

Wasserburg, G. J., Mazor, E., and Zartman, R. E., 1963. 'Isotopic and chemical composition of some terrestrial gases', In *Earth Science and Meteoritics*, J. Geiss and E. D. Goldberg (Eds.), North–Holland, Amsterdam, pp. 219–240.

von Weizäcker, C. G., 1937. 'Uber die Möglichkeit eines dualen β-Zerfalls von Kalium', *Physik. Zeitschr.*, **38**, 623.

The Oceans

The Evolution of Seawater

HEINRICH D. HOLLAND

Department of Geological Sciences, Harvard Univ., Mass. U.S.A.

I. Introduction to the Phanerozoic Ocean

We have not yet found samples of ancient seawater, nor are we ever likely to do so, and if we ever should come across such a sample, we would be hard pressed to prove that that is indeed what we had found. Our knowledge of the history of seawater is, therefore, based on our understanding the atmosphere–ocean–lithosphere system and on inferences drawn from the nature of ancient sediments and the products of their metamorphism.

The rate at which rivers deliver dissolved salts to the oceans is now quite well known. The time required for the present-day river flux to double the concentration of the various ions in seawater, if none were removed from the oceans, has been calculated repeatedly (see, for instance, Garrels and Perry, 1974); this turns out to be less than 10^3 years for some elements (Be, Al) and as much as 10^8 years for others (Cl, Br). In all cases the characteristic time is much less than the age of the Earth, and for most elements it is less than 10^7 years. The oceans have therefore acted as a system through which much more dissolved salt has passed than is present in the oceans today. For periods long compared with the characteristic time, the input of all ions must have roughly equalled the output, and the oceans as a system have tended to respond to changes in the input rate of an ion in times roughly comparable with that of the characteristic time. The characteristic time of calcium in the oceans is about 1×10^6 years. Drastic changes in the calcium concentration in the oceans during the past 1 m.y. are therefore unlikely, but the present concentration of calcium in ocean water is not necessarily similar to the Ca^{+2} concentration ten characteristic times, i.e. c. 10 m.y. ago. Since the characteristic time of most ions and molecules in seawater is geologically short, it follows that the composition of seawater in the not-so-distant past could have been rather different from that of the present day.

Intuitively one might expect, however, that changes in seawater composition have been modest. We see no drastic changes in the chemistry of the continents with time. Hence the composition of river water has probably remained roughly constant with time. On the other hand, we know of sizable temperature fluctuations in the past, the stand of the continents and their position relative to each other and to the poles have changed markedly with time, life evolved and finally populated the land, and the rate of heat generation within the Earth has decreased by a factor of roughly 3. The response of the atmosphere–ocean system to these changes is not easy to predict. Sillén (1961, 1967) proposed that the pH and the concentration of the major cations in seawater were essentially fixed by the chloride concentration and by equilibria between seawater and a number of common carbonate and silicate minerals. A search for proof that the proposed equilibria really control the chemistry of seawater has been somewhat disappointing, and it is very likely that kinetic models such as Broecker's (1971) will be more successful in explaining the operation of the oceans as a chemical system.

The growing doubts concerning the adequacy of Sillén's equilibrium model again raised questions regarding the possibility of extensive variations in seawater composition in the past. However, it has been shown (Holland 1972, 1974) that the constancy of the sequence of minerals in marine evaporites sets severe limits to excursions in seawater composition in the past. Concentrations of the major cations and anions in seawater more than a factor of 2 to 3 higher or lower than at present are unlikely. Unfortunately no marine evaporites older than late Precambrian have been discovered. Thus, even the rather wide limits imposed on the composition of seawater by the nature of marine evaporites are currently set for far less than the last quarter of Earth history.

II. The Precambrian Oceans

The fragmentary evidence which we presently possess regarding the composition of seawater in the Precambrian does not indicate that seawater then was very different from seawater today. Casts of what were probably gypsum and halite crystals in sedimentary rocks *c*. 2 b.y. old (Bell and Jackson, 1974) indicate that evaporation of seawater yielded common evaporite minerals. The presence of limestone in Precambrian sediments implies that seawater was approximately saturated with respect to calcite or aragonite then as now. The nature of many Precambrian dolomites suggests that penecontemporaneous dolomitization of calcite and aragonite took place but on a larger scale than in the recent past.

There are, however, some reasonably conspicuous differences between Precambrian and more recent sediments. There is no unanimity regarding their interpretation. This paper will show that some of the differences may be related to the progressive thermal decay of the Earth.

1. *Limestone, Dolomite and the Degassing History of the Earth*

The dolomite/calcite ratio is high in Precambrian carbonate sediments and appears to decrease progressively with time. Dolomite is relatively scarce in Cretaceous and Tertiary carbonates, and probably amounts to less than 1% of recent carbonates. Fig. 1 illustrates the time trend for the Mg/Ca ratio in carbonate rocks of the North American and Russian Platforms. The time trends for the other continents have not been as well documented but are almost certainly similar (see, for instance, Schidlowski, et al., 1975). A similar trend has been documented by Van Moort (1973) for the dolomite/calcite ratio in shaly rocks.

Two classes of explanations have been offered for these observations. The first includes those which call for the introduction of Mg and the removal of Ca long after sedimentation. In such schemes the smaller Mg/Ca ratio in Mesozoic and Cenozoic than in Precambrian and Paleozoic carbonates is interpreted in terms of the absence of dolomitization reactions which have already had time to affect the older carbonates. The second class of explanations invokes differences in the sediments during and/or shortly after sedimentation as the major cause for the observed age differences in the dolomite/calcite ratio of carbonates and of pelites. It is likely that most dolostones in the geologic record owe their origin to the percolation of hypersaline brines (see, for instance, Friedman and Sanders, 1967). Although dolomitization unrelated to the formation and early diagenesis of carbonate rocks is well documented, most dolomitization has probably occurred early in the history of carbonate sediments. If so, the observed trend in the dolomite/calcite ratio reflects real time variations in the operation of the ocean–atmosphere system.

The residence time of CO_2, Mg^{+2}, and Ca^{+2} in the ocean–atmosphere system is geologically short. The oceans and atmosphere therefore adjust their composition rapidly, probably on a time scale of $\leq 10^7$ years in response to changes in the supply of Mg^{+2}, Ca^{+2}, and CO_2. The systematics of this response are

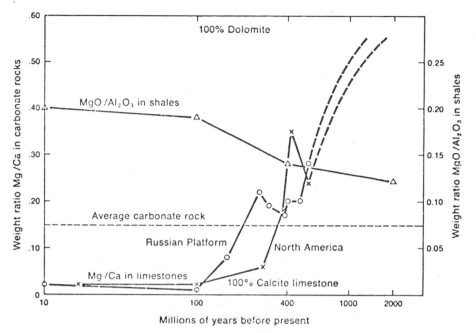

Fig. 1. Magnesium to calcium weight ratios in carbonate rocks from the Russian Platform and from North America as a function of age. Also plotted is the weight ratio MgO/Al_2O_3 for average shaly rocks as a function of their ages (Garrels and McKenzie, 1971)

outlined in Fig. 2. On a CO_2-free planet in which water is the major weathering agent Ca^{+2} and Mg^{+2} released during weathering are removed from the oceans as constituents of hydroxides and/or silicates. Clay minerals and zeolites are likely minerals in ocean sediments in a highly siliceous planet. Brucite and portlandite may appear as sediments from a silica-poor ocean.

If CO_2 is metered into the atmosphere of such a planet, calcite will appear as a sedimentary mineral when the CO_2 pressure in the atmosphere exceeds the equilibrium value for silicate–carbonate equilibria such as

$$6Ca_{0.166}Al_{2.33}Si_{3.67}O_{10}(OH)_2 + CO_2 + 8H_2O$$

Ca-montmorillonite

$$\rightleftharpoons CaCO_3 + 7Al_2Si_2O_5(OH)_4 + 8SiO_2$$

calcite   kaolinite   quartz

With gradually increasing rate of CO_2 input a progressively larger fraction of the Ca^{+2} released during weathering leaves the ocean as a constituent of $CaCO_3$. When the rate of CO_2 input exceeds the rate of Ca^{+2} release during

weathering, Mg^{+2} is also removed from the oceans as a constituent of a carbonate phase. Dolomite is the preferred Mg carbonate mineral; huntite and magnesite appear only sporadically. Siderite is also a potentially important CO_2 sink. Its stability field increases with increasing CO_2 pressure, but today virtually all iron is removed from the oceans as a constituent of oxide, sulphide, and silicate phases, and it is not clear that $FeCO_3$ has ever been a major CO_2 sink in marine sediments.

When the CO_2 input rate exceeds the combined release rate of Ca^{+2} and Mg^{+2}, carbonate and bicarbonate minerals of Na^+ and K^+ tend to form. Although this has happened in non-marine evaporite basins, the oceans have not been saturated with such phases during the Phanerozoic Era, and the minerals characteristic of the high pH–high carbonate facies of the Green River Formation and their metamorphic equivalents seem to be very rare in Precambrian sediments as well.

During weathering today and most of Earth history Mg^{+2} and Ca^{+2} released during weathering have come in part from the solution of

$$\frac{CO_2}{CaO + MgO + Na_2O}$$

Element	I	II	III	IV
Ca	sil.	carb. $\longrightarrow$		
Mg	sil.	sil.	carb. $\longrightarrow$	
Na	sil.	sil.	sil.	carb.

sil. : Element present in silicates

carb.: Element present partially or completely in carbonates

Fig. 2. The distribution of Ca, Mg and Na between silicate and carbonate phases as a function of the ratio between the CO_2 input rate to the atmosphere–ocean system and the sum of Ca, Mg and Na delivered to the oceans via river water

limestones and dolomite. Today roughly 80% of the Ca^{+2} and 50% of the Mg^{+2} in river water have come from this source. The redeposition of a quantity of limestone and dolomite equivalent to the quantity dissolved during weathering represents neither a gain nor a loss of CO_2 for the atmosphere–ocean system. The maintenance of a constant dolomite/limestone ratio in sediments therefore demands that the rate of CO_2 input to the atmosphere be such that Mg^{+2} and Ca^{+2} released from silicate weathering be removed as constituents of dolomite and limestone in the steady-state ratio. This state can be maintained in a system where earlier sediments are decarbonated during metamorphism and melting, and where the CO_2 released during this process is recombined at a constant rate with the Mg^{+2} and Ca^{+2} released during the weathering of the Mg- and Ca-silicate residues of metamorphism and magmatism. The rather imperfect record of Precambrian carbonates suggests that this type of steady state began very early and was maintained for a very long period of time. The carbonates of the c. 3·3 b.y. old Swaziland system consist of limestones,

dolomitic cherts, siliceous dolomites, and quartzose dolomitic limestones (Schidlowski et al., 1975). Analyses in Table 1 of the still older, c. 3·7 b.y. old metamorphosed sediments of the Isua area show that much of the CaO in the Ca-rich metapelites is balanced by CO_2. Two of the three calcitic schists also contain an abnormally high MgO/Al_2O_3 ratio, which suggests that dolomite was also present in these rocks prior to metamorphism. The abundance of dolomite in later Precambrian carbonates has already been mentioned.

The decrease and virtual disappearance of dolomite as a constituent of carbonate rocks during the late Phanerozoic indicates that this balance has not been maintained during the past several hundred million years. Three explanations can be offered:

1. The river input of Mg^{+2} may have been removed from the oceans by some mechanism other than incorporation in carbonate and silicate minerals;

2. The Mg/Ca ratio in silicate rocks undergoing weathering decreased markedly; and

Table 1. Chemical composition of supracrustal rocks from the Isua Area, West Greenland
(pers. comm., J. Allaart, 1975)

	17–30–19	17–29–99	17–29–77	17–31–64	17–31–53	17–31–40
SiO_2	55·10	46·72	57·75	52·68	47·40	44·00
TiO_2	0·60	1·50	1·97	1·22	1·22	0·89
Al_2O_3	16·69	21·08	21·31	13·02	14·41	10·51
Fe_2O_3	7·33	4·94	3·99	2·53	2·09	2·18
FeO	9·56	11·57	5·87	11·40	11·76	9·42
MnO	0·31	0·18	0·11	0·23	0·27	0·37
MgO	3·23	2·67	1·60	5·35	3·52	5·87
CaO	1·36	1·39	1·20	4·94	7·50	12·03
Na_2O	1·80	0·55	1·04	1·06	0·48	0·27
K_2O	2·45	5·86	2·99	3·89	4·60	3·62
P_2O_5	0·09	0·22	0·12	0·16	0·22	0·19
CO_2	—	—	—	2·05	4·90	9·45
H_2O^+	1·26	2·05	1·66	1·08	1·18	1·00
Total	99·78	98·73	99·60	99·65	99·64	99·86
$\dfrac{MgO \quad gm}{Al_2O_3 \quad gm}$	0·194	0·126	0·075	0·411	0·246	0·569
$\dfrac{CO_2 \, gm}{CaO \, gm}$	—	—	—	0·415	0·653	0·786

17–30–19 Graphitic garnet–staurolite–biotite schist.
17–29–99 Tourmaline bearing garnet–muscovite–biotite schist.
17–29–77 Garnet staurolite schist.
17–31–64 Hornblende–bearing biotite schist.
17–31–53 Carbon bearing garnet–biotite–schist.
17–31–40 Carbonate bearing biotite–garnet schist.

3. The rate of CO_2 input into the atmosphere–ocean system did not keep up with the rate of $Mg^{+2} + Ca^{+2}$ demand for the formation of dolomite.

A mechanism of the first type has been actively investigated during the past few years. Seawater cycling through oceanic crust loses virtually all of its Mg^{+2} and gains an equivalent quantity of Ca^+ (see, for instance, Mottl et al., 1974, Bonatti, 1975, Hajash, 1973). The annual rate of seawater cycling through the oceanic crust has been estimated on the basis of the observed heat flow deficit in mid-ocean ridge areas (Spooner and Fyfe, 1973); unfortunately the data base is not good, and the estimated rates of Mg^{+2} loss from the seawater to the oceanic crust are correspondingly uncertain. The loss rate could equal the rate of river input, but several mechanisms of Mg^{+2} removal into marine sediments are well documented (Drever, 1974), and it is unlikely that Mg^{+2} loss by seawater cycling through the oceanic crust exceeds 50% of the current river flux of Mg^{+2}. Even this percentage may be a strong upper limit. Fig. 1 shows the estimated course of the MgO/Al_2O_3 ratio in shales through geologic time. The basis of these averages is perhaps open to question, but if the rise in the MgO/Al_2O_3 ratio during the Phanerozoic is real, the data strongly favour the transfer of Mg from the carbonate reservoir to the silicate reservoir in shales rather than the exchange of Mg for Ca in the oceanic crust.

There is very little indication of a significant change in the MgO/CaO ratio in silicate rocks undergoing weathering. It seems more likely, therefore, that the decrease in the dolomite/calcite ratio during the Phanerozoic Era is due to a decrease in the ratio of CO_2 supply to $Mg^{+2} + Ca^{+2}$ demand. Two factors

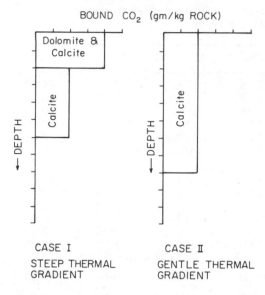

BOUND CO$_2$ (gm/kg ROCK)

CASE I
STEEP THERMAL
GRADIENT

CASE II
GENTLE THERMAL
GRADIENT

Fig. 3. A schematic view of the distribution of equal quantities of bound CO$_2$ in a crust with a steep geothermal gradient and in a crust with a gentle geothermal gradient

have decreased as well. Fig. 3 shows in a schematic way the effects of a decrease in mean geothermal gradient on the distribution of carbonate minerals in a crust of constant bound CO$_2$ content. When the geothermal gradient is so high that virtually all of the carbonate minerals are destroyed during metamorphism, bound CO$_2$ will be concentrated in the near-surface rocks. The CO$_2$/MgO + CaO mol ratio in near-surface rocks will therefore approach unity, and dolomite will be abundant. At lower geothermal gradients bound CO$_2$ will be distributed downward in the crust, the mol ratio CO$_2$/MgO + CaO in near-surface rocks will be lower, and dolomite will be a minor component of carbonate rocks.

A tendency in the direction of downward distribution of bound CO$_2$ with time is most likely. However, the complexity of the pattern of geothermal gradients and the variability in the nature of subduction zones and the subduction process are so considerable, that the connection between the sedimentary record of the Phanerozoic and any decrease in geothermal gradients must still be classed as highly speculative.

The elemental carbon content of sedimentary rocks offers a somewhat independent

could contribute to such a trend. The CO$_2$ degassing rate of the Earth has probably always consisted of a juvenile and a recycled component. The rate of juvenile degassing has almost certainly been decreasing with time. The rate of degassing of recycled CO$_2$ may

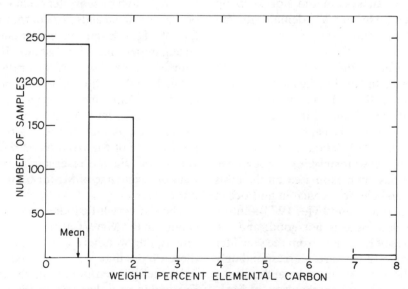

Fig. 4. The elemental carbon content of Archean shales from the Canadian Shield (Cameron and Jonasson, 1972)

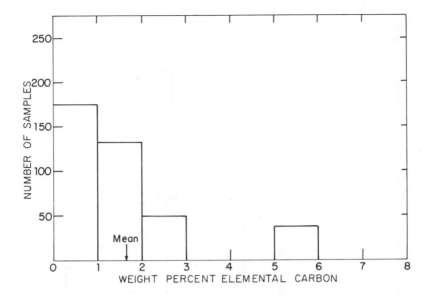

Fig. 5. The elemental carbon content of Proterozoic shales from the
Canadian Shield (Cameron and Jonasson, 1972)

measure of the CO_2 degassing rate of the
Earth. Schidlowski, et al. (1975) have shown
that the ratio of elemental carbon to carbonate
carbon in sedimentary rocks has remained
roughly constant during the last 3 b.y. Since
the release of Mg^{+2} and Ca^{+2} during weathering is proportional to the erosion rate, any
large changes in the ratio of the CO_2 input to
the $Mg^{+2} + Ca^{+2}$ demand should be visible in
the mean elemental carbon abundance of
sedimentary rocks. The geographic distribution of elemental carbon in sediments is, however, not uniform, and formidable sampling
problems stand in the way of obtaining believ-
able figures for modest changes in the mean $C°$
content of sedimentary rocks. $C°$ content of
Canadian Archean and Proterozoic shales
obtained from Cameron and Jonasson's
(1972) data in Fig. 4 and 5 shows that the mean
$C°$ content of Proterozoic shales is approximately twice that of the Archean shales.
Gehman's (1962) data in Fig. 6 for
Phanerozoic sediments and those of Ronov
and Migdisov (1971) indicate that the $C°$ content of Phanerozoic shales lies between 0·6
and 1·0%. These data rule out drastic changes
in the relationship of degassing rate to erosion
rate, but they do not rule out changes of the
magnitude required to explain the decrease in

the dolomite content of carbonates during the
Phanerozoic. Rather interestingly, the $S^=/C°$
ratio in shales seems to have decreased progressively with time (Cameron and Jonasson,
1972; Holland, 1973a).

2. *Changes in the Oxygen Content of the
    Atmosphere*

The oxygen content of the atmosphere
today is determined largely by the balance
between the rate of oxygen use during weathering and the rate of oxygen production due to
the burial of organic carbon, sulphide, and
ferrous oxide, largely in marine sediments
(see, for instance, Holland, 1973b).

Although the general properties of the control mechanism are well understood, the functional relationships between the rates of use,
the rate of production, and the P_{O_2} are not yet
well defined. Large-scale variation of P_{O_2} during the latter part of the Phanerozoic are ruled
out by the paleontologic record, and it is likely
that the present-day control mechanism has
operated quite efficiently for times long compared with the 4 m.y. characteristic time of
atmospheric oxygen.

The presence of apparently detrital grains of
uranite in a number of conglomerates more
than *c.* 1·9 b.y. old suggests that P_{O_2} in the

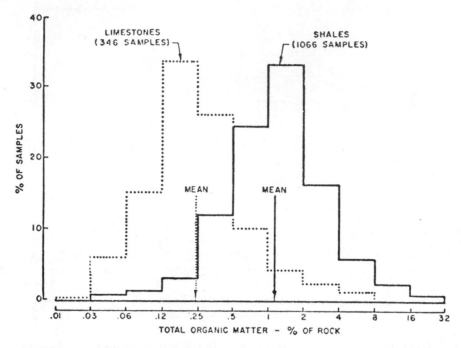

Fig. 6. The total organic content of Phanerozoic limestones and shales (Gehman, 1962); note that the organic content is 1·22 times the elemental carbon content of these sediments

Archean and early Proterozoic was considerably lower than at present. Extensive experiments by D. Grandstaff on the rate of oxidation and dissolution of uraninite have shown that for reasonable lengths of exposure of uraninite to the Precambrian atmosphere during weathering, transport, and deposition, uraninite is preserved only if $P_{O_2} \leq 4 \times 10^{-3}$ atm. The strongest evidence for a detrital origin of the uraninite in the Blind River (Canada) and the Witwatersrand and Dominion Reef (South Africa) conglomerates is their chemical composition, which is typical of a pegmatitic and granitic origin, and very different from the composition of hydrothermal uraninites and pitchblendes. Until a satisfactory mechanism is discovered for producing such uraninites diagenetically or hydrothermally, a detrital mode of origin seems very likely.

A much lower oxygen content of the atmosphere during the earlier Precambrian could be related to a number of variables. In the absence of organisms capable of liberating free oxygen during photosynthesis, free oxygen production would have been limited by the photodissociation of water in the upper atmosphere followed by the escape of hydrogen from the Earth. This process is thought to be quite inefficient, and it is unlikely that the quantity of iron oxidation and of hematite in early Precambrian sediments can be made to agree with this mode of oxygen production alone.

A more attractive alternative is the notion that the total rate of oxygen production via photosynthesis was limited by the availability of nutrients, and that the more rapid introduction of somewhat reducing volcanic gases limited P_{O_2} to values $\leq 2\%$ of the present atmospheric level. Unfortunately, neither the paleontologic nor the sedimentologic evidence for such a view are presently compelling.

III. Conclusion

The rate of heat generation by radioactive decay within the Earth has gradually decreased by roughly a factor of 3. Among the expected consequences of this decrease a number should have left an imprint on the operation of the ocean–atmosphere system and on the nature of the sedimentary record. The decrease in the dolomite/limestone ratio during the Phanerozoic can be interpreted rather readily in terms of a progressive decrease in the degassing rate and/or the mean geothermal gradient. The increase in atmospheric oxygen can be related to a progressive decrease in the rate of injection of reducing gases into the atmosphere. More data, however, are badly needed to confirm the hypothesis that the decrease in dolomite/calcite ratio and the rise of atmospheric oxygen are related to the slow thermal decay of the Earth.

References

Bell, R. T., and Jackson, G. D., 1974. 'Aphebian halite and sulfate indications in the Belcher Group, Northwest Territories', *Can. J. of Earth Sciences*, **11**, 722–728.

Bonatti, E., 1975. 'Metallogenesis at oceanic spreading centers', *Ann. Rev. of Planet. Sci.*, **3**, 401–432.

Broecker, W. S., 1971. 'A kinetic model for the chemical composition of seawater', *Quat. Res.*, **1**, 188–207.

Cameron, E. M., and Jonasson, T. R., 1972. 'Mercury in Precambrian shales of the Canadian Shield', *Geochim. Cosmochim Acta*, **36**, 985–1006.

Drever, J. I., 1974. 'The magnesium problem', Chapter 10 in Vol. 5 of *The Sea*, D. Goldberg (Ed.), Wiley–Interscience, New York.

Friedman, G. M., and Sanders, J. E., 1967. 'Origin and occurrence of dolostones', Chapter 6 in *Carbonate Rocks*, G. V. Chilingar, H. J. Bissell, and R. W. Fairbridge (Eds.), Elsevier, Amsterdam–London–New York.

Garrels, R. M., and Mackenzie, F. T., 1971. *Evolution of Sedimentary Rocks*. W. W. Norton, New York.

Garrels, R. M., and Perry, E. A., Jr., 1974. 'The cycling of carbon, sulfur, and oxygen through geologic time', In Vol. 5 of *the Sea*, D. Goldberg (Ed.), Wiley-Interscience, New York, 303–336.

Gehman, H. M., Jr., 1962. 'Organic matter in limestones', *Geochim. Cosmochim. Acta*, **26**, 867–884.

Hajash, A., Jr., 1975. 'Hydrothermal processes along mid-ocean ridges: an experimental investigation', *Ph.D. Dissertation*, Texas A & M University, College Station, Texas.

Holland, H. D., 1972. 'The geologic history of seawater—an attempt to solve the problem', *Geochim. Cosmochim. Acta*, **36**, 637–651.

Holland, H. D., 1973a. 'Systematics of the isotopic composition of sulfur during the Phanerozoic and its implications for atmospheric oxygen', *Geochim. Cosmochim. Acta*, **37**, 2605–2616.

Holland, H. D., 1973b. 'Ocean water nutrients, and atmospheric oxygen', Vol. 1 of *Proceedings of Symposium in Hydrogeochemistry and Biogeochemistry*, The Clarke Co., Washington.

Holland, H. D., 1974. 'Marine evaporites and the composition of seawater during the Phanerozoic', In *Studies in Paleo-Oceanography*, W. W. Hay (Ed.), Soc. of Econ. Paleont. and Mineral., Special Publication No. 20, 187–192.

Mottl, M., Corr, R. F., and Holland, H. D., 1974. 'Chemical exchange between seawater and mid-ocean ridge basalt during hydrothermal alteration: an experimental study' (Abs.), G. S. A. Ann. Meeting, 1974, *GAAPBC 6 (7)*, 879–880.

Ronov, A. B., and Migdisov, A. A., 1971. 'Geochemical history of the crystalline basement and the sedimentary cover of the Russian and North American platforms', *Sedimentology*, **16**, 137–185.

Schidlowski, M., Eichmann, R., and Junge, C. E., 1975. 'Precambrian sedimentary carbonates: carbon and oxygen isotope geochemistry and implications for the terrestrial oxygen budget', *Precambrian Research*, **2**, 1–69.

Sillén, L. G., 1961. 'The physical chemistry of seawater', In *Oceanography*, M. Sears (Ed.), A.A.A.S., Washington, D.C., 549–581.

Sillén, L. G., 1967. 'The ocean as a chemical system', *Science*, **156**, 1189–1197.

Spooner, E. T. C., and Fyfe, W. S., 1973. 'Sub-sea floor metamorphism, heat and mass transfer', *Contr. Mineral. and Petrol.*, **42**, 287–304.

Van Moort, J. C., 1973. 'The magnesium and calcium contents of sediments, especially pelites, as a function of age and degree of metamorphism', *Chem. Geol.*, **12**, 1–37.

$^{87}Sr/^{86}Sr$ Evolution of Seawater During Geologic History and its Significance as an Index of Crustal Evolution

JÁN VEIZER

Department of Geology, University of Ottawa, Ottawa, Ontario, Canada.

Introduction

An understanding of the formation of the continental crust is one of the basic tasks of geosciences. It is generally accepted that this crust was formed by a fractionation process from the upper mantle (cf. e.g. Ringwood, 1969; Taylor, 1964), but the details of this process and particularly the rates of crustal differentiation during geologic history are currently a focal point of considerable controversy. There is a broad consensus that the Archean–Proterozoic boundary marks probably the most important discontinuity of the preserved geologic record and that the Early Proterozoic crustal segments show considerable similarities to the Phanerozoic ones in their structural, petrogenic and sedimentary record (cf. e.g. Windley, 1973; Engel et al., 1974; Ronov, 1964; Veizer, in press and others). However, the nature of the continental crust in the Archean is widely disputed at present. The two opposing models could be summarized in their extremes as:

(a) Continental crust comparable in volume and chemical composition with the present one existed already early in the Archean. It is mostly assumed that this crust was thinner than at present, but covering most of the global surface;

(b) The Archean crust was mostly of the oceanic type and the felsic nuclei were of a subordinate importance only.

Consequently, the Archean–Proterozoic discontinuity would mark only a structural change, if the first alternative is accepted as a structural as well as compositional one in the case of the second alternative.

Since, on the one hand, the major portion of the present-day global surface is covered by crust of oceanic type and on the other hand even the oldest known crustal segment in Greenland (cf. Moorbath et al., 1972) has features similar to the continental crust, it is probable that the actual differentiation followed a path, which was somewhere between the two extremes. Considering the limited nature of the present-day constraints, we are probably not yet at a stage where we could state exactly the degree of validity of each model, but we may attempt to answer the question whether the actual differentiation path was closer to the one or the other above stated alternatives.

Present-day Cycle of Strontium in Near-surface Reservoirs

Fig. 1 summarizes the behaviour of strontium in near-surface environments. The fluxes among various reservoirs are not well known at present and also the $^{87}Sr/^{86}Sr$ ratios for the reservoirs are only approximations. Nevertheless, the Figure shows that strontium released from the continental crust and 'upper mantle' (including oceanic crust) accumulates in seawater. Because of its long residence time,

569

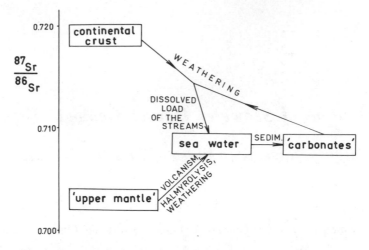

Fig. 1. Present-day cycle of strontium in near-surface reservoirs

strontium in seawater is distributed homogeneously and reflects a weighted average of the two inputs (cf. e.g. Peterman et al., 1970; Veizer and Compston, 1974).

The $^{87}Sr/^{86}Sr$ ratio of a reservoir is determined by the equation

$$(^{87}Sr/^{86}Sr)_t = (^{87}Sr/^{86}Sr)_0 + (^{87}Rb/^{86}Sr) \times (e^{\lambda t} - 1) \qquad (1)$$

where $(^{87}Sr/^{86}Sr)_t$ is the strontium isotopic ratio at time t; $(^{87}Sr/^{86}Sr)_0$ the initial ratio of the reservoir; $(^{87}4Rb/^{86}Sr)$ the Rb/Sr ratio of the reservoir; and λ the decay constant of ^{87}Rb. If the $(^{87}Sr/^{86}Sr)_0$ and t of the two input reservoirs are similar, the growth of radiogenic ^{87}Sr will depend on the Rb/Sr ratio only. Since the Rb/Sr ratio of the continental crust is higher, by about an order of magnitude, than the ratio of the 'upper mantle' (cf. e.g. Armstrong, 1968), *the addition of radiogenic ^{87}Sr*

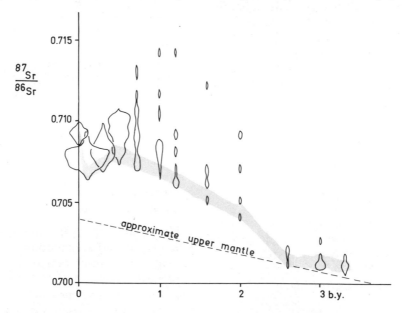

Fig. 2. Variations of $^{87}Sr/^{86}Sr$ ratios of sedimentary carbonate rocks during geologic history. Modified from Veizer and Compston (in prep.). The Bulawayan System plotted here at 3·0 b.y. may be only ~2·7 b.y. old.

into seawater will be a function of the evolution of the continental crust.

The output of strontium from seawater is via sedimentation, particularly of carbonate rocks. Thus, measuring the strontium isotopic composition of sedimentary carbonates during geologic history may help in the selection of the preferred model.

Variations and Significance of Seawater Strontium Isotopic Ratios During Geologic History

The variations in strontium isotopic composition of sedimentary carbonate rocks during geologic history were summarized by Veizer and Compston (in prep) and are presented in Fig. 2. As shown by those authors (Veizer and Compston, 1974), almost all secondary influences will tend to increase the $^{87}Sr/^{86}Sr$ ratios of the carbonates. Therefore the actual trend for seawater is defined by the low radiogenic part of the measured values. However, whether one takes only this low radiogenic part or the whole spread of values, it is clear that there is a sudden increase in $^{87}Sr/^{86}Sr$ ratios of seawater since ~ 2.5 b.y. ago and a decrease in the strontium isotopic ratios during the Phanerozoic. This would indicate a higher 'upper mantle' or lower continental crust contribution of strontium into seawater during the Archean and Phanerozoic. Another feature of the curve is that the Archean ratios are defining a trend close to the expected 'upper mantle' growth line. This would demand that the contribution of strontium from the continental crust into Archean oceans was of subordinate importance and was heavily outweighed by strontium of 'upper mantle' origin. It may be interpreted as indicating that *continental crust played only a subordinate role during the Archean.*

It is possible, although not likely, that the measured Archean sediments were formed only in local basins and not in the contemporaneous ocean. Thus the results would be of local significance only. However, if all 'local basins' derived their strontium from the 'upper mantle' input, as seems to be the case,

the above-suggested conclusion remains unaltered. This conclusion can be modified only if it is shown that the presently known Archean data are grossly unrepresentative.

Armstrong and Hein (1973) suggested that the recycling of continental crust through the mantle might have been faster during early stages of the Earth's history and consequently continental crust has been loosing its radiogenic components into the mantle. Even disregarding the difficulties with the subduction of the light non-refractory crust into deeper layers of the upper mantle (cf. e.g. Ringwood, 1969) and geochemical evidence against such a process from modern island arcs (cf. e.g. Faure and Powell, 1972 p. 41; Gill, 1974 and others), this does not seem to be a solution to the problem. In order to maintain $^{87}Sr/^{86}Sr$ ratio of seawater compatible with the 'upper mantle', the rate of recycling would have to be practically equal to the existing volume of the continental crust at any given moment during the Archean. This is difficult to conceive for any volume of the continental crust which is approaching its present-day value.

It seems, therefore, that the strontium isotopic data for seawater favour a model which assumes that the crust during the Archean was mostly of an oceanic or intermediate type and the continental crust played a subordinate role only. Consequently, the Archean–Proterozoic discontinuity marks not only a structural but also a compositional change. This is supported by the K_2O/Na_2O secular trends for igneous rocks and sediments compiled by Engel et al. (1974) (Fig. 3) as well as by additional sedimentological and geochemical data compiled by Veizer (in press), Ronov (1964), Ronov and Migdisov (1971) and others. There is surprisingly good correlation between the K_2O/Na_2O secular trends and the strontium isotopic trend for seawater. Both indicate that the crust was substantially less felsic in the Archean and they also demand a more mafic nature of the Phanerozoic crust as indicated by the 'trough' in both sets of data. In addition, the estimates of Engel et al. (1974) of the relative percentage of the 'continental crust and protocrust' show, that its proportion was low in the Archean and increased sharply at about 2·5 b.y. ago. Similarly the composition of this 'continental crust and protocrust' was rather mafic in the Archean as indicated not only by the K_2O/Na_2O but also by qz. monzonite/qz. diorite ratio. This remarkable agreement between both sets of data is hardly fortuitous. Those results would be also consistent with recent geological observations indicating three major tectonic regimes during geologic history: the essentially ensimatic 'green-stone belt' tectonic style of the Archean; mostly ensialic 'mobile belts' in the Proterozoic; and predominantly ensimatic 'plate tectonics' of the Phanerozoic (cf. e.g. Kröner et al., 1973; Engel et al., 1974).

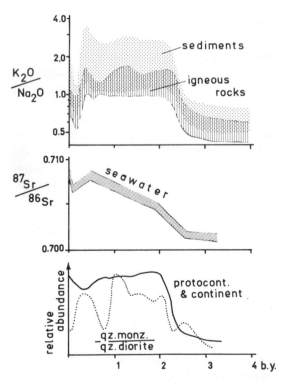

Fig. 3. Variations in K_2O/Na_2O ratios of sedimentary and igneous rocks; $^{87}Sr/^{86}Sr$ of 'seawater'; relative proportion of 'continental crust and protocrust'; and qz. monzonite/qz. diorite ratio during geologic history. Strontium isotope data from Veizer and Compston (in prep.) and the remaining information from Engel et al (1974).
Modified from Veizer and Compston (in prep.)

Model Evolution of the Continental Crust

Previous discussion indicated that the model favoured by present data is closer to the alternative with subordinate importance of the continental crust during the Archean than to the other extreme. However, it might be of interest to attempt a broad definition of the crustal fractionation rates during geologic history. The details of this discussion will be published elsewhere (Veizer, in press) and only a brief summary is included here.

Fig. 4 shows the present-day areal distribution of continental basement per 450 m.y. 'age' units. Although covering only about 2/3 of the continental surface, this is by far the most comprehensive compilation available at present. Therefore it may be assumed that the trend is representative of the whole continental crust. If the increase in area is proportional to the increase in volume, it might be possible to define the rate of fractionation of the continental crust.

The slope of decreasing area with increasing age in Fig. 4 is roughly linear for all time units, with the exception of the 1·8–2·7 b.y. interval. Since this scale is semilogarithmic, it would indicate that the 'straight line' follows an exponential function. Such an exponential slope can be generated by several processes and the basic models are the following (cf. Fig. 5):

(I) An exponentially increasing rate of continental crust formation towards present (e.g. Hurley and Rand, 1969);

(II) Total fractionation of the continental crust at an early stage of the Earth's history and afterwards only its recycling either through the mantle (Armstrong, 1968; Armstrong and Hein, 1973) or within the crust;

(III) Linear growth of the continental crust, combined with recycling, if the rate of recycling is faster than the rate of linear growth. If so, the recycling must be limited to recycling within the crust.

Models (I) and (II) are limiting cases and model (III) is between the two extremes.

It could be shown that, accepting the fractionation rates as indicated in Fig. 4, and cal-

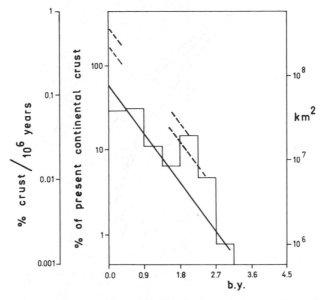

Fig. 4. Present-day areal distribution of the continental basement as a function of its geologic 'age'. Based on compilation of Hurley and Rand (1969)

culating Sr isotopic evolutionary curves for continental crust and 'upper mantle' in model (I), the results would not be compatible with existing estimates for either of those reservoirs. Therefore the slope in Fig. 4 is a residual slope caused by recycling.

In the case of model (II) the decrease in volume of continental crust for progressively older age units is proportional to the total volume of the continental crust. Thus

$$V_t = V_T\, e^{-kt} \qquad (2)$$

where V_t is the volume of the non-recycled crust after duration of cycling t; V_T equals total volume of the continental crust; and k is

the 'decay constant' defining the rate of recycling. Consequently the resulting cumulative slope will be linear on a semilogarithmic graph.

In the case of the model (III) situation V_T is not a constant but is increasing with t. Therefore, V_T at time t is

$$V_{T_t} = \int_0^t \frac{dV_T}{dt}\, dt \qquad (3)$$

Since the model assumes a linear increase in the volume of the continental crust and a constant rate of recycling

$$\frac{dV_F}{dt} - \frac{dV_R}{dt} = \frac{dV_T}{dt} \qquad (4)$$

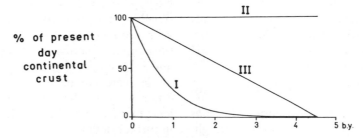

Fig. 5. Basic models of the volumetric growth of the continental crust during geologic history. The % on y axis are *cumulative* per cent

574

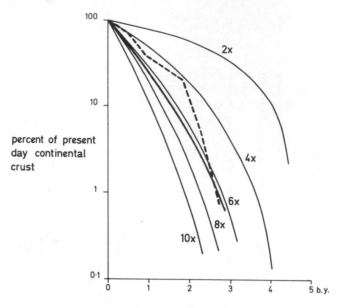

Fig. 6. *Cumulative* percentage of the continental crust older than t remaining after recycling in model (III) situation. The calculated model assumes linear growth of the continental crust since about 4·6 b.y. ago combined with recycling. The combined recycled and existing quantity of continental crust was 2, 4, 6, 8, 10 times of its present-day volume. The thick curved line represents the measured slope (cumulative curve of the 'straight line' in Fig. 4) and the dashed line the cumulative curve of the 'uncorrected' histogram in Fig. 4

where dV_F is the proportion of the crust formed per time unit dt, dV_R is the proportion recycled and dV_T is the proportion added. Since dV_R and dV_T are constants, dV_F must be constant as well. However, the ratio of

$$\frac{dV_R}{dt} \Big/ \int_0^t \frac{dV_T}{dt} dt \qquad (5)$$

is decreasing with increasing t and thus with the increasing volume of the continental crust. In other words, the older continental crust has been recycled (destroyed) faster than the younger one. If so, equation (2) will become binomial and expressed as

$$V_t = V_T(e^{-k_1 t} - e^{-k_2 t}), \quad \text{where } k_1 > k_2 \qquad (6)$$

and the resulting cumulative slopes will be as in Fig. 6. The cumulative curve of the 'straight line' in Fig. 4 seems to follow this distribution rather than the distribution required for model (II) situation.

It may be objected that the deviation of the measured cumulative curve from linearity is not large enough to warrant rejection of model (II). If this were the case, the rate of recycling required to accommodate the measured slope would be at least one order of magnitude greater than the recycling rates through the upper mantle acceptable on the basis of Rb–Sr systematics. Thus the major recycling must have been within the crust itself. If so, the Rb–Sr systematics would be consistent with such a model only if the total fractionation of the continental crust happened between ~3·0 and 2·0 b.y. ago. Considering the sensitivity of model calculations, this result is not very different from the final conclusion reached later in the text, and thus will not be discussed in further detail.

The projection of the 'straight line' on the ordinate in Fig. 4 indicates a fractionation rate of the continental crust (dV_F/dt) of ~0·13 volume % per 10^6 years. This divided by 6·5 (Fig. 6) gives ~0·02% per 10^6 years as the rate of constant addition (dV_T/dt). If the two excess periods between 2·7 and 1·8 b.y. ago were affected by the same binomial recycling, their rate of fractionation was 0·06 and 0·10% per 10^6 years, respectively. Adding all those fractions together would yield $(3700 \times 0·02 + 450 \times 0·10 + 450 \times 0·06 =)$ 146% of the present-day continental crust.

Considering the nature of constraints this might be an acceptable result. However, if the discussion in the previous section is correct, the Archean crust was predominantly of an oceanic or intermediate type and thus the calculated rate of fractionation of the continental crust would be valid since ~2·7 b.y. only. This will reduce our total to 108%, which is an encouraging result. The strontium isotopic model calculated on this basis satisfies all constraints and is also in good agreement with the trend for seawater (cf. Fig. 7).

Green (1975) concluded that for temperature gradients expected for the Archean from experimental studies on komatiite, basalt would be entirely within the garnet granulite stability field and thus could not sink into the mantle. Instead it woud be piled up against small continental nuclei, as present-day sediments do, and partially melt only if the thickness of this 'protocrust' reached about 25 km thus giving rise to granodioritic and granitic intrusives. Only with decrease in temperature gradients will basalt enter the stability field of eclogite and due to the subsequent formation of a density gradient trigger the subduction process.

If this contention is correct, the 'protocrust' of an oceanic or intermediate composition has been accumulating more or less linearly during geologic history (Fig. 6). However, due to the high temperature gradients during the Archean it could not have been reprocessed into stable continental crust (possibly through developed island arcs of the Andean type). This process was lagging behind until temperature gradients decreased to such a degree as to stabilize eclogite. If this happened at the end of the Archean it would trigger reprocessing of the accumulated 'protocrust' thus causing the most pronounced 'granitic phase' of the geologic history. The model calculations

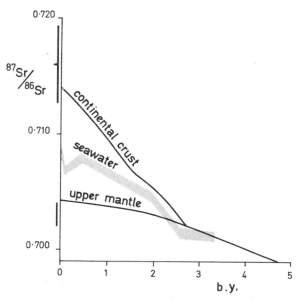

Fig. 7. Model strontium isotopic evolution of the continental crust and upper mantle. Rb and Sr budgets from Armstrong (1968). See text for further details

indicate that an equivalent of about 2/3 of the present-day areal extent of the upper continental crust ('granitic' layer) has been formed between 2·7 and 1·8 b.y. ago. *This would thus mark the emergence of stable cratons as the dominant structural phenomenon of the Earth crust.* During the subsequent 1·8 b.y. the accumulation of the 'protocrust' has been followed by its continual reprocessing into the stable continental crust.

The above conclusions seem to agree well with the new paleomagnetic (Piper et al., 1974; Irving and Lapointe, 1975), structural (cf. e.g. Shackleton, 1973) as well as sedimentological (cf. e.g. Ronov, 1964; Veizer, in press) data, which indicate the presence of stable platforms and intracratonic orogenic belts since the Early Proterozoic.

Consequences for Composition of Sedimentary Rocks

If the previous discussion is correct it will cause secular variations in proportions and composition of various sedimentary facies.

The total mass of sediments should grow proportionally with the growth of 'protocrust' and continental crust. The recycling of this mass will be probably similar to, or faster than, the recycling of the crust (cf. also Garrels and Mackenzie, 1971 Chapter 10). In addition, this recycling will be mostly cannibalistic and this will enhance mixing of the sedimentary mass.

Since during the Archean the source of the newly added material to sedimentary mass would be mostly the 'protocrust' of an oceanic or intermediate composition, the dominant facies should be graywackes and cherts from mechanical and chemical decomposition of alumosilicates. With emergence of the 'granitic' layer as the source of the new sedimentary material, the sedimentary sequences should be dominated by immature arkoses and mature sediments such as carbonates, sandstones, orthoquartzites, etc. This is indeed the case (cf. e.g. Strakhov, 1964; Ronov, 1964; Garrels and Mackenzie, 1971 Chapter 10; Cloud, 1972; Veizer, in press and others). In addition,

the chemical composition of clastic rocks should change accordingly. There should not be any remarkable chemical variations in composition of the Archean sediments due to more or less similar input. However, with emergence of the 'granitic' layer as the dominant source, the newly added mass of sediments will displace the average composition of clastic rocks more and more towards this felsic member. It would be unrealistic to expect secular trends for most of the major elements due to their small difference, say by a factor of two, between the two alternative sources. This is within the scatter caused by other factors. However, with a larger enrichment or depletion factor (e.g. trace elements), the trends should be observable with the present-day sampling and technical possiblities. Those predictions are well born out in Fig. 3 and for averaged values in Fig. 8 as well as by other

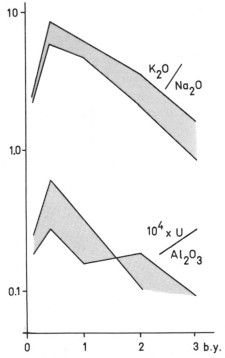

Fig. 8. K_2O/Na_2O and U/Al_2O_3 variations of the Russian Platform and North American shales and their partly metamorphosed equivalents (based on compilation of Ronov and Migdisov, 1971). Note the reverse of the trends for Phanerozoic, which corresponds with the 'trough' in Fig. 3

trends summarized, for example, by Ronov and Migdisov (1971), Veizer (in press) and others.

Conclusions

Considerations of the present-day 'age' distribution of the continental crust and strontium isotopic constraints indicate that the 'protocrust' of an oceanic or intermediate composition has been accumulating more or less linearly during geologic history. At the Archean–Proterozoic boundary this 'protocrust' has been reprocessed into stable continental crust, forming about 2/3 of the present-day 'granitic' layer during the 2·7–1·8 b.y.

interval. The subsequent accumulation of the 'protocrust' has been followed by its continuous reprocessing into stable continental crust. This evolution, if valid, should cause secular variations in proportions of various sedimentary facies and displace the average chemical composition of clastic sediments towards the felsic source member with decreasing age.

The reader should keep in mind that the above conclusions were based on model calculations and are therefore only as good as the basic constraints employed. How they compare with the real situation can be possibly judged from comparison with other papers in this volume.

References

Armstrong, R. L., 1968. 'A model for evolution of strontium and lead isotopes in a dynamic Earth', *Rev. Geophys.*, 6, 175–199.

Armstrong, R. L., and Hein, S. M., 1973. 'Computer simulation of Pb and Sr isotope evolution of the Earth's crust and upper mantle', *Geochim. Cosmochim. Acta*, **37**, 1–18.

Cloud, P. E., 1972. 'A working model of the primitive Earth', *Am. J. Sci.*, **272**, 537–548.

Engel, A. E. J., Itson, S. P., Engel, C. G., Stickney, D. M., and Cray, E. J., 1974. 'Crustal evolution and global tectonics: a petrogenic view', *Bull. Geol. Soc. Am.*, **85**, 843–858.

Faure, G., and Powell, J. L., 1972. *Strontium Isotope Geology*, Springer, 188 pp.

Garrels, R. M., and MacKenzie, F. T., 1971. *Evolution of sedimentary rocks*, Norton, 397 pp.

Gill, J. B., 1974. 'Role of underthrust oceanic crust in the genesis of a Fijian calc–alkaline suite', *Contrib. Miner. Petrol.*, **43**, 29–45.

Green, D. H., 1975. 'Genesis of Archean peridotitic magmas and constraints on Archean geothermal gradients and tectonics', *Geology*, **3**, 15–18.

Hurley, P. M., and Rand, J. R., 1969. 'Pre-drift continental nuclei', *Science*, **164**, 1229–1242.

Irving, E., and Lapointe, P. L., 1975. 'Paleomagnetism of Precambrian rocks of Laurentia', *Geoscience Canada*, **2**, 90–98.

Kröner, A., Anhausser, C. R., and Vajner, V., 1973. 'Neue Ergebnisse zur Evolution der präkambrischen Kruste in südlichen Africa', *Geol. Rundsch.*, **62**, 281–309.

Moorbath, S., O'Nions, R. K., Pankhurst, R. J., Gale, N. H., and McGregor, V. R., 1972.

'Further rubidium—strontium age determination of the very early Precambrian rocks of the Godthaab district, West Greenland', *Nature—Phys. Sci.*, **240**, 78–82.

Peterman, Z. E., Hedge, C. E., and Tourtelot, H. A., 1970. 'Isotopic composition of Sr in seawater throughout Phanerozoic time', *Geochim. Cosmochim. Acta*, **34**, 105–120.

Piper, J. D. A., 1974. 'Proterozoic crustal distribution, mobile belts and apparent polar movements', *Nature*, **251**, 381–384.

Ringwood, A. E., 1969. 'Composition and evolution of the upper mantle.' In P. J. Hart (Ed.), *The Earth's Crust and Upper Mantle*, Am. Geophys. Union, Geophys. Monogr., **13**, 1–17.

Ronov, A. B., 1964. 'Common tendencies in the chemical evolution of the Earth's crust, ocean and atmosphere', *Geochem. Int.*, **1**, 713–737.

Ronov, A. B., and Migdisov, A. A., 1971. 'Evolution of the chemical composition of the rocks in the shields and sediment cover of the Russian and North American Platforms', *Sedimentology*, **16**, 137–185.

Shackleton, R. M., 1973. 'Problems of the evolution of the continental crust', *Phil. Trans. Royal Soc. London. Math. Phys. Sci.*, **273**, 317–320.

Strakhov, N. M., 1964. 'Stages of development of the external geospheres and formation of sedimentary rocks in the history of the Earth', *Int. Geol. Rev.*, **6**, 1466–1482.

Taylor, S. R., 1964. 'Trace-element abundances and the chondritic earth model', *Geochim. Cosmochim. Acta*, **28**, 1989–1998.

578

Veizer, J., in press. Evolution of ores of sedimentary affiliation through geologic history; relations to the general tendencies in evolution of the crust, hydrosphere, atmosphere and biosphere.' In K. H. Wolf (Ed.), *Sedimentary Ore Deposits*, Elsevier.

Veizer, J., and Compston, W., 1974. $^{87}Sr/^{86}Sr$ composition of seawater during the Phanerozoic', *Geochim. Cosmochim. Acta*, **38**, 1461–1484.

Veizer, J., and Compston, W., in prep. '$^{87}Sr/^{86}Sr$ in Precambrian carbonates as an index of crustal evolution.'

Windley, B. F., 1973. 'Crustal development in the Precambrian', *Phil. Trans. Royal Soc. London.*, **A 273**, 321–341.

Basic Similarity of Archean to Subsequent Atmospheric and Hydrospheric Compositions as Evidenced in the Distributions of Sedimentary Carbon, Sulphur, Uranium and Iron

M. M. KIMBERLEY

Erindale College, University of Toronto, Mississauga, Ontario, L5L 1C6.

E. DIMROTH

Université du Québec, Chicoutimi, Québec.

Abstract

The distributions of sedimentary organic carbon, sulphur, uranium and iron are all related to atmospheric and hydrospheric component abundances, most directly to the partial pressure of atmospheric oxygen. There are many more similarities than differences among these sedimentary distributions through stratigraphically preserved Earth history and structural settings have had more obvious effects than organic evolution.

We know of no evidence which proves orders-of-magnitude differences between Middle Archean and subsequent atmospheric compositions, hydrospheric compositions, or total biomasses. Although there is no direct evidence for such, there may well have been a reducing atmosphere more than 3800 m.y. ago.

It is popularly believed that the Middle and Late Archean atmosphere contained practically no free oxygen (Cloud, 1972). This concept has had its roots in the three-fold division of the stratigraphic record based on abundance of macrofossils: Phanerozoic, Proterozoic and Archean. Lack of obvious Archean life has popularly been attributed to a hostile environment rich in toxic, reduced volcanic gases. Lack of Archean sulphates and red beds has similarly been attributed to peculiar atmospheric and hydrospheric compositions (Cloud, 1972).

The strongest support for the concept of an oxygen-poor Archean and Middle Precambrian atmosphere came with Holland's (1962) calculation of the maximum partial pressure of oxygen for uraninite stability, and his interpretation that the presumed placer deposits of the Witwatersrand, South Africa and Elliot Lake, Canada, could not have formed under a significantly oxygenic atmosphere. This has been followed by a variety of iron-formation genetic models which depend upon an oxygen-poor atmosphere (Lepp and Goldich, 1964; Cloud, 1973; Holland, 1973; Drever, 1974).

There is now substantial evidence against these interdependent concepts. Microorganisms have been reported from Middle Archean carbonaceous rocks of the Fig Tree

579

Group of the Swaziland Supergroup near Barberton, South Africa (Pflug, 1966) and possible eukaryotes older than 1800 m.y. have been found in stromatolitic carbonates of the Belcher Islands, Hudson Bay, Canada (Hofmann, 1974). Archean and Lower Proterozoic (Aphebian) mudrocks, e.g. argillites, shales, etc., sampled to date average 0·7 wt.% org. C and 1·6 wt.% org. C, respectively (Cameron and Jonasson, 1972).

This organic carbon either is the decay product of living matter or has formed by inorganic, radiative synthesis of hydrocarbons in the atmosphere (Lasaga, Holland and Dwyer, 1971). The latter possibility may be dismissed for the span of Earth history represented by sedimentary rocks because the resulting 'primordial oil slick' would have produced better developed bituminous gravels than those commonly found now on Mediterranean and other beaches, whereas Archean beach conglomerates contain no apparent clasts of bitumen (Dimroth et al., 1975).

Abundant Archean organic carbon is most likely a residual product of photosynthetic oxygen production. It is theoretically possible that the rate of this atmospheric oxygen production was low if most of the organic carbon in Archean sediments repeatedly survived weathering and resedimentation cycles (Garrels, Perry and Mackenzie, 1973, p. 1177). However, the distribution of organic carbon among Archean sedimentary facies is inconsistent with this possibility. The pattern of Archean carbon distribution does not differ in any obvious way from that of the Late Precambrian or Phanerozoic during which times single-cycle life processes have surely been the dominant control of distribution (Dimroth et al., 1975).

The average amount of organic carbon in Phanerozoic mudrocks is not markedly different, 0·5 wt.% (Ronov, 1958), from the Precambrian averages. If these statistics are accurate, the slight decrease may be due to organic degradation by bottom feeders and resulting increase in rate of oxidation. Isotopic ratios of organic carbon and carbonate carbon, a measure of net rate of organic carbon fixation, show no consistent time trend (Moore et al., 1974; Nagy et al., 1974). Even in the absence of photosynthesis, it is not certain that the partial pressure of atmospheric oxygen would be several orders of magnitude lower than at present, given the uncertainties in available calculations of the effect of photodissociation (Berkner and Marshall, 1965).

The present rate of atmospheric oxygen loss in the oxidation of volcanic gases is about two orders of magnitude less than the loss in oxidation of rocks undergoing weathering (H. D. Holland, pers. comm., 1971). Even if the rate of degassing were several times greater in the Archean than at present, it is unlikely that the rate of oxygen use would have approached that of modern weathering. If a lesser area of the Earth had been exposed to the atmosphere in the Archean, the rate of total oxygen consumption could have been significantly lower.

The distribution of sulphur in Archean and Proterozoic rocks is similar to that in Phanerozoic rocks of comparable type. Carbonaceous mudrocks of all ages are richly pyritic and basal sandstones of all clastic sequences are commonly cemented by pyrite. The sedimentary setting of pyrite in the 3100 m.y. old Dominion reef of South Africa (Nicolaysen et al., 1962) is comparable with that of pyrite in the Ordovician–Silurian basal Shawangunk Conglomerate of Pennsylvania and New Jersey, U.S.A. (Epstein and Epstein, 1972). Lower Proterozoic pyrite at Elliot Lake, Canada, commonly appears, microscopically, to be partially replacing quartz clasts.

The most obvious time trend in sulphur isotopic values is a decrease in scatter within individual pyritic deposits. The very low isotopic variation of some Lower Proterozoic pyritic bodies (Roscoe, 1969) and concomitant large sizes of commonly euhedral grains are probably the product of late-stage homogenization and crystal growth.

The lack of Archean red beds or sulphate deposits is hardly surprising given the relative lack of preserved shelf facies where subaerial oxidation or shallow-water chemical precipitation could take place. What Archean sulphates were deposited have probably been progressively lost to circulating groundwaters.

Concentration of silica, pyrite, uraninite, and gold between clasts in the Archean Dominion Reef of South Africa and in Middle Precambrian conglomerates and sandstones is interpreted to have been diagenetic (Kimberley, 1974a). The interstitial silica, pyrite, uraninite, and gold are considered to have precipitated from through-flowing groundwater as originally conceived by Du Toit (1953) for the Witwatersrand mineralizations. Lack of common heavy mineral concentration (Davidson, 1965) and broad sedimentological similarities to Tertiary mineralizations in the U.S.A. (Derry, 1960; Young, 1973; Brock and Pretorius, 1964) suggest that repeated cycles of dissolution from large volumes of exposed clastic sediment and pyroclastics were followed by precipitations along surfaces of contrasting permeabilities.

The principal observations commonly cited to counter such an epigenetic model are that ore is concentrated along the bottoms of Middle Precambrian host beds (Schidlowski, 1968), the thorium content of Middle Precambrian uraninite is high, despite thorium's insolubility in pure water, and detrital uraninite is presently found along the upper Indus River in Pakistan although it is unstable at an oxygen pressure more than about 10^{-21} atm. (Holland, 1962; Ramdohr, 1960; Zeschke, 1960).

These objections are not serious. Ore concentrations along the bottoms of host beds are commonly accompanied by lesser concentrations along the tops (Graton, 1930) where changes in permeability are typically less marked. Uraninite is also concentrated along the bottoms of host beds in the Permian (Ippolito, 1963). Thorium is generally a few times more abundant than uranium in clastic sediments and is soluble in organic solvents (Galkin et al., 1963). Abundant heavy hydrocarbon in Middle Precambrian ores (Schidlowski, 1967) records the probable former existence of such solvents. The observations that some carbonaceous forms in Witwatersrand ores appear to be algae partially replaced by thorian uraninite (Snyman, 1965; Hallbauer and Van Warmelo, 1974) also indicate thorium mobility. Detrital uraninite in the upper Indus river does not occur in an environment of potential preservation and is not being highly concentrated mechanically but does serve as a modern analogue of uranium readily removable by oxygenic groundwater and available for concentration along a favourable surface of precipitation (Davidson, 1964). Uraninite grains '. . . .which have the sphericity and polish suggestive of detrital grains' are found to be groundwater-precipitated replacements of detrital quartz near the base of the Triassic Chinle Formation in Arizona (Petersen, 1960, p. 138).

The relative constancy of uranium grade in ore beds throughout the Elliot Lake area of Canada is suggestive of a chemical control in mineralization (F. Joubin, pers. comm., 1975). The high uranium content of crystalline Archean source rocks is the probable main reason for uranium concentrations in the Lower Proterozoic. Tertiary mantles on uplifted, crystalline Precambrian rocks like the Shirley Basin of Wyoming (Harshman, 1972) are similarly rich in stratiform deposits of uraninite.

The most abundant Archean chemical sedimentary rocks are iron-formation and chert. Iron-formation, as the name is used here, is chemical sedimentary rock of any age which contains over 15 wt.% iron and is not largely composed of glauconite. Archean iron-formation is remarkably similar to younger iron-formation which is associated with similar terrigenous rocks. As in the case of younger iron-formations, there are two types of Archean iron-formations, shallow- (above wave base) and deep-water (below wave base) types, but the proportion of preserved Archean deep-water type is relatively high, commensurate with the high proportion of other deep-water Archean rocks. By contrast, shallow-water iron-formation constitutes something over 80% of all younger iron-formations. Continuous sedimentary sequences may include beds of both shallow- and deep-water types.

Unmetamorphosed iron-formations are usually readily distinguishable by field relationships and microscopic textures as to type. Shallow-water iron-formation of Archean to Late Pliocene age typically displays oolitic

582

texture which is very similar to that of Recent aragonitic oolite but which differs in microstructure from any other Recent chemical sediment. Oolitic texture has been widely reported from Archean iron-formations and ferriferous cherts of southern Africa (Ramsay, 1963; Wiles, 1957; Stowe, 1968). Associated shallow-water sedimentary structures include cross-bedding, ripple marks, and rill marks (Reimer, 1967; Viljoen and Viljoen, 1969; Mason, 1970). Eastern Canadian Archean iron-formation is characteristically non-oolitic, although pisolitic texture has been reported from the Lumby Lake area of Ontario (Woolverton, 1960).

The voluminous shallow-water iron-formation of Late Pliocene age, less than 5 m.y. old, which occurs around the south-western part of the Sea of Azov, U.S.S.R., must have formed under an atmosphere compositionally very similar to that at present. This deposit has been interpreted on geochemical and sedimentological evidence to have formed by diagenetic replacement of aragonitic oolite (Kimberley, 1974a,b). Overlying non-calcareous soil was presumably leached by organic acids of iron, silica, and aluminium which were carried down a few metres to the oolite and precipitated largely as the aluminous, ferrous septechlorite, chamosite.

An analogous process has recently been discovered occurring near Red Bay, north-western Andros Island, Bahamas (Kimberley, in press). Here the partially ferruginized oolite has banded structure and free silica, similar to the Middle Precambrian Gunflint, Labrador Trough, South African Transvaal, and other iron-formations. Paleogeographic reconstruction of the Transvaal iron-formation (Button, 1974) suggests a deltaic source of non-calcareous mud like that proposed for the Late Pliocene iron-formation (Kimberley, 1974a,b).

Deep-water iron-formation commonly found in Archean and less commonly in younger flysch sequences obviously is not the product of subaerial leaching. However, it has been proposed (Kimberley, 1975) that the mode of iron concentration has been closely related in that the primary sediment was largely aragonite, the source of iron was overlying, non-calcareous mud, and leaching occurred because of organic acid production during organic decay in that mud. The process would have occurred on a slope to which, during volcanically or tectonically active periods, fresh mud was supplied and leached mud eroded. The aragonite would have precipitated in supersaturated shallow water and been eroded to form resedimented aragonite-rich beds during volcanically and tectonically quiescent periods.

Exchange of calcium for iron and silica being leached a few centimetres above the carbonate could have occurred diffusively in deep water, like modern sulphate diffusion (Berner, 1964). Individual beds of deep-water iron-formation are quite thin, typically a few tens of centimetres at most, and diffusion could have operated at depths of burial less than those at which there would have been substantial upward pore fluid loss due to compaction. Any gravitationally induced inter-granular movement would have affected pore fluids and caused minor mixing between carbonate and clastic bed pore waters.

The main support for this hypothesis is the observation that both types of individual Archean, like all younger, iron-formation beds are characteristically overlain by carbonaceous mudrocks, e.g. shale, argillite, and phyllite, if the iron-formation itself has not been partially eroded. This mudrock is very seldom iron-poor. The hypothetical leached surficial mud has had a low preservation potential like that of the leached soil above shallow-water iron-formations. Ferruginized oolite beds exposed in the Pleistocene section at Morgan's Bluff, north-eastern Andros Island, are not overlain by black soil like that which is being leached during present-day ferruginization (Kimberley, in press). The organic carbon content of Canadian Archean mudrocks sampled to date is remarkably high, averaging 0.7 wt.% versus 1.6 wt.% for Lower Proterozoic mudrocks (Cameron and Jonasson, 1972) and 0.5 wt.% for Phanerozoic mudrocks (Ronov, 1958). The rate of production of organic acids was probably quite high in the Archean.

The occurrence of Archean stromatolites in the Bulawayan Series of Southern Rhodesia and at High Lake near Coronation Gulf, Canada (observation by J. F. Henderson, oral commun., 1974) provides evidence of Archean organically induced carbonate precipitation. In the absence of shelly fauna and red algae, the most probable form of carbonate precipitation would have been aragonite. Aragonite dissolves much more readily than magnesium-poor calcite in both meteoric water and undersaturated sea water (Kimberley, in press; Friedman, 1964) and would have been a highly reactive sediment.

Cherty, deep-water iron-formations are by no means restricted to the Archean. The thickest post-Paleozoic iron-formation, over 400 m thick, which occurs 130 km west of Inuvik, N.W.T., is of this type (Young, 1972). Like other deep-water iron-formations, this deposit contains abundant siderite. The Devonian Lahn-Dill iron-formations of Germany (Bottke, 1965) and the Jurassic Varcs deposit of Yugoslavia (Latal, 1952) are also of volcanic-associated, deep-water type. These two exhibit rare oolitic texture. Such Phanerozoic iron-formations do differ chemically from Archean counterparts, particularly in having higher phosphate contents. The reason for this is as yet unknown but may be related to a difference in chemical composition of Phanerozoic microorganisms. Archean banding may have been produced purely

diagenetically as it is today in the Bahamas (Kimberley, in press) or, in deep-water iron-formation, it may be partly a relict of primary lamination in the resedimented carbonate.

As similarly concluded from the distribution of sedimentary carbon, sulphur and uranium, Archean iron-formations are not interpreted here to indicate any 'orders-of-magnitude' differences between modern and ancient atmospheric or hydrospheric component abundances like those envisaged by most modern sedimentary geochemists (Holland, 1962; Garrels et al., 1973). The genetic processes favoured herein require no inorganic processes or inorganic solute abundances not presently found. The lack and possible absence of modern deep-water iron-formation is attributed to the relative abundance of organic calcite precipitation along existing shelf breaks which results in comparatively meagre aragonite supply to deep water. Carbonaceous Archean shales are here taken to record a high rate of Archean organic carbon burial and ample net photosynthetic oxygen supply to the atmosphere.

Uraniferous conglomerates similarly are best explained by an actualistic model. Comprehensive study of sedimentary uraniferous deposits of all ages reveals an apparent genetic link between Precambrian and Phanerozoic occurrences (Derry, 1960; Kimberley, 1974a). Groundwater concentration processes known to be presently forming such stratiform ores depend upon a high level of atmospheric oxygen.

References

Berkner, V. L., and Marshall, L. C., 1965. 'On the origin and rise of oxygen concentration in the Earth's atmosphere', *J. Atmos. Sci.*, **22**, 225–261.

Berner, R. A., 1964. 'An idealized model of dissolved sulfate distribution in recent sediments', *Geochim. Cosmochim. Acta*, **28**, 1497–1503.

Bottke, H., 1965. 'Die exhalativ-sedimentären devonischen Roteisensteinlagerstätten des Ostsauerlandes', *Beih. Geol. Jb.*, *Hannover*, No. 63.

Brock, B. B., and Pretorius, D. A., 1964. 'Rand basin sedimentation and tectonics', In *Geology of Some Ore Deposits of Southern Africa*, Vol. 1,

pp. 549–599, Johannesburg, Geol. Soc. South Africa.

Button, A., 1974. 'Iron formation as an end-member in carbonate sedimentary cycles in the Transvaal Supergroup, South Africa', *Econ. Geology research Unit, Witwatersrand Univ.*, *Circ. No. 89*, 10 pp.

Cameron, E. M., and Jonasson, T. R., 1972. 'Mercury in Precambrian shales of the Canadian Shield', *Geochim. Cosmochim. Acta*, **36**, 985–1006.

Cloud, P. E., 1972. 'A working model of the primitive Earth', *Am. J. Sci.*, **272**, 537–548.

584

Cloud, P. E., 1973. 'Paleoecological significance of the banded iron-formation', *Economic Geology*, **68**, 1135–1143.

Davidson, C. F., 1964. 'Uniformitarianism and ore genesis', *Min. Mag.*, **110**, 176–185; 244–253.

Davidson, C. F., 1965. 'The mode of origin of banket orebodies', *Inst. Mining and Metallurgy, Trans., B*, **74**, 319–338.

Derry, D. R., 1960. 'Evidence for the origin of the Blind River uranium deposits', *Econ. Geol.*, **55**, 906–927.

Dimroth, E., Coté, R., Provost, G., Rocheleau, M., Tassé, N., and Trudel, P., 1975. 'Third progress report on the stratigraphy, volcanology, sedimentology and tectonics of Rouyn-Noranda area', Quebec Dept. Natural Resources, *Open-file Report D.P. 300*, 56 pp., 2 Figs., 1 map.

Drever, J. I., 1974. 'Geochemical model for the origin of Precambrian banded iron-formations', *Geol. Soc. Amer. Bull.*, **85**, No. 7, 1099–1106.

Du Toit, A., (1953) *Geology of South Africa*, 3rd Edn., S. H. Haughton, (Ed.), Hafner, New York, p. 611.

Epstein, J. B., and Epstein, A. G., 1972. 'The Shawangunk Formation (Upper Ordovician (?) to Middle Silurian) in Eastern Pennsylvania', *U.S. Geol. Survey Prof. Paper 744*, p. 45.

Friedman, G. M., 1964. 'Early diagenesis and lithification in carbonate sediments', *Jour. Sed. Petrology*, **34**, 771–813.

Galkin, N. P., Maierov, A. A., and Veryatin, U. D., 1963. *'The Technology of the Treatment of Uranium Concentrates'*, R. W. Clarke, (Ed.), Macmillan, New York, p. 204.

Garrels, R. M., Perry, E. A., Jr., and Mackenzie, F. T., 1973. 'Genesis of Precambrian iron-formations and the development of atmospheric oxygen', *Econ. Geology*, **68**, 1173–1179.

Graton, L. C., 1930. 'Hydrothermal origin of the Rand gold deposits', *Econ. Geol.*, **25**, Supplement to No. 3, 185 pp.

Hallbauer, D. K., and Van Warmelo, K. T., 1974. 'Fossilized plants in thucholite from Precambrian rocks of the Witwatersrand, South Africa', *Precambrian Research*, **1**, 199–212.

Harshman, E. N., 1972. 'Geology and uranium deposits, Shirley Basin area, Wyoming', *U.S.G.S. Prof. Paper 745*, p. 82.

Hofmann, H. J., 1974. 'Mid-Precambrian pro-karyotes (?) from the Belcher Islands, Canada', *Nature*, **249**, No. 5452, 87–88.

Holland, H. D., 1962. 'Model for the Evolution of the Earth's Atmosphere', in *Petrologic Studies*, a volume in honour of A. F. Buddington, Geol. Soc. America, pp. 447–477.

Holland, H. D., 1973. 'The oceans: a possible source of iron in iron-formations', *Econ. Geol.*, **68**, 1169–1172.

Ippolito, F., 1963. Dieci anni di ricerca uranifera in Italia', *Giornale di Geologia, Annali del Museo Geologico di Bologna*, Ser. 2, **31**, 199–225.

Kimberley, M. M., 1974a. 'Origin of iron ore by diagenetic replacement of calcareous oolite', *Unpublished Ph.D. Thesis*, Princeton Univ., **1**, 345 pp.; **2**, 386 pp.

Kimberley, M. M., 1974b. 'Origin of iron ore by diagenetic replacement of calcareous oolite', *Nature*, **250**, 319–320.

Kimberley, M. M., 1975. 'Archean iron-formation and the early atmosphere', *1975 Geotraverse Symposium*, Univ. of Toronto, Canada, 5 pp.

Kimberley, M. M., 1975. 'Alteration and replacement of Pleistocene oolite by soil leachate on northwestern Andros Island, Bahamas: a modern analog of ironstone origin' (abstr.), Abstracts with programs, *North-Central Section, G.S.A.*, **7**, No. 6, 796–797.

Lasaga, A. C., Holland, H. D., and Dwyer, M. J., 1971. 'Primordial oil slick', *Science*, **174**, 53–55.

Latal, E., 1952. 'Die Eisenerzlagerstätten Jugoslaviens, in: *Symposium sur les Gisements de Fer du Monde*, Vol. 2, 19th Int. Geol. Congress, Algiers, p. 529–563.

Lepp, H., and Goldich, S. S., 1964. 'Origin of Precambrian iron formations', *Econ. Geol.*, **59**, 1025–1061.

Mason, R., 1970. 'Geology of the country between Francistown and Madinere, Northeastern Botswana', *Ph.D. Thesis*, Univ. Witwatersrand.

Moore, C. B., Lewis, C. F., and Kvenvolden, K. A., 1974. 'Carbon and sulfur in the Swaziland sequence', *Precambrian Research* **1**, 49–54.

Nagy, B., Kunen, S. M., Zumberge, J. E., Long, A., Moore, C. B., Lewis, C. F., Anhaeusser, C. R., and Pretorius, D. A., 1974. 'Carbon content and carbonate ^{13}C abundance in the Early Precambrian Swaziland sediments of South Africa', *Precambrian Research*, **1**, 43–48.

Nicolaysen, L. O., Burger, A. J., and Liebenberg, W. R., 1962. 'Evidence for the extreme age of certain minerals from the Dominion Reef conglomerates and the underlying granite in the western Transvaal', *Geochim. Cosmochim. Acta.*, **26**, 15–23.

Petersen, R. G., 1960. 'Detrital-appearing uraninite grains in the Shinarump member of the Chinle Formation in northern Arizona', *Econ. Geol.*, **55**, 138–149.

Pflug, H. D., 1966. 'Structural organic remains from the Fig Tree series of the Barberton Mountain Land', Economic Geol. Res. Unit, Witwatersrand Univ., Johannesburg, *Inform. Circ. No. 28.*

Ramdohr, P., 1960. *Die Erzmineralien und ihre Verwachsungen:* Berlin, Akademie–Verlag, 1089 pp.

Ramsay, J. G., 1963. 'Structural investigations in the Barberton Mountain Land', *Trans. Geol. Soc. South Africa*, **66**, 353–401.

Reimer, T. O., 1967. *Die geologie der Stolzburg Synklinale im Barberton Bergland (Transvaal-Südafrika)*, Diplomarbeit Geolog. Paleont. Institut, Johann Wolfgang Goethe Universität Anlage 32.

Ronov, A. B., 1958. 'Organic carbon in sedimentary rocks', *Geochemistry*, **5**, 510–536.

Roscoe, S. M., 1969. 'Huronian rocks and uraniferous conglomerates in the Canadian Shield', *Geol. Survey Canada, Paper 68–40*, 205 pp.

Schidlowski, A., 1967. 'Investigations of Pre-Cambrian thucholite', *Nature*, **216**, 560–563.

Schidlowski, A., 1968. 'The gold fraction from the Witwatersrand conglomerates from the Orange Free State Goldfield (South Africa)', *Mineralium Deposita*, **3**, 344–363.

Snyman, C. P., 1965. 'Possible biogenic structures in Witwatersrand thucholite', *Geol. Soc. South Africa, Trans.*, **68**, 225–235.

Stowe, C. W., 1968. 'The geology of the country south and west of Selukwe', *Rhodesia Geol. Survey Bull. 59*. 209 pp.

Viljoen, M. J., and Viljoen, R. P., 1969. 'The geological and geochemical significance of the upper formations of the Onverwacht Group', *Geol. Soc. South Africa, Spec. Pub. 2*, pp. 113–151.

Wiles, J. W., 1957. 'The geology of the eastern portion of the Hartley gold belt', *Rhodesia Geol. Survey, Bull. 44*, Pt. 1, 103 pp.

Woolverton, R. S., 1960. 'Geology of the Lumby Lake area', *Ont. Dept. Mines, Annual Rept.*, Vol. 69, Part 5, 52 pp.

Young, F. G., 1972. 'Cretaceous stratigraphy between Blow and Fish rivers, Yukon Territory', *G.S.C. Paper 72-1*. Part A, pp. 229–235.

Young, G. M. (Ed.), 1973. 'Huronian stratigraphy and sedimentation', Geol. Assoc. Canada, *Special paper 12*, 271 pp.

Zeschke, G., 1960. 'Transportation of uraninite in the Indus River, Pakistan', *Geol. Soc. South Africa Trans.*, **63**, 87–97.

Life Forms

Evidence of Archaean Life: A Brief Appraisal

J. WILLIAM SCHOPF

Department of Geology, University of California, Los Angeles, Los Angeles, California 90024.

Introduction

As is discussed elsewhere in this volume, it has become increasingly apparent in recent years that the Archaean may have been characterized by a tectono-sedimentary style decidedly different from that typical of later geologic time. In general, the Archaean appears to have been a time of unusually widespread volcanic and plutonic activity; rocks of this age tend to be geochemically primitive and sedimentary units are predominantly volcanogenic. In contrast, the Proterozoic (like the Phanerozoic) is characterized by typically cratonal sediments of the type deposited on submerged continental margins, and by less primitive igneous rocks. Interestingly, and perhaps significantly, this major geologic division is paralleled in the known record of Precambrian life. More than a score of microbiotas, assemblages comprised of literally thousands of individual fossils and containing a rich variety of taxa, are now known from the Precambrian (reviewed in Schopf, 1970, 1975). Without exception, however, and despite their widespread geographic distribution (occurring in Australia, Africa, North America, India, Europe and Soviet Siberia), such assemblages are restricted solely to the Proterozoic. As discussed below, and in marked contrast to the Proterozoic fossil record, however, evidence of Archaean life seems notably sparce. This apparent parallelism between the geologic and paleobiologic records presents an intriguing,

and as yet unsolved, problem, for it remains to be established whether it reflects a cause–effect relationship (i.e. the meagreness of the Archaean fossil record being a function of a sparcity of preserved, cratonal, fossiliferous facies) or whether the abrupt break in the known fossil record near the Archaean–Proterozoic 'boundary' reflects a major event in biological evolution (e.g. the origin and/or rapid diversification of a successful new stock of microorganisms). The solution to this and related problems hinges upon knowledge of the Archaean biota—what types of organisms existed during the Archaean? What can be said regarding their ecology, physiology and distribution? And how far into the geologic past can the fossil record be traced? In the following discussion are summarized available geochemical and paleobiological data bearing on these questions, data at present that are limited in quantity and subject to varying, and often conflicting, interpretation.

Evidence (?) of Archaean Life

Inorganic Geochemistry

It has been long suggested, and appears now to be widely accepted, that the Archaean atmosphere was devoid of all but trace amounts of free oxygen. Although consistent with an impressive array of both geologic and biologic data (Holland, 1962; Schopf, 1975, in press-a), this supposition nevertheless has seemed somewhat difficult to reconcile with the abundant occurrence of hematite and similarly oxidized mineral species in Archaean sediments, especially in ancient iron-formations, minerals that provide clear-cut evidence of syngenetic (whether subaerial or subaqueous), and commonly widespread, oxidation. In an effort to resolve this apparent conflict, Cloud (1974 and references cited therein) has constructed a novel, and assertedly speculative, hypothesis in which he suggests that there may have been an obligate link between the production of oxygen by primitive photosynthetic microorganisms and its consumption by reaction with ferrous iron during deposition of banded iron-formation. According to this view, such deposits constitute a necessary, if indirect, product of primitive photosynthesis. The hypothesis thus implies that oxygen-producing photoautotrophs were probably extant at least as early as 3750 m.y. ago, the approximate age of the oldest iron-formation now known (Moorbath et al., 1973). However, as is discussed in more detail elsewhere (Schopf, 1975), it seems evident that other, *non*-biologic sources of oxygen (e.g. the photodissociation of water vapour induced by ultraviolet light), sources capable of producing the trace amounts required for oxidation of ferrous minerals, were available on the primitive Earth; the hypothesized 'obligate' nature of the link between photoautotrophy and banded iron-formation is thus open to question. At present, therefore, the occurrence of banded iron-formation is probably best regarded as evidence consistent with, but not necessarily indicative of, the existence of oxygen-producing photoautotrophs (or, indeed, of the necessary existence of any type of biological activity).

Organic geochemistry

During the past decade, careful studies have demonstrated that many Archaean sediments contain organic matter, substances that could provide direct evidence of an Archaean biota; both extractable (e.g. amino acids, hydrocarbons, fatty acids, sugars, and porphyrins) and non-extractable (viz., 'kerogen') organic components have been detected (reviewed in Schopf, 1970; Kvenvolden, 1972; McKirdy, 1974). The extractable components are similar to (and in some cases indistinguishable from) biochemicals synthesized by extant living systems; they are thus of readily interpretable biologic significance. However, as has become increasingly apparent in recent years (cf. Smith et al., 1970), it is 'extremely difficult (if not impossible)' (McKirdy, 1974, p. 108) to establish that these components date from the time of sedimentation, rather than having been introduced (e.g. via ground waters) in the relatively recent geologic past. These components thus constitute suggestive, but inconclusive, evidence of Archaean life.

The insoluble carbonaceous components of Archaean sediments present a contrasting problem. With but few exceptions, this material—commonly comprising more than 98% of the total organic matter contained in ancient sediments—seems assuredly syngenetic with Archaean sedimentation. However, kerogen is a geochemically altered, chemically complex, polymeric material of variable but ill-defined composition. The precursor compounds from which it is formed and the diagenetic and geochemical processes involved in its formation and maturation have yet to be elucidated. Thus, to date, analytical studies of kerogen have provided insight into neither the physiology nor the biochemistry of Archaean life. Indeed, such studies have yet to provide conclusive evidence of the *existence* of Archaean life—criteria have yet to be developed that can be used to differentiate between kerogenous products of biologic, non-biologic, and combined (of both biotic and abiotic) origins.

Among geochemical data now available, perhaps the most suggestive of the existence of early life is that provided by analyses of the isotopic composition of kerogen in unmetamorphosed Archaean sediments. To date, more than fifty such analyses have been carried out (Oehler et al., 1972 and references cited therein; M. Schidlowski, pers. comm.). Results of these analyses indicate that the ^{13}C to ^{12}C ratio exhibited by Archaean organic matter (including samples dating back to about 3300 m.y.) is comparable with that of demonstrably biogenic carbon of younger geologic age—specifically, to ratios characteristic of organic material formed as a result of the fixation of carbon dioxide by autotrophic organisms. Further, this Archaean organic matter differs distinctly in isotopic composition from coexisting (Schopf et al., 1971) and younger carbon-bearing substances known to be of inorganic, non-biological origin. Although relatively little is known regarding the isotopic composition and mass distribution of carbon occurring in the various Archaean reservoirs, and although there can be little doubt that the Archaean carbon cycle—an anaerobic cycle perhaps dominated by an influx of abiotically fixed organic matter—was substantially different from that typical of later geologic time, it nevertheless seems a difficult matter to account for the isotopic composition of Archaean kerogen without the intervention of living systems. At present, the most plausible interpretation would appear to be that some and perhaps much of the carbonaceous material preserved in Archaean sediments (dating back to at least 3300 m.y. ago) is a result of biologic activity, material presumably derived in part from primitive (anaerobic, microscopic, prokaryotic), and perhaps bacterium-like, autotrophs.

'Microfossils'

Both microscopic (microfossils?) and megascopic (stromatolitic) objects of putative biologic origin have been reported from Archaean sediments. Although assessment of the possible biogenicity of these objects is a rather subjective matter—based in large measure on analogy with younger structures of well-established biologic origin—certain of the evidence currently available (that derived from studies of the stromatolitic structures) seems indicative of biologic activity.

Three varieties of microfossil-like objects have been reported from the Archaean: (1) rod-shaped (and possibly spheroidal) bacterium-like bodies; (2) filamentous, microscopic, thread-like structures; and (3) coccoidal, unicell-like microstructures. Unlike the other two categories of objects (which have been studied primarily in petrographic

thin sections), the bacterium-like bodies have been studied only by transmission electron microscopy. Unfortunately, however, the types of preparations examined (plastic or carbon surface replicas of fractured or acid-etched rock specimens and thin sections, or acid-resistant residues suspended on plastic-coated microscope grids) are readily subject to contamination in the laboratory, principally by airborne particulate matter. Moreover, various types of artifacts (e.g. folds, strands, wrinkles, bubbles, pits, etc.), some of which may look remarkably 'life-like', are common in such preparations. Thus, virtually all of the 'Precambrian microfossils' reported from such preparations appear to be neither Precambrian nor fossil; even the most convincing of these bodies are probably best regarded as merely suggestive, rather than compelling, evidence of Archaean life (for a more extensive discussion, see Schopf, 1975, pp. 235–240).

Similarly, the filamentous 'microfossils' thus far reported from the Archaean are less than convincing. Although several of the reported objects appear to be modern contaminants (detected in acid-resistant residues), the majority appear to be carbonaceous films and strands of 'accidental' form that have been distributed along bedding planes by normal sedimentary and diagenetic processes. In no case has the occurrence of cells been clearly demonstrated; and in no case have multiple examples been illustrated of the same filament morphology, forms that might reasonably be interpreted as representing a single taxonomic entity (cf. Schopf, 1975).

The spheroidal variety of Archaean microfossil-like objects presents a more difficult problem of interpretation. Microstructures of this type have been reported from several Archaean deposits, the oldest being the 3100 to 3300 m.y. old sediments of the Swaziland Sequence of South Africa. Numerous specimens have been detected in petrographic thin sections; the spheroids are demonstrably indigenous to Archaean-age sediments. Their possible biogenicity, however, is difficult to assess. They are of very simple morphology and, if true fossils, appear

592

commonly to be poorly preserved. Although they apparently occur in considerable abundance, they do not exhibit a degree of structural complexity (such as being demonstrably multicelled or having well-defined, regular, wall structure of surface ornamentation) that would make their biological interpretation wholly convincing. Moreover, they bear at least superficial resemblance to organic spheroids produced abiotically in laboratory experiments designed to simulate events that may have occurred on the primitive Earth and to the spheroidal 'organized elements' known to occur in some carbonaceous meteorites (Rossignol-Strick and Barghoorn, 1971). And, perhaps most perplexing, statistical analyses have shown that in population structure the Archaean spheroids are substantially more similar to such abiotically produced bodies than they are to modern microorganisms and to spheroidal microfossils of younger Precambrian age (Schopf, 1975, in press-b). Thus, while it is possible, and perhaps likely, that at least some of these Archaean fossil-like objects are actually microfossils (and while it is true that they are so regarded by many workers), the evidence as yet seems equivocal. Further studies are needed to define their true nature.

Stromatolites

Unlike the putative microfossils discussed above, megascopic structures—structures that in virtually all respects resemble Proterozoic stromatolites of assured biologic origin— appear to provide firm (albeit limited) evidence of Archaean life. The oldest of these structures, occurring in limestones of the Bulawayan Group of Rhodesia, are apparently between about 2900 and 3200 m.y. in age (Vail and Dodson, 1969). They are composed of closely spaced vertical columns comprised of stacked, hemispherical layers of calcareous and organic material (Schopf et al., 1971), a laminated organization essentially indistinguishable from that exhibited by younger stromatolites, both fossil and modern, that are known to have been formed by communities of photosynthetic microorganisms and a type of organization that seems difficult to explain as having resulted from solely inorganic (e.g. sedimentary or accretionary processes. Their biologic (algal and/or bacterial) origin seems further evidenced by results of carbon isotopic analyses (Schopf et al., 1971). Thus, although neither the Bulawayan stromatolites nor younger Archaean stromatolites known from two calcareous rock units in Canada (Hofmann, 1971; P. Hoffman, pers. comm.) are known to contain structurally preserved microfossils, all appear to be of biologic origin. At present, however, little is known regarding the organisms—presumed to have been primitive, microscopic and photosynthetic—that were responsible for their formation.

Summary

As is evident from the foregoing discussion, few firm conclusions can be drawn regarding Archaean life. The occurrence of stromatolites in the Archaean, and the carbon isotopic composition of Archaean organic matter, both augur well for the existence of an Archaean biota. Moreover, the presence of relatively abundant, diverse, and morphologically complex microorganisms in deposits of early Proterozoic age seems certain to evidence a prior episode (of unknown duration) of Archaean evolution. None of these lines of evidence, however, provide more than the most limited of insight into the nature and antiquity of early life. At present, and despite the numerous studies of recent years, the Archaean biota remains very much a mystery.

Acknowledgement

Research leading to the preparation of this report has been supported by NSF Grant GB 37257 (Systematic Biology Program) and by NASA Grant NGR 05-007-407; travel costs connected with my participation in the NATO Advanced Study Institute were also defrayed by the NASA grant.

References

Cloud, P. E., 1974. 'Evolution of ecosystems', *Am. Scientist*, **62**, 54–66.

Hofmann, H. J., 1971. 'Precambrian fossils, pseudofossils and problematica in Canada', *Geol. Surv. Can. Bull.*, **189**, 146 pp.

Holland, H. D., 1962. 'Model for the evolution of the Earth's atmosphere', *Petrologic Studies, A Volume to Honor A. F. Buddington* (Ed.), A. E. J. Engel et al. (Geol. Soc. Am., New York), 447–477.

Kvenvolden, K. A., 1972. 'Organic geochemistry of early Precambrian sediments', *Proc. XXIV Internat. Geol. Cong., Montreal*, Sect. 1, Precambrian Geology, 31–41.

McKirdy, D. M., 1974. 'Organic geochemistry in Precambrian research', *Precambrian Res.*, **1**, 75–137.

Moorbath, S., O'Nions, R. K., and Pankhurst, R. J., 1973. Early Archaean age for the Isua Iron Formation, West Greenland', *Nature*, **245**, 138–139.

Oehler, D. Z., Schopf, J. W., and Kvenvolden, K. A., 1972. 'Carbon isotopic studies of organic matter in Precambrian rocks', *Science*, **175**, 1246–1248.

Rossignol-Strick, M., and Barghoorn, E. S., 1971. 'Extraterrestrial abiogenic organization of organic matter: the hollow spheres of the Orgueil meteorite', *Space Life Sci.*, **3**, 89–107.

Schopf, J. W., 1970. 'Precambrian microorganisms and evolutionary events prior to the origin of vascular plants', *Biol. Rev. Cambridge Phil. Soc.*, **45**, 319–352.

Schopf, J. W., 1975. 'Precambrian paleobiology: problems and perspectives', *Ann. Rev. Earth Planet. Sci.*, **3**, 212–249.

Schopf, J. W., in press-a. 'Paleobiology of the Precambrian: the age of blue-green algae', In Dobzhansky, T., et al. (Eds.), *Evolutionary Biology*, Plenum, New York, Vol. 7, 1–43.

Schopf, J. W., in press-b. 'Are the oldest "fossils," fossils?', *Origins of Life*.

Schopf, J. W., Oehler, D. Z., Horodyski, R. J., and Kvenvolden, K. A., 1971. 'Biogenicity and significance of the oldest known stromatolites', *J. Paleontol.*, **45**, 477–485.

Smith, J. W., Schopf, J. W., and Kaplan, I. R., 1970. 'Extractable organic matter in Precambrian cherts', *Geochim. Cosmochim. Acta*, **34**, 659–675.

Vail, J. R., and Dodson, M. H., 1969. 'Geochronology of Rhodesia', *Trans. Geol. Soc. S. Afr.*, **72**, 79–113.

Micropalaeontological Evidence from the Onverwacht Group, South Africa

M. D. MUIR, and P. R. GRANT

Department of Geology, Royal School of Mines, Prince Consort Road, London SW7 2BP, England.

Introduction

Because of the very great age of the sediments ($3 \cdot 355 \times 10^9$ years; Hurley et al., 1972), the biogenicity or not of the microscopic carbonaceous structures from the rocks of the Onverwacht group, of the Barberton Mountain Land, South Africa has long been a controversial matter. The first reports (Engel et al., 1968; Nagy et al., 1968; Nagy and Nagy, 1969) were cautious and used the term microstructures rather than microfossils. The microstructures did not have a narrow size range, and were of simple morphology. Controversy also centred on the interpretation of carbon stable isotope data from different parts of the succession (Oehler et al., 1972), but despite the disputed biogenicity of the microstructures, there is abundant evidence for biological activity in the Onverwacht Group. The cherts are rich in organic carbon (Moore et al., 1973, give values of between $0 \cdot 04$ and $1 \cdot 95\%$) but most of this is not preserved in a morphologically recognizable form. The amorphous organic matter is distributed through the sediments in fine laminations, similar to those produced by blue–green algae in shallow marine and lacustrine environments today (Muir and Hall, 1974). When such laminations become domed upwards, they develop into domal or columnar stromatolites such as have been described from the Archaean greenstone belts of Rhodesia (Huntsman Dolomite, Schopf et al., 1971) and Canada (Steep Rock Lake, Walcott, 1912). Thus, evidence for biological activity is not rare in greenstone belts, and the present study presents micropalaeontological evidence for the biogenicity and environment of deposition of the Onverwacht Group sediments.

Materials and Methods

The present study is based on black cherts given to the writers by J. Brooks, who was sent the samples by M. Viljoen, and by D. O. Hall. The cherts were collected from the Theespruit, Hooggenoeg, Kromberg, and Swartkoppie Formations of the Onverwacht Group (Viljoen and Viljoen, 1969, p. 13). However, the present discussion is based on new results from samples from the Kromberg Formation. The authors are also indebted to C. Anhaeusser for advice on localities of interest visited during a field study by one of us (M. D. M.).

The materials have been examined in petrological thin sections and in the scanning electron microscope. Measurements have only been made in the light microscope, and where structures were too small to permit accurate direct measurement, specimens were photographed. Then the negatives were enlarged in a photographic enlarger to give a magnification of ×4000, and the images measured using a gridded overlay accurate to ± 0·2 mm. For the photography, the microscope ·vas focused carefully on the largest diameter of each of the structures of interest. Single cells, cells in pairs, cells in stromatolites, cells in tissue, and filaments were measured, and for comparison, similar measurements were made on samples from the 1·4 × 10⁹ year old Amelia Dolomite from the McArthur Group, N.T. Australia.

Results

Unicells

In some samples, small unicells are very common. They occur set in structureless organic matter that lies along the bedding planes. In some cases (Figs. 1, 2 and 3), the unicells are domed upwards. The degree of doming can be acute, as in Fig. 1, or more gentle, as in Figs. 2 and 3 (the bedding is parallel to the long side of these three micrographs). In some cases the unicells appear to be randomly oriented (Fig. 4), and in others they lie along the surfaces of the bedding (Fig. 5). The unicells in the domed structures range in size from 1·6 to 5·0 μm, with a mean maximum diameter (m.m.d.) of 4 μm. This peak position is different from that of the random or flat-lying unicells which range in size from 1·3 to 5·0 μm, with a m.m.d. of 3·0 μm, see Text-Fig. 1.

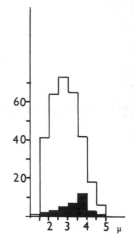

Text-Fig. 1. Showing the differences in size distribution between domal spheroids (solid histogram), and randomly distributed spheroids (outline histogram). Vertical axis is numbers of specimens measured

In order to test the consistency of the size distributions of the very small unicells, counts were made on the populations of two slides (Text-Fig. 2). In both cases the pattern of distribution is similar, with a main and an ancillary peak at approximately twice the mode of the main peak. Truncation occurs toward the lower end of the size range due to measuring difficulties.

The normality of the distribution of the 'main peaks' was tested by the computer programme SNORT (Preston, 1969). Sample 2 was found to be normal, whilst sample 1 was not (see Text Figs. 2 and 3), and in consequence parametric statistical comparative

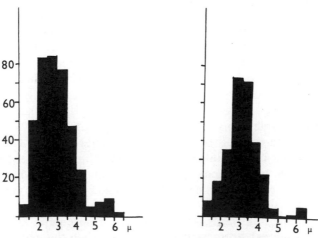

Text-Fig. 2. Comparison between the size distribution of populations from two slides (GNV 1, left; ONV 2, right). Vertical axis is numbers of specimens measured

(a)

Sample	No. measured	Mean diam. (μm)	Variance	Normality test				
				R_A	Sk	K	K-S	χ
ONV 2	269	3·1	0·55	A	A	A	A	R
ONV 1	407	2·9	0·58	R	R	R	A	R
ONV 1 BP1	137	3·1	0·66	R	R	R	A	R
ONV 1 BP2	126	2·7	0·50	R	A	R	A	R
ONV 1 BP3	138	3·0	0·50	R	A	R	A	R

R_A = Range, Sk = Skewness, K = Kurtosis, χ = chi-square of 1·5 grouping. R = Reject at 5% level. A = Accept that statistic falls within range expected by variation of a gaussian distribution at 5% level. For discussion of statistical techniques, see Preston (1969).

(b)

ONV 1	R			
BP 1				
BP 2			R	
BP 3			A	R
	ONV 2	ONV 1	BP 1	BP 2

Kolmogorov–Smirnov non-parametric comparison, critical level, 5%. Other symbols are the same as Text-Fig. 3(a).

Text-Fig. 3. Summary statistics of distributions of maximum diameters of organic spheroids from Kromberg Cherts

techniques could not be employed. However, the similarity of the variance within the populations made non-parametric testing feasible. A Kolmogorov–Smirnov non-parametric comparison was made between the samples, and the hypothesis that the samples were the same at the 5% level of confidence was rejected. Because the thin section was cut perpendicular to the bedding, the distributions of spheres from three individual bedding planes in sample 1 were able to be compared (Text-figs. 3 and 4). None of these distributions is normal, and only BP1 and BP3 are considered similar by the Kolmogorov–Smirnov non-parametric text at the 5% level.

The ancillary peaks are caused by some slightly larger unicells being interspersed among the smaller ones. They have thicker walls, which appear to be rather porous, and their m.m.d. is approximately 6·0 μm. There are also some considerably larger unicells, occurring in small clusters in depressions on the bedding planes. These have a m.m.d. of about 10·0 μm, and range from 8·0 to 12·5 μm, are very thick-walled, and are often black and opaque as a result of diagenetic or metamorphic recrystallization of the chert matrix (Figs. 10 and 11).

Small *paired unicells* are not uncommon in the Onverwacht. In the present study, 28 pairs were found, and the individuals measured parallel to the long axis of the pair (which is always at right-angles to the line of division between the individuals). Their m.m.d. is approximately 1·5 μm, ranging from 1·0 to 4·0 μm. They are distinctly smaller than the single unicells. A comparison with the paired individuals and single unicells from the Amelia Dolomite shows a similar relationship (Text-Fig. 5). Typical pairs are illustrated in Figs. 6–8, while Fig. 9 shows a pair that are just slightly separated.

In some thin sections, numerous examples of apparently *multicellular tissue* have been

598

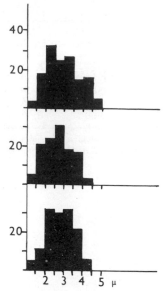

Text-Fig. 4. Size distribution of small spheroids in three bedding planes. Vertical axis is numbers of specimens measured: BP1 is at the top, and BP3 at the bottom

15 and 16 show the smallest of these. They are chains of small cells (1·5 to 3·5 μm m.m.d.) which occur generally at right-angles to the direction of the bedding. They follow a fairly sinuous course and are often closely associated with the small single unicells.

A larger version of the 'chain of cells' type of filament also occurs (Figs. 17 and 18). The cells in these chains vary from 9·5 to 12 μm in length and from 6·5 to 10·0 μm in breadth. They are thick-walled, opaque, and occur lying either parallel to the bedding or at a low angle to it, and the chains follow a straight or slightly sinuous course. The maximum length, so far found, of one of these filaments is more than 110 μm.

observed (Figs. 12–14). The cells in this tissue are slightly larger, on average, than the single unicells, around 4·0 μm, ranging from 2·0 to 6·0 μm (Text-Fig. 6). Preservation of the material is poor, and tends to be inconsistent even within a single area.

In addition to the cell types described above, a variety of *filament* types occurs. Figs.

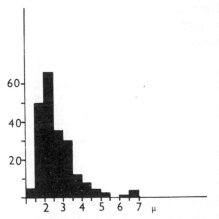

Text-Fig. 6. Size distribution of cells occurring in multicellular tissue. Vertical axis is number of cells counted

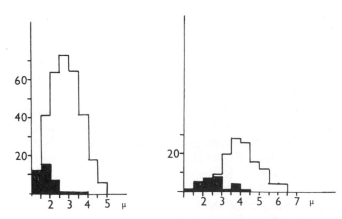

Text-Fig. 5. Size distributions of paired individuals (solid histograms), and single unicells (outline histograms). Onverwacht sample on the left; Amelia Dolomite sample on the right. Vertical axis is numbers counted

Non-segmented filaments are also present (Figs. 19 to 22). Their state of preservation is variable, their average diameter is approximately $4 \cdot 0 \, \mu m$, and the longest specimen so far discovered is approximately $100 \, \mu m$. This particular example has a rather pointed termination (top of Fig. 19). Most of the non-segmented filaments occur lying at a slight angle to the bedding (see Fig. 23, for a low power view of the specimen in Fig. 19). The bedding runs parallel to the long side of micrographs 19 to 23. The filament in Fig. 21 describes two almost right-angled bends.

Discussion

The assemblage described above represents a mixture of several species, all of which lived in intimate association. The relationships of the microfossils with the bedding is closely comparable with that observed in many later Precambrian microbiotas. In younger stromatolitic black cherts (e.g. the Amelia Dolomite assemblages (Croxford et al., 1973; Muir, 1974)), small cocci frequently delineate a bedding surface of domal stromatolites. The small domed arcs of unicells (Figs. 1–3) are very similar to this and the size distributions of both the single cells and the paired individuals from the Onverwacht and the Amelia Dolomite are alike. Furthermore, the resemblance of the amorphous organic matrix of the unicells to the mucilaginous matrix of the Amelia Dolomite unicells is striking. The paired cells from both assemblages resemble chroococcacea algae in the process of dividing. In both cases, the mean size of the pair-individuals is smaller than that of closely associated single unicells, as would be expected in cells that had just undergone division.

In younger Precambrian cyanophyte assemblages, distinct differences can be observed between the appearance of the cell wall and the mucilaginous sheath that surrounds it. The cell wall is usually thin and hyaline, while the sheath is more diffuse and granular in texture. Sheaths abandoned in life because of threat of burial in sediment are common in the Amelia Dolomite assemb-

lages. In the Onverwacht material, the slightly larger unicells of about $6 \, \mu m$ diameter (Fig. 5), some of the non-septate filaments (Fig. 22) and even parts of the multicellular tissue (Fig. 14) may represent such abandoned sheaths. Furthermore, the termination of the filament in Fig. 19 resembles a typically cyanophycean pointed terminal cell, and the morphologies and dimensions of the fossils are consistent with their being interpreted as blue–green algae or flexibacteria. The earliest reports (Engel et al., 1968; Nagy and Nagy, 1968, 1969) described only spheroidal bodies and doubted their biogenicity because of their simple morphology, and wide size range. A wide range of morphologies has now been reported (Brooks et al., 1973; Muir and Hall, 1974), and the present work has demonstrated that the reason for the wide spread of the size range is that the basic assumption that only one population, or type of microfossil, occurs in the samples. From our results, there are several statistically distinct populations of small spheroids. Even the morphologically similar populations from three adjacent bedding planes show size distribution variations which can be interpreted as reflecting different physico-chemical environmental conditions at the times of deposition of the beds. These minor differences are expected, and not commented on, in younger Precambrian microfossil assemblages. In modern monospecific populations, such variations are often seasonally controlled (Braarud and Nordli, 1963), and in a fossil, probably polyspecific assemblage, they must be the product both of seasonal, physico-chemical changes and the number and the number of types present.

However, comparison with younger Precambrian and Recent assemblages is rendered difficult because the state of preservation of the material is poor. The colour of the organic matter is very dark brown or black, and crystalline graphite occurs sporadically throughout the Onverwacht Group (Karkhanis, pers. comm.). The effect of metamorphism is very variable and dependent on many factors: because the Onverwacht succession is largely volcanic, the effect of intrusives and extrusives is likely to be important locally, although the

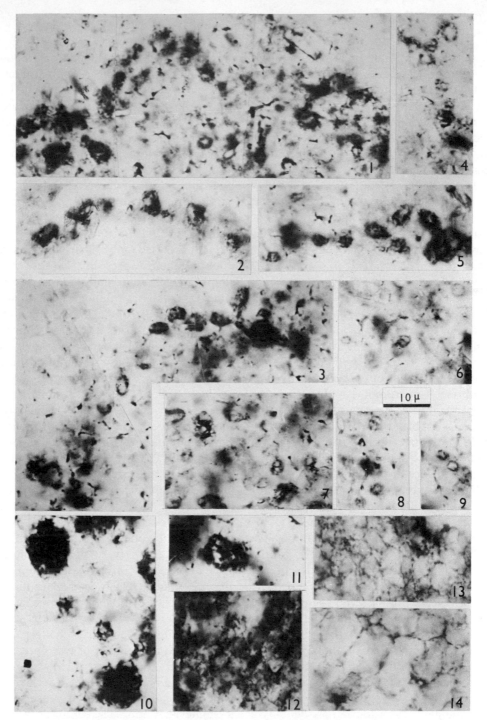

Figs. 1–3. Cells following the laminations of domal stromatolites. Fig. 1, ONV 5/20: K1A, 101·2, 13·6; Fig. 2, ONV 14/5: K1A1, 112·6, 9·3; Fig. 3, ONV 5/20: K1A1, 112·7, 9·6

Fig. 4. Random arrangement of cells in mucilage. ONV 6/4, K2, 103·5, 14·1

Fig. 5. Cells lying parallel to bedding. ONV 7/2: K1A, 110·2, 4·5

Figs. 6–8. Paired cells. Fig. 6, ONV 15/29: K1A, 112·7, 9·8. Fig. 7, 0NV 16/3; K1A, 112·8, 11·3; Fig. 8, ONV 15/6: K1A, 112·6, 10·0

Fig. 9. Two small cells that appear to have just separated. ONV 8/30; K1A, 111·2, 6·5

Figs. 10–11. Large cells. Fig. 10, ONV 11/16; K1A, 105·6, 17·2. Fig. 11. ONV 11/17; K1A, 105·6, 17·4

Figs. 12–14. Multicellular tissue. Fig. 12, ONV 13/4; K3, 106·5, 10·5; Fig. 13, ONV 13/8, K3, 106·5, 18·2; Fig. 14, ONV 13/6, K3, 106·0, 13·5

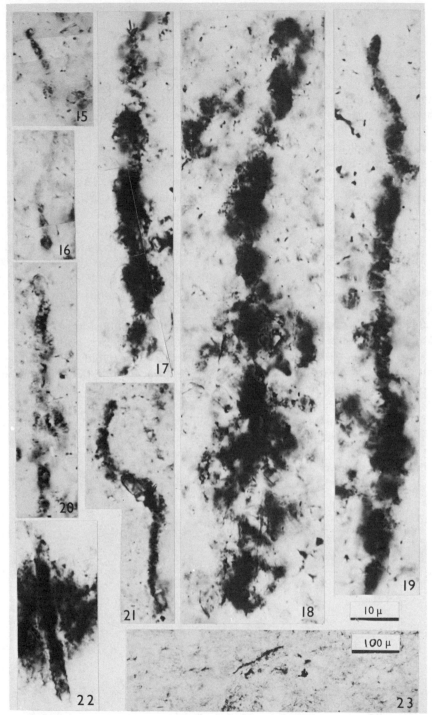

Figs. 15–16. Chains of small cells. Fig. 15, ONV 15/18, K1A, 112·7, 10·6; Fig.
16, ONV 16/27; K1A, 112·8, 12·8

Figs. 17–18. Chains of large cells. Fig. 17, ONV 11/21. K1A1, 103·8, 19·8. Fig.
18, ONV 14/9, KłA1, 112·6, 9·3

Figs. 19–23. Non-septate filaments. Fig. 19, ONV 11/26; K1A, 103·2, 19·2; Fig.
20, ONV 5/8; K1A1, 98·2, 9·2; Fig. 21, ONV 6/10, K1A, 103·6, 13·6: Fig. 22,
ONV 13/21; K1A 104·3, 10·5; Fig. 23, ONV 11/32 (as Fig. 19—low power
view)

In all cases, the term ONV refers to a negative number, the numbers prefixed with
'K' are slide numbers, and the remaining numbers are microscope stage coordi-
nates of microscopes belonging to the Dept. of Geology, Royal School of Mines,
London.

602

degree of metamorphism is controlled by the proximity of a sample to the base of a lava flow, the homogeneity of the sediment type and the degree of lithification that had occurred before extrusion. The samples described here contained large crystals of biotite and chlorite indicating a greenschist metamorphic facies. Most Onverwacht sediments have suffered even more metamorphism (Viljoen and Viljoen, 1969). This has affected the chemistry of the organic matter to the extent that aromatization is commoner in Onverwacht samples than in younger kerogens (Engel et al., 1968). Brooks and Shaw (1972) showed that increasing metamorphism produced an increase in the aromaticity.

Oehler et al. (1972) examined the stable carbon isotope ratios for a small number of samples from different formations of the Onverwacht Group. Their results indicated that $\delta\,^{13}C_{PDB}$ values were anomalously heavy in samples from the Theespruit Formation below the Middle Marker Horizon (see Viljoen and Viljoen, 1969, for a summary of the stratigraphy), and suggested that these values may have indicated either an abiogenic origin

for the organic matter, or that the kerogen may have represented the remains of non-photosynthetic organisms. Silverman (1964), and Calvin (1969) had suggested that relatively heavy C-isotope values were to be expected where kerogen had suffered metamorphism. Such a mechanism was invoked by Brooks et al. (1973) to explain the Theespruit values. The matter remained controversial (Dungworth and Schwartz, 1974) until McKirdy and Powell (1974) showed that the $\delta\,^{13}C_{PDB}$ value of kerogen could be directly related to its H/C ratio, a measure of metamophism. The results of McKirdy and Powell (1974) have been combined with those of Dungworth and Schwartz (1974) and Brooks (pers. comm.) in Text-Fig. 7. There is an obvious relationship between the $\delta\,^{13}C_{PDB}$ and the H/C values, with no break in the sequence between older and younger samples. Indeed, one sample that we could not plot in Text-Fig. 7, because its value was too low (−10·8) comes from the Late Proterozoic Skillogalee Dolomite of South Australia, which contains undisputed biogenic structures (Schopf and Barghoorn, 1969).

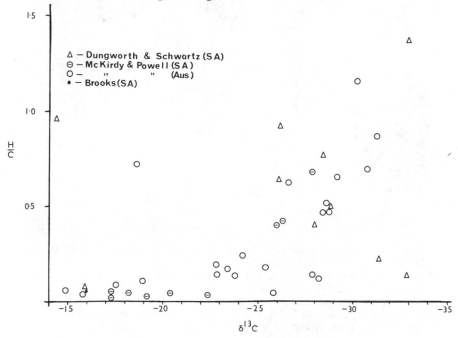

Text-Fig. 7. Relationship of $\delta\,^{13}C$ values with H/C ratios for a number of Precambrian rocks. S.A. = South Africa, Aus = Australia. Data from various sources

Thus, the $\delta^{13}C_{PDB}$ values from Precambrian kerogens can be used to indicate metamorphic grade as well as the biogenicity of the original material. In fact, modern marine algae have $\delta^{13}C_{PDB}$ values of between -10 and -20 (Brooks, 1971) which are very similar to Oehler et al.'s (1972) 'abiogenic' values from the Theespruit Formation. Precambrian kerogens, with values lighter than -25, are, in fact, anomalously light.

While the biogenicity of the structurally preserved organic matter was in doubt, no inferences could be drawn regarding the environment of deposition of the Onverwacht Group. However, the microfossils described here show a variety of forms comparable with those displayed by cyanophyte assemblages from younger Precambrian stromatolitic black chert assemblages. Furthermore, the commonest form of kerogen in the Onverwacht cherts is in fine, algal mat type laminations, very similar to those produced by blue–green algae, or even under hot spring conditions by photosynthetic bacteria (Walter et al., 1972). There is no direct method of determining whether the laminations are cyanophycean or bacterial in origin, but some indirect arguments can be amassed. Firstly, the Onverwacht Group consists predominantly of volcanic rocks and suitable conditions for the formation of bacterial laminae in geyserite could have existed.

On the other hand, the range of morphologies, and expecially the multicellular tissues, suggest a cyanophyte level of organization. Blue–green algae live in the photic zone, i.e. in shallow water or littoral type environments, to have sufficient light for photosynthesis, and those morphologically similar to our microfossils are sediment dwellers which supports our conclusion that the Kromberg Formation sediments were laid down in shallow water. Viljoen and Viljoen (1969) figure shallow water sedimentary structures, and quote van den Berg, from an unpublished account, as reporting cross-bedded oolitic cherts. (Interestingly, van den Berg apparently contends that the chert is secondary, having replaced pre-existing detrital feldspar and carbonate.) Thus the sedimentology and palaeontology appear to be in agreement as to the nature of the depositional environment.

References

Braarud, T., and Nordli, E., 1963. 'Reproduction and size variation in *Halosphaera viridis* of Northern Waters', *Nytt Mag. Bot.*, **10**, 131–136.

Brooks, J., 1971. 'Some chemical and geochemical studies on sporopollenin', In Brooks, J., Grant, P. R., Muir, M., van Gijzel, P., and Shaw, G. (Eds.), *Sporopollenin*, Academic Press, London and New York, 351–407.

Brooks, J., Muir, M., and Shaw, G., 1973. 'Chemistry and morphology of Precambrian microorganisms', *Nature*, **244**, 5413, 215–217.

Brooks, J., and Shaw, G., 1972. 'Geochemistry of Sporopollenin', *Chemical Geology*, **10**, 69–88.

Calvin, M., 1969. *Chemical Evolution*, Oxford University Press, 1–268.

Croxford, N. J. W., Janecek, J., Muir, M. D., and Plumb, K. A., 1973. 'Micro-organisms of Carpentarian (Precambrian) age from the Amelia Dolomite, McArthur group, Northern Territory, Australia', *Nature*, **245**, 5419, 28–30.

Dungworth, G., and Schwartz, A., 1974. 'Organic matter and trace elements in Precambrian rocks from South Africa', *Chemical Geology*, **14**, 167–172.

Engel, A. E. J., Nagy, B., Nagy, L. A., Engel, C. G., Kremp, G. O. W., and Drew, C., 1968. 'Algal-like fossils in the Onverwacht Series, South Africa. Oldest recognised life-like forms on earth', *Science*, **161**, 1005–1008.

Hurley, P. M., Pinson, W. H., Nagy, B., and Teska, T. M., 1972. 'Ancient age of the Middle Marker Horizon, Onverwacht Group, Swaziland Sequence, South Africa', *Earth Planet. Science Lett.*, **14**, 360–366.

Kvenvolden, K., 1974. 'Natural evidence for chemical and early biological evolution', *Origins of Life*, **5**, 71–86.

McKirdy, D. M., and Powell, T. G., 1974. 'Metamorphic alteration of carbon isotopic composition in ancient sedimentary organic matter: new evidence from Australia and South Africa', *Geology*, December 1974, 591–595.

Moore, C. B., Lewis, C. F., and Kvenvolden, K. A., 1973. 'Carbon and sulfur in the Swaziland sequence', *Precambrian Research*, **1**, 49–54.

Muir, M. D., 1974. 'Microfossils from the Middle Precambrian McArthur Group, Northern Territory, Australia', *Origins of Life*, **5**, 105–118.

604

Muir, M. D., and Hall, D. O., 1974. 'Diverse microfossils in Precambrian Onverwacht Group rocks of South Africa', *Nature*, **242**, 5482, 376–378.

Nagy, B., and Nagy, L. A., 1969. 'Early Precambrian Onverwacht microstructures. Possibly the oldest fossils on Earth?', *Nature*, **223**, 1226–1228.

Nagy, B., Nagy, L. A., Bitz, M. C., Engel, C. G., and Engel, A. E. J., 'Investigations of the Precambrian Onverwacht sedimentary rocks in South Africa', (Abst.), *4th Internat. Meeting on Organic Geochemistry*, Amsterdam 23.

Oehler, D. Z., Schopf, J. W., and Kvenvolden, K. A., 1972. 'Carbon isotopic studies of organic matter in Precambrian rocks', *Science*, **175**, 1246–1248.

Preston, D. A., 1969. 'Fortran IV program for sample normality tests', *Kansas Geol. Survey Computer Contribution*, 1–20.

Schopf, J. W., and Barghoorn, E. S., 1969. 'Microorganisms from the Late Precambrian of South Australia', *J. Paleontology*, **43**, 111–118.

Schopf, J. W., Oehler, D. Z., Horodyski, R. J., and Kvenvolden, K. A., 1971. 'Biogenicity and significance of the oldest known stromatolites', *J. Paleontology*, **45**, 477–485.

Silverman, S. R., 1964. 'Investigations of petroleum origin and evolution mechanisms by carbon isotope studies', In Graig, H., Miller, S. L., and Wasserburg, G. J. (Eds.), *Isotopic and Cosmic Chemistry*, North–Holland, Amsterdam, 92–102.

Viljoen, M. J., and Viljoen, R. P., 1969. 'An introduction to the geology of the Barberton granite-greenstone terrain. 9–28. & The geological and geochemical significance of the upper formations of the Onverwacht Group'. 113–153, In *Upper Mantle Project*, Geol. Soc. South Africa, Spec. Publ., 2.

Walcott, C. D., 1912. 'Notes on fossils from limestone of Steep Rock Series, Ontario', *Geol. Surv. Canada, Mem 28*, 16–23.

Walter, M. R., Bauld, J., and Brock, T. D., 1972. 'Siliceous algal and bacterial stromatolites in hot spring and geyser effluents of Yellowstone National Park', *Science*, **178**, 402–405.

Author Index

References to citations in the cursive text are shown in ordinary type. Numerals in italic refer to detailed citation of the original reference. Numerals in bold refer to authors of articles.

610

Griffin, W. L.   56, 70, *72*, 161, *163*, 385, *386*
Griffiths, D. H.   135, *144*
Griggs, D. T.   370, *372*
Grikurov, G. E.   87, *94*
Grögler, N.   *75*
Grossman, L.   14, *17*, 24, 26, 27, *29*, *30*, 200, *201*, 369, *372*, 539, 540, 541, *545*
Grout, F. F.   299, *302*
Grove, T. L.   *75*
Groves, D. I.   **303**
Gruner, J. W.   *302*
Guazzone, G.   141, *144*
Gulson, B.   *415*
Gunn, B. M.   282, 283, *286*, **389**, 390, 391, 392, 393, 394, 395, 396, 397, 398, 400, 401, *402*, *403*
Gunthorpe, R. J.   **303**, 476

Haas, J. L., Jr.   462, *465*
Hackman, B. D.   134, *145*
Hahn-Weinheimer, P.   49, *52*
Hajash, A., Jr.   *402*, 563, *567*
Hall, D. H.   *155*, 343, *346*
Hall, D. O.   595, 599, *604*
Hall, H. T.   23, 24, 25, 26, *30*
Hallbauer, D. K.   581, *584*
Hallberg, J. A.   215, 220, *222*, 258, 259, 263, 266, 268, 269, *274*, *276*, 305, 311, *313*, 378, *386*
Halpern, M.   136, *145*
Hamilton, D. L.   218, *222*
Hamilton, W.   138, *145*
Hanks, T. C.   22, 27, *29*, *30*, 34, 35, *52*, 92, *94*, 367, 369, *373*, 539, *545*
Hanson, G. N.   59, 66, 67, *72*, *74*, 133, 140, 141, 142, *143*, 258, *273*, 299, 300, 301, *301*, *302*, 351, 352, 353, 357, *358*, *359*, *360*, 378, 382, *386*, 407, 409, 411, *415*, 421, *423*, 476, *478*, 487, *488*
Hardy, J. P.   *545*
Hari Narain, *see* Narain, H.
Härme, M.   484, *488*
Harre, W.   *498*
Harris, A. L.   *478*
Harris, A. W.   13, 14, *17*, *18*
Harris, F. S.   *52*
Harris, N. B. W.   *95*
Harrison, N. M.   56, 67, *76*, 265, *274*, 420, 421, 422, *424*, 457, *465*
Harshman, E. N.   581, *584*
Hart, S. R.   39, 40, *52*, 56, 66, 67, 68, 69, 70, *72*, 115, *128*, 132, 133, 140, *144*, *145*, 258, *274*, 353, *359*, 414, *416*, 419, 422, *424*
Hartman, L. W.   62, 460, 462, *465*
Hartmann, W. K.   7, *17*, 68, *72*
Hasan, Z.   409, *417*
Haskin, L. A.   59, *73*, *75*, *424*
Haskin, M. A.   *424*
Hassenforder, B.   *94*

Hattori, H.   419, *424*
Hausen, D. M.   *313*
Havens, R. G.   *221*, *252*
Hawkesworth, C. J.   105, 132, *145*, 318, 320, *321*, 323, *330*, 351, 352, 353, *359*
Hayashi, C.   541, *545*
Hays, J. F.   *75*
Hayslip, D. L.   421, *424*
Head, J. W.   12, *17*, *18*
Hebeda, E. H.   *521*
Hedge, C. E.   22, *30*, *53*, 56, 66, 70, *72*, *74*, *75*, 84, *94*, *302*, *346*, 352, 353, *359*, 360, 409, *416*, *424*, *509*, *577*
Heier, K. S.   **159**, 159, 160, 161, 162, *163*, *164*, 200, *201*, 229, *235*, 296, *298*, 356, 357, *359*, 367, *373*, 378, *386*, 407, 413, 414, *416*, 484, *488*, *555*
Heimlich, R. A.   *424*, 501, 502, 504, 505, 506, *509*, *510*
Hein, S. M.   46, *51*, 56, *72*, 571, 572, *577*
Heinrich, E. W.   500, 504, *509*
Helmke, P. A.   *73*
Hempkins, W. B.   459, *465*
Henderson, J. F.   583
Henley, R. W.   455, 462, 463, *465*
Henriksen, N.   109, *110*, 433, *439*
Henriksen, S.   *535*
Hensen, B. J.   365, *373*
Hepworth, J. V.   317, *321*, 494, 496, *498*
Herd, R. K.   *176*, *223*, 234, *235*, *387*, 447, *453*
Hermann, A. G.   419, *424*
Herrin, E.   367, *373*
Herz, N.   89, *94*
Hess, H. H.   47, *52*, 503, *510*
Hesstvedt, E.   527, *535*
Hickman, M. H.   85, *94*, 320, *321*, 323, *330*, 352, 353, *359*
Hietanen, A.   301, *302*
Higgins, A. K.   214, *222*
Higuchi, H.   *18*
Hildreth, R. A.   *358*, *510*
Hill, J.   *346*
Hill, P. G.   *74*
Hills, F. A.   506, 507, 508, *509*, *510*
Hills, J. G.   541, *545*
Himmelberg, G. R.   372, *373*
Hitchen, C. S.   495, *498*
Hjartarson, H.   *535*
Hodges, C. A.   *18*
Hodges, F. N.   *17*
Hodgson, G. W.   529, *535*
Hoffman, J. H.   *302*
Hoffman, P.   592
Hofmann, H. J.   580, *584*, 592
Hogan, L.   541, *545*, 553, *555*
Holdaway, M. J.   364, *373*
Holden, J. C.   79, 87, *94*
Holland, H. D.   35, *52*, 526, *535*, 539, 543, *545*, 547, *555*, **559**, 560, 565, *567*, 579, 580, 581, 583, *584*, 589, *593*

618